A FLORA *of* SONOMA COUNTY

A FLORA *of*

SONOMA COUNTY

Manual of the Flowering Plants and Ferns of Sonoma County, California

CATHERINE BEST, JOHN THOMAS HOWELL,
WALTER & IRJA KNIGHT, *and* MARY WELLS

First Edition
1996

California Native Plant Society

Cover photograph and design by Beth Hansen

Printed and bound in the United States.
ISBN: 0-943460-29-8 (hardcover)
ISBN: 0-943460-28-x (softcover)

Library of Congress Catalog Card Number: 96-086104

Front cover: Coast live oak (*Quercus agrifolia*) grace many hillsides throughout Sonoma County. Photograph by Beth Hansen.
Back cover, top: Valley oak (*Quercus lobata*) and Oregon oak (*Q. garryana*) in flooded pastureland south of Trenton in the Laguna de Santa Rosa shown following a rainstorm. January 1990. Photograph by Walter Knight.
Back cover, bottom, and page 1: Sonoma sunshine (*Blennosperma bakeri*) grows in a number of vernal pools in Sonoma County and has become rare by habitat loss. June 1996. Photograph by Reny Parker.

Redwood (*Sequoia sempervirens*), Douglas-fir (*Pseudotsuga menziesii*), and coast live-oak (*Quercus agrifolia*) growing in the Willow Creek watershed. June 1995. Photograph by Ken Wilson.

Redwood (*Sequoia sempervirens*), madrone (*Arbutus menziesii*), and coast live-oak (*Quercus agrifolia*) grow along Austin Creek. May 1993. Photograph by Jim Finn.

CONTENTS

Valley oak (*Quercus lobata*) growing in the Laguna de Santa Rosa. October 1995. Photograph by Ken Wilson.

A forest of foothill pine (*Pinus sabiniana*) to the west from Pine Flat Road at about 2300 ft. elevation. In Sonoma County this species usually grows scattered as individuals, often on serpentine soils. January 1990. Photograph by Walter Knight.

ACKNOWLEDGMENTS

This flora was created with the assistance of many persons. We give special thanks to the Frank Almeda, Bruce Bartholomew and the staff of the Botany Department at the California Academy of Sciences, San Francisco, where much of the research was done for this project. The Academy collections include the Dudley Herbarium, which was formerly at Stanford University. The staff at the University of California and Jepson herbaria, Berkeley, gave the authors access to their research facilities, and the late Lawrence R. Heckard encouraged us to proceed with the flora in its present style. We thank the contributors to the various herbaria who collected plants from all parts of the county and spent thousands of hours in the preparation of the materials to be filed therein. Stephen W. Edwards, director, Regional Park Botanic Garden, Berkeley, CA, provided encouragement to bring the project to a successful conclusion. Gilbert Muth, Professor of Botany, Biology Department, Pacific Union College, Angwin, CA, provided a data base printout of the plants collected in Sonoma County that are on deposit at Pacific Union's herbarium.

Charles F. Quibell, Sonoma State University, Rohnert Park, CA, furnished herbarium facilities, and assisted in a detailed analysis of parts of this flora. Rolf Erickson and Terry Wright, Sonoma State University, offered consultation on geological aspects. Nancy Harrison, Santa Rosa Junior College, provided access to the herbarium, and to her card files. Reid Moran submitted critique and suggestions dealing with all phases of the project. Glenn Keator monitored the work during part of its later stages. Wayne Roderick reported many uncollected plant sites.

Sally deBecker, Pacific Gas and Electric Co., San Ramon, CA, made available herbarium specimens which were collected in the geothermal areas of Sonoma County. Robert Powell, Department of Agronomy and Range Science, University of California, Davis, CA, critiqued the project in its early stages of development. Douglas Stone, Martinez, CA, gave valuable information regarding his Sonoma County collections, especially in the geothermal area, which are filed in various northern California herbaria. Charles A. Patterson, El Cerrito, CA, transported us in his 4-wheel drive vehicle over relatively inaccessible areas of Rincon Ridge, and furnished plant lists of prospective sewage disposal sites. The late Peter Rubtzoff made available a considerable part of his card file covering his wetland collections. Roxanne Bittman, Sacramento, CA, provided a computer diskette with status of rare and endangered plants. Peter G. Connors, Bodega Marine Laboratory, University of California, revised the plant list of the Bodega Bay area.

For making suggestions in selected groups, we thank Mary E. Barkworth, Ron Coleman, W. J. Crins, Alva Day, the late James C. Hickman, Judith Jernstedt, Daryl Koutnik, Reid Moran, Randall Morgan, Elizabeth Neese, Martin Ray, David M. Thompson, Michael C. Vasey and the late Lawrence R. Heckard.

Others who made suggestions for the improvement of, or made contributions to, the manuscript include David Amme, Doug Barbe, Vanette Bunyan, Joe Callizo, Greg de Nevers, Ed Deering, Elizabeth McClintock, James T. Vale, H. Keith Wagnon, Jack and Betty Guggolz.

Dieter H. Wilken and David J. Keil reviewed the manuscript and made numerous valuable suggestions to improve it.

The California Native Plant Society acknowledges with gratitude financial support for this publication from the Milo Baker Chapter of CNPS and from Jack and Betty Guggolz, long-time CNPS friends and Sonoma County residents.

Goldfields (*Lasthenia californica*) and yellow bush lupine (*Lupinus arboreus*) carpet the coastal terraces at Salt Point Park in spring. May 1996. Photograph by Reny Parker.

Elsanjo Lake at Annadel State Park with clumps of bulrush (*Scirpus* sp.) emerging from the lake. June 1993. Photograph by Jim Finn.

PREFACE

California is home to a rich and diverse flora with over 30% of all vascular plants occurring in North America. Our knowledge of this botanical treasure trove has grown dramatically through the preparation of floristic studies whose focus has ranged from small politically bounded areas to comprehensive accounts of the entire state. Every flora, whether it is of a local nature or more comprehensive in scope, is a progress report that records native species as well as those that have become naturalized through the modification of natural plan communities by human activities.

The preparation of local floras, especially in California, has had great appeal for both the amateur and professional botanist. The late John Thomas Howell (1903-1994), long-time Curator of Botany at the California Academy of Sciences, was a staunch proponent of California floristics. He felt strongly that botanists should be encouraged to collect, study and write about the plants of small geographic areas. This, as he was always quick to point out, is an endeavor that can produce important contributions to the knowledge of any major flora. Inspired by his example and willingness to always lend a helping hand, numerous amateur and professional botanists have taken an active role in the production of local California floras. The Sonoma Flora is yet another example of Howell's legacy.

Active work on this flora started in 1980 when Howell invited Catherine Best and Mary Wells to join the effort. They were subsequently joined in 1987 by Walter and Irja Knight who also had a long-standing interest in the flora of Sonoma County. Although Tom Howell had planned to orchestrate preparation of this flora, advancing age and ill health precluded his active involvement throughout the course of the project. Consequently, completion of the flora fell to the other authors, the Knights, Kay Best and Mary Wells. Walter Knight assumed the time consuming task of entering all the data into the computer and creating several versions of the flora to be proofed and field-checked. The authors are all grateful for the expertise and generous time given by many caring individuals. Foremost among those who responded to requests for assistance are Dieter Wilken and David Keil, who critically reviewed scientific content and offered many helpful suggestions. Barbara Leitner contributed her general editing expertise to the entire manuscript. Stephen Lowens spent long hours at the computer entering the tedious corrections that each review brought to light, and Beth Hansen provided her meticulous design skill to the project. Publication of this volume would have been much more difficult if it had not been for Phyllis Faber who worked tirelessly and harmoniously with the authors and managed production of this flora with her usual finesse It is to the conscientious dedication of the authors and the cooperation of this industrious team that we owe a debt of gratitude for the appearance of what will surely be a welcome and useful account of the vascular plants of Sonoma County, California.

Frank Almeda
California Academy of Sciences
San Francisco, CA 94118
June, 1996

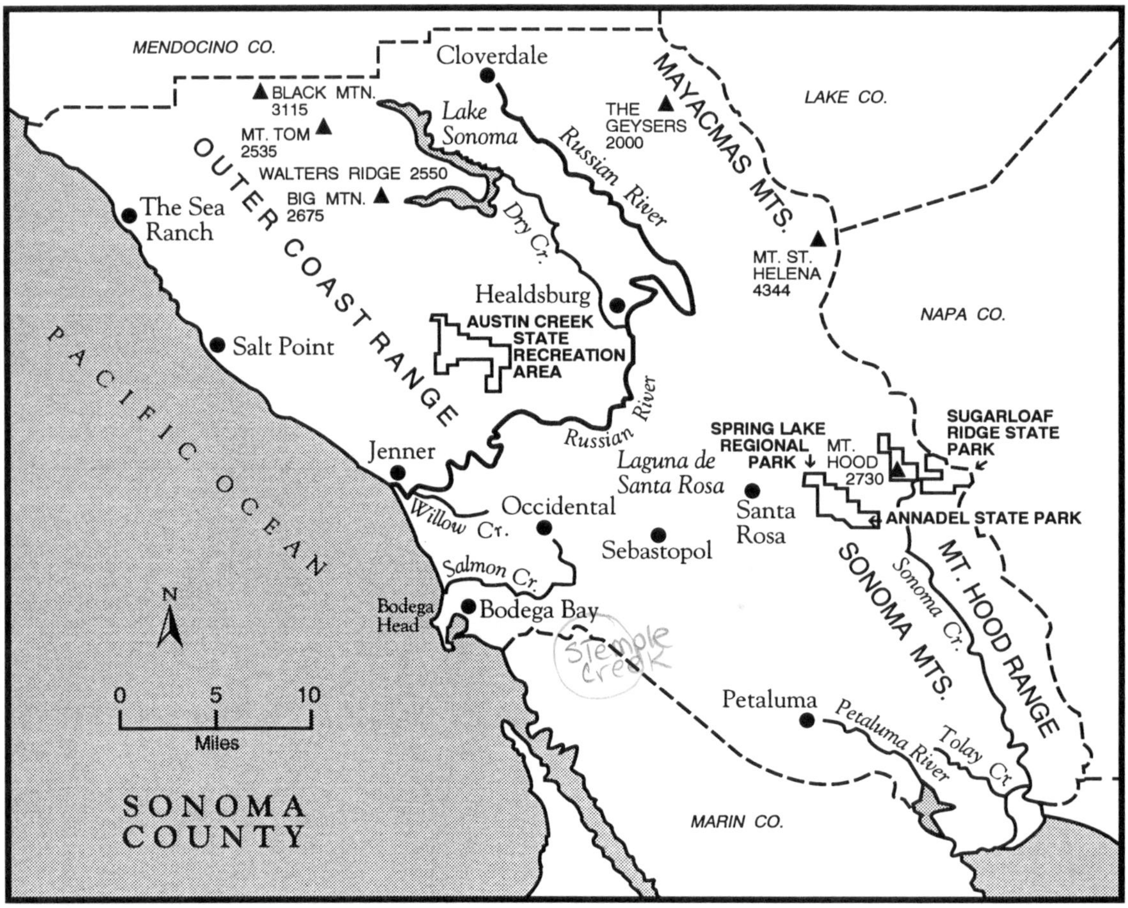

Sonoma County map by Paula Nelson.

INTRODUCTION

This work contains a list of the native and naturalized vascular plants of Sonoma County, California. No algae, lichens or mosses are included.

A concentrated attempt has been made to provide a reference which will be of assistance to students of botany, planning agencies, garden club members, and developers, yes, and even botanists, in their attempts at understanding the flora and patterns of vegetation of the county.

This introduction covers the general characteristics of the flora of Sonoma County as well as providing a mini-history of botanical science in the county. Particularly notable in this section is the listing (on page 10) of the status of rare and endangered plants as identified in the California Native Plant Society's *Inventory of Rare and Endangered Vascular Plants of California*, 1994.

TOPOGRAPHY

The topography of Sonoma County is dominated by a series of low-lying mountain ranges, generally below 3000 feet but reaching a maximum of 4495 feet. These ranges along with dividing valleys and river drainages define the major physical features of the county.

Mountain Ranges: The Mayacmas Mountain dominate the northeastern section of the county and reach their highest elevation at 4495 feet on the southwest flank of Cobb Mountain. The summit is at 4722 feet located in Lake county. Mt. St. Helena is eleven air miles south-southeast of Cobb Mountain with its lookout located in Sonoma County.

The Mt. Hood Range, elevation 2730 feet, lies along the west flank of the Mayacmas Mountains. Its western slopes face the Kenwood-Oakmont area; its eastern slopes face the York Creek-Sulphur Creek area of Napa County.

The Sonoma Mountains extend from Bennett Valley to Sears Point, attaining their highest elevation at 2295 feet. The range separates the Petaluma and Sonoma valleys.

The northern part of the county bordering Mendocino County is rugged and virtually inaccessible. Major peaks include Mt. Tom 2535 feet, Black Mountain 3115 feet, Big Mountain 2675 feet and Walters Ridge 2550 feet. The western section of the county is part of the Outer Coast Range.

Intervening Valleys and Drainages: The Russian River watershed is the largest drainage in the county. Flowing southward from Cloverdale, the river flows westward southwest of Healdsburg, ultimately reaching the Pacific Ocean at Jenner. The segment of the Russian River drainage system that includes the flatlands southwest, west and northwest of Santa Rosa as far as Cunningham, Sebastopol and Graton, is listed as "Laguna de Santa Rosa" or as "Western Santa Rosa Plain."

The Petaluma River is tidal up to Petaluma. Its freshwater continuation is known as Willow Brook, and extends almost to Penngrove. Sonoma Creek and the Petaluma River both discharge into San Pablo Bay. The lower part of Sonoma Creek, along with a maze of sloughs in the area of Skaggs Island, is also tideal. Sonoma Creek drains the Kenwood region between the Mayacmas and Mt. Hood ranges on the east and the Sonoma Mountains on the west.

Dry Creek and Warm Springs Creek are impounded by Warm Springs Dam to form Lake Sonoma, which collects its water from western segments of northern Sonoma County. Lake Sonoma inundated the site of the former town of Skaggs Springs.

Secondary drainages are: (1) Tolay Creek, located between the Petaluma River and Sonoma Creek near Sears Point Raceway, draining into San Pablo Bay, and (2) San Antonio Creek, along the Marin-Sonoma county line, into the lower part of the Petaluma River. Drainages into the Pacific Ocean are: (1) Stemple Creek, with its source near Pepper and Meacham roads in the Stony Point district, (2) Estero Americano, along the Marin-Sonoma county line including the drainage of Valley Ford and Bloomfield areas, (3) Salmon Creek, its source near Occidental, meandering near the towns of Bodega and Freestone and entering the ocean north of Bodega Harbor, and (4) the South and Wheatfield forks of the Gualala River, draining the northwestern part of the county.

CLIMATE

The weather along the Russian River and in the Petaluma and Sonoma valleys is generally mild with only a few days reaching a temperature of 100° F during the summer. A winter low of around 25° F (rarely lower) may be expected away from the coast. Coastal areas are usually much cooler than in the interior during the daytime throughout the year, and the low temperatures during the winter season are much milder than those in the interior.

Along the coast from April until early September, fog persists much of the time in morning hours and usually burns off before noon. This condition occurs to a lesser degree in the Russian River and Petaluma valleys. During September, October and November many of the coastal areas are almost fog-free. During the fog season, Stemple Creek sometimes serves as a corridor to funnel the evening fog eastward along Railroad Avenue until it comes in contact with, and collects against, the Sonoma Mountains southeast of Cotati. The fog then diffuses to the north and south so that Petaluma, Cotati and Rohnert Park receive fog indirectly rather than directly from the ocean. Sonoma Valley, located to the east, receives less fog.

During most of the year trade winds off the Pacific Ocean tend to cool western and central areas of Sonoma County, except in September when warmer northeasterly breezes with low humidity sometimes prevail. During the rainy season winds are frequently from the south and southeast. Generally between Stemple Creek and the Marin County line in the Two Rock-Stony Point area, strong winds prevail particularly in the afternoon during the spring months and well into the summer. Petaluma is the region affected the most and residents of the southern part of the county refer to the condition as the "Petaluma wind tunnel."

The rainy season is, for the most part, from October until April. Twenty to thirty-five inches of precipitation may be expected over most of the county. The *Petaluma Argus-Courier* reports that the average annual rainfall in the Petaluma area is about 29.3 inches. The *San Francisco Chronicle* lists Santa Rosa's average at 29.88 inches. The *Press Democrat* reports average rainfall at Rohnert Park as 29.76 inches. In the Cazadero area, rainfall is normally heavier, averaging 60 to 90 inches.

During three winters in the last few decades, close to 40 inches of rain fell in central and southern Sonoma County. Up until the winter of 1991-1992 the whole state had suffered six years of drought.

SERPENTINE vs. SERPENTINITE

The mineral serpentine possesses unique qualities that result in concentrations of specialized and often rare plants. Because serpentine tends to occur frequently in Sonoma County, we present the following discussion of this mineral taken from *The Four Seasons,* 8(2):20-27, Origin of Serpentinites, by Stephen W. Edwards (1988). In part: "If you pick up a chunk of the rock, it is *serpentinite*. The rock consists mainly of varying proportions of the serpentine group of minerals: lizardite, chrysotile, and/or antigorite. These minerals *alone* are serpentine, and the rock consisting mainly of them *can* be called serpentine because it is dominated by them. However, nearly everywhere the rock contains minor amounts of such minerals as brucite, talc, magnetite, and various carbonates, including aragonite and calcite. Often chromite or magnesite is present. Serpentinite is a metamorphic rock, and it may also contain olivines, pyroxenes, and other minerals remnant from the igneous (solidified from magma) rock that was metamorphosed. There are also whole suites of minerals that arise during metamorphism (some are noted above) and in the process of weathering. So the rock itself is quite complex, and that is why most geologists refer to it as *serpentinite* rather than serpentine. It is more than just serpentine minerals."

"Outcrops of the most intensely deformed serpentinite are greenish gray, the color of weathered bronze. The material is cut by innumerable curved slip-surfaces that are braided together to form layers, and these in turn are braided together on a larger scale. These slip surfaces curve around the scattered blocks of unsheared, feltlike serpentinite that is commonly dark green or black. Some extensively sheared, severely altered material is almost white, but scattered through it are unsheared blocks of serpentinite that weather to a rusty red. Sheared material also may be honey-colored or otherwise light-colored, tending to have a waxy luster and to be translucent." —*F. G. Wells* (1951).

Some interesting areas where plants growing on serpentine occur:

(1) Around Occidental along Stoetz Lane, Harrison Grade, Morelli Lane, Facendini Lane, Occidental Road, junction of Joy and Bittner roads, and the old quarry near Camp Meeker.

(2) A large region in the Austin Creek State Recreation Area and northwest to The Butcherknife, Red Slide, Red Oat Ridge and The Cedars.

(3) The geothermal area at the end of Pine Flat Road covering thousands of acres between Socrates Mine and The Geysers (2000 feet) on the west-facing slopes of Cobb Mountain.

(4) A small outcrop on the north face of Geyser Peak (sometimes called Sulphur Peak) near the summit at 3427 ft.

(5) An expanse in the upper reaches of Knights Valley between Foote Creek and Maacama Creek, extending northward to the southern segment of Ida Clayton Road.

(6) Bradford Mountain, a large area which is accessible from West Dry Creek Road.

(7) Between Pythian Road and the summit of Mt. Hood.

(8) On the south flank of Red Mountain (2530 feet) in Sugarloaf Ridge State Park. Another outcrop near here is at the upper end of Adobe Canyon about ½ mile past the parking lot.

(9) At the entrance of Pepperwood Preserve on Franz Valley Road.

(10) Southwest of Lake Sonoma along Stewarts Point-Skaggs Springs Road near junction to overlook.

At sites 3, 4 and 5 above, the serpentine endemic *Arctostaphylos viscida* ssp. *pulchella* is found in dense populations. Conspicuous in some of the other areas are the endemics *Ceanothus jepsonii*, *Quercus durata*, *Cupressus sargentii* and *Rhamnus tomentella*.

PLANT ASSOCIATIONS

Redwood Forest. Small groves of redwood (*Sequoia sempervirens*) are scattered through canyons along the coast north of the Russian River. One of the best populations is in the Armstrong Redwoods State Reserve. Small stands extend along the Russian River to Trenton and south to the Vine Hill area. There are several groves along the Bohemian Highway, Willow Creek Road and Coleman Valley Road. Excellent populations of various numbers and densities appear at Joy Woods, Bohemian Grove, Pepperwood, Ida Clayton Road, Sugarloaf Ridge, Cooper's Grove, Taylor Mountain and Fitch Mountain. Mingling with redwood trees are California laurel (*Umbellularia californica*) and tanbark oak (*Lithocarpus densiflorus*). The latter two species form a more conspicuous part of the forest where redwoods were cut for lumber in past decades. In the early 1960s, the California Department of Transportation planted some small groves of redwoods along US 101 between Petaluma and Windsor.

Douglas-fir Forest. There are dense stands of Douglas-firs (*Pseudotsuga menziesii*) in the Russian River drainage system from Mill Creek southward to Bodega Highway and Salmon Creek Road. Other forests are along Rincon Ridge and Mark West Creek east of US 101. Scattered populations occur through the

Sonoma Mountains and Sugarloaf Ridge. Lesser stands are found throughout the northern section of the county. There are small, isolated occurrences of western hemlock (*Tsuga heterophylla*) in the Gualala River drainage and grand fir (*Abies grandis*) reaches its southern limit at Willow Creek Road. Douglas-fir comes in contact and combines with stands of coast redwood.

Tanbark Oak-Madrone Forest. As the redwood forest ascends the canyon sides, it gradually diminishes and the Tanbark oak-madrone (*Lithocarpus densiflorus-Arbutus menziesii*) association becomes more prevalent. Sometimes abundant are canyon live oak (*Quercus chrysolepis*), coast live oak (*Q. agrifolia*), giant chinquapin (*Chrysolepis chrysophylla*) and California laurel (*Umbellularia californica*).

Oak-Buckeye Forest. This association consists mainly of black oak (*Quercus kelloggii*), Oregon oak (*Q. garryana*), valley oak (*Q. lobata*) and buckeye (*Aesculus californica*), and extends over most of the county. Some of the accompanying shrubs include hillside gooseberry (*Ribes californicum*), cream bush (*Holodiscus discolor*), elderberry (*Sambucus mexicana*), coyote brush (*Baccharis pilularis*) and snowberry (*Symphoricarpos albus* var. *laevigatus*).

Pine Forests. Scattered throughout the northern half of the county is ponderosa pine (*Pinus ponderosa*), generally in association with Douglas-fir. A sizeable sugar pine (*P. lambertiana*) population occurs in the Gualala River region at the inner edge of a redwood forest. Scattered individuals occur near the head of Austin Creek and at about 1800 ft. elevation in the Mayacmas Range along the Sonoma-Lake county boundary north of Mt. St. Helena. Foothill pine (P. *sabiniana*) frequents the foothills in association with blue oak (*Quercus douglasii*). A large foothill pine forest appears northwest of Pine Flat Road at about 1800 feet elevation. Although foothill pines grow off serpentine in the Sierran foothills, in Sonoma County they are usually on serpentine. Knobcone pine (*P. attenuata*) is in the Cavedale and Trinity roads area and in the geothermal region on the western face of Cobb Mountain. The trees of this species are frequently found in groves of approximately all the same height since the seeds of this closed-cone pine germinate readily after a fire. However at Pepperwood twenty knobcone pines occur as scattered individuals only and not in the typical monospecific stands. Stephen W. Edwards reports the existence of a plantation of about 50 Coulter pines (*P. coulteri*) ¼ mile from the summit of the Cavedale-Trinity Road loop. Also some trees have been planted east of Ida Clayton Road on the west flank of Mt. St.Helena. The Coulter pines are not native to the area.

Among the closed cone species, Bishop pine (*Pinus muricata*) is common on coastal bluffs and terraces north of the Russian River, the next occurrence to the south being on Point Reyes Peninsula. In association are Monterey pine (*Pinus radiata*) and Monterey cypress (*Cupressus macrocarpa*) which have been introduced as windbreaks long ago from the Monterey Peninsula.

At Salt Point State Park in a barrens directly behind the frontal forest on the higher area east of State Route 1 is the southernmost-known station of both pygmy pine (*Pinus contorta* ssp. *bolanderi*) and Mendocino or pygmy cypress (*Cupressus goveniana* ssp. *pigmaea*).

Chaparral. Restricted to exposed and relatively dry slopes of the interior sections of the county, its best development, (sometimes on serpentine), frequently called "hard chaparral," may be seen in Hooker, Stuart, Nuns and Adobe canyons and along Cavedale, Trinity, Los Alamos, St. Helena and Franz Valley roads, as well as in various other areas to the east of State Route 12. An easily accessible area to see well-developed chaparral while hiking is in Annadel State Park near Santa Rosa. Also a small area is on the west slope of Taylor Mountain. Large patches are scattered throughout the geothermal area on the western segment of Cobb Mountain and in the Geyser Peak-Black Mountain area. Some principal constituents are manzanita (*Arctostaphylos* spp.), *Ceanothus* spp., chamise (*Adenostoma fasciculatum*), toyon (*Heteromeles arbutifolia*), mountain mahogany (*Cercocarpus betuloides*), chaparral pea (*Pickeringia montana*), silk-tassel bush (*Garrya* spp.) and pitcher sage (*Lepechinia calycina*).

Coastal Brush. Found on maritime bluffs and mesas of interior hills and canyons. Dense thickets can act as "soft chaparral." The principal species are the dominant coyote brush (*Baccharis pilularis*), poison oak (*Toxicodendron diversilobum*), California blackberry (*Rubus ursinus*), red-flowering currant (*Ribes sanguineum* var. *glutinosum*), ninebark (*Physocarpus capitatus*), sticky monkey flower (*Mimulus aurantiacus*) and California sagebrush (*Artemisia californica*).

Coastal Grassland. There are two recognizable types of coastal grassland, based on their location, character of the soil, amount of moisture and possibly some other factors (Howell 1949). Coastal slopes have a more turflike development than similar habitat in the interior areas. Most native species fall into the category of bunchgrasses. These were prevalent before the advent of alien species. The aliens proliferated during land cultivation and conversion to pasture areas. Some of the readily found native grasses in coastal areas are: pacific hair grass (*Deschampsia cespitosa* ssp. *holciformis*), pacific reed grass (*Calamagrostis nutkaensis*), California fescue (*Festuca californica*), red fescue (*F. rubra*), and California oat grass (*Danthonia californica*). Common introductions are six-weeks fescue (*Vulpia bromoides*), foxtail fescue (*V. myuros*), little quaking grass (*Briza minor*), silver hairgrass (*Aira caryophyllea*), soft chess (*Bromus hordeaceus*) and velvet grass (*Holcus lanatus*).

Valley and Hill Grassland. Many species are found throughout the grassy areas from coastal hills to interior slopes and valleys. Some grasses characteristic of these regions are: foothill needlegrass (*Nassella lepida*), purple needlegrass (*N. pulchra*), Pacific blue grass (*Poa secunda*), blue bunchgrass (*Festuca idahoensis*), perennial ryegrass (*Lolium perenne*), big squirreltail (*Elymus multisetus*), and beardless wild-rye (*Leymus triticoides*), besides a host of other natives and introductions.

Vernal Pool. Vernal pools are characterized by a variable aquatic and terrestrial regime: long-term or periodic inundation during the growing season, desiccation during summer and early fall and a flora dominated by native annual species adapted to a variable aquatic and terrestrial regime. Physical factors distinguishing vernal pool types include class of subsoil restricting layer (e. g. hardpan), soil type, and hydrology, which help create unique conditions with respect to soil and water chemistry and a seasonal timing of frequency and duration of inundation. Since pools are isolated from each other, a distinctive flora evolves that is often unique.

In decades past the principal area where this association occurred in Sonoma County was around Laguna de Santa Rosa, Sonoma County Airport and Windsor. Industrial parks and residential development have eliminated most of the vernal pools in this region and around parts of Sonoma Valley. Some vernal pools still exist in Todd Road Preserve. Habitats for Sebastopol meadowfoam (*Limnanthes vinculans*), Burke's goldfields (*Lasthenia burkei*), Sonoma sunshine (*Blennosperma bakeri*), semaphore grass (*Pleuropogon californicus*) and *Downingia* spp. have been reduced at an alarming rate.

Freshwater Marsh. Considerable research on the flora of fresh water marshes by Peter Rubtzoff has resulted in efforts to preserve this valuable but scarce county resource. As a result, large areas of the Laguna de Santa Rosa have been preserved or are in the process of gaining that status. Common species are narrow-leaved cattail (*Typha angustifolia*), southern cattail (*T. domingensis*), broad-leaved cattail (*T. latifolia*), bur-reed (*Sparganium erectum* ssp. *stoloniferum*), floating-leaved pondweed (*Potamogeton natans*), long-leaved pondweed (*P. nodosus*), small pondweed (*P. pusillus*), water hemlock (*Cicuta douglasii*), ditch-carrot (*Oenanthe sarmentosa*), common water-plantain (*Alisma plantago-aquatica*), common sedge (*Carex densa*), soft-stem tule (*Scirpus tabernaemontani*) and broad-leaved arrowhead (*Sagittaria latifolia*).

Pitkin and Perry marshes are still somewhat intact, whereas Cunningham Marsh has been reduced in size by encroaching agricultural activity. Pitkin Marsh lily (*Lilium pardalinum* ssp. *pitkinense*) and Pitkin Marsh Indian paintbrush (*Castilleja uliginosa*) sometimes are not seen for several years. In fact the latter plant may be extinct. Although Pitkin Marsh still has some good habitat, much of the marsh has been altered by nearby extensive cultivation (vineyards, Christmas tree farms and pastures) and a six- to seven-year drought during the late 1980s and early 1990s which has affected groundwater availability in parts of the marsh.

Salt Marsh. San Pablo Bay Area, especially in the Sears Point region, contains many acres of salt marsh. Species diversity in this plant community is very low. The flatlands between San Pablo Bay and Petaluma are the most representative of this plant community in the county. The perennial Petaluma popcorn-flower (*Plagiobothrys mollis* var. *vestitus*) has not been seen since Congdon collected it in Petaluma in 1880, and it must be presumed extinct. Along the coast there are brackish marshes at Bodega Bay and Salmon Creek. Because these habitats are oozy, odoriferous and insectiferous, they have not received as much attention as have the freshwater marshes. Also fewer endemics occur in salt marsh habitats. Common plants to be seen in the Petaluma basin are alkali heath (*Frankenia salina*), prairie bulrush (*Scirpus maritimus*), marsh grindelia (*Grindelia hirsutula*), fleshy jaumea (*Jaumea carnosa*) and pickleweed (*Salicornia virginica*). The area suffering most from industrial development is between Petaluma and Lakeville.

Riparian. This vegetation is well-developed where rivers and creeks sustain water flow throughout the year. Some common species are big-leaf maple (*Acer macrophyllum*), box elder (*A. negundo* ssp. *californicum*) and Oregon ash (*Fraxinus latifolia*). Red alder (*Alnus rubra*) is on the stream edges at or near the coast and white alder (*Alnus rhombifolia*) on the watercourses away from the coast. Willows (*Salix exigua*, *S. lasiolepis*, *S. lucida* ssp. *lasiandra* and *S. laevigata*) are common along watercourses. A good place to study willows is at Hacienda just downstream from the bridge across the Russian River where six species grow within a small area. They may all be seen from the same spot.

Dunes and Beaches. Dunes are well-developed at Bodega Dunes County Campground, the mouth of Salmon Creek, Wright's Beach and Goat Rock Beach, with smaller ones in between. Plants characterizing this association are yellow bush lupine (*Lupinus arboreus*), beach sand-verbena (*Abronia umbellata*), yellow sand verbena (*A. latifolia*), beach strawberry (*Fragaria chiloensis*), sand mat (*Cardionema ramosissimum*), wild buckwheat (*Eriogonum latifolium*), the ice plants (*Carpobrotus chilensis* and *C. edulis*), and the prolific, introduced dune grass (*Ammophila arenaria*).

SOME EARLY COLLECTORS AND THE LOCALITIES THEY VISITED

Archibald Menzies — Bodega Point (1792).

Johann Friedrich Eschscholtz — naturalist on Kotzebue's second expedition on the ship *Predpriiatiie*, visited the mission at Sonoma and then made the overland trip to the coastal Russian settlement at Bodega Bay (fall of 1824).

David Douglas — Sonoma, collected type of *Pogogyne parviflora* Bentham (1831).

E. Voznesenskii — collected 346 specimens, including some duplicates, representing 214 species and varieties with the printed information, "California boreal." Eight specimens were collected by Kuprianov, governor of Russian America at that time. (Russian America around Fort Ross and as far east as Mt. St. Helena.) Some duplicates of the plant collections are on deposit at California Academy of Sciences herbarium (1840-1841). (See J. T. Howell, *Leaflets of Western Botany,* A collection of California plants, 2:17-20, [1937]).

Karl Theodore Hartweg — Sonoma area (1846).

John M. Bigelow — Santa Rosa-Mark West Creek (1854).

William H. Brewer — botanist for the Whitney Geological Survey, collected in Petaluma, Sonoma and Santa Rosa (Apr. 1862), and from Petaluma to Tomales via Two Rock (Nov. 1862).

John Torrey — Sonoma Valley (1865).

Henry N. Bolander — also collecting for the Whitney Geological Survey, visited Petaluma and the Russian River Valley northward to Cloverdale (1864-1866).

Hans Hermann Behr — Sonoma area (during the 1860s).

Albert Kellogg and W. G. W. Harford — countywide (1868-1870).

Volney Rattan — Russian River area and in Sonoma Valley (1877-1885).

Joseph W. Congdon, lawyer, amateur botanist, resident of Petaluma — various parts of the county (1880-1895).

Marcus E. Jones — throughout the county (1882). He returned in the 1930s to collect again.

Frederic Theodore Bioletti — Mt. Hood area (1891-1893), sometimes accompanied by Chas. A. Michener.

Katharine Curran Layne Brandegee — countywide (mid-1880s and 1890s).

Edward Lee Greene — throughout central and eastern part of county (1888-1894).

J. Burtt Davy — extensively (1890s).

William Russel Dudley — southern Sonoma County (1893).

F. G. Krauss — Knights Valley and Sonoma Valley areas (1893-1895).

Alice Eastwood — Mt. Hood area (1893) and later other parts of the county (until the 1940s). Her collections from Bodega Point refer to Bodega Head.

Frank Bice — Felta Creek area southwest of Healdsburg (spring of 1895).

William Albert Setchell — around Cazadero (1897).

William W. Carruth — various parts of the county (1899-1907).

Helen D. Geis — Sulphur Creek area (1902).

Amos A. Heller — various areas of the county (1902-1903), sometimes accompanied by H. F. Brown.

Milo S. Baker — most parts of the county (1899-1961). He collected over 13,000 specimens, and they are mostly deposited at Sonoma State University along with his field records.

As we consider Milo Baker to be the patriarchal botanist of the county, we have chosen to reprint this thumbnail sketch of his life which was originally published in *Leaflets of Western Botany,* 8:85, (1957), written by Mr. and Mrs. H. C. Cantelow:

"BAKER, MILO SAMUEL. Educator, botanist; born in Strawberry Point, Iowa, 19 July 1868, now resides near Kenwood, Sonoma Co., CA. Lifelong student of the genus *Viola*; a pioneer explorer of Modoc Co., CA.; for many years professor of biology in Santa Rosa Junior College; brought to the attention of botanists generally the remarkable Pitkin Marsh, Sonoma Co., CA. He has been most generous in supplying specimens to specialists and has donated many specimens, including isotypes of new taxa based on his collections, to the California Academy of Sciences and other institutions.

"He received his B. S. degree at University of California, Berkeley, in 1895 and his A. M. degree at Stanford University in 1929. Baker's first collections were from Modoc County in 1893, where he collected *Cupressus bakeri* Jepson. He died in 1961."

LOCAL PLANT COLLECTIONS AND LISTS

The terms The Sea Ranch, Quad., Salt Point, Willow Creek, Bodega Head, Todd Road Preserve, Fairfield Osborn, Bouverie, Sugarloaf Ridge, Annadel, Spring Lake, Spring Lake Village, Pepperwood or Lake Sonoma area are used as collection sites. It is meant that the specimen was selected from one of the following lists:

The Sea Ranch designates collections from original lists published by Hektner & Foin in 1977 and 1978 that include only plants west of Hwy. 1. Vouchers are on deposit at DAV. Collections made by Mary P.

Wells (mostly 1981-1985) include the entire development known as "The Sea Ranch." Material found only east of Hwy. 1 is designated as *Wells* and is deposited at CAS, except for some she observed (acc. Wells) or some for which she had photographs. Included in The Sea Ranch list is a compilation of grasses by Hektner and Wells (1981-1983). Reference is also made to species listed only in Mahaffey (1990).

Quad. indicates results of Soil-Vegetation Survey 1947-1964 designated by quadrangles; see map in the map section at the back of the book. Specimens are on file at DAV.

Salt Point designates plants found at Salt Point including parts of Miller Creek, Gerstle Cove and Stump Beach. The list was prepared by Roger Raiche and the JEPS staff in the 1980s. Some contributions are by Mary P. Wells, for which specimens are on deposit at CAS.

Willow Creek incorporates determinations made by Elaine Weinreb, 1979, for an environmental study for the benefit of the California State Department of Parks and Recreation. It includes an area alongside Willow Creek from its confluence with the Russian River to a point about 1½ miles upstream and the high ridges on either side of the stream.

Bodega Head designates collections from a published work by Barbour in 1970 and an addition in 1972. In 1973 *The Coastal Ecology of Bodega Head* by M. G. Barbour, R. B. Craig, F. R. Drysdale and M. T. Ghiselin was published by U. C. Press. In these works, the Bodega Head area includes not only Bodega Head but the area from Bodega Bay village to Salmon Creek west and southwest of Hwy. 1. In 1990 and 1991 collections were made and filed in the BML herbarium by Peter G. Connors. Some specimens from the Barbour study are on deposit at DAV and others at BML.

Todd Road Preserve is a 75 acre preserve located southwest of Santa Rosa. Most of the collections were made by David W. Self and/or Dennis Taniguchi for an environmental biology project at Sonoma State University. Specimens were collected between 1979 and 1982 and most of them are on deposit at NCC. Specimens collected in 1986 by John Thomas Howell and Knight are designated "Todd Road Preserve (H&K)" and are on deposit at CAS.

Fairfield Osborn Preserve is located in the western foothills of the Sonoma Mountains east of Sonoma State University. Larry Serpa did the original list (S) with some vouchers at NCC during 1979-1980, and a subsequent list was prepared by Howell and associates (H) in the early 1980s. Some vouchers are deposited at CAS. Plants on both lists are indicated by (*).

Bouverie Audubon Preserve is located in Sonoma Valley east of State Route 12 near Agua Caliente. Phyllis Ellman prepared the list (1982-1984), and subsequently it was added to by observations of Howell and associates. Some vouchers are deposited at CAS.

Sugarloaf Ridge State Park is south of Mt. Hood and north of Trinity Road. The list was prepared by K. M. Stocking mostly during 1965-1970. Many of the specimens are on file at NCC.

Annadel State Park is just east of Spring Lake Regional Park in the eastern part of Santa Rosa. The list was prepared basically by Tim Tunison and added to later by Tom Wainwright. The lists were concluded in 1979.

Spring Lake Regional Park is in the eastern section of Santa Rosa. Keith Wagnon began collecting there during 1988 and the list was developed principally from his collections. Vouchers are on file at NCC.

Spring Lake Village is on 5555 Montgomery Drive in Santa Rosa. The list evolved from observations and collections by Keith Wagnon and James Vale together or individually during 1988 and subsequently.

Pepperwood Ranch Natural Preserve is a 3117 acre property owned by the California Academy of Sciences on Franz Valley Road. The list was prepared by Greg de Nevers in 1985, and he has added to it since. Vouchers are on file at CAS.

Lake Sonoma area is at the eastern end of the Stewarts Point-Skaggs Springs Road. The list was prepared by Jack Guggolz in the late 1970s and early 1980s and there are vouchers for most of the specimens at NCC. Some collections are from areas which have since been inundated by Lake Sonoma.

PALEOBOTANICAL ASPECTS

Principal sources: *Axelrod* (1944); *Chaney* (1951).

A series of volcanic eruptions during the Cenozoic Era (0.01 to 66.4 million years ago) scattered ash over a considerable part of Sonoma County. Deposits of this sort are highly favorable for the preservation of a life record. One of the finest examples in the world of an ancient forest is preserved at the Petrified Forest. Enormous trunks of redwoods (*Sequoia*) lie where they fell. Also there are winged seeds of four other conifers: fir (*Abies*), spruce (*Picea*), Douglas-fir (*Pseudotsuga*) and hemlock (*Tsuga*). Other redwood associates are represented including wax myrtle (*Myrica*), red alder (*Alnus*), chinquapin (*Castanopsis*), tan-oak (*Lithocarpus*), California laurel (*Umbellularia*), *Rhododendron* and huckleberry (*Vaccinium*). These plant remains indicate that climatic conditions during the upper Pliocene time were much like those of today, and that the forest was like the one which now extends from the Oregon line southward along the coast to Point Sur. Some genera, which include chestnut (*Castanea*), elm (*Ulmus*) and red bay (*Persea*), while no longer extant here, are known in eastern United States. Water chestnut (*Trapa*), an aquatic plant, is now native only in China.

Excellent paleobotanical sites besides the Petrified Forest are Neer's Hill (Moving Hill) east of Old Redwood Highway (Mendocino Ave.) in the northern part of Santa Rosa, and Matanzas Creek east of Mt. Taylor between Petaluma Hill Road and Bennett Valley Road. The Sonoma paleobotanical flora is represented by slightly more than 600 specimens, but owing to the large number of species this is considered insufficient to make an accurate statement on the relative abundance of the various species. However, judging from various data, the following species are most abundant at the three localities:

PETRIFIED FOREST	NEER'S HILL	MATANZAS CREEK
Ilex sonomensis	*Abies sonomensis*	*Platanus paucidentata*
Persea coalingensis	*Persea coalingensis*	*Populus alexanderi*
Pseudotsuga sonomensis	*Quercus douglasoides*	*Populus pliotremuloides*
Quercus bockéei	*Quercus lakevillensis*	
Quercus lakevillensis	*Sequoia langsdorfii*	
Sequoia langsdorfii	*Trapa americana*	

SONOMA COUNTY ENDEMICS

Betty L. Guggolz

Sonoma County is floristically rich, its vegetation diversity created, at least in part, by a broad range in altitude from sea level to 4000 feet, a wide variety of soils, a Mediterranean climate and an abundance of habitat types. Several of these habitats contain taxa that are endemic to Sonoma County, meaning that they do not occur naturally anywhere else in the world. Approximately 26% of Sonoma County's 1416 plant species, subspecies and varieties listed as rare and endangered in the CNPS "Inventory" are endemic to this county.

Approximately 42% of Sonoma County's plant endemics are edaphic (soil related) endemics, occurring in serpentine or on serpentine-related soils. The lack of nutrients, presence of heavy metals, sparse soil cover and high erodibility factor discourage the growth of many species, leaving sufficient open space for plants that are serpentine-tolerant but unable to compete in heavily vegetated areas. How each serpentine-tolerant plant species adapts to the elements of serpentine, whether by screening out the toxics or selectively absorbing the needed nutrients, is one small part of a complex subject that needs more study to develop a better understanding of California's serpentine flora. This need is further emphasized by the recent discoveries of three new serpentine endemic taxa in the county.

The 18 Sonoma County endemics include:

Arctostaphylos bakeri ssp. *bakeri*
Arctostaphylos bakeri ssp. *sublaevis*
Arctostaphylos densiflora
Arctostaphylos stanfordiana ssp. *decumbens*
Blennosperma bakeri
Calochortus raichei
Carex albida
Castilleja uliginosa
Clarkia imbricata
Cordylanthus tenuis ssp. *capillaris*
Dichanthelium acuminatum
Erigeron serpentinus
Lilium pardalinum ssp. *pitkinense*
Plagiobothrys mollis var. *vestitus*
Sidalcea oregana ssp. *valida*
Streptanthus glandulosus var. *hoffmanii*
Streptanthus morrisonii ssp. *hirtiflorus*
Streptanthus morrisonii ssp. *morrisonii*

STATUS OF LISTED SONOMA COUNTY PLANT SPECIES

Betty L. Guggolz

In the California Native Plant Society's *Inventory of Rare and Endangered Vascular Plants of California*, (1994), Sonoma County ranks second among California's 58 counties in state-listed species and seventh in those that are rare or endangered (Lists 1 and 2). The following lists contain the current status of the 122 plants included in the *Inventory's* fifth edition.

C1 = Category 1: Candidate taxa for which information is sufficient for federal listing.
C2 = Category 2: Candidate taxa for which information is not yet sufficient for listing.
C3c = Category 3: Taxa previously considered as a candidate, but currently unable to qualify.
FE = Federally listed as endangered.
PE = Proposed endangered.
CE = State listed as endangered.
CR = State listed as rare.
CT = State listed as threatened.
CEQA = California Environmental Quality Act.
CEQA? = May or may not be protected under CEQA.
+ See note in *Inventory*
* Although listed by the California Natural Diversity Data Base as occurring in Sonoma County, no precise documented location information is available for these taxa.

NAME	STATE	FED.
List 1A: Plants Presumed Extinct in California:		
Castilleja uliginosa	CE	C2
Plagiobothrys mollis var. *vestitus*	CEQA	C2
List 1B: Plants Rare, Threatened or Endangered in California and Elsewhere:		
**Abronia umbellata* ssp. *brevifolia*	CEQA	C2
Agrostis blasdalei	CEQA	C2
Alopecurus aequalis var. *sonomensis*	CEQA	C2
Arctostaphylos bakeri ssp. *bakeri*	CR	C3c
Arctostaphylos bakeri ssp. *sublaevis*	CEQA	
Arctostaphylos canescens ssp. *sonomensis*	CEQA	
Arctostaphylos densiflora	CE	C2
Arctostaphylos stanfordiana ssp. *decumbens*	CEQA	
Astragalus clarianus	CT	C1
Astragalus tener var. *tener*	CEQA	
Blennosperma bakeri	CE	FE
Calochortus raichei	CEQA	C2
Campanula californica	CEQA	C2
Carex albida	CE	C1
Ceanothus confusus	CEQA	C2
Ceanothus divergens	CEQA	C2

Ceanothus foliosus var. *vineatus*	CEQA	C2
Ceanothus sonomensis	CEQA	C2
Chlorogalum pomeridianum var. *minus*	CEQA	
Chorizanthe cuspidata var. *cuspidata*	CEQA	C2
Chorizanthe cuspidata var. *villosa*	CEQA	
Chorizanthe valida	CE	FE
Clarkia imbricata	CE	C1
**Collinsia corymbosa*	CEQA	
Cordylanthus maritimus ssp. *palustris*	CEQA	C2
Cordylanthus mollis ssp. *mollis*	CR	C1
Cordylanthus tenuis ssp. *capillaris*	CR	PE
Cupressus goveniana ssp. *pigmaea*	CEQA	C2
Delphinium bakeri	CR	C1
Delphinium luteum	CR	C1
Dichanthelium lanuginosum var. *thermale*	CE	C2
Dirca occidentalis	CEQA	
Erigeron angustatus	CEQA	
Erigeron serpentinus	CEQA	
Erigeron supplex	CEQA	C2
Eriogonum nervulosum	CEQA	C2
Fritillaria liliacea	CEQA	C2
Hesperolinon bicarpellatum	CEQA	C2
Horkelia tenuiloba	CEQA	
Lasthenia burkei	CE	FE
Layia septentrionalis	CEQA	
Legenere limosa	CEQA	C2
Lessingia arachnoidea	CEQA	C2
Lilium maritimum	CEQA	C1
Lilium pardalinum ssp. *pitkinense*	CE	C1
Limnanthes vinculans	CE	FE
Lupinus sericatus	CEQA	C3c
Lupinus tidestromii	CE	FE
Monardella villosa ssp. *globosa*	CEQA	
Navarretia leucocephala ssp. *bakeri*	CEQA	
Navarretia leucocephala ssp. *plieantha*	CE	C1
Penstemon newberryi var. *sonomensis*	CEQA	
Pleuropogon hooverianus	CR	C2
Potentilla hickmanii	CE	C1
Rhynchospora californica	CEQA	C2
Sidalcea calycosa ssp. *rhizomata*	CEQA	
Sidalcea hickmanii ssp. *viridis*	CEQA	C2
Sidalcea oregana ssp. *valida*	CE	C1
Streptanthus brachiatus ssp. *brachiatus*	CEQA	C1
Streptanthus brachiatus ssp. *hoffmanii*	CEQA	C1
Streptanthus glandulosus ssp. *secundus* var. *hoffmanii*	CEQA	C2
Streptanthus morrisonii ssp. *elatus*	CEQA	C1
Streptanthus morrisonii ssp. *hirtiflorus*	CEQA	C1
Streptanthus morrisonii ssp. *kruckebergii*	CEQA	C2
Streptanthus morrisonii ssp. *morrisonii*	CEQA	C2
+*Suaeda californica*	CEQA	PE
Tracyina rostrata	CEQA	C3c
Trifolium amoenum	CEQA	C2

List 2: Plants Rare, Threatened or Endangered in California, but More Common Elsewhere.

Calamagrostis crassiglumis	CEQA	C2
Carex californica	CEQA	
Carex comosa	CEQA	
Downingia pusilla	CEQA	C3c
Rhynchospora globularis var. *globularis*	CEQA	

List 3: Plants About Which We Need More Information - A Review List

Cardamine pachystigma var. *dissectifolia*	CEQA?	
Erigeron bioletti	CEQA?	
**Eriogonum luteolum* var. *caninum*	CEQA?	C3c
Hemizonia congesta ssp. *leucocephala*	CEQA?	
Lessingia hololeuca	CEQA?	
Lupinus arboreus var. *eximius*	CEQA?	C2
Pogogyne douglasii ssp. *parviflora*	CEQA?	C3c
Polygonum marinense	CEQA?	C2

List 4: Plants of Limited Distribution - A Watch List:

Antirrhinum virga	CEQA?	
Arabis blepharophylla	CEQA?	C3c
Arctostaphylos hispidula	CEQA?	C3c
Asclepias solanoana	CEQA?	
Astragalus breweri	CEQA?	
Astragalus rattanii var. *rattanii*	CEQA?	
Calamagrostis bolanderi	CEQA?	
Calamagrostis ophiditis	CEQA?	
Calandrinia breweri	CEQA?	
Calyptridium quadripetalum	CEQA?	
Calystegia collina ssp. *oxyphylla*	CEQA?	C2
Ceanothus gloriosus var. *gloriosus*	CEQA?	
Cirsium andrewsii	CEQA?	
Cordylanthus tenuis ssp. *brunneus*	CEQA?	
Cypripedium californicum	CEQA?	C3c
Cypripedium montanum	CEQA?	C3c
Eleocharis parvula	CEQA?	
Elymus californicus	CEQA?	C3c
Eriogonum ternatum	CEQA?	
Erysimum franciscanum	CEQA?	C2
Erythronium helenae	CEQA?	
Grindelia stricta var. *angustifolia*	CEQA?	
Helianthus exilis	CEQA?	C3c
Hesperevax sparsiflora var. *brevifolia*	CEQA?	
Lilium rubescens	CEQA?	
**Linanthus grandiflorus*	CEQA?	
Lomatium repostum	CEQA?	
Madia nutans	CEQA?	
Micropus amphibolus	CEQA?	
Monardella undulata	CEQA?	
Monardella viridis ssp. *viridis*	CEQA?	
Orobanche valida ssp. *howellii*	CEQA?	
Perideridia gairdneri ssp. *gairdneri*	CEQA?	C2
Piperia candida	CEQA?	
Pityopus californicus	CEQA?	C3c

Ranunculus lobbii	CEQA?	
Rhynchospora alba	CEQA?	
Ribes victoris	CEQA?	
Veratrum fimbriatum	CEQA?	C3c

REFERENCES for this Section

Bass, R. E., and A. I. Herson. 1992. Successful CEQA Compliance: a Step-by-step Approach. Solano Press Books, Point Arena, CA.

Kruckeberg, A. R. 1984. California Serpentines: Flora, Vegetation, Geology, Soils, and Management Problems. U. C. Press, Berkeley, CA.

Ornduff, R. 1969. *Limnanthes vinculans*, a new California Endemic. Brittonia 21: No. 1, pp. 11-14.

Skinner, M. W. and B. Pavlik. 1994. Inventory of Rare and Endangered Vascular Plants of California, California Native Plant Society Special Publication No. 1 (5th Edition).

U. S. Dept. of Interior, Fish and Wildlife Service. 1993. 50 CFR part 17, Plant Taxa for Listing as Endangered or Threatened Species. Notice of Review, pp. 51144-51190.

WEEDS

Noxious List. A significant element of the flora of Sonoma County consists of known introduced species, some of which have been classified by the California Department of Food and Agriculture as noxious. The following are the noxious weeds occurring in Sonoma County:

Acaena novae-zelandiae
Aegilops triuncialis
Allium paniculatum var. paniculatum
Allium vineale
Araujia sericifera
Cardaria chalepensis
Cardaria draba
Carduus pycnocephalus
Carduus tenuiflorus
Carthamus baeticus
Carthamus lanatus
Carthamus leucocaulos
Centaurea calcitrapa
Centaurea iberica
Centaurea solstitialis
Cirsium arvense
Cortaderia jubata
Crupina vulgaris
Cynara cardunculus
Cytisus scoparius
Elytrigia repens
Euphorbia esula
Euphorbia oblongata
Gaura drummondii
Genista monspessulana
Hydrilla verticillata
Hypericum perforatum
Imperata brevifolia
Lepidium latifolium
Lythrum salicaria
Nothoscordum inodorum
Onopordum acanthium
Pennisetum clandestimun
Polygonum amphibium var. emersum
Polygonum cuspidatum
Polygonum sachalinense
Senecio jacobaea
Senecio mikanioides
Solanum carolinense
Solanum elaeagnifolium
Taeniatherum caput-medusae
Ulex europaeus

EARLY WEED RECORDS

Adobe Brick Analysis. Some of the earliest records of weedy introductions have come from analyses of the contents of traditional adobe bricks. Investigators sampled bricks from many historic structures throughout California and the Southwest, including some buildings in Sonoma County. Following are the names of the dated buildings studied and lists of the plants found:

The Sonoma Mission at Sonoma, founded in 1824, and Vallejo's Old Adobe east of Petaluma, erected 1834-1845, are constructed of brick consisting of heavy loam mixed with sand and coarse gravel. Plant remains (especially seed) embedded in the sun-dried, adobe bricks have been subjected to critical botanical analysis; Hendry and Kelly (1925), Hendry (1931) and Hendry and Bellue (1936).

From the Sonoma building plant material taken from brick fragments piled in rear of the building and taken from inside walls during restoration in early 1920s include:

Pisum sativum	Garden pea
Triticum compactum humboldtii	Little club wheat
Triticum vulgare albidum	Beardless wheat
Hordeum vulgare pallidum	Coast barley
Sonchus asper	Sow thistle
Scirpus sp.	Tule
Ranunculus sp.	Buttercup
Hemizonia sp.	Tarweed

From the Vallejo Old Adobe (bricks from partition walls in north wing):

Erodium sp.	Filaree
Trifolium olivaceum = *T. albopurpureum var. olivaceum*	Native clover
Bromus sp.	Brome grass
Polygonum aviculare = *P. arenastrum*	Knotweed
Rumex crispus	Curly dock
Dondia californica = *Suaeda c.*	Sea blite
Elymus triticoides = *Leymus t.*	Wild-rye
Triticum compactum humboldtii	Little club wheat
Triticum vulgare graecum	Propo wheat
Avena sterilis byzantiana	Oat
Avena sativa	Oat

Russian Weed Collections. Other introduced plants which often are weedy were collected in 1840-41 by E. Voznesenskii in "boreal California" presumably between Fort Ross and Mt. St. Helena, (*Leaflets of Western Botany, Howell, 2:2, pp. 17-20,* [1937]):

Anagallis arvensis	Pimpernel
Erodium cicutarium	Red-stemmed filaree
Erodium moschatum	White-stemmed filaree
Physalis philadelphica	Tomatillo
Poa annua	Annual blue grass
Raphanus sativus	Wild radish
Silene gallica	Windmill pink

Plant names in this section are those which appeared in the papers by Hendy, *et al.* Many since have been reduced to synonyms and are shown where possible. See comment about *Suaeda* in the Catalog of Plants.

ABOUT THIS BOOK

Families and species are listed alphabetically under the major groups. Names mainly follow *The Jepson Manual* (1993). Names of authors are not abbreviated but spelled out. Names of the introduced plants are starred (★), and the native region is given. Common names are shown when they are well-established and widely used.

Synonyms are at the end of the entry for the taxon. Repeated generic names are abbreviated to the first letter, as *Tetragona expansa* to *T. expansa*. Repeated epithets also are abbreviated to the first letter unless the ending is different; then they are spelled out. Thus *Rhus trilobata* to *R. t.* var.; but *Cymopteris terebinthinus* and *Pteryxia terebinthina*.

After a brief statement of the frequency and the type of habitat for each taxon, we list representative known localities in the county. Some of these localities are based on cited herbarium specimens. Others, such as The Sea Ranch and Salt Point, are based on local lists (see p. 7 for details). Others are based on statements of authorities, as "acc. Guggolz (1991)." Still others are from citations in botanical literature such as "*acc. Munz* (1959)." Not all herbarium specimens have been listed.

Peter Rubtzoff made collections in the 1950s and later which are annotated "Bennett Mtn. Lake." The lake is now listed on contemporary maps as "Ledson Marsh" in Annadel State Park. Another small body of water within the park which is mentioned occasionally as a collection site is Lake Ilsanjo. Maps of marshes throughout southern Sonoma County are found on p. 331.

Spelling of place names: Mayacmas Mts. is in accordance with topographic maps; however some other maps show the spelling as Mayacamas Mts. The former name was chosen for this flora. Various maps disagree on usage of apostrophes in place names. Our preference is: Cooper's Grove, Duncan's Landing, Duncans Mills, Knights Valley, Nuns Canyon, Stewarts Point and Wright's Beach. Herbarium labels have indicated Hood's Peak, Mt. Hood and Hood Mtn. Herein Mt. Hood is used.

Skaggs Springs, frequently cited, no longer exists. The town was removed during the construction of Warm Springs Dam and the site is now covered by Lake Sonoma.

A map of geothermal areas in Sonoma County, found on p. 333, shows the location of such places as Caldwell Ranch, Caldwell Pines, Geyser Rock, Lincoln Rock, Bear Creek and Squaw Creek.

Rincon Summit refers to the high point on Calistoga Road east of Rincon Valley. Rincon Ridge and Rincon Valley are in northeastern and eastern Santa Rosa respectively. Sonoma Mtn. Road refers to the road on the AAA map designated as Sonoma Mtn. Road #2.

Herbarium specimens are cited by collector (name *italicized*) and year date, as *Howell* (1934). (Note again that names of collectors are distinguished by being italicized, wheras names of authorities who contributed collection sites are not. Authors found in the bibliography are in italics preceded by acc.) The specimens cited are mostly in the herbarium at the California Academy of Sciences (CAS), in which case the herbarium is not cited. All other herbaria are cited using standard abbreviations from *Index Herbariorum* (Holmgren, et al., 1990): UC, University of California, Berkeley, CA; CM, Carnegie Museum of Natural History, Pittsburgh, PA., etc., (see p. 17).

Type localities are mostly derived from the card file of the California Academy of Sciences developed by the late Lewis Rose. Also some type localities have been selected from *A Flora of California* by W. L. Jepson (1909-1943), and from publications relating to new plant species and new combinations. Types designated from Fort Ross or Russian Settlement many times were collected as seeds by commercial traders from Russia. They would take the seed to the St. Petersburg Botanic Garden. The emerging plant would be used as the type and in many instances was described by Fischer & Meyer. Many of these seed collections were made between 1835 and 1845.

Collections in herbaria marked only "Mt. St. Helena" without the designation of the county have not been included herein because the mountain extends into three counties. If it is not known where a specimen is filed, the herbarium indication is (U).

The collection sites cited are followed by the year collected instead of a collection number as in most regional floras. The latter is the traditional method of preparing a flora, but it is felt that the year collected is much more significant and interesting to an average student who is not preparing a paper for publication or a monograph. The frequency of occurrence in the Catalog of Plants described by "common, occasional, etc." applies only to Sonoma County.

The metric system is used in plant measurements. Elevations and depths are noted in feet. Elevations pertain to the range of the *entire* distribution of the plant - not solely Sonoma County - and are followed by the flowering period of each plant.

Keys have been partially adapted by Walter Knight (unless otherwise specified) from those in Abrams vol. 1 (1923); Abrams vol. 2 (1944); Abrams vol. 3 (1951); Abrams and Ferris vol. 4 (1960); Wolf and Wagener (1948); Correll (1950); Hitchcock (1950); Howell (1949); Mason (1957); Munz (1959); Hitchcock and Cronquist (1961); Thomas (1961); Tutin *et al.* (1964-1980); McMinn & Maino (1967); Munz Supplement (1968); Smith & Sawyer (1981); Lellinger (1985); Beauchamp (1986); Hickman (1989); Whitson (1991); and Hickman (1993).

ABBREVIATIONS:

acc. - according to
ave., aves.- avenue, avenues
blvd. - boulevard
ca - about
can. - canyon
cf. - compares favorably
cm - centimeters
cr. - creek
dm - decimeters
dist. - district
elev. - elevation
emend. - adjusted description
ext. - extension
f. - filial; pertaining to the generation following the parental
jct. - junction
hwy. - highway
m - meters
mm - millimeter/millimeters
mtn. - mountain
mts. - mountains
nw., wsw., etc. - directions
reg. - regional
var. - variety; plural - vars.
sp. - species; plural - spp.
ssp. - subspecies; plural - sspp.
yds. - yards

× - multiplication sign, meaning "times," or indicating hybridization; also hand lens power, e.g., 10×
= - equal to; + - or more
/ - per
° - degree of compoundness, branching, or angle (e.g., 1° [primary], 2° [secondary]); angular degree
< - less than; > - more than
<< - much less than; >> - much more than
± - more or less
♂ - staminate; ♀ - pistillate
Headers: (F) Ferns and fern allies; (G) Gymnosperms; (D) Dicots; (M) Monocots
★ - Plants not native to Sonoma County

ABBREVIATIONS OF HERBARIA

AHUC = University of California, Department of Agronomy and Range Science, Davis, CA.
BML = Bodega Marine Laboratory, University of California, (Davis affiliation).
CAS = California Academy of Sciences, Botany Department, San Francisco, CA.
CDA = State of California, California Food & Agriculture, Sacramento, CA.
CM = Carnegie Museum of Natural History, Pittsburgh, PA.
COLO = University of Colorado, Boulder, CO.
DAV = University of California, Davis, CA.
DS = Dudley Herbarium, California Academy of Sciences, Botany Department, San Francisco, CA (formerly at Stanford University).
GH = Gray Herbarium, Harvard University, Cambridge, MA.
HSC = Humboldt State University, Arcata, CA.
JEPS = Jepson Herbarium, University of California, Berkeley, CA.
K = Kew Gardens, England.
LE = Herbarium, St. Petersburg, Russia.
MIN = University of Minnesota, St. Paul, MN.
MO = Missouri Botanical Garden, St. Louis, MO.
NCC = designates collections deposited at Sonoma State University at Rohnert Park, CA. Some years ago the Milo Baker Herbarium from Santa Rosa Junior College was consolidated with the Rohnert Park facility. In the past, the acronym has appeared as ROPA.
ND = University of Notre Dame, Notre Dame, IN.
NY = New York Botanical Garden, Bronx, NY.
OBI = California Polytechnic State University, San Luis Obispo, CA.
PG&E = collections from the geothermal area of the Mayacmas Mts. on deposit at the Pacific Gas and Electric Company herbarium at San Ramon, CA.
PUA = Pacific Union College, Angwin, CA.
RSA = Rancho Santa Ana Botanic Garden, Claremont, CA.
SACT = California State University, Sacramento, CA.
SBBG = Santa Barbara Botanic Garden, Santa Barbara, CA.
SFSU = San Francisco State University, San Francisco, CA.
SRJC = Santa Rosa Junior College, Santa Rosa, CA.
UC = University of California, Berkeley, CA.
UCR = University of California, Riverside, CA.
UCSB = University of California, Santa Barbara, CA.
US = U. S. National Herbarium, Smithsonian Institution, Washington, DC.
UT = University of Utah, Salt Lake City, UT.
WTU = University of Washington, Seattle, WA.
U = Unknown herbarium.

With no herbarium designation in the Catalog of Plants, the specimen is on deposit at California Academy of Sciences, San Francisco, CA. It may be either a part of the Academy collection or of the Dudley Herbarium of Stanford University.

TABULATION OF THE FLORA

	Families	Genera	Native Taxa	Introduced Taxa	County Endemics
Ferns and Fern Allies	11	20	41	1	0
Gymnosperms	4	8	14	3	0
Angiosperm, Dicots	100	492	1016	383	14
Angiosperm, Monocots	21	145	321	142	4
Totals	136	665	1392	529	18

The *Sonoma Flora* includes descriptions of 136 families, 665 genera, 1392 native taxa and 529 introduced taxa. Native taxa are those believed to have occurred in California before the arrival of Europeans. Introduced taxa have been introduced from other parts of the world and have persisted or in some cases have naturalized and become aggressively competitive. The percentage of introduced taxa in this flora is 27.5%. There are 18 species that are endemic and occur only in Sonoma County.

LIST OF LARGEST GENERA

Number of specific and infraspecific taxa in each genus

Carex	39	Arctostaphylos	17	Clarkia	13
Trifolium	38	Ranunculus	15	Mimulus	13
Lupinus	21	Lathyrus	14	Polygonum	18
Lotus	20	Navarretia	14	Bromus	17
Juncus	21				

MAJOR GROUPS

(Keys by Walter Knight except where noted otherwise.)

1. Plants without seeds or flowers, reproducing by 1-celled spores borne in sporangia . . . . . . (p 19) Ferns and fern allies
1: Plants with seeds usually produced by cones or flowers
   2. Seeds produced in cones or at the ends of short twigs . . . . . . (p 27) Gymnosperms
   2: Seeds enclosed in mature pistils; plants bearing true flowers . . . . . . (p 31) Angiosperms
      3. Cotyledons 2; leaves mostly netted-veined; flowers usually 4-5-merous . . . . . . (p 31) Dicots
      3: Cotyledons 1; leaves mostly parallel-veined; flowers usually 3 and 6-merous . . . . (p 248) Monocots

FERNS AND FERN ALLIES

1. Plants neither fernlike nor free-floating aquatics, the leaves mostly minute or ± scalelike (except in *Isoetes)*, never circinate; fertile leaves often clustered in conelike strobili.
   2. Leaves all basal . . . . . . *(p 22) Isoetaceae*
   2: Leaves cauline
      3. Stems hollow, jointed, leaves in whorls, separated by leafless internodes . . . . . . *(p 22) Equisetaceae*
      3: Stems solid, leafy throughout . . . . . . *(p 25) Selaginellaceae*
1: Plants with circinate, fernlike, or 4-leaf clover-like leaves, or free-floating aquatics with small overlapping leaves
   4. Plants not particularly fernlike, but floating aquatics or creeping on mud; bearing 2 kinds of spores in round to ovoid bony sporocarps near the base of leaf-stalks or on the underside of branches
      5. Leaves circinate, petioled, the blades 4-leaf cloverlike, (lacking on *Pilularia)*; plants mostly creeping on and rooting in the mud . . . . . . *(p 22) Marsileaceae*
      5: Leaves not circinate, sessile, overlapping; plants mostly floating on water . . . . *(p 19) Azollaceae*
   4: Plants fernlike, terrestrial or on mud, not floating; bearing 1 kind of spore in sporangia
      6. Sporangia large, globular, without an annulus, borne in a panicle from the base of green blade which appears lateral; rootstock almost none; frond generally one, seeming to arise from cluster of fleshy roots . . . . . . *(p 23) Ophioglossaceae*
      6: Sporangia minute, stalked, with an annulus of thick-walled cells on 1 side and produced on the underside of the leaves; rootstock developed; fronds > 1; roots not fleshy
         7. Sori dorsal upon the veins, separate, not marginal
            8. Indusia inflated, bladderlike . . . . . . *(p 20) Dryopteridaceae*
            8: Indusia, if present, not so
               9. Sori round to oval
                  10. Stipes continuous with the rhizome; blades 1-3-pinnate; indusia present . . . . . . *(p 20) Dryopteridaceae*
                  10: Stipes jointed to the rhizome; blades pinnatifid or pinnatisect; indusia wanting . . . . . . *(p 23) Polypodiaceae*
               9: Sori oblong or linear to lunate or horseshoe-shaped
                  11. Venation netted, sori large . . . . . . *(p 20) Blechnaceae*
                  11: Venation not netted; sori small . . . . . . *(p 20) Dryopteridaceae*
         7: Sori marginal or submarginal (borne at or near the apices of the veins) or in a few cases the sporangia decurrent on the veins or completely covering them
            12. Sporangia following the veins throughout . . . . . . *(p 24) Pteridaceae*
            12: Sporangia borne at or near the apices of the veins
               13. Plants with fronds to 150 cm, coarse; sporangia (very difficult to see) borne on a veinlike receptacle connecting the ends of the veins; indusium double, the inner one minute, concealed . . . . . . *(p 20) Dennstaedtiaceae*
               13: Plants with fronds to 60 (80) cm; sporangia not borne on a special receptacle; indusia (if any) single . . . . . . *(p 24) Pteridaceae*

AZOLLACEAE. MOSQUITO OR WATER FERN FAMILY

Azolla filiculoides Lamarck. MOSQUITO FERN. AMERICAN WATER FERN. Rare, or rarely reported, quiet ponds: The Sea Ranch, *Wells* (1984); Button Ranch on Walker Road (U. C. property) between Petaluma-Valley Ford Road and Pepper Road, acc. Connors (1992); locally common in Laguna de Santa Rosa (when there is water in it), acc. Harrison (1992); Fairfield Osborn (S); Willow Brook, *Knight* (1986); Annadel; Spring Lake; Pepperwood. < 5300 ft.

BLECHNACEAE

1. Blade deeply pinnately lobed to 1-pinnate with unlobed 1 leaflets; leaves of 2 kinds, fertile and sterile markedly dissimilar; sori 2 per 1 leaflet, linear, > 10 x longer than wide, 1 along each side of 1 leaflet midrib . *Blechnum*
1: Blade 1-pinnate with deeply pinnately lobed 1 leaflets; fertile and sterile leaves all ± alike; sori many per lobe, oblong, 2-4 x longer than wide, end-to-end along each side of lobe midrib, some oblong to linear along 1 leaflet midrib as well . . . . *Woodwardia*

Blechnum spicant (Linnaeus) J. Smith. DEER FERN. Occasional, coast and coastal hills, moist or wet ground along streams and about marshes: Timber Ridge Road at The Sea Ranch, *Wells* (1981); along road n. of Plantation, acc. Baker; Salt Point; Willow Creek. The Pacific Coast plant has been treated differently from the European type by some workers, but most regard it as the same species. <10500 ft. -*B. s. ssp. nipponicum* (Kunze) Löve & Löve.

Woodwardia fimbriata J. E. Smith. WESTERN CHAIN FERN. Locally common, moist or wet soil along streams, marshy ground: The Sea Ranch; Salt Point; ca 8 mi. from Cazadero on King Ridge Road, common for ca ¼ mi., acc. Roderick (1980); Bodega Head on road to parking lot, acc. Best (1991); ca 1 mi. nw. of Camp Meeker, acc. J. & B. Guggolz (1991); Pitkin Marsh, *Rubtzoff* (1951); Fairfield Osborn (H); Bouverie; Cooper's Grove on Sonoma Mtn. Road, acc. Knight (1989); Sugarloaf Ridge; Annadel; Gates Road off Calistoga Road, acc. Wells et. al. (1984); Pepperwood; Ida Clayton Road, acc. Knight (1989); ca 2000 ft. elev. Anderson Trail on Black Mtn. between Geyser Peak and upper end of Pine Flat Road, acc. J. & B. Guggolz (1992). <7600 ft.

DENNSTAEDTIACEAE

Pteridium aquilinum (Linnaeus) Kuhn var. **pubescens** Underwood. BRACKEN. Common, grassy hills, brushy canyons, open woods: The Sea Ranch, *J. H. Thomas* (1976); Salt Point; Willow Creek; Bodega Head; Duncans Mills Marsh, *Rubtzoff* (1957); Pitkin Marsh, *Rubtzoff* (1951); Crane Can., *Knight* (1963); Fairfield Osborn (H); Tresch Ranch, Walker Road near Two Rock, acc. J. & B. Guggolz (1992); w. of Sears Point Raceway in Tolay Cr. drainage, acc. Patterson (1990); Bouverie; Sugarloaf Ridge; Mt. Hood, *Knight* (1964); Annadel; Spring Lake; Pepperwood; chaparral area sw. of Bradford Mtn. 1½ mi. w. of Dry Cr., acc. J. & B. Guggolz (1992); Lake Sonoma area; ca 2000 ft. elev. Anderson Trail on Black Mtn. between Geyser Peak and upper end of Pine Flat Road, acc. J. & B. Guggolz (1992). <10600 ft.

DRYOPTERIDACEAE

1. Indusium inflated, bladderlike . . . . *Cystopteris*
1: Indusium, if present, not so
    2. Sori oblong or linear to lunate or horseshoe-shaped . . . . *Athyrium*
    2: Sori round to oval
        3. Indusium roundish-reniform, attached at its sinus . . . . *Dryopteris*
        3: Indusium orbicular, centrally peltate . . . . *Polystichum*

Athyrium filix-femina (Linnaeus) Roth var. **cyclosorum** Ruprecht. WESTERN LADY FERN. Fairly common, marshes, coastal hills, wet ground, shade or partial shade: The Sea Ranch; Salt Point; Duncans Mills Marsh, *Rubtzoff* (1957); Willow Creek; Willow Cr. Marsh, *Rubtzoff* (1959); lower Salmon Cr., *Rubtzoff* (1962); 7 ft. elev. Bodega Head, *Clifton* (1980 PUA); Pitkin Marsh, *Rubtzoff* (1951); Laguna de Santa Rosa near Llano Road, *Rubtzoff* (1958); Bouverie; Annadel. < 10500 ft. *A. filix-femina* var. *sitchense* Ruprecht. *A. filix-femina* var. *californicum* Butters.

Cystopteris fragilis (Linnaeus) Bernhardi. BRITTLE FERN, FRAGILE FERN. Occasional, streamsides; moist, rocky canyon slopes, sun or shade: Bouverie; Sonoma Cr., acc. Baker; n. of Lawndale Road near Kenwood, *Quibell* (1982 NCC); foot of Mt. Hood, *Heller* (1902); Annadel; Pepperwood; Skaggs Springs, *Hauver* (1915); Lake Sonoma area; 1500 ft. elev. w. flank Mt. St. Helena, *Grummer & Hemphill* (1981 PUA). 160-12600 ft.

Dryopteris

1. Leaves ± 3-pinnate, segments deeply pinnately lobed . . . . . *D. expansa*
1: Leaves 1-2-pinnate, segments deeply pinnately lobed or not . . . . . *D. arguta*

Dryopteris arguta (Kaulfuss) Maxon. COASTAL WOOD FERN. Common, wooded summer-dry slopes: 6-8 mi. e. of Stewarts Point, acc. Roderick (1989); Salt Point; Cazadero, *Graff* (1926); Fairfield Osborn (H); Button Ranch on Walker Road (U. C. property) between Petaluma-Valley Ford Road and Pepper Road, acc. Patterson (1990); Petaluma, *Congdon* (1880); Bouverie; Sugarloaf Ridge; Annadel; Santa Rosa Cr. e. of Santa Rosa, *Heller* (1902); Spring Lake; Pepperwood; Fitch Mtn., *Rubtzoff* (1952); Mill Cr. Road, *Knight* (1964); chaparral area sw. of Bradford Mtn. 1½ mi. w. of Dry Cr., acc. J. & B. Guggolz (1992); Cloverdale, *Rattan* (1877). < 8250 ft.

Dryopteris expansa (Presl) Fraser-Jenkins & Jermy. SPREADING WOOD FERN. Rare, coastal woods: Hwy. 1 four mi. s. of Stewarts Point, acc. Baker; Duncans Mills, *Davy* (1896); Pepperwood. Blades basically 3 x pinnate. < 1700 ft. *D. austriaca* (Jacquin) Woynar. *D. dilatata* (Hoffman) A. Gray, misapplied.

Polystichum

1. Fronds once-pinnate, pinnae variously serrate to evenly incised, never pinnately lobed to pinnatifid
   2. Pinnae narrow-lanceolate . . . . . *P. munitum*
   2: Pinnae obliquely imbricate . . . . . *P. imbricans*
1: Fronds more compound, with at least the lower pinnae pinnately lobed or divided at the base
   3. Pinnae mostly pinnate, the lowest pinnules again divided or cleft . . . . . *P. dudleyi*
   3: Pinnae divided or lobed, but not totally pinnate . . . . . *P. californicum*

Polystichum californicum (D. C. Eaton) Diels. CALIFORNIA SHIELD FERN. Occasional, hills, deep moist shaded canyons: 500 ft. elev. Ward Cr. in Cazadero quad., *Clifton* (1977 PUA); Sweetwater Springs Road, acc. Knight & Edwards (1986); 1 mi. from top of Sonoma Mtn. Road, *Lowry* (1983 SFSU); Bouverie, *Knight & Ellman* (1986); Gates Road off Calistoga Road, *Wells et. al.* (1984); Hood Mtn. Regional Park, acc. Edwards (1988); Annadel; Mill Cr. Road, *Knight & Edwards* (1986); Pepperwood; Lake Sonoma area; first large rocky outcrop encountered from Jimtown end of road, acc. Edwards (1992). < 2700 ft. This species is a tetraploid hybrid between *P. dudleyi* and *P. munitum.* It backcrosses with both parents, acc. *Lellinger,* p. 276 (1985,).

Polystichum dudleyi Maxon. DUDLEY SHIELD FERN. Rare, deep canyons: near bridge on Gualala River near Annapolis, acc. Baker; Cazadero, acc. D. H. Wagner; Bouverie, *Knight & Ellman* (1986); Mill Cr. Road, *Knight & Edwards* (1986 NCC); Skaggs Springs, *Hawver* (1915). < 1700 ft.

Polystichum imbricans (D. C. Eaton) D. H. Wagner. Rock Sword Fern. Rarely detected or reported, rocky ground, usually open dry places, rock crevices, generally at higher elevations than *P. munitum:* near summit of Bennett Valley Road, acc. Baker; Geyser Valley, acc. D. H. Wagner. 600-8300 ft. -*P. m.* var. *I.* (D. C. Eaton) Maxon.

Polystichum munitum (Kaulfuss) Presl. WESTERN SWORD FERN. Common, wooded slopes and canyons: The Sea Ranch; Salt Point; Cazadero, *Congdon* (1890), *Graff* (1926); Willow Creek; Bodega Head; Guerneville, *Rattan* (1877); Pitkin Marsh, *Rubtzoff* (1952); Fairfield Osborn (S); Tresch Ranch, Walker Road near Two Rock, acc. J. & B. Guggolz (1992); Bouverie; Sugarloaf Ridge; Cooper's Grove on Sonoma Mtn. Road, acc. Knight (1990); Mt. Hood, *Knight* (1964); Annadel; Spring Lake; Pepperwood; Franz Valley, *Hemphill* (1936 PUA). < 5300 ft.

EQUISETACEAE. HORSETAIL FAMILY

Equisetum. (Principally adapted from key by *R. F. Hoover* [1970]).

1. Erect shoots of 2 distinct sorts: one unbranched, pale and ending in a strobilus (conelike cluster of sporophylls); the other with whorls of slender branches, bright green and without strobilus
   2. Sterile stems mostly > 5 mm thick (except near tip), over 30 cm tall at maturity; strobilus 4-10 cm long . . . . *E. telmateia* ssp. *braunii*
   2: Sterile stems 3 mm thick or less, rarely >30 cm tall at maturity; strobilus ca 2-3 cm long . . . *E. arvense*

1: Erect shoots ± alike, simple or unbranched, all bright green
   3. Main erect shoots with ca 15-25 longitudinal ridges, dying down to base in winter . . . . . . *E. laevigatum*
   3: Main erect shoots with ca 25-42 longitudinal ridges, evergreen at least in lower portion
      4. Shoots in lower part remaining green in winter, with ca 25-34 ridges . . . . *E. ferrisii*
      4: Shoots wholly green in winter with 35-42 ridges . . . . *E. hyemale ssp. affine*

Equisetum arvense Linnaeus. COMMON HORSETAIL. Uncommon, moist or wet soil along streams, alluvial flats, shaded banks: The Sea Ranch; Salt Point; Willow Creek; Bodega Head; Copeland Cr. on Sonoma State Univ. campus, *Self* (1979 NCC); near Kenwood, acc. Baker. < 10000 ft.

Equisetum x **ferrisii** Clute emend. Hauke. FERRIS' SCOURING RUSH. Moist, deep soil, thickets or open woods: Sonoma State Univ. campus, *Howell* (1987); Pepperwood. -*E. hyemale* ssp. *affine* x *E. laevigatum.*

Equisetum hyemale Linnaeus ssp. **affine** (Engelmann) Calder & R. H. Taylor. GIANT SCOURING RUSH. Uncommon, colonies, moist brushy places along streams: Salt Point; Goat Rock State Beach, *Wells* (1981); Bodega Head; Fairfield Osborn (S). < 10000 ft.

Equisetum laevigatum A. Braun. NARROW HORSETAIL. Relatively common, dry, often disturbed places, open or in partial shade: railroad right-of-way at Santa Rosa High School (now Junior College), acc. Baker; Spring Lake; s. of College Ave. in Laguna de Santa Rosa drainage, *Harrison* (1974 NCC); Pepperwood; Lake Sonoma area; ca 2000 ft. elev. Anderson Trail on Black Mtn. between Geyser Peak and upper end of Pine Flat Road, acc. J. & B. Guggolz (1992); 2900 ft. elev. w. flank Mt. St. Helena, *Grummer* (1979 PUA). < 10000 ft.

Equisetum telmateia Ehrhart ssp. **braunii** (Milde) R. L. Hauke. GIANT HORSETAIL. Common, moist soil, woods and thickets, frequently near springs and along streams: The Sea Ranch, *J. H. Thomas* (1976); Stewarts Point, *Porter* (1967); Salt Point; Goat Rock Beach, acc. Wells (1983); Bodega Head; Duncans Mills Marsh, *Rubtzoff* (1957); ca 1 mi. nw. of Camp Meeker, *J. & B. Guggolz* (1991 NCC); Forestville Marsh, *Rubtzoff* (1955); Guerneville Marsh, *Rubtzoff* (1954); Fairfield Osborn (H); Pepperwood; 2190 ft. elev. Squaw Cr., acc. Clifton (1984). < 3300 ft.

ISOETACEAE. QUILLWORT FAMILY

Isoetes

1. Translucent membrane covering < 75% of sporangium . . . . *I. howellii*
1: Translucent membrane covering > 75% of sporangium . . . . *I. nuttallii*

Isoetes howellii Engelmann. HOWELL'S QUILLWORT. Occasional, swampy places, shallow ponds that dry during summer: near Wright School, acc. Baker; laguna e. of Sebastopol, *Howell* (1937); Todd Road Preserve; near Kenwood, *Howell* (1935); Bennett Valley Road summit, *Howell* (1934); Spring Lake; Laguna de Santa Rosa drainage near Guerneville Road at Piner Road and Todd Road, *Harrison* (1978 NCC); Pepperwood. Usually < 4600 ft.

Isoetes nuttallii Engelmann. NUTTALL'S QUILLWORT. Occasional, wet or moist soil, grassland, along ephemeral brooks: Pitkin Marsh, *Howell* (1937); Cunningham Marsh, *Howell* (1947); Fairfield Osborn (S); Adobe Can. above Kenwood Springs, acc. Baker. Usually < 4600 ft.

MARSILEACEAE. PEPPERWORT FAMILY

1. Leaves 4-foliolate, petioled; sporocarps ovoid; plants hairy . . . . *Marsilea*
1: Leaves filiform, lacking a definite blade; sporocarps globose; plants glabrous . . . . *Pilularia*

Marsilea vestita Hooker & Greville. CLOVER-FERN. Rare, muddy banks, edge of ponds, vernal areas, swale bottoms: Laguna de Santa Rosa near Roseland Cr. flood control channel, *Harrison* (1978 NCC); Laguna de Santa Rosa between Hall Road and River Road, acc. Rubtzoff (1978). < 7200 ft. -*M. mucronata* A. Braun.

Pilularia americana A. Braun. AMERICAN PILLWORT. Rare, heavy soil, vernal pools; s. of Hwy. 12 in Laguna de Santa Rosa drainage, *Harrison* (1978 NCC); jct. Starr Road and Old Redwood Hwy, acc. J. & B. Guggolz (1995). < 5000 ft.

OPHIOGLOSSACEAE. ADDER'S-TONGUE FAMILY

Botrichium multifidum (S. Gmelin) Ruprecht. GRAPE-FERN. Rare, wet meadows, brushy flats: The Sea Ranch; Salt Point; Pitkin Marsh, *Howell* (1945), Rubtzoff (1951). < 9300 ft. -*B. silaifolium* Presl. *B. m.* (Gmelin) Ruprecht var. *intermedium* (D. C. Eaton) Farwell.

POLYPODIACEAE

Polypodium

1. Blades leathery; rhizome glaucous; sori 3-4 mm broad . . . . *P. scouleri*

1: Blades not leathery; rhizome not glaucous; sori 1-3 mm broad,rarely larger

  2. Blade shape ± deltate to ovate; frond aspect upright; lower blade surface not noticeably glaucous; high percentage of pinnae with veins netted; sori shape elongate-oval-round, not raised . . . . *P. californicum*

  2: Blade shape lanceolate-ovate, oblong-ovate; frond aspect ± drooping, arching; lower blade surface slightly glaucous or glaucous; pinnae with veins less netted; sori raised

    3. Frond aspect ± drooping; lower blade surface slightly glaucous; pinnae with veins less netted than *P. californicum;* sori oval . . . . *P. calirhiza*

    3: Frond aspect drooping, arching; lower blade surface glaucous; pinnae with veins not netted; sori shape round,rarely ± oval . . . . *P. glycyrrhiza*

Polypodium californicum Kaulfuss. CALIFORNIA POLYPODY. Common, rocks, shaded canyons, open hills, sometimes epiphytic: 98 ft. elev. Fort Ross, *Clifton* (1981 PUA); Willow Creek; Fairfield Osborn (S); Button Ranch on Walker Road (U. C. property) between Petaluma-Valley Ford Road and Pepper Road, acc. Patterson (1990); 1½ mi. n. of Sonoma, *Baker* (1944); Bouverie; Sugarloaf Ridge; Spring Lake; Pepperwood; Lake Sonoma area; ca 2000 ft. elev. Anderson Trail on Black Mtn. between Geyser Peak and upper end of Pine Flat Road, acc. J. & B. Guggolz (1992). < 5000 ft.

Polypodium calirhiza S. Whitmore & A. R. Smith. Probably common but as yet poorly collected, sometimes epiphytic: Bodega, base of e. cliffs, *Whitmore* (1980 UCSB); Adobe Can., *Michener & Bioletti* (1892 UC). < 4000 ft.

Polypodium calirhiza x **P. glycyrrhiza**. Occasional, canyons, shaded rocks, logs, sometimes epiphytic: Cazadero, *Graff* (1926); Pine Flat Road, *Knight* (1963).

Polypodium glycyrrhiza D. C. Eaton. LICORICE FERN. Common, canyons, shaded rocks, logs, sometimes epiphytic: Miller Cr. in Salt Point State Park; ca 1 mi. nw. of Camp Meeker, acc. J. & B. Guggolz (1991); Rodger's Cr. 1 mi. e. of Old Adobe Road, 4 mi. s. of Petaluma, *Buckett* (1959 NCC); Lichau Road n. of Penngrove, *Stocking* (1965 NCC), *Tanforan* (1968 NCC); St. Helena Road 5½ mi. from Calistoga Road, *Edwards* (1986 NCC). < 2000 ft.

Polypodium scouleri Hooker & Greville. LEATHER LEAF FERN. Occasional, along coast or short distance inland, rocks, tree trunks, epiphytic: The Sea Ranch, acc. Wells (1985); ravine at e. edge of Hwy. 1 n. of main entrance to Salt Point growing on Douglas-fir trunk, acc. Roderick (1989); Redwood Can. 1 mi. n. of Fort Ross, *Howell* (1930); on rocks along road to Goat Rock Beach from Hwy. 1, acc. Roderick (1972); Bodega Head; near Larkspur Rock at s. end of Bodega Bay, acc. Baker. < 2000 ft.

PTERIDACEAE

1. Sporangia following the veins throughout . . . . *Pentagramma*
1: Sporangia borne at or near the apices of the veins special receptacle; indusia (if any) single
  2. Sporangia borne on the under side of sharply reflexed membranous lobes . . . . *Adiantum*
  2: Sporangia not borne on the back of reflexed lobes
    3. Lower leaf surface with scales . . . . *Cheilanthes*
    3: Leaf lower surface without scales
      4: Sterile leaf segments ±sessile . . . . *Aspidotis*
      4: Sterile leaf segments (and fertile) stalked . . . . *Pellaea*

Adiantum

1. Leaf ± palmate-pinnate . . . . *A. aleuticum*
1: Leaf 2-3-pinnate . . . . *A. jordanii*

Adiantum aleuticum (Ruprecht) C. A. Paris. FIVE-FINGER FERN. Rare, deep shaded ravines: The Sea Ranch, *Wells* (1984); 1 mi. e. of Stewarts Point, acc. Roderick (1989); Salt Point; 3-4 mi. up Willow Cr. from Bridgehaven, acc. Roderick (1965); Guerneville, *Rattan* (1877); De Turk and Graham canyons near Glen Ellen, acc. Baker; Annadel; Pepperwood; 2700 ft. elev. w. flank Mt. St. Helena, *Grummer* (1980 PUA). < 11200 ft. -*A. pedatum* Linnaeus var. *a.* Ruprecht.

Adiantum jordanii C. Mueller. CALIFORNIA MAIDENHAIR FERN. Occasional, shaded rocky slopes: Salt Point at Miller Cr., *Wells* (1984); ca 2-3 mi. w. of Cazadero acc. Roderick (1987); ca 1 mi. nw. of Camp Meeker, *J. & B. Guggolz* (1991 NCC); Fairfield Osborn (S); Bouverie; Sugarloaf Ridge; Annadel; Spring Lake; Pepperwood; Mill Cr. Road w. of Healdsburg, *Knight* (1964); Skaggs Springs, *Hawver* (1916); Pine Flat Road, *Knight* (1963); Lake Sonoma area; property bordered by Chianti and Dutcher Cr. roads w. of US 101 and w. of Geyserville, acc. J. & B. Guggolz (1991); ca 2000 ft. elev. Anderson Trail on Black Mtn. between Geyser Peak and upper end of Pine Flat Road, acc. J. & B. Guggolz (1992). < ca 4000 ft.

Adiantum x tracyi C. C. Hall. Cultivated by Hugh B. Logan in his garden in Inverness, Marin Co., from plants originally collected near the S. Fork Gualala River and close to Guerneville. The hybrid has also been collected in Humboldt and Marin counties. -*A. jordanii* x *A. aleuticum.*

Aspidotis

1. Sporangia continuous along both sides of segment midvein; false indusia with many shallow, regular teeth; leaf segments linear, ± entire . . . . *A. densa*
1: Sporangia not continuous along both sides of segment midvein; false indusia with few-many deep, irregular teeth; leaf segments lanceolate to triangular
  2. False indusia ± as wide as long, free, 1-2 (5) per segment, with few coarse, irregular teeth or ± entire . . . . *A. californica*
  2: False indusia ± as wide as or wider than long, usually fused at bases, 3-5(7) per segment, with many coarse, irregular teeth or lobes . . . . *A. carlotta-halliae*

Aspidotis californica (Hooker) Copeland. CALIFORNIA LACE FERN. Rare, rocky places: hills e. of Sonoma Valley: Bouverie; e. of Agua Caliente in Sonoma Valley, *Howell* (1943); near Maplewood in Adobe Can., acc. Baker; Lake Sonoma area; 2700 ft. elev. w. flank Mt. St. Helena, *Grummer* (1979 PUA). 60-4500 ft. -*Cheilanthes c.* (Hooker) Mettenius.

Aspidotis x carlotta-halliae (Wagner & Gilbert) Lellinger. This is a rare fertile species derived from a hybrid between *A. Californica* and *A. densa.* Known only from the coast of California. There is no local information regarding this fern.

Aspidotis densa (Brackenridge) Lellinger. SERPENTINE FERN, INDIAN'S DREAM, CLIFF BRAKE. Common, rocky slopes, generally on serpentine: outcrop near jct. of Joy Road and Bittner Road between Bodega and Occidental, *Ferris* & Lorraine (1961); ca 1 mi. nw. of Camp Meeker, acc. J. & B. Guggolz (1991); hills near Guerneville, *Branscomb* & Roberts (1958); Morelli Lane near Occidental, *Knight* (1963); Adobe Can., acc. Baker; Pepperwood; Lake Sonoma area; ca 2000 ft. elev. Anderson Trail on Black Mtn. between Geyser Peak and upper end of Pine Flat Road, acc. J. & B. Guggolz (1992); Socrates Mine at end of Pine Flat Road, *LaPré & Dolan* (1987 UCR); Geysers geothermal area; *Quibell* (1980 NCC). 300-11299 ft. -*Onychium densum* Brackenridge.

Cheilanthes

1. Blades frequently 2-pinnate; segments mostly oblong, densely tomentose beneath . . . . *C. gracillima*
1: Blades 3-4-pinnate, segments usually roundish to oval,imbricate-paleaceous beneath
   2. Segments with no scales above, scales beneath large, much exceeding segments, whitish to light brown . . . . *C. covillei*
   2: Segments with few minute pale stellate scales above, scales beneath smaller, more numerous, darker, many reduced to entangled cilia . . . . *C. intertexta*

Cheilanthes covillei Maxon. Uncommon, rock crevices: 2080 ft. elev. Bothe-Napa State Park, *Grummer* (1981 PUA); 3100 ft. elev. w. flank of Mt. St. Helena, *Grummer* (1979 PUA). 1900-8000 ft.

Cheilanthes gracillima D. C. Eaton. LIP FERN. LACE FERN. Rare, rocky edges, crevices: Mt. St. Helena summit, *B. R. Jackson* (1924), *Howell* (1928), *Grummer* (1978 PUA); 3698 ft. elev. Cobb Mtn., *Jokerst* (1985 PUA). 1300-10600 ft.

Cheilanthes intertexta (Maxon) Maxon. COASTAL LIP FERN. Rare, rock crevices: Bouverie; Pine Flat Road, *Knight* (1964). 900-9300 ft.

Pellaea

1. Stipes 8-40 cm long, adaxially flattened, straw-colored or light brown to gray; laminae lanceolate to nearly triangular 8-40 cm long, 3-18 cm wide, mostly 3-pinnate, the ultimate segments elliptical, sessile to short-stalked, subarticulate, truncate to subcordate at base, not mucronate at apex . . . . *P. andromedifolia*
1: Stipes 4-20 cm long, adaxially flattened or very shallowly grooved, brown, reddish-brown, or purplish-black; laminae ovate-lanceolate to rarely suboblong 4-30 cm long, 1¼-11cm wide, 2- or 3-pinnate, the pinnules simple or ternate, the pinnules or ultimate segments linear 3-5 (-7) mm long, sessile, not articulate, strongly underrolled, obtuse to truncate at base, mucronate at apex . . . . *P. mucronata*

Pellaea andromedifolia (Kaulfuss) Fée. COFFEE FERN. Dry, rocky slopes, brush or woods: Fairfield Osborn (H); between Enterprise Road and Cooper's Grove near summit of Sonoma Mtn. Road, *Lowry* (1983 SFSU); Bouverie; Sugarloaf Ridge; Annadel; Spring Lake; Pepperwood; Fitch Mtn., *Rubtzoff* (1952); Lake Sonoma area; Skaggs Springs, *Hawver* (1915). 100-6000 ft.

Pellaea mucronata (D. C. Eaton) D. C. Eaton. BIRD'S-FOOT FERN. Occasional, dry, rocky, sunny slopes: ca 1 mi. nw. of Camp Meeker, acc. J. & B. Guggolz (1991); Bouverie; Cavedale Road in Hooker Can., *Knight* (1964); Sugarloaf Ridge; Annadel; Pepperwood; Fitch Mtn., *Rubtzoff* (1952); Lake Sonoma area; 2400 ft. elev. Little Geysers, *Malloch* (1977 PG&E). 70-8000 ft.

Pentagramma triangularis (Kaulfuss) Yatskievych, Windham & Wollenweber. GOLDBACK FERN. Common, summerdried slopes, partial shade: The Sea Ranch, acc. Wells (1985); Salt Point; Bodega Dunes, *Connors* (1991 BML); Fairfield Osborn (H); Bouverie; Sugarloaf Ridge; Annadel; Spring Lake; Pepperwood; between Knights Valley and Franz Valley, *Wiggins* (1932); Fitch Mtn., *Rubtzoff* (1952); Lake Sonoma area; Alexander Valley quad.; Pine Flat Road, *Knight* (1963); Ida Clayton Road quad. < 7600 ft. -*Pityrogramma t.* (Kaulfuss) Maxon.

SELAGINELLACEAE. SPIKE-MOSS FAMILY

Selaginella

1. Main stems above ground erect, rooting only at or near stem-base; rhizome conspicuous at base of erect stems . . . . *S. bigelovii*
1: Main stems above ground prostrate, pendent or spreading, rooting at branch forks; rhizome 0 or inconspicuous
   2. Foliage ± firm . . . . *S. wallacei*
   2: Foliage soft, mosslike . . . . *S. kraussiana*

Selaginella bigelovii L. Underwood. Uncommon, dry rocky banks: Larkspur Rock at s. end of Bodega Bay, *Eastwood & Howell* (1939); Bouverie; Sugarloaf Ridge; Annadel; rock crevices near Santa Rosa Cr., acc. Baker; Pepperwood; Lake Sonoma area; ca 2000 ft. elev. Anderson Trail on Black Mtn. between Geyser Peak and upper end of Pine Flat Road, acc. J. & B. Guggolz (1992). <6600 ft.

★ **Selaginella kraussiana** (Kunze) A. Braun. Naturalized: Griffith Woods in se. Vine Hill District, *Knight* (1965). Native of S. Africa. Ca 200 ft.

Selaginella wallacei Hieronymus. WALLACE MOSS-FERN. Shaded or partially shaded rocks: Possibly incorrectly reported from around Bouverie area. < 5600 ft.

GYMNOSPERMS

1. Seeds borne in woody cones
   2. Subtending bracts present, ± free from cone scales, (often concealed by them); leaves needlelike . . . . *(p 28) Pinaceae*
   2: Subtending bracts not conspicuous, fused to cone scales; leaves scalelike, subulate, or needlelike
      3. Leaf arrangement opposite or whorled; leaves scalelike or subulate . . . . *(p 27) Cupressaceae*
      3: Leaf arrangement alternate; leaves needlelike, subulate or scalelike *(p 30) Taxodiaceae*
1: Seed solitary at tip of short twig, partly or completely enclosed by subtending aril . . . . . . *(p 29) Taxaceae*

CUPRESSACEAE. CYPRESS FAMILY

1. Cones and their scales oblong, imbricated, maturing the 1st year; seeds 2 to each scale . . . . . . *Calocedrus*
1: Cones subglobose, their scales shield- or wedge-shaped, maturing in 1 or 2 years; seeds few to many on each scale . . . . *Cupressus*

Calocedrus decurrens (Torrey) Florin. INCENSE-CEDAR. Known from the e. slope of the Mayacmas Range in Napa and Lake counties (Cobb Mtn., Mt. St. Helena [*Sudworth* 1908]), but no Sonoma Co. record. To be watched for along streams in canyons, but only on serpentine. 600-8000 ft. Apr.-May. —*Libocedrus d.* Torrey.

Cupressus:

1. Dorsal surface of mature leaves usually bearing a gland or pit, which may be inactive, but generally exudes a clear resin which turns gray or whitish upon drying; foliage gray, glaucous gray-green, or dull, dusty-green, but never bright green . . . . *C. macnabiana*
1: Dorsal surface of mature leaves usually without an apparent gland or pit, sometimes present as an inconspicuous closed pit, or rarely a few are active and exude resin; foliage either bright green, dark green, or rarely dusty-green but never a real gray
   2. Branchlets rather thick and harsh; ♂ cones with central scales bearing 6 to 8 or even 10 pollen sacs; ♀ cones 25-35 mm long, mainly longer than broad, the scales 8 to 12; seeds rich, dark, shiny brown . . . *C. macrocarpa*
   2: Branchlets usually thin, not harsh; ♂ cones with central scales bearing not >6 pollen sacs; ♀ cones mostly usually 20 mm or less in diameter, nearly spherical, the scales 6 to 20; seeds dull brown, slightly glaucous, or dull to shiny black
      3. Tree with long, slender leader, and when in good soil growing to 30 m in height, or dwarfed in sterile soil and producing ♀ cones when < 1 m high; foliage a deep, dark, dull, blackish-green; seeds shiny black or dull brown . . . . *C. goveniana* ssp. *pigmaea*
      3: Tree with a compact pyramidal crown, or if sparse, sometimes irregular, but without long whiplike leader, rarely if ever, > 10 m. high; foliage dull, dusty green, or rarely a little glaucous, seeds glaucous, sometimes very conspicuously so . . . . *C. sargentii*

Cupressus goveniana Gordon ssp. **pigmaea** (Lemmon) J. Bartel. MENDOCINO CYPRESS, PYGMY CYPRESS. Rare, slopes and flats, southernmost known station: Salt Point. 100-1000 ft. —*C. pygmaea* (Lemmon) Sargent.

Cupressus macnabiana A. Murray. MACNAB CYPRESS. Rare, areas of serpentine and rhyolite tuff: The Cedars, *Dolan & LaPré* (1987 UCR), acc. *Kruckeberg* [1984]; Cavedale Road at head of Hooker Can. the southernmost station, acc. *Griffin & Critchfield*, Map 23, (1972), *Knight* (1964); Pine Flat Road, *Knight* (1963); Geysers geothermal area, *B. Coate* (1981); 3160 ft. elev. Socrates Mine, *Clifton* (1984 PUA); 2400 ft. elev. w. margin of Little Geysers, acc. Crane (1980), acc. Stone (1983). 900-3000 ft.

★ **Cupressus macrocarpa** Gordon. MONTEREY CYPRESS. Volunteers in the vicinity of planted trees may be expected in disturbed soil near coast: The Sea Ranch, *Wells* (1984), *J. H. Thomas* (1976); Bodega Head. Indigenous in San Mateo, Monterey and San Luis Obispo counties. < 100 ft.

Cupressus sargentii Jepson. SARGENT CYPRESS. Uncommon, serpentine chaparral: near mouth of Gualala River (Stewarts Point quad.); Annapolis, Fort Ross, Cazadero, Duncans Mills and Arched Rock quads.; Harrison Grade Road e. of Occidental, acc. Knight (1989); Morelli Lane e. of Occidental, *Knight* (1963); past Johnson Ranch at end of Pythian Road on Mt. Hood, *Knight* (1964); 2400 ft. elev. Mt. Hood, *Hemphill* (1979 PUA); Tombs Cr. quad.; 2400-3100 ft. elev. on slope e. of Little Sulphur Cr. in geothermal area, *Raiche* (1991 JEPS); Pine Flat Road, *Knight* (1963). 650-3300 ft.

PINACEAE. PINE FAMILY

1. Cones erect, disintegrating on tree; mature scales falling individually; leaves notched at apex . . . . . . *Abies*
1: Cones pendulous or reflexed, persistent on branches or falling to ground intact; scales persistent
  2. Leaves needlelike in fascicles 2-5; cones maturing after 1st year . . . . *Pinus*
  2: Leaves neither needlelike nor in fascicles; cones maturing the 1st year
    3. Bracts of cone-scales not exserted . . . . *Tsuga*
    3: Bracts of cone-scales obviously exserted . . . . *Pseudotsuga*

Abies grandis (Douglas) Lindley. GRAND FIR. Occasional, groves of coastal influence: near mouth of Gualala River (Stewarts Point quad.); Annapolis quad.; Salt Point; Fort Ross quad.; Willow Cr. Road about halfway between Bridgehaven and Coleman Valley Road, *Knight*. <2500 ft. Apr.-May.

Pinus: (Key ± acc. *Kozloff & Beidleman* [1994]).
1. Needles in bundles of 2
  2. Cones up to ca 9 cm long, usually closed at maturity; needles usually 10-12 cm long . . . . . *P. muricata*
  2: Cones < 5 cm long, open at maturity; needles mostly 3-6 cm long
    3. Tree low 3-10(-16) m tall . . . . *P. contorta* ssp. *contorta*
    3: Tree canelike dwarf 6-15 dm high . . . . *P. c.* ssp. *bolanderi*
1: Needles in bundles of 3 or 5
  4. Needles in bundles of 5 (cones up to 40 cm long, open at maturity) . . . . *P. lambertiana*
  4: Needles in bundles of 3
    5. Cones usually at least 15 cm long (open at maturity)
      6. Cones mostly 25-30 cm long; attachment of cones off-center at base; needles green . . *P. coulteri*
      6: Cones mostly 15-20 cm long; attachment of cones centered at base; needles gray-green *P. sabiniana*
    5: Cones rarely up to 15 cm long
      7. Needles commonly 12-20 cm long; cones open at maturity . . . . *P. ponderosa*
      7: Needles usually < 12 cm long; cones opening if exposed to fire or prolonged drying
        8. Cones ca 2 × long as wide . . . . *P. attenuata*
        8: Cones < 2 × long as wide . . . . *P. radiata*

Pinus attenuata Lemmon. KNOBCONE PINE. Locally common, dry rocky ridges and slopes: Bouverie; Cavedale Road, acc. Knight (1990); 1½ mi. from Hwy. 12, Trinity Road, *Ruckert* (1960);Sugarloaf Ridge; Mt. Hood, *Knight* (1964); Pepperwood; Alexander Valley quad.; 3-4 mi. up Ida Clayton Road from Kellogg, ca a 25-year-old forest, acc. Roderick (1989); 4298 ft. elev. Mt. St. Helena lookout, *Grummer* (1978 PUA); summit of Pine Flat Road, *Knight* (1963); ca 2000 ft. elev. Anderson Trail on Black Mtn. between Geyser Peak and upper end of Pine Flat Road, acc. J. & B. Guggolz (1992); PG&E geothermal units 13 and 18, acc. Quibell (1980); Little Geysers, acc. Quibell, (1980); along road between PG&E units 14 and 18, acc. Crane (1980); ¼ mi. nw. of Eureka Mine in geothermal area, acc. Crane (1980). < 6500 ft. Mar.-May.

Pinus contorta Loudon ssp. **bolanderi** (Parlatore) Critchfield. PYGMY PINE. Uncommon, pine barrens (pygmy forest): Salt Point. <850 ft. May.

Pinus contorta ssp. **contorta**. SHORE PINE, BEACH PINE. Coastal trees native s. to the vicinity of Manchester, Mendocino Co.; local records are of cultivated or perhaps naturalized trees: The Sea Ranch near Annapolis Road, *Wells* (1985). < 550 ft. May.

★ **Pinus coulteri** D. Don. COULTER PINE. Introduced: plantation of ca 50 trees ¼ mi. from summit of Cavedale-Trinity Road loop, acc. S. W. Edwards (1990); a few trees planted e. of Ida Clayton Road on w. flank of Mt. St. Helena, acc. Knight (1991). Native of sw. California.

Pinus lambertiana Douglas. SUGAR PINE. "A large sugar pine population occurs at the inner edge of the redwood forest in the Gualala River region" in nw. Sonoma Co. and adjacent Mendocino Co., acc. *Griffin & Critchfield*, p. 27, (1972): e. of The Sea Ranch beyond n. end of Timber Ridge Road, *Wells* (1981); many specimens near Annapolis most of which have been harvested, acc. Wells (1985), "near the head of Austin Cr." acc. (*Erythea* 4:152 [1896]); Tombs Cr. quad.; Mayacmas Range between Sonoma and Lake counties s. to Mt. St. Helena (where first reported by Alice Eastwood, [*Zoe* 5:36. 1900]); 4298 ft. elev. Mt. St. Helena lookout, *Grummer* (1978 PUA). < 10200 ft. May-June.

Pinus muricata D. Don. BISHOP PINE. Common, coastal bluffs and terraces n. of Russian River: Rockpile Cr. quad.; The Sea Ranch Timber Preserve, (depauperate), acc. Wells (1983); Annapolis and

Stewarts Point quads.; stressed forest in upper terrace Salt Point State Park, acc. Edwards (1990); Bodega Head (probably planted). < 1000 ft. Mar.-Apr.

Pinus ponderosa Lawson. WESTERN YELLOW PINE, PONDEROSA PINE. Occasional, forested flats, hills, mountains: Smith Cr. across from Northwood (Guernewood Park) on Russian River, *McMurphy* (1908); Pepperwood; Tombs Cr. quad.; ridge e. side of Dry Cr. Valley below Warm Springs Dam, acc. Quibell (1991); Alexander Valley and Ida Clayton Road quads.; 4298 ft. elev. Mt. St. Helena lookout, *Grummer* (1978 PUA); Pine Flat Road, *Knight* (1964); Socrates Mine quad.; PG&E geothermal unit 18, acc. Quibell (1980); 3300 ft. elev. crest of Mayacmas Mts. at Caldwell Pines, acc. Stone (1989). 500-7300 ft. May-June.

★ **Pinus radiata** D. Don. MONTEREY PINE. Introduced: planted as windbreaks in coastal areas, acc. Knight (1991). Native of central California coast.

Pinus sabiniana Douglas. DIGGER PINE, FOOTHILL PINE. Common, hills, dry slopes: Annadel; Spring Lake; Lake Sonoma, Geyserville, Alexander Valley and Ida Clayton Road quads.; ca 2000 ft. elev. Anderson Trail on Black Mtn. between Geyser Peak and upper end of Pine Flat Road, acc. J. & B. Guggolz (1992); PG&E geothermal unit 13, acc. Quibell (1981); 2600 ft. elev. Coldwater Cr. Can. ½ airmile sw. of Lake Co. line at Caldwell Pines, acc. Stone (1989); Big Sulphur Cr. and Cloverdale quads. 470-4700 ft. Apr.-May.

Pseudotsuga menziesii (Mirbel) Franco. DOUGLAS-FIR. Commonest coniferous tree in Sonoma Co.: all quads.; Salt Point; Willow Creek; Bodega Dunes, *Connors* (1991 BML); Sugarloaf Ridge; chaparral area sw. of Bradford Mtn 1½ mi. w. of Dry Cr., acc. J. & B. Guggolz (1992); Lake Sonoma area; 4298 ft. elev. Mt. St. Helena lookout, *Grummer* (1978 PUA); ca 2000 ft. elev. Anderson Trail on Black Mtn. between Geyser Peak and upper end of Pine Flat Road, acc. J. & B. Guggolz (1992); PG&E geothermal unit 13, acc. Quibell (1981). < 7200 ft. —*P. taxifolia* (Lambert) Britton.

Tsuga heterophylla (Rafinesque) Sargent. WESTERN HEMLOCK. Rare, Gualala River drainage in coastal woods, probably the southernmost distributional limits: 3½ mi. se. of Gualala, *B. F. Smith* (1962); valley crossing at Gualala River between The Sea Ranch and Annapolis, *Knight* (1964); between Cazadero and Plantation, acc. Baker. < 2400 ft.

TAXACEAE. YEW FAMILY

1. Aril reddish, open at top, maturing with seed in 2 seasons; leaf < 25 mm, spine or bristle tip 0 or < 0.5 mm . . . . *Taxus*

1: Aril green or purplish, closed at top, maturing with seed in 1 season; leaf > 25 mm, spine or bristle tip 1-1.5 mm . . . . *Torreya*

Taxus brevifolia Nuttall. PACIFIC YEW. Very rare or rarely detected on coastal slopes in wooded shaded canyons; the only Sonoma County record is that of Baker: "Gualala Canyon near Annapolis bridge." 35-5000 ft. Apr.-May. Pacific yew is widespread in Mendocino County and local at two stations in Marin County. A cancer treatment based on the chemical content of the bark is now actively pursued, and in some quarters, it is felt that the population will diminish considerably until a synthetic similar material is developed.

Torreya californica Torrey. CALIFORNIA NUTMEG. Locally common, wooded canyons, generally moist ground or near streams: e. of Stewarts Point on Stewarts Point-Skaggs Springs Road, *Raven* (1958); Guerneville, *Congdon* (1880); Camp Meeker, *Howell* (1924); Hooker Cr. s. of Cavedale Road, *Raven* (1954); Pepperwood, *Almeda & Breedlove* (1983); Sugarloaf Hill between Mt. St. Helena lookout and Ida Clayton Road, acc. Knight (1964); Lake Sonoma area; between Geyser Peak and Black Mtn., *McMurphy* (1909); Geysers geothermal area (relatively frequent), *Quibell* (1979). 100-6500 ft. Apr.-May.

TAXODIACEAE. BALD CYPRESS FAMILY

Sequoia sempervirens (D. Don) Endlicher. COAST REDWOOD. Common, coastal hills, mountains; occasional, moist canyons in the interior: The Sea Ranch; Villa Grande 1 mi. nw. of Monte Rio, *Taber* (1960); near Trenton along Russian River, *Heller* (1902); near Camp Meeker, *Howell* (1924); near end of Warrington Road off Petaluma Hill Road, acc. Roderick (1958); near end of Hamilton Road off Adobe Road, grove visible from Adobe Road, acc. Roderick (1980); Bouverie; up Trinity Road from Hwy. 12, *Ruckert* (1960); Sugarloaf Ridge; near Santa Rosa, *Congdon* (1880); Pepperwood; Knights Can. (probably near Knights Valley), *Klauss* (1893); chaparral area sw. of Bradford Mtn 1½ mi. w. of Dry Cr., acc. J. & B. Guggolz (1992); Lake Sonoma area. This is the state tree of California.

ANGIOSPERMS
DICOTS

1. Petals wanting or not evident (occasionally simulated by the petal-like calyx, as in *Sterculiaceae* [*Fremontodendron*] and *Nyctaginaceae*)
    2. Woody, well-developed shrubs and trees . . . . (p. 31) **Group 1**
    2: Herbaceous, occasionally suffrutescent . . . . (p. 32) **Group 2**
1: Petals present, evident
    3. Corolla of separate petals
        4. Stamens many, > twice as many as petals (p. 33) **Group 3**
        4: Stamens 10 or fewer, not > twice as many as petals (p. 33) **Group 4**
    3: Corolla of ± connate petals, as if growing together . . . . (p. 35) **Group 5**

A family designation might appear in two different groups if a character is sufficiently borderline to fit into each group.

Group 1. Woody plants lacking petals, or petals not evident

1. Leaves compound, blade divided into separate parts
    2. Plants woody vines climbing by twisting petioles (sepals petal-like) . . . *(p 200) [Clematis] Ranunculaceae*
    2: Plants trees or shrubs, not climbing
        3. Leaves alternate
            4. Fruit a nut . . . . *(p 159) Juglandaceae*
            4: Fruit a legume . . . . *(p 126) [Acacia, Albizia] Fabaceae*
        3: Leaves opposite
            5. Leaflets 3-5; samaras double, pubescent . . . . *(p 38) [Acer negundo] Aceraceae*
            5: Leaflets 3-7; samaras single, glabrous . . . . *(p 174) [Fraxinus] Oleaceae*
1: Leaves simple, blade entire, toothed or lobed
    6. Parasitic on tree and shrub branches and disconnected from the ground . . . . *(p 246) Viscaceae*
    6: Growing conventionally on the ground
        7. Plants climbing vines with large pipe-shaped flowers . . . . *(p 48) [Aristolochia] Aristolochiaceae*
        7: Plants not climbing or vinelike
            8. Flowers (at least ♂) in catkins or in catkin-like spikes
                9. Leaves alternate
                    10. Calyx present in ♂ flowers
                        11. Fruit a nut enveloped in a leafy involucre, or many nutlets in a cone-like catkin . . . . *(p 87) Betulaceae*
                        11: Fruit an acorn with a basal cup, or a spiny bur enveloping 1-3 nuts . *(p 147) Fagaceae*
                    10: Calyx none, ♂ flowers subtended by bracts
                        12. Fruit a capsule, seeds with a coma, ♀ flowers in catkins . . . . . *(p 220) Salicaceae*
                        12: Fruit 1-seeded without a coma; ♀ flowers solitary in axil of scales . *(p 173) Myricaceae*
                9: Leaves opposite (catkins drooping) . . . . *(p 151) Garryaceae*
            8: Flowers not in catkins
                13. Plants normally trees with an obvious main trunk
                    14. Leaves entire, lanceolate to ± ovate, aromatic when crushed
                        15. Ovary superior, maturing as a drupe; stamens in 3 series of 3 each, not vividly colored . . . . *(p 166) Lauraceae*
                        15: Ovary inferior, maturing into a woody capsule, stamens many, white to yellow or red . . . . *(p 173) [Eucalyptus] Myrtaceae*
                    14: Leaves toothed or lobed, ovate to roundish, not aromatic . . . . *(p 242) Ulmaceae*
                13: Plants normally shrubs with several stems from near base
                        16. Flowers 1-4 cm broad
                            17. Flowers mostly 2.5-4 cm across, leaves stellate-pubescent . *(p 241) Sterculiaceae*
                            17: Flowers barely 1 cm across; leaves not stellate-pubescent *(p 241) Thymelaeaceae*
                        16: Flowers < 1 cm
                            18. Fruit a utricle or an achene
                                19. Fruit a utricle . . . . *(p 111) [Suaeda] Chenopodiaceae*
                                19: Fruit an achene
                                    20. Style 1; ovary surrounded by a floral-tube, plumose in

fruit . . . . *(p 210) [Cercocarpus] Rosaceae*
20: Styles 2-3, ovary not surrounded by a floral-tube; not plumose in fruit . . . . *(p 190) [Eriogonum fasciculatum] Polygonaceae*
18: Fruit a drupe . . . . *(p 205) [Rhamnus] Rhamnaceae*

Group 2. Herbaceous plants without petals, or petals not evident

1. Plants aquatic ± submerged, or, on wet mud as the water evaporates
2. Leaves mostly dissected, whorled
3. Ovary superior; leaves 3-times forked . . . . *(p 111) Ceratophyllaceae*
3: Ovary inferior; leaves ± pinnatifid . . . . *(p 155) [Myriophyllum] Haloragaceae*
2: Leaves entire, frequently opposite and often clustered into terminal rosettes
4. Stamen 1
5. Ovary superior, splitting into 4 pieces when ripe; leaves 3-8 mm . . . . . *(p 101) Callitrichaceae*
5: Ovary inferior; fruit indehiscent; leaves 10-35 mm . . . . *(p 155) [Hippuris] Haloragaceae*
4: Stamens 4-5
6. Ovary superior . . . . *(p 199) [Glaux] Primulaceae*
6: Ovary inferior . . . . *(p 174) [Ludwigia palustris] Onagraceae*
1: Plants terrestrial, occasionally growing in ± wet sites
7. Ovary superior
8. Perianth lacking entirely
9. Flowers perfect, borne in a spike . . . . *(p 86) [Achlys] Berberidaceae*
9: Flowers imperfect, ♀ flowers ± stalked . . . . *(p 125) Euphorbiaceae*
8: Perianth present as a calyx, frequently petaloid
10. Pistils numerous . . . . *(p 200) [Anemone, Caltha, Thalictrum] Ranunculaceae*
10: Pistils 1 or 2
11. Flowers perigynous, ovary inside a hypanthium
12. Leaves alternate; hypanthium urn-shaped [*Sanguisorba*], or ± urn-shaped [*Aphanes*], or ± cone-shaped [*Acaena*] . . . . *(p 210) Rosaceae*
12: Leaves opposite; perianth petal-like . . . . *(p 173) [Abronia] Nyctaginaceae*
11: Flowers hypogynous; the ovary not so enclosed
13. Style simple
14. Calyx 5-6-parted; stamens 6-10, ♀ flowers lack perianth . . . . *(p 125) [Eremocarpus] Euphorbiaceae*
14: Calyx mostly 4-parted; stamens 4-5
15. Leaves mostly < 8 mm long . . . . *(p 242) [Soleirolia] Urticaceae*
15: Leaves larger
16. Fruit an achene *(p 242) [Urtica, Hesperocnide, Parietaria] Urticaceae*
16: Fruit usually a capsule [sometimes *Athysanus, Coronopus, Rorippa*] . . . . *(p 92) Brassicaceae*
13: Styles > 1
17. Ovary 1-loculed (must be observed under magnification)
18. Ovule or seed solitary; fruit mostly an achene or utricle; leaves frequently alternate
19. Flowers borne in a tubular or campanulate involucre . . . . *(p 190) [Chorizanthe, Eriogonum, Lastarriaea] Polygonaceae*
19: Flowers not borne in an involucre
20. Leaves with obvious stipular sheath above each node . . . . *(p 190) [Fagopyrum, Polygonum] Polygonaceae*
20: Leaves lacking such a sheath
21. Calyx mostly 6-cleft or of 6 sepals . . . . *(p 190) [Pterostegia, Rumex] Polygonaceae*
21: Calyx lobes or sepals 1, 4 or 5
22. Bracts subtending flowers, not scarious . . . . *(p 111) Chenopodiaceae*
22: Bracts subtending flowers scarious; plants not mealy, scurfy or fleshy . . . . *(p 39) Amaranthaceae*
18: Ovules or seeds > 1; fruit a capsule; leaves opposite
23. Leaves of each pair usually unequal; (sepals are ± petal-like on *Mollugo*) . . . . *(p 38) Aizoaceae*
23: Leaves of each pair usually ca = ... *(p 106) Caryophyllaceae*

17: Ovary > 1-loculed
24. Leaves whorled or opposite ..... *(p 38) [Glynus, Cypselea] Aizoaceae*
24: Leaves alternate .... *(p 125) Euphorbiaceae*
7: Ovary ± inferior
25. Flowers imperfect .... *(p 119) Datiscaceae*
25: Flowers perfect
26. Sepals 3, leaves 2 per stem; fruit a fleshy capsule ...... *(p 48) [Asarum] Aristolochiaceae*
26: Sepals 5, leaves many; fruit nutlike, angled, with 2-5 horns .. *(p 38) [Tetragonia] Aizoaceae*

Group 3. Plants with separate petals; stamens more than twice as many as petals

1. Ovary superior
2. Plants aquatic .... *(p 173) Nymphaeaceae*
2: Plants terrestrial, if in wet areas, not having floating leaves
3. Plants trees or shrubs
4. Stipules obvious or ± so
5. Stamens joined into a tube around the pistil; leaves palmately veined, blades rounded, often stellate-pubescent .... *(p 170) [Malvastrum] Malvaceae*
5: Stamens not united into a tube; leaves various, not stellate-pubescent .... *(p 210) Rosaceae*
4: Stipules none
6. Flowers red .... *(p 102) Calycanthaceae*
6: Flowers white or yellow
7. Petals 5; ovaries 5 .... *(p 210) Holodiscus] Rosaceae*
7: Petals 4 or 6; ovary 1 .... *(p 181) Papaveraceae*
3: Plants herbaceous, or woody at base only; not definite shrubs
8. Sepals 2, sometimes connate to tip (number of sepals in *Eschscholzia* really not evident), rarely 3-4
9. Plants fleshy; sepals persistent ..... *(p 196) [Portulaca, Calandrinia ciliata] Portulacaceae*
9: Plants not fleshy; sepals caducous at anthesis .... *(p 181) [Eschscholzia] Papaveraceae*
8: Sepals usually > 2
10. Sepals 3, rarely 2 .... *(p 181) [Meconella linearis, Platystemon] Papaveraceae*
10: Sepals 4-8 (rarely 2 or 3 on *Resedaceae*, rarely 3 on *[Carpobrotus] Aizoaceae*)
11. Leaves all in rosettes, fleshy .... *(p 196) [Lewisia] Portulacaceae*
11: Leaves otherwise
12. Stamens joined into a tube around the pistil .... *(p 170) Malvaceae*
12: Stamens otherwise
13. Flowers irregular, in racemes
14. Pistil 1, compound; maturing ovary at summit .. *(p 205) Resedaceae*
14: Pistils 3 or more, simple; maturing ovaries not at summit .... *(p 200) Ranunculaceae*
13: Flowers regular, solitary or in cymes or panicles
15. Stamens attached to receptacle
16. Leaves opposite .... *[Hypericum] (p 113) Clusiaceae*
16: Leaves alternate
17. Ovary 3 ±-celled .... *(p 113) Cistaceae*
17: Ovaries 1-celled .... *(p 200) Ranunculaceae*
15: Stamens attached to hypanthium .... *(p 210) [Fragaria, Geum, Potentilla] Rosaceae*
1: Ovary at least partly inferior
18. Petals numerous; plant succulent .... *(p 38) [Carpobrotus] Aizoaceae*
18: Petals few; plants not succulent
19. Plants shrubs; fruit fleshy *(p 210) [Amelanchier, Cotoneaster, Malus, Pyracantha, Pyrus] Rosaceae*
19: Plants herbs; fruit a capsule .... *(p 169) Loasaceae*

Group 4. Plants with separate petals; stamens not more than twice as many as petals

1. Pistils > 1, nearly or quite separate
2. Plants fleshy, at least the leaves so; (petals on *Parvisedum* slightly fused 116at base) *(p 116) Crassulaceae*
2: Plants not succulent
3. Trees; leaves ill-smelling when crushed; fruit a samara .... *(p 238) Simaroubaceae*
3: Herbs, leaves not ill-smelling when crushed; fruit otherwise
4. Leaves simple, entire, all basal; receptacle elongating in fruit . *(p 200) [Myosurus] Ranunculaceae*
4: Leaves lobed or compound, basal and cauline; receptacle not elongate in fruit

5. Stamens attached to hypanthium . . . . *(p 210) [Horkelia] Rosaceae*
5: Stamens attached to receptacle . . . . *(p 167) Limnanthaceae*
1: Pistil 1, of 1 or more ± united carpels
6. Plants climbing by means of tendrils; flowers regular . . . . *(p 246) Vitaceae*
6: Plants not climbing by tendrils; flowers regular or irregular
7. Styles 2-8, mostly separate to near base
8. Plants woody, well-developed trees or shrubs
9. Ovary superior
10. Leaves alternate; fruit a dry berrylike drupe . . . . *(p 40) Anacardiaceae*
10: Leaves opposite; fruit a samara . . . . *(p 38) [Acer macrophyllum] Aceraceae*
9: Ovary seeming ± inferior
11. Leaves alternate, with stipules
12. Leaves ± oblong, pinnately veined; stamens 10; fruit a pome
. . . . *(p 210) [Heteromeles] Rosaceae*
12: Leaves mostly palmately lobed; stamens 5; fruit a berry . *(p 154) Grossulariaceae*
11: Leaves opposite, without stipules (a subshrub) . . . *(p 183) [Whipplea] Philadelphaceae*
8: Plants herbaceous
13. Plants submerged aquatic or on wet mud flats and rooting at the nodes
14. Leaves ± finely dissected: submersed whorled, emergent opposite
. . . . *(p 155) [Myriophyllum] Haloragaceae*
14: Leaves entire, opposite . . . . *(p 120) Elatinaceae*
13: Plants terrestrial
15. Ovary superior
16. Leaves modified for insect catching, with stout glandular hairs on upper surface
. . . . *(p 119) Droseraceae*
16: Leaves not insectivorous
17. Leaves compound with 3 leaflets . . . . *(p 180) Oxalidaceae*
17: Leaves simple
18. Leaves clustered or in rosettes
19. Inflorescence an umbel, raceme or panicle (but not 1-sided)
. . . . *(p 196) Portulacaceae*
19: Inflorescence a panicle with 1-sided spikelike clusters
. . . . *(p 184) [Limonium] Plumbaginaceae*
18: Leaves otherwise
20. Leaves alternate
21. Plants ± fleshy; sepals 2
(p 196) [Calandrinia ciliata, Montia linearis] Portulacaceae
21: Plants not fleshy; sepals 4-5
22. Leaves linear to lanceolate . . . . *(p 167) Linaceae*
22: Leaves ovate, lobed, cordate
. . . . . . *(p 222) [Tiarella, Tolmiea] Saxifragaceae*
20: Leaves opposite, fascicled or whorled
23. Leaves revolute . . . . *(p 150) Frankeniaceae*
23: Leaves not revolute . . . . *(p 106) Caryophyllaceae*
15: Ovary ± inferior (*Saxifraga* occasionally more superior in fruit)
24. Flowers in panicles, usually ± scapose, *[Boykinia, Heuchera, Lithophragma, Tellima]* — [*Parnassia* and *Saxifraga* ovary superior to ± inferior]
. . . . *(p 222) Saxifragaceae*
24: Flowers in umbels, heads or spreading panicles
25. Flowers mostly in umbels, solitary to panicled; styles 5; fruit a berry or drupe . . . . *(p 48) Araliaceae*
25: Flowers mostly in compound umbels; styles 2; fruit 2 dry 1-seeded halves that separate from each other but usually remain attached for some time to a central axis
26. Herbage with hairs stellate . . . . . . *(p 40) [Bowlesia] Apiaceae*
26: Herbage without stellate hairs [all other genera] *(p 40) Apiaceae*
7: Style 1, occasionally divided toward the apex
27. Ovary inferior
28. Herbs or subshrubs . . . . *(p 174) Onagraceae*
28: Shrubs or trees
29. Stamens opposite petals . . . . *(p 205) Rhamnaceae*
29: Stamens alternate with petals . . . . *(p 116) Cornaceae*

27: Ovary superior, occasionally seeming inferior because enveloped in but not adnate to floral tube
  30. Plants well-developed trees or shrubs
    31. Flowers irregular, the petals not all alike
      32. Ovary 3-loculed; fruit a leathery capsule ...... *(p 156) Hippocastanaceae*
      32: Ovary 1-loculed; fruit a legume .... *(p 126) Fabaceae*
    31: Flowers regular, the petals basically similar
      33. Leaves compound
        34. Foliage very aromatic; (subshrub) .... *(p 220) Rutaceae*
        34: Foliage not aromatic
          35. Leaves opposite; fruit a samara *(p 174) [Fraxinus dipetala] Oleaceae*
          35: Leaves alternate; fruit not a samara *(p 86) [Berberis] Berberidaceae*
      33: Leaves simple
        36. Stamens 4-5
          37. Stamens opposite the petals .... *(p 205) [Ceanothus*, most *Rhamnus] Rhamnaceae*
          37: Stamens alternate with the petals .... *(p 110) Celastraceae*
        36: Stamens 10
          38. Leaves mostly 3-6 cm long .... *(p 120) Ericaceae*
          38: Leaves not > 1 cm long; (sometimes up to 15 stamens) .... *(p 210) [Adenostoma] Rosaceae*
  30: Plants herbaceous
    39. Petals 6 in 2 circles .... *(p 86) [Vancouveria] Berberidaceae*
    39: Petals otherwise
      40. Sepals 2 or 3
        41. Leaves deeply dissected; petals pouched at base .... *(p 181) [Dicentra] Papaveraceae*
        41: Leaves entire; petals free *(p 181) [Meconella californica] Papaveraceae*
      40: Sepals > 3
        42. Flowers nearly or quite regular
          43. Leaves compound .... *(p 247) Zygophyllaceae*
          43: Leaves simple and entire or divided, but not truly compound, (except rhizomal leaves and occasionally cauline leaves on *Cardamine* in *Brassicaceae*)
            44. Ovary appearing inferior, really superior and free in the floral tube; flowers 4-6 merous .... *(p 169) Lythraceae*
            44: Ovary appearing superior
              45. Sepals 4; filaments 6, 4, rarely 2, separate .... *(p 92) Brassicaceae*
              45: Sepals mostly 5; filaments 5, 8 or 10, often joined at base
                46. Mature carpels dividing from each other as 1-2-seeded units .... *(p 152) Geraniaceae*
                46: Mature carpels forming a many-seeded capsule .... *(p 120) Ericaceae*
        42: Flowers absolutely irregular, the petals dissimilar
          47. Leaves compound .... *(p 126) Fabaceae*
          47: Leaves simple
            48. Leaves not peltate
              49. Sepals nearly alike .... *(p 244) Violaceae*
              49: Two sepals enlarged as wings .. *(p 190) Polygalaceae*
            48: Leaves peltate .... *(p 241) Tropaeolaceae*

Group 5. Plants with petals ± connate, or appearing so

1. Ovary superior
  2. Stamens > *5*
    3. Corolla urn-shaped, bell-shaped, funnel-shaped, or with petals obviously united
      4. Plants mycoparasites, or parasites .... *(p 120) [Monotropoideae] Ericaceae*
      4: Plants not mycoparasites or parasites ... *(p 120) [Arbutus, Arclostaphylos, Gaultheria, Vaccinium, sometimes Rhododendron] Ericaceae*
    3: Corolla not urn-shaped, etc., or tubular, the petals united near the base only
      5. Pistils 4-5, plants succulent; petals only slightly fused at base on *Parvisedum*, petals ± free on *Sedum spathulifolium* .... *(p 116) [Crassula, Dudleya] Crassulaceae*

5: Pistil 1; plants not succulent
6. Corolla regular
7. Fruit a legume . . . . *(p 126) [Acacia, Albizia] Fabaceae*
7: Fruit not a legume
8. Petals free but fused at the base to the filament tube and falling together as a unit . . . . *(p 170) Malvaceae*
8: Petals more obviously connate or appearing so
9. Leaves simple . . . . *(p 196) [Calyptridium] Portulacaceae*
9: Leaves with leaflets 3 . . . . *(p 180) Oxalidaceae*
6: Corolla irregular
10. Leaves simple; petals 3, appearing to be 5 because of 2 petaloid sepals *(p 190) Polygalaceae*
10: Leaves dissected; petals 4 in 2 dissimilar pairs; sepals 2 *(p 181) [Dicentra] Papaveraceae*
2: Stamens not > 5
11. Plants parasitic, lacking chlorophyll
12. Plants delicate twining or trailing vines; disconnected from ground at maturity *(p 118) Cuscutaceae*
12: Plants not twining or trailing vines; root-parasites . . . . *(p 179) Orobanchacea*
11: Plants not or only partially parasitic and having chlorophyll
13. Corolla regular
14. Pistils 2; plants with milky juice
15. Inflorescence with flowers solitary or in cymes . . . . *(p 47) Apocynaceae*
15: Inflorescence umbel-like . . . . *(p 49) Asclepiadaceae*
14: Pistil 1; plants without milky juice
16. Ovary 1-loculed; plants ± fleshy
17. Plants prostrate, corolla none but calyx plaited, salverform, and colored like a corolla; style and stigma 1; fruit dry, hard . . . . *(p 173)* [See Group 2, 12:, *Abronia*] *Nyctaginaceae*
17: Plants vines; corolla with petals ± fused at base to form a shallow cup; styles 3, sometimes ± fused; fruit fleshy or papery, enclosed in a perianth . . . . *(p 86) Basellaceae*
16: Ovary and fruit otherwise; plants not fleshy
18. Petals shallowly connate, stamens opposite corolla lobes
19. Style 1; fruit a several-seeded capsule . . . . *(p 199) [Dodecatheon, Anagallis] Primulaceae*
19: Styles 5; fruit 1-seeded, membranous *(p 184) [Armeria] Plumbaginaceae*
18: Petals not shallowly connate, stamens alternate with corolla lobes, or didynamous
20. Foliage with translucent glands; twig tips sticky . . *(p 172) Myoporaceae*
20: Foliage otherwise
21. Corolla with small, dry-scarious, veinless lobes . . *(p 183) Plantaginaceae*
21: Corolla not so
22. Inflorescence a cyme; ovary 4-loculed
23. Leaves convex, stiffly dentate, with apical hard bristle; cyme axillary . . . . *(p 48) Aquifoliaceae*
23: Leaves plane, not stiffly dentate, without apical hard bristle; cyme terminal or flowers scattered, sometimes a raceme or panicle . . . . *(p 88) Boraginaceae*
22: Inflorescence not a cyme (except partly in *Hydrophyllaceae);* ovary 1- 2- or 3-loculed
24. Anthers opening by terminal pores . . . . *(p 120) [Rhododendron] Ericaceae*
24: Anthers otherwise
25. Ovary 3-loculed; style 3-cleft . . *(p 185) Polemoniaceae*
25: Ovary 1-2-celled; style not 3-cleft
26. Calyx 4-5 toothed or -cleft; style 1
27. Ovary 1-loculed; stigmas 2 . . . . *(p 151) Gentianaceae*
27: Ovary 2-loculed; stigma usually 1

28. Leaves opposite
29. Leaves with close white tomentum beneath . . . . *(p 101) Buddlejaceae*
29: Leaves densely silvery-scaly beneath . . . . *(p 174) [Olea] Oleaceae*
28: Leaves alternate . . . . *(p 239) Solanaceae*
26: Calyx of 5 distinct sepals, or sepals joined only at very base; styles 2 or 1
30. Plants twining or trailing, sometimes prostrate to erect; infl. not coiled . . . . *(p 114) Convolvulaceae*
30: Plants erect or diffuse; inflorescence often coiled . . . . *(p 156) Hydrophyllaceae*
13: Corolla ± irregular
31. Fruit of 2-4 nutlets
32. Ovary 4-lobed, style usually arising from base of ovary lobes; fruit 4 nutlets . . . . *(p 160) Lamiaceae*
32: Ovary entire or shallowly lobed, style arising from tip of ovary; fruit 2-4 nutlets . . . . *(p 243) Verbenaceae*
31: Fruit otherwise
33. Fruit an achene in the calyx . . . . *(p 183) Phrymaceae*
33: Fruit a capsule with > 1 seed
34. Ovary 1-loculed
35. Plants aquatic, often with alternate green branch systems (dissected leaves); corolla spurred; bladders borne on stem or "leaves" trap small organisms; stamens 2 . . . . *(p 166) Lentibulariaceae*
35: Plants terrestrial, leaves simple; corolla gibbous; stamens 4, plus one vestigial . . . . *(p 172) Martyniaceae*
34: Ovary 2-loculed . . . . *(p 225) Scrophulariaceae*
1: Ovary inferior or partly so
36. Stamens distinct
37. Leaves alternate . . . . . *(p 102) [Campanula, Githopsis, Heterocodon, Triodanus] Campanulaceae*
37: Leaves opposite or whorled
38. Stamens 1-3; flowers irregular; ovary 1-celled . . . . *(p 243) Valerianaceae*
38: Stamens 4-5, infrequently 2; flowers regular or irregular
39. Ovary 1-celled; flowers in dense involucrate heads . . . . *(p 119) Dipsacaceae*
39: Ovary 1-5-celled; flowers not in dense heads, not involucrate
40. Flowers regular; leaves whorled; ovary 2-celled; plants herbaceous; stems often with retrorse hairs . . . . *(p 218) Rubiaceae*
40: Flowers regular or irregular; leaves opposite; ovary 1-5-celled; plants woody; stems without retrorse hairs
41. Ovary mostly 2-5-loculed . . . . *(p 104) [Lonicera, Sambucus, Symphoricarpos] Caprifoliaceae*
41: Ovary 1-loculed . . . . *(p 104) [Viburnum] Caprifoliaceae*
36: Stamens united into a tube around the style
42. Flowers not in heads
43. Plants with tendrils, vinelike; leaves palmately veined
44. Plant with sharp stiff translucent harsh hairs, corolla large, yellow-orange . . . . *(p 117) [Cucurbita] Cucurbitaceae*
44: Plant ± scabrous or hairy, becoming glabrous; corolla small, white or cream-colored to yellowish-green . . . . *(p 117) [Marah] Cucurbitaceae*
43. Plants without tendrils. not vinelike; leaves pinnately veined . . . . *(p 102) [Downingia, Legenere, Lobelia, Nemcladus] Campanulaceae*
42: Flowers in involucrate heads . . . . *(p 50) Asteraceae*

ACERACEAE. MAPLE FAMILY

Acer

1. Leaf pinnately compound, leaflets 3-5; petals 0 . . . . *A. negundo* var. *californicum*
1: Leaf simple, palmately lobed; petals present . . . . *A. macrophyllum*

Acer macrophyllum Pursh. BIG-LEAF MAPLE. Common, streams, damp canyons: acc. Joseph White, resident unit ranger at Sugarloaf Ridge State Park, the largest big-leaf maples in California grow there, (acc. *S. F. Chronicle, Sun. Punch Sec., p. 6, 6/30/91*). Fort Ross, Cazadero, Duncans Mills and Arched Rock quads.; Neeley Hill area s. of Guerneville, *Rubtzoff* (1953); Russian River at Forest Hills, *Rubtzoff* (1980); Fairfield Osborn (H); 1000 ft. elev. Trinity Road, *Kawahara* (1968); Kenwood, *Nelson* (1933); hills se. of Santa Rosa, *Heller & Brown* (1902); Spring Lake Park, *Maxwell* (1990); Calistoga Road between St. Helena Road and Porter Cr. Road, *Knight* (1963); Spring Lake; Tombs Cr. quad.; 350-400 ft. elev. Old Skaggs Springs Road, *Utech, Sato & Ohara* (1990 CM); Lake Sonoma area; ca 2000 ft. elev. Anderson Trail on Black Mtn. between Geyser Peak and upper end of Pine Flat Road, acc. J. & B. Guggolz (1992); 2600 ft. elev. Coldwater Cr. Can. near Caldwell Pines, acc. Stone (1989). < 5000 ft. Apr.-May.

Acer negundo Linnaeus var. **californicum** (Torrey & A. Gray) Sargent. BOX ELDER. Occasional, in valleys, along streams: Freezeout Road near Duncans Mills, *Wells* (1987); ca 1 mi. nw. of Camp Meeker, acc. J. & B. Guggolz (1991); Vacation Beach near Guerneville, *Rubtzoff* (1953); Forest Hills across Russian River from Hacienda, *Rubtzoff* (1970); Pitkin Marsh, *Rubtzoff* (1957); Fairfield Osborn (*); Sonoma Mtn. Road near Enterprise Road, *Knight* (1964); Bouverie; Pepperwood; Lake Sonoma area; Geyser Peak, *Knight & Edwards* (1986); w. flank Mt. St. Helena, acc. Grummer & Hemphill (1983); Cloverdale quad. < 6000 ft. Mar.-Apr.

AIZOACEAE. CARPET-WEED FAMILY
[Sometimes MOLLUGINACEAE, and TETRAGONIACEAE, in part]

1. Ovary superior; petals none
    2. Leaves verticillate; calyx-tube lacking or nearly so; capsule loculicidal
        3. Plants glabrous; stamens 3-5; seeds without a coiled aril . . . . *Mollugo*
        3: Plants pubescent; stamens 5-10; seeds with a coiled aril . . . . *Glinus*
    2: Leaves opposite; calyx-tube evident; capsule circumscissile . . . . *Cypselea*
1: Ovary inferior
    4. Leaves flat; petals none . . . . *Tetragonia*
    4: Leaves thick, fleshy; petals many . . . . *Carpobrotus*

Carpobrotus

1. Leaves straight, not serrulate, 3-5 cm long; flowers rose-magenta, 3-5 cm broad . . . . *C. chilensis*
1: Leaves curved, serrulate on lower angle, ultimately being 7-10 cm long; flowers yellow or purple, 8-10 cm broad. . . . . *C. edulis*

★ **Carpobrotus chilensis** (Molina) N. E. Brown. SEA FIG. Common, dunes, coastal flats: The Sea Ranch; Salt Point; Bodega Head. < 350 ft. Apr.-September. Native of S. America. —*C. aequilaterus* (Haworth) N. E. Brown. *Mesembryanthemum chilense* Molina. Acc. Connors at Bodega Marine Laboratory, a suspected hybrid grows with both parents: *C. chilensis* and *C. edulis*.

★ **Carpobrotus edulis** (Linnaeus) N. E. Brown. HOTTENTOT FIG. Widely planted as a groundcover and soilbinder, along the coast, naturalized: The Sea Ranch, *J. H. Thomas* (1976); Salt Point; Bodega Head. <350 ft. Apr.-Oct. Native of S. Africa. —*Mesembryanthemum edule* Linnaeus.

★ **Cypselea humifusa** Turpin. Inconspicuous, probably overlooked, desiccated pond bottoms, reservoirs, streams: lake in Jack London State Park, *Wagnon* (1991 NCC); pond on River Road 3 mi. w. of Fulton, *Rubtzoff* (1969); Laguna de Santa Rosa ne. of Graton, *Rubtzoff* (1958); Bennett Mtn. Lake, *Rubtzoff* (1966). < 3300 ft. July-Oct. Introduced from W. Indies.

★ **Glinus lotoides** Linnaeus. Uncommon, pond margins, strands: Laguna de Santa Rosa, *Rubtzoff* (1966); pond on Russian River Road 3 mi. w. of Fulton, *Rubtzoff* (1969). < 3300 ft. June-Nov. Native of Europe.

★ **Mollugo verticillata** Linnaeus. INDIAN CHICKWEED. Uncommon, low places that are wet in the spring, along streams on strands and bottoms: Russian River beaches near Guerneville, *Rubtzoff* (1961); Fulton, *McMurphy* (1906). < 3300 ft. May-Nov. Perhaps introduced from tropical America.

★ **Tetragonia tetragonioides** (Pallas) Kuntze. NEW ZEALAND SPINACH. Infrequent, coastal, naturalized, disturbed and alkaline places: Gualala Point Reg. Park, *Knight* (1984); mouth of Salmon Cr., acc. Baker (1935); Bodega Harbor, *Connors* (1991 BML). < 300 ft. Apr.-September. Native of Australasia. —*T. expansa* Murray.

AMARANTHACEAE. AMARANTH FAMILY

Amaranthus

1. Fruit indeshiscent, usually inflated, thick-walled; plant prostrate . . . . . *A. deflexus*
1: Fruit circumscissile, thin-walled, plant prostrate to erect
  2. Inflorescence clusters axillary, throughout plant, not clearly terminal
    3. Stems ascending to erect, plant spheric; ♀ sepals 3, 0.7-2.2 mm . . . . . *A. albus*
    3: Stems prostrate; ♀ sepals (1) 3-5, 1-3.5 mm
      4. Seed 1.2-1.7 mm diameter; ♀ sepals 4-5, outer green along midrib . . . . . *A. blitoides*
      4: Seed 0.8-1.1 mm diameter; ♀ sepals 1-3, white and scabrous throughout *A. californicus*
  2: Inflorescence clearly terminal, sometimes also axillary
    5. Inner ♀ sepals spoonlike to obovate, tips obtuse to notched, usually irregularly toothed *A. retroflexus*
    5: Inner ♀ sepals narrow, occasionally spoonlike, tips acute to short-pointed
      6. ♀ bracts usually 2-3.5 mm; ♂ sepals and stamens 3-5; inflorescence of many short, crowded lateral spikes . . . . . *A. hybridus*
      6: ♀ bracts 4.5-8 mm; ♂ sepals and stamens 3-4; inflorescence usually of 1 long terminal spike and few lateral spikes . . . . . *A. powellii*

★ **Amaranthus albus** Linnaeus. TUMBLEWEED. Occasional, weed, waste and cultivated areas: Todd Road Preserve; Copeland Cr. at Sonoma State Univ. Campus, *Self* (1979 NCC); near Old Adobe e. of Petaluma, *Knight* (1963); 1 mi. sw. of Sonoma in cultivated ground, *Howell* (1957); 1 mi. n. of Sears Point, *Baker* (1948); Bennett Valley, *Knight* (1963); Pepperwood. < 7200 ft. June-Oct. Native of tropical America.

Amaranthus blitoides S. Watson. Occasional, native weed, cultivated ground, waste places: Vacation Beach, *Rubtzoff* (1953); Forest Hills across the river from Hacienda, *Rubtzoff* (1961); Willow Brook, *Knight* (1986); 1 mi. from US 101 on Sulphur Cr. Road, *Knight* (1963). < 7000 ft. July-Nov. —*A. graecizans* of authors, not Linnaeus.

Amaranthus californicus (Moquin) S. Watson. Occasional, dried mud, damp flats: Laguna de Santa Rosa n. of Cunningham, *Rubtzoff* (1958); Laguna de Santa Rosa n. of Sebastopol near Occidental Road, *Rubtzoff* (1958); 3 mi. w. of Fulton, *Rubtzoff* (1969); wet place at cemetery along Magnolia Ave. near Petaluma, acc. Knight (1964). < 9000 ft. July-Oct.

★ **Amaranthus deflexus** Linnaeus. Occasional, weed, along streets, gardens, waste places: lawn of Occidental Community Center, *Best & Wells* (1983); Santa Rosa Jr. Coll. campus at n. entrance to science building, *Baker* (1943); corner Marshall and Magnolia aves. near Petaluma, *Knight* (1963); Spring Lake. < 2250 ft. May-Nov. Naturalized from the Old World.

★ **Amaranthus hybridus** Linnaeus. Occasional, semicosmopolitan weed, cultivated ground, waste places: laguna n. of Sebastopol at Occidental Road crossing, *Howell* (1946); 1 mi. n. of Sebastopol, *Tryner* (1939 NCC); near Cunningham Marsh, *Baker* (1946 NCC). < 1000 ft. June-Nov. Native of Eurasia, N. Africa.

Amaranthus powellii S. Watson. Locally abundant, waste places: Vacation Beach near Guerneville, *Rubtzoff* (1956); Armstrong Woods Road 1 block from Main St., Guerneville in gutter crack, *Knight* (1990 NCC); w. of Sebastopol on Cherry Ridge near Grandview, *Best* (1983); laguna n. of Sebastopol, *Howell* (1946); Laguna de Santa Rosa, *Rubtzoff* (1958); Todd Road Preserve; Bennett Valley, *Knight* (1963); Ducker Cr. dike in Rincon Valley, *Wells* (1985). < 3000 ft. July-Oct.

★ **Amaranthus retroflexus** Linnaeus. REDROOT. Common, waste places, gardens, orchards: Copeland Cr. between Snyder Lane and County Club Drive in Rohnert Park, *Self* (1979 NCC); s. of Cotati firehouse, Stocking (1965 NCC); Sonoma State Univ. campus on Copeland Cr., *Self* (1979 NCC); Petaluma Hill Road 1 mi. n. of Sonoma State Univ., *Stocking* (1965 NCC); Wiggins Hill w. of Petaluma in cultivated field, *Knight* (1963); along Willow Brook between Old Redwood Hwy. and Corona Road, *Knight & Edwards* (1986); near Kenwood, *Baker* (1947 NCC); Bennett Valley, *Knight* (1963); Pepperwood. < 7600 ft. June-Nov. Native of tropical America.

ANACARDIACEAE. SUMAC FAMILY

1. Flowers whitish, in loose axillary panicles; fruit glabrous, whitish; leaves subglabrous, shining; branches stiff,erect, or sometimes climbing . . . . *Toxicodendron*
1: Flowers yellowish, in sessile spikes; fruit hairy, red; leaves usually pubescent; branches tending to be arched . . . . *Rhus*

Rhus trilobata Torrey & A. Gray. SQUAW BUSH. Rare, seepage areas: property bordered by Chianti and Dutcher Cr. roads w. of US 101 and w. of Geyserville, acc. J. & B. Guggolz (1991); 2720 ft. elev. Caldwell Ranch, acc. Clifton (1984); 3200 ft. elev. Caldwell Pines, acc. Clifton (1984). < 7000 ft. Mar.-Apr. —*R. t.* var. *quinata* Jepson.

Toxicodendron diversilobum (Torrey & A. Gray) E. Greene. POISON OAK. Common, ubiquitous, thickets, wooded slopes, variable habit: The Sea Ranch; Annapolis and Stewarts Point quads.; Salt Point; Willow Creek; Arched Rock quad.; Bodega Head; w. of Santa Rosa in Laguna de Santa Rosa drainage, *Harrison* (1978 NCC); Fairfield Osborn (*); Button Ranch on Walker Road (U. C. property) between Petaluma-Valley Ford Road and Pepper Road, acc. Patterson (1990); Bouverie; Sugarloaf Ridge; Spring Lake; along Santa Rosa Cr. in e. Santa Rosa, *Wagnon* (1989 NCC); Pepperwood; chaparral area sw. of Bradford Mtn. 1½ mi. w. of Dry Cr., acc. J. & B. Guggolz (1992); 350-400 ft. elev. Old Skaggs Springs Road, *Utech, Sato & Ohara* (1990 CM); Lake Sonoma area; ca 2000 ft. elev. Anderson Trail on Black Mtn. between Geyser Peak and upper end of Pine Flat Road, acc. J. & B. Guggolz (1992); Geyserville, Ida Clayton Road, Socrates Mine, Geothermal, Big Sulphur Cr. and Cloverdale quads. Causes an uncomfortable dermatitis for human beings. < 6000 ft. June-Nov. —*Rhus diversiloba* Torrey & A. Gray. Acc. Margot Patterson Doss, a poison-oak plant the size of a tree, estimated to be several hundred years old, is living in Sugarloaf Ridge State Park (*S. F. Chronicle, Sun. Punch Sec. p. 6, 6/30/91*).

APIACEAE [UMBELLIFERAE]. PARSLEY OR CARROT FAMILY

1. Inflorescence capitate . . . . *Eryngium*
1: Inflorescence umbellate
  2. Ovary and fruit with prickles, bristles, scales or tubercles
    3. Ovary and fruit armed with hooked bristles, tubercles or spines
      4. Perennial or biennial; flowers perfect and ♂ . . . . *Sanicula*
      4: Annual; flowers all perfect
        5. Plants glabrous; leaf-divisions ± elongate, filiform . . . . *Apiastrum*
        5: Plants ± pubescent; leaf-divisions shorter
          6. Involucre of obvious foliaceous bracts; leaves 3-4-pinnatisect; fruit bristly on ribs only . . . . *Yabea*
          6: Involucre none or of linear bracts; leaves pinnate to 3-4-pinnatisect; fruit bristly or tubercled throughout
            7. Fruit not beaked; bractlets of involucel > pedicels . . . . *Torilis*
            7: Fruit beaked; bractlets of involucel < pedicels . . . . *Anthriscus*
    3: Ovary and fruit armed with non-hooked bristles, tubercles or spines
      8. Fruit several x longer than broad; oil-tubes absent or obscure
        9. Fruit with a beak longer than the body . . . . *Scandix*
        9: Fruit not beaked or if so, the beak shorter than the body . . . . *Osmorhiza*
      8: Fruit not > twice as long as broad; oil-tubes evident . . . . *Daucus*
  2: Ovary and fruit not bearing prickles, bristles, scales or tubercles
    10. Leaves simple
      11. Ovary and fruit glabrous; leaves glabrous

12. Leaves with a definite roundish to ovate blade . . . . *Hydrocotyle*
12: Leaves reduced to hollow cylindrical jointed phyllodes . . . . *Lilaeopsis*
11: Ovary and fruit covered with stellate hairs; leaves ± stellate-pubescent . . . . *Bowlesia*
10: Leaves variously compound
13. Ribs of the fruit not prominently winged; fruit terete in cross-section or flattened ± laterally
14. Flowers yellow
15. Involucel absent; leaf-divisions filiform; plants with anise odor . . . . *Foeniculum*
15: Involucel present; leaf-divisions linear to ovate; plants lacking anise odor *Tauschia*
14: Flowers not yellow
16. Annual or perennial with celery odor . . . . *Apium*
16: Perennial or biennial; without celery odor, plants usually caulescent
17. Involucral bracts 3-parted to the middle into filiform divisions, closely reflexed . . . . *Ammi*
17: Involucral bracts entire or toothed, spreading, rarely reflexed, occasionally none
18. Stems purple-blotched; oil-tubes none or obscure; leaves decompound into small divsions . . . . *Conium*
18: Stems not purple-blotched; oil-tubes present; leaves pinnately or ternate-pinnately divided into mostly larger divisions
19. The leaves all once-pinnate
20. Stylopodium depressed; oil-tubes 1-3 in the intervals; ribs of fruit corky, equal . . . . *Sium*
20: Stylopodium conical; oil-tubes solitary in the intervals; ribs of fruit filiform, pericarp forming a continuous corky covering . . . . *Berula*
19: The leaves pinnately or ternate-pinnately divided or the upper once-pinnate
21. Stylopodium prominent; ribs not corky; dry ground and meadows
22. Leaf-divisions few, mostly entire; fruit-ribs filiform . . . . *Perideridia*
22: Leaf-divisions many, incised or serrate; fruit-ribs prominent to ± winged . . . . *Ligusticum*
21: Stylopodium none or low; ribs corky; marshes and stream banks
23. Styles 1/5 to 1/3 as long as the fruit; fruit broadly ovoid to roundish . . . . *Cicuta*
23: Styles ca ½ as long as fruit; fruit subcylindric . *Oenanthe*
13: Ribs of the fruit (all or some of them) winged; fruit ± dorsally compressed
24. Lateral ribs winged, dorsal ribs filiform
25. Corollas all alike; oil-tubes mostly as long as the fruit . . . . *Lomatium*
25: Corollas of marginal flowers of umbel larger than of central; oil tubes reaching only halfway to base of fruit . . . . *Heracleum*
24: Lateral, dorsal and intermediate ribs winged or prominent
26. Stems leafy, rather tall . . . . *Angelica*
26: Stems scarcely or not developed above ground
27. Leaf-divisions broad, 5-30 mm wide; maritime . . . . *Glehnia*
27: Leaf-divisions mostly < 5 mm wide; interior . . . . *Cymopterus*

★ **Ammi majus** Linnaeus. BISHOP'S WEED. Rare, weed, fields, waste places: w. of Stony Point Road in Laguna de Santa Rosa, *Harrison* (1976 NCC); within 100 yds. of nw. corner Manor Lane and Adobe Road near Petaluma Adobe, acc. Knight (1989); 5.3 mi. s. of El Verano, *Bacigalupi* (1955); Bennett Valley Road, *Rubtzoff* (1975). Usually < 3300 ft. May-July. Native of Eurasia.

Angelica

1. Plant 3-8 dm tall; umbel shape in mature fruit spherical . . . . *A. hendersonii*
1: Plant 6-18 dm tall; umbel shapes various
2. Umbel shape flat-topped or bowl-shaped; fruits glabrous . . . . *A. californica*
2: Umbel shape spherical; fruits scabrous to tomentose . . . . *A. tomentosa*

Angelica californica Jepson. Occasional, mostly in sandstone, shale or volcanics: Crane Can., *Stocking* (1963 NCC); Trinity Road, *Eastwood & Howell* (1938); Bennett Valley, *Baker* (1931 NCC); Fitch Mtn., *Rubtzoff* (1952); Lake Sonoma area; Cloverdale, *M. Jones* (1931). 50-5000 ft. June-Aug.

Angelica hendersonii Coulter & Rose. Common, coastal bluffs, flats, mostly in sandy-humus soil: The Sea Ranch; Salt Point; 1 mi. n. of Fort Ross, *Howell* (1930); Bodega Head. < 1000 ft. June-July. —*A. tomentosa* var. *h.* DiTomaso.

Angelica tomentosa S. Watson. Common, ± moist, shaded places, usually on serpentine: n. of Duvoul Cr. n. of Camp Meeker, *Edwards* (1988); Harrison Grade Road, *Knight* (1963); Bouverie; Trinity Grade, *Eastwood & Howell* (1939); Pepperwood; between Petrified Forest and Mark West Springs, *Abrams* (1916); Lake Sonoma area; Alder Glen Springs near Cloverdale, *Eastwood & Howell* (1938). 500-7800 ft. June-Aug.

★ **Anthriscus caucalis** Bieberstein. BUR-CHERVIL. Common, waste places: The Sea Ranch, *Wells* (1985); Bodega Head; Annadel, *Wells* (1984); Pepperwood, *Breedlove* (1981); Lake Sonoma Area; Sulphur Cr., *Baker* (1932 NCC). Naturalized from Europe. < 1500 ft. Apr.-June. —*A. scandicina* (Weber) Mansfeld. *A. neglecta* Boissier & Reuter var. *scandix* (Scopoli) Hylander.

Apiastrum angustifolium Nuttall. WILD-CELERY. Occasional, gravelly or loose soil, usually on slopes: Bodega Bay, *Heller & Brown* (1902); burned chaparral on s. slope of Mt. Hood, *Baker* (1947 NCC); Santa Rosa Cr. Can., *Baker* (1930 NCC); Pepperwood; Petrified Forest Road, *Baker* (1940 NCC); 5 mi. w. of Healdsburg, *Baker* (1928 NCC). < 5000 ft. Mar.-Apr.

★ **Apium graveolens** Linnaeus. CELERY. Rare, damp areas: Salt Point, *Wagnon & Vale* (1988 NCC); Duncans Mills, *M. Jones* (1882); Bodega Harbor Marsh, *Rubtzoff* (1955). Usually < 3300 ft. May-July. Naturalized from Europe.

Berula erecta (Hudson) Coville. WATER-PARSNIP. Common, sluggish water, marshes: Bodega Bay, *Best* (1986); Pitkin Marsh, *Rubtzoff* (1951 NCC); Perry Marsh, *Howell* (1945); 2 mi. n. of Sebastopol on Gravenstein Hwy., *Baker* (1945 NCC); Lichau Road (now near Fairfield Osborn), *Rubtzoff* (1964). < 6000 ft. July-Oct.

Bowlesia incana Ruiz López & Pavón. Occasional, partially shaded slopes: Petaluma Hill Road, *Koch* (1939 NCC); Petaluma, *Bolander* (1865 JEPS). 50-4000 ft. Mar.-May.

Cicuta douglasii (de Candolle) Coulter & Rose. WATER HEMLOCK. Occasional, wet soil, low marshy places, around springs: Bodega Harbor Marsh, *Rubtzoff* (1955); Pitkin Marsh, *Howell & Stacey* (1936); Forestville Marsh, *Rubtzoff* (1955); Laguna de Santa Rosa e. of Graton, *Rubtzoff* (1958); Laguna de Santa Rosa e. of Cunningham, *Rubtzoff* (1958); Cooper's Grove on Sonoma Mtn. Road, *Baker* (1928 NCC). < 8000 ft. June-Sept

★ **Conium maculatum** Linnaeus. POISON HEMLOCK. Very common; low, waste places: The Sea Ranch, *Wells* (1985); Willow Creek; marsh in dunes of Bodega Marine Lab. dormitory, *Rubtzoff* (1970); Laguna de Santa Rosa, *Rubtzoff* (1958); Todd Road Preserve; Lichau Road (now near Fairfield Osborn), *Rubtzoff* (1968); Wiggins Hill along Bodega Ave. w. of Petaluma, *Knight* (1963); Copeland Cr., Rohnert Park, *Self* (1979 NCC), *E. Murray* (1975 NCC); Button Ranch on Walker Road (U. C. property) between Petaluma-Valley Ford Road and Pepper Road, acc. Patterson (1990); w. of Sears Point Raceway in Tolay Cr. drainage, acc. Patterson (1990); Glen Ellen, *Baker* (1928 NCC); Sugarloaf Ridge; Spring Lake; Pepperwood. Usually < 3300 ft. Apr.-July. Naturalized from Europe.

Cymopterus terebinthinus (Hooker) M. E. Jones. Infrequent, brushy areas, serpentine barrens: Coleman Valley Road ca 6 mi. w. of Occidental, *Best* (1981); Bohemian Hwy. ca 3½ mi. n. of Occidental, *Best* (1981); Harrison Grade Road, *Best* (1981); Fred McMurray Ranch sw. of Healdsburg, *F. W. Hoffman* (1950 NCC); ridge on n. side of Dry Cr. Valley, *Raiche* (1985); Geyserville quad. 400-11000 ft. May-June. —*Pteryxia terebinthina* (Hooker) Coulter & Rose.

Daucus

1. Biennial, bract segments elongate, narrowly linear to ± threadlike; central flower of umbel usually purple; fruit widest at middle . . . . *D. carota*

1: Annual, bract segments short, linear to lanceolate; central flower of umbel white; fruit widest below middle . . . . *D. pusillus*

★ **Daucus carota** Linnaeus. WILD CARROT, QUEEN ANNE'S LACE. Common, low waste places:The Sea Ranch, acc. *Mahaffey* (1990); Salt Point; Bodega Head; ca 1 mi. nw. of Camp Meeker, acc. J. & B. Guggolz (1991); Occidental Marsh, *Rubtzoff* (1957); Fife Cr. Marsh w. of Guerneville, *Rubtzoff* (1962); Guerneville Marsh, *Rubtzoff* (1954); field in flatlands n. of Freestone on Bohemian Hwy., acc. Knight (1990); near Valley Ford, *Wells* (1981); Royal Oaks Mobile Home Park in Petaluma, *Knight* (1985); Wiggins Hill w. of Petaluma on Bodega Highway, *Knight* (1963); Hope Valley near Kenwood, *Baker* (1957 NCC); Gates Road just off Calistoga Road, acc. Wells et al. (1984); Pepperwood. < 4000 ft. May-Sept. Native of Eurasia.

Daucus pusillus Michaux. RATTLESNAKE WEED. Common, grassland:The Sea Ranch; Horseshoe Cove, *Knight* (1963); Salt Point; mouth of Russian River, *Best* (1981); Bodega Head; Fairfield Osborn (H); Sobre Vista Road w. of Arnold Drive near El Verano, *Rubtzoff* (1971); Bouverie; Mt. Hood, *Baker* (1947 NCC); 4¼ mi. w. of Glen Ellen on Bennett Valley Road, Rubtzoff (1971); field s. of Santa Rosa, *D. Campbell* (1966 NCC); Santa Rosa Cr., *Baker* (1930 NCC); Pepperwood; Lake Sonoma area; Geyserville and Ida Clayton Road quads.; ca 2000 ft. elev. Anderson Trail on Black Mtn. between Geyser Peak and upper end of Pine Flat Road, acc. J. & B. Guggolz (1992). < 5000 ft. Apr.-June.

Eryngium

1. Bracts and bractlets very rigid, margins usually entire, prominently thickened . . . . *E. armatum*
1: Bracts and bractlets ± flexible, margin usually sharply toothed, not thickened . . . . *E. aristulatum*

Eryngium aristulatum Jepson. Occasional, summer areas of vernal pools, salt marshes: Laguna de Santa Rosa n. of Sebastopol, *Rubtzoff* (1958); Atascadero Cr. Marsh, *Rubtzoff* (1958); Todd Road Preserve; Hessel Road near Gravenstein Hwy., acc. J. & B. Guggolz (1991); Petaluma, *Elmer* (1903); 4¼ mi. w. of Glen Ellen on Bennett Valley Road, *Rubtzoff* (1971); dry margin of Coney Lake (Ledson Marsh) near Kenwood, *Baker* (1950 NCC); 850 ft. elev. Shiloh Ranch in valley of small tributary of Mark West Cr. drainage 2.3 airmiles ne. of jct. of Shiloh and Faught roads, 0.6 mi. wsw. of Mark West Springs, acc. Stone (1985); along Calistoga Road about 200 yds. from Hwy. 12, *Knight* (1964); Pepperwood. < 3300 ft. May-Aug. —*E. oblanceolatum* Coulter & Rose, type locality: Sonoma Valley, *Torrey* (1865 U).

Eryngium armatum (S. Watson) Coulter & Rose. Occasional, where water seeps or stands in spring, fields: The Sea Ranch; Stump Beach in Salt Point State Park, *Wells* (1983); Salt Point, *Best* (1983); Fort Ross, *Howell* (1930); Goat Rock Beach, *Best* (1984); 4 mi. n. of Bodega Bay, *Howell* (1930); 1 mi. e. of Bodega Bay, *Howell* (1930); Blank Road near Gravenstein Hwy., acc. J. & B. Guggolz (1991); Pitkin Marsh, *Baker* (1946), *Rubtzoff* (1951); Cunningham Marsh, *Howell* (1946); Todd Road Preserve; 8 mi. w. of Petaluma on Bodega Hwy., *Howell* (1946); Petaluma, *Congdon* (1880), *Elmer* (1903); Black Point Road between Sonoma and Petaluma, *Howell* (1930). <700 ft. May-Aug.

★ **Foeniculum vulgare** Miller. FENNEL. Very common, obvious roadside weed of summer and autumn:The Sea Ranch; Bodega Head; Todd Road Preserve; Copeland Cr., Rohnert Park, *Self* (1979 NCC); Fairfield Osborn (S); Tresch Ranch, Walker Road near Two Rock, acc. J. & B. Guggolz (1992); w. of Sears Point Raceway in Tolay Cr. drainage, acc. Patterson (1990); Bouverie; Sugarloaf Ridge; Spring Lake; n. boundary of Santa Rosa Jr. Coll., *E. M. Hill* (1937 NCC); Pepperwood; Healdsburg, *Heller* (1902); Lake Sonoma area. < 1500 ft. May-Sept. Naturalized from the Mediterranean region.

Glehnia littoralis (A. Gray) Miquel ssp. **leiocarpa** (Mathias) Hultén. Rare, coastal strand: Salt Point. ± 0 ft. May-June. —*G. leiocarpa* Mathias.

Heracleum lanatum Michaux. COW-PARSNIP. Common, inland, damp gullies and ravines; near coast, moist hillsides brushy or open wooded areas:The Sea Ranch; Willow Creek; Bodega Head; lower Salmon Cr. Marsh, *Rubtzoff* (1962); upper Green Valley Marsh, *Rubtzoff* (1974); Laguna de Santa Rosa near Llano Road, *Rubtzoff* (1958); Fairfield Osborn (S); Rainsville Road near Stony Point Road, *Ott* (1934); near Cooper's Grove on Sonoma Mtn. Road, *Utech & Knight* (1986 CM); Mark West Cr. w. of Fulton, *Rubtzoff* (1970); Pepperwood; Lake Sonoma area. < 8500 ft. Apr.-July. —*H. maximum* Bartram.

Hydrocotyle

1. Leaves round-reniform, not peltate . . . . *H. ranunculoides*
1: Leaves peltate
   2. Inflorescence an interrupted spike . . . . *H. verticillata* var. *verticillata*
   2: Inflorescence with spike mostly simple . . . . *H. v.* var. *triradiata*

Hydrocotyle ranunculoides Linnaeus f. MARSH PENNYWORT. Occasional, swampy ground, shallow freshwater ponds, frequently floating: mouth of Salmon Cr., *Bacigalupi* (1928); Bodega Head; Perry Marsh, *Howell* (1945); Forestville Marsh, *Rubtzoff* (1955); Fairfield Osborn (S); 5 ¼ mi. ne. of Penngrove near Lichau Road, *Rubtzoff* (1969); bed of lagoon on W. College Ave. w. of Santa Rosa, *Baker* (1928 NCC). < 5000 ft. Mar.-Aug.

Hydrocotyle verticillata Thunberg var. **triradiata** (A. Richard) Fernald. Rare, swampy or wet areas, low meadows, borders of marshes: Santa Rosa, *Baker* (JEPS); Channel Drive e. of Annadel State Park, *Rubtzoff & Harrison* (1976). < 350 ft. Apr.-Sept. —*H. prolifera* Kellogg.

Hydrocotyle verticillata var. **verticillata**. Occasional, swampy ground, shallow freshwater ponds: dunes near Bodega Marine Laboratory dormitory, *Rubtzoff* (1971); Perry Marsh, *Howell* (1945); Laguna de Santa Rosa n. of Sebastopol, *Rubtzoff* (1958); Kenwood Marsh, *Rubtzoff* (1974); Spring Lake. < 350 ft. Apr.-Sept. —*H. v.* var. *cuneata* (Coulter & Rose) Jepson.

Ligusticum apiifolium (Nuttall) A. Gray. LOVAGE. Rare, shaded banks and meadows: The Sea Ranch, acc. Wells (1985); Salt Point; 1 mi. e. of Bodega Bay on road to Bodega, *Howell* (1930); Bennett Valley Road, *Baker* (1931 NCC); Mark West Springs Road 1 mi. w. of Mark West Springs, *Knight* (1989 NCC). < 6000 ft. June-July.

Lilaeopsis occidentalis Coulter & Rose. Uncommon, marshes, brackish flats: lower Salmon Cr., *Rubtzoff* (1973); Bodega Head Marsh, *Rubtzoff* (1954); marsh in dunes s. of Bodega Marine Lab. dormitory, *Rubtzoff* (1971). ± 0 ft. June-Aug.

Lomatium

1. Peduncles conspicuously inflated and swollen at the apex . . . . *L. nudicaule*
1: Peduncles not obviously swollen at the apex
  2. Fruit ± deeply emarginate at each end, the wings distinct on each side of body; leaflets mostly broad in outline . . . . *L. repostum*
  2: Fruit not emarginate or barely so, the wings ± joined above and below the body; leaflets mostly narrow
    3. Leaves with mostly few or large divisions, ternately or pinnately divided, divisions mostly remote . . . . *L. californicum*
    3: Leaves decompound, dissected into many small divisions
      4. Ovaries and young fruit variously pubescent or roughened . . . . *L. dasycarpum*
      4: Ovaries and fruit glabrous
        5. Bractlets of involucel obovate, sometimes connate
          6. Plants often with several cauline leaves; wings of fruit broader than body, dorsal ribs obsolete . . . . *L. utriculatum*
          6: Plants with not > 1 stem leaf; wings of fruit narrower than body, or if broader, dorsal ribs evident. . . . . *L. caruifolium*
        5: Bractlets filiform to lance-linear, rarely oblance-acuminate . . . . *L. macrocarpum*

Lomatium californicum (Torrey & A. Gray) Mathias & Constance. Occasional, open plains, slopes: n. of Occidental, *Best* (1981); Lafferty Ranch on Sonoma Mtn., *Arnold et al.* (1971); Sonoma region, *Baker* (1930 NCC), *Stocking* (1964 NCC); Bennett Valley Road, *Baker* (1930 NCC); Pepperwood; Lake Sonoma area; Geyser Peak, *Knight & Edwards* (1986). 500-6000 ft. Apr.-June.

Lomatium caruifolium (Hooker & Arnott) Coulter & Rose. Occasional, meadows, moist grassy hills: Los Alamos Can., *Baker* (1928 NCC); jct. Shiloh Road and US 101, acc. Guggolz (1989); 2 mi. n. of Windsor, *Eastwood & Howell* (1936); several mi. n. of Windsor, *Baker* (1940 NCC); 4 mi. s. of Geyserville, *Constance & Rollins* (1942); summit of Cloverdale grade, *Eastwood & Howell* (1936); Lake Sonoma area, acc. P. Gilbert (1989). 200-2200 ft. Mar.-May.

Lomatium dasycarpum (Torrey & A. Gray) Coulter & Rose. Common, dry ridges: area of serpentine on Stewarts Point-Skaggs Springs Road, *J. & B. Guggolz* (1977 NCC); Black Point Road between Sonoma and Petaluma, *Howell* (1930); Bennett Valley, *Baker* (1926 NCC); Spring Lake; Lake Sonoma area; Geyser Peak, *Knight & Edwards* (1986). < 5200 ft. Mar.-June.

Lomatium macrocarpum (Torrey & A. Gray) J. Coulter & Rose. Occasional, dry stony areas: Petaluma, *Congdon* (1880); Bennett Valley, *Leitholt* (1893); the principal hog-parsnip on the serpentine at Chalk Hill Winery, Windsor, *Harrison* (1989 SRJC); Geyser Peak, *Knight & Edwards* (1986). 200-10000 ft. Apr.-June.

Lomatium nudicale (Pursh) Coulter & Rose. Occasional, open slopes: 1 mi. from top of Sonoma Mtn., *Lowry* (1983 SFSU); Sonoma Valley Reg. Park, *Harrison* (1975 SRJC); Hwy. 12 across and just n. of Nuns Can. Road, *J. & B. Guggolz* (1990 NCC); Bennett Valley, *Baker* (1929 NCC); Park Trail Drive off Summerfield Road before development, *Harrison* (1980 SRJC). 100-6500 ft. Apr.-June.

Lomatium repostum (Jepson) Mathias. Common, brushy places, chaparral: hills east of Agua Caliente (near Bouverie), *Howell* (1966); Cavedale Road near McNab Cypress, *Quibell* (1988 NCC); 700 ft. elev. in burned chaparral on s. slope of Mt. Hood, *Baker* (1947); Lawndale Road, Kenwood, *Harrison* (1978 SRJC); PG&E transmission tower site on Los Alamos Road, *Harrison* (1982 SRJC); Rincon Ridge, *Stebbins* (1966); Quail Court, Santa Rosa, *Harrison* (1983 SRJC); Pepperwood; hills above Calistoga, *Eastwood* (1923); s. of Lake Sonoma, *J. & B. Guggolz* (1990 JEPS); Ida Clayton Road at summit, *Harrison* (1976 NCC); serpentine on Geysers-Healdsburg Road, *Harrison* (1979 SRJC); Pine Mtn. Road, *Knight* (1985); 3230 ft. elev. along fire trail ca ½ mi. w. of Rattlesnake Fire Trail in recent burn of geothermal area, *Raiche* (1990 JEPS); ca 2000 ft. elev. Anderson Trail on Black Mtn. between Geyser Peak and upper end of Pine Flat Road, acc. J. & B. Guggolz (1992); BLM personnel reported observing plant in The Geysers area, acc.*Kruckeberg* (1984). 300-2700 ft. Apr.-May.

Lomatium utriculatum (Torrey & A. Gray) Coulter & Rose. Occasional, grassland: Devil Cr., a tributary of E. Austin Cr., on serpentine, *F. W. Hoffman* (1950 NCC); Larkspur Rock at Bodega Bay, *Baker* (1935 NCC); 1 mi. e. of Bodega Bay on road to Bodega, *Howell* (1930); Dutch Bill Cr. near Camp Meeker, *Stocking* (1964 NCC); Fairfield Osborn (*); Bennett Valley, *Baker* (1936 NCC); Pepperwood; Lake Sonoma area. < 5500 ft. Feb.-May.

Oenanthe sarmentosa J. S. Presl. DITCH-CARROT. Common, shallow water of marshes and ponds, wet springy places: 2 mi. s. of Gualala on Hwy. 1, *Knight* (1964); 7 mi. s. of Gualala, *Thomas* (1976); Stewarts Point, *Abrams* (1920); Salt Point; Duncans Mills Marsh, *Rubtzoff* (1955); Willow Creek; mouth of Salmon Cr., *Ferris* (1937); Guerneville Marsh, *Rubtzoff* (1955); Pitkin Marsh, *Rubtzoff* (1951 NCC); marsh 2 mi. n. of Sebastopol, *Baker* (1945 NCC); Llano Road w. of Rohnert Park, *Howell* (1986); Bloomfield, *Congdon* (1880). <3300 ft. June-Oct.

Osmorhiza chilensis Hooker & Arnott. SWEET CICELY. Common, brushy or wooded slopes: The Sea Ranch, *Wells*(1984); Willow Cr. Road, *Utech* (1988 CM); Guerneville, *Rattan* (1877); Fairfield Osborn (*); hills near Rainsville Road w. of Petaluma, *Ott* (1934 NCC); Bouverie; Sugarloaf Ridge; Gates Road off Calistoga Road near Humbug Cr., acc. Wells et al. (1984); Petrified Forest, *Heller* (1902); Fitch Mtn., *Rubtzoff* (1952); Colverdale, *Rattan* (1877); Lake Sonoma area; between Jimtown & Mercuryville on Geysers Road, *Utech & Knight* (1966 CM); 2400 ft. elev. Little Geysers, *Malloch* (1977 PG&E). 300-9000 ft. Apr.-July.

Perideridia

1. Styles short, usually < 1 mm long; plants from a dense fascicle of usually numerous (5-20) narrow, fibrous or slightly thickened monostelic roots
    2. Leaflets linear 1-6 mm broad; rays 10-20; petioles 4-7 cm long . . . . *P. kelloggii*
    2: Leaflets ovate 5-25 mm broad; rays usually 20-25; lower petioles 6-25 cm long . . . . *P. howellii*
1: Styles elongate, 1-2.5 mm long; plants from one or a cluster of several (2-5) swollen, polystelic roots
    3. Leaves simply pinnate with 3-5 pinnae, the lower pinnae occasionally deeply lobed or ternately dissected
        4. Tuberous roots 2-3-clustered; petals 5- or 7-veined . . . . *P. gairdneri* ssp. *borealis*
        4: Tuberous roots single; petals 1-veined . . . . *P. gairdneri* ssp. *gairdneri*
    3: Leaves bipinnately or tripinnately dissected
        5. Involucre inconspicuous or obsolete, ca 10 bracts . . . . *P. oregana*
        5: Involucre obvious at anthesis, the bracts 6-10 . . . . *P. bolanderi*

Perideridia bolanderi (A. Gray) Nelson & J. F. Macbride. Uncommon, meadows, slopes: Wilson Ranch on Starr Road nw. of county airport, *J. & B. Guggolz* (1988 JEPS). 3300-6600 ft. June-Aug.

Perideridia gairdneri (Hooker & Arnott) Mathias ssp. **borealis** Chuang & Constance. Rare, open areas: Starr Road w. of Windsor, *Wilson* (1989 NCC). < 10000 ft. June-July.

Perideridia gairdneri ssp. **gairdneri**. YAMPAH. Locally common but declining, threatened by loss of habitat; adobe flats, grassland, under pines along the coast and outer Coast Ranges: Hwy. 1 between Stewarts Point and Fort Ross, *Baker* (1931 NCC); Pitkin Marsh, *Rubtzoff* (1951, 1967); Laguna de Santa Rosa e. of Graton between Occidental and Guerneville roads, *Rubtzoff* (1966); Laguna de Santa Rosa e. of Sebastopol, *Rubtzoff*(1958); Atascadero Cr. Marsh, *Rubtzoff*(1962, 1966), *Constance, Chuang, Ornduff*

& Rubtzoff (1965 UC); Cunningham Marsh, *Howell* (1946), *J. & B. Guggolz* (1991); Todd Road Preserve; near Todd Road in Laguna de Santa Rosa drainage, *Harrison* (1978 NCC); near Kenwood, *J. & B. Guggolz* (1984 JEPS). < 1200 ft. June-July.

Perideridia howellii (Coulter & Rose) Mathias. Uncommon, damp areas: Freestone Marsh, *Rubtzoff* (1962 UC). 300-5000 ft. July-Aug.

Perideridia kelloggii (A. Gray) Mathias. Common, open grasslands, adobe flats, outcrops of serpentine: Morelli Lane 1 mi. ne. of Occidental, *Best* (1981); next drainage n. of Duvoul Cr. n. of Camp Meeker, *Edwards* (1986); Perry Marsh, *Howell* (1945); 2 mi. n. of Sebastopol on Gravenstein Hwy., *Baker* (1945 NCC); s. of Hwy. 12 w. of Santa Rosa in Laguna de Santa Rosa drainage, *Harrison* (1978 NCC); Fairfield Osborn (*); Sonoma Valley, *Torrey* (1865 CM); Santa Rosa Road, *Congdon* (1880); Joost Ranch on Calistoga Road, *Knight* (1963); 4 mi. n. of Cloverdale, *Chuang* (1963 UC); The Geysers, *M. Jones* (1931); Lake Sonoma area. < 3300 ft. July-Aug.

Perideridia oregana (S.Watson) Mathias. Uncommon, open flats, stream beds, grassy or gravelly slopes, oak woodlands, edge of open pine forests, outcrops of serpentine: 1½ mi. nw. of The Geysers, *Constance, Chuang & Quibell* (1967 UC). 1000-6600 ft. July-Aug.

Sanicula

1. Basal leaves ternately or palmately divided, rarely entire
    2. Involucel-bractlets much longer than the umbellets, obvious; plants ± prostrate, with foliage greenish-yellow at anthesis . . . . *S. arctopoides*
    2: Involucel-bractlets shorter than the umbellets, inconspicuous; plants erect, with foliage other than greenish-yellow
        3. Stem solitary, erect 3-8 dm high; leaves near base round-cordate in outline, palmately and deeply 3-5-lobed into broad obovate segments . . . . *S. crassicaulis*
        3: Stems few, branched from near base 1-5 dm high; leaves ovate in outline 3-lobed or -parted, the divisions toothed or pinnately parted with bristle-tipped teeth; cauline leaves rather remote, reduced upward . . . . *S. laciniata*

1: Basal leaves pinnatifid or pinnately or ternate-pinnately dissected or compound
    4. Leaves once or twice pinnatifid with a toothed rachis . . . . *S. bipinnatifida*
    4: Leaves 1- to 3-pinnate or pinnately or ternate-pinnately decompound, without a margined or toothed rachis
        5. Stems from fusiform taproots . . . . *S. bipinnata*
        5: Stems from globose or irregular tuberous roots . . . . *S. tuberosa*

Sanicula arctopoides Hooker & Arnott. FOOTSTEPS OF SPRING, YELLOW MATS. Infrequent, usually in areas of coastal influence: The Sea Ranch; 1 mi. n. of Stewarts Point, *Knight* (1964); Salt Point; near Jenner, *Reed* (1939 NCC); Wright's Beach, *Hastings* (1940 NCC); Bodega Head, acc. Connors (1991); Larkspur Rock near Bodega Bay, *Ott* (1934 NCC); Button Ranch on Walker Road (U.C. property) between Petaluma-Valley Ford Road and Pepper Road, acc. Connors (1992). < 1000 ft. Mar.-May.

Sanicula bipinnata Hooker & Arnott. POISON SANICLE. Occasional, shady or open slopes: Sugarloaf Ridge; Alexander Valley quad.; Sulphur Cr. Road, *Baker* (1934 NCC). 70-3300 ft. Apr.-May.

Sanicula bipinnatifida Hooker. PURPLE SANICLE. Frequent, open slopes, brushy areas: The Sea Ranch, *Wells* (1984); Fort Ross quad.; Willow Creek; near Todd Road in Laguna de Santa Rosa drainage, *Harrison* (1978 NCC); Todd Road Preserve; Fairfield Osborn (*); Button Ranch on Walker Road (U. C. property) between Petaluma-Valley Ford Road and Pepper Road, acc. Patterson (1990); Petaluma cemetery, *Ott* (1934 NCC); Bouverie; Pepperwood; Healdsburg quad.; Lake Sonoma area; property bordered by Chianti and Dutcher Cr. roads w. of US 101 and w. of Geyserville, acc. J. & B. Guggolz (1991); Alexander Valley and Ida Clayton Road quads. 70-4200 ft. Mar.-May.

Sanicula crassicaulis de Candolle. GAMBLE WEED. Frequent, grassland, brushy areas: The Sea Ranch; Stewarts Point quad.; Salt Point; Fort Ross and Cazadero quads.; 1 mi. e. of Bodega Bay on road to Bodega, *Howell* (1930); Willow Cr. Road, *Utech* (1988 CM); w. of Santa Rosa in Laguna de Santa Rosa drainage, *Harrison* (1978 NCC); Fairfield Osborn (*); Bouverie; Bennett Mtn. Lake, *Rubtzoff* (1963); Gates Road off Calistoga Road, acc. Wells et al. (1984); Pepperwood; hills between Santa Rosa and Calistoga, *C. R. Bell* (1952 NCC); 350-400 ft. elev. Old Skaggs Springs Road, *Utech, Sato & Ohara* (1990 CM); Lake Sonoma area; property bordered by Chianti and Dutcher Cr. roads w. of US 101 and w. of Geyserville, acc. J. & B. Guggolz (1991); Alexander Valley quad. < 3300 ft. Mar.-May.

Sanicula laciniata Hooker & Arnott. COAST SANICLE. Quite common, grassland, brushy areas: The Sea Ranch, *Wells* (1984); the island on Devil Cr., a tributary of E. Austin Cr., on serpentine, *F. W. Hoffman* (1950 NCC); Willow Creek; ca 1 mi. nw. of Camp Meeker, acc. J. & B. Guggolz (1991);

Bouverie; Sugarloaf Ridge; Bennett Mtn. Lake, *Rubtzoff* (1963); Adobe Can., *Baker* (1927 NCC); Mayacmas Mts., *Applegate* (1934 NCC); Gates Road off Calistoga Road, acc. Wells et al. (1984); Pepperwood; Petrified Forest, *Eastwood* (1915); chaparral area sw. of Bradford Mtn 1½ mi. w. of Dry Cr., acc. J. & B. Guggolz (1992); Sonoma Lake area. 100-3000 ft. Mar.-May.

Sanicula tuberosa Torrey. Infrequent, wooded to open slopes: Adobe Can., *Baker* (1927 NCC); Pepperwood, *de Nevers* (1988). 300-9000 ft. Mar.-July.

★ **Scandix pecten-veneris** Linnaeus. SHEPHERD'S NEEDLE. Common, waste places: Willow Creek; upper Green Valley Marsh area nw. of Graton, *Rubtzoff* (1974); 2 sites w. and sw. of Santa Rosa in Laguna de Santa Rosa drainage, *Harrison* (1978 NCC); Todd Road Preserve; Roberts Road e. of Petaluma Hill Road, *Stocking* (1963 NCC); Fairfield Osborn (H); 1½ mi. nw. of Petaluma, *Ott* (1934 NCC); w. of Sears Point Raceway in Tolay Cr. drainage, acc. Patterson (1990); e. foothills of Sonoma Mts. on Sobre Vista Road 9 mi. w. of Arnold Drive, *Rubtzoff* (1971); Bouverie; Sugarloaf Ridge; Green Hill Drive in Santa Rosa, *Stocking* (1965 NCC); Pepperwood; Lake Sonoma area; Ida Clayton Road quad. 50-3300 ft. Apr.-June. Naturalized from Eurasia.

Sium suave Walter. WATER-PARSNIP. Infrequent, wet places: Laguna de Santa Rosa e. of Graton, *Rubtzoff* (1960); Atascadero Cr. Marsh, *Rubtzoff* (1958); Laguna de Santa Rosa e. of Sebastopol, *Rubtzoff* (1960); Laguna de Santa Rosa near Trenton, *Rubtzoff* (1960); swamp near Healdsburg, *Rattan* (1877). < 6600 ft. July-Aug.

Tauschia kelloggii (A. Gray) J. F. Macbride. Rare, chaparral, mixed evergreen forest, yellow pine forest, sometimes on serpentine: Pepperwood. 150-6000 ft. Apr.-June.

Torilis

1. Plant erect; peduncle > leaf; umbel open, not headlike; outer, inner fruit-halves equally prickly . . *T. arvensis*
1: Plant spreading; peduncle < leaf; umbel dense, headlike; outer fruit ½ prickly, inner only tubercled *T. nodosa*

★ **Torilis arvensis** (Hudson) Link. HEDGE-PARSLEY. Occasional, waste places: The Sea Ranch, *Wells* (1985); Jenner, *Leary & Sherfey* (1961); Fairfield Osborn (H); Button Ranch on Walker Road (U. C. property) between Petaluma-Valley Ford Road and Pepper Road, acc. Connors (1992); 2 mi. s. of Old Adobe e. of Petaluma, *Winter* (1932 NCC); Adobe Can., *Baker* (1950 NCC); Annadel, *Wells* (1984); Spring Lake; Pepperwood, *Breedlove* (1980); near Napa Co. line on St. Helena Road, *Wagnon & Vale* (1988 NCC); chaparral area sw. of Bradford Mtn. 1½ mi. w. of Dry Cr., acc. J. & B. Guggolz (1992); overlook at Lake Sonoma, *Best* (1981); confluence of Galloway and Dry creeks, *Howell* (1967); Canyon Road between Lake Sonoma and Geyserville, *Thomas* (1976); Cherry Cr. Road w. of Cloverdale, *Knight & Edwards* (1986); ca 2000 ft. elev. Anderson Trail on Black Mtn. between Geyser Peak and upper end of Pine Flat Road, acc. J. & B. Guggolz (1992). 125-6000 ft. Apr.-July. Native of Europe. —*T. japonica* (Houttuyn) de Candolle.

★ **Torilis nodosa** (Linnaeus) Gaertner. KNOTTED HEDGE-PARSLEY. Fairly common, open hills, waste places: Bodega Head quad.; Forest Hills 5½ mi. e. of Guerneville, *Rubtzoff* (1968); Fairfield Osborn (H); along Sonoma Cr. at base of Mt. Hood, *Heller* (1902); Franz Valley Road ca ¾ mi. from Mark West Road, *Ferris & Lorraine* (1948); Pepperwood; Healdsburg and Geyserville quads. < 5000 ft. Apr.-June. Naturalized from Europe.

Yabea microcarpa (Hooker & Arnott) Koso-Poljansky. HEDGE-PARSLEY. Occasional, sandy or rocky soils: Willow Brook n. of Petaluma, *Knight* (1985 NCC); Kenwood, *Baker* (1927 NCC); Pepperwood; Dry Cr. Valley w. of Healdsburg, *Baker* (1930 NCC). < 3300 ft. Apr.-June. —*Caucalis m.* Hooker & Arnott.

APOCYNACEAE. DOGBANE OR INDIAN HEMP FAMILY

1. Plant trailing; flowers blue . . . . *Vinca*
1: Plant not trailing; flowers white usually with pinkish veins . . . . *Apocynum*

Apocynum

1. Plant diffusely branched; leaves drooping or spreading; corolla 4-6 mm . . . . *A. androsaemifolium*
1: Plant ± stiffly erect; leaves ascending; corolla 2.5-5 mm . . . . *A. cannabinum*

Apocynum androsaemifolium Linnaeus. BITTER DOGBANE. Occasional, grassy areas, edge of thickets: near Sebastopol, *Howell* (1931); Bennett Valley, acc. Baker; Pepperwood; e. base Fitch Mtn., *Rubtzoff* (1952); Lake Sonoma area, 3200 ft. elev. Caldwell Pines, acc. Clifton (1984). 650-8000 ft. June-Aug. —*A. pumilum* (A. Gray) E. Greene.

Apocynum cannabinum Linnaeus. INDIAN HEMP. Rare, wooded slopes, flatlands: Santa Rosa, *Heller* (1902); off College Ave. behind Borden Villa tract, *Potter* (1986 SRJC); Coffey Lane, nw. Santa Rosa, *Knight* (1963); large population in a field next to the Quik Stop on Ninth Ave. a couple of blocks e. of Lincoln School, Santa Rosa, acc. Harrison, (1992); probably between Santa Rosa and Mt. St. Helena, *Voznesenskii* (1840-1841 LE). < 6000 ft. June-Aug. —*A.* var. *glaberrimum* A. de Candolle.

★ **Vinca major** Linnaeus. GREATER PERIWINKLE. Escaping from cultivation, becoming naturalized, moist shaded places, or in sun near coast: The Sea Ranch; Salt Point; within a mi. of w. end of Bay Hill Road from Hwy. 1, acc. Knight (1991); Bodega Head; near Freestone, *Rubtzoff* (1974); Bouverie; Lafferty Ranch on Sonoma Mtn., acc. Arnold et al. (1971); Enterprise Road between Bennett Valley and Sonoma Mtn. roads, acc. Knight (1990); Santa Rosa, *Heller & Brown* (1902); Pepperwood. Various altitudes. Mar.-July. Native of the Mediterranean region.

AQUIFOLIACEAE. HOLLY FAMILY

★ **Ilex aquifolium** Linnaeus. ENGLISH HOLLY. Seedling from nearby parents: Royal Oak Mobile Home Park, N. McDowell Blvd., Petaluma, acc. Knight (1991). 'Big Bull' ornamental ♂ with large, nearly smooth-edged leaves. < 700 ft. Native of Europe, w. Asia to China.

ARALIACEAE. GINSENG FAMILY

1. Plant erect; leaf pinnately compound . . . . *Aralia*
1: Plant a woody vine; leaf simple, lobed to ± entire . . . . *Hedera*

Aralia californica S. Watson. ELK CLOVER, SPIKENARD. Infrequent, somewhat shaded and moist spots: Salt Point; ½ mi. w. of Monte Rio, *Stocking* (1964 NCC); ca 1 mi. nw. of Camp Meeker, acc. J. & B. Guggolz (1991); Fairfield Osborn (S); Cooper's Grove on Sonoma Mtn. Road, acc. Knight (1990); Bouverie; Pepperwood; Skaggs Springs Road, acc. Van Soelen (1988); Lake Sonoma area; 1500 ft. elev. w. flank Mt. St. Helena, acc. Grummer & Hemphill (1983). < 6600 ft. June-Aug.

★ **Hedera helix** Linnaeus. ENGLISH IVY. Common, weed: 100 ft. elev. 1 mi. s. of Fort Ross, *Knight* (1984); Blank Road near Gravenstein Hwy., acc. J. & B. Guggolz (1991); Enterprise Road between Bennett Valley and Sonoma Mtn. roads, acc. Knight (1990); Sonoma Mtn. Road, acc. Knight (1990); Gates Road off of Calistoga Road, acc. Wells et al. (1984); along Santa Rosa Cr. in e. Santa Rosa, *Wagnon* (1989 NCC). < 3300 ft. Native of Eurasia.

ARISTOLOCHIACEAE. PIPEVINE FAMILY

1. Plants woody climbing vines; flowers irregular . . . . *Aristolochia*
1: Plants low herbs; flowers regular . . . . *Asarum*

Aristolochia californica Torrey. DUTCHMAN'S PIPE, PIPEVINE. Common, wooded or brushy areas, stream banks: Neeley Hill area near Guerneville, *Rubtzoff* (1951); Forest Hills 5½ mi e. of Guerneville, *Rubtzoff* (1986, 1987); Russian River opposite Vacation Beach, *Sargent* (1922), *Rubtzoff* (1951); Bouverie;

Sonoma Cr. at base of Mt. Hood, *Heller & Brown* (1902); Sugarloaf Ridge; Santa Rosa, *Howell* (1934); Leslie Road near Mark West Springs, acc. Knight (1990); Pepperwood; Healdsburg quad. < 2500 ft. Jan.-Apr.

Asarum caudatum Lindley. WILD-GINGER. Occasional, moist soil, shade: The Sea Ranch; Duncans Mills Marsh, *Rubtzoff* (1955); Willow Creek; Camp Meeker, *Howell* (1924); Green Valley Road w. of Graton, *Rubtzoff* (1955); Pitkin Marsh, *Rubtzoff* (1951); Forestville Marsh, *Rubtzoff* (1955); w. of Sebastopol, *Heller & Brown* (1902); Adobe Can. near Kenwood Springs, acc. Baker; Ida Clayton Road, *Knight* (1963), *Quibell* (1979 NCC). < 7200 ft. May-July.

ASCLEPIADACEAE. MILKWEED FAMILY

1. Stems not twining . . . . *Asclepias*
1: Stems twining . . . . *Araujia*

★ **Araujia sericifera** Brotero. BLADDER-FLOWER. Rare, disturbed places, noxious weed: acc. *Barbe* (1991). 325-1300 ft. Native of S. America.

Asclepias

1. Corona hoods without hornlike projections
    2. Stems strongly flattened . . . . *A. solanoana*
    2: Stems terete . . . . *A. cordifolia*
1: Corona heads with hornlike projections
    3. Hoods 2-3 × as long as the stamens and stigma; filament-column almost lacking . . . . *A. speciosa*
    3: Hoods ca as long as stamens and stigma; filament-column well-developed
        4. Leaves narrow, 1 cm or less wide; pedicels erect in fruit . . . . *A. fascicularis*
        4: Leaves broader, mostly 2-7 cm wide; pedicels deflexed in fruit . . . . *A. eriocarpa*

Asclepias cordifolia (Bentham) Jepson. PURPLE MILKWEED. Occasional, wooded or open slopes: Franz Valley Road, *Barnhart* (1981 SRJC); near Dry Cr., *Cantelow* (1934); forest at Lake Sonoma site, *Harrison* (1973 SRJC); confluence of Dry and Dutcher crs., *Mullin* (1933 NCC); Lake Sonoma area; ca 2000 ft. elev. Anderson Trail on Black Mtn. between Geyser Peak and upper end of Pine Flat Road, acc. J. & B. Guggolz (1992). 150-6600 ft. May-July.

Asclepias eriocarpa Bentham. INDIAN MILKWEED, KOTOLO. Uncommon, dry barren areas: Knights Valley, acc. *Jeps. Fl. of Calif. vol. 3,* (1939). 600-6500 ft. June-Aug.

Asclepias fascicularis Decaisne. NARROW-LEAF MILKWEED. Common, meadows, fields: 2 sites w. of Santa Rosa in Laguna de Santa Rosa drainage, *Harrison* (1978 NCC); Todd Road Preserve; Fairfield Osborn (S); laguna in Chileno Valley, *Rubtzoff* (1975); Casa Grande Road se. of Petaluma, *Knight (1963);* Ducker Cr., Rincon Valley, *Wells* (1985); corner Benicia Drive and Montecito Blvd., Rincon Valley, *Wells* (1987); Spring Lake; Santa Rosa, *Heller* (1902); Pepperwood; Knights Valley, *Breedlove* (1980); Healdsburg, *Rattan* (1877); Lake Sonoma area; Asti, *Knight* (1963); 2200 ft. elev. Squaw Cr., acc. Clifton (1984). 175-7200 ft. June-Sept. —*A. mexicana*, misapplied.

Asclepias solanoana Woodson. PROSTRATE MILKWEED, SOLANOA. Occasional, outcrops of serpentine, streamsides: Layton Mine at Austin Cr., *F. Hoffman* (1949 UC), acc. *Kruckeberg* (1984); 800 ft. elev. The Cedars n. of Cazadero at headwaters of Big Austin Cr., *Bacigalupi* (1956 JEPS), *Raiche* (1982); near jct. of Socrates Mine Road with Pine Flat Road, *H. Sharsmith* (1952), acc. S. & J. Edwards (1989). 750-5400 ft. June. —*Solanoa purpurascens* (A. Gray) E. Greene.

Asclepias speciosa Torrey. SHOWY MILKWEED. Occasional, dry stony and rocky places: near Dry Cr., *Cantelow* (1934); Lake Sonoma area; Pine Flat Road, *W., I., & T. Knight* with *S. Edwards* (1985); 2800 ft. elev. Caldwell Ranch, acc. Clifton (1984). < 6500 ft. May-July.

ASTERACEAE [COMPOSITAE]. SUNFLOWER FAMILY

Key ± acc. D. J. Keil, *Hickman* (1993)

Key to Groups:

1. Ligules (strap-shaped, petal-like corollas) 0
   2. Heads disciform (flowers of 2 kinds [♀ or sterile and bisexual or ♂], in same or different heads) **Group 1**
   2: Heads discoid (flowers bisexual, usually all fertile [sometimes outermost enlarged, ± bilateral])
      3. Receptacle bearing chaff (scalelike bractlets) or stiff hairs among flowers
         4. Receptacle chaffy . . . . **Group 2**
         4: Receptacle bristly . . . . **Group 3**
      3: Receptacle ± naked (sometimes bearing minute scales or short hairs among flowers)
         5. Pappus 0 or only a low crown . . . . **Group 4**
         5: Pappus well-developed
            6. Pappus of bristles (sometimes with an additional series of shorter bristles or scales) . **Group 5**
            6: Pappus of flat, ± membranous scales or stiff, ± needlelike awns . . . . **Group 6**

1: Ligules present
   7. Heads ligulate or of 2-lipped disk flowers; sap usually milky . . . . **Group 7**
   7: Heads radiate; flowers of 2 kinds (ray flowers ♀ or sterile; disk flowers usually bisexual); sap usually clear
      8. Receptacle bearing chaff scales among disk flowers (sometimes only in a ring between ray and disk flowers)
         9. Phyllaries in 1 series, each subtending a ray flower . . . . **Group 8**
         9: Phyllaries in 2+ series, not all subtending ray flowers . . . . **Group 9**
      8: Receptacle naked or bearing minute scales or hairs (rarely stiff bristles) among flowers
         10. Pappus 0 or only a low crown . . . . **Group 10**
         10: Pappus well-developed on ray or disk flowers (or both)
            11. Pappus of flat, ± membranous scales or stiff, ± needlelike awns . . . . **Group 11**
            11: Pappus of bristles (sometimes with an additional series of shorter bristles or scales)
               12. Ligules white to purple . . . . **Group 12**
               12: Ligules yellow to orange or red . . . . **Group 13**

Group 1: Heads disciform; flowers of 2 kinds

1. ♀ and ♂ flowers in different heads
   2. ♀ and ♂ heads on same plant (monoecious)
      3. Stems armed with 3-branched spines . . . . *Xanthium spinosum*
      3: Stems unarmed
         4. ♂ heads in long terminal inflorescence; burs usually 3-10 mm . . . . *Ambrosia*
         4: ♂ heads congested; burs often > 20 mm . . . . *Xanthium strumarium*
   2: ♀ and ♂ heads on different plants (dioecious)
      5. Shrubs with sticky leaves, not woolly . . . . *Baccharis*
      5: Herbs not sticky, plants white-woolly . . . . *Anaphalis*

1: ♀ or sterile flowers in same heads with ♂ or bisexual flowers
   6. Outer flowers without corollas; heads not embedded in wool
      7. Heads long-peduncled; outer fruits stalked . . . . *Cotula*
      7: Heads sessile (at stem forks); fruits sessile, spine-tipped . . . . *Soliva*
   6: Outer flowers with (sometimes narrowly cylindric) corollas; heads sometimes embedded in wool
      8. Outer flowers sterile (corollas sometimes expanded, ± bilateral); heads not very small, not embedded in wool
         9. Fertile flowers 1-2 per head; basal leaves entire or toothed, cauline pinnately lobed . . . . *Crupina*
         9: Fertile flowers many per head; leaves spiny . . . . *Cnicus*
      8: Outer flowers ♀ (corollas often narrowly cylindric); heads sometimes very small, embedded in wool
         10. Phyllaries (or outermost bracts of head) papery, membranous or scarious, sometimes green only below middle or in narrow, central band; heads often embedded in wool
            11. Phyllaries usually many, overlapping, inner > flowers; receptacle naked; ♀ flowers not subtended by bracts
               12. Pappus 0 or a minute crown; herbage glabrous to densely ± short-hairy . . *Artemisia*
               12: Pappus of bristles
                  13. Plant from creeping rhizome; herbage white-woolly . . . . *Anaphalis*

13: Plant from taproot or caudex; usually ± woolly or tomentose . . . *Gnaphalium*

11: Phyllaries 0 or 6-8, not overlapping, < flowers; outer receptacle chaffy; some or all ♀ flowers individually subtended by bracts

14. Disk flowers bisexual, pappus often of > 12 bristles, ± exserted; inner ♀ flowers with pappus . . . . *Filago*

14: Disk flowers ♂, pappus 0 or 1-12 bristles, included; ♀ pappus 0

15. Chaff scales subtending disk flowers (disk chaff) enlarged, rigid, ± spreading, very different from scales subtending ♀ flowers (♀ chaff)

16. Disk chaff scale tips usually spinelike, strongly hooked inward; ♀ chaff scales closed, strongly 3-veined, tips usually strongly scarious-winged; receptacle not bristly, usually widest at tip . . . . *Ancistrocarphus*

16: Disk chaff scale tips erect or spreading, ± flat to folded, not spinelike, obtuse; ♂ chaff scales open, concave, veins obscure, tips not or barely scarious-winged; receptacle bristly, widest at base . . . . . . *Hesperevax*

15: Disk chaff 0 or scales ± gradually reduced, scarious, erect

17. Leaves usually opposite; chaff scales net-veined, inner surface with scarious wing that is hidden in head;pappus 0 . . . . *Psilocarphus*

17: Leaves alternate or seeming whorled; chaff scales parallel-veined, scarious wing visible in head; pappus usually of 1-5 bristles . . . . . . *Micropus*

10: Phyllaries green usually throughout, sometimes scarious-margined; heads usually not embedded in wool

18. Pappus well-developed, when present

19. Pappus of flattened scales or barbed awns (pappus 0 on *Lasthenia glabrata*); leaves usually opposite throughout . . . . *Lasthenia*

19: Pappus of bristles; leaves opposite or alternate

20. Main phyllaries in 1 series, equal . . . . *Erechtites*

20: Main phyllaries in 2+ series

21. Stems threadlike; ♀ flowers 1-5 . . . . *Pentachaeta*

21: Stems stouter; ♀ flowers several-many

22. Heads ± flat, buttonlike . . . . *Erigeron*

22: Heads usually cylindric or bell-shaped . . . . *Conyza*

18. Pappus 0 or reduced to a minute crown

23. Phyllary margins widely scarious or transparent

24. Delicate annual; ♀ corollas threadlike . . . . *Pentachaeta*

24: Perennial; ♀ corollas short, ± wide

25. Leaves pinnate . . . . *Tanacetum*

25: Leaves entire to ± lobed . . . . *Artemisia*

23: Phyllary margins usually not scarious or transparent

26. Ovary of disk flowers much reduced; style tip truncate or tack-shaped

27. Leaves widely triangular, usually ± basal or cauline below middle, green above, white below; inflorescence open; corollas white; fruit glandular . . . . *Adenocaulon*

27: Leaves linear to narrowly ovate, surfaces ± alike, usually green to gray on one or both surfaces; inflorescence ± dense; corollas ± yellowish white; fruit not glandular . . . . *Artemisia*

26: Ovary of disk flowers well-developed; style tip ± branched

28. Plant glandular, ± strongly scented . . . . *Madia*

28: Plant not glandular, not strongly scented . . . . *Pentachaeta*

Group 2: Heads discoid; receptacle chaffy

1. Leaf margins spiny . . . . *Carthamus*

1: Leaf margins not spiny

2. Pappus (on fertile fruit only) of small scales and stiff bristles . . . . *Crupina*

2: Pappus 0 . . . . *Santolina*

Group 3:Heads discoid; receptacle bristly

1. Leaves not spiny

2. Phyllaries spine-tipped

3. Phyllary spines hooked at tip . . . . *Arctium*

3: Phyllary spines straight . . . . *Centaurea*

2: Phyllaries not spine-tipped . . . . *Acroptilon*

1: Leaves spiny
- 4. Corollas yellow to orange-red
  - 5. Pappus of many, unequal, narrow scales in several series (sometimes 0), usually 0 on outer fruits; flowers all fertile; fruit 4-angled . . . *Carthamus*
  - 5: Pappus of 20 stiff bristles or awns in 2 series, on all fruits; outer flowers sterile (corolla 3-lobed; ovary vestigial); fruit cylindric, 20-ribbed . . . *Cnicus*
- 4: Corollas white to blue, red, or purple
  - 6. Pappus of long-plumose bristles
    - 7. Largest leaves toothed to deeply lobed; involucre body 1-6 cm diameter; receptacle usually not fleshy; phyllaries linear to ovate . . . *Cirsium*
    - 7: Largest leaves often ± compound, involucre body 3-15 cm diameter; receptacle fleshy; phyllaries ovate . . . *Cynara*
  - 6: Pappus of ± rough or barbed bristles or slender scales
    - 8. Stems spiny-winged . . . *Carduus*
    - 8: Stems not spiny-winged
      - 9. Leaves not blotched along veins; heads not long-peduncled; pappus of free, slender scales . . . *Carthamus leucocalos*
      - 9: Leaves white-blotched along veins; heads long-peduncled; pappus of many long bristles, deciduous in a ring . . . *Silybum*

Group 4: Heads discoid; receptacle naked; pappus 0 or reduced to crown

1. Shrubs; receptacle flat-topped or weakly convex; plant woolly, ± tufted . . . *Artemisia*

1: Annuals; receptacle conic to spheric; plants glabrous or puberulent, not tufted . . . *Chamomilla*

Group 5: Heads discoid; receptacle naked; pappus of bristles

1. Shrubs or subshrubs
- 2. Corollas white or cream-colored to pink- or red-tinged
  - 3. Ovaries very reduced, sterile; style included . . . *Baccharis*
  - 3: Ovaries forming fruits; style often exerted . . . *Brickellia*
- 2: Corollas yellow to orange
  - 4. Phyllaries in 5 ± distinct vertical ranks . . . *Chrysothamnus nauseosus* ssp. *albicaulis*
  - 4: Phyllaries not in distinct vertical ranks
    - 5. Stems ± tomentose . . . *Chrysothamnus parryi*
    - 5: Stems glabrous or ± lightly puberulent . . . *Ericameria*

1: Herbs
- 6. Corollas white to purple
  - 7. Leaves and phyllaries spiny . . . *Onopordum*
  - 7: Leaves and phyllaries not spiny
    - 8. Annual; outer flowers often bilateral, ± raylike . . . *Lessingia*
    - 8: Perennial; flowers all radial
      - 9. Herbage glabrous, sticky; leaves entire or finely toothed; involucre 3.5-5 mm — corollas white . . . *Baccharis douglasii*
      - 9: Herbage puberulent to loosely tomentose; leaves sharply toothed; involucre 10-15 mm; corollas cream-colored . . . *Brickellia*
- 6: Corollas yellow to orange
  - 10. Phyllaries in 2-several series, unequal, often strongly graduated
    - 11. Pappus bristles flat, readily deciduous; involucre gummy . . . *Grindelia*
    - 11: Pappus bristles ± cylindric, ± persistent; involucre usually not gummy
      - 12. Annual; outer flowers often bilateral, ± raylike . . . *Lessingia*
      - 12: Perennial; flowers radial
        - 13. Leaves linear to narrowly oblong; disk wide, buttonlike . . . *Erigeron*
        - 13: Leaves linear to ovate; disk ± narrow
          - 14. Hairs smooth; various habitats . . . *Aster*
          - 14: Hairs minutely knobby (20×); usually gravelly places, sand bars . . . *Heterotheca oregona*
  - 10: Phyllaries in 1-2 series, ± equal
    - 15. Cauline leaves opposite . . . *Arnica*
    - 15: Cauline leaves alternate
      - 16. Annual or short-lived perennial from slender taproot . . . *Senecio*
      - 16: Perennial, often from caudex or rhizome

17. Leaves palmately veined and lobed
  18. Erect herb . . . . *Cacaliopsis*
  18: Twining vine . . . . *Senecio mikanioides*
17: Leaves pinnately veined (or lateral veins obscure), entire to pinnately lobed
  19. Leaves densely white-tomentose below; phyllaries not black-tipped . . *Luina*
  19: Leaves glabrous to loosely tomentose below; phyllaries often black-tipped . . . . *Senecio*

Group 6: Heads discoid; receptacle naked; pappus of scales or awns

1. Heads ± sessile, sometimes dense 2° clusters . . . . *Eriophyllum*
1: Heads peduncled
  2. Phyllaries in 2-many overlapping series, outer << inner
    3. Involucre gummy; pappus awns readily deciduous . . . . *Grindelia*
    3: Involucre hairy or glandular; pappas scales often ± dissected into bristles . . . . *Lessingia*
  2: Phyllaries in 1-3 series, subequal, weakly overlapping or not, outer ± = inner
    4. Heads ± spheric; leaves decurrent . . . . *Helenium*
    4: Heads hemispheric; leaves not decurrent . . . . *Chaenactis*

Group 7: Heads ligulate; ligules 5-lobed; sap usually milky

1. Receptacle with long chaff scales among flowers . . . . *Hypochaeris*
1: Receptacle usually naked (sometimes short bristly or minutely scaled)
  2. Pappus none . . . . *Rhagadiolus*
  2: Pappus present
    3. Corollas blue . . . . *Cichorium*
    3: Corollas yellow or white to purple
      4. Pappus of scales, stiff awns, or awn-tipped scales, not plumose
        5. Outer fruits enfolded by phyllaries, pappus scales short, deltate; inner fruits with pappus scales long, narrow . . . . *Hedypnois*
        5: Outer fruits usually not enfolded by phyllaries; pappus of outer and inner fruits ± alike
          6. Outer phyllaries long, linear; pappus scales stiff, bristlelike; uncommon alien . . . . *Tolpis*
          6: Outer phyllaries < or = inner, usually wide (if linear, short); pappus scales few-many, bristle-tipped; native
            7. Annual; head erect; involucre glabrous, outer phyllaries not < ¼ inner; pappus scales 5, lanceolate, silvery bristle short, smooth, from notched scale tip . . . . *Uropappus*
            7: Annual or perennial; head ± nodding in bud; involucre glabrous to hairy, outer phyllaries often < ¼ inner; pappus scales 1-35, variously colored, tip entire to unevenly cut, bristle from tip smooth to plumose
              8. Perennial from fleshy taproot(s); ligules >> involucre; pappus scales 5-35 *Microseris*
              8: Annual; ligules ± = involucre; pappus scales 5 (often < 5 in *Microseris douglasii*)
                9. Head usually nodding in bud; pappus scales usually deltate to ovate (if lanceolate, tapered to bristle); leaves basal . . . . *Microseris*
                9: Head not strongly nodding; pappus scales narrowly lanceolate, the tip irregularly cut; cauline leaves sometimes present . . . . *Stebbinsoseris*
      4. Pappus of simple bristles or plumose
        10. Pappus of plumose bristles or awns
          11. Pappus bristles very unequal . . . . *Leontodon*
          11: Pappus bristles ± =
            12. Leaves linear to lanceolate, grasslike, entire; involucre 2.5-5 cm
              13. Perennial; phyllary series several; fruit beak 0 . . . . *Scorzonera*
              13: Biennial; phyllary series 1, fruit beak > body . . . . *Tragopogon*
            12: Leaves lanceolate to elliptic (or scalelike), larger leaves ± toothed or lobed; involucre 0.5-2 cm
              14. Corollas white to pink or lavender; native
                15. Fruit beaked; corollas white or cream-colored, sometimes red-veined . . . . *Rafinesquia*
                15: Fruit beak 0; corollas pink to pale lavender . . . . . . *Stephanomeria*
              14: Corollas yellow; alien
                16. Leaves in basal rosette; heads solitary on scapose peduncles, outer phyllaries inconspicuous . . . . *Leontodon*
                16: Leaves cauline; heads 1-many, not on scapose peduncles . . . . *Picris*

10: Pappus of many, smooth or barbed, stiff or soft bristles
 17. Fruit ± compressed
  18. Fruit beaked (beak sometimes short, thick) . . . . *Lactuca*
  18: Fruit beak 0 . . . . *Sonchus*
 17: Fruit not compressed
  19. Fruit cylindric, slender (not beaked) . . . . *Hieracium*
  19: Fruit tapered at both ends
   20. Pappus bristles readily deciduous (not beaked) . . . . *Malacothrix*
   20: Pappus bristles persistent
    21. Stem branched, leafy (leaves often reduced); beak 0 . . . . . . *Crepis*
    21: Stem unbranched, leafless (except some annual *Agoseris*); beak slender, >> body
     22. Phyllaries in 2 series, outer reflexed, inner erect, linear *Taraxacum*
     22: Phyllaries subequal or overlapping in 2-4 series, inner often elongated in fruit . . . . *Agoseris*

Group 8: Heads radiate; phyllaries in 1 series, subtending ray flowers; receptacle chaffy (at least between ray and disk flowers)

Phyllary margin folded, ± clasping or enclosing ray ovaries

1. Disk pappus 0
 2. Phyllary margins not adjacent or overlapping, ± clasping fruit, but not completely enclosing fruit
  3. Upper leaves and phyllaries tipped by sessile, open-pit glands . . . . *Holocarpha*
  3: Upper leaves or phyllary glands 0 or stalked
   4. Upper leaves and phyllaries without tack-shaped glands; lower leaves often toothed or lobed . . . . *Hemizonia*
   4: Upper leaves and phyllaries with tack-shaped glands; lower leaves usually entire . . . *Calycadenia*
 2: Phyllary margins adjacent or overlapping, completely enclosing fruit
  5. Perennial from long fleshy rhizome (disk pappus bristles 0-5) . . . . *Holozonia*
  5: Annual
   6. Ray fruit compressed side-to-side, 3-angled, or obovoid; phyllary margins folded around angled or rounded ovary face; lower leaves usually entire . . . . *Madia*
   6: Ray fruit compressed front-to-back; phyllary margins folded across flat ovary face; lower leaves usually toothed or lobed
    7. Disk flowers 6, ♂; heads usually closing by mid-day . . . . *Lagophylla*
    7: Disk flowers 12-100+, usually fruiting; heads open throughout day . . . . *Layia*
1: Disk pappus present
 8. Disk pappus bristles or scales plumose
  9. Perennial . . . . *Madia*
  9: Annual . . . . *Layia*
 8: Disk pappus bristles or scales not plumose, sometimes minutely fringed
  10. Phyllary margins clasping ovary, not overlapping, partly enclosing fruit
   11. Upper leaves and or phyllaries tipped by tack-shaped glands; ligules usually deeply and palmately lobed, lobes spreading . . . . *Calycadenia*
   11: Upper leaves and phyllaries without tack-shaped glands; ligules usually shallowly 2-3-lobed, lobes ± parallel . . . . *Hemizonia*
  10: Phyllary margins overlapping around ovary, completely enclosing fruit
   12. Perennial from fleshy rhizome; (disk pappus bristles 0-5) . . . . *Holozonia*
   12: Annual
    13. Pappus of 10 blunt, shiny, white scales in 2 series, inner 6-11 mm; fruit 10 ribbed, ± cylindric . . . . *Achyrachaena*
    13: Pappus of bristles or scales < 5 mm; fruit not ribbed . . . . *Layia*

Group 9: Heads radiate; phyllary series 2+; receptacle chaffy ± throughout

1. Chaff scales folded around disk ovaries
 2. Ray flowers fruiting (style present, ovary well-developed) . . . . *Wyethia*
 2: Ray flowers sterile (ovary vestigial, style usually 0)
  3. Chaff scales 0-3-lobed . . . . *Helianthus*
  3: Chaff scales entire, obtuse, tip hairy . . . . *Helianthella*
1: Chaff scales flat, linear to ovate, not folded around disk ovaries
 4. Ray corollas white to purple

5. Disk corollas white; leaves aromatic . . . . *Achillea*

5: Disk corollas yellow

6. Leaves linear, entire; (ray corollas pale yellow, often purple-tinged) . . . . *Rigiopappus*

6: Leaves pinnately lobed to compound

7. Base of corolla narrow . . . . *Anthemis*

7: Base of corolla wide, enclosing fruit tip . . . . *Chamaemelum*

4: Ray corollas yellow to orange-red (sometimes multi-colored)

8. Phyllaries ± alike, in 2+ series; (ray corollas pale yellow, often purple-tinged) . . . . . . *Rigiopappus*

8: Phyllaries in 2 very different series (inner ± membranous)

9. Pappus scales 2 . . . . *Coreopsis*

9: Pappus awns 2-4 . . . . *Bidens*

Group 10: Heads radiate; receptacle naked; pappus 0 or low crown

1. Leaves opposite, at least below

2. Upper leaves alternate

3. Ray corollas without small lobe opposite ligule . . . . *Eriophyllum*

3: Ray corollas with small lobe opposite ligule . . . . *Monolopia*

2: Upper leaves opposite

4. Phyllaries in 2+ series, widely overlapping; perennial, fleshy . . . . *Jaumea*

4: Phyllaries in 1-2 series, barely overlapping; annual or perennial, usually not fleshy . . . . . *Lasthenia*

1: Leaves alternate throughout

5. Ligules 1 mm or less

6. Leaves finely dissected; phyllaries unequal, overlapping in several series, scarious-margined *Tanacetum*

6: Leaves linear, entire; phyllaries in 2 series . . . . *Rigiopappus*

5: Ligules > 1 mm

7. Ray corollas white to purple (sometimes multi-colored)

8. Phyllaries unequal in 2+ series, usually conspicuously overlapping, scarious-margined

9. Heads solitary on long peduncles; leaves serrate or pinnately lobed . . . . *Leucanthemum*

9: Heads usually many on short peduncles; leaves 2-3-pinnately dissected *Tanacetum parthenium*

8: Phyllaries ± =, in 1-2 series, usually little overlapping, usually not scarious-margined

10. Ligules 25-40 mm; coarse perennial . . . . *Osteospermum*

10: Ligules < 10 mm; annual or slender perennial

11. Lower leaves pinnately lobed . . . . *Blemmosperma*

11: Leaves entire

12. Perennial; lawn weed . . . . *Bellis*

12: Annual; dry habitats . . . . *Pentachaeta*

7: Ligules yellow to orange

13. Phyllaries unequal, overlapping in several series

14. Phyllary margins conspicuously scarious

15. Heads solitary or in open, leafy cymes; outer fruits 3-angled . . . . . *Chrysanthemum*

15: Heads in dense, flat-topped clusters; fruits cylindric . . . . *Tanacetum*

14: Phyllary margins not or barely scarious . . . . *Arctotheca*

13: Phyllaries ± =, in 1-3 series, slightly overlapping or not

16. Ray fruit prickly, knobby, or papillate on back; phyllary margins usually narrowly scarious

17. Involucre 5-8 mm . . . . *Blennosperma*

17: Involucre 10-20 mm . . . . *Calendula*

16: Ray fruit smooth on back; phyllary margins usually not scarious

18. Heads ± spheric; phyllaries spreading or reflexed . . . . *Helenium*

18: Heads cylindric to hemispheric; phyllaries erect or ascending

19. Plants green, ± glabrous . . . . *Blennosperma nanum*

19: Plants ± gray-tomentose

20. Ray corollas with small lobe opposite ligule . . . . *Monolopia*

20: Ray corollas without small lobe opposite ligule . . . . *Eriophyllum*

Group 11: Heads radiate; receptacle usually naked; pappus of scales or awns

1. Ligules white to purple

2. Perennial . . . . *Tanecetum parthenium*

2: Annual . . . . *Rigiopappus*

1: Ligules yellow to red (sometimes multi-colored)

3. Phyllaries in several series, strongly graduated

4. Leaves tomentose below; stems creeping, herbaceous . . . . . *Arctotheca*
4: Leaves glabrous to short-hairy; stems erect, woody below
5. Phyllary tips spreading to recurved; involucre gummy; pappus of 2-8 slender deciduous awns; leaves oblong, lanceolate, ovate . . . . . *Grindelia*
5: Phyllary tips ± appressed; involucre not or barely gummy; pappus of 1-2 series of finely toothed, white or yellowish scales usually ½ fruit length or much reduced; leaves linear . . . . . *Gutierrezia*
3: Phyllaries in 1-3 series, ± =, not strongly graduated
6. Leaves usually opposite throughout . . . . . *Lasthenia*
6: Leaves usually alternate (sometimes opposite below)
7. Pappus of rigid, tapered awns; leaves linear, entire . . . . . *Rigiopappus*
7: Pappus of ± thin scales; leaves various, entire to lobed
8. Phyllaries in 1 series . . . . . *Eriophyllum*
8: Phyllaries in 2-3 ± = series . . . . . *Helenium*

Group 12: Heads radiate; ligules white to purple; receptacle naked; pappus of bristles

1. Leaves palmately lobed or dentate (basal) . . . . . *Petasites*
1: Leaves entire to pinnately lobed
2. Ligules inconspicuous, barely surpassing disk or not
3. Resin-filled phyllary vein 1 . . . . . *Conyza canadensis*
3: Phyllaries without resin-filled veins
4. Pappus single; phyllaries often ± thickened near tip . . . . . *Aster*
4: Pappus double, outer of short bristles or scales; phyllaries not thickened near tip
5. Disk wide, flat; disk flowers many; fruit not beaked . . . . . *Erigeron*
5: Disk narrow; disk flowers 15-25; fruit short-beaked . . . . . *Tracyina*
2: Ligules conspicuous; usually much exceeding disk
6. Pappus bristles < 10
7. Fruit beakless . . . . . *Pentachaeta*
7: Fruit short-beaked . . . . . *Tracyina*
6: Pappus bristles usually > 20
8. Main phyllaries ± = in 1-3 series
9. Pappus usually double, outer of short bristles or scales, inner of soft bristles; rays usually < 2 mm wide . . . . . *Erigeron*
9: Pappus single; rays often > 2 mm wide
10. Main phyllaries subequal, in 2-3 series . . . . . *Aster*
10: Main phyllaries =, in 1 series (often with very short outer phyllaries) . . . . . *Senecio*
8: Main phyllaries unequal, overlapping in 2-several series
11. Subshrub . . . . . *Lessingia filaginifolia*
11: Herbs
12. Ray flowers sterile (style 0, ovary ± vestigial) . . . . . *Lessingia*
12: Ray flowers fruiting, style present . . . . . *Erigeron*

Group 13: Heads radiate; ligules yellow to red; receptacle naked; pappus of bristles

1. Leaves opposite at least below
2. Trailing, fleshy; pappus bristles 1-5; saline habitats . . . . . *Jaumea*
2: Erect, not fleshy; pappus bristles many; inland . . . . . *Arnica*
1: Leaves alternate throughout
3. Annual or biennial
4. Ray pappus 0 . . . . . *Heterotheca grandiflora*
4: Ray (and disk) fruit with pappus
5. Main phyllaries ± =, in 1-3 series
6. Fruit compressed . . . . . *Pentachaeta*
6: Fruit not compressed . . . . . *Senecio*
5: Main phyllaries unequal, in 2-several series
7. Pappus bristles flat, deciduous; involucre gummy . . . . . *Grindelia*
7: Pappus bristles ± cylindric, usually persistent; involucre not gummy
8. Fruit short-beaked, phyllaries deciduous when head dry . . . . . *Tracyina*
8: Fruit not beaked; phyllaries persistent on old heads . . . . . *Pentachaeta*
3: Perennial to shrubs
9. Shrubs or subshrubs
10. Main phyllaries in 1 series, = (often a few, often much shorter, outer phyllaries present) *Senecio*

10: Main phyllaries in 2-7 series, often graduated, often unequal
 11. Pappus bristles flattened; phyllaries ± ovate; involucre gummy . . . . *Grindelia*
 11: Pappus bristles ± cylindric; phyllaries usually linear to narrowly lanceolate . . *Ericameria*
9: Herbs
 12. Main phyllaries in 1 series, = (often with few, very short, outer phyllaries) . . . . *Senecio*
 12: Phyllaries in 2+ series, subequal to strongly graduated
 13. Disk pappus double, outer of very short bristles or scales
 14. Stem simple, erect; phyllaries ± =; fruit 2-ribbed . . . . *Erigeron*
 14: Stem often branched, prostrate to erect; phyllaries ± unequal . . . . . . *Heterotheca*
 13. Disk pappus single
 15. Pappus bristles flat, deciduous; involucre strongly gummy, especially in bud *Grindelia*
 15: Pappus bristles ± cylindric, usually persistent; involucre not strongly gummy
 16. Leaves toothed or lobed . . . . *Solidago*
 16: Leaves entire
 17. Heads small, clustered at branch tips; inflorescence leafy, paniclelike; stem erect, from rhizome; fruit ± 1 mm . . . . *Euthamia*
 17: Heads 1-several, small to large; inflorescence various; stem from caudex or rhizome; fruit usually << 1 mm . . . . *Solidago*

Achillea millefolium Linnaeus. COASTAL YARROW. Common, tomentose plant of maritime situations, valleys, hills: The Sea Ranch; Salt Point, *Wells* (1982); Cazadero and Duncans Mills quads.; Willow Creek; 1025 ft. elev. Coleman Valley Road, acc. Clifton (1984); near mouth of Salmon Cr., *Ferris* (1937); Bodega Bay dunes, *Ferris* (1932); Pitkin Marsh, *Rubtzoff* (1951), *Howell* (1938); Perry Marsh, *Howell* (1945); Stoetz Lane ne. of Occidental, *Utech & Knight* (1986 CM); Sebastopol, *Dows* (1907); Todd Road Preserve; Tresch Ranch, Walker Road near Two Rock, acc. J. & B. Guggolz (1992); Stage Gulch Road near Lakeville, *Kawahara* (1968); Bouverie; hills near Santa Rosa, *Heller* (1902); Spring Lake Village; Spring Lake; top of Fitch Mtn. near Healdsburg, *Rubtzoff* (1952); Pepperwood; chaparral area sw. of Bradford Mtn. 1½ mi. w. of Dry Cr., acc. J. & B. Guggolz (1992); ca 2000 ft. elev. Anderson Trail on Black Mtn. between Geyser Peak and upper end of Pine Flat Road, acc. J. & B. Guggolz (1992); 2040 ft. elev. Squaw Cr., acc. Clifton (1984); 3200 ft. elev. Caldwell Ranch, acc. Clifton (1984). < 11500 ft. Mar.-Aug. —*A. m.* Linnaeus var. *arenicola* (Heller) Nobs. *A. borealis* Bongard ssp. *a.* (Heller) Keck. *A. b.* ssp. *californica* (Pollard) Keck. *A. a.* A. A. Heller, type locality: sand hills at upper end of Bodega Bay, *Heller* (1902).

Achyrachaena mollis Schauer. BLOW-WIVES. Common, hill and valley grassland: 2 sites w. of Santa Rosa in Laguna de Santa Rosa drainage, *Harrison* (1978 NCC); E. Cotati Ave., *Murray* (1975 NCC); Wilfred Road in Cotati Valley, *Quibell* (1990 JEPS); Fairfield Osborn (H); Agua Caliente, *Lehman* (1975 NCC); Santa Rosa; Pepperwood; Lake Sonoma area; Ida Clayton Road quad.; 2040 ft. elev. Squaw Cr., acc. Clifton (1984). < 3200 ft. Apr.-May.

★ **Acroptilon repens** (Linnaeus) de Candolle. RUSSIAN KNAPWEED. Rare, ruderal: Santa Rosa area, acc. *Barbe* (1991). Native of Europe. < 6200 ft. May-Sept. —*Centaurea repens* Linnaeus.

Adenocaulon bicolor Hooker. SILVER ARROW. Common, moist areas, shaded canyons: Cazadero, *Bacigalupi* (1926); Stewarts Point, *Best* (1981); Annapolis and Fort Ross quads.; ca 1 mi. nw. of Camp Meeker, acc. J. & B. Guggolz (1991); Guerneville, *Rattan* (1877); Fairfield Osborn (*); 6.3 mi. up Cavedale Road, *Knight* (1984); along Sonoma Cr. at foot of Mt. Hood, *Heller* (1902); Pepperwood; Lake Sonoma; Cloverdale, *Rattan* (1877). < 7000 ft. June-Aug.

Agoseris

1. Annual; achene-beak 2-3 × as long as achene-body; involucre 12-18 mm high . . . . *A. heterophylla*
1: Perennial; achene-beak variable; involucre mostly higher
 2. Beak < twice as long as body of achene
 3. Beak slightly longer than achene; herbage thinly hirsute . . . . *A. apargioides* var. *apargioides*
 3: Beak ± ½ as long as achene; herbage with some crisped hairs, glabrate . . . . *A. a.* var. *eastwoodiae*
 2: Beak 2-4 x as long as achene-body
 4. Achene-body truncate at apex, abruptly beaked; leaf-segments retrorse; phyllaries in 2 quite unlike sets . . . . *A. retrorsa*
 4: Achene-body tapering at apex, gradually beaked; leaf-segments not retrorse; phyllaries mostly in 3 sets. . . . . *A. grandiflora*

Agoseris apargioides (Lessing) E. Greene var. **apargioides**. SEASIDE AGOSERIS. Infrequent, grassy hillsides: Bodega Head, *Eastwood* (1915), *Barbour* (1968 BML); Petaluma, *Harford* (1870); Santa Rosa, *Heller* (1902). < 1700 ft. Apr.-Aug. —*A. hirsuta* E. Greene. *Macrorhyncus harfordii* Kellogg, type locality: Petaluma, *Harford* (1870 U).

Agoseris apargioides var. **eastwoodiae** (Fedde) Munz. Occasional, coastal dunes, maritime bluffs, beaches: Goat Rock State Beach, *Wells* (1981); mouth of Salmon Cr., *Bacigalupi* (1928); cliffs at upper end of Bodega Bay, *Ferris* (1932). < 170 ft. Apr.-Aug. —*A. maritima* Eastwood, type locality: Bodega Point, *Eastwood* (1903). *A. e.* Fedde.

Agoseris grandiflora (Nuttall) E. Greene. GRAND MOUNTAIN DANDELION. Common, dry or rather moist places: Cazadero, *Bacigalupi* (1926); Stewarts Point, *Best* (1981); Annapolis and Fort Ross quads.; ca 1 mi. nw. of Camp Meeker, acc. J. & B. Guggolz (1991); Guerneville, *Rattan* (1877); Fairfield Osborn (*); 6.3 mi. up Cavedale Road, *Knight* (1984); Sugarloaf Ridge; along Sonoma Cr. at foot of Mt. Hood, *Heller* (1902); Pepperwood; Lake Sonoma quad.; Cloverdale, *Rattan* (1877). < 7000 ft. May-July. —*A. plebeia* (E. Greene) E. Greene.

Agoseris heterophylla (Nuttall) E. Greene var. **heterophylla**. ANNUAL MOUNTAIN DANDELION. Common, open, dry, often weedy situations: Cazadero quad.; Pitkin Marsh, *McCormick* (1933); Fairfield Osborn (H); Pepperwood; Lake Sonoma area, Geyser Peak, *Knight & Edwards* (1986); Ida Clayton Road quad.; 3200 ft. elev. Socrates Mine, acc. Clifton (1984); 4100 ft. sw.-facing slope Cobb Mtn., *Cooney* (1977 PG&E); 2040 ft. elev. Squaw Cr., acc. Clifton (1984). Var. **kymapleura** E. Greene differs in that the ribs of outer achenes are corky-thickened and undulate: Pepperwood. < 6600 ft. Apr.-July.

Agoseris retrorsa (Bentham) E. Greene. SPEARLEAF MOUNTAIN DANDELION. Dry slopes and ridges: Units 20 and 21, *Malloch* (1977 PG&E); 2190 ft. elev. Squaw Cr.; 2440 ft. elev. Caldwell Ranch; 3000 ft. elev. Caldwell Pines, all acc. Clifton (1984). 1300-9200 ft. May-Aug.

Ambrosia

1. Fruiting involucre unarmed or with a single row of short prickles below the single beak; leaves once-pinnatifid into coarse lobes, greenish, rough-pubescent; stem simple 5-12 dm high . . . . *A. psilostachya*
1: Fruiting involucre armed with several rows of spines below the 1-4 beaks; stems radiating, procumbent branching, forming loose mats 1-3 m across
   2. Leaves simple, blade ovate, rhombic or oval-oblanceolate, silvery-canescent with silky hairs *A. chamissonis*
   2: Leaves 1-3 × pinnatifid into oblong or obovate segments, less silvery . . . . *A. bipinnatifida*

Ambrosia bipinnatifida (Nuttall) E. Greene. BEACH BUR. Common, dunes deep beach sand: Grows with next taxon. < 85 ft. Mar.-Sept. —*A. chamissonis* var. *pinnatisecta* Lessing.

Ambrosia chamissonis (Lessing) E. Greene. BEACH-BUR. Dunes, deep sand of beaches. Leaf forms intergrade with previous: The Sea Ranch, *Wells* (1981); Stump Beach at Salt Point State Park, *Wells* (1982); mouth of Salmon Cr., *Bacigalupi* (1928); Spud Point Marina at Bodega Harbor, *Clifton* (1980 PUA). < 85 ft. July-Nov. *Franseria c.* Lessing.

Ambrosia psilostachya de Candolle. WESTERN RAGWEED. Occasional, low fields: Pitkin Marsh, *Rubtzoff* (1951 NCC); Fairfield Osborn (*); near Santa Rosa, *Bellue* (1938 NCC). < 3300 ft. July-Nov. —*A. p.* de Candolle var. *californica* (Rydberg) S. F. Blake, currently not recognized.

Anaphalis margaritacea (Linnaeus) Bentham & Hooker f. PEARLY EVERLASTING. Occasional, brushy moist slopes: The Sea Ranch, *Wells* (1981); Bodega areas, *Voznesenskii* (1840-1841 LE); Shell Beach, *Harrison* (1981 SRJC); Bodega Head, *Eastwood* (1915), *Harrison* (1981 SRJC); Lake Sonoma area; ca 2000 ft. elev. Anderson Trail on Black Mtn. between Geyser Peak and upper end of Pine Flat Road, acc. J. & B. Guggolz (1992). < 10300 ft. June-Aug.

Ancistrocarphus filagineus A. Gray. WOOLLY FISHHOOKS. Occasional, brushy places: Santa Rosa Cr. Can., *Baker* (1930 NCC); Pepperwood; w. of Dry Cr., *Baker* (1934 NCC); Jensen's Ranch at Dorman Cr. 4 mi. W. Of Healdsburg, *Mason* (1934); Lake Sonoma area. < 6200 ft. Apr.-May. —*Stylocline filaginea* (A. Gray) A. Gray.

Anthemis

1. Bracts of receptacle present only toward the middle; ribs of the achenes tuberculate; herbage scented *A. cotula*
1: Receptacle bracteate throughout; ribs of achenes smooth; herbage unscented . . . . *A. arvensis*

★ **Anthemis arvensis** Linnaeus. CORN CHAMOMILE. Uncommon, waste places: Spring Lake. < 3399 ft, Native of Europe.

★ **Anthemis cotula** Linnaeus. MAYWEED, STINKWEED, DOG-FENNEL. Widespread, valleys and hills, pastures, near habitations, along roads: The Sea Ranch; Freestone, *Huntington* (1987 NCC); Willow Cr., *Knight* (1984); Todd Road Preserve; Copeland Cr. at Sonoma State Univ. campus, *Self* (1979 NCC); Button Ranch on Walker Road (U. C. property) between Petaluma-Valley Ford Road and Pepper Road, acc. Patterson (1990); w. of Sears Point Raceway in Tolay Cr. drainage, acc. Patterson (1990); Spring Lake; Santa Rosa, *Heller* (1902), Pepperwood; Lake Sonoma area. < 6600 ft. Apr.-Aug. Naturalized from the Old World.

★ **Arctium minus** (Hill) Bernhardi. BURDOCK. Rare, widely scattered in state: mouth of Gualala River, *Baker* (1920 NCC). < 4000 ft. June-Oct. Native of Europe.

★ **Arctotheca calendula** (Linnaeus) Levyns. CAPEWEED. Vegetatively reproducing at entrance to Bodega Dunes State Park and Salmon Cr. Park headquarters, acc. L. Davis (1990); Korbel Winery near Russian River, acc. Wagnon (1990). Becoming more invasive. < 900 ft. Introduced into California from s. Africa as an ornamental ground cover in the 1960s.

Arnica discoidea Bentham. RAYLESS ARNICA. Occasional, open woodlands: Neeley Hills se. of Guerneville, *Rubtzoff* (1951; Trinity Grade, *Eastwood & Howell* (1939); Lake Sonoma area; s. of the Geysers, *McMurphy* (1909); Cherry Cr. Road, *Knight & Edwards* (1986); 3300 ft. elev. Squaw Cr., acc. Clifton (1983). 700-5000 ft. Stewarts Point-Skaggs Springs Road between Lake Sonoma and Annapolis Road, *Vale & Wagnon* (1988 NCC); Lake Sonoma area; 3080 ft. elev. Socrates Mine, acc. Clifton (1984). 700-5000 ft. May-Aug.

Artemisia

1. Plants definitely woody at base or above
   2. Plant a shrub 6-15 dm high, herbage strigulose; involucre 2-3 mm high . . . . *A. californica*
   2: Plant suffrutescent, 4-7 dm high; herbage silky-tomentose; involucre 3.5-4.5 mm high . *A. pycnocephala*

1: Plants not woody at base
   3. Involucre campanulate 2-3 mm wide, ± tomentose; ray-flowers 6-10; disk-flowers 10-25 *A. douglasiana*
   3: Involucre terete or narrow-ovoid, < 2 mm wide, yellow-green and shining; ray-flowers 3-7; disk-flowers 2-8 . . . . *A. suksdorfii*

Artemisia californica Lessing. CALIFORNIA SAGEBRUSH. Common, dry coastal bluffs, canyon sides, rocky ridges: near Fort Ross, *Eastwood & Howell* (1937); Red Hill Road toward Pt. Reyes Sta. from Petaluma, *Baker* (NCC 1934); 2 mi. s. of Sonoma, *Matthau* (1964). < 2800 ft. Aug.-Dec.

Artemisia douglasiana Besser. MUGWORT. Common, places wet in the spring: The Sea Ranch; Gerstle Cove at Salt Point, *Wells* (1982); Bodega Bay, *King* (1980); Bodega, *Rose* (1962); Pitkin Marsh, *Rubtzoff* (1951); w. of Santa Rosa in Laguna de Santa Rosa drainage, *Harrison* (1978 NCC); Copeland Cr. on Sonoma State Univ. campus, *Self* (1979 NCC); Fairfield Osborn (*); Tresch Ranch, Walker Road near Two Rock, acc. J. & B. Guggolz (1992); w. of Sears Point Raceway, Tolay Cr. drainage, acc. Patterson (1990); Bouverie; Los Alamos Road, *Knight* (1963); Bennett Valley, *Knight* (1963); Spring Lake; Pepperwood; Lake Sonoma area; 2720 ft. elev. Caldwell Ranch, acc. Clifton (1984). < 7200 ft. June-Dec.

Artemisia pycnocephala de Candolle. COASTAL SAGEWORT. Common, ocean beaches, dunes: mouth of Russian River, *Breedlove* (1961); Goat Rock Beach, *Wells* (1987); Wright's Beach, acc. Connors (1992); Bodega Head, *Eastwood* (1915). < 660 ft. June-Aug.

Artemisia suksdorfii Piper. COASTAL MUGWORT. Uncommon, damp coastal ravines: Salt Point; along road just n. of Jenner, *Clausen* (1939); coast near Jenner, *Bailey* (1930 NCC). Probable southern limit. < 1000 ft. May-Aug.

Aster

1. Annual . . . . *A. subulatus* var. *ligulatus*

1: Perennial
   2. Phyllaries often with purple tip or margin; disk-corollas with tube = or > the limb; stems 2-6 dm high . . . . *A. radulinus*
   2: Phyllaries occasionally purple-tipped, but not purple-margined; disk-corollas with tube distinctly < the limb; stems 3-10 dm high . . . . *A. chilensis*

Aster chilensis Nees. Common, usually in moist soil: The Sea Ranch; Bodega Bay, *Barbour* (1976 BML); 36 ft. elev. Bodega Harbor area, *Clifton* (1980 PUA); Cunningham Marsh, *Howell* (1946); Sebastopol, *Abrams* (1907); 2 sites s. and sw. of Santa Rosa in Laguna de Santa Rosa drainage, *Harrison* (1978 NCC); Wiggins Hill 6 mi. w. of Petaluma, *Knight* (1963); Petaluma, *Congdon* (1880); 0.3 mi. from San Antonio Road on I Street Ext. s. of Petaluma; *Knight* (1963); 1 mi. n. on Hwy. 121 from Hwy. 37 (jct. of Sears Point and Black Point cutoffs) *Baker* (1948); along Sonoma Cr. towards Glen Ellen 3 mi. from Kenwood, *Baker* (1932); Bennett Valley, *Knight* (1963); Coffey Lane, Santa Rosa, *Knight* (1963). < 1700 ft. June-Aug. —*A. sonomensis* E. Greene, type locality: in open fields of the Sonoma Valley in low subsaline ground, currently not recognized.

Aster radulinus A. Gray. BROAD-LEAVED ASTER. Occasional, colonies favoring loose soil: Hidden Acres Road between Grange Road and Sonoma Mtn. Road, *Quibell* (1990 NCC); Alta Vista Lane on Rincon Ridge, Santa Rosa, *Knight* (1984); Pepperwood; chaparral area sw. of Bradford Mtn. 1½ mi. w. of Dry Cr., acc. J. & B. Guggolz (1992); Pine Mtn. Road, *Knight* (1963); Geyser Peak, *Knight & Edwards* (1986); 2440 ft. elev. Caldwell Ranch, acc. Clifton (1984). 400-5000 ft. July-Oct.

Aster subulatus Michaux var. **ligulatus** Shinners. SLIM ASTER. Uncommon, on low wet ground: Spring Lake. < 6600 ft. July-Oct. —*A. exilis* S. Elliott.

Baccharis

1. Leaves cuneate to oblong, 1-nerved or with a pair of weak lateral veins; heads partly or wholly in sessile or pedunculate axillary glomerules . . . . *B. pilularis*

1: Leaves linear-lanceolate to ovate, usually rather strongly triple-nerved; heads in small or large panicles at tips of stems or branches

  2. Stems essentially herbaceous; leaves ovate to lanceolate; phyllaries thin; fertile achenes hirtellous . . . . *B. douglasii*

  2: Stems woody; leaves chiefly linear or linear-lanceolate; phyllaries thick and firm; fertile achenes glabrous . . . . *B. salicifolia*

Baccharis douglasii de Candolle. Uncommon, damp thickets, salt marshes: 310 ft. elev. 1 mi. nnw. of Camp Meeker where Grub Cr. enters reservoir, *J. & B. Guggolz* (1991 NCC); Dutch Bill Cr. at n. edge of Occidental, *Best & Wells* (1983); Pitkin Marsh, *Rubtzoff* (1953); Perry Marsh, *Howell* (1930); Fairfield Osborn (H); Pepperwood. < 2500 ft. July-Oct.

Baccharis pilularis de Candolle. COYOTE BRUSH. Common, hills, valley areas, open flats: The Sea Ranch; Salt Point; Russian Gulch n. of Jenner, *McKenzie* (1980); Duncans Mills, *M. Jones* (1882); Willow Creek; Bodega Head and dunes, acc. Connors (1991); Todd Road Preserve; Fairfield Osborn (*); Schollenberger Park, Petaluma, *Knight* (1963); Tresch Ranch, Walker Road near Two Rock, acc. J. & B. Guggolz (1992); along US 101 between Petaluma and Marin Co. line, *Howell* (1981); w. of Sears Point Raceway in Tolay Cr. drainage, acc. Patterson (1990); Bouverie; Sugarloaf Ridge; Calistoga Road between St. Helena Road and Porter Cr. Road, *Knight* (1963); Spring Lake; Pepperwood; Lake Sonoma area; ca 2000 ft. elev. Anderson Trail on Black Mtn. between Geyser Peak and upper end of Pine Flat Road, acc. J. & B. Guggolz (1992); 2720 ft. elev. Caldwell Ranch, acc. Clifton (1984). < 5000 ft. Aug.-Dec. —*B. p.* var. *consanguinea* (de Candolle) Kuntze, currently not recognized.

Baccharis salicifolia (Ruiz López & Pavón) Persoon. MULE FAT, SEEP-WILLOW, WATER-WALLY. Common, wet areas, damp thickets, salt marshes: 249 ft. elev. nr. Salmon Cr., *Clifton* (1981 PUA); ca 1 mi nw. of Camp Meeker, acc. J. & B. Guggolz (1991); Spring Lake; Pepperwood; Lake Sonoma area. < 4100 ft. Apr.-Oct. —*B. viminea* de Candolle. *B. glutinosa* Persoon.

★ **Bellis perennis** Linnaeus. ENGLISH DAISY. Common, moist grassy places: Gualala River, *Abrams* (1920); The Sea Ranch, *Wells* (1985); Stewarts Point, *Parks* (1925); 49 ft. elev. Gerstle Cove at Salt Point State Park, *Clifton* (1981 PUA); parking area at Stump Beach in Salt Point State Park, acc. Knight in 1990; Duncans Mills quad.; Willow Creek; Guerneville, *Syme* (1917). < 670 ft. Apr.-Sept. Naturalized from Europe.

Bidens

1. Plants perennial, erect or decumbent; leaves simple; rays 1.5-3 cm long . . . . *B. laevis*

1: Plants annual, erect; leaves compound; rays inconspicuous . . . . *B. frondosa*

Bidens frondosa Linnaeus. STICKTIGHT. Uncommon, marshy places: Laguna de Santa Rosa at crossing of Occidental Road, *Howell* (1946); Spring Lake, *Knight* (1984). Usually < 5000 ft. Aug.-Oct.

Bidens laevis (Linnaeus) Britton, Sterns & Poggenburg. BUR-MARIGOLD. Uncommon, marshy places: Pitkin Marsh, *Rubtzoff* (1952 NCC). < 1100 ft. Aug.-Nov.

Blennosperma

1. Lower leaves 7-15-lobed, the lobes mostly 5 mm or less long . . . . *B. nanum*
1: Lower leaves entire or 3-lobed, the lobes 10-15 mm long . . . . *B. bakeri*

Blennosperma bakeri Heiser. SONOMA SUNSHINE. Once locally abundant, now rare, occurring only in Laguna de Santa Rosa and the Sonoma area: w. outskirts of Sonoma in "hog wallow" ca ¼ mi. s. of Napa St. in field on e. side of street, *Baker* (type 1946 UC), *Howell* (1949), *Rose* (1949); Montclair Park subdivision on Newcomb St. in s. Sonoma, acc. *Santa Rosa Press Democrat* 5/29/91; thinly scattered over an area of 40-50 acres 4.6 mi. s. of El Verano on Arnold Drive, *Roderick* (1961), *Ornduff & Heckard* (1964); Glen Ellen in vernal pool, *Barnhart* (1982 SRJC); Fisher property off Ludwig Ave. sw. of Santa Rosa, *Harrison* (1983 SRJC); Llano Road and Roseland Cr., *Harrison* (1979 SRJC); jct. Piner and Fulton roads, *Harrison* (1975 SRJC). Habitats have been decimated by development and agriculture until only a relatively few remain. < 350 ft. Mar.-Apr.

Blennosperma nanum (Hooker) S. F. Blake. Common, scattered, in sunny, wet, low flats: w. of Santa Rosa in Laguna de Santa Rosa drainage, *Harrison* (1978 NCC); ½ mi. s. of Warfield Dist. near jct. of Hwy. 12 and Trinity Grade, *Baker* (1952 NCC); Sonoma Hwy. near old Dunbar School, *Baker* (1946 NCC); Annadel; Los Alamos Road, *Knight* (1963); Spring Lake; near Windsor, *Baker* (1952). < 5200 ft. Feb.-Apr.

Brickellia californica (Torrey & A. Gray) A. Gray. Occasional, dry slopes: Los Alamos Road, *Knight* (1963); along Santa Rosa Cr., e. Santa Rosa, *Wagnon* (1989 NCC); Pepperwood. <9000 ft. Aug.-Oct.

Cacaliopsis nardosmia (A. Gray) A. Gray. SILVER CROWN. Perhaps southern limit of distribution. Uncommon, brushy or wooded slopes: geothermal area near power station B, line 7, *Quibell* (1980 NCC); geysers Unit 13, *Quibell* (1981 NCC). 500-6200 ft. Apr.-May. —*Cacalis n.* A. Gray, type locality: near The Geysers in Sonoma County, *Bolander* (1868 U).

★ **Calendula arvensis** Linnaeus. FIELD-MARIGOLD. Occasional, orchards, vineyards, fields, populations fairly dense and conspicuous: Sonoma, *Bacigalupi,* (1946); just e. of Schellville, *Howell* (1954); just e. of Sonoma, Howell (1946), *Bacigalupi* (1946 JEPS); roadcut at Fetters Hot Springs, *Howell* (1959); Ducker Cr. in Rincon Valley, Santa Rosa, *Wells* (1988); just off Farmers Lane n. of Bennett Valley Road, *Wells* (1989); along W. College Ave. in Santa Rosa in Laguna de Santa Rosa drainage, *Harrison* (1978 NCC). < 700 ft. Mar.-Apr. Naturalized from Eurasia.

Calycadenia

1. Ray-achenes rugose; stems glabrous throughout or scabrid toward the summit; ligules yellow, occasionally with small red spot at base . . . . *C. truncata*
1: Ray-achenes smooth; stems pubescent, at least above, herbage scabrous; ligules white
   2. Stem simple or openly branched; heads medium large, crowded and often glomerate; ray flowers 2-8 . . . . *C. multiglandulosa*
   2: Stem densely and often dichotomously branched above; heads small, never glomerate; ray-flowers usually 1 . . . . *C. pauciflora*

Calycadenia multiglandulosa de Candolle. STICKY ROSIN WEED. Locally common, serpentine: Salt Point; 2½ mi. n. of Occidental on open hillside, *Keck* (1933); 1024 ft. elev. Coleman Valley Road, *Clifton* (1984 PUA); between Sebastopol and Occidental along hwy., *Baker* (1930). 100-3300 ft. May-Oct.

Calycadenia pauciflora A. Gray. Rare, exposed, rocky, dry slopes: Pepperwood; serpentine on Ida Clayton Road, *J. & B. Guggolz* (1989 DAV); serpentine at Chalk Hill Winery, Windsor, *Harrison* (1989 SRJC); Mill Cr. Road beyond schoolhouse, *Baker* (1930); The Geysers, *Rattan* (1877), *M. Jones* (1931), *Keck* (1944). 200-3300 ft. May-Sept.

Calycadenia truncata de Candolle. ROSIN WEED. Occasional, dry, sunny areas: Adobe Can., *Baker* (1932 NCC); Sonoma Cr. Can. near Kenwood Springs, *Howell* (1932); Los Alamos Road, *Knight* (1963); Pepperwood; along hwy. between Calistoga and Santa Rosa, *Baker* (1933 NCC). < 5000 ft. June-Oct.

Carduus

1. Flower-heads 2.5-4 cm in diameter, solitary or few-clustered, flowers conspicuously purplish-pink . . . *C. acanthoides*
1: Flower-heads 1.5-2.5 cm in length, heads clustered, flowers ± inconspicuously evident
    2. Flower-heads 1.5-2 cm long, usually numerous and densely clustered at the ends of stems and branches, frequently > 5 (5-20) in a cluster . . . *C. tenuiflorus*
    2: Flower heads 2-2.5 cm long, usually 2-5-clustered at the ends of stems and branches, or occasionally solitary . . . *C. pycnocephalus*

★ **Carduus acanthoides** Linnaeus. PLUMELESS THISTLE. Uncommon, opening of Douglas-fir forest: Pepperwood, *de Nevers* (1991). < 4200 ft. Native of e. Europe.

★ **Carduus pycnocephalus** Linnaeus. ITALIAN THISTLE. Common, waste areas: The Sea Ranch; Salt Point; Bodega Head; Freestone, *Huntington* (1987 NCC); Hessel Road near Gravenstein Hwy., acc. J. & B. Guggolz (1991); Blank Road near Gravenstein Hwy., acc. J. & B. Guggolz (1991); Todd Road Preserve; Fairfield Osborn (H); Penngrove, *Knight* (1965); Tresch Ranch, Walker Road near Two Rock, acc. J. & B. Guggolz (1992); w. of Sears Point Raceway in Tolay Cr. drainage, acc. Patterson (1990); Bouverie; Pepperwood; Lake Sonoma area; between Jimtown and Mercuryville on Geysers Road, *Utech & Knight* (1986 CM); 3200 ft. elev. Socrates Mine, acc. Clifton (1984); ca 2000 ft. elev. Anderson Trail on Black Mtn. between Geyser Peak and upper end of Pine Flat Road, acc. J. & B. Guggolz (1992); 3300 ft. elev. Squaw Rock, acc. Clifton (1983); 3600 ft. elev. Geyser Rock, acc. Clifton (1983). < 4000 ft. May-July. Naturalized from Mediterranean region.

★ **Carduus tenuiflorus** Curtis. SLENDER-FLOWERED THISTLE. Common, waste areas: hwy. a short distance n. of Stewarts Point, acc. Baker; Fort Ross, Cazadero and Duncans Mills quads.; Bodega Harbor Marsh, *Rubtzoff* (1955); Bodega Bay, *Sauffer* (1973); Todd Road Preserve, *Howell* (1987); Geyserville, Alexander Valley and Ida Clayton Road quads. < 3300 ft. May-July. Native of s. Europe.

Carthamus

1. Corolla yellow or whitish
    2. Heads broadly ovoid, phyllaries shorter or little longer than flowers . . . *C. lanatus*
    2: Heads narrowly ovoid or oblong, the outer phyllaries attenuate and much longer than flowers . . . *C. baeticus*
1: Corolla pale violet . . . *C. leucocaulos*

★ **Carthamus baeticus** (Boissier & Reuter) Nyman. SMOOTH DISTAFF THISTLE. Uncommon, waste places: Crane Can., *Ferlatte* (1973 NCC); Willow Brook n. of Petaluma, *Knight & Wagnon* (1988 NCC); Spring Lake; near Geyserville, acc. *Barbe* (1991). —*C. lanatus* ssp. *b.* (Boissier & Reuter) Nyman. *Kentrophyllum b.* Boissier & Reuter. < 1700 ft. July-Aug. Naturalized from Mediterranean region.

★ **Carthamus lanatus** Linnaeus. WOOLLY DISTAFF THISTLE. Occasional, waste places: Crane Can., *Ferlatte* (1973 NCC); Pepperwood; confluence of Dry and Galloway crs., *Howell* (1967); 10 locations throughout the county, acc. *Barbe* (1991). < 3600 ft. July-Aug. Naturalized from Old World.

★ **Carthamus leucocaulos** Sibthorp & Smith. WHITESTEM DISTAFF THISTLE. Locally abundant, waste places: Crane Can., *Ferlatte* (1973 NCC); both slopes of the dam of a reservoir ca 1 mi. w. of the Asti turnoff of US 101, *Fuller* (1969). < 1000 ft. Naturalized from Old World.

Centaurea

1. Flowers purplish or pinkish; stems wingless
    2. Plants perennial; achenes with definite pappus . . . *C. iberica*
    2: Plants annual or biennial; achenes without pappus or with merely a vestige . . . *C. calcitrapa*
1: Flowers yellow; stems winged with decurrent leaf-bases
    3. Spines slender, purplish, 1 cm long or less; plants branched mostly above the base; corolla glandular . . . *C. melitensis*
    3: Spines stout, yellow, 1-2 cm long; plants branched from base; corolla without glands . . . *C. solstitialis*

★ **Centaurea calcitrapa** Linnaeus. PURPLE STAR-THISTLE. Common, waste places; dense populations throughout the county, acc. *Barbe* (1991): San Antonio Road and US 101, *Knight* (1963); e. side Gravenstein Hwy. opposite Hessel Road near top of hill, *J. & B. Guggolz* (1990 JEPS); Copeland Cr. on Sonoma State Univ. campus, *Self* (1979 NCC); Spring Lake; Pepperwood. Usually < 3300 ft. July-Oct. Naturalized from Old World.

★ **Centaurea iberica** Sprengel. IBERIAN STAR-THISTLE. Uncommon, occurring locally: Santa Rosa, *Bremmer* (1932); Pepperwood; 2 locations in geothermal area between Mt. St. Helena and Pine Mtn., acc. *Barbe* (1991). < 1000 ft. Aug.-Nov. Native of Eurasia.
★ **Centaurea melitensis** Linnaeus. NAPA STAR-THISTLE, TOCALOTE. Common, cultivated and waste ground, roadsides, fields: The Sea Ranch; Fairfield Osborn (H); w. of Sears Point Raceway in Tolay Cr. drainage, acc. Patterson (1990); Spring Lake; Pepperwood; Fitch Mtn., *Rubtzoff* (1952); Geyserville quad.; Pine Flat Road, *Knight* (1963); 2040 ft. elev. Squaw Cr., acc. Clifton (1984). <7200 ft. May-June. Naturalized from Mediterranean region.
★ **Centaurea solstitialis** Linnaeus. BARNABY'S THISTLE, YELLOW STAR-THISTLE. Common, roadsides, waste ground: Bodega Head; Stony Point Road, *Filippi* (1965 NCC); Copeland Cr. between Sonoma State Univ. and Snyder Lane, *Self* (1979 NCC); w. of Sears Point Raceway in Tolay Cr. drainage, acc. Patterson (1990); railroad bridge over Petaluma River, *Knight* (1963); Pepperwood; chaparral area sw. of Bradford Mtn. 1½ mi. w. of Dry Cr., acc. J. & B. Guggolz (1992); Lake Sonoma area; along ridge between Socrates Mine and Mercuryville in geothermal area, *Dolan* & LaPré (1987 UCR); 3200 ft. elev. Socrates Mine, acc. Clifton (1984); ca 2000 ft. elev. Anderson Trail on Black Mtn. between Geyser Peak and upper end of Pine Flat Road, acc. J. & B. Guggolz (1992); 2800 ft. elev. Bear Cr. Can., acc. Clifton (1984). < 4200 ft. May.-0ct. Naturalized from Eurasia.

Chaenactis glabriuscula de Candolle. YELLOW PINCUSHION. Locally occasional, rock outcrops: 3600 ft. elev. Geyser Rock, acc. Clifton (1983); 3150 ft. elev. upper Squaw Cr. drainage 0.8 air-miles nw. of Geyser Rock, acc. Stone (1983); 3619 ft. elev. Anderson Cr. 1 mi. sw. of Cobb Mtn. summit, acc. Clifton (1984). < 5200 ft. May-June.

★ **Chamaemelum fuscatum** (Brotero) Carvalho. DUSKY-CHAMOMILE. Locally abundant, vineyards: roadside US 101 n. of Cloverdale, *Baker* (1940 NCC); ½ mi. s. of Santa Rosa, *Parsons* (1983 SFSU); Lake Sonoma area; 2 mi. s. of Asti, *Munz* (1952 NCC), *Howell* (1966, 1974), *Raven* (1963), *Baker* (1940 NCC); Asti, acc. Knight (1991). < 500 ft. Mar.-Apr. Naturalized from Mediterranean region. —*Anthemis fuscata* Brotero.

Chamomilla

1. Pappus-crown minute, entire, achenes with 2 glandular lines which = length of achene but not into the minute pappus-crown; mature heads usually 6-9 mm high . . . . *C. suaveolens*

1: Pappus-crown evident with 2 short lobes or teeth, these bearing oblanceolate or elliptic glands which scarcely extend on to the body of the achene; mature heads usually 10-11 mm high . . . . *C. occidentalis*

Chamomilla occidentalis (E. Greene) Rydberg. Uncommon, open ground inundated in winter: Laguna de Santa Rosa e. of Graton, *Rubtzoff* (1974); Laguna de Santa Rosa n. of Sebastopol, *Rubtzoff* (1958); Chileno Valley, *Rubtzoff* (1970); 5 mi. n. of Sears Point, *Rubtzoff* (1970). < 8000 ft. May-Aug. —*Matricaria o.* E. Greene.
★ **Chamomilla suaveolens** (Pursh) Rydberg. PINEAPPLE WEED. Common, weedy, waste places: The Sea Ranch; upper Green Valley Marsh nw. of Graton, *Rubtzoff* (1974); Todd Road Preserve; Fairfield Osborn; Tresch Ranch, Walker Road near Two Rock. acc. J. & B. Guggolz (1992); Chileno Valley, *Rubtzoff* (1964); 2 mi. n. of Adobe Road and Sonoma Mtn. Road, *Lowry* (1983 SFSU); very common in Petaluma area, acc. Knight (1990); Button Ranch on Walker Road (U. C. property) between Petaluma-Valley Ford Road and Pepper Road, acc. Patterson (1990); w. of Sears Point Raceway in Tolay Cr. drainage, acc. Patterson (1990); Sugarloaf Ridge; Spring Lake; 98 ft. elev. Bailhache Ave. along Russian River near Healdsburg, *Pitcher* (1975 PUA); Lake Sonoma area; between Jimtown and Mercuryville on Geysers Road, *Utech & Knight* (1986 CM). < 8000 ft. May-Aug. Native in w. N. America but probably introduced into northern California. —*Matricaria matricarioides* (Lessing) Porter.

★ **Chrysanthemum segetum** Linnaeus. CORN CHRYSANTHEMUM. Common, coastal, fields, waste places: 1½ mi. s. of Salmon Cr., *Knight* (1963); Bodega Head; 4 mi. e. of Valley Ford near Bloomfield, *Howell* (1930). Naturalized from Mediterranean region. < 700 ft. Apr.-Aug.

Chrysothamnus

1. Plants 5-20 dm high; foliage gray or white with ± copious tomentum, leaves 1 (-4) mm wide; inflorescence mostly cymose; phyllaries obtuse to moderately attenuate . . . . *C. nauseosus* ssp. *albicaulis*

1: Plants up to 5 dm high; foliage green, scabro-glandular, leaves 4-8 mm wide; inflorescence mostly racemose or spicate; phyllaries very attenuate . . . . *C. parryi* ssp. *latior*

Chrysothamnus nauseosus (Pallas) Britton ssp. **albicaulis** (Nuttall) Hall & Clements. RUBBER RABBITBRUSH. Uncommon, rock outcrops: 3600 ft. elev. Geyser Rock, acc. Clifton (1983). 150-10500 ft. Late summer.

Chrysothamnus parryi (A. Gray) E. Greene ssp. **latior** Hall & Clements. PARRY RABBITBRUSH. Uncommon, forest openings: 3200 ft. elev. Caldwell Pines, acc. Clifton (1984). 2200-6300 ft. July-Sept.

★ **Cichorium intybus** Linnaeus. CHICORY. Widespread, roadsides, waste places: The Sea Ranch; Salt Point; Two Rock, *Congdon* (1880); Lakeville Road, *Howell* (1930); Black Point Road, *Howell* (1930); False Lake Meadow and Ledson Marsh in Annadel, acc. Harrison with Rubtzoff. < 5000 ft. June-Oct. Naturalized from Mediterranean region.

Cirsium

1. Flowers unisexual, the plants dioecious; heads 1-1.5 cm high . . . . *C. arvense*

1: Flowers bisexual; heads > 1.5 cm high

  2. Stem spinose-winged; leaves scabrous-hispid above . . . . *C. vulgare*

  2: Stem not spinose-winged; leaves not scabrous-hispid above

    3. Phyllaries (outer, occasionally middle) ± pectinate-spinose on margins, middle phyllaries spreading . . . . *C. andrewsii*

    3: Phyllaries not pectinate on the margins, rarely the outer phyllaries spiny-margined in *C. remotifolium* and *C. douglasii* var. *breweri*; uppermost spiny-margined leaves occasionally closely subtend the head and appear to be phyllaries

      4. Back of at least some of the phyllaries with a glandular spot on ridge . . *C. douglasii* var. *breweri*

      4: Back of phyllaries with no glandular spot on ridge

        5. Corolla throat gradually narrowed downward into the tube, corolla without a distinct separation between throat and tube; flowers generally straight in the head, pinkish, purplish or reddish (or white in albinos); pubescence arachnoid or inconspicuously villous-arachnoid

          6. Heads often as broad as long or broader; phyllaries usually straight, almost acicular above appressed base, embellished with a fine, grayish-white webbing; corolla 2.5-3.5 cm long, scarcely exceeding involucre . . . . *C. occidentale* var. *occidentale*

          6: Heads usually longer than broad; phyllaries usually deflexed, outward or inward, linear to ovatish but not acicular; corolla to 4 cm long, usually exceeding the involucre

            7. Plants thinly to thickly arachnoid-tomentose, rarely white-tomentose; phyllaries often subglabrate . . . . *C. o.* var. *venustum*

            7: Plants snowy-white with dense, feltlike tomentum; phyllaries white-arachnoid-tomentose except for the glabrous yellow spine . . . . *C. o.* var. *candidissimum*

        5: Corolla-throat ± distinctly constricted at the base, the throat and tube usually distinct; flowers straight in head or often the outermost widely spreading, whitish, lavender, pink, rose or sometimes reddish; pubescence, at least on stem, consisting of crisp-puberulent and arachnoid hairs

          8. Phyllaries, at least the inner ones, ending in a ± expanded chartaceous appendage with the margins often fimbriate, lacerate, spinulose-denticulate, or crisped, phyllaries occasionally little expanded in intermediate hybrids . . . . *C. remotifolium*

          8: Phyllaries without an expanded, scarious or chartaceous appendage

            9. Phyllaries densely and conspicuously pubescent . . . . *C. brevistylum*

            9: Phyllaries subglabrous to thinly arachnoid . . . . *C. quercetorum*

Cirsium andrewsii A. Gray. FRANCISCAN THISTLE. Uncommon, wet or marshy ground along streams or around seepages: San Mateo Co. to Sonoma Co., acc *Munz* (1959). < 330 ft. June-July.

★ **Cirsium arvense** (Linnaeus) Scopoli. CANADA THISTLE. Uncommon, damp places: Willow Cr. Marsh, *Rubtzoff* (1955); near Caldwell Pines, acc. *Barbe* (1991). < 6000 ft. June-Sept. Native of Europe.

Cirsium brevistylum Cronquist. INDIAN THISTLE. Occasional, brushy slopes, coastal influence: The Sea Ranch; 1 mi. n. of Fort Ross, *Howell* (1930); 8 mi. n. of Bodega Bay, *Howell* (1930); 2 mi. e. of town of Bodega, *Howell* (1930). < 1700 ft. Apr.-Aug.

Cirsium douglasii de Candolle var. **breweri** (A. Gray) Keil & C. Turner. SWAMP THISTLE. Common, often on serpentine, wet places: The Sea Ranch; 6½ mi. s. of Gualala, *Eastwood & Howell* (1937); near Stewarts Point, *Abrams* (1920); Salt Point; 5.7 mi. n. of Fort Ross, *Cantelow* (1941); Bay Flat Road near Spud Point Marina at Bodega Harbor, *Clifton* (1984 PUA); Bodega Point, *Eastwood* (1915); Bodega Bay, *Ripley & Barneby* (1945); Mt. Hood, *Baker* (1926); Santa Rosa Cr. e. of Santa Rosa, *Heller* (1902); Sausel Cr. n. of Alexander Valley, *Hoover* (1966). —*C. b.* A. Gray. *C. canescens* Petrak. Sometimes forming possible hybrid with *C. quercetorum*: 6 mi. n. of Bodega Bay, *Eastwood & Howell* (1939); serpentine seep at lower elev. of Pine Flat Road, *W., I. & T. Knight* with *S. Edwards* (1986). < 7200 ft. June-Aug.

Cirsium occidentale (Nuttall) Jepson var. **candidissimum** (E. Greene) J. F. Macbride. SNOWY THISTLE. Uncommon: Bay Road 2 mi. w. of Bodega, *Ewan* (1935). < 6500 ft. June-Sept. —*C. pastoris* J. T. Howell.

Cirsium occidentale var. **occidentale**. COBWEBBY THISTLE. Occasional, usually in sandy places: The Sea Ranch, acc. Wells (1985); Horseshoe Cove, *Knight* (1963); Bodega Head, *J. Taylor* (1941), *Cantelow* (1941), *Howell* (1955); 2 mi. w. of Bodega, *Ewan* (1935); Fairfield Osborn (S); Lake Sonoma area; between Jimtown and Mercuryville on Geysers Road, *Utech & Knight* (1986 CM); ca 2000 ft. elev. Anderson Trail on Black Mtn. between Geyser Peak and upper end of Pine Flat Road, acc. J. & B. Guggolz (1992). < 2100 ft. May-July. —*C. coulteri* Harvey & Gray.

Cirsium occidentale var. **venustum** (E. Greene) Jepson. VENUS THISTLE. Occasional, rocky, clayey or sandy soil of grassland and brush: St. Helena Road 2 mi. from Napa Co. line, *Knight* (1964); Pepperwood; Fitch Mtn., *Rubtzoff* (1952); Skaggs Springs, *Hawver* (1915); 3200 ft. elev. Socrates Mine, acc. Clifton (1984); ca 2000 ft. elev. Anderson Trail on Black Mtn. between Geyser Peak and upper end of Pine Flat Road, acc. J. & B. Guggolz (1992); 2040 ft. elev. Squaw Cr., acc. Clifton (1984); 3619 ft. elev. Anderson Cr. 1 mi. sw. Cobb Mtn. summit, acc. Clifton (1984). < 12000 ft. Apr.-July. —*C. proteanum* J. T. Howell.

Cirsium quercetorum (A. Gray) Jepson. BROWNIE THISTLE. Locally common, mostly near coast: The Sea Ranch; 6 mi. n. of Bodega Bay, *Eastwood & Howell* (1937); 1 mi. e. of Bodega Bay on road to Bodega, *Howell & Stacey* (1930); Bodega Head, *Connors* (1991); Stony Point, *Howell* (1938). < 1000 ft. May-July. —*C. Walkerianum.* Petrak.

Cirsium remotifolium (Hooker) de Candolle. Uncommon: Bohemian Hwy. 2½ mi. from Camp Meeker en route to Monte Rio, *Howell* (1936); e. of Vine Hill School, *Baker* (1946 NCC); Bennett Valley Road in woods, *Baker* (1931 NCC); Pepperwood; 2 mi. e. of Petrified Forest, *Howell* (1937); Lake Sonoma area; along Cherry Cr. Road w. of Cloverdale, *Knight & Edwards* (1986); Alder Springs n. of Cloverdale, *Eastwood & Howell* (1938); 2040 ft. elev. Squaw Cr., *Clifton* (1984 PUA). < 6500 ft. June-Sept. —*C. callilepis* (E. Greene) Jepson var. *c.*, *C. c.* var. *pseudocarlinioides* (Petrak) J. T. Howell.

★ **Cirsium vulgare** (Savi) Tenore. BULL THISTLE. Common, weed, waste places:The Sea Ranch, *Wells* (1981); Salt Point; Willow Creek; 8½ mi. n. of Bodega Bay *Howell* (1930); Bodega Head; Freestone, *Huntington* (1987 NCC); Cunningham Marsh, *Howell* (1946); 3 sites w. and sw. of Santa Rosa at Laguna de Santa Rosa drainage, *Harrison* (1978 NCC); Todd Road Preserve (H&K); Fairfield Osborn (S); e. end of Railroad Ave. between Davis Ave. and Petaluma Hill Road, acc. Knight (1990); Button Ranch on Walker Road (U. C. property) between Petaluma-Valley Ford Road and Pepper Road, acc. Patterson (1990); w. of Sears Point Raceway in Tolay Cr. drainage, acc. Patterson (1990); Spring Lake; Pepperwood; Geyserville and Ida Clayton Road quads.; ca 2000 ft. elev. Anderson Trail on Black Mtn. between Geyser Peak and upper end of Pine Flat Road, acc. J. & B. Guggolz (1992). < 7600 ft. June-Sept. Native of Europe.

★ **Cnicus benedictus** Linnaeus. BLESSED THISTLE. Uncommon, weed, waste places: near Forestville, *Huffine* (1925 NCC); Vine Hill, *Huffine* (1932 NCC); Healdsburg, *Rattan* (1877). < 2700 ft. Apr.-July. Native of Europe.

Conyza

1. Midvein of phyllary brown, resin-filled . . . . *C. canadensis*
1: Midvein of phyllary green to purple, not resin-filled
  2. Central stem overtopped by lateral branches; pressed heads usually ± 1 cm wide; dry, reflexed phyllaries whitish or dull brown inside . . . . *C. bonariensis*
  2: Central stem > lateral branches; pressed heads usually < 1 cm wide; dry, reflexed phyllaries reddish-brown inside . . . . *C. floribunda*

★ **Conyza bonariensis** (Linnaeus) Cronquist. Frequent, sidewalk cracks, waste places: Petaluma Blvd. S., acc. Knight (1990); Magnolia Ave. w. of Petaluma, *Knight* (1963); e. of Farmers Lane, Santa Rosa, *Wells* (1989); Spring Lake; Pepperwood. Usually < 3300 ft. June-Aug. Native to S. America. —*Erigeron b.* Linnaeus.

Conyza canadensis (Linnaeus) Cronquist. HORSEWEED. Common, waste areas: Bodega Head; upper Pitkin Marsh, *Rubtzoff* (1951); Occidental, *Best & Wells* (1983); 2 mi. from Freestone on Bodega Hwy., *Best* (1983); Todd Road Preserve; Copeland Cr. on Sonoma State Univ. campus, *Self* (1979 NCC); foot of Jefferson St. at river in Petaluma, *Knight* (1963); Spring Lake. Usually < 6600 ft. June-Sept. —*Erigeron c.* Linnaeus.

★ **Conyza floribunda** Kunth. Rare, widely naturalized but scarce in the county: 3 blocks w. of Laguna de Santa Rosa Cr., Sebastopol, *Wells* (1990). < 330 ft.June-Sept. Sometimes considered = *C. bonariensis*. Native of tropical America.

★ **Coreopsis lanceolata** Linnaeus. Garden escape: The Sea Ranch, acc. Wells (1983). < 1700 ft. May-July. Native of e. U. S.

Cotula

1. Plant annual; leaves finely pinnate; heads 2-5 mm broad . . . . . *C. australis*
1: Plant perennial; leaves entire or pinnatifid; heads 8-10 mm broad . . . . . *C. coronopifolia*

★ **Cotula australis** (Sieber) Hooker f. Occasional, weed of gardens, waste places: n. of Pitkin Marsh in Trenton region, *Hoffing* (1940 NCC); Adobe Can. in garden, *Baker* (1956 NCC); Pepperwood. < 850 ft. Introduced from Australia.

★ **Cotula coronopifolia** Linnaeus. BRASS BUTTONS. Common, wet ground, low fields: The Sea Ranch; Horseshoe Cove, *Knight* (1963); Salt Point; Bodega Head; Fairfield Osborn, *Quibell* (1972 NCC); near US 101 bridge at Petaluma River, *Knight* (1963); Annadel; 3 sites w. and sw. of Santa Rosa in Laguna de Santa Rosa drainage, *Harrison* (1978); Pepperwood; 98 ft. elev. Bailache Ave. along Russian River near Healdsburg, *Pitcher* (1975 PUA). < 1000 ft. Mar.-Dec. Naturalized from S. Africa

Crepis occidentalis Nuttall. HAWKSBEARD. Uncommon, dry stony areas: 4200 ft. elev. at Units 20 and 21, *Malloch* (1977 PG&E), *Cooney* (1977 PG&E). 2700-9000 ft. June-Aug.

★ **Crupina vulgaris** Cassini. BEARDED CREEPER. Invasive, open places; a potentially serious range weed: Bennett Valley Golf Course, *Breedlove* (1976), *L. Davis* (1991 NCC); upper side of Canyon Trail ca ½ mi. uphill from Spring Cr. bridge at Annadel, *Wagnon & Vale* (1989 NCC). ± 800 ft. Native of the Old World.

Cynara

1. Leaves prominently spiny; phyllaries spinose-tipped . . . . . *C. cardunculus*
1: Leaves scarcely spiny; phyllaries unarmed . . . . . *C. scolymus*

★ **Cynara cardunculus** Linnaeus. CARDOON, ARTICHOKE THISTLE. Occasional, pastureland: n. of jct. of Walker and Pepper roads e. of Two Rock, acc. Patterson (1990); w. of Sears Point Raceway in Tolay Cr. drainage, acc. Patterson (1990); US 101 just n. of Washington St. exit, Petaluma, acc. Best (1981); near Freestone and Schellville, acc. *Barbe* (1991). < 1700 ft. May-July. Native of Mediterranean region.

★ **Cynara scolymus** Linnaeus. ARTICHOKE. Rare: acc. J. T. Howell, *Leaf. of West. Bot., 9:2 p. 28,* (1959), mentioned from Sonoma Co. < 1700 ft. May-July. Native of Mediterranean region.

Erechtites

1. Leaves ± lobed or pinnatifid; heads scarcely calyculate at base . . . . . *E. glomerata*
1: Leaves finely and sharply dentate, not at all lobed or pinnatifid; heads markedly calyculate . . . . . *E. minima*

★ **Erechtites glomerata** (Poiret) de Candolle. AUSTRALIAN FIREWEED. Common, invasive weed, waste places: The Sea Ranch; 50 ft. elev. Gerstle Cove at Salt Point State Park, *Clifton* (1981 PUA); Bodega Head. < 1700 ft. June-Aug. Native of Australasia. —*E. arguta* (A. Richard) de Candolle. *Senecio a.* A. Richard. *S. glomeratus* Desfontaines.

★ **Erechtites minima** (Poiret) de Candolle. AUSTRALIAN FIREWEED. Common, invasive weed, waste places: The Sea Ranch, *J. H. Thomas* (1976); Salt Point; mouth of Salmon Cr., *Ferris* (1937); 7 ft. elev. Bodega Head, *Clifton* (1980 PUA); Cunningham Marsh, *Howell* (1946); Fairfield Osborn (*); Wiggins Hill w. of Petaluma, *Knight* (1963).Usually < 1700 ft. July-Sept. Native of Australasia. —*E. prenanthoides* (A. Richard) de Candolle. *Senecio minimus* (Poiret) de Candolle.

Ericameria

1. Plants to 1.5 m high; ray-flowers present; leaves filiform 4-12 mm long, scarcely exceeding the axillary fascicles; heads ± paniculate . . . . . *E. ericoides*

1: Plants to 3 m high; ray flowers absent; leaves linear to filiform 3-6 cm long; heads in terminal cymes or cymose panicles . . . . . *E. arborescens*

Ericameria arborescens (A. Gray) E. Greene. GOLDEN-FLEECE. Common, chaparral: Bouverie; Cavedale Road, *Knight* (1984); Joost Ranch on Calistoga Road; *Knight* (1963); ca ½ mi. n. of Gates Road near Calistoga Road, acc. J. & B. Guggolz (1992); Pepperwood; Franz Valley School Road, acc. Lovell (1978); Harold Smith Ranch on Pine Mtn. Road, *Knight* (1963). < 4000 ft. Aug.-Nov. —*Haplopappus a.* (A. Gray) H. M. Hall.

Ericameria ericoides (Lessing) Jepson. MOCK HEATHER. Uncommon, coastal slopes, dunes: Bodega dunes, acc. Connors (1991); Ranch Road just n. of Bodega Bay on Nattkemper property, acc. Knight (1967); 16 ft. elev. Bay Flat Road near Spud Point Marina at Bodega Harbor, *Clifton* (1980 PUA). < 1700 ft. Aug.-Nov. —*Haplopappus e.* (Lessing) Hooker & Arnott.

Erigeron (Key ± acc. G. L. Nesom, *Hickman* [1993]).

1. Heads discoid
   2. Basal leaves present at flowering; cauline leaves gradually reduced . . . . . *E. supplex*
   2: Basal leaves 0 at flowering; cauline leaves usually evenly sized and spaced
      3. Glandless stem hairs spreading
         4. Stems 30-90 cm, erect; leaves 20-40 mm . . . . . *E. bioletti*
         4: Stems 10-30 cm, decumbent to ascending; leaves 10-25 mm . . . . . *E. petrophyllus*
      3: Glandless stem hairs 0-sparse, appressed
         5. Stems 30-90 cm, many, from woody caudex . . . . . *E. angustatus*
         5: Stems 8-20(30) cm, arising singly from slender, rhizomelike caudex branches . . . . . *E. reductus* var. *angustus*

1: Heads radiate
   6. Phyllaries strongly graded in 3-5 series . . . . . *E. serpentinus*
   6: Phyllaries ± =
      7. Heads 6-15 mm diameter, ray flowers 150-400 . . . . . *E. philadelphicus*
      7: Heads 15-35 mm diameter, ray flowers 80-165 . . . . . *E. glaucus*

Erigeron angustatus E. Greene. Occasional, serpentine areas, often in loose rocks, usually in brushy vegetation: near Cazadero, *Cooke* (1943 UC); Dutch Bill Cr.at Camp Meeker, *F. W. Hoffman* (1947 UC); between Occidental and Graton, *Howell* (1936); 2 mi. e. of Occidental, *Mason & Grant* (1946 UC). < 500 ft.

Erigeron biolettii E. Greene. Infrequent, dry rocky places: Duncans Mills, *M. Jones* (1882); near Guernewood Park, *H. M. Hall* (1925); Russian River below Guerneville, *Davy* (1896 UC); dry hills near Agua Caliente, Sonoma Valley, *Jepson* (1933 JEPS); Hope Valley at summit above Coleman's Ranch near Kenwood, *Baker* (1950 NCC); Adobe Can., *Howell* (1932); 2600 ft. elev. Mt. Hood, *Baker* (1948), *Biolettii* (1893 ND, type); St. Helena Road, *Knight* (1963); Mountain Home Ranch Road near Petrified Forest, *Stocking* (1960 NCC). < 3600 ft. July-Aug. —*E. inornatus* var. *b.* (E. Greene) Jepson.

Erigeron glaucus Ker Gawler. SEASIDE DAISY. Common, coastal strands, bluffs: The Sea Ranch; Horseshoe Cove, *Knight* (1963); Salt Point; Goat Rock State Beach, *Wells* (1981); 56 ft. elev. near Salmon Cr., *Reeve* (1980 PUA); Bodega Head; 52 ft. elev. Bodega Bay, *Escobar* (1974 PUA), *Marxmiller* (1975 PUA). < 75 ft. Apr.-Aug.

Erigeron petrophilus E. Greene. Uncommon, rock outcrops: 3000 ft. elev. w. flank Mt. St. Helena, *Grummer* (1980 PUA); 3120 ft. elev. Bear Cr. Can., acc. Clifton (1984); 3619 ft. elev. Anderson Cr. 1 mi. sw. of Cobb Mtn. summit, acc. Clifton (1984). 1700-6600 ft. July-Sept.

Erigeron philadelphicus Linnaeus. PHILADELPHIA DAISY. Rare, moist, open or protected grassy areas: Pitkin Marsh, *Quibell* (1987 NCC); w. side of Los Guilicos Valley, *Baker* (1950 NCC); s. slope of Black Mtn. near spring, *J. & B. Guggolz* (1992 UC NCC). < 4000 ft. Apr.-July.

Erigeron reductus (Cronquist) G. L. Nesom var. **angustatus** (A. Gray) G. L. Nesom. RAYLESS DAISY. Occasional, rocky outcrops: Monte Rio, *Campbell* (1920); 2 mi. e. of Occidental, *Mason & Grant* (1946); 3 mi. n. of Occidental on Bohemian Hwy., *Best* (1981); 1025 ft. elev. Coleman Valley Road, *Clifton* (1984 PUA); between Occidental and Graton, *Howell* (1936); Sugarloaf Ridge; Pepperwood; 3280 ft. elev. Socrates Mine, acc. Clifton (1984); unnamed ridge on w. side of Dry Cr. Valley, n. of Crane Cr., s. of Bradford Mtn., *Raiche* (1985); 3200 ft. elev. Caldwell Pines, acc. Clifton (1984). 300-4800 ft. July-Aug. —*E. inornatus* (A. Gray) A. Gray var. *a.* A. Gray.

Erigeron serpentinus G.L. Nesom. SERPENTINE DAISY. Rare, serpentine soil, seeps: Layton Mine at Austin Cr., *F. W. Hoffman* (1947 UC, type); above Austin Cr. below Layton Mine buildings at The Cedars, *Raiche* (1983 JEPS). 200-2200 ft.

Erigeron supplex A. Gray. SUPPLE DAISY. Rare, near coast: near N. Fire Station at The Sea Ranch, *Wells* (1985); 1.7 mi. n. of Stewarts Point, *Vestal* (1937). < 175 ft. May-June.

Eriophyllum

1. Heads 1.5-3 cm broad, 1-5+ on long peduncles
   2. Stems not strongly woody at base; achenes short 2.5-3 mm long; pappus present . . . . . *E. lanatum* var. *achillaeoides*
   2: Stems branching from a woody base; achenes 3-4 mm long; pappus reduced to a crown of erose teeth . . . . . *E. l.* var. *arachnoideum*

1: Heads 1.5 cm or less broad, several to many in sessile clusters or in short, pedunculate, branching clusters
   3. Ultimate branches bearing inflorescences stout; phyllaries 8-12, scarcely overlapping; heads 5-15+, peduncles < 1 cm . . . . . *E. staechadifolium*
   3: Ultimate branches bearing inflorescences slender; phyllaries 4-6, overlapping; heads 3-30+, peduncles 0-25 cm . . . . . *E. confertiflorum*

Eriophyllum confertiflorum (de Candolle) A. Gray. GOLDEN-YARROW. Common, brushy slopes: between Bodega Bay and Valley Ford, *Knight* (1964); Bouverie; ¾ mi. e. of jct. of Sonoma Mtn. Road and Bennett Valley Road near Glen Ellen, *Baker* (1934 NCC); Santa Rosa Cr. Road, *Baker* (1928 NCC); Lake Sonoma area. < 10000 ft. Apr.-Aug. —*E. c.* var. *discoideum* E. Greene, type locality: Sonoma County.

Eriophyllum lanatum (Pursh) J. Forbes var. **achillaeoides** (de Candolle) Jepson. COMMON WOOLLY SUNFLOWER. Common, dry slopes, brushy places: jct. Bittner and Joy roads near Occidental, *Ferris & Lorraine* (1961); Stoetz Lane e. of Occidental, *Utech & Knight* (1986 CM); Tresch Ranch, Walker Road near Two Rock, acc. J. & B. Guggolz (1992); between Sonoma Cr. and Mt. Hood, *Heller* (1902); between Mark West Springs and Petrified Forest, *Abrams* (1916); Pepperwood; Knights Valley, *Krauss* (1893); Petrified Forest, *Eastwood* (1915); near Healdsburg, *Constance* (1933); Healdsburg, *Rattan* (1877); Fitch Mtn., *Rubtzoff* (1952); head of Dohrman Cr. 4 mi. w. of Healdsburg on Jensen Ranch, *Mason* (1934); 350-400 ft. elev. Old Skaggs Springs Road, *Utech, Sato & Ohara* (CM 1990); Lake Sonoma area; along ridge between Socrates Mine and Mercuryville in geothermal area, *Dolan & LaPré* (1987 UCR); ca 2000 ft. elev. Anderson Trail on Black Mtn. between Geyser Peak and upper end of Pine Flat Road, acc. J. & B. Guggolz (1992); 2 mi. n. of Cloverdale, *Beane & James* (1949); 1½ mi. nw. of Cloverdale on Boonville Road, *Breedlove* (1962). < 4200 ft. May-July.

Eriophyllum lanatum var. **arachnoideum** (Fischer & Avé-Lallemant) Jepson. WOOLLY SUNFLOWER. Common, coastal bluffs, shady banks: The Sea Ranch, *Wells* (1981); 1 mi. n. of Fort Ross, *Howell* (1930); Russian Gulch n. of Jenner, *Constance* (1935); 5.1 mi. s. of Jenner on Hwy. 1, *Howell* (1930); Bodega Bay, *Heller & Brown* (1902); Bodega Point, *Eastwood* (1901, 1915); Cazadero, *Bacigalupi* (1926); Duncans Mills, *M. Jones* (1882); Coleman Valley Road, *J. & B. Guggolz* (1990 JEPS); Neeley Hill area s. of Guerneville, *Rubtzoff* (1953); 1024 ft. elev. Coleman Valley Road, *Clifton* (1984 PUA); Harrison Grade Road, *Baker* (1935); Bennett Valley, *Leitholt* (1893); Sugarloaf Ridge; 2 mi. w. of Petrified Forest, *Rose* (1933), *Rose & Howell* (1947); St. Helena Road, *Knight* (1963); head of Dohrman Cr. 4 mi. w. of Healdsburg on Jensen Ranch, *Mason* (1934); Fitch Mtn., *Rubtzoff* (1952); chaparral area sw. of Bradford Mtn. 1½ mi. w. of Dry Cr., acc. J. & B. Guggolz (1992); Pine Flat Road, *Knight* (1963); ca 2000 ft. elev. Anderson Trail on Black Mtn. between Geyser Peak and upper end of Pine Flat Road, acc. J. & B. Guggolz (1992); Geyser Peak, *Knight & Edwards* (1986); 3700 ft. elev. Cobb Mtn., *Jokerst* (1985 PUA);

2040 and 3060 ft. elev. Squaw Cr., acc. Clifton (1983, 1984); 3000 ft. elev. Caldwell Pines, acc. Clifton (1984). < 4000 ft. Apr.-Aug. —*Bahia arachnoidea* Fischer & Avé-Lallemant, type locality: Fort Ross Russian Colony.

Eriophyllum staechadifolium Lagasca. LIZARD-TAIL, SEASIDE WOOLY SUNFLOWER. Common, coastal bluffs, beaches: The Sea Ranch; Horseshoe Cove, *Knight* (1963); Salt Point; Goat Rock State Beach, *Wells* (1981); 49 ft. elev. Bodega Head, *Clifton* (1977 PUA); Pinnacle Rock Trail in Bodega Harbour development, *Huntington* (1989 NCC). < 350 ft. Mar.-Sept. —*E. s.* var. *artemisiaefolium* (Lessing) J. F. Macbride, currently not recognized.

Euthamia occidentalis Nuttall. WESTERN GOLDENROD. Occasional, moist ground: 200 ft. elev. Willow Cr. Road, *Knight* (1984); near bridge on Magnolia Ave. 0.3 mi. from Thompson Lane w. of Petaluma, *Knight* (1963); Pepperwood. < 7800 ft. July-Nov. —*Solidago o.* (Nuttall) Torrey & A. Gray.

Filago

1. Longest heads 5-7 mm, yellowish, in dense, spheric groups of 12-20, largest groups 11-15 mm; receptacle 5-15 × as long as wide, narrowly cylindric; middle chaff-scales acuminate, ± awned, persistent, obscurely winged . . . . . *F. pyramidata*
1: Longest heads 3-4.5 mm, grayish, loose ± hemispheric groups of 1-10, all groups < 11 mm; receptacle 0.5-2 × as long as wide, expanded at tip; middle chaff-scales narrowly obtuse to acute, each falling with a fruit, scarious- winged
    2. Uppermost leaves much longer than the heads . . . . . *F. gallica*
    2: Uppermost leaves scarcely or not longer than the heads . . . . . *F. californica*

Filago californica Nuttall. Occasional, grassland, brushy areas: Salt Point; sw. slope Mt. Hood, *Baker* (1947 NCC); Mt. Hood near Kenwood, *Mason* (1947 NCC); hills near Boundary Ranch w. of Dry Cr. Valley, *Baker* (1934 NCC). < 6200 ft. Mar.-June.

★ **Filago gallica** Linnaeus. Uncommon, waste places: Bodega dunes, acc. Connors (1991); 3 mi. e. of Valley Ford, *Howell* (1930); w. of Santa Rosa in Laguna de Santa Rosa drainage, *Harrison* (1978 NCC); Fairfield Osborn (H); Button Ranch on Walker Road (U. C. property) between Petaluma-Valley Ford Road and Pepper Road, acc. Connors (1992); Bouverie; Spring Lake; Pepperwood; Lake Sonoma area; Ida Clayton Road quad.; ca 2000 ft. elev. Anderson Trail on Black Mtn. between Geyser Peak and upper end of Pine Flat Road, acc. J. & B. Guggolz (1992); Geyser Peak, *Knight & Edwards* (1986); 2860 ft. elev. Pine Flat Road, acc. Clifton (1984); 3200 ft. elev. Socrates Mine, acc. Clifton (1984); 2440 ft. elev. Caldwell Ranch, acc. Clifton (1984). < 4700 ft. Apr.-June. —*Logfia g.* (Linnaeus) Casson & Germain. Naturalized from Europe.

★ **Filago pyramidata** Linnaeus. Uncommon, waste areas: hills e. of Geyserville on road to The Geysers, *Howell* (1976); Geyser Peak, *Knight & Edwards* (1986). 300-1700 ft. Native of Mediterranean region. —*F. vulgaris* Lamarck, misapplied.

Gnaphalium

1. Inflorescence paniculate or cymose-paniculate or corymbose, the branches spreading or much congested, the solitary or glomerate heads not leafy-bracted
    2. Leaves persistently woolly above as well as below, grayish or white
        3. Annuals; heads 150-230-flowered
            4. Phyllary-tips almost hyaline, whitish faintly tinged with brown; heads 3-3.5 mm high; corollas reddish at tip . . . . . *G. luteo-album*
            4: Phyllary-tips scarious, yellowish or straw-colored; heads 4-6 mm high; corolla yellow . . *G. stramineum*
        3: Perennials, herbaceous or ± woody below; heads 35-90-flowered . . . . *G. canescens* ssp. *beneolens*
    2: Leaves green and glandular above, or if at all floccose, thinly so and usually becoming glabrate
        5. Heads campanulate to globose; phyllaries pearly white except in extreme age, few outer pointed . . . . . *G. californicum*
        5: Heads turbinate or narrowly ovate; phyllaries pink or tinged with pink, ± rounded to pointed . . . . . *G. ramosissimum*
1: Inflorescence spiciform or capitate and leafy-bracted, or of leafy-bracted glomerules at the tips of the branches, the leaves surpassing the heads
    6. Lower leaf-surface closely white-pannose, the subappressed hairs tightly enmeshed; involucre densely woolly at base only 4-6 mm high; phyllaries with greenish base, light brown or often purplish . . . *G. purpureum*
    6: Lower leaf-surface loosely villose-lanate; involucre nearly buried in wool 3-4 mm high; phyllaries greenish-centered below . . . . . *G. palustre*

Gnaphalium californicum de Candolle. Common, moist brushy slopes, coastal headlands: The Sea Ranch; Bodega Point, *Eastwood* (1915); forested hills se. of Guerneville, *Rubtzoff* (1952); Calistoga Road at Joost Ranch, *Knight* (1963); Fitch Mtn., *Rubtzoff* (1952); Lake Sonoma area; Sulphur Cr. Road 1 mi. from US 101 near Cloverdale, *Knight* (1963). < 6200 ft. Jan.-July.

Gnaphalium canescens de Candolle ssp. **beneolens** (Davidson) Stebbins & Keil. EVERLASTING. CUDWEED. Occasional, brushy slopes: Calistoga Road at Joost Ranch, *Knight* (1963); 5 mi. w. of Calistoga, *Krauss* (1893); Pepperwood. < 4000 ft. July-Nov. A specimen collected by *Knight* (1963) at offramp of US 101 and Railroad Ave. appears to be a hybrid: *G. c.* ssp. *beneolens* × *G. californicum*. Another specimen of the latter was collected on Los Alamos Road, *Knight* (1963).

★ **Gnaphalium luteo-album** Linnaeus. Common, coastal hills, roadsides, flat areas: sandbar in S. Fork Gualala River, *Wells* (1984); near n. end of the Sea Ranch, *J. H. Thomas* (1976); Stump Beach in Salt Point State Park, *Wells* (1982); edge of marsh at mouth of Salmon Cr., *Ferris* (1937); Bodega Head; Bodega Point, *Eastwood* (1915); Bodega Bay, *Heller* (1932); Guerneville, *Rattan* (1903); Pitkin Marsh, *Rubtzoff* (1951); Sebastopol, *Dows* (1907); Copeland Cr. on Sonoma State Univ. campus, *Self* (1979); Button Ranch on Walker Road (U. C. property) between Petaluma-Valley Ford Road and Pepper Road, acc. Patterson (1990); Willow Brook n. near Petaluma, *Howell & Knight* (1987); Rincon Valley, *Wells* (1985); Annadel, *Wells* (1984); Bechtel cottage in Pepperwood, *Howell et al.* (1982). < 6800 ft. All year. Introduced from the Old World.

Gnaphalium palustre Nuttall. Occasional, damp banks, stream beds: between Occidental and Graton, *Howell & Stacey* (1936); Pitkin Marsh, *Rubtzoff* (1951); 3 sites w. of Santa Rosa in Laguna de Santa Rosa drainage, *Harrison* (1978); Two Rock, *Howell* (1936); Santa Rosa, *Heller* (1902). < 9000 ft. May-Oct.

Gnaphalium purpureum Linnaeus. Common, dry, often disturbed or waste places: The Sea Ranch; 7 mi. s. of Gualala, *J. H. Thomas* (1976); Stewarts Point quad.; s. end of main parking lot at Salt Point State Park, *Wells* (1982); Horseshoe Cove, *Knight* (1963); 1 mi. n. of Salmon Cr. on Hwy. 1, *Keck & Clausen* (1935); ca 1 mi. nw. of Camp Meeker, acc. J. & B. Guggolz (1991); Guerneville, *Rattan* (1877); Pitkin Marsh, *Rubtzoff* (1951); Cunningham Marsh, *Howell* (1947); w. of Santa Rosa in Laguna de Santa Rosa drainage, *Harrison* (1978 NCC); Bouverie; Pepperwood; Alexander Valley quad. < 4000 ft. Apr.-July.

Gnaphalium ramosissimus Nutall. Rare in flat dry areas: Russian Gulch n. of Jenner, *Harrison* (1989 SRJC); Bodega Bay just e. of Highway 1, *Harrison* (1987 SRCJ); Stoltz Lane 2 mi ne. of Occidental, *Best* (1984). < 2100 ft. Mar.-Sept.

Gnaphalium stramineum Kunth. Common, maritime slopes, brushy and wooded hills, open flats: The Sea Ranch, acc. Stone (1980); Stewarts Point quad.; Salt Point, *Wells* (1982); Horseshoe Cove, *Knight* (1963); Goat Rock State Beach, *Wells* (1981); Sonoma Coast State Beach s. of Coleman Beach, *Frankel* (1964); Bodega Head; 2 sites w. of Santa Rosa in Laguna de Santa Rosa, *Harrison* (1978 NCC); in ditch on Magnolia Ave. w. of Petaluma, *Knight* (1963); Howarth Park in Santa Rosa, *W., I. & T. Knight* (1984); Pepperwood; 2720 ft. elev. Caldwell Ranch, acc. Clifton (1984). < 6600 ft. June-Oct. —*G. chilense* Spengel.

Grindelia

1. Plant a subshrub, 10-15 dm, tidal and/or saline areas . . . . *G. stricta var. angustifolium*
1: Plant parts not woody, < < shorter, areas not tidal and/or saline
   2. Tips of phyllaries erect or spreading, some gradually curved but not sharply reflexed . . . . . *G. hirsutula*
   2: Tips of phyllaries (at least of some middle and outer ones) sharply reflexed or looped
      3. Plants coastal or near coastal ± succulent; mostly procumbent to decumbent . . . . *G. stricta* var. *platyphylla*
      3: Plants interior non-succulent; stems ± erect . . . . *G. camporum*

Grindelia camporum E. Greene. GUM PLANT. Common, dry habitats: Tomales Road near Marin-Sonoma county line, *Knight* (1963); between Cotati and Santa Rosa, *Eastwood* (1921); 6 mi. from Sonoma on road to Napa, *Howell* (1932); Sonoma, *Rose* (1940); Pepperwood; Mark West Cr. n. of Santa Rosa, *Heller* (1902); between Petrified Forest and Mark West, *Abrams* (1916); 4 mi. from Petrified Forest on road to Knights Valley, *Howell* (1929); Asti (s. of Cloverdale), *Knight* (1963); 3200 ft. elev. Socrates Mine, acc. Clifton (1984); The Geysers, *M. Jones* (1931). < 4700 ft. May-Oct.

Grindelia hirsutula Hooker & Arnott var. **hirsutula**. GUM PLANT. Common, dry slopes: 1024 ft. elev. Coleman Valley Road, *Clifton* (1984 PUA); ca 1 mi. nw. of Camp Meeker, *J. & B. Guggolz* (1991 UC); 8 mi. w. of Petaluma on Bodega Hwy., *Howell* (1930); Black Point Road, *Howell* (1930); banks of

Petaluma River, *Arnold* (1963 NCC); 4 mi. from Santa Rosa on road to Sonoma, *Howell* (1929); Bennett Valley, *Gutermute* (1932 NCC); Pepperwood. < 5700 ft. May-Oct.

Grindelia stricta de Candolle var. **angustifolium** (A. Gray) M. A. Lane. Common, brackish areas: along Petaluma River intermittently from Petaluma to mouth of river, acc. Knight (1994). < 3000 ft. Aug.-Oct. —*G. humilis* Hooker & Arnott misapplied.

Grindelia stricta de Candolle var. **platyphylla** (E. Greene) M. A. Lane. Common, coastal salt marshes, maritime bluffs, sometimes on serpentine: The Sea Ranch; Salt Point; near Goat Rock Beach, *Best* (1982); Coleman Valley Road ca 7 mi. W. Of Occidental, *Best* (1981); Duncan's Landing between Salmon Cr. mouth and Jenner, *Wells* (1983); 4 mi. n. of Bodega Bay, *Howell* (1930); 2 mi. n. of Bodega Bay, *Munz* (1936), *King* (1980); w. side of Bodega Bay, *Best* (1986); Bodega Head; 30 ft. elev. Bodega, *Clifton* (1980 PUA); Pinnacle Gulch Trail in Bodega Harbour development, *Huntington* (1989 NCC); Coleman Valley Road ca 7 mi. w. of Occidental, *Best* (1981). < 700 ft. June-Sept. —*G. s.* ssp. *venulosa* (Jepson) Keck. *G. arenicola* Steyermark.

Gutierrezia californica (de Candolle) Torrey & Gray. CALIFORNIA MATCHWEED, SNAKEWEED. Rare, dry hills, flats: Sonoma, *Behr* (1860s U). 300-1000 ft. May-Oct. —*G. bracteata* Abrams.

★ **Hedypnois cretica** (Linnaeus) Dumont de Courset. CRETE WEED. Common, widely naturalized weed: mouth of Salmon Cr., *Baker* (1930 NCC), *Howell* (1931); Lichau Road ca 1 mi. e. of Roberts Road which begins from Petaluma Hill Road, *Clopton* (1973 NCC); Annadel Heights, e. Santa Rosa, *Harrison* (1989 SRJC); Burma and Live Oak trails in Annadel, acc. Harrison (1992); Spring Lake; Mark West Cr. near schoolhouse, *Baker* (1927 NCC); Bechtel cottage at Pepperwood, *Howell et al.* (1982). < 2700 ft. Apr.-May. Native of Mediterranean region.

Helenium

1. Rays inconspicuous, much shorter than the disk; stems openly branched above .... *H. puberulum*
1: Rays conspicuous, equalling to much extending the disk; stems simple or few-branched above
  2. Peduncles stout and enlarged (flared) at summit; disk hemispheric 2-3 cm wide .... *H. bolanderi*
  2: Peduncles slender; disk subglobose 1.5-2 cm wide .... *H. bigelovii*

Helenium bigelovii A. Gray. SNEEZEWEED. Uncommon, moist meadowy areas: Pitkin Marsh, *Howell & Stacey* (1936); swamps near Santa Rosa Cr., *Bigelow* (type, 1854 U); 3160 and 3200 ft. elev. Socrates Mine, acc. Clifton (1984). ± 0-10000 ft. June-Aug.

Helenium bolanderi A. Gray. Uncommon, coastal swamps, bluffs, moist meadows: The Sea Ranch; Cunningham Marsh, *Howell* (1947). < 330 ft. June-Aug.

Helenium puberulum de Candolle. Common, moist meadows, marshes: The Sea Ranch; Stewarts Point; Stewarts Point-Skaggs Springs Road within valley of the Wheatfield Fork Gualala River, *Wells* (1980); Salt Point; Willow Cr. Road 2½ mi. from Bridgehaven, *Knight* (1963); 7 ft. elev. Bodega Head, *Clifton* (1980 PUA); ca 1 mi. nw. of Camp Meeker, acc. J. & B. Guggolz (1991); Button Ranch on Walker Road (U. C. property) between Petaluma-Valley Ford Road and Pepper Road, acc. Connors (1992); Bouverie; Spring Lake; Pepperwood; Lake Sonoma area. < 4000 ft. June-Sept.

Helianthella californica A. Gray. CALIFORNIA HELIANTHELLA. Occasional, grassy slopes, chaparral: Sonoma Mtn. Road just w. of Enterprise Road, acc. Quibell (1991); 3 mi. e. of Sonoma on Wood Valley Road, *Best* (1981); Bouverie; 1½ mi. up Cavedale Road from Hwy. 12, *Utech & Knight* (1986 CM); 3 mi. up Trinity Road, acc. Knight (1989); Pepperwood; Dry Cr. area w. of Healdsburg, *Raiche* (1985); ca 2000 ft. elev. Anderson Trail on Black Mtn. between Geyser Peak and upper end of Pine Flat Road, acc. J. & B. Guggolz (1992); Geyser Peak, *Knight & Edwards* (1986). < 3500 ft. Apr.-June.

Helianthus

1. Stems 0.5-4 m tall, usually rough-hispid; leaves usually cordate at base .... *H. annuus*
1: Stems 0.3-1.3 m tall, rough-scabrous to hirsute-villous; leaves cuneate to truncate at base ...... *H. exilis*

Helianthus annuus Linnaeus. COMMON SUNFLOWER. Uncommon, waste places and roadsides: 300 ft. elev. ¼ mi. e. of Ash Cr. bridge on Geysers Road, *Andreae* (1992 NCC). < 6500 ft. Feb.-Oct.

Helianthus exilis A. Gray. SERPENTINE SUNFLOWER. Difficult to distinguish from *H. bolanderi* A. Gray, which has lance-acuminate involucral bracts. Rare, vernally wet seeps, serpentine: 3160 ft. elev. Socrates Mine, *Clifton* (1984 PUA). < 4000 ft. July-Nov.

Hemizonia

1. Leaves rigidly spine-tipped
    2. Disk pappus 0
        3. Leaves and bracts scabrous . . . . . *H. pungens* ssp. *pungens*
        3: Leaves and bracts glabrous . . . . . *H. p.* ssp. *maritima*
    2: Disk pappus present
        4. Chaff scales obtuse, soft-hairy; anthers yellow to black . . . . . *H. fitchii*
        4: Chaff scales obtuse to acute, ± resinous-thickened, not long soft-hairy; disk flower anthers yellow
            5. Herbage glandular and long-hairy . . . . . *H. parryi* ssp. *parryi*
            5: Herbage not glandular, variously hairy . . . . . *H. p.* ssp. *rudis*
1: Leaves not spine-tipped
    6. Ray achenes beakless; ligules white to yellow; chaff scales scattered throughout, ± fused at base
        7. Corollas ± yellow
            8. Flowering July-Nov.; 2-7 dm high; leaves inconspicuous at time of flowering, very short, crowded into a basal rosette; densely glandular throughout . . . . . *H. congesta* ssp. *congesta*
            8: Flowering Apr.-July; 1-3 (-5) dm high; leaves prominent at time of flowering, elongated, not crowded into a basal rosette
                9. Leaves not grasslike, deep green; herbage glandular throughout . . . . . *H. multicaulis* ssp. *multicaulis*
                9: Leaves grasslike, usually silvery sericeous; herbage glandular above . . . . *H. m.* ssp. *vernalis*
        7: Corollas white
            10. Phyllary tips usually >> body . . . . . *H. congesta* ssp. *leucocephala*
            10: Phyllary tips usually < body
                11. Heads usually ± sessile along inflorescence branches, in clusters . . *H. c.* ssp. *clevelandii*
                11: Heads usually peduncled, ± terminal on inflorescence branches . . . *H. c.* ssp. *luzulifolia*
    6: Ray achenes with short, off-center beak; ligules yellow; chaff scales in 1 ring between ray and disk flowers, free or ± fused . . . . . *H. corymbosa*

Hemizonia congesta de Candolle ssp. **clevelandii** (E. Greene) Babcock & H. M. Hall. Common, dry interior grassy hills: between Fulton and Sebastopol, *McMurphy* (1908); Spring Lake; foothills e. of Fulton, *McMurphy* (1908); Blue Bell Lane of Coddingtown in Santa Rosa, *Knight* (1963); 1 mi. s. of Sonoma, *McMurphy* (1909); 1 mi. w. of Santa Rosa en route to Fulton, *McMurphy* (1909); Santa Rosa, *McMurphy* (1917); between Petrified Forest and Mark West Springs, *Abrams* (1916); 4 mi. w. of Petrified Forest, *Rose* (1933); Pepperwood; 2 mi. n. of Healdsburg, *Rose* (1935); Lytton Station between Healdsburg and Cloverdale, *McMurphy* (1917); 3200 ft. elev. Socrates Mine, acc. Clifton (1984); 3600 ft. elev. Geyser Rock, *Clifton* (1983 PUA); hills near Cloverdale, *McMurphy* (1909). < 4700 ft. June-Oct. —*H. cl.* E. Greene.

Hemizonia congesta ssp. **congesta**. HAYFIELD TARWEED. Common, grassy valleys, hillsides, not along immediate coast: lagoon at Sebastopol, *Heller* (1902); Leddy Station between Santa Rosa and Sebastopol, *McMurphy* (1909); Hwy. 116 and Hessel Road, *J. & B. Guggolz* (1990 JEPS); near Two Rock, *Quibell* (1988 NCC); Petaluma Hill Road 4½ mi. s. of Santa Rosa, *McMurphy* (1909); Fairfield Osborn (S); 5 mi. ne. of Penngrove, *Rose* (1964); 7½ mi. nw. of Petaluma, *H. M. Hall* (1916); roadside near Petaluma, *McMurphy* (1917); jct. of Adobe and Casa Grande roads near Petaluma, *Hoover* (1969); between Petaluma and Sonoma, *McMurphy* (1909); s. of Petaluma, *H. M. Hall* (1930); 8 mi. e. of Petaluma en route to Sonoma, *Heusi* (1930); Petaluma Cr. at bridge on Black Point cutoff, *Raven* (1955); Spring Lake; Santa Rosa, *Abrams* (1907), *Dudley* (1896); s. of Santa Rosa, *Clausen* (1932); 4½ mi. s. of Santa Rosa, *McMurphy* (1909); 1 mi. w. of Santa Rosa en route to Fulton, *McMurphy* (1909); Pepperwood; Lake Sonoma area. < 1000 ft. July-Nov. —*H. lutescens* (E. Greene) Keck.

Hemizonia congesta ssp. **leucocephala** (B. D. Tanowitz) Keil. Common, open fields: 2 mi. nw. of Camp Meeker, *J. & B. Guggolz* (1991 NCC UC); Perry Marsh, *Howell* (1945); 3 sites w. of Santa Rosa in Laguna de Santa Rosa drainage, *Harrison* (1978 NCC); Todd Road Preserve; 8 mi. w. of Petaluma on Bodega Road, *Howell* (1930); Chileno Valley, *J. & B. Guggolz* (1990 DAV); Los Alamos Road, *Knight* (1963); 1200 ft. elev. Calistoga Road, *Clifton* (1981 PUA); Spring Lake; 2 mi. n. of Windsor, *Eastwood & Howell* (1936); n. of Arata Lane, Windsor, *J. & B. Guggolz* (1991 DAV); Pepperwood; Lake Sonoma area; Harold Smith Ranch on Pine Mtn. Road, *Knight* (1963). < 1200 ft. May-Oct. —*H. c.* de Candolle.

Hemizonia congesta ssp. **luzulifolia** Babcock & H. M. Hall. HAYFIELD TARWEED. Common, hills, valleys: Todd Road Preserve; Fairfield Osborn (*); 6½ mi. ne. of Penngrove, *Rose* (1964); 2.3 mi. sw. of Schellville en route to Napa, *Keck* (1933); Glen Ellen, *Howell* (1932); 5.9 mi. up Cavedale Road, *Knight* (1984); Mt. Hood, *Knight* (1964); Los Alamos Road, *Knight* (1963); Mark West Cr. on Calistoga Road near jct. of St. Helena Road, *Hoover* (1966); Pepperwood; Knights Valley, *Tracy* (1930); n. of Healdsburg, *H. M. Hall* (1924); Lake Sonoma area; Sausal Cr. n. of Alexander Valley, *Hoover* (1966); 1 mi. e. of Asti, *J. & B. Guggolz* (1990 JEPS); along ridge between Socrates Mine and Mercuryville in geothermal area, *Dolan & LaPré* (1987 UCR); Caldwell Pines, acc. Stone (1989). < 3300 ft. Apr.-June. —*H. l.* de Candolle ssp. *l., H. l.* ssp. *rudis* (Bentham) Keck.

Hemizonia corymbosa (de Candolle) Torrey & A. Gray. Uncommon, mostly open areas: The Sea Ranch; Salmon Cr., acc. Connors (1991); Lohrman Lane w. of Petaluma, *Knight* (1965); between Napa jct. and Petaluma (probably Lakeville), *McMurphy* (1910). < 1000 ft. May-Oct.

Hemizonia fitchii A. Gray. FITCH'S SPIKEWEED. Common, dry hills, plains: Guerneville, *Rattan* (1877); 1 mi. e. of Schellville en route to Napa, *Keck* (1933); Spring Lake; near Santa Rosa, *Knight* (1965); Pepperwood; Mill Cr. Road, *Baker* (1925, 1930 NCC); near Healdsburg, *Baker* (1930), *Rose* (1935); Lytton, n. of Healdsburg, *McMurphy* (1917); 6 mi. n. of Geyserville, *Howell* (1930); Caldwell Pines, acc. Stone (1989). 100-3300 ft. May-Nov. —*Centromadia f.* (A. Gray) E. Greene.

Hemizonia multicaulis Hooker & Arnott ssp. **multicaulis**. SEASIDE TARWEED. Occasional, seaward bluffs along immediate coast: The Sea Ranch; 8 mi s. of Fort Ross, *Howell* (1930); 1 mi. n. of Jenner, *Best* (1981); n. of Jenner, *Ripley & Barneby* (1945); ½ mi. s. of Jenner, *Keck* (1933); 1 mi. n. of mouth of Salmon Cr., *Keck & Clausen* (1935); 4 mi. n. of Bodega Bay, *Howell* (1930), *Cantelow* (1941); Bodega Bay, *H. M. Hall* (1931). < 300 ft. May-July. —*H. congesta* ssp. *c.*

Hemizonia multicaulis ssp. **vernalis** Keck. KECK'S TARWEED. Occasional, away from immediate coast, serpentine: Monte Rio, *Campbell* (1920); 2½ mi. n. of Occidental, *Keck* (1933); Morelli Lane 2 mi. n. of Occidental, *Best* (1981); Harrison Grade Road between Occidental and Graton, *Utech & Knight* (1986 CM). < 700 ft. Apr.-June. —*H. congesta* ssp. *c.*

Hemizonia parryi E. Greene ssp. **parryi**. PAPPOSE SPIKEWEED. Occasional, low, often alkaline fields: Willow Brook n of Petaluma, *Howell & Knight* (1987); 2.3 mi. sw. of Schellville en route to Napa, *Keck* (1933); 5 mi. e. of Mark West Springs on Petrified Forest Road, *Eastwood & Howell* (1938). < 750 ft. June-Oct. —*Centromadia p.* (E. Greene) E. Greene.

Hemizonia parryi ssp. **rudis** (E. Greene) Keck. Uncommon, fields: hard-packed road surface in very saline area on lower Tubbs Island ¼ mi. from San Pablo Bay, *Knight* (1985). < 400 ft. June-Oct. —*Centromadia r.* E. Greene.

Hemizonia pungens (Hooker & Arnott) Torrey & A. Gray ssp. **maritima** (E. Greene) Keck. Uncommon, grassy embankment: Keokuk St. near St. Vincent's school, Petaluma, *Knight* (1963). < 200 ft. Apr.-Oct. —*Centromadia m. E.* Greene.

Hemizonia pungens ssp. **pungens**. COMMON SPIKEWEED. Occasional, dry flat areas; Willow Cr. Road at 200 ft. elev., *Knight* (1984); w. of Petaluma, *H. M. Hall* (1931); adobe flat on Madison St., Petaluma, *Koch* (1938 NCC); summit of hills between Petaluma and Sonoma, *McMurphy* (1917); s. of Piner Road e. of Peterson Lane, n. of Cassidy Court (now occupied by housing tract), Santa Rosa, *Harrison* (1978 NCC). < 1700 ft. Apr.-Oct. —*Centromadia p.* (Hooker & Arnott) E. Greene.

Hesperevax

1. Plants 0.5-3 or 4 cm high; longest leaves 6-12 (14) mm, ± round . . . . *H. sparsiflora* var. *brevifolia*
1: Plants 3-10 cm high; longest leaves (10) 13-30 mm, ± obovate . . . . *H. s.* var. *sparsiflora*

Hesperevax sparsiflora (A. Gray) E. Greene var. **brevifolia** (A. Gray) J. Morefield. SHORT-LEAVED EVAX. Uncommon, open areas: 7 mi. s. of Gualala (The Sea Ranch), *J. H. Thomas* (1976). < 700 ft. Mar.-May. —*Evax s.* (A. Gray) Jepson var. *b.* (A. Gray) Jepson, in part.

Hesperevax sparsiflora var. **sparsiflora**. Common, compacted soil, open or brushy hills: The Sea Ranch; Salt Point, *Vale & Wagnon* (1988 NCC); Bodega Head, acc. Connors (1991); near Todd Road sw. of Santa Rosa, *Harrison* (1978 NCC); Fairfield Osborn (*); Petaluma, *Eastwood* (1921); along Dry Cr. Road ca 3 mi. past Warm Springs Road, *Breedlove* (1976); Pepperwood; Lake Sonoma area. < 3000 ft. Mar.-May. —*Evax s.*

Heterotheca

1. Ray flowers 0; outer pappus bristles very few . . . . *H. oregona*
   2. Leaves on branches ± (sub)glabrous, ± densely glandular . . . . *H. o.* var. *oregona*
   2: Leaves on branches ± densely hairy, sparsely glandular . . . . *H. o.* var. *rudis*

1: Ray flowers present; outer pappus bristles usually many
   3. Plant annual to short-lived perennial; ray fruits ± glabrous; ray pappus 0 . . . . *H. grandiflora*
   3: Plants perennial; ray fruits hairy; ray pappus present
      4. Lower stem sparsely woolly, upper leaves oblanceolate, little reduced, flat, strigose to long-woolly . . . . *H. sessiliflora* ssp. *bolanderi*
      4: Lower stems bristly, upper leaves elliptic to lanceolate, usually much reduced, sometimes wavy-margined, strigose and bristly . . . . *H. s.* ssp. *echioides*

★ **Heterotheca grandiflora** Nuttall. TELEGRAPH WEED. Occasional, waste places: along highway at Northwood near Guerneville, *F. W. Hoffman* (1943). Usually < 1000 ft. Jan.-Dec. Presumed to be native of s. California.

Heterotheca oregona (Nuttall) Shinners var. **oregona**. Occasional, gravelly places, sand bars: along railroad tracks just after leaving Rincon Valley toward Santa Rosa, *Howell* (1926); Pepperwood; Lake Sonoma area; 2 mi. from jct. of Healdsburg-Geysers Road on Sulphur Cr. Road towards Cloverdale, acc. Vale (1991); 1 mi. from US 101 on Sulphur Cr. Road, *Knight* (1963). < 3300 ft. June-Sept. —*Chrysopsis o.* (Nuttall) A. Gray var. *o.*

Heterotheca oregona var. **rudis** (E. Greene) Semple. Rare, sand bar: S. Fork Gualala River near Hot Spot on The Sea Ranch, *Wells* (1984). < 3300 ft. July-Oct. —*Chrysopsis o.* (Nuttall) A. Gray var. *r.* (E. Greene) Jepson.

Heterotheca sessiliflora (Nuttall) Shinners ssp. **bolanderi** (A. Gray) Semple. GOLDENASTER. Occasional, open areas: The Sea Ranch; Salt Point; Two Rock, *Congdon* (1880); Lakeville Road, *Howell* (1930); Black Point Road, *Howell* (1930); False Lake Meadow and Ledson Marsh in Annadel, acc. Harrison with Rubtzoff. < 500 ft. June-Nov. —*Chrysopsis villosa* (Pursh) Nuttall var. *b.* (A. Gray) Jepson.

Heterotheca sessiliflora ssp. **echioides** (Bentham) Semple. < 3700 ft. July-November. —*Chrysopsis echioides* Bentham, type locality: "Bodegas." *C. villosa* var. *e.* (Bentham) A. Gray.

Hieracium albiflorum Hooker. WHITE-FLOWERED HAWKWEED. Common, dry open wooded slopes: The Sea Ranch, acc. *Mahaffey* (1990); Stewarts Point-Skaggs Springs Road 3 mi. e. of Stewarts Point, *Best* (1981); Salt Point, Fort Ross and Duncans Mills quads; ca 1 mi. nw. of Camp Meeker, acc. J. & B. Guggolz (1991); Fairfield Osborn (H); 1 mi. from Old Adobe e. of Petaluma, *Koch* (1939); Bouverie; Annadel, *Wells* (1984); Gates Road off Calistoga Road, acc. Wells et al. (1984); Pepperwood; chaparral area sw. of Bradford Mtn. 1½ mi. w. of Dry Cr., acc. J. & B. Guggolz (1992); Lake Sonoma area; Alexander Valley quad.; between Jimtown and Mercuryville on Geysers Road, *Utech & Knight* (1986 CM); ca 2000 ft. elev. Anderson Trail on Black Mtn. between Geyser Peak and upper end of Pine Flat Road, acc. J. & B. Guggolz (1992); 2200 ft. elev. near geysers Unit 18, *Malloch* (1977 PG&E); 2040 ft. elev. Squaw Cr., acc. Clifton (1984); 2440 ft. elev. Caldwell Ranch, acc. Clifton (1984). < 9500 ft. June-Aug.

Holocarpha virgata (A. Gray) Keck. TARWEED. Occasional, fields: 4 sites w. of Santa Rosa in Laguna de Santa Rosa drainage, *Harrison* (1978 NCC); Lafferty Ranch on Sonoma Mtn., Arnold et al., (1971); between Petaluma and Napa jct. (probably near Arnold Drive and Hwy. 121), *McMurphy* (1910); s. side Guerneville Hwy. ca 6 mi. from Santa Rosa, *Baker* (1945 NCC); s. side Hall Road w. of Piezzi Road ese. of Graton, *J. & B. Guggolz* (1989 NCC); Sugarloaf Ridge. < 2700 ft. June.-Nov.

Holozonia filipes (Hooker & Arnott) E. Greene. Uncommon, gravelly creek beds, pond margins: near Guerneville Road in Laguna de Santa Rosa drainage, *Harrison* (1978 NCC); w. of El Verano on Grove St., *Best* (1980); dry margin of Coney Lake (Ledson Marsh) at Kenwood, *Baker* (1954 NCC); Fulton, *McMurphy* (1906). < 500 ft. June-Oct.

Hypochaeris (Incorrectly spelled **Hypochoeris.**)

1. Plants annual, glabrous; leaves usually not deeply lobed; ligules scarcely > involucre . . . . *H. glabra*

1: Plants perennial, pubescent to hispid; leaves pinnatifid; ligules < than involucre . . . . *H. radicata*

★ **Hypochaeris glabra** Linnaeus. SMOOTH CAT'S EAR. Common weed, fields, waste places: The Sea Ranch; Salt Point; Fort Ross, Cazadero, Duncans Mills, Arched Rock, Bodega Head and Sebastopol quads.; Todd Road Preserve (H&K); many sites w. of Santa Rosa in Laguna de Santa Rosa drainage, *Harrison* (1978 NCC); Fairfield Osborn (*); Bouverie; Sugarloaf Ridge; Piner Road w. of Santa Rosa, *J. & B. Guggolz* (1990 JEPS); Pepperwood; Healdsburg, Geyserville and Ida Clayton Road quads.; 3100 ft. elev. w. flank Mt. St. Helena, *Grummer* (1980 PUA); 3200 ft. elevation Socrates Mine, acc. Clifton (1984); ca 2000 ft. elev. Anderson Trail on Black Mtn. between Geyser Peak and upper end of Pine Flat Road, acc. J. & B. Guggolz (1992); 2400 ft. elev. Little Geysers, *Malloch* (1977 PG&E); 2040 ft. elev. Squaw Cr., acc. Clifton (1984). < 4000 ft. Mar.-June. Naturalized from Europe.
★ **Hypochaeris radicata** Linnaeus. HAIRY CAT'S EAR. Common weed, lawns, waste places: The Sea Ranch, *J. H. Thomas* (1976); Horseshoe Cove, *Knight* (1963); Salt Point; Fort Ross and Arched Rock quads.; many sites w. of Santa Rosa in Laguna de Santa Rosa drainage, *Harrison* (1978 NCC); Bodega Head, *Clifton* (1980 PUA); Freestone, *Huntington* (1987 NCC); Pitkin Marsh, *Rubtzoff* (1951); Todd Road Preserve (H&K); Fairfield Osborn (S); Bouverie; Sugarloaf Ridge; Gates Road off Calistoga Road, acc. Wells et al. (1985); Bennett Valley, *Heller* (1902); Pepperwood; ca 2000 ft. elev. Anderson Trail on Black Mtn. between Geyser Peak and upper end of Pine Flat Road, acc. J. & B. Guggolz (1992). <1700 ft. May-Nov. Naturalized from Europe.

Jaumea carnosa (Lessing) A. Gray. Locally common, saline places: Lower Salmon Cr. Marsh, *Rubtzoff* (1962); Bodega Head at base of cliffs in marsh, *Clifton* (1980 PUA); near railroad tracks near Petaluma marina, acc. Knight (1990); mouth of Petaluma River, acc. Knight (1990); Sears Point Hwy., *Baker* (1946 NCC). < 20 ft. May-Oct.

Lactuca

1. Leaves linear-lanceolate or pinnatifid with linear-lanceolate lobes, the margins entire or remotely denticulate; panicles piciform, with short ascending branches; flowers ca 8-10 . . . . *L. saligna*
1: Leaves oblong or elliptic in outline, the margins obviously spinulose-denticulate, sometimes entire; panicle open, with widely spreading branches; flowers 14-20 . . . . *L. serriola*

★ **Lactuca saligna** Linnaeus. WILLOW LETTUCE. Common weed, fields, waste places:Occidental Community Center, *Best & Wells* (1983); Guerneville en route to Armstrong Grove, *Bravo* (1982); w. of Santa Rosa in Laguna de Santa Rosa drainage, *Harrison* (1978 NCC); Todd Road Preserve, *Howell* (1987); along Copeland Cr. on Sonoma State Univ. campus, *Self* (1979 NCC); Kentucky St. sidewalk crack in Petaluma, *Knight* (1963); weedy garden 1 mi. sw. of Sonoma, *Howell* (1957); Los Alamos Road, *Knight* (1963); Spring Lake; Pepperwood; 2000 ft. elev. e. of Unit 18, *Malloch* (1977 PG&E). < 2700 ft. Aug.-Nov. Naturalized from Europe.
★ **Lactuca serriola** Linnaeus. PRICKLY LETTUCE. Common weed, waste places: Skillman Ln. nw. of Petaluma, *Knight* (1963); e. end of Railroad Ave. between Petaluma Hill Road and Davis Ln., acc. Knight (1990); flower bed at Petaluma Jr. High School, acc. Knight (1991); bed of Ducker Cr., Rincon Valley, *Wells* (1985); between high school and junior collecge on railroad track spur, Santa Rosa, *M. A. Thomas* (1934). < 6600 ft. May-Sept. Native of Europe. —*L. s.* var. *integrata* Grenier & Godron, currently not recognized.

Lagophylla ramosissima Nuttall. Occasional, open places: Pitkin Marsh, *Baker* (1928 NCC); Fairfield Osborn (H); Kenwood Springs, *Baker* (1934 NCC); along highway ½ mi. n. of Warfield near jct. Hwy.12 and Trinity Road, *Baker* (1934 NCC); Pepperwood; Geysers Road, *Harrison* (1979 SRJC); Big Sulphur Cr., *Harrison* (1987 SRJC). < 5200 ft. May-Oct.

Lasthenia (Key ± acc. R. Ornduff, *Hickman* [1993]).

1. Phyllaries partly fused
   2. Pappus present . . . . *L. glaberrima*
   2: Pappus 0 . . . . *L. glabrata*
1: Phyllaries free
   3. Plants with all leaves ± entire
      4. Plants annual; coastal or inland . . . . *L. californica*
      4: Plants perennial; coastal
         5. Stem erect, simple to few-branched; roots fleshy; leaves usually < 2 mm wide . . . . *L. macrantha* ssp. *bakeri*

5: Stem ± decumbent, usually branched; roots fibrous; leaves usually > 2 mm wide . . . . . . *L. m.* ssp. *macrantha*

3: Plants with middle cauline leaves usually pinnately divided

6. Fruit > 2 mm

7. Pappus awns > 4; ray corollas < 3 mm . . . . . . *L. maritima*

7: Pappus awns 2-3; ray corollas > 4 mm . . . . . . *L. minor*

6: Fruit < 1.5 mm

8. Pappus usually of 1 long awn and many short scales . . . . . . *L. burkei*

8: Pappus usually of 3 or more long awns mixed with several short scales . . . . . . *L. fremontii*

Lasthenia burkei (E. Greene) E. Greene. BURKE'S GOLDFIELDS. Habitats have been decimated by development and agriculture until only a relatively few populations remain. Species now very restricted to a small area in Laguna de Santa Rosa and Windsor. With the exception of two locations in Lake County, it is not known to occur outside of Sonoma County. < 1700 ft. Apr.-May.

Lasthenia californica Lindley. GOLDFIELDS. Common, open areas, fields: The Sea Ranch; ½ mi. n. of Moon Rocks campground at Salt Point State Park, *Huntington* (1989 NCC); Willow Creek; Bodega Head; Coleman Valley Road, *J. & B. Guggolz* (1990 JEPS); Stoetz Lane between Occidental and Graton, *Utech & Knight* (1986 CM); Todd Road Preserve; Fairfield Osborn (H); Hwy. 12 between Kenwood and Trinity Road, acc. Knight (1989); Sugarloaf Ridge; 4¼ mi. w. of Glen Ellen on Bennett Valley Road, *Rubtzoff & Froebe* (1971); Pepperwood; Lake Sonoma area; Alexander Valley and Ida Clayton Road quads.; 2400 ft. elev. Little Geysers, *Malloch* (1977 PG&E). < 5000 ft. Mar.-May. —*L. chrysostoma* (Fischer & C. Meyer) E. Greene. *Baeria c.* Fischer & C. Meyer, type locality: vicinity of Bodega Bay.

Lasthenia fremontii (A. Gray) E. Greene. Uncommon, fields: 1 mi. n. of Windsor, *Howell* (1943); 2 mi. n. of Windsor, *Eastwood & Howell* (1936). < 2300 ft. Apr.-May. —*Baeria f.* (Torrey) A. Gray.

Lasthenia glaberrima A. de Candolle. Common, damp clay soils: Bodega Point, *Eastwood* (1917); upper Stemple Cr. Marsh, *Rubtzoff & Peñalosa* (1961); Laguna de Santa Rosa e. of Graton, *Rubtzoff* (1963); Laguna de Santa Rosa ne. of Graton at end of Piner Road, *Rubtzoff* (1966); Laguna de Santa Rosa n. of Cunningham near Todd Road, *Rubtzoff* (1958); Todd Road 1¼ mi. from Hwy. 116 and 0.6 mi. from Llano Road, *Rubtzoff* (1958), *Howell & Knight* (1986); across road from Todd Road Preserve, *Sorrie* (1978); Fairfield Osborn (*); 6½ mi. ne. of Penngrove on Lichau Road, *Rubtzoff* (1968); laguna in Chileno Valley, *Rubtzoff* (1972); 5 mi. n. of Sears Point, *Rubtzoff* (1970); 4.7 mi. s. of El Verano on Arnold Drive, *Raven* (1963); near Kenwood, *Howell* (1935); 4¼ mi. w. of Glen Ellen on Bennett Valley Road, *Rubtzoff & Froebe* (1972); Spring Lake; Wright School, *Baker* (1926 NCC); Santa Rosa, *Eastwood* (1921); 3 mi. s. of Healdsburg, *Beane & James* (1949). < 3000 ft. Apr.-June.

Lasthenia glabrata Lindley. Occasional, open areas: Eardley Ave. 0.1 mi. n. of jct. with W. College Ave, Santa Rosa, *Harrison* (1978 NCC); Petaluma, *Howell* (1933); Santa Rosa, *Eastwood* (1921); near Geyserville, *Baker* (1929 NCC). < 3300 ft. Mar.-May.

Lasthenia macrantha (A. Gray) E. Greene ssp. **bakeri** (J. T. Howell) Ornduff. Uncommon, damp areas: marsh on Gravenstein Hwy. s. of Molina near Sebastopol, *Baker* (1939 NCC). < 1700 ft. May-June. —*Baeria m.* (A. Gray) A. Gray var. *b.* (J. T. Howell) Keck.

Lasthenia macrantha ssp. **macrantha**. Occasional, maritime grassy areas: 49 ft. elev. Gerstle Cove at Salt Point State Park, *Clifton* (1981 PUA); between Salmon Cr. and Wright's Beach, *Baker* (1938 NCC); 8 mi. n. of Bodega Bay, *Howell* (1931); near beach n. of Bodega Bay, *Baker* (1926 NCC); Larkspur Rock just e. of Bodega Bay, *Baker* (1931 NCC). < 1700 ft. Mar.-Aug. —*Baeria m.* (A. Gray) A. Gray.

Lasthenia maritima (A. Gray) M. C. Vasey. BIRDROCK GOLDFIELDS, MARITIME GOLDFIELDS. Rare, immediate coast: Gualala Point rock, and mouth of Russian River 3 rocks s. of Peninsula Bluff, acc. Michael C. Vasey, *Madroño 32:131-142, July* (1985); Salt Point, *Vale & Wagnon* (1988 NCC). < 400 ft. Mar.-May. —*Baeria minor* (de Candolle) Ferris ssp. *maritima* (A. Gray) Ferris. *L. minor* ssp. *maritima* (A. Gray) Ornduff.

Lasthenia minor (A. de Candolle) Ornduff. Uncommon, stream banks, coastal bluffs: Portuguese Beach State Park, *J. & B. Guggolz* (1991 NCC); mouth of Salmon Cr. 2 mi. n. of Bodega Bay, *Howell* (1931); 1 mi. n. of Bodega Bay, *Baker* (1928 NCC); Bodega Head. < 2300 ft. Mar.-Apr. —*Baeria m.* (de Candolle) Ferris.

Layia

1. Bracts of receptacle subtending each disk-flower; herbage eglandular; phyllaries with prominent pustulate processes at base of hairs
   2. Stem erect; leaves not succulent, basal pinnately parted into linear or oblong blunt lobes; tips of phyllaries not dilated or rounded . . . . *L. chrysanthemoides* ssp. *chrysanthemoides*
   2: Stems prostrate, the central peduncle horizontal; leaves succulent, the lower with broad rounded lobes; tips of phyllaries dilated and rounded . . . . *L. c.* ssp. *maritima*
1: Bracts of receptacle limited to a ring between ray- and disk-flowers; herbage glandular; the phyllaries pubescent but not as in previous 1.
   3. Ligule yellow with white tip (rarely yellow throughout); pappus commonly merely scabrid, if plumose, also with inner wool, of 18-30 bristles; anthers black; tips of phyllaries longer than basal portion
      4. Stems prostrate or decumbent, stout, succulent, 1-3 dm long; herbage green; basal and lower cauline leaves dentate to pinnatifid with salient, rotund, short lobes; phyllary tips rounded . . . . *L. platyglossa* ssp. *platyglossa*
      4: Stems erect, less succulent, to 5 cm high; the herbage a grayer green; leaf-lobes narrower and longer; phyllary-tips not wider or rounded . . . . *L. p.* ssp. *campestris*
   3: Ligule yellow or white but not as previous 3. (except occasionally in *L. gaillardioides*); pappus plumose, with inner woolly hairs only in *L. septentrionalis*; tips of phyllaries shorter than basal portion
      5. Anthers yellow; involucre not urceolate; stems not dark-dotted; pappus white . . . . *L. septentrionalis*
      5: Anthers black; involucre urceolate; stems dark-dotted; pappus rufous . . . . *L. gaillardioides*

Layia chrysanthemoides (de Candolle) A. Gray ssp. **chrysanthemoides**. SMOOTH TIDY-TIPS. Occasional, vernally moist ground: 8 mi. n. of Bodega Bay, *Howell* (1931); w. of Santa Rosa in Laguna de Santa Rosa drainage, *Harrison* (1978 NCC); Todd Road Preserve; Laguna de Santa Rosa floodplain w. of Llano Road, acc. J. & B. Guggolz (1985); Chileno Valley, *Rubtzoff* (1964); Lynch Road 3 mi. e. of Adobe Road near Petaluma, *Quibell* (1979 NCC); 9 mi. sw. of Sonoma on road to San Rafael, *Howell* (1931); near Boyes Hot Springs 2 mi. n. of Sonoma, *Howell* (1931); Lake Sonoma area; property bordered by Chianti and Dutcher Cr. roads w. of US 101 and w. of Geyserville, acc. J. & B. Guggolz (1991). < 2600 ft. Mar.-May.

Layia chrysanthemoides ssp. **maritima** Keck. Uncommon, coastal bluffs: 2½ mi. n. of Salmon Cr., *Baker* (1932 NCC); 6.2 mi. s. of Jenner, *Clausen* (1935 type UC). < 300 ft. May.

Layia gaillardioides (Hooker & Arnott) de Candolle. Occasional, grassy places: between Fort Ross and Jenner, *Baker* (1932 NCC); beach near Salmon Cr., *Hanson* (1928); s. of Bodega Bay, *Baker* (1930 NCC); Tresch Ranch on Walker Road near Two Rock, *J. & B. Guggolz* (1989 JEPS). < 4000 ft. Apr.-June.

Layia platyglossa (Fischer & C. Meyer) A.Gray ssp. **campestris** Keck. Uncommon, grassy openings, slopes: Pepperwood. < 6600 ft. Mar.-May.

Layia platyglossa ssp. **platyglossa**. TIDY-TIPS. Common, coastal bluffs, grassy flats:The Sea Ranch; Stewarts Point, *Baker* (1932 NCC), *Ott* (1933 NCC); Gerstle Cove at Salt Point State Park, *Clifton* (1981 PUA); near Fort Ross, *Heller* (1913), *Howell* (1930); Wright's Beach, *Baker* (1932 NCC); near mouth of Salmon Cr., *Baker* (1928, 1932 NCC); Bodega Head. < 600 ft. May-June. —*Callichroa p.* Fischer & C. Meyer., type locality: Fort Ross.

Layia septentrionalis Keck. Rare, cliffs, bluff tops — sandy soils or serpentine: 3600 ft. elev. Geyser Rock, *Clifton* (1983 PUA). < 4000 ft. Apr.-May.

★ **Leontodon taraxacoides** (Villars) Mérat. HAWKBIT. Uncommon, weed, open hills, fields: Pygmy Forest at The Sea Ranch Timber Preserve, *Wells* (1983); near n. Fire Station at The Sea Ranch, *Wells* (1984); 2.3 mi. s. of Stewarts Point, *Frankel* (1964); Stump Beach at Salt Point State Park, *Wells* (1982); Pitkin Marsh, *Howell* (1945), *Rubtzoff* (1957); 2 mi. n. of Sebastopol, *Baker* (1945 NCC); near Cotati firehouse, *Jordan* (1965 NCC). < 3300 ft. May-Aug. Native of Europe. —*L. leysseri* (Wallroth) G. Beck.

Lessingia

1. Ray flowers present . . . . *L. filaginifolia* var. *californica*
1: Ray flowers 0
   2. Phyllaries glabrous or puberulent . . . . *L. ramulosa*
   2: Phyllaries woolly or tomentose
      3. Herbage without glands

4. Pappus < fruit; style appendages short-triangular 0.2-0.4 mm; basal leaves usually fallen before flowering, glabrous if present . . . . *L. arachnoidea*
4: Pappus = or > fruit; style appendages tapered 0.6-0.8 mm; basal leaves present at flowering, tomentose . . . . *L. hololeuca*
3: Herbage with glands . . . . *L. leptoclada*

Lessingia arachnoidea E. Greene. Rare, grassy slopes: n. of Duvoul Cr. n. of Camp Meeker, *Edwards* (1987); 2 mi. nw. of Camp Meeker, *J. & B. Guggolz* (1991 UC); Sweetwater Springs Road on serpentine near mine, *Knight & Edwards* (1986). < 1000 ft. July-Sept. —*L. hololeuca* var. *a.* (E. Greene) J. T. Howell.

Lessingia filaginifolia (Hooker & Arnott) M. A. Lane var. **californica** (de Candolle) M. A. Lane. CALIFORNIA-ASTER. Occasional, coastal hills and mesas: n. end of Timber Ridge Road and 2 other locations at The Sea Ranch, *Wells* (1984); near Stewarts Point, *Abrams* (1920); Gerstle Cove, Salt Point State Park, *Wells* (1982); Stillwater Cove below Salt Point, *Harrison* (1985 SRJC). < 3300 ft. July-Oct. —*Corethrogyne c.* de Candolle var. *obovata* (Bentham) Kuntze. *C. o.* Bentham, type locality, "Bodegas."

Lessingia hololeuca E. Greene. Occasional, grassland: near mouth of Russian River, *Dudley* (1893); Stoetz Lane 2 mi. ne. of Occidental, *Best* (1981); Petaluma, *Congdon* (1880); hills e. of Santa Rosa Valley, *E. Greene* (type 1888 UC). < 1000 ft. June-Oct.

Lessingia leptoclada A. Gray. SIERRA LESSINGIA. Rare, open slopes and valleys, serpentinized soil; previously reported only from Sierra Nevada: between Camp Meeker and Bohemian Grove off Bohemian Hwy., *J. & B. Guggolz* (1992 NCC). < 6200 ft. July-Oct. Has been previously identified as *L. arachnoidea*; however *L. a.* is eglandular. This is minutely crystalline-glandular throughout.

Lessingia ramulosa A. Gray. Occasional, open ground, clay soils: ca 1 mi. nw. of Camp Meeker, *J. & B. Guggolz* (1991 UC); serpentine at Grub Cr. near Camp Meeker, *Harrison* (1977 SRJC); between Occidental and Graton, *Howell & Stacey* (1936); Wagnon Ranch 4 mi. w. of Sebastopol, *Tryner* (1938); near Sonoma, *Hartweg* (1846); Windsor Jr. High on athletic field, *Harrison* (1977 SRJC); serpentine at Chalk Hill Winery, Windsor, *Harrison* (1989 SRJC). < 3300 ft. July-Oct. —*L. bicolor* E. Greene, type locality: Santa Rosa, *Heller* (1902 U).

Leucanthemum

1. Heads 7-9 cm diameter; leaves ± regularly toothed; middle stem leaves > 5 cm . . . . *L. maximum*
1: Heads 3-7 cm diameter; leaves irregularly pinnately lobed ± halfway to midrib; middle stem leaves < 5 cm . . . . *L. vulgare*

★ **Leucanthemum maximum** (Ramond) de Candolle. Presumed to be one of the parents of Shasta Daisy. Uncommon, coastal strand: Salt Point, *Wells* (1982). < 700 ft. June-Aug. Closely related to *L. vulgare*. Native of Europe. —*Chrysanthemum m.* Ramond.

★ **Leucanthemum vulgare** Lamarck. OX-EYE DAISY. Along coastal area; in a wide range of habitats and common as a roadside and field weed: The Sea Ranch on Schooner Drive, acc. Wells. < 6600 ft. June-Aug. Naturalized from Europe. —*Chrysanthemum l.* Linnaeus.

Luina hypoleuca Bentham. Rare, cliffs, bluff sides: 3800 ft. elev. w. flank Mt. St. Helena, *Grummer* (1980 PUA); 2400 ft. elev. w. flank Mt. St. Helena, *Grummer* (1981 PUA). < 7000 ft. Apr.-June. Type locality: near The Geysers.

Madia

1. Pappus present
   2. Plants tall perennials of dense woods; leaves opposite well up the stem, those of rosette wider *M. madioides*
   2: Plants low vernal annuals; leaves alternate above first basal pairs, those of the rosette narrow . . . *M. nutans*
1: Pappus none
   3. Disk-flowers more than 1, pubescent; phyllary with empty flat tip
      4. Disk-achenes sterile; receptacle pubescent; ray-flowers conspicuous . . . . *M. elegans*
      4: Disk-achenes fertile; receptacle glabrous; ray-flowers inconspicuous
         5. Achenes obovoid, black and shining . . . . *M. anomala*
         5: Achenes strongly laterally compressed, dull
            6. Glandular above the middle, slender and flexuous, heads sessile to long-peduncled . . . . *M. gracilis*
            6: Glandular well down toward base; stem stout; heads sessile to short-peduncled . . . . *M. sativa*
   3: Disk-flowers 1 (or 2), glabrous, phyllary embracing the achene and furrowed with longitudinal grooves to tip . . . . *M. exigua*

Madia anomala E. Greene. Occasional, chaparral, grassy slopes: ca 1 mi. nw. of Camp Meeker, acc. J. & B. Guggolz (1991); Lafferty Ranch on Sonoma Mtn., acc. Arnold et al. (1971); hills n. of Sonoma, *Rechinger & Howell* (1953); hills e. of Agua Caliente at edge of Sonoma Valley, *Howell* (1959); 2 mi. s. of Mark West Springs, *Yates* (1934 UC); near Mark West Springs, *Eastwood & Howell* (1938); Pepperwood; Lake Sonoma area; ca 2000 ft. elev. Anderson Trail on Black Mtn. between Geyser Peak and upper end of Pine Flat Road, acc. J. & B. Guggolz (1992); 4300 ft. elev. geysers Units 20 and 21, *Cooney* (1977 PG&E). < 4500 ft. May-June.

Madia elegans Lindley ssp. **elegans**. COMMON MADIA. Common, flatlands: ca 1 mi. nw. of Camp Meeker, acc. J. & B. Guggolz (1991); Pitkin Ranch, *Howell* (1945); Sebastopol, *Dows* (1907); near jct. of Todd Road and US 101, *Knight* (1963); Pepperwood; 2900 ft. elev. w. flank Mt. St. Helena, *Grummer* (1979 PUA); 300 ft. e. of geysers Unit 18, *Malloch* (1977 PG&E); 3600 ft. elev. Anderson Cr. 1 mi. sw. of Cobb Mtn. summit, acc. Clifton (1984); mountains near The Geysers, *McMurphy* (1909); hills near Cloverdale, *McMurphy* (1909). < 8000 ft. June-Aug. Ssp. **densiflora** (E. Greene) Keck differs as stem is stout up to 2½ m high compared to 8 dm high; rare, serpentine slopes: 3160 and 3200 ft. elev. Socrates Mine, *Clifton* (1984 PUA). < 3300 ft. Aug.-Nov. Ssp. **vernalis** Keck differs from other subspecies as has large heads which remain open longer through the day: Fort Ross colony, *Voznesenskii* (1841 LE); 2 mi. w. of Monte Rio on road to Jenner, *Clausen* (1935). < 3300 ft. Mar.-June.

Madia exigua (Smith) A. Gray. LITTLE TARWEED. Common, open areas: Stewarts Point quad.; ca 1 mi. nw. of Camp Meeker, acc. J. & B. Guggolz (1991); Stoetz Lane ne. of Occidental, *Best* (1981), *Utech & Knight* (1986 CM); Harrison Grade Road 1.3 mi. from jct. w. of Green Valley Road, *Ferris & Lorraine* (1961); Valley Ford, *Knight & Utech* (1986); Annadel, *Wells* (1984); Pepperwood; 0.3 mi. n. of entrance to Petrified Forest, *Clausen* (1935); Fitch Mtn., *Rubtzoff* (1952); Lake Sonoma area; Ida Clayton Road quad.; ca 2000 ft. elev. Anderson Trail on Black Mtn. between Geyser Peak and upper end of Pine Flat Road, acc. J. & B. Guggolz (1992); 2440 ft. elev. Caldwell Ranch, acc. Clifton (1984); 3000 ft. elev. Caldwell Pines, acc. Clifton (1984). < 8300 ft. May-July.

Madia gracilis (Smith) Keck. GUMWEED MADIA, SLENDER TARWEED. Common, open areas, wooded hillsides: Skaggs Springs Road 6 mi. e. of Stewarts Point, *Best* (1981); ca 1 mi. nw. of Camp Meeker, acc. J. & B. Guggolz (1991); Fairfield Osborn (H); 2 mi. n. of Sonoma, *Howell* (1929); Bouverie; Spring Lake; Mark West Springs, *Eastwood & Howell* (1938); between Calistoga and Santa Rosa, *Thomas* (1966); near Petrified Forest, *Heller* (1902), *Abrams* (1916), *Clausen* (1935), *Howell* (1937); Pepperwood, *Breedlove* (1983); Fitch Mtn., *Rubtzoff* (1952); between Skaggs Springs and Geyserville, *J. H. Thomas* (1976); Lake Sonoma area; Alexander Valley and Ida Clayton Road quads.; 2900 ft. elev. w. flank of Mt. St. Helena, *Grummer* (1979 PUA); 3120 ft. elev. Bear Cr. Can., acc. Clifton (1984); 2440 ft. elev. Caldwell Ranch, acc. Clifton (1984); 3000 ft. elev. Caldwell Pines, acc. Clifton (1984). < 8000 ft. Apr.-Aug.

Madia madioides (Nuttall) E. Greene. WOODLAND TARWEED. Common, moist coniferous woods: near South Fork Gualala River at The Sea Ranch, *Wells* (1981); Annapolis and Stewarts Point quads.; Salt Point; Fort Ross quad.; Skaggs Springs Road 3 mi. e. of Stewarts Point, *Best* (1981); Cazadero, *Bacigalupi* (1926); Neeley Hills se. of Guerneville, *Rubtzoff* (1954); Guerneville, *Rattan* (1877); Fairfield Osborn (H); Bouverie; along Sonoma Cr. at foot of Mt. Hood, *Heller* (1902); Bennett Valley, *Leitholt* (1893); Rincon Valley, *Wells* (1984); 2 mi. w. of Chalk Hill Road on Hwy. 128, *J. & B. Guggolz* (1989 NCC); 4 mi. ne. of Mark West Springs, *H. M. Hall* (1931); Pepperwood; between Jimtown and Mercuryville on Geysers Road, *Utech & Knight* (1986 CM); 2190 ft. elev. Squaw Cr., acc. Clifton (1984). < 4500 ft. May-Sept.

Madia nutans (E. Greene) Keck. Uncommon, open areas: Adobe Can., *Clausen* (1935); Mt. Hood, *Bioletti* (1893); Spring Lake; Petrified Forest, *Clausen* (1935); Knights Valley, *Krauss* (1895). 250-3200 ft. Mar.-May. —*Callichroa n.* E. Greene, type locality: mountains of Sonoma County, also Mt. St. Helena.

★ **Madia sativa** Molina. Chile Tarweed. COAST TARWEED. Common, appearing weedy: The Sea Ranch, *Wells* (1981); s. end of The Sea Ranch, *J. H. Thomas* (1976); 6½ mi. n. of Jenner, *Keck* (1932); n. of Jenner, *Ripley & Barneby* (1945); 1 mi. n. of Salmon Cr., *Keck & Clausen* (1935); 1 mi. e. of Bodega Bay on road to Bodega, *Howell* 1930); Bodega Head, *Clifton* (1980 PUA); near Occidental Community Center, *Best & Wells* (1981); Pitkin Marsh, *Rubtzoff* (1951); 6 mi. w. of Petaluma along road to Tomales, *Heusi* (1930); near Petaluma, *H. M. Hall* (1917); Live Oak Road w. of Petaluma, *Knight* (1963); Warrington Road off Petaluma Hill Road, acc. Knight (1963); old Redwood Hwy. and Railroad Ave. near Cotati, *Knight* (1963); 2 mi. s. of Kenwood, *Baker* (1932 NCC); s. base of Mt. Hood on Sonoma Cr., *Howell* (1932); Adobe Can., *Baker* (1932); 1½ mi. sw. of Kenwood on lower road to Glen Ellen, *Howell* (1932); Spring Lake; Leddy electric railroad station between Santa Rosa and Sebastopol, *McMurphy* (1909). Between Sebastopol and Santa Rosa, possibly a hybrid with *M. elegans*, *Abrams & Luther Burbank* (1907). <3250 ft. May-Oct. Native of Chile and Argentina. —*M. capitata* Nuttall.

Malacothrix floccifera (de Candolle) S. F. Blake. Occasional, sandy and rocky places: Westminster Woods, Camp Meeker on serpentine, *Harrison* (1978 SRJC); quarry on Bohemian Hwy. on serpentine, *Harrison* (1980 SRJC); first roadcut on Franz Valley Road from Mark West Springs Road, *de Nevers* (1981); Pepperwood; 3200 ft. elev. Socrates Mine, acc. Clifton (1984); s. slope of Black Mtn. *J. & B. Guggolz* (1992 UC NCC); 3120 ft. elev. Bear Can., acc. Clifton (1984); Cloverdale, *M. Jones* (1931). < 6600 ft. Apr.-June.

Micropus

1. Phyllaries membranaceous; ♀ flowers 9-11 . . . . *M. amphibolus*
1: Phyllaries much thicker; ♀ flowers 5-6 . . . . *M. californicus*

Micropus amphibolus A. Gray. MT. DIABLO COTTONWEED. Occasional, shallow soil, rocky areas: grassy slope near rocks 3 mi. se. of Sonoma, *Howell* (1965); hills e. of Agua Caliente, *Howell* (1948); Lafferty Ranch on Sonoma Mtn., acc. Arnold et al. (1970); Dorman Can. 4 mi. w. of Healdsburg, *Mason* (1934); Geyser Peak, *Knight & Edwards* (1986). 150-2700 ft. Apr.-May. —*Stylocline amphibola* (A. Gray) J. T. Howell.

Micropus californicus Fischer & C. Meyer. SLENDER COTTONWEED. Occasional, low dry open areas: 1024 ft. elev. Coleman Valley Road, acc. Clifton (1984); ca 1 mi. nw. of Camp Meeker, acc. J. & B. Guggolz (1991); mouth of Salmon Cr., *Howell* (1931); Port Bodega, type (1835); Fairfield Osborn (*); Button Ranch on Walker Road (U. C. property) between Petaluma-Valley Ford Road and Pepper Road, acc. Connors (1992); s. of El Rose Dr. and w. of D Street in grassland, Petaluma, *B. Anderson* (1987); Bouverie; Spring Lake; Pepperwood; chaparral area sw. of Bradford Mtn. 1½ mi. w. of Dry Cr., acc. J. & B. Guggolz (1992); Lake Sonoma area; property bordered by Chianti and Dutcher Cr. roads w. of US 101 w. of Geyserville, acc. J. & B. Guggolz (1991); between Jimtown and Mercuryville on Geysers Road, *Utech & Knight* (1986 CM); 2040 ft. elev. Squaw Cr., acc. Clifton (1984). < 5600 ft. Apr.-June. —*Gnaphalodes californica* (Fischer & C. Meyer) E. Greene.

Microseris

1. Plants perennial; flowers > > involucre; pappus parts often > 5
    2. Outer phyllaries narrowly ovate or wider, generally glabrous . . . . *M. laciniata*
    2: Outer phyllaries linear to lanceolate, often mealy and blackish hairy . . . . *M. paludosa*
1: Plants annual; flowers ± = involucre; pappus parts < 6
    3. Pappus scales curved only at base, midrib linear, thicker at the base . . . . *M. bigelovii*
    3: Pappus scales ± curved throughout, midrib evenly tapered from thick base . . . . *M. douglasii*

Microseris bigelovii (A. Gray) Schultz Bipontinus. Infrequent, coastal bluffs and flats: s. part of The Sea Ranch, *J. H. Thomas* (1976); Fort Ross, *Chisaki & Kamb* (1952); mouth of Salmon Cr., *Howell* (1931); Bodega Head Marine Lab. property, *Connors* (1993 BML). < 300 ft. Apr.-June. —*M. obtusata* E. Greene, type locality: Bodega Point, *Eastwood* (1899).

Microseris douglasii (de Candolle) Schultz Bipontinus. Common, grassy areas: Sebastopol quad.; 2 sites w. and sw. of Santa Rosa in Laguna de Santa Rosa drainage, *Harrison* (1978 NCC); Fairfield Osborn (H); Lafferty Ranch on Sonoma Mtn., acc. Arnold et al. (1971); 1 mi. w. of Mark West bridge, *Baker* (1936 NCC); Pepperwood; hwy. between Geyserville and Cloverdale, *Baker* (1929 NCC); Sulphur Cr., *Baker* (1936 NCC); Lake Sonoma area. < 3300 ft. Mar.-Apr.

Microseris laciniata (Hooker) Schultz Bipontinus. Occasional, woody or meadowy areas: e. of Sebastopol, *Howell* (1937); Los Guilicos Valley, *Bioletti* (1982 UC); 11 mi. ne. of Santa Rosa, *Eastwood & Howell* (1936); Cloverdale, *Bolander* (1864); n. of Cloverdale, *Rattan* (1880); Alder Glen Springs on Cloverdale Grade to Hopland, *Eastwood & Howell* (1938). < 6600 ft. May-July. —*Scorzonella maxima* Bioletti, type locality: Los Guilicos Valley, *Bioletti* (1892 UC); *S. procera* E. Greene. *Calais l.* A. Gray, type locality: Mark West Cr., *Bolander* (1864 UC).

Microseris paludosa (E. Greene) J. T. Howell. Occasional, grassy and wooded areas: near Todd Road in Laguna de Santa Rosa drainage, *Harrison* (1978 NCC); 7 mi. n. of Petaluma on Stony Point Road, *Howell & Stacey* (1937); vernal pool 2 mi. nw. of Windsor by Old Redwood Hwy., *de Nevers* (1981); 2 mi n. of Windsor, *Baker* (1934 NCC); Pepperwood. < 1000 ft. May-June.

Monolopia major de Candolle. May be extinct in county; rolling hills and plains: Petaluma, *Congdon* (1880). < 3500 ft. Mar.-May.

★ **Onopordum acanthium** Linnaeus. SCOTCH THISTLE.Waste places: adjacent to Freestone area, acc. *Barbe* (1991). < 5200 ft. Native of Europe.

★ **Osteospermum fruticosum** (Linnaeus) Norlindh. FREEWAY DAISY. Escape from cultivation: Spring Lake. < 700 ft. Apr.-June. Native of Africa.

Pentachaeta

1. Stem simple or with few erect branches; disk-corollas dilated at throat, constricted at orifice ..... *P. exilis*
1: Stem diffusely branched; disk-corollas narrowly linear ..... *P. alsinoides*

Pentachaeta alsinoides E. Greene. Uncommon, grassy slopes of foothills: Santa Rosa, *Eastwood* (1921); 4 mi. w. of Mark West Springs, *Eastwood & Howell* (1938); Lake Sonoma area. < 1100 ft. Apr.-May. —*Chaetopappa a.* (E. Greene) Keck.

Pentachaeta exilis (A. Gray) A. Gray. Uncommon, grassy slopes of foothills: Lake Sonoma area; 2440 ft. elev. Caldwell Ranch, acc. Clifton (1984); Cloverdale, *Rattan* (1865, 1867). < 3100 ft. Apr.-May. —*Chaetopappa e.* (A. Gray) Keck.

Petasites frigidus (Linnaeus) Fries var. **palmatus** (Aiton) Cronquist. COLTSFOOT. Locally common, deep shade, wooded areas: The Sea Ranch; Salt Point; Fort Ross gulch beside hwy., *A. S. M. McKenzie* (1975); 1 mi. from Monte Rio on Moscow Road, *Knight* (1990 NCC); 4.85 mi. from end of Mill Cr. Road, *Knight & Edwards* (1986).< 1350 ft. Mar.-Apr. —*P. palmatus* (Aiton) A. Gray.

★ **Picris echioides** Linnaeus. BRISTLY OX-TONGUE. Common, well-established weed, waste places: Bodega Head; Todd Road Preserve (H&K); between Cotati and Santa Rosa, *Eastwood* (1921); along Copeland Cr. on Sonoma State Univ. campus, *Self* (1979 NCC); Fairfield Osborn (S); Button Ranch on Walker Road (U. C. property) between Petaluma-Valley Ford Road and Pepper Road, acc. Patterson (1990); w. of Sears Point Raceway in Tolay Cr. drainage, acc. Patterson (1990); Rincon Valley, *Wells (*1984, 1985); Spring Lake, *Wells* (1987); Blue Bell Lane of Coddingtown in Santa Rosa, *Knight* (1963); Pepperwood; 2420 ft. elev. Caldwell Ranch, acc. Clifton (1984). < 2500 ft. June-Dec. Native of Mediterranean region.

Psilocarphus

1. Well-developed receptacular bracts ca 3 mm long at maturity
   2. Largest head 9-14 mm; receptacle deeply lobed; chaff scales narrowly cylindric with length > 3 × width ..... *P. brevissimus* var. *multiflorus*
   2: Largest head 6-9 mm; receptacle entire or shallowly lobed; chaff scales obovoid with length < 3 × width ..... *P. b.* var. *brevissimus*
1: Well-developed receptacular bracts ca 2 mm long at maturity
   3. Leaves linear or linear-oblanceolate, mostly 6-12 × as long as wide; achenes narrowly oblong or elliptic-oblong ..... *P. oregonus*
   3: Leaves spatulate, oblong or oblanceolate, mostly 1.5-6 × as long as wide; achenes broadly oblanceolate to narrowly obovate ..... *P. tenellus*

Psilocarphus brevissimus Nuttall var. **brevissimus**. DWARF WOOLLY-HEADS. Occasional, dried beds of vernal pools: Bodega dunes, *Connors* (1992 BML); near Sonoma Valley Regional Park, *Best* (1990); Bennett Valley Road, *Baker* (1928 NCC); Santa Rosa, *Congdon* (1880); Lake Sonoma area. < 8300 ft. Apr.-June.

Psilocarphus brevissimus var. **multiflorus** Cronquist. DWARF WOOLLY-MARBLES. Reported in *Munz* (1959) "sites of vernal pools about San Francisco Bay." 30-1700 ft. May-June. There is no known local information about this being in the county.

Psilocarphus oregonus Nuttall. OREGON WOOLLY-HEADS. Uncommon, dried vernal pools: near Todd Road sw. of Santa Rosa in Laguna de Santa Rosa drainage, *Harrison* (1978 NCC); Fairfield Osborn (H). 35-5000 ft. May-June.

Psilocarphus tenellus Nuttall.Common, dried vernal pools, hard-packed ground: The Sea Ranch; 3 mi. e. of Valley Ford on road to Petaluma, *Howell* (1930); many sites w. of Santa Rosa in Laguna de Santa Rosa drainage, *Harrison* (1978 NCC); Fairfield Osborn (H); Spring Lake; Pepperwood. 35-6600 ft. Apr.-June.

Rafinesquia californica Nuttall. CALIFORNIA CHICORY. Shrubby slopes, open woods: Lake Sonoma. < 5000 ft. Apr.-July.

★ **Rhagadiolius stellatus** Gaertner. Uncommon weed of disturbed places: Sugarloaf Ridge State Park, acc. Bowcutt (1993). <600 ft. Native of Europe.

Rigiopappus leptocladus A. Gray. Uncommon, dry hard ground: Miller Cr. Road just beyond schoolhouse, *Butler* (1929 NCC); along trail to Geyser Peak from near Mercuryville, *Knight & Edwards* (1986); Ida Clayton Road quad. < 7700 ft. Apr.-June.

★ **Santolina chamaecyparissus** Linnaeus. LAVENDER-COTTON. Uncommon, introduction: Pitkin Marsh, *Rubtzoff* (1952). < 1700 ft. May-June. Native of Mediterranean region.

★ **Scorzonera hispanica** Linnaeus. VIPER'S GRASS, SPANISH SALSIFY. Rare, escape: around Calistoga and Knights Valley, *Eastwood* (1900). < 1000 ft. Native of Europe.

Senecio

1. Plants perennial
  2. Stems climbing; leaves ivy-like . . . . . *S. mikanioides*
  2: Stems not climbing; leaves not ivy-like
    3. Leaves pinnatifid into narrowly linear segments or occasionally entire and linear; plants ± woody at base . . . . . *S. flaccidus* var. *douglasii*
    3: Leaves not narrowly linear or divided into linear segments; plants largely herbaceous
      4. Leaves scattered along the stems, only slightly or gradually reduced up the stem . . . *S. jacobaea*
      4: Leaves largely basal, the cauline rapidly reduced up the stem
        5. Stems mostly solitary, simple 3-20 dm long, arising from a short, erect, short-lived crown and with numerous, fibrous roots
          6. Herbage ± villous with multicellular hairs, later glabrate
            7. Rays 5-12; involucre mostly 8-12 mm high . . . . . *S. integerrimus* var. *major*
            7: Rays 0-2, involucre 5-8 mm high . . . . . *S. aronicoides*
          6: Herbage subglabrous from the beginning . . . . . *S. hydrophilus*
        5: Stems few to several 1-8 dm tall, arising from running rootstocks or short woody caudex
          8. Ligules orange; stems with slender running rootstocks 1-2 (-3) dm long; basal leaves ± tufted, roundish 2-5 cm long, ± dentate, often purplish, especially beneath, on rather long, slender petioles . . . . . *S. greenei*
          8: Ligules yellow; stems few-several 3-8 dm tall, short woody caudex; leaves ± dissected, lyrate to sharply toothed . . . . . *S. eurycephalus*

1: Plants annual
  9. Bracteoles at base of involucre well-developed, black-tipped
    10. Heads radiate; plants sticky . . . . . *S. elegans*
    10: Heads discoid; plants not sticky . . . . . *S. vulgaris*
  9: Bracteoles at base of involucre none or poorly developed, not black-tipped . . . . . *S. sylvaticus*

Senecio aronicoides de Candolle. CALIFORNIA BUTTERWEED. Occasional, oak woodland, forest openings: 3 mi. w. of Occidental on Coleman Valley Road, *Stocking* (1969 NCC); Pepperwood; 3300 ft. elev. Squaw Cr., acc. Clifton (1983); 2440 ft. elev. Caldwell Ranch, acc. Clifton (1984); 3000 ft. elev. Caldwell Pines, acc. Clifton (1984). < 8000 ft. Apr.-July.

★ **Senecio elegans** Linnaeus. PURPLE RAGWORT. Uncommon, escaped ornamental: Bodega Dunes State Park, *Connors* (1993 BML). < 330 ft. May-July. Native of S. Africa.

Senecio eurycephalus A. Gray. Uncommon, rock outcrops, forest openings: 3600 ft. Geyser Rock, acc. Clifton (1983); 3200 ft. elev. Caldwell Pines, acc. Clifton (1984). 1000-5600 ft. Apr.-June.

Senecio flaccidus Lessing var. **douglasii** B. Turner & T. Barkley. Uncommon, dry stream beds: ca 5 mi. n. of Cloverdale on US 101, *Baker* (1946 NCC). < 5400 ft. June-Oct. —*S. douglasii* de Candolle.

Senecio greenei A. Gray. Locally common, serpentine, wooded or brushy slopes: 4000 ft. elev. w. flank Mt. St. Helena, *Grummer* (1980 PUA); woods near The Geysers, *E. Greene* (type U); along ridge between Socrates Mine and Mercuryville in geothermal area, *Dolan & LaPré* (1987 UCR); 3290 ft. elev. Socrates Mine, acc. Clifton (1984); 2880 ft. elev. locally plentiful on Rattlesnake Fire Trail ¾ mi. airline ese. from Mercuryville, *Raiche* (1991 JEPS); 2420 ft. elev. along Rattlesnake Fire Trail in geothermal area, locally common on serpentine, *Raiche* (1990 JEPS); near Pine Mtn. fire road ca 2 mi. w. of its origin on Western Mines Road, acc. Raiche (1984); near jct. of Socrates Mine Road with Pine Flat-Middletown Road in Mayacmas Mts., *H. K. Sharsmith* (1952 NCC); 3060 ft. Squaw Cr., acc. Clifton (1983). 1200-5000 ft. May-June.

Senecio hydrophilus Nuttall. SWEET MARSH RAGWORT. Uncommon, damp areas: Geyser Peak, *Knight & Edwards* (1986). < 8000 ft. May-Aug.

Senecio integerrimus Nuttall var. **major** (A. Gray) Cronquist. ONE-STEM BUTTERWEED. Uncommon, slopes, woods: Healdsburg region, *Butler* (1936 NCC); ne. portion of Jenson Ranch w. of Dry Cr., *Baker* (1936 NCC). Has sometimes been identified as *S. laetiflorus* E. Greene. 450-12000 ft. May-Aug.

★ **Senecio jacobaea** Linnaeus. TANSY RAGWORT. Uncommon, waste places, pastures: just s. of Long Meadow Drive on Hwy. 1, The Sea Ranch, *Wells* (1984). < 5000 ft. July-Sept. Introduced from Europe.

★ **Senecio mikanioides** Walpers. GERMAN-IVY. Common, coastal areas especially, rampant growing vine which sometimes completely covers and kills its supporting trees or makes mats along the ground: s. side of creek below Timber Cove, *Wells* (1985); between old barn near lodge and the sea at The Sea Ranch, *Wells* (1981); Hwy. 1 at turnoff to Fisk Mill Cove n. of Salt Point, acc. Knight (1990); just n. of Fort Ross along Hwy. 1, acc. Knight (1990); Hwy. 1 at Jenner, *Knight* (1990 NCC); Bodega Head, acc. Best (1988); Pepperwood. < 700 ft. Dec.-Mar. Native of S. Africa.

★ **Senecio sylvaticus** Linnaeus. Uncommon, dunes: Coleman Valley Marsh, *Rubtzoff* (1969); 3 ft. elev. Bodega Head, *Clifton* (1980 PUA). < 650 ft. June-Sept. Native of Europe.

★ **Senecio vulgaris** Linnaeus. COMMON GROUNDSEL. Common, weed, flower beds, waste places: s. end of The Sea Ranch along coastal bluffs, *J. H. Thomas* (1976); Bodega Head; Russian River near Trenton, *Heller & Brown* (1902); along Copeland Cr. on Sonoma State Univ. campus, *Self* (1979 NCC); Blomquist Ranch near Penngrove, *Knight* (1963); Button Ranch on Walker Road (U. C. property) between Petaluma-Valley Ford Road and Pepper Road, acc. Patterson (1990); w. of Sears Point Raceway in Tolay Cr. drainage, acc. Patterson (1990); Bouverie; Sugarloaf Ridge; Rincon Valley, *Wells* (1982); Spring Lake Village; Pepperwood; Lake Sonoma area; 3600 ft. elev. Geyser Rock, acc. Clifton (1983). < 5000 ft. Most months. Naturalized from Eurasia.

★ **Silybum marianum** (Linnaeus) Gaertner. MILK THISTLE. Common, waste places, pastures: The Sea Ranch; Salt Point; Willow Creek; Bodega Head; Todd Road Preserve; Rohnert Park, *O'Toole* (1977 NCC); Crane Can., *Stocking* (1963); near Cooper's Grove on Sonoma Mtn. Road, *Utech & Knight* (1986 CM); Tresch Ranch, Walker Road near Two Rock, acc. J. & B. Guggolz (1992); w. of Sears Point Raceway in Tolay Cr. drainage, acc. Patterson (1990); Sugarloaf Ridge; Spring Lake; Lake Sonoma area; ca 2000 ft. elev. Anderson Trail on Black Mtn. between Geyser Peak and upper end of Pine Flat Road, acc. J. & B. Guggolz (1992). < 2700 ft. May-July. Naturalized from Mediterranean region.

Solidago

1. Plants with well-developed, creeping, herbaceous rhizomes; stems ± equally leafy, the lowest leaves not prominently different from the upper and deciduous in age.
   2. Leaves densely puberulent on both faces, the middle and upper usually elliptic and entire . *S. californica*
   2: Leaves puberulous mostly on nerves beneath or usually glabrous, the middle and upper lanceolate, mostly sharply toothed . . . . *S. canadensis*

1: Plants with rather short woody rhizomes or a branched caudex; stems with lower and basal petiolate leaves much larger than the upper, reduced, sessile ones . . . . *S. spathulata*

Solidago californica Nuttall. CALIFORNIA GOLDENROD. Common, forest openings, clearings, fields: s. of Duncan's Landing, *Wells* (1983); Bodega Head, *Knight* (1963); Bodega Point, *Eastwood* (1915); Pitkin Marsh, *Rubtzoff* (1951); Sebastopol, *Dows* (1907); Petaluma, *Congdon* (1880); Santa Rosa Cr. e. of Santa Rosa, *Heller* (1902); Pepperwood; near jct. Porter Cr. Road and Calistoga Road, *Knight* (1963); 2000 ft. elev. e. of geysers Unit 18, *Malloch* (1977 PG&E). < 8000 ft. July-Oct.

Solidago canadensis Linnaeus ssp. **elongata** (Nuttall) Keck. CANADA GOLDENROD. Occasional, brushy areas: Vine Hill Region, acc. J. B. Roof (1972); Perry Marsh, *Howell* (1945); Cunningham Marsh 1 mi. w. of Cunningham, *Howell* (1946); along Canfield Road s. of Hwy. 116, acc. J. & B. Guggolz (1992); jct. San Antonio Road and D St. Ext. s. of Petaluma, *Knight* (1963); chaparral area sw. of Bradford Mtn. 1½ mi. w. of Dry Cr., acc. J. & B. Guggolz (1992). < 9300 ft. May-Sept.

Solidago spathulata de Candolle. COAST GOLDENROD. Occasional, usually on sandy coastal hills, dunes: The Sea Ranch; Salt Point; Perry Marsh, *Howell* (1945); Cunningham Marsh ca 1 mi. w. of Cunningham, *Howell* (1946). < 400 ft. May-Nov.

★ **Soliva sessilis** Ruiz López & Pavón. Occasional, weed, paths, roadsides, disturbed places:The Sea Ranch, *Wells* (1981); Duncans Mills quad.; top of bluff n. of Salmon Cr., *Baker* (1930 NCC); mouth of Salmon Cr., *Howell* (1931); Bodega Head quad.; Watson School on Bodega Hwy. w. of Freestone, *Best* (1983); Noldecke Ranch near Sebastopol, *Cantelow* (1939); Todd Road Preserve; Fairfield Osborn (H); Buena Vista Winery near Sonoma, *Best* (1984); Bouverie; Spring Lake; around Santa Rosa, *Brunk* (1933 NCC); Pepperwood, *Breedlove* (1983); Skylane Blvd. between Shiloh Road and Airport Blvd., *Knight* (1989 NCC); Fitch Mtn., *Rubtzoff* (1952); 2040 ft. elev. Squaw Cr., acc. Clifton (1984). < 2500 ft. Apr.-July. —*S. daucifolia* Nuttall. *S. pterosperma* (Jussieu) Lessing. Native of S. Am.

Sonchus

1. Basal auricles of the leaves straight, acute; achenes striate, transversely wrinkled . . . . *S. oleraceus*
1: Basal auricles of the leaves rounded; achenes 3-nerved on each side, otherwise smooth . . . . *S. asper*

★ **Sonchus asper** (Linnaeus) Hill. PRICKLY SOW THISTLE. Very common, weed, roadsides, disturbed places: The Sea Ranch; Bodega Head; Todd Road Preserve (H&K); along Copeland Cr. on Sonoma State Univ. campus, *Self* (1979 NCC); Fairfield Osborn (S); near slough draining Kenwood Marsh, *Howell* (1932); Gates Road off Calistoga Road, acc. Wells et al. (1984); Spring Lake; Pepperwood; ca 2000 ft. elev. Anderson Trail on Black Mtn. between Geyser Peak and upper end of Pine Flat Road, acc. J. & B. Guggolz (1992); 2400 ft. elev. Little Geysers near Unit 18, *Malloch* (1977 PG&E). < 6400 ft. Most months. Naturalized from Eurasia.

★ **Sonchus oleraceus** Linnaeus. COMMON SOW THISTLE. Common, weed, roadsides, fields, waste places: The Sea Ranch; Salt Point; Willow Creek; Bodega Head; Goat Rock State Beach, acc. Wells (1981); Bodega Head; Todd Road Preserve; along Copeland Cr. on Sonoma State Univ. campus, *Self* (1979 NCC); Crane Can., *Knight* (1963); Tresch Ranch, Walker Road near Two Rock, acc. J. & B. Guggolz (1992); Bouverie; Sugarloaf Ridge; Rincon Valley, *Wells* (1984); Spring Lake Village, Pepperwood, *Breedlove* (1983); Lake Sonoma area; 3290 ft. elev. Socrates Mine, acc. Clifton (1984). < 5000 ft. Most months. Naturalized from Europe.

Stebbinsoseris heterocarpa (Nuttall) Chambers. Occasional, rocky to clay soils, usually ± inland: Camp Meeker serpentine, *Logan* (1977 SRJC); 1500 ft. elev. on Cavedale Road, *Bufford, Wood, Bartholomew & Knight* (1987); ca 2000 ft. elev. Anderson Trail on Black Mtn. between Geyser Peak and upper end of Pine Flat Road, acc. J. & B. Guggolz (1992). < 3300 ft. Mar.-May. Derived from *Microseris douglasii x Unopappus.*

Stephanomeria

1. Plants mostly < 5 dm high; sides of fruit not grooved; pappus plumose only above, naked or merely hirsutulous below the middle, the base thickened or paleaceous . . . . *S. exigua* ssp. *coronaria*
1: Plants mostly 5-30 dm high; sides of fruit each with a narrow, longitudinal groove; pappus-bristles plumose almost throughout, the base scarcely thickened . . . . *S. virgata*

Stephanomeria exigua Nuttall ssp. **coronaria** (E. Greene) Gottlieb. Uncommon, low dry areas: Pepperwood. < 6600 ft. May-Sept. —*S. e.* var. *c.* (E. Greene) Jepson.

Stephanomeria virgata Bentham. Occasional, deserted fields, disturbed areas: Camp Meeker, *Harrison* (1977 SRJC); Bouverie; Lake Sonoma area; Sulphur Cr. 3 mi. e. of Cloverdale, *Knight* (1963). < 6700 ft. July-Oct.

Tanacetum

1. Rays not evident, flowers yellow; leaves bipinnatifid . . . . *T. camphoratum*
1: Rays white, disk flowers yellow; leaves broad-ovate to oblong-ovate, lower ones pinnate and then pinnatisect, the upper ones deeply pinnately parted . . . . *T. parthenium*

Tanacetum camphoratum Lessing. DUNE TANSY. Acc. *Baker* (1954) as occurring in "sand dunes at Bodega Bay," but there is no information concerning any collection there. < 100 ft. June-Sept. —*T. douglasii* de Candolle, misapplied.

★ **Tanacetum parthenium** (Linnaeus) Schultz Bipontinus. FEVERFEW. An old garden favorite: Spring Lake, *Harrison* (1986). Usually < 850 ft. June-Sept. Native of Europe. —*Chrysanthemum p.* (Linnaeus) Bernardi.

★ **Taraxacum officinale** Wiggers. COMMON DANDELION. Very common, widely scattered, weed: Pitkin Marsh, *Rubtzoff* (1952); Fairfield Osborn (H); Bouverie; Sugarloaf Ridge; Rincon Valley, *Wells* (1984); Spring Lake; Santa Rosa, *Raven* (1950); Pepperwood; Lake Sonoma area. < 11000 ft. Most months. Native of Europe. —*T. laevigatum* (Willdenow) de Candolle.

★ **Tolpis barbata** (Linnaeus) Gaertner. Rare weed, sunny slopes: Gilman Cr. watershed, Austin Cr. State Rec. Area, *Wagnon* (1994 NCC). Pepperwood, *de Nevers* (1995). < 1600 ft. Native of s. Europe. —*T. umbellata* Bertoloni.

Tracyina rostrata S. F. Blake. BEAKED TRACYINA. Rare, dry grassy slopes: 600 ft. elev. confluence of Galloway and Dry crs. ca 8 air-miles w. of Cloverdale, *Howell* (1967). < 1700 ft. May-June.

Tragopogon

1. Outer ligules yellow . . . . *T. dubius*
1: Outer ligules purplish . . . . *T. porrifolius*

★ **Tragopogon dubius** Scopoli. GOAT'S BEARD. Rare, weed: Spring Lake. < 9000 ft. May-July. Native of Europe.

★ **Tragopogon porrifolius** Linnaeus. SALSIFY, OYSTER PLANT. Occasional, weed, fields, waste places: 1 mi. from top of Sonoma Mtn. Road, *Lowry* (1983 SFSU); field near Charles St. in Cotati, *Murray* (1975 NCC); Sonoma State Univ. near Bridgit Drive, *Stocking* (1963 NCC); Fairfield Osborn (S); vacant lot on Douglas St. in Petaluma, *Gutermute* (1932 NCC); 397 ft. elev. Porter Cr. Road, *Brantley* (1970 PUA); Pepperwood. < 6000 ft. Apr.-June. Naturalized from Europe.

Uropappus lindleyi (de Candolle) Nuttall. SILVER PUFFS. Occasional, open woods, grassy places: Pepperwood, *Breedlove* (1983); Lake Sonoma area; 1280 ft. elev. Ida Clayton Road, *Clifton* (1984 PUA); 3620 ft. elev. Anderson Cr. 1 mi. sw. summit of Cobb Mtn., acc. Clifton (1984); 2040 ft. elev. Squaw Cr., acc. Clifton (1984). < 6000 ft. Apr.-June. —*Microseris l.* (de Candolle) A. Gray. *M. linearifolia* (de Candolle) Schultz Bipontinus.

Wyethia

1. Outer phyllaries foliaceous, greatly exceeding the disk
   2. Plants shining, sparsely short-pilose to glabrous; achenes 10-12 mm long . . . . . . *W. glabra*
   2: Plants densely white-tomentose, becoming glabrate in age; achenes 12-15 mm long . . . . *W. helenioides*
1: Outer phyllaries narrow, bractlike, or, if foliaceous, scarcely exceeding the disk . . . . *W. angustifolia*

Wyethia angustifolia de Candolle) Nuttall. NARROW-LEAVED MULE EARS. Common, open grassy hillsides: The Sea Ranch, *Wells* (1985); Stewarts Point near milepost 48.32, *Wells* (1985); Salt Point; Bodega Bay, *Barbour* (1976 BML); Neeley Hills area s. of Guerneville, *Rubtzoff* (1953); Todd Road Preserve (H&K); Fairfield Osborn (S); n. end Enterprise Road near Bennett Valley Road, *Quibell* (1982 NCC); ½ mi. up Cavedale Road, *Thorne* (1965); Sugarloaf Ridge; Pepperwood, *Breedlove* (1983); road between Petrified Forest and Mark West Springs, *Abrams* (1916); Petrified Forest, *Eastwood* (1915); 11 mi. n. of Santa Rosa, *Eastwood & Howell* (1936); 2 mi. n. of Windsor, *Cantelow* (1962); chaparral area sw. of Bradford Mtn. 1½ mi. w. of Dry Cr., acc. J. & B. Guggolz (1992); Skaggs Springs, *Hawver & Parsons* (1915); Lake Sonoma area; Canyon Road near Geyserville, *Breedlove* (1962); 3200 ft. elev. Socrates Mine, acc. Clifton (1984); 2040 ft. elev. Squaw Cr., acc. Clifton (1984). < 6600 ft. Apr.-July.

Wyethia glabra A. Gray. COAST MULE EARS. Common, dry foothills: The Sea Ranch, acc. Wells (1985); Skaggs Springs Road e. of Stewarts Point, acc. Wells (1984); w. of Graton, *L. L. Davis* (1975 NCC); Grange Road (continuation of Crane Can. Road), *Quibell* (1969 NCC); Fairfield Osborn (*); Tresch Ranch, Walker Road near Two Rock, acc. J. & B. Guggolz (1992); Petaluma, *Congdon* (1880); El Verano, *Griffin* (1922); n. end Enterprise Road near Bennett Valley Road, *Quibell* (1982 NCC); Sugarloaf Ridge; Holland Drive near Bennett Valley, *Stocking* (1964 NCC); Rincon Ridge, *Knight & Edwards* (1986); Spring Lake; Lake Sonoma area; ca 2000 ft. elev. Anderson Trail on Black Mtn. between Geyser Peak and upper end of Pine Flat Road, acc. J. & B. Guggolz (1992); 2400 ft. elev. Caldwell Ranch, acc. Clifton (1984); Cloverdale, *Breedlove* (1962). < 3000 ft. Mar.-May.

Wyethia helenioides (de Candolle) Nuttall. GRAY MULE EARS. Uncommon, forest openings: Stewarts Point-Skaggs Springs Road near top of ridge, *Quibell* (1986 NCC); 2760 ft. elev. Caldwell Ranch, acc. Clifton (1984); 3000 ft. and 3200 ft. elev. Caldwell Pines, acc. Clifton (1984). < 6600 ft. Mar.-July.

Xanthium

1. Leaves sessile or shortly petiolate, white or gray beneath with 1-, 2-, or 3-fid spines at base . . . *X. spinosum*
1: Leaves long-petiolate, green beneath without spines at base of petiole . . . . *X. strumarium*

★ **Xanthium spinosum** Linnaeus. SPINY COCKLEBUR. Common, weed, pastures, waste places: Wiggins Hill w. of Petaluma, *Knight* (1963); Todd Road Preserve; Laguna de Santa Rosa floodplain w. of Llano Road, acc. J. & B. Guggolz (1985); Fairfield Osborn (S); ¼ mile e. off US 101 near railroad tracks in Rohnert Park, *Self* (1979 NCC); fields around edge of Sonoma State Univ. campus, *Stocking* (1963 NCC); Spring Lake Village; Santa Rosa Jr. Coll. campus, *H. Thomas* (1934 NCC); Pepperwood; Lake Sonoma area. < 3000 ft. July-Oct. Native of Europe.

Xanthium strumarium Linnaeus. COCKLEBUR. Common, cosmopolitan weed, pastures, waste places: laguna n. of Sebastopol at Occidental Road crossing, *Howell* (1946); jct. US 101 and Todd Road, *Knight* (1963); Todd Road Preserve (H&K); along Copeland Cr. on Sonoma State Univ. campus, *Self* (1979 NCC); Pepperwood; Lake Sonoma. < 1700 ft. July-Oct. —*X. s.* var. *canadense* (Miller) Torrey & A. Gray. Worldwide.

BASELLACEAE. BASELLA FAMILY

★ **Anredera cordifolia** (Tenore) Steenis. MIGNONETTE VINE. Escape in a flower bed at Friends House in Rincon Valley, *Wells* (1985). < 1700 ft. Native of tropical and subtropical America.

BERBERIDACEAE. BARBERRY FAMILY

1. Plants shrubs with pinnate, spiny-toothed leaves . . . . *Berberis*
1: Plants perennial herbs; leaves basal, mostly ternate
   2. Leaves with 3 sessile leaflets; sepals and petals none . . . . *Achlys*
   2: Leaves once or twice ternate; petiolulate; sepals and petals present . . . . *Vancouveria*

Achlys californica I. Fukuda & H. G. Baker. VANILLA LEAF, DEER FOOT. Rather rare, coastal woods, often deep shade: The Sea Ranch, acc. Wells (1985); Annapolis, *Sutliffe* (1920). < 4500 ft. Apr.-June.

Berberis

1. Filaments with a pair of recurved teeth near the apex; racemes short, from small terminal or lateral buds; bud-scales few, deciduous, small; leaflets 3-13, pinnately veined
   2. Foliage not very dense; leaflets plane 3-5 or 7, the margins with many, long, slender spines *B. aquifolium*
   2: Foliage mostly forming a dense terminal fascicle; leaflets usually 7-13, rather crowded on the rachis, ± bubbly- undulate, shallowly sinuate-dentate, the numerous teeth prickly, the lowest pair close to base of petiole . . . . *B. pinnata*
1: Filaments without teeth; racemes elongated, loose, solitary or few from a terminal bud; bud scales large, persistent; leaflets 11-21, ± palmately veined . . . . *B. nervosa*

Berberis aquifolium (Pursh) Nuttall. OREGON-GRAPE. Uncommon, dry wooded or open slopes, sometimes on serpentine: Pine Flat area, *Harrison* (1989 SRJC). 1200-6900 ft. Mar.-May. —*B. piperiana* (Abrams) McMinn. *Mahonia a.* Nuttall.

Berberis nervosa Pursh. LONGLEAF MAHONIA. Occasional, shaded slopes, hills: Stewarts Point, acc. Jepson *Fl. Calif. 1:550* (1909); Salt Point; Camp Meeker, *Howell* (1924); 2400 ft. elev. ½ mi. from county line in Black Mtn. area, *J. & B. Guggolz* (1990 NCC); 2500 ft. elev. w. flank Mt. St. Helena, *Grummer* (1980 PUA). < 6600 ft. Apr.-June. —*Mahonia n.* (Pursh) Nuttall.

Berberis pinnata Lagasca. CALIFORNIA BARBERRY. Occasional, rocky slopes: Bodega Bay, *Heller & Brown* (1902); Bodega Head, acc. Connors (1990); Salmon Cr. Road w. of town of Bodega, *Knight* (1963); 2720 ft. elev. Caldwell Ranch, *Clifton* (1984 PUA). < 4200 ft. Mar.-May.

Vancouveria

1. Leaves deciduous, leaflets thin; panicles glabrous . . . . *V. hexandra*
1: Leaves persistent; leaflets coriaceous; panicles glandular . . . . *V. planipetala*

Vancouveria hexandra (Hooker) Morren & Decaisne. NORTHERN VANCOUVERIA. Rare, deep shade, coniferous woods: Plantation Road near Hwy. 1, acc. Baker, the only record known from Sonoma County, probably the southernmost distributional limit. < 6000 ft. May-July.

Vancouveria planipetala Calloni. INSIDE-OUT FLOWER. Occasional, moist coniferous woods: The Sea Ranch, acc. Wells (1985), acc. *Mahaffey* (1990); Stewarts Point Skaggs Springs Road just e. of Annapolis bridge, acc. Quibell (1991); Salt Point; Cazadero, *Congdon* (1892), *Bacigalupi* (1926); ca 1 mi. nw. of Camp Meeker, acc. J. & B. Guggolz (1991); Guerneville, *Rattan* (1877), *Congdon* (1880), *Rubtzoff* (1953); Green Valley Road between Graton Road and Pocket Can. Road, *Baker* (1899); Pepperwood; w. of Calistoga near Humbug Cr., acc. Wells (1984); Skaggs Springs, *Hawver* (1915); Lake Sonoma area. < 4300 ft. May-June.

BETULACEAE. BIRCH FAMILY

1. ♂ flowers solitary in axil of each bract, without calyx . . . . *Corylus*
1: ♂ flowers 3-6 together in the axils of each bract, with calyx . . . . *Alnus*

Alnus

1. Leaves cordate at base . . . . *A. cordata*
1: Leaves not cordate at base
   2. Leaf margin tightly rolled under; coastal tree . . . . *A. rubra*
   2: Leaf margin usually ± flat; lowland or montane tree . . . . *A. rhombifolia*

★ **Alnus cordata** (Loiseleur) Duby. ITALIAN ALDER. Regeneration from landscape-planted trees: 240 ft. elev. Spring Lake Park, Santa Rosa, *Wagnon* (1991 NCC). < 1000 ft. Native of Italy and Corsica.

Alnus rhombifolia Nuttall. WHITE ALDER. Common, live streams, hills away from coast: The Sea Ranch, acc. Wells (1985); along Matanzas Cr. between Grange and Sonoma Mtn. roads, acc. Quibell (1991); Fairfield Osborn (*); Alta Vista Ranch n. of Sonoma; Bouverie; Sonoma Cr. near Glen Ellen, acc. Baker; Spring Lake; Pepperwood; Ida Clayton Road quad.; Mill Cr. w. of Healdsburg, acc. Knight (1986); ca 2000 ft. elev. Anderson Trail on Black Mtn. between Geyser Peak and upper end of Pine Flat Road, acc. J. & B. Guggolz (1992); 2840 ft. elev. Bear Cr. Can., acc. Clifton (1984); 2600 ft. elev. Coldwater Cr. Can. ½ airmile sw. of Lake Co. line at Caldwell Pines, acc. Stone (1989). < 8000 ft. Jan.-Apr.

Alnus rubra Bongard. RED ALDER, OREGON ALDER. Common, coastal streams, swampy ground: The Sea Ranch ½ mi. s. of Gualala River, *Madison, Wells & Rhyne* (1985); Annapolis and Stewarts Point quads.; 8 mi. s. of Fort Ross, *Howell* (1934); Duncans Mills Marsh, *Rubtzoff* (1957, 1959); Willow Creek; Willow Cr. Marsh, *Rubtzoff* (1959); lower Salmon Cr. Marsh; *Rubtzoff* (1957, 1962); Guerneville Marsh, *Rubtzoff* (1954); Vacation Beach near Guerneville, *Rubtzoff* (1951); 350-400 ft. elev. Old Skaggs Springs Road, *Utech, Sato & Ohara* (1990 CM). < 3300 ft. Mar.-Apr. —*A. oregona* Nuttall.

Corylus cornuta Marshall var. **californica** (A. de Candolle) W. Sharp. HAZELNUT. Common, wooded areas: The Sea Ranch, acc. Wells (1985); Salt Point; Annapolis, Cazadero and Duncans Mills quads.; Willow Creek; Willow Cr. Road, *Utech* (1988 CM); Crane Can. Road, *Knight* (1963); Hidden Oaks Road between Grange and Sonoma Mtn. roads, acc. Quibell (1991); Tresch Ranch, Walker Road near Two Rock, acc. J. & B. Guggolz (1992); Sugarloaf Ridge; Spring Lake; Ida Clayton quad.; Fitch Mtn., *Rubtzoff* (1952); Mill Cr. Road, *Knight* (1964); Lake Sonoma area; Geyserville quad.; 1500 ft. elev. w. flank Mt. St. Helena, acc. Grummer & Hemphill (1983); Cloverdale quad. < 7200 ft. Jan.-Apr. —*C. c.* ssp. *c.* (A. de Candolle) E. Murray.

BORAGINACEAE. BORAGE FAMILY

1. Flowers rotate . . . . *Borago*
1: Flowers tubular, salverform or funnelform
 2. Style borne on summit of the fruit on pericarp, falling away with the nutlets; stigma discoid, usually surmounted with a short sterile appendage . . . . *Heliotropium*
 2: Style borne between the lobes of the fruit (nutlets) and attached directly and independently to the gynobase or receptacle; stigma capitate, lacking a sterile apical appendage
  3. Nutlets widely spreading in fruit, armed with barbed (or hooked) prickles
   4. Nutlets flat, armed on the margins with hooked bristles; flowers white; annual . . . . . *Pectocarya*
   4: Nutlets subglobose, armed totally with barbed prickles; flowers mostly blue; perennial . . . . *Cynoglossum*
  3: Nutlets erect, often not armed with prickles
   5. Nutlets attached, often by the base, to the flattish or convex receptacle . . . . *Myosotis*
   5: Nutlets attached ± laterally to an elongate or conical receptacle
    6. Corolla bright blue, bilateral . . . . *Echium*
    6: Corolla white or yellow or orange, radial
     7. Flowers white
      8. Nutlets attached laterally along a usually partly opened and generally forked groove or slit in pericarp . . . . *Cryptantha*
      8: Nutlets lacking a distinct ventral groove or slit in the pericarp, this replaced by an elevated keel and ± conspicuous attachment scar
       9. Leaves opposite in basal area . . . . *[Allocarya] Plagiobothrys*
       9: Leaves totally alternate . . . . *Plagiobothrys*
     7: Flowers yellow or orange . . . . *Amsinckia*

Amsinckia

1. Back of nutlets tessellate, like a very small cobblestone pavement; corolla-tube 20-nerved below stamens; calyx-lobes unequal in width and at least 2 of them fused for some length; (possibly in the county) *A. douglasiana*
1: Back of nutlets not tessellate, the roughenings not crowded to give pavementlike appearance; corolla-tube 10-nerved below stamens; calyx-lobes various
 2. Calyx with the 2 axial lobes united at least to middle (except in the early flowers); leaves mostly erose-dentate; nutlets dark, 1.5-2 mm long . . . . *A. spectabilis*
 2: Calyx with lobes all distinct; leaf-margins usually entire; nutlets usually gray or brown, mostly 2.5-3.5 mm long
  3. Corolla-throat constricted and nearly closed by intruding hairy saccate processes; stamens inserted evenly below the constriction . . . . *A. lycopsoides*
  3: Corolla-throat open and glabrous, stamens ± irregularly inserted in throat
   4. Corolla yellow, pale, 5-7 mm long, 2-3 mm broad, the tube almost or quite included in calyx; style 2.5-3 mm long; plant usually cinereous, the stems finely strigulose as well as spreading-hirsute . . . . *A. menziesii* var. *menziesii*
   4: Corolla orange-yellow 7-20 mm long, the tube distinctly exserted; style 5-18 mm long; stems hispid, but scarcely if at all finely strigose . . . . *A. m.* var. *intermedia*

Amsinckia lycopsoides Lehmann. BUGLOSS FIDDLENECK. Rare, usually in damp ground: Lake Sonoma area; Diamond Mtn. Road (just over county line, extension of Sharpe Road) between Petrified Forest Road and St. Helena Road, acc. *Muth* (1984). < 1300 ft. Apr.-June.

Amsinckia menziesii (Lehmann) Nelson & J. F. Macbride var. **intermedia** (Fischer & C. Meyer) Ganders. COMMON FIDDLENECK, RANCHER'S FIREWEED. Widespread, common, along roads, grassy fields: The Sea Ranch, acc. Wells (1983); Bodega Bay and Russian River area (type 1834 — collected by Russian commercial traders); n. end of Bodega Bay, *Howell* (1930); Fairfield Osborn (S); n. of Penngrove on Old Redwood Highway, *Best* (1981); Tresch Ranch, Walker Road near Two Rock, acc. J. & B. Guggolz (1992); w. of Sears Point Raceway in Tolay Cr. drainage, acc. Patterson (1990); n. of Sonoma near Boyes Hot Springs, *Howell* (1931); Bouverie; Annadel; Sugarloaf Ridge; Spring Lake Village; Pepperwood; Lake Sonoma area. < 5600 ft. Mar.-June. —*A. intermedia* Fischer & C. Meyer. *A. acuticarpa* Suksdorf, type locality: Lake Sonoma area. < 5600 ft. Mar.-June. —*A.* intermedia Fischer & C. Meyer. *A. acuticarpa* Suksdorf, type locality: Santa Rosa, *Eastwood* (1921). *A. communis* Suksdorf, type locality: Petaluma,

Eastwood (1921). *A. viridifolia* Suksdorf, type locality: Petaluma, *Eastwood* (1921). *A. hispidula* Suksdorf, type locality: Petrified Forest, *Eastwood* (1915).

Amsinckia menziesii var. **menziesii**. MENZIES' FIDDLENECK. Occasional, dry grassy areas: Chileno Valley Road near San Antonio Cr., *Knight* (1964); Fairfield Osborn (H); hills e. of Agua Caliente n. of Sonoma, *Howell* (1948); Bouverie; Spring Lake; Park Trail Drive near Howarth Park, *Harrison* (1981); Pepperwood; 2040 ft. elev. Squaw Cr., acc. Clifton (1984). < 5600 ft. Apr.-June. —*A. helleri* Brand.

Amsinckia spectabilis Fischer & C. Meyer. SEASIDE FIDDLENECK. Occasional, coastal dunes, coastal terraces: Horseshoe Cove, *Knight* (1963); 56 ft. elev. near Salmon Cr., *Reeve* (1980 PUA); entrance to Bodega Marine Reserve, *Knight & Best* (1991). < 330 ft Mar.-June. —*A. latifolia* Suksdorf, type locality: Bodega Head, *K. Brandegee* (1905 UC). *A. nigricans* Brand, type locality: Bodega Bay, *Heller* (1902).

★ **Borago officinalis** Linnaeus. COMMON BORAGE. Rare, escape: Fort Ross-Russian River-Bodega areas, *Voznesenskii* (1840-1841); Portuguese Beach s. of Wright's Beach, *Sanborn* (1958); Salmon Cr. (town), acc. Knight (1964); Spring Lake Village. < 100 ft. Native of Europe.

Cryptantha

1. Nutlets roughened or at least one of them so . . . . *C. muricata*
1: Nutlets smooth and shining, not roughened
  2. Calyx armed with pale, encrusted, spreading, hooked or curved hairs . . . . *C. flaccida*
  2: Calyx armed with straight, ± tawny, unencrusted hairs
    3. Nutlets with eccentric groove, 1 side of nutlet on lower surface appearing as if somewhat deformed . . . . *C. affinis*
    3: Nutlets with a centrally placed groove
      4. Dorsal side of nutlets low-convex or flat, alike in form and size . . . . *C. torreyana*
      4: Dorsal side of nutlets rounded-convex, the lateral angles rounded or obtuse
        5. Style < half as long as nutlets; fruiting calyx 1.5-2 mm long . . . . *C. microstachys*
        5: Style almost as long as nutlets; fruiting calyx longer
          6. Spikes bracteate, dense; nutlets mostly 4 . . . . *C. leiocarpa*
          6: Spikes naked, or the lowest flowers sometimes bracteate; nutlets solitary . . . *C. hispidula*

Cryptantha affinis (A. Gray) E. Greene. Uncommon, dry slopes: 3100 ft. elev. w. flank Mt. St. Helena, *Grummer* (1979 PUA). 2600-10000 ft. June-Aug.

Cryptantha flaccida (Lehmann) E. Greene. Occasional, dry flats, slopes, sometimes on serpentine: serpentine talus at Westminster Woods near Camp Meeker, *Harrison* (1981 SRJC); Santa Rosa, *Rattan* (1877); Pepperwood; Petrified Forest, *Eastwood* (1915); Skaggs Springs, *Heller* (1940); 3600 ft. elev. Geyser Rock, acc. Clifton (1983). < 6600 ft. Apr.-June.

Cryptantha hispidula Brand. Rare, slightly serpentinized area: 3500 ft. elev. in upper Cobb Cr. drainage area, acc. Stone (1989). 600-3600 ft. Apr.-June.

Cryptantha leiocarpa (Fischer & C. Meyer) E. Greene. COAST CRYPTANTHA. Occasional, sandy areas, coastal: Goat Rock State Park, *Best* (1980); Wright's Beach, *Baker* (1934 NCC); near mouth of Salmon Cr., *Baker* (1930 NCC); Bodega Head, *K. Brandegee* (probably in mid-1880s). < 670 ft. Apr.-June. —*Echinospermum leiocarpum* Fischer & C. Meyer, type locality: Fort Ross, Russian traders (1830s).

Cryptantha microstachys (A. Gray) E. Greene. TEJON CRYPTANTHA. Uncommon, dry openings and chaparral: Oakville to Glen Ellen, *Raven* (1951); 3100 ft. elev. w. flank Mt. St. Helena, *Grummer* (1979 PUA). 600-5600 ft. Apr.-June.

Cryptantha muricata (Hooker & Arnott) Nelson & J. F. Macbride. JONES' CRYPTANTHA. Locally common: Mt. Hood, *Bioletti* (1893); sw. slopes of Mt. Hood near base at Sugarloaf Ridge, *Baker* (1947); ca 800 ft. elev. w. slope Mt. Hood on Rose Butte, *Baker* (1947 NCC); Hope Valley near Kenwood, *Baker* (1928 NCC); Santa Rosa Cr. Can., *Baker* (1930 NCC); St. Helena Road near county line, *Baker* (1942 NCC); on ridge w. of Calistoga on w. slope ca ½ mi. S. of Santa Rosa-Calistoga Hwy, *Baker* (1940 NCC) ca 2000 ft. elev. Anderson Trail on Black Mtn. between Geyser Peak and upper end of Pine Flat Road, acc. J. & B. Guggolz (1992); 2040 ft. elev. Squaw Cr., acc. Clifton (1984). 500-9000 ft. Apr.-June. —*C. m.* var. *jonesii* (A. Gray) I. M. Johnston.

Cryptantha torreyana (A. Gray) E. Greene. Occasional, partial shade, open slopes: road to Armstrong Grove, *Gossin* (1933 NCC); Santa Rosa Cr. Can., *Baker* (1932); Petrified Forest, *Eastwood* (1915); 2860 ft. elev. Pine Flat Road, acc. Clifton (1984); 3200 ft. elev. Socrates Mine, acc. Clifton (1984); ca 2000 ft. elev. Anderson Trail on Black Mtn. between Geyser Peak and upper end of Pine Flat Road, acc. J. & B. Guggolz (1992); 3300 ft. elev. Squaw Cr., acc. Clifton (1983). 50-7000 ft. May-Aug.

Cynoglossum grande Lehmann. HOUND'S TONGUE. Common, edge of woods, semishaded areas: The Sea Ranch, acc. Wells (1983); Cazadero quad.; ca 1 mi. nw. of Camp Meeker, acc. J. & B. Guggolz (1991); 1025 ft. elev. Coleman Valley Road, acc. Clifton (1984); Fairfield Osborn (*); Tresch Ranch, Walker Road near Two Rock, acc. J. & B. Guggolz (1992); Bouverie; Kenwood, *Broderson* (1985 NCC); 3 mi. nw. of Sebastopol, *Hagg* (1934 NCC); 1 mi. from top of Sonoma Mtn. Road, *Lowry* (1983 SFSU), *Knight* (1964); Sugarloaf Ridge; Spring Lake; Pepperwood; chaparral area sw. of Bradford Mtn. 1½ mi. w. of Dry Cr., acc. J. & B. Guggolz (1992); property bordered by Chianti and Dutcher Cr. roads w. of US 101, w. of Geyserville, acc. J. & B. Guggolz (1991); between Jimtown and Mercuryville on Geysers Road, *Utech & Knight* (1986 CM); Geyser Peak, *Knight & Edwards* (1986); Lake Sonoma area; 2450 ft. elev. Caldwell Ranch, acc. Clifton (1984); 3000 ft. elev. Caldwell Pines, acc. Clifton (1984). < 5000 ft. Mar.-June.

★ **Echium plantagineum** Linnaeus. VIPER'S BUGLOSS. Uncommon, grassland: Pepperwood. <1000 ft. May-June. Native of s. Europe.

Heliotropium curassavicum Linnaeus. SALT HELIOTROPE. Occasional, disturbed and alkaline areas, river banks: Salmon Cr. dunes, *Connors* (1991 BML); Laguna de Santa Rosa e. of Graton at end of Hall Road, *Rubtzoff* (1973); laguna in Chileno Valley, *Rubtzoff* (1971); Healdsburg in sand along Russian River, *Heller* (1902), *Millard* (1969 PUA); Russian River near Geyserville, *Baker* (1923 NCC). < 7000 ft. Mar.-Oct. —*H. curassavicum* ssp. *oculatum* (Heller) Thorne. *H. spathulatum* Rydberg ssp. *oculatum* (Heller) Ewan. *H. spathulatum* var. *oculatum* (Heller) Johnston. *H. oculatum* Heller, type locality: sand along Russian River at Healdsburg, *Heller* (1902 U).

Myosotis

1. Corolla-limb 1-5 mm broad; fruiting pedicels < to << calyx . . . . *M. discolor*
1: Corolla-limb 5-10 mm broad; fruiting pedicels > or = calyx . . . . *M. latifolia*

★ **Myosotis discolor** Persoon. BLUE SCORPION-GRASS. Common, waste places, roadsides, fields: Pilot's Reach at The Sea Ranch, *Wells* (1985); Salt Point; Willow Cr. Marsh, *Rubtzoff* (1959); Arched Rock quad.; lower Salmon Cr. Marsh, *Rubtzoff* (1962); 70 ft. elev. near Salmon Cr., *Clifton* (1984 PUA); Austin Cr. Marsh, *Rubtzoff* (1955); upper Green Valley Marsh nw. of Graton, *Rubtzoff* (1974); Occidental Marsh, *Rubtzoff* (1957); Freestone Marsh, *Rubtzoff* (1957); Forestville Marsh, *Rubtzoff* (1969); near Vine Hill, *Crum* (1940); Pitkin Marsh, *Baker* (1932), *Rubtzoff* (1951); 1 mi. n. of Bodega Hwy. along Furlong Road, *Wagnon* (1939 NCC); Pepperwood. < 4400 ft. Apr.-July. Naturalized from Europe. —*M. versicolor* (Persoon) Smith.

★ **Myosotis latifolia** Poiret. FORGET-ME-NOT. Common, naturalized locally, moist shady woods: The Sea Ranch, acc. *Mahaffey* (1990); Annapolis Road e. of The Sea Ranch, *Wells* (1983); 98 ft. elev. Fort Ross, *Clifton* (1981 PUA); near Camp Meeker, *Howell* (1924), *Hagg* (1934 NCC); Coffee Lane off of Occidental Road, *Wells* (1986); Laguna de Santa Rosa area e. of Graton, *Rubtzoff* (1960); Pepperwood; Gates Road off Calistoga Road, *Wells* (1984); Fitch Mtn., *Rubtzoff* (1952). < 1000 ft. Feb.-July. Native of nw. Africa —*M. sylvatica* (Ehrhart) Hoffmann, misapplied.

Pectocarya

1. Calyx radial, sepals ± =; lower cauline leaves opposite . . . . *P. pusilla*
1: Calyx bilateral, upper 2 sepals > others; lower cauline leaves mostly alternate . . . . *P. penicillata*

Pectocarya penicillata (Hooker & Arnott) A. de Candolle. WINGED PECTOCARYA. Occasional, gravelly or sandy soils: Spring Mtn. Road ca 16 mi. from Santa Rosa, *Baker* (1930). 300-7000 ft. Mar.-May.

Pectocarya pusilla (A. de Candolle) A. Gray. LITTLE PECTOCARYA. Uncommon, sandy or gravelly places: Royal Oaks Mobile Home Park, Petaluma, *Knight* (1987 NCC); Pepperwood; Sulphur Cr., *Baker* (1932 NCC). < 6000 ft. Apr.-June.

Plagiobothrys

1. Plants with lower leaves opposite, not forming a rosette; nutlets attached by a scar or groove, not carunculate, mostly erect; corolla-tube often yellowish within . . . . [Section *Allocarya*]
   2. Plants perennial, coarse . . . . *P. mollis* var. *vestitus*
   2: Plants annual, slender-stemmed

3. Scar large, ¼-½ the length of the nutlet, deeply excavated, broadly flanged, armed along the keels and between with many barbed appendages . . . . *P. greenei*
3: Scar small, not much if at all excavated, 1/5 the length of nutlet or less, no barbed appendages
  4. Nutlet attachment basal, frequently substipitate; calyx-lobes strongly ribbed
    5. Plants prostrate; calyx-lobes connivent and together directed off toward 1 side of flower . . . . *P. leptocladus*
    5: Plants erect or ascending; calyx-lobes spreading equally
      6. Corolla 5-12 mm broad . . . . *P. stipitatus* var. *stipitatus*
      6: Corolla ca 2.5 mm broad . . . . *P. s.* var. *micranthus*
  4: Nutlet attachment lateral to obliquely basal; calyx-ribs only rarely thickened
    7. Scar linear or nearly so, usually borne on a narrow knifelike attachment . . . . . . *P. undulatus*
    7: Scar broad, the attachment not at all knifelike, or if so, the nutlets rather asymmetric
      8. Ventral keel not in a groove . . . . *P. bracteatus*
      8: Ventral keel in a conspicuous groove, at least below the middle
        9. Nutlets densely tuberculate, the pericarp thick and rather bony . . . . *P. reticulatus* var. *rossianorum*
        9: Nutlets sparsely if at all tuberculate, the pericarp thin . . . . *P. r.* var. *reticulatus*

1: Plants with leaves all alternate; nutlets attached above the base to a caruncle or thickened scar, oblique or incurved; corolla white throughout . . . . [Section *Plagiobothrys*]
  10. Calyx circumscissile in fruit, < 4 mm long; lobes usually connivent over fruit; nutlets usually 1-3 . . . . *P. nothofulvus*
  10: Calyx not circumscissile, or if so, the strongly accrescent calyx > 4 mm long; calyx-lobes erect or spreading; nutlets usually 4
    11. Nutlets with a conspicuous annular caruncle and 2.3-3.3 mm long . . . . *P. fulvus*
    11: Nutlets with a solid caruncle and < 2.3 mm long
      12. Transverse dorsal crests of nutlets very narrow and sharp, with median keel enclosing polygonal granulate areolas . . . . *P. canescens*
      12: Transverse dorsal crests of nutlets very low and broad, separated only by low lineate ridges . . . . *P. tenellus*

Plagiobothrys bracteatus (T. J. Howell) I. M. Johnston. Frequent, moist areas: many sites w. and sw. of Santa Rosa in Laguna de Santa Rosa, *Harrison* (1978 NCC); 2 mi. w. of US 101 at 4220 Langner Ave. near Rohnert Park, *B. Keller* (1990); Petaluma Hill Road, *Koch* (1939 NCC); Fairfield Osborn (H); e. of Petaluma, *Howell* (1933); Petaluma, *Eastwood* (1921); Sobre Vista Road between Fetters Hot Springs and Eldridge in Sonoma Valley, *Raven* (1950); Bennett Valley, *Baker* (1935); Spring Lake; near Wright School w. of Santa Rosa, *Baker* (1934); Santa Rosa, *Eastwood* (1921); Pepperwood. < 5300 ft. Apr.-June. —*Allocarya bracteata* T. J. Howell.

Plagiobothrys canescens Bentham. Uncommon, gravelly adobe of plains, hillsides: vernal flats, Cotati, *Raven* (1950); Lake Sonoma area, site now inundated. < 5800 ft. Mar.-May.

Plagiobothrys fulvus (Hooker & Arnott) I. M. Johnston. FULVOUS POPCORN FLOWER. Occasional, openings of valleys and foothills: Fort Ross quad.; w. of Santa Rosa in Laguna de Santa Rosa drainage, *Harrison* (1978 NCC); 6 mi. from Santa Rosa on road to Sonoma, *Howell* (1945); near Warfield (former community e. of Glen Ellen on Hwy. 12), *Koch* (1940 NCC); Alexander Valley quad. < 1700 ft. Mar.-May. —*P. f.* var. *campestris* (E. Greene) I. M. Johnston, not clearly distinct. *P. c.* E. Greene.

Plagiobothrys greenei (A. Gray) I. M. Johnston. Occasional, lowlands on bottom of desiccated winter pools: low field sw. of Sonoma, *Howell* (1949); summit of Bennett Valley Road from Glen Ellen to Santa Rosa, *Howell* (1934); Bennett Valley Road at vernal pool, *Baker* (1934 NCC); Spring Lake; w. of Santa Rosa and s. of Hwy. 12 in Laguna de Santa Rosa drainage, *Harrison* (1978 NCC). < 3000 ft. Mar.-May. —*Allocarya g.* (A. Gray) E. Greene.

Plagiobothrys leptocladus (E. Greene) I. M. Johnston. Infrequent, usually heavy alkaline soils, vernal pools: The Sea Ranch, *J. H. Thomas* (1976); 2 mi. w. of US 101 at 4220 Langner Ave. near Rohnert Park, *B. Keller* (1990 NCC). < 8200 ft. Mar.-May. —*Allocarya leptoclada* E. Greene.

Plagiobothrys mollis (A. Gray) I. M. Johnston var. **vestitus** (E. Greene) I. M. Johnston. PETALUMA POPCORN FLOWER. Habitat unknown. Probably < 200 ft. June-July. —*Allocarya vestita* E. Greene, type locality: Petaluma, *Congdon* (1880). Presumed extinct, *Jepson*.

Plagiobothrys nothofulvus (A. Gray) A. Gray. RUSTY POPCORN FLOWER. Common, grassy hillsides, fields: Bodega Bay; 3 sites w. of Santa Rosa in Laguna de Santa Rosa drainage, *Harrison* (1978 NCC); 1/8 mi. w. of Stony Point Road on Meacham Road, *Quibell* (1970 NCC); near cemetery in Petaluma, *Ott* (1934 NCC); Fairfield Osborn (*); Bouverie; Sugarloaf Ridge; Spring Lake; 3 mi. w. of Santa Rosa, *Hagg*

(1934 NCC); Pepperwood; Lake Sonoma area; 2240 ft. elev. Squaw Cr., acc. Clifton (1984). < 2700 ft. Mar.-May.

Plagiobothrys reticulatus (Piper) I. M. Johnston var. **reticulatus**. Infrequent, moist places, fields: The Sea Ranch, *Wells* (1981); Bodega Bay, *Armstrong & Hotle* (1934 NCC); Willow Cr., *Howell* (1961); Pepperwood. Usually < 1000 ft. May-July. —*Allocarya reticulata* Piper.

Plagiobothrys reticulatus var. **rossianorum** I. M. Johnston. Occasional, moist places, open areas: Schoolhouse Beach between Jenner and Salmon Cr., *Stocking* (1968 NCC); Cazadero, *Congdon* (1880); Bodega Bay, *K. Brandegee* (1905); Stoetz Lane near Occidental, *Best* (1984); lagoon e. of Sebastopol, *Howell & Stacey* (1937); 4 mi. e. of Valley Ford, *Howell* (1930). < 1000 ft. May-June. —*Allocarya californica* (Fernald & J. F. Macbride) E. Greene. *Myosotis californica* Fischer & C. Meyer, type locality: Fort Ross, *Jepson, Fl. of CA, 3:363* (1943) which explains complicated situation of types from the Russian colony. *P. diffusus* (E. Greene) I. M. Johnston, indistinct.

Plagiobothrys stipitatus (E. Greene) I. M. Johnston var. **micranthus** (Piper) I. M. Johnston. Occasional, grassy areas, wet flats or meadows: Schoolhouse Beach between Jenner and Salmon Cr., *Stocking* (1967 NCC); Todd Road Preserve; Madison St., Petaluma, *Koch* (1938 NCC); 2420 ft. elev. Caldwell Ranch, *Clifton* (1984 PUA). < 5000 ft. Apr.-July. —*Allocarya stipitata* E. Greene ssp. *micrantha* Piper.

Plagiobothrys stipitatus var. **stipitatus**. STIPITATE ALLOCARYA. Common, heavy soils, low places: 3 sites w. and sw. of Santa Rosa in Laguna de Santa Rosa, *Harrison* (1978 NCC); Gravenstein Hwy. between Cotati and Sebastopol, *Hanagrass & Anderson* (1934 NCC); just e. of Petaluma, *Howell* (1933); end of Reclamation Road (off Black Point Hwy.), *Weaver* (1975 NCC); Schellville, *Cantelow* (1939); near salt marsh n. of Sears Point, *Keck* (1935 GH); Bennett Valley Road ½ mi. e. of fairgrounds, *Baker* (1935 NCC); w. of hwy. ca 1 mi. n. of Mark West Cr. bridge (Old Redwood Hwy.), *Baker* (1934 NCC). < 1600 ft. Mar.-May. —*Allocarya stipitata* E. Greene ssp. *stipitata*.

Plagiobothrys tenellus (Nuttall) A. Gray. SLENDER POPCORN FLOWER. Common, meadows, grassy slopes, sometimes disturbed areas: Bodega Head; w. of Santa Rosa in Laguna de Santa Rosa, *Harrison* (1978 NCC); Sonoma Mtn., *Baker* (1925 NCC); Hope Valley near Kenwood, *Baker* (1928 NCC); Bennett Valley, *Conners & Mahoney* (1930 NCC); Santa Rosa Cr., *Heller & Brown* (1902); Pepperwood, *Breedlove* (1983); Spring Mtn. Road, *Baker* (1945 NCC); Cloverdale-Geysers Road, *Baker* (1934 NCC). < 5600 ft. Mar.-May.

Plagiobothrys undulatus (Piper) I. M. Johnston. COAST POPCORN FLOWER. Occasional, mud flats of vernal situations: 5 sites w. and sw. of Santa Rosa in Laguna de Santa Rosa drainage, *Harrison* (1978 NCC); vicinity of Wright School (Price Road near Wright Road) near pond, *Baker* (1934 NCC); 2 mi. w. of US 101 at 4220 Langner Ave. near Rohnert Park, *B. Keller* (1990 NCC); Bennett Valley Road, *Baker* (1935 NCC); marsh e. of the lagoon, Sebastopol, *Howell & Stacey* (1937). < 1300 ft. Mar.-June. —*Allocarya undulata* Piper.

BRASSICACEAE [CRUCIFERAE]. MUSTARD FAMILY

GROUPS:

(Key ± acc. Reed C. Rollins. *Hickman* [1993])

1. Fruit usually at least 4 × longer than wide, usually linear . . . . **Group 1**
1: Fruit < 4 × longer than wide, not linear
  2. Fruit round to ± flat in the X-section and occasionally inflated . . . . **Group 2**
  2: Fruit flat, not inflated
    3. Fruit 1-seeded, 1-chambered, indehiscent . . . . **Group 3**
    3: Fruit usually 2+-seeded, 2-chambered, usually dehiscent
      4. Fruit flat perpendicular to septum . . . . **Group 4**
      4: Fruit flat parallel to septum . . . . **Group 5**

GROUP 1:

1. Leaf hairs branched (occasionally also simple)
  2. Leaf deeply lobed or compound; fruit < 3 cm . . . . *Descurainia*
  2: Leaf entire to usually shallowly lobed, fruit frequently > 3 cm
    3. Fruit flat, not inflated
      4. Hairs appressed, forked to multibranched; petals usually > 15 mm, blade much wider than claw . . . . *Erysimum*
      4: Hairs appressed and not, simple to multibranched; petals < 15 mm, blade little wider than claw

. . . . *Arabis*
3: Fruit ± flat to ± 4-sided or round in X-section, sometimes inflated
5. Flowers showy; petals 14-29 mm long . . . . *Erysimum*
5: Flowers not showy; petals 4-10 mm long; upper stem leaves clasping . . . . *Arabis glabra*
1: Leaf hairs 0 or simple
6. Calyx ± urn-shaped, nearly closed at full flower, often brightly colored; petals usually strap-shaped . . . . *Streptanthus*
6: Calyx not urn-shaped, fully opening, usually greenish; petals usually clawed or spoon shaped, sometimes linear to obovate
7. Fruit not opening by valves . . . . *Raphanus*
7: Fruit opening by valves
8. Fruit beak prominent, usually nearly as wide as valves
9. Seeds 2 rows per chamber; petals usually prominently dark-veined . . . . *Eruca*
9: Seeds 1 row per chamber; petals not dark-veined
10. Fruit valves with 1 midvein . . . . *Brassica rapa*
10: Fruit valves with 3-7 veins, especially when young
11. Fruit appressed to stem, beak base abruptly swollen . . . . *Hirschfeldia*
11: Fruit not appressed to stem, beak base usually not abruptly swollen . . . . *Sinapis*
8: Fruit beak usually 0 or obscure (except (*Brassica*), narrower than the valves
12. Fruit usually ± flat . . . . *Cardamine*
12: Fruit ± 4-sided to round in X-section
13. Seeds 2 rows per chamber . . . . *Rorippa nasturtium-aquaticum*
13: Seeds 1 row per chamber
14. Cauline leaves clasping stem
15. Fruit beak > 5 mm; seed spheric . . . . *Brassica nigra*
15: Fruit beak 0 or obscure, any remaining style < 3 mm; seed not spheric . . . . *Barbarea*
14: Cauline leaves not clasping stem
16. Fruit ascending to reflexed 1-7 cm; petals 3-6 mm, white, pale yellow, pinkish . . . . *Guillenia lasiophylla*
16: Fruit erect to ascending 0.8-10 cm; petals 1.5-10 mm, pale yellow . . . . *Sisymbrium*

GROUP 2:
1. Leaf usually lobed to compound, sometimes entire
2. Fruit transversely jointed into 2 segments; leaf very fleshy; dunes . . . . *Cakile*
2: Fruit not transversely jointed; leaf not very fleshy; inland . . . . *Rorippa*
1: Leaf entire to shallowly lobed . . . . *Cardaria*

GROUP 3:
1. Fruit 2-2.5 mm wide, hooked-hairy, wing 0 . . . . *Athysanus*
1: Fruit usually > 3 mm wide, hooked hairs 0, wing obvious, often perforated . . . . *Thysanocarpus*

GROUP 4:
1. Inflorescence axillary; fruit of 2 nutletlike halves, wrinkled . . . . *Coronopus*
1: Inflorescence often terminal; fruit not of nutletlike halves, not wrinkled
2. Seed 1 per chamber . . . . *Lepidium*
2: Seeds 2+ per chamber
3. Upper cauline leaves ± clasping stem
4. Fruit obtriangular-obcordate, corners acute; basal leaves usually lobed to deeply dissected *Capsella*
4: Fruit ± obovate to round, corners rounded; basal leaves entire to dentate . . . . *Thlaspi*
3: Upper cauline leaves 0 or few, sessile . . . . *Teesdalia*

GROUP 5:
1. Leaves basal . . . . *Draba*
1: Leaves basal and cauline
2. Flower and fruit > 10 mm wide; fruit stalk above receptacle 7-12 mm . . . . *Lunaria*
2: Flower and fruit < 5 mm wide; fruit stalk not above receptacle . . . . *Lobularia*

Arabis
1. Plants cespitose from a woody branched caudex; stems several . . . . *A. breweri*
1: Plants not cespitose; stems 1-few
2. Sepals yellowish; petals narrow, yellowish-white . . . . *A. glabra*
2: Sepals purplish; petals broadly spatulate, rose purple . . . . *A. blepharophylla*

Arabis blepharophylla Hooker & Arnott. COAST ROCK CRESS. Occasional, coastal rocky places: Bodega Head; Bodega Bay area, *Heller & Brown* (1902); Estero Americano on Marin Co. side ½ mi. from coast, acc. Connors (1990). < 1650 ft. Feb.-Apr.

Arabis breweri S. Watson. BREWER'S ROCK CRESS. Occasional, dry rocky summits and slopes: Mt. Hood, *Heller & Brown* (1902); Lake Sonoma area; Geyser Peak, *Knight & Edwards* (1986); 3100 ft. elev. w. flank Mt. St. Helena, *Grummer* (1979 PUA); 3800 ft. elev. w. flank Mt. St. Helena, *Grummer* (1980 PUA); 1100 ft. elev. Anderson Cr. 1 mi. sw. summit of Cobb Mtn., acc. Clifton (1984); Geysers PG&E facility, *Quibell* (1977 NCC); 3600 ft. elev. Geyser Rock, acc. Clifton (1983). 1000-10500 ft. Mar.-July.

Arabis glabra (Linnaeus) Bentham. TOWER MUSTARD. Occasional, shaded mountains, canyons: Cazadero, *Bacigalupi* (1928); Bodega Bay; Mt. Hood, *Heller & Brown* (1902); between Fort Ross and Mt. St. Helena, *Voznesenskii* (1840-1841 LE); 2 mi. w. of Mark West Springs, *Eastwood & Howell* (1938); Pepperwood; ca 2000 ft. elev. Anderson Trail on Black Mtn. between Geyser Peak and upper end of Pine Flat Road, acc. J. & B. Guggolz (1992); 3300 ft. elev. Squaw Cr., acc. Clifton (1983). < 7900 ft. Mar.-July.

Athysanus pusillus (Hooker) E. Greene. Occasional, grassy areas: Harrison (1992); Howarth Park, Santa Rosa, acc. Harrison (1992); Pepperwood; Saddle Mtn. off Calistoga Road, *Harrison* (1978 SRJC); Lake Sonoma area. < 6000 ft. Mar.-June.

Barbarea

1. Basal leaves simple or with 2-4 small lateral leaflets; petals pale yellow 3-5 mm long; siliques 2-3.5 cm long . . . . *B. orthoceras*
1: Basal leaves with 10-20 lateral leaflets, petals deep yellow 6-8 mm long; siliques 4-8 cm long . . . . *B. verna*

Barbarea orthoceras Ledebour. Common, gullies, creek beds, fields: Willow Creek; Bodega Head; near Graton, *Baker* (1921 NCC); road bank near Sonoma Cr., *Baker* (1930 NCC); 850 ft. elev. Shiloh Ranch in valley of small tributary in Mark West Cr. drainage, 2.3 airmiles ne. of jct. of Shiloh and Faught roads, 0.6 airmiles wsw. of Mark West Springs, acc. Stone (1985); Lake Sonoma area; 3 mi. s. of Cloverdale, *Winters* (1935 NCC); 2700 ft. elev. w. flank Mt. St. Helena, *Grummer* (1979 PUA); The Geysers, *Quibell* (1988 NCC). < 11000 ft. May- Sept. —*B. orthoceras* var. *dolichocarpa* Fernald, currently not recognized.

★ **Barbarea verna** (Miller) Ascherson. EARLY WINTER CRESS, SCURVY GRASS. Uncommon, weed, waste places: Bohemian Hwy., *Best* (1983); laguna e. of Sebastopol, *Howell* (1951). < 1650 ft Apr.-July. Native of Europe.

Brassica

1. Upper stem-leaves not clasping . . . . *B. nigra*
1: Upper stem-leaves with clasping base . . . . *B. rapa*

★ **Brassica nigra** (Linnaeus) Koch. BLACK MUSTARD. Common, grain fields, grassy slopes, waste places: The Sea Ranch; Bodega Head; Todd Road Preserve; Lafferty Ranch on Sonoma Mtn., acc. Arnold et al. (1971); Sugarloaf Ridge; Rincon Valley, *Wells* (1984); Spring Lake; Pepperwood; Lake Sonoma area. < 5000 ft. Apr.-July. Naturalized from Europe.

★ **Brassica rapa** Linnaeus. FIELD MUSTARD. One of the commonest spring wildflowers: Salt Point; Willow Creek; Bodega Head; jct. Davis Lane and E. Railroad Ave. ne. of Penngrove, acc. Knight (1991); n. of Penngrove, *Howell* (1981); Tresch Ranch, Walker Road near Two Rock, acc. J. & B. Guggolz (1992); Bouverie; Rincon Valley, *Wells* (1984); Spring Lake; near Mark West Springs, *Emmel* (1971); Pepperwood; Lake Sonoma area. (A strain of this species with low erucic acid, var. *sylvestris* [Linnaeus] Janchen, developed in Canada is the source of the present, popular, low cholesterol, low polyunsaturated fat spread called *canola*. It is an acronym with CAN designating Canada, O oil and LA low acid). < 5000 ft. Jan.- May. Naturalized from Europe. —*B. campestris* Linnaeus.

★ **Cakile maritima** Scopoli. SEA ROCKET. Locally common, beach sand: The Sea Ranch; Stump Beach at Salt Point; Goat Rock State Beach, *Wells* (1981); Bodega Head; 10 ft. elev. Bay Flat Road near Spud Point Marina at Bodega Harbor, *Clifton* (1984 PUA). < 170 ft. June-Nov. Naturalized from Europe.

★ **Capsella bursa-pastoris** (Linnaeus) Medikus. SHEPHERD'S PURSE. Common, weed: The Sea Ranch; Todd Road Preserve; Fairfield Osborn (S); Tresch Ranch, Walker Road near Two Rock, acc. J. & B. Guggolz (1992); Jess St., Petaluma, *Knight* (1963); w. of Sears Point Raceway in Tolay Cr. drainage, acc. Patterson (1990); Bouverie; Sugarloaf Ridge; Spring Lake; Pepperwood; Lake Sonoma area. < 8000 ft. All months. Naturalized from Europe.

Cardamine

1. Plants not having rhizomal leaves; petals white; siliques erect . . . . *C. oligosperma*
1: Plants with rhizomal leaves; petals pale rose to white, pink; siliques not erect
  2. Rhizomatous leaves usually simple
    3. Siliques 2-4 mm wide; rhizomatous leaves coarsely toothed at apex; petals pink . . . . *C. pachystigma* var. *dissectifolia*
    3: Siliques 1-2 mm wide; rhizomatous leaves with whole margin sinuate; petals pale rose to white
      4. Margin of rhizomatous leaves deeply sinuate; cauline leaves mostly compound . . . . *C. californica* var. *sinuata*
      4: Margin of rhizomatous leaves crisped, undulate; cauline leaves mostly simple *C. c.* var. *cardiophylla*
  2: Rhizomatous leaves typically compound, mostly pinnately 3-7-foliolate . . . . *C. c.* var. *californica*

Cardamine californica (Torrey & A. Gray) E. Greene var. **californica**. MILK MAIDS, TOOTHWORT. Common, shady slopes, banks: The Sea Ranch; Salt Point; Bohemian Hwy. 3.6 mi. nw. of Occidental, *Huntington* (1987 NCC); Willow Creek; among redwoods w. of Sebastopol, *Heller & Brown* (1902); Fairfield Osborn (*); Cooper's Grove, *Baker* (1924 NCC); Sonoma Mtn. Road near Enterprise Road, *Knight* (1964); 2 mi. nw. of Petaluma, *Ott* (1934 NCC); Bouverie; Sugarloaf Ridge; Warfield at jct. of Hwy. 12 and Trinity Road, *Clausen* (1933); near Los Guilicos, *Baker* (1950 NCC); Spring Lake; Green Hill Drive, Santa Rosa, *Stocking* (1964 NCC); near Healdsburg, *R. H. Thompson* (1880), *Abrams* (1918); chaparral area sw. of Bradford Mtn. 1½ mi. w. of Dry Cr., acc. J. & B. Guggolz (1992); Skaggs Springs, *Hawver* (1915); Lake Sonoma area. < 4000 ft. Feb.-May. —*Dentaria c.* Nuttall. *D. integrifolia* Nuttall var. *c.* (Nuttall) Jepson. *Dentaria c. Torrey & A. Gray. D. integrifolia* Nuttall var. *c.* (Torrey & A. Gray) Jepson.

Cardamine californica (Torrey & A. Gray) E. Greene var. **cardiophylla** (E. Greene) Rollins. Uncommon, cliffs, bluff sides: 4199 ft. elev. w. flank Mt. St. Helena, *Grummer* (1980 PUA). < 4300 ft. Mar.-May. —*Dentaria c.* (Torrey & A. Gray) var. *c.* (E. Greene) Detling.

Cardamine californica var. **sinuata** (E. Greene) O. Schulz. Uncommon, brush and forest openings: 3280 ft. elev. Socrates Mine, acc. Clifton (1984); 3200 ft. elev. Caldwell Pines, acc. Clifton (1984). < 1000 ft. Mar.-May. —*Dentaria c.* var. *s.* (E. Greene) Detling.

Cardamine oligosperma Torrey & A. Gray. BITTER-CRESS. Common, weedy: The Sea Ranch, acc. Wells (1981); Bodega Head; Duncans Mills Marsh, *Rubtzoff* (1955, 1957, 1964); Willow Cr. Marsh, *Rubtzoff* (1959); Vacation Beach near Guerneville, *Rubtzoff* (1960); Ross Marsh (near Ross Road sse. of Forestville), *Rubtzoff* (1959); Fairfield Osborn (S); Royal Oaks Mobile Home Park, acc. Knight (1990); Bouverie; Gates Road off Calistoga Road, acc. Wells et al. (1984); Annadel; Spring Lake Village; between Benicia Drive and Bambi Lane, Santa Rosa, *Wells* (1990); near Santa Rosa, *Heller & Brown* (1902); Pepperwood; Lake Sonoma area. < 3500 ft. Mar.-July.

Cardamine pachystigma (S. Watson) Rollins var. **dissectifolia** (Detling) Rollins. DISECTED-LEAF TOOTHWORT. Rare, loose unconsolidated talus, scree, often on serpentine: along the Kingcaid (Burmah) Road in upper Cobb Cr. drainage (milepost C5.3) ca 1.3 mi. w. of the summit of Cobb Mtn., acc. Stone (1983); acc. botanical survey report for the proposed Geothermal Public Power Line (WESCO. 1986), prepared for Sacramento Municipal Utility Dist. et al.; reported by Jokerst in same area. < 3000 ft. Mar.-Apr. —*Dentaria p.* S. Watson var. *dissectifolia* Detling.

Cardaria

1. Fruit widely ovate or widely obovate to ± round, not narrowed at septum . . . . *C. chalepensis*
1: Fruit widely cordate or widely ovate, usually narrowed at septum . . . . *C. draba*

★ **Cardaria chalepensis** (Linnaeus) Handel-Mazzetti. LENS-PODDED HOARY CRESS. Rare, waste places: near Stewarts Point, and e. of Schellville, acc. *Barbe* (1991); Pepperwood. < 5000 ft. Mar.-June. Native of se. Asia. —*C. draba* (Linnaeus) Desvaux var. *c.* (Linnaeus) O. Schulz.

★ **Cardaria draba** (Linnaeus) Desvaux. HEART-PODDED HOARY CRESS. Rare, disturbed areas: now in the county, acc. *Barbe* (1991). < 4000 ft. Mar.-July. Native of Eurasia.

★ **Coronopus didymus** (Linnaeus) Smith. WART CRESS. Common, sidewalk cracks, salt marshes, waste places: Horseshoe Cove, *Knight* (1963); Bodega Harbor area, *Clifton* (1980 PUA); Rincon Valley, *Wells* (1985); Spring Lake; Santa Rosa, *Rattan* (1877); Pepperwood. < 6700 ft. Mar.-July. Naturalized from Eurasia.

★ **Descurainia sophia** (Linnaeus) Webb. TANSY MUSTARD. Uncommon, weed, waste places: Royal Oaks Mobile Home Park, Petaluma, *Knight* (1987 NCC); Spring Lake. < 8400 ft. May-Aug. Naturalized from Europe.

Draba verna Linnaeus. WHITLOW-GRASS. Infrequent, sterile, gravelly soil: 1 mi. n. of Mark West Lodge off Leslie Road, *Van Soelen* (1977 NCC); Pepperwood; Lake Sonoma area. < 6600 ft. Feb.-May.

★ **Eruca vesicaria** (Linnaeus) Cavanilles ssp. **sativa** (Miller) Thellung. GARDEN-ROCKET. Rare, waste places e. of Sebastopol, *Winter* (1935 NCC). < 1300 ft. May-July. Native of Europe. —*E. sativa* Miller.

Erysimum

1. Petals bright orange, occasionally yellow; pods markedly quadrangular . . . . *E. capitatum*
1: Petals pale yellow, yellow to cream; pods not quadrangular
  2. Plants fleshy . . . . *E. menziesii* ssp. *concinnum*
  2: Plants not fleshy . . . . *E. franciscanum*

Erysimum capitatum (Douglas) E. Greene. WESTERN WALLFLOWER. Occasional, stony, dry places: Steep hillside on River Road near Hilton, *Mayer* (1968 NCC); 2 mi. e. of Sonoma, *McNally* (1970 NCC); Lake Sonoma area; 2400 ft. elev. w. flank Mt. St. Helena, *Grummer* (1981 PUA); 3280 ft. elev. Socrates Mine, acc. Clifton (1984). ± 13000 ft. Mar.-July.

Erysimum franciscanum G. Rossbach. SAN FRANCISCO WALLFLOWER. Uncommon, dunes: s. side mouth of Russian River, *Best* (1981); Goat Rock State Park, *Mayer* (1968 NCC), *Best* (1981); Bodega Bay, *Heller & Brown* (1902). < 1700 ft. Mar.-June.

Erysimum menziesii (Hooker) Wettstein ssp. **concinnum** (Eastwood) R. A. Price. Rare, rocky slopes of oceanic headlands: Bodega Point, *Eastwood* (1899, 1900). < 1000 ft. Mar.-May. —*E. c.* Eastwood.

Guillenia lasiophylla (Hooker & Arnott) E. Greene. CALIFORNIA MUSTARD. Locally common, grassland, sometimes recent burns: 8½ mi. n. of Bodega Bay, *Howell* (1930); Bodega Dunes, *Connors* (1992 BML); Sugarloaf Ridge; Mt. Hood, *Baker* (1947 NCC); 2720 ft. elev. w. flank Mt. St. Helena, *Grummer* (1980 PUA); 3080 ft. elev. Socrates Mine, acc. Clifton (1984); 3300 ft. elev. upper Squaw Cr. drainage 1.2 mi. Nw. Of Geyser Rock, acc. Stone (1983). < 8200 ft. Mar.-June. —*Thelypodium lasiophyllum* (Hooker & Arnott) E. Greene. *Caulanthus lasiophyllus* (Hooker & Arnott) Payson var. *inalienus* (B. L. Robinson) Payson.

★ **Hirschfeldia incana** (Linnaeus) Lagrèze-Fossat. SUMMER MUSTARD. Common, weed, waste places: The Sea Ranch, *Wells* (1972); overlook at mouth of Russian River, *Wells* (1985); Goat Rock State Beach, *Wells* (1981); Copeland Cr. on Sonoma State Univ. campus, *Self* (1979 NCC); w. of Sears Point Raceway in Tolay Cr. drainage, acc. Patterson (1990); Magnolia Ave. w. of Petaluma, *Knight* (1963); Sugarloaf Ridge; Spring Lake. < 5300 ft. May-Oct. Naturalized from Europe. —*Brassica geniculata* (Desfontaines) J. Ball.

Lepidium

1. Leaves perfoliate or with clasping bases (at least some of cauline)
  2. Fruits inflated and with obviously winged margins; cauline leaves sagittate at base . . . . . . *L. campestre*
  2: Fruits not greatly inflated, not winged; cauline leaves perfoliate . . . . *L. perfoliatum*
1: Leaves not perfoliate nor with clasping base
  3. Silicles notched at apex but not or slightly winged
    4. Pedicels terete or only slightly flattened
      5. Stems glabrous or nearly so; stem-leaves 1-4 cm broad, entire or at most dentate; stamens 6 . . . . *L. latifolium*
      5: Stems pubescent or stem-leaves < 1 cm broad, or both; stamens mostly 2 or 4 . . . . . *L. strictum*
    4: Pedicels strongly flattened . . . . *L. nitidum*
  3: Silicles plainly winged at apex with 2 prominent divergent lobes or teeth

6. Pedicels slightly flattened but slender; petals 0.5 mm long to lacking; silicles glabrous 2.5-3.5 mm long . . . . *L. oxycarpum*
6: Pedicels very wide and flat; petals 2-4 mm long; silicles glabrous to pubescent 5.5-7 mm long . . *L. latipes*

★ **Lepidium campestre** (Linnaeus) R. Brown. COW-CRESS. Disturbed and waste places: Pepperwood. 800-6500 ft. May-July. Introduced from Europe.

★ **Lepidium latifolium** Linnaeus. SLOUGH MUSTARD. Very common, weed, flood plains, marshes, usually in saline areas: along Petaluma River from turning basin to Marina Ave., Petaluma, *Knight* (1963); under n. end of elevated portion US 101 adjacent to Lakeville St., Petaluma, acc. Knight (1991); Gilardi's Resort on Lakeville Hwy., *Knight* (1986); jct. at Frémont Drive and Broadway near Schellville, *Howell* (1950); River Road 3 mi. w. of Fulton, *Rubtzoff* (1967). < 6000 ft. June-Aug. Native of Eurasia.

Lepidium latipes Hooker. DWARF PEPPERGRASS. Uncommon, alkaline flats, vernal areas: Petaluma, *Eastwood* (1921); near jct. I St. Extension and San Antonio Road, *Follette & Williams* (1984); 5 mi. e. of Mark West Springs, *Eastwood & Howell* (1938). < 2700 ft. Mar.-May.

Lepidium nitidum Nuttall. SHINING PEPPERGRASS. Common, open places: Bodega Point, *Eastwood* (1915); ca 1 mi. nw. of Camp Meeker, acc. J. & B. Guggolz (1991); w. of Santa Rosa in Laguna de Santa Rosa drainage, *Harrison* (1978 NCC); Todd Road Preserve; Fairfield Osborn (H); Tresch Ranch, Walker Road near Two Rock, acc. J. & B. Guggolz (1992); Bouverie; Sugarloaf Ridge; Sonoma Cr. at e. base Mt. Hood, *Heller & Brown* (1902); Spring Lake; Santa Rosa, *Heller* (1902), *Baker* (1928); Pepperwood; Lake Sonoma area. < 5000 ft. Feb.-May.

Lepidium oxycarpum Torrey & A. Gray. Uncommon, saline flats: just e. of Petaluma, *Howell* (1933); 9 mi. sw. of Sonoma on road to San Rafael, *Howell* (1931); 2 mi. n. of Sears Point, *Keck* (1935). < 2100 ft. Mar.-May.

★ **Lepidium perfoliatum** Linnaeus. SHIELD-CRESS. Rare, waste places: Sonoma, *Kuhn* (JEPS). < 5000 ft. Mar.-June. Native of Europe.

★ **Lepidium strictum** (S. Watson) Rattan. Occasional, paths, compacted soil: Bodega Head, *Best* (1987); Watson School on Bodega Hwy. ca 3 mi. s. of Freestone, *Best* (1981); Fairfield Osborn (H); Bouverie; Spring Lake. < 1000 ft. Mar.-May. Naturalized from S. America.

★ **Lobularia maritima** (Linnaeus) Desvaux. SWEET ALYSSUM. Occasional, weed, gardens: Sonoma State Univ. campus, *Caruso* (1970 NCC); Royal Oaks Mobile Home Park, acc. Knight (1990); 2420 ft. elev. Caldwell Ranch, *Clifton* (1984 PUA). < 2500 ft. Much of year. Naturalized from Europe. —*Alyssum maritimum* (Linnaeus) Lamarck.

★ **Lunaria annua** Linnaeus. MOONWORT, HONESTY. Occasional, garden escape < 700 ft. Late spring-summer. Native of Europe.

Raphanus

1. Siliques not longitudinally grooved, 2-3 seeded; flowers mostly purple or white . . . . *R. sativus*
1: Siliques longitudinally grooved, 4-6-seeded; flowers yellow . . . . *R. raphanistrum*

★ **Raphanus raphanistrum** Linnaeus. JOINTED CHARLOCK. Infrequent, poorly collected, weed, waste places: Canfield Road, Stony Point area, acc. Knight (1993). < 2600 ft. Apr.-June. Naturalized from Europe.

★ **Raphanus sativus** Linnaeus. WILD RADISH. Very common, weed, waste places: The Sea Ranch, *J. H. Thomas* (1976); ca ½ mi. n. of Stewarts Point, *Wells* (1985); Fort Ross-Russian River- Bodega areas, *Voznesenskii* (1840-1841 LE); Willow Creek; Bodega Head; Sebastopol, *Dows* (1907); Todd Road Preserve; Tresch Ranch, Walker Road near Two Rock, acc. J. & B. Guggolz (1992); between San Antonio Cr. and Petaluma along Hwy. 101, *Howell* (1981); w. of Sears Point Raceway in Tolay Cr. drainage, acc. Patterson (1990); Bouverie; Spring Lake; Pepperwood; Lake Sonoma area. < 3300 ft. Feb.-July. Naturalized from Europe.

Rorippa

1. Plants perennial; petals white; cauline leaves 1-pinnate; plant in aquatic to very wet places; rooting at the nodes or not . . . . *R. nasturtium-aquaticum*
1: Plants annual or biennial; petals yellow; cauline leaves simple, entire to deeply pinnately lobed, rarely 1-pinnate; plant often in wet but usually not aquatic places; rarely, if ever, rooting at nodes
    2. Fruit ± as long as wide 1-2.5 mm . . . . *R. sphaerocarpa*

2: Fruit usually > 2 × longer than wide 1.5-15 mm
3. Fruit oblong to ± round, often ± narrowed near middle, usually > 1.5 mm wide . . . . *R. palustris* var. *occidentalis*
3: Fruit linear to oblong, not narrowed near middle, < 1.5 mm wide . . . . *R. curvisiliqua*

Rorippa curvisiliqua (Hooker) Britton. YELLOW CRESS. Common, vernal areas, moist ravine: Austin Cr. marsh, *Rubtzoff* (1954); Bodega Head; 1 mi. se. of Bloomfield, *Rubtzoff* (1962); upper Stemple Cr., *Rubtzoff* (1961); Vacation Beach near Guerneville, *Rubtzoff* (1953); Forest Hills near Hacienda, *Rubtzoff* (1968); Atascadero Cr., *Rubtzoff* (1958); Cunningham Marsh, *Howell* (1947); Laguna de Santa Rosa near Occidental Road, *Rubtzoff* (1958); Todd Road Preserve; Fairfield Osborn (S); near Agua Caliente in Sonoma Valley, *Rubtzoff* (1973); Rincon Valley, *Wells* (1984); Santa Rosa Cr. e. of Santa Rosa, *Heller* (1902); Spring Lake; Pepperwood; 2 mi. nw. of Windsor on Old Redwood Hwy., *de Nevers* (1981). < 3000 ft. Apr.-Sept. —*Radicula curvisiliqua* (Hooker) E. Greene.

★ **Rorippa nasturtium-aquaticum** (Linnaeus) Hayek. WATER CRESS. Common, wet banks, stream sides, quiet water: The Sea Ranch; Salt Point (near ranger's residence); Horseshoe Cove, *Knight* (1963); 1 mi. n. of jct. of Myers Grade and Hwy. 1, *Huntington* (1989 NCC); Willow Creek; ca 1 mi. nw. of Camp Meeker, acc. J. & B. Guggolz (1991); Fairfield Osborn (S); Copeland Cr. tributaries at Fairfield Osborn, acc. Quibell (1991); 1½ mi. from top of Sonoma Mtn. Road, *Lowry* (1983 SFSU); Lafferty Ranch on Sonoma Mtn., acc. Arnold et al. (1971); Tresch Ranch, Walker Road near Two Rock, acc. J. & B. Guggolz (1992); Sugarloaf Ridge; along Santa Rosa Cr. in e. Santa Rosa, acc. Wagnon (1989); Pepperwood; Lake Sonoma area; 2190 ft. elev. Squaw Cr., acc. Clifton (1984). < 9000 ft. Mar.-Nov. Naturalized from Europe. —*Nasturtium officinale* R. Brown.

Rorippa palustris (Linnaeus) Besser var. **occidentalis** (S. Watson) Rollins. Occasional, wet places: Parktrail Drive off Summerfield Road, Santa Rosa, before development, *Harrison* (1980 UC); geysers Unit 13, *Quibell* (1981 NCC). < 6600 ft. Apr.-July. —*R. pacifica* Howell. *R. islandica* (Oeder) Borbás var. *occidentalis* (S. Watson) Butters & Abbe.

Rorippa sphaerocarpa (A. Gray) Britton. Uncommon, damp areas: Sonoma Cr. bottom near Agua Caliente Road, *Rubtzoff* (1973). < 8500 ft. June-Sept. —*R. obtusa* (Nuttall) Britton. *R. truncata* (Jepson) Stuckey.

★ **Sinapis arvensis** Linnaeus. CHARLOCK. Common, weed, roadsides, waste ground: Todd Road Preserve; Copeland Cr. on Sonoma State Univ. campus, *Self* (1979 NCC); E. Cotati Ave., *Howell* (1982); n. of Penngrove, *Howell* (1981); Madison St. near Edith St., Petaluma, *Koch* (1939 NCC); Baker property in Adobe Can., *Baker* (1934 NCC). < 1700 ft. Mar.-Oct. Naturalized from Eurasia. —*Brassica kaber* (de Candolle) Wheeler var. *k.*, *B. k.* var. *pinnatifida* (Stokes) Wheeler.

★ **Sisymbrium officinale** Linnaeus. HEDGE MUSTARD. Common, weed, waste places: Bodega Head; Pitkin Marsh, *Rubtzoff* (1951); w. of Santa Rosa in Laguna de Santa Rosa drainage, *Harrison* (1978 NCC); Fairfield Osborn (H); Tresch Ranch, Walker Road near Two Rock, acc. J. & B. Guggolz (1992); w. of Sears Point Raceway in Tolay Cr. drainage, acc. Patterson (1990); Bouverie; Gates Road off Calistoga Road, acc. Wells et al. (1984); Spring Lake; Pepperwood; between Jimtown and Mercuryville on Geysers Road, *Utech & Knight* (1986 CM). < 7300 ft. Apr.-July. Naturalized from Europe.

Streptanthus (Key ± acc. Buck, Taylor and Kruckeberg, *Hickman* [1993]).
1. Plant with filaments free to base; anthers of longest (upper) stamen pair > ½ greater than others
2. Plant a subshrub; sepals 8-10 mm . . . . *S. tortuosus* var. *suffrutescens*
2: Plant a biennial (annual to weak perennial); woody base 0 or weak; sepals < 8 mm . . *S. t.* var. *tortuosus*
1: Plant with longest (upper) filament pair ± fused or closely adherent, their anthers < < others
3. Plants biennial (non-flower rosettes present within flower population); cauline leaves usually gradually reduced upwards; calyx ± biradial; herbage glabrous
4. Plant usually 5-12 dm, branched only above; petals whitish or brownish, lower not or faintly darker than upper, lacking purple except occasionally on veins of upper . . . . *S. morrisonii* ssp. *m.*
4: Plant usually 2.5-5 dm, usually branched throughout; lower petals usually ± purple, upper whitish or with less purple
5. Leaves purple-mottled above, upper cauline leaves 2-4 × longer than wide
6. Sepals yellowish, nearly glabrous . . . . *S. m.* ssp. *elatus*
6: Sepals purple, densely hairy . . . . *S. m.* ssp. *hirtiflorus*

5: Leaves unmottled above, upper cauline leaves 1-2 × longer than wide; sepals yellowish, nearly glabrous
7. Sepals glabrous, rose-purple; upper fused filaments orange . . . . *S. brachiatus* ssp. *b.*
7: Sepals usually hairy, yellow or dark purple; upper fused filaments yellow *S. b.* ssp. *hoffmanii*
3: Plants annual; cauline leaves usually abruptly reduced upward (or calyx bilateral and herbage usually ± bristly below)
8. Calyx ± biradial; lower petals usually purple, upper lighter; herbage glabrous; fruit ± narrowed between seeds
9. Middle cauline leaves ± linear, barely or not clasping; basal leaves ± not rosetted; stem usually > 1.5 dm . . . . *S. barbiger*
9: Middle cauline leaves usually narrowly lanceolate or wider, deeply clasping; basal leaves usually ± rosetted
10. Middle cauline leaves ± narrowly lanceolate, basal leaves often coarsely and sharply dentate . . . . *S. batrachopus*
10: Middle cauline leaves widely lanceolate or wider; basal leaves usually not coarsely and sharply dentate, longest leaves usually > 3 cm . . . . *S. breweri*
8: Calyx ± bilateral; all 4 petals uniformly colored or upper darker; herbage usually ± bristly below; fruit not narrowed between seeds
11. Inflorescence ± 1-sided
12. Sepals whitish or greenish-yellow to reddish-purple; petals whitish to purplish . . . . *S. glandulosus* ssp. *secundus* var. *s.*
12: Sepals and petals otherwise
13. Sepals rose; petals ± purplish . . . . *S. g.* ssp. *s.* var. *hoffmanii*
13: Sepals white to greenish-yellow without purple; petals whitish . . . . *S. g.* ssp. *s.* var. *sonomensis*
11: Inflorescence not 1-sided; sepals usually lilac-lavender to dark purple, or greenish-white . . . . *S. g.* ssp. *g.* var. *g.*

Streptanthus barbiger E. Greene. Locally common, ridges of serpentine: 1400 ft. elev. The Cedars at Big Austin Cr., *Raiche* (1984); Harrison Grade Road, *Baker* (1947 NCC); Lake Sonoma area; e. side of Old Stewarts Point-Skaggs Springs Road (near Skaggs Springs), *Guggolz* (1977 NCC); 3300 ft. elev. upper Cobb Cr. drainage sw. of Geyser Rock, acc. Stone with Clifton (1983); 3150 ft. elev. upper Squaw Cr. drainage 0.8 airmiles nw. of Geyser Rock, *Stone* (1983 JEPS); 650 ft. elev. Hot Springs Road w. of Cloverdale, *Raiche* (1984). Acc. Hickman (1993), intergrades with *S. batrachopus*. 650-3500 ft. May-July.

Streptanthus brachiatus F. W. Hoffman ssp. **brachiatus**. SOCRATES MINE JEWELFLOWER. Locally common, roadcuts, tailings, cut slopes, open chaparral, serpentine: 1½ mi. se. of jct. Rattlesnake Fire Trail and Healdsburg-Geysers Road on PG&E Transmission Line Road, acc. Roland (1979); wnw. of Harness Camp < 1 mi. n. of n. branch of Little Sulphur Cr., acc. Roland (1979); 3000 ft. elev. Sonoma-Lake county line near Contact Mine e. of Pine Flat, *Kruckeberg & Hoffman* (type 1949 UC); near jct. of Socrates Mine and Pine Flat roads, *H. Sharsmith* (1952 NCC); a transmission line tower ca ½ mi. n. of Socrates Mine, *Crane* (1977 UCR); open chaparral ca ¼ mi. e. of transmission tower, *Crane* (1977 UCR); main barrens near Socrates Mine which are mostly disturbed by mining, *Dolan & LaPré* (1987 UCR); small natural barren surface to n. of Socrates Mine which has been cut by a ridgeline access road, *Dolan & LaPré* (1987 UCR). Also Dolan & LaPré (1987) report a historic locality on a ridge of open chaparral on serpentine just w. of Mercuryville. These locations are all in or near the geothermal area. 1800-3000 ft. May-July.

Streptanthus brachiatus ssp. **hoffmanii** R. Dolan & L. LaPré. FREED'S JEWELFLOWER. Rare: endemic to Sulphur Cr. drainage on Lake-Sonoma county line and se. to Mt. St. Helena (1985), *Dolan & LaPré* (1987). < 4000 ft. May-July.

Streptanthus breweri A. Gray. Uncommon, openings in brush: 3060 ft. elev. Squaw Cr., acc. Clifton (1983). 800-5500 ft. May-July.

Streptanthus glandulosus Hooker ssp. **glandulosus** var. **glandulosus**. JEWELFLOWER. Occasional, rocky ridges, dry frequently loose disturbed soil: Red Slide n. of Cazadero (purple flowers), *Dolan & LaPré* (1986 UCR); Dorr Cabin in Austin Cr. area (yellow flowers), *Dolan & LaPré* (1986 UCR); Santa Rosa Can. to n. of Mt. Hood, *Howell* (1931); Mt. Hood, *Bioletti* (1889 UC); 9½ mi. from St. Helena on Spring Mtn. Road, *Howell* (1931); Lake Sonoma area (3 locations by Guggolz); Socrates Mine at the end of Pine Flat Road, *Dolan & LaPré* (1986 UCR); ca 2000 ft. elev. Anderson Trail on Black Mtn. between Geyser Peak and upper end of Pine Flat Road, acc. J. & B. Guggolz (1992); 3620 ft. elev. Anderson Cr. 1 mi. sw. of summit of Cobb Mtn., acc. Clifton (1984); 3600 ft. elev. Geyser Rock, acc. Clifton (1983);

3300 ft. elev. Squaw Cr., acc. Clifton (1983). 50-4500 ft. Apr.-May. —*S. bioletti* E. Greene, type locality: Mt. Hood, *Bioletti* (1893 UC).

Streptanthus glandulosus ssp. **secundus** (E. Greene) Kruckeberg var. **hoffmanii** Kruckeberg. Uncommon, rocky places: edge of redwood grove at Russian Gulch 8 mi. s. of Fort Ross, *Constance* (type 1938 UC); Russian Gulch ± e. of 1st bend in road after crossing stream n. of Jenner, *Hoffman* (1948 UC); Red Slide at headwaters of E. Austin Cr., *Hoffman* (1948 UC); 1400 ft. elev. The Cedars at Big Austin Cr., acc. Raiche (1984). <1500 ft. Mar.-May.

Streptanthus glandulosus ssp. **secundus** var. **secundus**. Locally common, rocky places: Cazadero, *Graff* (1925); Stoetz Lane ne. of Occidental, *Utech & Knight* (1986 CM); 4 mi. from Petrified Forest on road to Knights Valley, *Howell* (1929); near Petrified Forest, *Howell* (1937); Fitch Mtn., *Rubtzoff* (1964); 3100 ft. elev. w. flank Mt. St. Helena, *Grummer* (1979 PUA). < 4500 ft. Mar.-May.

Streptanthus glandulosus ssp. **secundus** var. **sonomensis** Kruckeberg. Uncommon, usually on serpentine: Bohemian Hwy. 3.6 mi. nw. of Occidental, *Huntington* (1986 NCC); Great Eastern Quicksilver Mine near Guerneville, *Hoffman* (type 1948 UC); Lake Sonoma area 6 collections from different places; Kelly Road 0.2 mi. w. of jct. with Rockpile Road, *J. & B. Guggolz* (1986 NCC); 1280 ft. elev. Ida Clayton Road, *Clifton* (1974 PUA). < 1500 ft. Mar-May. Perhaps indistinct from var. *secundus*.

Streptanthus morrisonii F. W. Hoffman ssp. **elatus** F. W. Hoffman. THREE PEAKS JEWELFLOWER. Rare, serpentine barren: 1 specimen on s.-facing slope, 1 specimen on n.-facing slope, Sonoma-Lake co. line w. of Middletown off Western Mine Road, *Raiche* (1984). < 300 ft. Spring-early summer.

Streptanthus morrisonii ssp. **hirtiflorus** F. W. Hoffman. DORR'S CABIN JEWELFLOWER. Uncommon, confined to two small ridges of barrens of serpentine located in the center of a vast outcrop of greywacke: headwaters of E. Austin Cr. area above Dorr Cabin, *Hoffman* (type 1948 UC), *Dolan & LaPré* (1986 UCR); 1800 ft. elev. King Ridge, The Cedars, acc. Raiche (1984); Big Austin Cr., The Cedars, acc. Raiche (1984). <2700 ft. Spring-early summer.

Streptanthus morrisonii ssp. **morrisonii**. MORRISON'S JEWELFLOWER. Uncommon, among large rocks, boulders and talus material of serpentine: The Cedars, Hoffman acc. *Kruckeberg* (1984); headwaters of Big Austin Cr. at Layton Chromite Mine, *Hoffman* (type 1946 UC); Red Slide se. of The Cedars, *Dolan & LaPré* (1986 UCR). 400-3300 ft. Spring-early summer.

Streptanthus tortuosus Kellogg var. **suffrutescens** (E. Greene) Jepson. Rare, dry rocky areas: Mt. Hood, *Bioletti* (type 1893 UC), *Baker* (1948 NCC); Caldwell Pines, acc. Stone (1989). 0-3600 ft. Apr-May.

Streptanthus tortuosus var. **tortuosus**. MOUNTAIN JEWELFLOWER. Uncommon, ridge tops: 4298 ft. elev. near Mt. St. Helena lookout, *Grummer* (1978 PUA); 3000 ft. elev. upper Bear Cr. drainage 0.4 air-miles sse. of co. line at Caldwell Pines, acc. Stone (1984). 1000-7000 ft. May-Aug.

★ **Teesdalia coronopifolia** (Bergeret) Thellung. Rare, open areas: Grange Road just n. of Quenza/Perraco roads, *Quibell* (1975 UC); Annadel, *Best* (1990); vacant lot at sw. part of jct. of Fulton and Piner roads in Santa Rosa, *Quibell* (1977 NCC); s. side of Alton Lane ca ½ mi. w. of Piner Road, *Quibell* (1990 NCC); Alton Lane near Fulton, *Best* (1990). < 300 ft. Native of s. Europe. Possible first record of the species in N. America.

★ **Thlaspi arvense** Linnaeus. PENNY-CRESS, FAN-WEED. Rare: Royal Oaks Mobile Home Park, Petaluma, *Knight* (1987 NCC); Rincon Valley, *Wells* (1985). < 1700 ft. May-Aug. Native of Europe.

Thysanocarpus

1\. Fruiting pedicels straight or recurved at tip only; silicles 8-10 mm wide, the wing with narrow almost filiform rays . . . . *T. radians*

1: Fruiting pedicels ± recurved their whole length; silicles 3-7 mm wide, the wing with broad or no rays

  2\. Silicles 3-4 mm wide, the wing usually not perforate . . . . *T. curvipes* var. *c.*

  2: Silicles 5-6 mm wide, the wing often perforate . . . . *T. c.* var. *elegans*

Thysanocarpus curvipes Hooker var. **curvipes**. LACEPOD, FRINGEPOD. Occasional, grassland: Fairfield Osborn (S); Tresch Ranch, Walker Road near Two Rock, acc. J. & B. Guggolz (1992); 2 mi. from Sonoma between Sonoma and Boyes Hot Springs, *Howell* (1929); Sonoma Valley Reg. Park e. of Glen Ellen on s. side of paved trail midway between e. and w. boundaries, *Quibell* (1992 NCC); Sugarloaf Ridge; Los Alamos Can., *Baker* (1933 NCC), *Koch* (1938 NCC); Spring Lake; Pepperwood, *Breedlove & Almeda* (1983); Lake Sonoma area; Geyser Peak, *Knight & Edwards* (1986); 2400 ft. elev. Little

Geysers, *Malloch* (1977 PG&E); 2040 ft. elev. Squaw Cr., acc. Clifton (1984). < 6000 ft. Mar.-May. —*T. pulchellus* Fischer & C. Meyer, type locality: middle California. *T. curvipes* var. *involutus* E. Greene, type locality: Sonoma Co. (probably in vicinity of Mt. Hood), *Bioletti* (1893 UC).

Thysanocarpus curvipes var. **elegans** (Fischer & C. Meyer) Robinson. Uncommon, grassland: Lake Sonoma area; Bouverie; 797 ft. elev. Alexander Valley, *Bennett* (1971 PUA). Sometimes considered indistinct from species. < 1000 ft. Mar.-May.

Thysanocarpus radians Bentham. SPOKEPOD. Occasional, grassy or brushy slopes: w. of Santa Rosa in Laguna de Santa Rosa drainage, *Harrison* (1978 NCC); near Boyes Hot Springs 2 mi. n. of Sonoma, *Howell* (1931); summit of Bennett Valley Road from Glen Ellen to Santa Rosa, *Howell* (1934); Kenwood, *Baker* (1927 NCC); cemetery near Windsor, *Koch* (1938 NCC); < 4000 ft. Mar.-May.

BUDDLEJACEAE. BUDDLEJA FAMILY

★ **Buddleja davidii** Franchet. BUTTERFLY BUSH, SUMMER LILAC. Uncommon, probably escaped from a garden: Hwy. 116 along Russian River ½ mi. e. of Monte Rio, *Best* (1984). < 700 ft. Native of China.

CALLITRICHACEAE. WATER-STARWORT FAMILY

Callitriche

1. Plants with ♀ flowers on distinct pedicel >> fruit; leaves linear to linear-spatulate 5-12 mm joined at base by tiny winged ridges; fruits ca 1 mm, wing-margined . . . . *C. marginata*
1: Plants with ♀ flowers sessile or with pedicel scarcely ¼ length of fruit; fruit winged only near the top if at all
  2. Leaves all linear, 1-nerved, 5-20 mm, bright lustrous green, not wing-joined at base; floral bracts lacking . . . . *C. hermaphroditica*
  2: Leaves varied, upper ones often broadened and several-nerved and joined at base by winglike ridge; floral bracts usually present
    3. Carpel faces marked with tiny pits in rather regular longitudinal lines; fruit usually ±wing-margined at top, ca 1 mm, 1/5-1/3 longer than broad; submerged leaves linear 5-20 mm, retuse, vein thickened but not protruding at tip . . . . *C. verna*
    3: Carpel faces marked with pits irregularly patterned, not in rows; fruit scarcely at all winged, usually ca as broad as long
      4. Fruit usually widest above middle and ± obovate in outline; submerged leaves bidentate but with midvein barely perceptibly thickened at tip; emergent leaves often > 5 mm broad; stems mostly 5-40 cm . . . . *C. heterophylla* var. *bolanderi*
      4: Fruit usually subrotund or oblong in outline; submerged leaves with midvein much-thickened and slightly protruding at tip; leaves all < 5 mm broad; stems scarcely 5 cm, usually submerged, when emergent leaves not strongly dimorphic . . . . *C. trochlearis*

Callitriche hermaphroditica Linnaeus. NORTHERN WATER-STARWORT. Occasional, submerged, still water, streams, lakes: Pitkin Marsh, *Rubtzoff* (1952 NCC); close-by Aurora School near modern Two Rock Coast Guard Training Center, (barely in Marin County), *Howell* (1946). < ± 6600 ft. Apr.- Aug. —*C. autumnalis* Linnaeus var. *a*. (A plant growing w. of Santa Rosa has been named *C. autumnalis* Linnaeus var. *bicarpellaris* Fenley, [type JEPS], leaves sometimes 3-cleft; bracts conspicuous, obovate, orbicular or subquadrate; nutlets reniform, broadly winged: Jepson, *Fl. of CA, 2:436*, (1936).

Callitriche heterophylla Pursh var. **bolanderi** (Hegelmaier) Fassett. BOLANDER'S WATER-STARWORT. Occasional, wet ground, quiet water: 2 mi. e. of Stewarts Point, *Nobs & Smith* (1949); Coleman Marsh, *Rubtzoff* (1964); upper Stemple Cr. Marsh, *Rubtzoff* (1961); Bodega Head; Guerneville Marsh, *Rubtzoff* (1955); Atascadero Cr. Marsh, *Rubtzoff* (1958); Pitkin Marsh, *Rubtzoff* (1951); Cunningham Marsh, *Howell* (1947); Hamilton Ranch near Old Adobe e. of Petaluma, *Koch* (1939); Kenwood, *Bioletti* (1893 UC); Bennett Mtn. Lake, *Rubtzoff* (1955). > ± 6600 ft. Apr.-Aug. —*C. heterophylla* ssp. *bolanderi* (Hegelmaier) Calder & Taylor.

Callitriche marginata Torrey. CALIFORNIA WATER-STARWORT. Occasional, drying mud, vernal areas: Duncans Mills Marsh, *Rubtzoff* (1957); 2 sites w. of Santa Rosa in Laguna de Santa Rosa drainage, *Harrison* (1978 NCC); Todd Road Preserve; 1600 ft. elev. Lichau Road ne. of Penngrove, *Rubtzoff* (1955); Schellville, acc. *Jepson's Fl. of CA, Vol. II,* (1936); w. outskirts of Sonoma, *Baker* (1952 NCC); Santa

Rosa,*Eastwood* (1921); muddy places along Mark West Cr., *Bigelow* (type 1854 U). < ± 5000 ft. Mar.-May.

Callitriche trochlearis Fassett. Fairly common, mud, shallow water: Cazadero, *Davy* (1896 UC); Bodega Bay Marsh, *Rubtzoff* (1962); Freestone Marsh, *Rubtzoff* (1962); upper Stemple Cr. Marsh, *Peñalosa* (1961); Freestone Marsh, *Rubtzoff* (1962); Pitkin Marsh, *Rubtzoff* (1952); Laguna de Santa Rosa n. of Sebastopol near Occidental Road, *Rubtzoff* (1958); Santa Rosa, *Eastwood* (1921); 115 ft. elev. Sonoma Valley near Eldridge, *Jepson* (1914 JEPS); Pepperwood; US 101 just s. of Asti, *Raven* (1963). < ± 1700 ft. Mar.-June.

Callitriche verna Linnaeus. VERNAL WATER-STARWORT. Occasional, mud, shallow water: Willow Cr. Marsh, *Rubtzoff* (1959); Occidental Marsh, *Rubtzoff* (1962); Freestone Marsh, *Rubtzoff* (1962); Guerneville Marsh, *Rubtzoff* (1955); Guerneville, *Rattan* (1877); Laguna de Santa Rosa e. of Sebastopol, *Rubtzoff* (1955); Button Ranch on Walker Road (U. C. property) between Petaluma-Valley Ford Road and Pepper Road, acc. Connors (1992); Fairfield Osborn; 1 mi. from Old Adobe e. of Petaluma, *Koch* (1939 NCC); s. of Kenwood opposite railroad station, *Baker* (1935 NCC); near Kenwood, *Howell* (1935); Spring Lake; Dutcher Cr. Road, w. of Geyserville, *J. & B. Guggolz*. < 13000 ft. May-Aug. —*C. stenocarpa* Hegelmaier.

CALYCANTHACEAE. CALYCANTHUS FAMILY

Calycanthus occidentalis Hooker & Arnott. SPICEBUSH, SWEET-SHRUB. Common, moist places: The Sea Ranch, s. end of Timber Ridge Road, acc. Wells (1983); Annapolis, *Sutliffe* (1920); Duncans Mills, *M. Jones* (1882); Bohemian Hwy. 3 mi. n. of Occidental, *Best* (1983); Guerneville, *Rattan* (1877); Russian River at Forest Hills, *Rubtzoff* (1980, 1987); between Vacation Beach and e. Guernewood Park, *Rubtzoff* (1959); Fife Cr. Marsh at Guerneville, *Rubtzoff* (1960); Green Valley Road 0.8 mi. e. of Harrison Grade Road jct., *Ferris & Lorraine* (1961); Bouverie; Sugarloaf Ridge; Gates Road off Calistoga Road, acc. Wells et al. (1984); Spring Lake; along Santa Rosa Cr. in e. Santa Rosa, acc. Wagnon (1989); n. and ne. slopes of Fitch Mtn., *Rubtzoff* (1952); head of Dorman Cr., tributary of Dry Cr., *Mason* (1934); Skaggs Springs, *Parsons* (1915); Lake Sonoma area; ca 2000 ft. elev. Anderson Trail on Black Mtn. between Geyser Peak and upper end of Pine Flat Road, acc. J. & B. Guggolz (1992); 1500 ft. elev. w. flank Mt. St. Helena, acc. Grummer & Hemphill (1983); 2680 ft. elev. Pine Flat Road, acc. Clifton (1984); The Geysers, *McMurphy* (1901). Usually < 5000 ft. Apr.-Aug.

CAMPANULACEAE. BELLFLOWER FAMILY
[including LOBELIACEAE]

1. Corolla regular; anthers and filaments distinct
  2. Capsule dehiscent on the side; leaves broad to roundish
    3. Capsule opening by lateral openings formed by uplifting of small lids
      4. At least the terminal flowers pedicelled; all flowers normally opening and expanding; ovary short and broad;annual or perennial . . . . . *Campanula*
      4: At least the terminal and upper flowers sessile; lower flowers mostly cleistogamous with vestigial corollas, upper expanded; ovary ± elongate; annuals . . . . . *Triodanis*
    3: Capsule opening by irregular fissures . . . . . *Heterocodon*
  2: Capsule dehiscent at the apex; leaves narrow, oblong . . . . . *Githopsis*
1: Corolla irregular
  5. Anthers all alike, distinct; flowers minute, inconspicuous . . . . . *Nemacladus*
  5: Anthers united into a tube, 2 shorter than the others; flowers mostly showier
    6. Flowers sessile in the axils of foliaceous bracts, the ovary linear and simulating a pedicel; capsule elastically dehiscent by long slits on the sides . . . . . *Downingia*
    6: Flowers pedicelled; capsule round to fusiform, indehiscent or opening by valves
      7. Corolla wanting or not > 4 mm long, cleft nearly to the base on the upper side; fruit with 1 cavity . . . . . *Legenere*
      7: Corolla 6-40 mm long; capsule bilocular . . . . . *Lobelia*

Campanula

1. Plants perennial
   2. Style longer than and well-exerted from corolla; leaves serrate; corolla bright blue, the lobes 2-3 × as long as tube . . . . *C. prenanthoides*
   2: Style not longer than corolla; leaves crenulate; corolla pale blue, campanulate, lobes ca = tube . . . . *C. californica*

1: Plants annual
   3. Leaves 2-3 × as long as broad, oblong-ovate . . . . *C. angustiflora*
   3: Leaves > 3 × as long as broad, never oblong-ovate . . . . *C. griffinii*

Campanula angustiflora Eastwood. Occasional, burns, chaparral, sometimes on serpentine, disturbed places: Camp Meeker, *Harrison* (1977); ca 1 mi. nw. of Camp Meeker, acc. J. & B. Guggolz (1991); near foot of Trinity Road, *Baker* (1938 NCC); w. slope of Mt. Hood, *Baker* (1947 NCC); summit of St. Helena Road, *Morin* (1978). 100-2100 ft. May-June.

Campanula californica (Kellogg) A. A. Heller. SWAMP HAREBELL. Occasional, freshwater swamp, usually coastal: The Sea Ranch; Salt Point; Duncans Mills Marsh, *Rubtzoff* (1967); Pitkin Marsh, *Rubtzoff* (1952); near e. end of Templeton Road, Pitkin Marsh, *Bacigalupi* (1967 JEPS); Perry Marsh, *Howell* (1945). < 500 ft. June-Sept.

Campanula griffinii N. Morin. Uncommon, wooded areas, serpentine: 2¼ mi. se. of Monte Rio (probably Bohemian Hwy.), *Hoover* (1941). < 2100 ft. —*C. angustiflora* Eastwood var. *exilis* J. T. Howell.

Campanula prenanthoides Durand. CALIFORNIA HAREBELL. Common, dryish wooded areas, poorly collected: Camp Meeker, *Harrison* (1977 SRJC); Pepperwood; Lake Sonoma area. < 6600 ft. June-Sept.

Downingia

1. Corolla 2-4 mm long, scarcely or not exceeding the sepals; lower lip of corolla ascending . . . . . *D. pusilla*

1: Corolla definitely > sepals, its lower lip forming a sharp angle with the corolla-tube
   2. Dorsal sinus of corolla (between the lobes of upper lip) usually produced backward and outward into short hornlike process . . . . *D. ornatissima*
   2: Dorsal sinus of corolla without projecting horn or fold
      3. Ovary and capsule with a longitudinal septum to which the ovules are attached
         4. Lips of the corolla with 1 or more purple spots
            5. Anthers 2.5-3.5 mm exerted from the corolla tube . . . . *D. pulchella*
            5: Anthers < 2.5 mm included in the corolla tube . . . . *D. concolor*
         4. Lower lip of corolla without any purple areas . . . . *D. cuspidata*
      3: Ovary and capsule lacking a central septum, the ovules borne in 2 lines on the ovary wall . . . *D. yina*

Downingia concolor E. Greene. Very common, ditches, vernal pools: lagoon e. of Sebastopol, *Howell & Stacey* (1937); Wilfred Station n. of Cotati, acc. Jepson (1921); nw. corner Snyder Lane and Southwest Blvd. in Rohnert Park, *Caruso* (1979 NCC); vernal pool in Rohnert Park, *Quibell* (1978 NCC); in ditch along N. McDowell Blvd. near railroad crossing in Petaluma, acc. Knight (1987); near El Verano, *Cantelow* (1936); Arnold Drive ½ mi. from jct. of Hwy. 116, *Best* (1983); Sonoma, *Bioletti* (1891), *Rose* (1940); sw. of Sonoma on Schellville Road, *Cantelow* (1938); 6½ mi. n. of Sonoma on Hwy. 12, *Crampton* (1966); Spring Lake; Southwest Community Park near Hearn Ave., Santa Rosa, *Alimonti* (1988 NCC); Santa Rosa, *Heller & Brown* (1902), *Eastwood* (1921); Piner Road w. of Santa Rosa, *J. & B. Guggolz* (1990 JEPS); Pepperwood; near Windsor, *Carruth* (1901); 2 mi. nw. of Windsor by Old Redwood Hwy., *de Nevers* (1981); 2 mi. n. of Windsor, *Eastwood & Howell* (1936); 3 mi. s. of Healdsburg, *Beane & James* (1949). < 4700 ft. Apr-July. —*Bolelia concolor* E. Greene.

Downingia cuspidata Jepson. Occasional, vernal situations, clay soil: near Guerneville Hwy. in Laguna de Santa Rosa drainage, *Harrison* (1978 NCC); 4.7 mi. s. of El Verano on Arnold Drive, *Raven* (1963); False Lake Meadow in Annadel, *Baldwin* (1980 SRJC); vernal area s. of Shiloh Road e. of Hwy. 101, acc. J. & B. Guggolz (1988, population since extirpated). —*Bolelia c.* E. Greene, type locality: Los Guilicos Valley, *Bioletti* (1893 UC). < 1850 ft. Mar.-June.

Downingia ornatissima E. Greene. Uncommon, heavy clay mud: Pepperwood. < ± 700 ft. Apr.-May. —*Bolelia o.* E. Greene.

Downingia pulchella (Lindley) Torrey. Uncommon, muddy places, vernal pools, ditches: 2 mi se. of the jct. of Hwy 116 and Hwy 112 e. of Arnold Drive near Sonoma Creek, *Connors* (1996 BML) < 900 ft. May-June.

Downingia pusilla (G. Don) Torrey. Locally common, ditches, vernal areas: 4.4 mi. n. of Sears Point, *Weiler* (1960); near Sonoma Valley Reg. Park, *Best* (1990), *Quibell & Moran* (1982 NCC); summit of Bennett Valley Road from Glen Ellen to Santa Rosa, *Howell* 1934); Piner Elementary School, *Brunn* (1982 SRJC); Herb Road, Windsor, *Harrison* (1987 SRJC). < ± 500 ft. Also native in Chile. —*D. humilis* E. Greene. *Bolelia h.* E. Greene, type locality: moist plain in Sonoma Co., *Bioletti* (1892 ND).

Downingia yina Applegate. Rare, vernal areas: s. of Rohnert Park Expressway, w. of Country Club Drive and e. of railroad tracks, Rohnert Park, *C. H. & C. L. Quibell* (1978 NCC). There is a rumor that this species once grew at the Sonoma Co. Airport between Santa Rosa and Windsor. < 6000 ft. July-Aug.

Githopsis

1. Corolla 5-15 mm long; capsule fusiform, ± strongly ribbed . . . . *G. specularioides*
1: Corolla 3-5 mm long; capsule linear, slightly enlarged upward, not ribbed . . . . *G. diffusa*

Githopsis diffusa A. Gray ssp. **robusta** N. Morin. Uncommon, open places: Kenwood Springs, Adobe Canyon, *Baker* (1926 NCC). < 6600 ft. Apr.-June.

Githopsis specularioides Nuttall. Occasional, open places, burns: ca 1 mi. nw. of Camp Meeker, acc. J. & B. Guggolz (1991); Bouverie; Annadel; Pepperwood; Lake Sonoma area; 3100 ft. elev. w. flank Mt. St. Helena, *Grummer* (1979 PUA); 2860 ft. elev. Pine Flat Road, acc. Clifton (1984); White Rock Ranch near Geysers Road, *J. & B. Guggolz* (1992 NCC); Little Geysers, *Malloch* (1977 PG&E); Squaw Cr., acc. Clifton (1984); 3000 ft. elev. Caldwell Pines, acc. Clifton (1984). < 4700 ft. May-June.

Heterocodon rariflorum Nuttall. Occasional, damp grassy or shaded areas, vernal pools: Pitkin Marsh, *Rubtzoff* (1952); 6½ mi. ne. of Penngrove on Lichau Road, *Rubtzoff* (1969); Macartney Woods, Adobe Can. area, *Baker* (1927); above Kenwood Springs, Adobe Can., *Baker* (1926); Adobe Can., *Clausen* (1935); 4¼ mi. w. of Glen Ellen on Bennett Valley Road, *Rubtzoff, Froebe & Harrison* (1974); Bennett Mtn. Lake, *Rubtzoff* (1955); Pepperwood; Lake Sonoma area; 2700 ft. elev. w. flank Mt. St. Helena, *Grummer* (1979 PUA); 4000 ft. elev. Socrates Mine, acc. Clifton (1984). < 7600 ft. Apr.-July.

Legenere limosa (E. Greene) McVaugh. Uncommon, vernal pools: ca 1400 ft. elev. 6½ mi. ne. of Penngrove on Lichau Road, *Rubtzoff* (1969). < 1500 ft. May-June. —*Howellia l.* E. Greene.

Lobelia erinus Linnaeus. Escape from cultivation: Friends House in Rincon Valley, *Wells* (1985). < 350 ft. July-Oct. Native of S. Africa.

Nemacladus capillaris E. Greene. THREADPLANT. Rare, dry slopes, burns: road to Dunlevy's off Geysers-Healdsburg Road, *Harrison* (1979 SRJC); 2700 ft. elev. on w. flank Mt. St. Helena, *Grummer* (1979 PUA). 1300-7000 ft. May-July.

Triodanis biflora (Ruiz López & Pavón) E. Greene. VENUS LOOKING-GLASS. Uncommon, grassland, disturbed, burned areas: Button Ranch on Walker Road (U.C. property) between Petaluma-Valley Ford Road and Pepper Road, *Connors* (1992 BML); Summit Ranch w. of Calistoga near Santa Rosa Hwy., acc. *Baker* (1954). < 6600 ft. Apr.-June.

CAPRIFOLIACEAE. HONEYSUCKLE FAMILY

1. Corolla rotate or urn-shaped
  2. Leaves pinnate; drupe 3-5-seeded . . . . *Sambucus*
  2: Leaves simple; drupe 1-seeded . . . . *Viburnum*
1: Corolla tubular or campanulate
  3. Corolla campanulate, short, regular or nearly so; berry white . . . . *Symphoricarpos*
  3: Corolla tubular or campanulate, ± irregular; berry red or black . . . . *Lonicera*

Lonicera

1. Flowers borne in pairs on a common peduncle in the leaf-axils; leaves deciduous
  2. Plants native shrubs; flowers 1-2 cm long; corolla yellow, tinged with red *L. involucrata* var. *ledebourii*
  2: Plants introduced vines; flowers 3-4 cm long; corolla white changing to yellow . . . . *L. japonica*

1: Flowers borne in whorls and sessile, the whorls solitary or more commonly in terminal spikes or panicles; leaves mostly evergreen
  3. Corollas 3-5 cm long, the slender tube not at all gibbous . . . . *L. etrusca*
  3: Corollas 2-2.5 cm long, tube gibbous
    4. Inflorescence glandular-pubescent; corolla glandular-pubescent outside, reddish-purple outside, yellow inside . . . . *L. hispidula* var. *vacillans*
    4: Inflorescence glabrous; corolla glabrous throughout, yellowish, frequently with a reddish tinge . . . . *L. interrupta*

★ **Lonicera etrusca** Santi. HONEYSUCKLE. Probably an escape from cultivation: Bouverie. Introduced from Mediterranean region. < 1000 ft.

Lonicera hispidula Douglas var. **vacillans** A. Gray. HONEYSUCKLE. Common, streams, wooded slopes: The Sea Ranch, *Wells* (1985); Annapolis, *Sutliffe* (1920); Salt Point; Salmon Cr. dunes, *Connors* (1991 BML); Pitkin Marsh, *Rubtzoff* (1952); Crane Can., *Knight* (1963); Fairfield Osborn (S); Button Ranch on Walker Road (U. C. property) between Petaluma-Valley Ford Road and Pepper Road, acc. Patterson (1990); Bouverie; along Sonoma Cr. at foot of Mt. Hood, *Heller* (1902); Spring Lake; along Santa Rosa Cr. in e. Santa Rosa, acc. Wagnon (1989); Santa Rosa, *Knight* (1963); Pepperwood, *Breedlove* (1980); Petrified Forest, *Eastwood* (1915); Fitch Mtn., *Rubtzoff* (1952); chaparral area sw. of Bradford Mtn. 1½ mi. w. of Dry Cr., acc. J. & B. Guggolz (1992); Lake Sonoma area; ca 2000 ft. elev. Anderson Trail on Black Mtn. between Geyser Peak and upper end of Pine Flat Road, acc. J. & B. Guggolz (1992); Geyser Peak, *Knight & Edwards* (1986). < 3700 ft. Apr.-July. —*L. hispidula* var. *californica* Jepson.

Lonicera interrupta Bentham. CHAPARRAL HONEYSUCKLE. Uncommon, dry ridges, slopes: Sebastopol quad.; 3150 ft. w. flank Mt. St. Helena, *Grummer* (1979 PUA). 500-6000 ft.

Lonicera involucrata (Richardson) Banks var. **ledebourii** (Eschscholtz) Jepson. TWINBERRY. Common, moist places, streams: The Sea Ranch, *Wells* (1984); Salt Point, *Wells* (1982); n. of Fort Ross, *Cantelow* (1940); Russian Gulch n. of Jenner, *McKenzie* (1981); Bridgehaven near jct. of Willow Cr. Road and Hwy. 1 along Russian River, *Utech & Knight* (1984); near Duncans Mills, *M. Jones* (1882); Bodega dunes, *Connors* (1990 BML); along stream s. of Occidental, *McMinn* (1922); Pitkin Marsh, *Rubtzoff* (1951); Cooper's Grove on Sonoma Mtn. Road, acc. Knight (1990); Santa Rosa Cr. e. of Santa Rosa, *Heller* (1902); along Mark West Cr. near Mark West Lodge, *de Nevers* (1981); Pepperwood. < 5000 ft. Mar.-June. —*L. l.* Eschscholtz.

★ **Lonicera japonica** Thunberg. JAPANESE HONEYSUCKLE. Occasional, escape from cultivation: Gualala Point Reg. Park, *Knight* (1986); upper Pitkin Marsh, *Rubtzoff* (1952); w. of Snyder Lane, Rohnert Park, *Caruso* (1979 NCC). Usually < 3300 ft. Spring-summer. Native of Asia.

Sambucus

1. Inflorescence flat topped; fruit blue . . . . *S. mexicana*
1: Inflorescence subglobose to pyramidal; fruit red . . . . *S. racemosa*

Sambucus mexicana C. Presl. BLUE ELDERBERRY. Common, open places: Roberts Road e. of Petaluma Hill Road, acc. Quibell (1991); Fairfield Osborn (S); Sugarloaf Ridge; Howarth Park in Santa Rosa, *Knight* (1964); Pepperwood; Petrified Forest, *Phelps* (1932); Fitch Mtn., *Rubtzoff* (1952); between Jimtown and Mercuryville, Geysers Road, *Utech & Knight* (1986 CM); Pine Mtn. Road ne. of Cloverdale, *Knight* (1963); ca 2000 ft. elev. Anderson Trail on Black Mtn. between Geyser Peak and upper end of Pine Flat Road, acc. J. & B. Guggolz (1992). < 10000 ft. Mar.-Sept. —*S. caerulea* Rafinesque.

Sambucus racemosa Linnaeus. RED ELDERBERRY. Common, damp woods, flats, low elevations: The Sea Ranch; Salt Point; Willow Creek; ca 5 mi. e. of Bridgehaven, Willow Cr. Road, *Utech & Knight* (1984); between Duncans Mills and Monte Rio, Hwy. 116, acc. Knight (1990). < 6000 ft. Mar.-July. —*S. callicarpa* E. Greene.

Symphoricarpos

1. Shrubs erect, generally 1-2 m tall; leaves ± glabrous above, often subcoriaceous; young twigs glabrous to minutely puberulent; fruit 10-15 mm in diameter . . . . *S. albus* var. *laevigatus*
1: Shrubs mostly trailing, commonly < 5 dm tall; leaves pubescent above, thin; young twigs pubescent; fruit 4-6 mm in diameter . . . . *S. mollis*

Symphoricarpos albus (Linnaeus) S. F. Blake var. **laevigatus** (Fernald) S. F. Blake. SNOWBERRY. Common, banks, flats, canyons, streams, forests: Willow Creek; Sebastopol, *Dows* (1907); Crane Can., *Knight* (1963); Fairfield Osborn (H); Bouverie; Sugarloaf Ridge; n. of Santa Rosa, *Heller* (1902); Pepperwood; Fitch Mtn., *Rubtzoff* (1952); w. of Cloverdale, *M. Jones* (1931). < 4000 ft. May-July. —*S. rivularis* Suksdorf.

Symphoricarpos mollis Nuttall. CREEPING SNOWBERRY. Common, shaded slopes, chaparral, forests: The Sea Ranch; Duncans Mills quad.; Fairfield Osborn (H); Button Ranch on Walker Road (U. C. property) between Petaluma-Valley Ford Road and Pepper Road, acc. Connors (1992); Pepperwood; 1500 ft. elev. w. flank Mt. St. Helena, acc. Grummer & Hemphill (1983); 3200 ft. elev. Caldwell Pines, acc. Clifton (1984). < 10000 ft. Apr.-June.

Viburnum ellipticum Hooker. Rare, chaparral, forest: Adobe Can., *Baker* (1931); 1 mi. n. of end of Leslie Road off Mark West Springs Road, *Mock* (1984 SRJC); Pepperwood (3 sites). 400-4900 ft. May-June.

CARYOPHYLLACEAE. PINK FAMILY

1. Fruit a utricle, petals 0 or < 0.6 mm, scalelike; stamens arising from hypanthium rim; stipules present (except *Scleranthus*), scarious
  2. Stipules 0; hypanthium in fruit very hard . . . . *Scleranthus*
  2: Stipules present; hypanthium in fruit herbaceous to ± hard
    3. Awn very stout, spine-tipped, 1.5-4 mm; flower densely woolly . . . . *Cardionema*
    3: Awn threadlike to ± stout, not especially spiny, 0.5-1.5 mm; flower with hairs ± straight or tips coiled . . . . *Paronychia*

1: Fruit a capsule; petals present, sometimes 0; stamens from ovary base or on disk around ovary; stipules 0; (except on couplet 7.)
  4. Sepals fused, tube prominent, lobes or teeth < tube; petals long-clawed, usually appendaged near junction with claw
    5. Styles 3 (4); fruit valves or teeth 3, 6 or 10 . . . . *Silene*
    5: Styles 2; fruit valves usually 4
      6. Flower subtended by involucral bracts . . . . *Petrorhagia*
      6: Flower with 0 involucral bracts; (stamens 10; inflorence open; pedicels 5-40+ mm; calyx in flower 10-17 mm, 5- angled or 5-keeled; petal appendages 0) . . . . *Vaccaria*
  4: Sepals ± free; petals not long-clawed, not appendaged at junction with claw
    7. Stipules 0.5-11 mm, ovate to bristlelike, scarious
      8. Leaf-blade oblanceolate to obovate; plant glabrous; styles 0.1-0.3 mm, fused below, 3-branched . . . . *Polycarpon*
      8: Leaf-blade awl-like to linear; plant glandular-hairy, especially inflorescence (except some *Spergularia*); styles usually 0.3-3 mm, free, 3-5 (if shorter, then stipules bristlelike)
        9. Leaves ± whorled, 10-30 per node; styles, fruit valves 5 . . . . *Spergula*
        9: Leaves opposite, often with axillary leaf-clusters; styles, fruit valves 3 . . . . *Spergularia*
    7: Stipules 0
      10. Petals 2-lobed, often ± to base (sometimes 0 in *Stellaria, Cerastium*)
        11. Fruit cylindric, often curved, tip 10-toothed . . . . *Cerastium*
        11: Fruit spheric to ± ovoid, tip not toothed, valves 6
          12. Petals 2-lobed, lobes < 1/3 petal length; rhizomes with tuberlike thickenings 3-12 mm diameter; plant glandular-hairy, especially below; fruit spheric, valves rolled under; seeds 1-2 . . . . *Pseudostellaria*
          12: Petals 2-lobed > half to base, sometimes 0 petals; rhizomes not enlarged; plant not glandular; fruit ± ovoid,valves not rolled under; seeds several-many . . . . *Stellaria*
      10: Petals entire, irregularly toothed, or ± notched (sometimes 0 petals in *Minuartia, Sagina*)
        13. Fruit valves 8; petals 4; plant erect; leaf blades ± linear, rigid, acute . . . . . . *Moenchia*
        13: Fruit valves 3-6; petals 5 (if 4 then plant not erect; leaf blade not linear, rigid and acute)
          14. Styles 5, sometimes 4, alternate with sepals; fruit valves 5, sometimes 4 . . . *Sagina*
          14: Styles 3, opposite sepals; fruit valves 3 or 6
            15. Ovary sutures 3; fruit valves 3 . . . . *Minuartia*
            15: Ovary sutures 6; fruit valves 6 . . . . *Arenaria*

★ **Arenaria serpyllifolia** Linnaeus. SANDWORT. Uncommon, weed, lawns, moist places: Baker's garden, Adobe Can., *Howell* (1935). 500-6000 ft. Apr.-June. Naturalized from Europe.

Cardionema ramosissimum (J. A. Weinmann) Nelson & J. F. Macbride. SAND MAT. Locally common, coast, sandy places: The Sea Ranch; Horseshoe Cove, *Knight* (1963); Salt Point; Bodega Head. < 500 ft. Apr.-Aug. Disjunct to Chile.

Cerastium

1. Plants perennial, with prostrate or creeping basal branches
   2. Petals 2-3 × as long as sepals; basal branches becoming dry and withered and with many axillary fascicles; their leaves linear to oblong, not hirsute . . . . *C. arvense*
   2: Petals ca as long as sepals; basal branches green, with few or no axillary fascicles, leaves oblong, hirsute . . . . *C. fontanum* ssp. *vulgare*

1: Plants annual, without basal, persistent, sterile branches; petals lacking or ca = the sepals . . *C. glomeratum*

Cerastium arvense Linnaeus. FIELD CHICKWEED. Common, grassland, damp places: The Sea Ranch; Salt Point; ½ mi. n. of Fort Ross, *Morey* (1940); Cazadero, *Bacigalupi* (1926); Coleman Valley Road, 4 mi. w. of Occidental, *Best* (1981), *J. & B. Guggolz* (1990 JEPS); Bodega Bay, *Heller* (1902); Bodega Point, *Eastwood* (1915); Bodega Head; Bouverie; Spring Lake Village. < 8200 ft. Feb.-Aug.

★ **Cerastium fontanum** Baumgarten ssp. **vulgare** (Hartman) Greuter & Burdet. Locally common weed: many sites w. and sw. of Santa Rosa in Laguna de Santa Rosa, *Harrison* (1978 NCC). Naturalized from Eurasia. < 7300 ft. Mar.-Aug. —*C. vulgatum* Linnaeus, misapplied.

★ **Cerastium glomeratum** Thuillier. MOUSE-EAR CHICKWEED. Common, weed, roadsides, gardens, disturbed places: n. of Fort Ross on Hwy. 1, *Cantelow* (1940); Cazadero and Arched Rock quads.; ca 2 mi. e. of Occidental, *Best* (1982); lower Pitkin Marsh, *Rubtzoff* (1952); Button Ranch on Walker Road between Petaluma-Valley Ford Road and Pepper Road, acc. Patterson (1990); w. of Sears Point Raceway in Tolay Cr. drainage, acc. Patterson (1990); Bouverie; Cavedale Road, *Raven* (1954); Sugarloaf Ridge; Santa Rosa, *Eastwood* (1921); Pepperwood; Lake Sonoma area; ca 2000 ft. elev. Anderson Trail on Black Mtn. between Geyser Peak and upper end of Pine Flat Road, acc. J. & B. Guggolz (1992). < 5300 ft. Feb.-May. Naturalized from Europe. —*C. viscosum* Linnaeus, misapplied.

Minuartia

1. Petals shorter than sepals; leaves narrowly lanceolate 2-5 cm long . . . . *M. pusilla*

1: Petals longer than sepals, or at least =
   2. Leaves ± lanceolate, 2-5 mm long . . . . *M. californica*
   2: Leaves filiform, 7-40 mm long . . . . *M. douglasii*

Minuartia californica (A. Gray) Mattfeld. Uncommon, disintegrated rock, grassy areas: hills e. of Agua Caliente along Sonoma Valley, *Howell* (1948); Pepperwood; near Windsor, *Heller & Brown* (1902). < 5000 ft. Mar.-June. —*Arenaria c.* (A. Gray) Brewer. *A. breviolia* Maguire var. *c.* A. Gray, type locality: in valleys and among high ridges near Sonoma.

Minuartia douglasii (Torrey & A. Gray) Mattfeld. Occasional, sterile dry places: jct. of Bittner and Joy roads near Occidental, *Ferris & Lorraine* (1961); 2.7 mi. e. of Occidental, *Rubtzoff* (1957); Hessel Road near Gravenstein Hwy., acc. J. & B. Guggolz (1991); Pepperwood; Skaggs Springs, *Parsons* (1915); Lake Sonoma area; Geyserville, *Rodda* (1929); along ridge between Socrates Mine and Mercuryville, geothermal area, *Dolan & LaPré* (1986 UCR); ca 2000 ft. elev. Anderson Trail on Black Mtn. between Geyser Peak and upper end of Pine Flat Road, acc. J. & B. Guggolz (1992); 2400 ft. elev. Little Geysers, *Rubtzoff & Malloch* (1977 PG&E). 300-6000 ft. Apr.-June. —*Arenaria douglasii* Fenzl.

Minuartia pusilla (S. Watson) Mattfeld. Rare, open slopes, dry woods: Adobe Can., *Baker* (1927 NCC); Lake Sonoma area. 350-8000 ft. Apr.-July. —*Arenaria p.* S. Watson.

★ **Moenchia erecta** (Linneaus) Gaertner, Meyer & Scherbius ssp. **erecta**. Uncommon, disturbed areas: The Sea Ranch, acc. Mahaffey (1996). < 600 ft. Native of SW Europe.

★ **Paronychia franciscana** Eastwood. Uncommon, coastal influence: in grass above Larkspur Rock e. of Bodega Bay, *Baker* (1930 NCC); Bodega Bay behind elementary school, *Connors* (1992 BML). < 850 ft. Apr.-June. Native to Chile.

Petrorhagia

1. Leaf sheath length ± = width, 1-2 mm; petals truncate or shallowly notched . . . . *P. prolifera*
1: Leaf sheath length 1.5-3 x width, 3-9mm; petals obcordate to 2 lobed . . . . *P. nanteuilii*

★ **Petrorhagia nanteuilii** (Burnat) P. Ball & Heywood. PROLIFEROUS PINK. Uncommon, open areas: Big Austin Cr. at the base of "The Butcher Knife," *Bacigalupi et al.* (1956 DAV), *R. K. Rabeler, Sida 11:31,* (1985); 4.4 mi. n. of Cazadero, King Ridge Road around corner from Big Austin Cr. bridge, *Raiche* (1980 NCC). < 1500 ft. Native of Old World.

★ **Petrorhagia prolifera** (Linnaeus) P. Ball & Heywood. Occasional, grassland, rocky places: The Sea Ranch on sandbar on S. Fork of Gualala River near Hot Spot, *Wells* (1984); Fairfield Osborn (S); acc. *R. K. Rabeler, Sida 11:38,* (1985), acc. *Ellman* (1978); Lafferty Ranch on Sonoma Mtn., acc. Arnold et al. (1971); Westerbecke Ranch w. of El Verano, acc. Best (1980); Annadel, *Wagnon* (1989 NCC); Spring Lake; Pepperwood; Lake Sonoma area; property bordered by Chianti and Dutcher Cr. roads w. of US 101 and w. of Geyserville, acc. J. & B. Guggolz (1991); ca 2000 ft. elev. Anderson Trail on Black Mtn. between Geyser Peak and upper end of Pine Flat Road, acc. J. & B. Guggolz (1992); 2040 ft. elev. Squaw Cr., acc. Clifton (1984). < 2100 ft. May. Native of Old World. —*Tunica p.* (Linnaeus) Scopoli.

★ **Polycarpon tetraphyllum** (Linnaeus) Linnaeus. FOUR-LEAVED ALLSEED. Common, weedy, sidewalk cracks, flower beds, dunes: Salt Point; dunes on s. side of mouth of Russian River, *Best* (1982); Bodega dunes, *Connors* (1991 BML); Bodega Head, acc. Connors (1991); Royal Oaks Mobile Home Park in Petaluma, acc. Knight (1991); Spring Lake Village; Pepperwood, *Breedlove* (1983). < 1400 ft. May-July. Native of Europe.

Pseudostellaria jamesiana (Torrey) W. A. Weber & R. L. Hartman. Uncommon, meadows, damp places, usually serpentine: 1 mi. nw. of Camp Meeker, acc. J. & B. Guggolz (1991); Coleman Valley Road, *Cummings* (1970 NCC). 700-9000 ft. May-July. —*Stellaria j.* Torrey. *Arenaria j.* (Torrey) Shinners.

Sagina

1. Plants annual, often glandular-hairy on pedicels or sepals or both; sterile basal rosettes 0; stem threadlike; pedicels ± straight
   2. Upper leaves minutely ciliate near base; sepals 4 (5); petals 0 or rarely 4, minute, < sepals . *S. apetala*
   2: Upper leaves glabrous; sepals usually 5; petals usually 5, ± = sepals . . . *S. decumbens* ssp. *occidentalis*
1: Plants perennial, glabrous; sterile basal rosettes often present; stem stouter, not threadlike; pedicels often recurved below flowers
   3. Sepals 4 (5), spreading to ascending after fruit dehiscence; petals ¼-½ × sepals or occasionally 0 . . . . *S. procumbens*
   3: Sepals 5, ± appressed after fruit dehiscence; petals 3/4-1 × sepals . . . . . . *S. maxima* ssp. *crassicaulis*

★ **Sagina apetala** Arduino. DWARF PEARLWORT. Uncommon, inconspicuous weed, hard-packed soil: Bodega Dunes, *Connors* (1991 BML); Bouverie; Spring Lake Village; Pepperwood. < 2300 ft. Apr.-June. Naturalized from Europe. —*S. a.* var. *barbata* Fenzl.

Sagina decumbens (Elliott) Torrey & A. Gray ssp. **occidentalis** (S. Watson) G. Crow. WESTERN PEARLWORT. Occasional, hard-packed or damp ground: Stoetz Lane ne. of Occidental, *Best* (1982); Pitkin Marsh, *Rubtzoff* (1952); w. of Santa Rosa in Laguna de Santa Rosa drainage, *Harrison* (1978 NCC); Santa Rosa, *Eastwood* (1921); 2400 ft. elev. Little Geysers, *Malloch* (1977 PG&E). < 6600 ft. Mar.-June. —*S. occidentalis* S. Watson.

Sagina maxima A. Gray ssp. **crassicaulis** (S. Watson) G. Crow. BIG PEARLWORT. Rare, near dunes: bluff at Salt Point, *Wagnon* (1989 NCC); 10 ft. elev. Bodega Head, *Clifton* (1981 PUA). < 100 ft. June-Dec. —*S. c.* S. Watson var. *littoralis* (Hultén) Hultén.

Sagina procumbens Linnaeus. ARCTIC PEARLWORT. Occasional, moist places: Horseshoe Cove, *Knight* (1963); Santa Rosa, *Wells* (1985); Pepperwood, *Breedlove* (1983). < 900 ft. May-Sept.

★ **Scleranthus annuus** Linnaeus. KNAWEL. Rare: Pepperwood; Alexander Valley quad. 500-4200 ft. Mar.-June. Native of Europe.

Silene

1. Plants annual
   2. Calyx with 20-30 well-developed ribs . . . . . *S. multinervia*
   2: Calyx generally with 10 ribs, if more, the ribs indistinct . . . . . *S. gallica*

1: Plants perennial
   3. Corolla bright red . . . . . *S. californica*
   3: Corolla white or pink to purplish
      4. Stems 1-4, from a heavy crown; basal leaves 15-30 mm wide . . . . . *S. scouleri* ssp. *grandis*
      4: Stems more numerous; basal leaves 2-12 mm wide . . . . . *S. verecunda* ssp. *platyota*

Silene californica Durand. INDIAN PINK. Common, dry brushy areas: Bohemian Hwy. 3.8 mi. nw. of Occidental, *Huntington* (1986 NCC); Bouverie; Sugarloaf Ridge; Joost Ranch on Calistoga Road between Alpine and Porter Cr. roads, acc. Knight (1989); Pepperwood; chaparral area sw. of Bradford Mtn. 1½ mi. w. of Dry Cr., acc. J. & B. Guggolz (1992); Lake Sonoma area; Ida Clayton Road quad.; ca 2000 ft. elev. Anderson Trail on Black Mtn. between Geyser Peak and upper end of Pine Flat Road, acc. J. & B. Guggolz (1992); Geyser Peak, *Knight & Edwards* (1986); 3100 ft. elev. w. flank Mt. St. Helena, *Grummer* (1979 PUA). < 7300 ft. Mar.-Aug.

★ **Silene gallica** Linnaeus. WINDMILL PINK. Common, disturbed places: The Sea Ranch; Salt Point; Fort Ross, Cazadero and Duncans Mills quads.; Willow Creek; Bodega Marine Reserve, *Connors* (1991 BML); Bodega Head; Pinnacle Gulch Trail in Bodega Harbour development, *Huntington* (1989 NCC); 2 sites w. of Santa Rosa in Laguna de Santa Rosa, *Harrison* (1978 NCC); Fairfield Osborn (H); Bouverie; 1½ mi. up Cavedale Road from Hwy. 12, *Utech & Knight* (1986 CM); Sugarloaf Ridge; Spring Lake; Pepperwood; Lake Sonoma area; Ida Clayton Road quad.; ca 2000 ft. elev. Anderson Trail on Black Mtn. between Geyser Peak and upper end of Pine Flat Road, acc. J. & B. Guggolz (1992); Geyser Peak, acc. Knight & Edwards (1986); 2400 ft. elev. Little Geysers, *Malloch* (1977 PG&E); 2040 ft. elev. Squaw Cr., acc. Clifton (1984). < 3300 ft. Feb.-June. Naturalized from Europe.

Silene multinervia S. Watson. Rare, burns, disturbed places: w. slope Mt. Hood, *Baker* (1947 NCC). < 6600 ft. Apr.-May.

Silene scouleri Hooker ssp. **grandis** (Eastwood) C. L. Hitchcock & Maguire. Occasional, coastal grassland: The Sea Ranch, acc. Wells (1983); bluffs at Wright's Beach, *F. W. Hoffman* (1951); cliff at Bodega Bay, *Baker* (1931); Bodega Bay at Bodega Harbor, *Harrison* (1976 SRJC). < 1000 ft. May-Aug. —*S. s.* var. *pacifica* (Eastwood) C. L. Hitchcock. *S. grandis* Eastwood, type locality: Bodega Point, *Eastwood* (1900).

Silene verecunda S. Watson ssp. **platyota** (S. Watson) C. L. Hitchcock & Maguire. Uncommon, openings in brush: 3800 ft. elev. w. flank Mt. St. Helena, *Grummer* (1980 PUA); 3280 ft. elev. Socrates Mine, acc. Clifton (1984). 1000-11000 ft. June-Aug.

★ **Spergula arvensis** Linnaeus. SPURREY, STICKWORT, STARWORT. Common, weedy, fields, orchards: The Sea Ranch; Bodega Head; 5 sites w. of Santa Rosa in Laguna de Santa Rosa, *Harrison* (1978 NCC); Todd Road Preserve; Magnolia and Marshall Aves. w. of Petaluma, *Knight* (1963); Bouverie; Rincon Valley, *Wells* (1984); Spring Lake; Pepperwood. < 700 ft. Most months. Native of Europe.

Spergularia

1. Leaves densely fascicled; stipules lance-acuminate; plants perennial, (occasionally annual in *S. rubra*)
   2. Sepals mostly 5-10 mm long; seeds 0.7-0.9 mm long; styles 0.6-3 mm long . . . . . *S. macrotheca*
   2: Sepals mostly 3-5 mm long; seeds 0.4-0.6 mm long; styles 0.2-0.8 mm long
      3. Stem below the inflorescence mostly glabrous; capsule equaling calyx; seeds not winged, reticulate . . . . . *S. rubra*
      3: Stem below the inflorescence mostly glandular-pubescent; capsule longer than calyx; seeds usually winged,papillate . . . . . *S. villosa*

1: Leaves not densely fascicled; stipules mostly deltoid, not long-acuminate; annual
   4. Stamens 6-10 . . . . . *S. bocconii*
   4: Stamens 2-5 . . . . . *S. marina*

★ **Spergularia bocconii** (Scheele) Merino. SAND-SPURREY. Rare, dunes, waste places: Goat Rock State Park, *Best* (1982). < 1300 ft. Apr.-Sept. Naturalized from Europe.

Spergularia macrotheca (Hornemann) Heynhold. STICKY SAND-SPURREY. Common, rock outcrops, coastal bluffs, saline marshes: The Sea Ranch, *Knight* (1986); Horseshoe Cove, *Knight* (1963); Salt Point; Bodega Head; Bodega Harbor salt marsh, *Clifton* (1980 PUA); Petaluma, *Eastwood* (1921). <850 ft. Most of year.

Spergularia marina (Linnaeus) Grisebach. COASTAL SAND-SPURREY. Locally common, alkaline places: Salt Point; Bodega Bay, *Rossbach* (1956); Petaluma, *Knight* (1963). < 2300 ft. Mar.-Sept.
★ **Spergularia rubra** Linnaeus) J. & C. Presl. Sand-Spurrey. Common, weed, hard-packed soil, sidewalk cracks, disturbed places: Horseshoe Cove, *Knight* (1963); Salt Point State Park, *Wells* (1982); Timber Cove n. of Fort Ross, *McKenzie* (1980); 6 m. n. of Bodega Bay, *Howell* (1937); w. of Santa Rosa in Laguna de Santa Rosa drainage, *Harrison* (1978 NCC); Fairfield Osborn (H); Spring Lake; Santa Rosa, *Eastwood & Howell* (1936); Pepperwood; Lake Sonoma area. < 8000 ft. Apr.-Sept. Naturalized from Europe.
★ **Spergularia villosa** (Persoon) Cambessèdes. SAND-SPURREY. Occasional, sandy or hard-packed soil: Kentucky St., Petaluma, *Knight* (1963); Gilardi's Resort near Lakeville, e. of Petaluma, *Knight* (1984). < 1500 ft. Apr.-July. Introduced from S. America

Stellaria

1. Plants annual
    2. Internodes with longitudinal line of hairs; leaves ovate . . . . . *S. media*
    2: Internodes lacking line of hairs; upper leaves lance-linear . . . . . *S. nitens*
1: Plants perennial
    3. Plants glandular-puberulent . . . . . *S. littoralis*
    3: Plants not glandular . . . . . *S. borealis* ssp. *sitchana*

Stellaria borealis Bigelow ssp. **sitchana** (Steudel) Piper. Uncommon, wet places: The Sea Ranch; Perry Marsh, *Howell* (1945). < 7000 ft. June-July. —*S. sitchana* Steudel var. *bongardiana* (Fernald) Hultén.

Stellaria littoralis Torrey. Uncommon, moist dunes: adjacent to Salmon Cr. *Baker* (1930 NCC); Bodega Point, *Eastwood* (1915); Bodega Head and Dunes, *Connors* (1991 BML). < 135 ft. Mar.-July.
★ **Stellaria media** (Linnaeus) Villars. COMMON CHICKWEED. Very common, weed, gardens, disturbed areas: The Sea Ranch; Horseshoe Cove, *Knight* (1963); Kolmer Gulch between Jenner and The Sea Ranch, *McKenzie* (1981); Willow Creek; Bodega Head; many sites w. and sw. of Santa Rosa in Laguna de Santa Rosa drainage, *Harrison* (1978 NCC); Todd Road Preserve (H&K); Fairfield Osborn (S); Tresch Ranch, Walker Road near Two Rock, acc. J. & B. Guggolz (1992); corner Magnolia and Marshall aves. w. of Petaluma, *Knight* (1963); Bouverie; Sugarloaf Ridge; Gates Road off Calistoga Road, acc. Wells et al. (1984); Spring Lake Village; Pepperwood; Lake Sonoma area; Geyserville quad.; 2040 ft. elev. Squaw Cr., acc. Clifton (1984). < 4200 ft. Feb.-Sept. Naturalized from Eurasia.

Stellaria nitens Nuttall. SHINING CHICKWEED. Occasional, inconspicuous, grassy places: Stoetz Lane ne. of Occidental, *Utech & Knight* (1986 CM); Sugarloaf Ridge; Pepperwood, *Breedlove* (1983); The Geysers, *Quibell* (1980 NCC); PG&E geysers Unit 13, *Quibell* (1981 NCC); 2040 ft. elev. Squaw Cr., acc. Clifton (1984). < 5000 ft. Mar.-June.

★ **Vaccaria hispanica** (Miller) Rauschert. COW-HERB, COCKLE. Rare, weed of grainfields: Sonoma, *Jepson, Fl. of CA, 1:500,* (1922), *Kuhn (1914)*; mountains near Calistoga, *Krauss* (1894). < 9500 ft. May-Aug. Native of Europe. —*V. segetalis* (Necker) Garcke.

CELASTRACEAE. STAFF-TREE FAMILY

1. Shrubs large erect; leaves petioled; seed completely enclosed in a fleshy red aril . . . . . *Euonymus*
1: Shrubs small prostrate; leaves subsessile; seed with a whitish aril at base . . . . . *Paxistima*

Euonymus occidentalis Torrey. WESTERN BURNING BUSH, PAWNBROKER BUSH. Occasional, moist shaded places, coastal hills, canyons: Salt Point State Park, *Tunison* (1977 SRJC); 2 mi. n. of Fort Ross, *Mason* (1928); Duncans Mills Marsh, *Rubtzoff* (1960); Guerneville, *Rattan* (1877); Monte Rio, *McMinn* (1922); 5 mi. w. of Sebastopol, *Wagnon* (1939); Mill Cr. near Cooper's Ranch, *Baker* (1929). < 5300 ft. Apr.-June.

Paxistima myrsinites (Pursh) Rafinesque. OREGON BOXWOOD. Uncommon, open or shaded wooded slopes: Salt Point, acc. Quibell (1991); Camp Meeker, *R. Brown* (1984); Annadel; Dry Cr. Valley (now Lake Sonoma), *Baker* (1934); Dorman Cr. off of Dry Cr., *Mason* (1934); Cherry Cr. Road, w. of Cloverdale, *Knight & Edwards* (1986). < 6600 ft. May-July.

CERATOPHYLLACEAE. HORNWORT FAMILY

Ceratophyllum demersum Linnaeus. HORNWORT. Occasional, ponds, slow streams: Duncans Mills, *M. E. Jones* (1882); dune pond near Bodega Marine Lab., *Rubtzoff* (1971); Atascadero Cr. Marsh w. of Sebastopol, *Rubtzoff* (1956); Button Ranch on Walker Road (U.C. property) between Petaluma-Valley Ford Road and Pepper Road, acc. Connors (1992); Fairfield Osborn (S); Pepperwood. Usually < 6600 ft. June-Aug.

CHENOPODIACEAE. GOOSEFOOT FAMILY

1. Leaves foliaceous, flattened, not particularly fleshy or scaly
   2. Flowers perfect, occasionally only ♀, all with calyx and not enclosed in a pair of bracts
      3. Calyx indurate at base in age, with ovary partly sunk . . . . *Beta*
      3: Calyx not indurate; with ovary superior . . . . *Chenopodium*
   2: Flowers imperfect, the ♀ enclosed in 2 accrescent bractlets . . . . *Atriplex*
1: Leaves not or scarcely flattened but fleshy and sublinear, or scaly and spiny
   4. Leaves scalelike; stems and branches fleshy . . . . *Salicornia*
   4: Leaves not scalelike, but fleshy and sublinear, or spiny-tipped
      5. Leaves not tipped with spine or bristlelike hair
         6. Fruiting calyx transversely winged . . . . *Kochia*
         6: Fruiting calyx not transversely winged . . . . *Suaeda*
      5: Leaves tipped with spine or bristly hair . . . . *Salsola*

Atriplex

1. Plants perennial
   2. Fruiting bracts ± free, ± compressed, ± thin; ♂ and ♀ flowers usually in same clusters *A. californica*
   2: Fruiting bracts fused to middle or above, thick, spongy-fleshy; ♀ inflorescence usually below ♂
      3. Fruiting bracts fleshy, reddish 3-6 mm . . . . *A. semibaccata*
      3: Fruiting bracts spongy, gray to white 4-12 mm . . . . *A. leucophylla*
1: Plants annual
   4. Leaves usually green on both surfaces, glabrous to sparsely powdery or fine-scaly
      5. Fruiting bract margin narrow, minutely dentate . . . . *A. triangularis*
      5: Fruiting bract margin wide, entire . . . . *A. patula* var. *obtusa*
   4: Leaves white to gray, densely and finely scaly, especially below
      6. Fruiting bracts usually widest below middle, rhombic or ovate . . . . *A. rosea*
      6: Fruiting bracts, including stalk, usually widest at or above middle, diamond-shaped . . . *A. suberecta*

Atriplex californica Moquin. Uncommon, coastal bluffs: Salt Point; Bodega Marine Reserve salt marsh, *Connors* (1991 BML); Bodega Point, *Eastwood* (1915). < 200 ft. Apr.-Nov.

Atriplex leucophylla (Moquin) D. Dietrich. Uncommon, beaches: The Sea Ranch; salt marsh at entrance to Bodega Marine Reserve at Bodega Bay, *Edwards* (1976). < 200 ft. Apr.-Oct.

Atriplex patula Linnaeus var. **obtusa** (Chamisso) C. Hitchcock. SPEAR ORACLE. Uncommon, along shore: Bodega Head. < 200 ft. Aug.-Nov. —*A. angustifolia* var. *o.* Chamisso. *A. gmelinii* C. Meyer.

★ **Atriplex rosea** Linnaeus. TUMBLING ORACLE. Uncommon, alkaline areas: Petaluma River near Lakeville St., Petaluma, *Knight* (1963). < 8000 ft. July-Oct. Naturalized from Eurasia.

★ **Atriplex semibaccata** R. Brown. AUSTRALIAN SALTBUSH. Uncommon, along roads, waste places: Petaluma River at bridge on Black Point Road, *Raven* (1955); roadcut on Hwy. 37 ca 1 mi. from Sears Point jct., *Knight* (1964). < 3300 ft. Apr.-Dec. Naturalized from Australia.

★ **Atriplex suberecta** Verdoorn. Uncommon, pastureland, waste places: Willow Brook between Old Redwood Hwy. and Corona Road, *Knight & Edwards* (1984), *Knight* (1987). < 1000 ft.Native of Australia.

Atriplex triangularis Willdenow. SPEARSCALE. Common, usually moist saline areas: The Sea Ranch; Bodega Head; near Todd Road in Laguna de Santa Rosa drainage, *Harrison* (1978 NCC); Llano Road 0.6 mi. from Gravenstein Hwy., *Howell & Knight* (1986 NCC); flood plain s. of Petaluma, *Knight* (1984); Spring Lake. < 1000 ft. June-Nov. —*A. hastata* Linnaeus.

★ **Beta vulgaris** Linnaeus. BEET. Uncommon, escape from gardens: Petaluma, *Jepson* (JEPS). < 700 ft. July-Oct. Native of Europe.

Chenopodium

1. Plants ± glandular-pubescent or resinous-glandular, especially about the calyx, not farinose
  2. Flowers in glomerules, these in capitate clusters or small spikes
    3. Glomerules small, headlike, all axillary; seed vertical . . . . *C. pumilio*
    3: Glomerules in short spikes, the upper in panicles; seed mostly horizontal
      4. Stems and branches loosely white-villous; spikes quite leafless . . . . *C. ambrosioides* var. *vagans*
      4: Stems and branches smooth; spikes leafy or intermixed with leaves . . . . *C. a.* var. *ambrosioides*
  2: Flowers solitary in small cymes, these spreading-recurved and in elongated panicles . . . . *C. botrys*
1: Plants mostly farinose, not glandular-pubescent, but occasionally with nonglandular hairs
  5. Seeds horizontal, or almost entirely so
    6. Leaves shining on upper surface, ± rhombic; only younger parts farinose, plant branched from base . . . . *C. murale*
    6: Leaves dull on upper surface, various in shape
      7. Pericarp closely adherent to seed and removable with difficulty; at least lower leaves conspicuously sinuate- dentate
        8. Pericarp and seed-surface essentially smooth
          9. Seeds chiefly 1.2-1.5 mm broad; sepals largely covering fruit . . . . *C. album*
          9: Seeds 0.9-1.2 mm broad; sepals exposing mature fruit . . . . *C. strictum*
        8: Pericarp and seed-surface alveolate . . . . *C. berlandieri*
      7: Pericarp free from or easily removed from seed, or if adherent, leaves rhombic- to rounded-ovate . . . . *C. vulvaria*
  5: Seeds ± vertical
    10. Plants annual, glabrous; stems up to 6 dm long . . . . *C. rubrum*
    10: Plants perennial, farinose on upper parts; stems 3-8 dm long or more . . . . *C. californicum*

★ **Chenopodium album** Linnaeus. LAMB'S QUARTERS, PIGWEED. Common, waste places: The Sea Ranch; Bodega Bay; Jonive and Cherry Ridge roads near Freestone, *Best* (1983); Occidental Road and Gravenstein Hwy., *Best* (1983); w. of Santa Rosa in Laguna de Santa Rosa drainage, *Harrison* (1978 NCC); Bodega Ave. w. of Petaluma, *Knight* (1963); Kenwood Marsh, *Howell* (1932); Rincon Valley, *Wells* (1984); Spring Lake Village; 7½ mi. from Santa Rosa on road to Petrified Forest, *Howell* (1929). < 6000 ft. June-Oct. Naturalized from Europe.

★ **Chenopodium ambrosioides** Linnaeus var. **ambrosioides**. MEXICAN TEA. Common, waste places: Forest Hills near Hacienda, *Rubtzoff* (1965); Pitkin Marsh, *Rubtzoff* (1952); Laguna de Santa Rosa ne. of Graton, *Rubtzoff* (1958); Todd Road Preserve; Copeland Cr. on Sonoma State Univ. campus, *Self* (1979 NCC); Willow Brook n. of Petaluma, *Knight & Edwards* (1984); Spring Lake; bed of Santa Rosa Cr. near Santa Rosa, *Baker* (1928). < 4700 ft. June-Dec. Naturalized from tropical America.

Chenopodium ambrosioides var. **vagans** (Standley) J. T. Howell. Uncommon, disturbed areas: Bodega Head. < 4700 ft. June-Dec. Naturalized from S. America.

Chenopodium berlandieri Moquin. PITSEED GOOSEFOOT. Common, ± waste places: Vacation Beach 2 mi. below Guerneville, *Rubtzoff* (1961); Fife Cr. Marsh at Guerneville, *Rubtzoff* (1961); Odd Fellows Park near Hilton on Russian River, *Howell* (1938); 5½ mi. e. of Guerneville at Forest Hills, *Rubtzoff* (1965); Laguna de Santa Rosa ne. of Graton, *Rubtzoff* (1962). Var. *zschackei* (Murray) Murray, often confused with *C. album*. < 9000 ft. June-Oct.

★ **Chenopodium botrys** Linnaeus. JERUSALEM-OAK. Occasional, weed, sandy areas: sandbar in S. Fork of Gualala River at Hot Spot on The Sea Ranch, *Wells* (1984); Forest Hills near Hacienda, *Rubtzoff* (1965); Spring Lake; Hidden Valley area of Santa Rosa, *Wells* (1985); Pepperwood; Fitch Mtn. *Rubtzoff* (1952). < 7000 ft. June-Oct. Naturalized from Eurasia.

Chenopodium californicum (S. Watson) S. Watson. Occasional, dry plains, slopes: Bodega Head; Bodega Point, *Eastwood* (1915); Kenwood, acc. Baker (1935). < 6600 ft. Mar.-June.

★ **Chenopodium murale** Linnaeus. Common, weed, gardens, orchards, waste areas: mouth of Salmon Cr., acc. Baker (1935); Laguna de Santa Rosa near Occidental Road, *Rubtzoff* (1958); Buena Vista Winery near Sonoma, *Best* (1984); Rincon Valley, *Wells* (1984); Pepperwood. < 9700 ft. Most of the year, especially in the spring. Naturalized from Europe.

★ **Chenopodium pumilio** R. Brown. Occasional, weed, stream bottoms: Forest Hills across river from Hacienda, *Rubtzoff* (1961); 2290 ft. elev. Socrates Mine, acc. Clifton (1984). < 9900 ft. June-Sept. Naturalized from Australia.

Chenopodium rubrum Linnaeus. Uncommon, weed, waste places: near Kenwood in Adobe Can., *Baker* (1941 NCC). Some forms may be introduced. < 3700 ft. July-Oct.
★ **Chenopodium strictum** Roth var. **glaucophyllum** (Aellen) Wahl. Rare, low areas: San Antonio Cr. near US 101, *Best* (1980). < 5700 ft. Aug.-Oct. Native of e. U.S.
★ **Chenopodium vulvaria** Linnaeus. Rare, waste places: Chileno Valley, *Tennon* (1973); at county line on US 101 at San Antonio Cr., acc. Best (1985). < 4700 ft. June-Oct. Native of Eurasia.

★ **Kochia scoparia** (Linnaeus) Schrader. SUMMER-CYPRESS. Rare, waste places: Pepperwood, *de Nevers* (1992). < 4800 ft. Aug.-Oct. Introduced from Eurasia.

Salicornia

1. Plants annual; central flower of a node usually higher than lateral flowers . . . . *S. bigelovii*
1: Plants perennial from rhizomes; flowers of a node usually at same level . . . . *S. virginica*

Salicornia bigelovii Torrey. GLASSWORT. Rare, alkaline flats: hwy. n. of Sears Point, *Baker* (1946 NCC). < 65 ft. July-Nov.
Salicornia virginica Linnaeus. PICKLEWEED. Common, alkaline flats: Bodega Head; Bodega Harbor, *Clifton* (1980 PUA); Jefferson St., Petaluma, *Knight* (1963); Gilardi's Resort near Lakeville, *Knight* (1984). < 350 ft. Aug.-Nov. —*S. pacifica* Standley.

★ **Salsola tragus** Linnaeus. RUSSIAN THISTLE, TUMBLEWEED. Occasional, weed, disturbed ground: Schellville at the railroad tracks, *Best* (1986); Pepperwood. < 9000 ft. July-Oct. Native of Eurasia. —*S. kali* Linnaeus var. *tenuifolia* Tausch. *S. ruthenica* Iljin. *S. iberica* Sennen & Pau. *S. australis* R. Brown. All misapplied.

Suaeda californica S. Watson. SEA-BLITE, SEEP-WEED. Margins of coastal salt marshes. A component of the adobe brick at Vallejo's Old Adobe e. of Petaluma, acc. *Hendry & Kelly* (1925). It may have been a member of the botanical community along the Petaluma River between Lakeville and Petaluma in the early 19th century. < 15 ft. July-Oct.

CISTACEAE. ROCK-ROSE FAMILY

Helianthemum scoparium Nuttall. PEAK RUSH-ROSE, ROCK-ROSE. Occasional, rocky ridges, dry slopes: The Sea Ranch, acc. Wells (1983); Summerhome Park on Russian River, *de Nevers* (1983); Bouverie; sw. slope Mt. Hood, *Baker* (1947 NCC); Mountain Home Ranch Road off Petrified Forest Road, *Stocking* (1966 NCC); 1150 ft. elev. and higher to summit of Bradford Mtn. in Dry Cr. drainage area on fire trail, acc. J. & B. Guggolz & W. Knight (1992). < 5000 ft. Mar.-May.

CLUSIACEAE. ST. JOHN'S WORT FAMILY. [GUTTIFERAE or HYPERICACEAE]

Hypericum

1. Petals < or = sepals; styles ± 1 mm; placentas parietal . . . . *H. anagalloides*
1: Petals >> sepals; styles 3-10 mm; placentas axile
  2. Leaf linear to lanceolate, usually folded, tip acute; stems many, from woody caudex . . . . *H. concinnum*
  2: Leaf ± elliptic or oblong, flat, obtuse; stems few from taproot or rhizome
    3. Leaf elliptic; sterile axillary branches usually < 2 cm; fruit 3-lobed . . . . *H. formosum* var. *scouleri*
    3: Leaf narrowly oblong; sterile axillary branches usually 2-10 cm; fruit unlobed . . . . . . *H. perforatum*

Hypericum anagalloides Chamisso & Schlechtendal. TINKER'S PENNY. Common, marshy conditions: The Sea Ranch; 1 mi. s. of Gualala on Hwy. 1, *Wolf* (1927); Kruse Ranch Road 1 mi. w. of Plantation, *Hastings* (1940 NCC); Salt Point; Forestville Marsh, *Rubtzoff* (1954, 1955); Pitkin Marsh, *Baker* (1935 NCC), *Rubtzoff* (1951); Laguna de Santa Rosa ne. of Graton near Laguna Road, *Rubtzoff* (1960, 1961); Cunningham Marsh, *Howell* (1946); Button Ranch on Walker Road (U. C. property) between Petaluma-Valley Ford Road and Pepper Road, acc. Connors (1992); Macartney Woods on Mt. Hood, *Baker* (1927

NCC); Pepperwood; Lake Sonoma area; marsh near highest summit of Pine Mtn. Road ne. of Cloverdale, *Knight* (1963). < 10000 ft. June-Aug.

Hypericum concinnum Bentham. GOLD-WIRE. Common, dry brushy slopes: ca 1 mi. nw. of Camp Meeker, acc. J. & B. Guggolz (1991); Bouverie; Gates Road off Calistoga Road, acc. Wells et al. (1984); Pepperwood; Mountain Home Road near Petrified Forest, *Stocking* (1966 NCC); chaparral area sw. of Bradford Mtn. 1½ mi. w. of Dry Cr., acc. J. & B. Guggolz (1992); Skaggs Springs Road near Lake Sonoma, *Best* (1981); Lake Sonoma area; ca 2000 ft. elev. Anderson Trail on Black Mtn. between Geyser Peak and upper end of Pine Flat Road, acc. J. & B. Guggolz (1992); Geyser Peak, *Knight & Edwards* (1986 CM); probably en route to Mt. St. Helena from Fort Ross, *Voznesenskii* (1840-1841 LE); Geyserville, Alexander Valley and Ida Clayton Road quads.; Sonoma Geysers, *Brewer* [U]. < 3300 ft. May-July.

Hypericum formosum Humboldt, Bonpland & Kunth var. **scouleri** (Hooker) J. Coulter. Infrequent, wet meadows: Pitkin Marsh, *Rubtzoff* (1951); The Geysers, *Rattan* (1887). < 5000 ft. June-Aug. —*H. scouleri* Hooker.

★ **Hypericum perforatum** Linnaeus. KLAMATH WEED. Common, waste places: Stewarts Point-Skaggs Springs Road at the Annapolis turnoff, *Best* (1981); Salt Point; ca 1 mi. nw. of Camp Meeker, acc. J. & B. Guggolz (1991); Fife Cr. Marsh near Guerneville, *Rubtzoff* (1962); Russian River near Guerneville, *Rubtzoff* (1956); n. of Sebastopol on Hwy. 116 near Occidental Road, *Best* (1981); Annadel, *Wells* (1984); Pepperwood; w. of Porter Cr. Road, *Stocking,* (1960 NCC); Asti turnoff from US 101 on e. side of freeway, acc. Knight (1991); near Cloverdale, *Breedlove* (1962). < 5000 ft. June-Sept. Native of Europe.

CONVOLVULACEAE. MORNING-GLORY FAMILY

1. Styles 2, distinct or at least partly so; flowers < 1 cm long
  2. Ovary deeply 2-lobed; creeping, matted herb . . . . . *Dichondra*
  2: Ovary entire; erect herb . . . . . *Cressa*
1: Style entire, or cleft at apex only; flowers > 1 cm long
  3. Stigma oblong, ± cylindrical, the stigmatic area and style distinct; ovary with incomplete septum *Calystegia*
  3: Stigma linear, the stigmatic area and the style ± continuous; ovary with a complete septum . *Convolvulus*

Calystegia

1. Calyx enclosed or closely subtended by a pair of large sepal-like bracts
  2. Corolla rose-purplish . . . . . *C. soldanella*
  2: Corolla white to cream, occasionally pinkish in age
    3. Corolla 4-5 cm long; plant stemless or nearly so
      4. Plant pilose-pubescent; pedicels much shorter than petioles . . . . . *C. subacaulis*
      4: Plant ± tomentose; pedicels ca as long as petioles . . . . . *C. collina* ssp. *oxyphylla*
    3: Corolla 2.5-3.5 cm long; plants usually with stems 1-4 dm long; variously pubescent *C. malacophylla*
1: Calyx subtended by more remote, entire bracts that are not like sepals
  5. Plants puberulent to pubescent . . . . . *C. occidentalis*
  5: Plants glabrous
    6. Leaves sharp-pointed; flowers ochroleucous ca 3 cm long . . . . . *C. purpurata* ssp. *purpurata*
    6: Leaf-lobes obtuse or rounded; corolla white ca 4-5 cm long . . . . . *C. p.* ssp. *saxicola*

Calystegia collina (E. Greene) Brummitt ssp. **oxyphylla** Brummitt. MT. ST. HELENA MORNING-GLORY. Locally common, brushy and forest openings: summit between Mark West Valley and Franz Valley, *Ferris & Lorraine* (1948); The Geysers, *Cantelow* (1938); Pepperwood; Pine Flat Road 7.9 mi. from jct. with Hwy. 128 and 1 mi. sw. of Contact Mine, *J. & B. Guggolz* (1990 JEPS); 3200 and 3280 ft. elev. Socrates Mine, acc. Clifton (1984); ca 2000 ft. elev. Anderson Trail on Black Mtn. between Geyser Peak and upper end of Pine Flat Road, acc. J. & B. Guggolz (1992); 2440 ft. Caldwell Ranch, acc. Clifton (1984). < 3300 ft. Apr.-June. —*Convolvulus malacophyllus* E. Greene ssp. *collinus* (E. Greene) Abrams.

Calystegia malacophylla (E. Greene) Munz. Occasional, dry steep slopes: near Mt. Bradford in Dry Cr. drainage area, acc. Edwards (1986); ca 2000 ft. elev. Anderson Trail on Black Mtn. between Geyser Peak and upper end of Pine Flat Road, acc. J. & B. Guggolz (1992); Geyser Peak, acc. Knight & Edwards (1986); Geysers Unit 18, *Quibell* (1981 NCC). 1000-11600 ft. June-Aug. —*Convolvulus malacophyllus* E. Greene.

Calystegia occidentalis (A. Gray) Brummitt. Occasional, open brushy slopes, ridges: Salt Point; ca 1 mi. nw. of Camp Meeker, acc. J. & B. Guggolz (1991); Bodega Head; s. of Two Rock Military Station, *J. & B. Guggolz* (1990 JEPS); Bouverie; 1½ mi. up Cavedale Road from Hwy. 12, *Utech & Knight* (1986 CM); Sugarloaf Ridge; Pepperwood; Lake Sonoma area; 12 mi. from Calistoga on Alexander Valley Road, *Howell* (1934); ca 2000 ft. elev. Anderson Trail on Black Mtn. between Geyser Peak and the upper end of Pine Flat Road, acc. J. & B. Guggolz (1992); Sulphur Cr. Can., *Baker* (1932 NCC). < 4000 ft. May-July. —*C. polymorphus* E. Greene. *Convolvulus occidentalis* A. Gray.

Calystegia purpurata (E. Greene) Brummitt ssp. **purpurata**. Occasional, grassland, burned chaparral: Adobe Can., *Baker* (1934 NCC); s. slope Mt. Hood, *Baker* (1947 NCC); Green Hill Drive in Santa Rosa, *Stocking* (1960 NCC); 2860 ft. elev. Pine Flat Road, acc. Clifton (1984); 3600 ft. elev. Geyser Rock, acc. Clifton (1983); 2040 ft. Squaw Cr., acc. Clifton (1984); 3000 ft. elev. Caldwell Pines, acc. Clifton (1984). < 3300 ft. May-July. —*C. purpuratus* ssp. *solanensis* (Jepson) Brummitt. *Convolvulus luteolus* A. Gray var. *solanensis* Jepson.

Calystegia purpurata ssp. **saxicola** (Eastwood) Brummitt. Occasional, coastal: The Sea Ranch; 49 ft. elev. Gerstle Cove at Salt Point State Park, *Clifton* (1981 PUA); 8½ mi. n. of Bodega Bay, *Howell* (1930); Bodega Head. <350 ft. May-Aug. —*Convolvulus occidentalis* A. Gray var. *saxicola* (Eastwood) J. T. Howell. *Convolvulus saxicola* Eastwood, type locality: Bodega Point, *Eastwood* (1901). *Convolvulus purpuratus* E. Greene var. *s.* Jepson.

Calystegia soldanella (Linnaeus) R. Brown. BEACH MORNING-GLORY. Occasional, dunes: The Sea Ranch; Salt Point; mouth of Salmon Cr. *Howell* (1931); Doran Beach near Bodega Harbour development, *Huntington* (1989 NCC). < 165 ft. Apr.-Aug. —*Convolvulus s.* Linnaeus.

Calystegia subacaulis Hooker & Arnott. HILL MORNING-GLORY. Common, dry open areas: ca 1 mi. nw. of Camp Meeker, *J. & B. Guggolz* (1991 NCC); near Vine Hill School ca 5 mi. n. of Sebastopol, *Howell* (1955); w. of Santa Rosa in Laguna de Santa Rosa drainage, *Harrison* (1978 NCC); Todd Road Preserve (H&K); Lafferty Ranch on Sonoma Mtn., acc. Arnold et al. (1971); chaparral area sw. of Bradford Mtn. 1½ mi. w. of Dry Cr., acc. J. & B. Guggolz (1992); Lake Sonoma area; property bordered by Chianti and Dutcher Cr. roads w. of US 101 and w. of Geyserville, acc. J. & B. Guggolz (1991); serpentine on Pine Flat Road just e. of Deer Cr. crossing, *J. & B. Guggolz* (1989 NCC); ca 2000 ft. elev. Anderson Trail on Black Mtn. between Geyser Peak and upper end of Pine Flat Road, acc. J. & B. Guggolz (1992). < 2100 ft. Apr.-June. —*Convolvulus subacaulis* (Hooker & Arnott) E. Greene.

★ **Convolvulus arvensis** Linnaeus. BINDWEED, ORCHARD MORNING GLORY. Common, deep-rooted troublesome weed, orchards, fields, waste places: Duncans Mills quad.; Bodega Head; Freestone, *Huntington* (1987); Guerneville Marsh, *Rubtzoff* (1954); Pitkin Marsh, *Rubtzoff* (1951); Laguna de Santa Rosa near Occidental Road n. of Sebastopol, *Rubtzoff* (1958); W. College Ave., Santa Rosa, *Howell* (1945); Todd Road Preserve; Fairfield Osborn (S); Sugarloaf Ridge; Spring Lake; 11 mi. n. of Santa Rosa, *Eastwood & Howell* (1936); Pepperwood; Lake Sonoma area. < 5000 ft. May-Oct. Native of Europe.

Cressa truxillensis Kunth var. **vallicola** (Heller) Munz. ALKALI WEED. Uncommon, saline places: near Gilardi's Resort near Lakeville, *Knight* (1963); 5 mi. n. of Sears Point, *Rubtzoff* (1970). < 4000 ft. May-Oct.

Dichondra donelliana Tharp & M. C. Johnston. Occasional, coastal prairie, open brushy or grassy areas: Salt Point; 5 mi. n. of Fort Ross, *W. P. & A. G. Vestal* (1947); Joy Road near jct. of Bittner Road, *Wagnon* (1989 NCC); Bodega Head, *Best* (1988); 5 mi. w. of Sebastopol near Bodega Hwy. on Wagnon Road, *Wagnon* (1940 NCC); Spring Lake. < 1000 ft. Mar.-June. Acc. *McClintock* et al., *Fl. of San Bruno Mtns., pp. 101-102*, (1991), *Dichondra* is a poorly understood genus of about 14 species, occurring mostly in N. and S. America, New Zealand and Australia. It is best known in California by the cultivated lawn substitute that appears to be *D. micrantha* Urban but generally has been called *D. repens* J. R. & G. Forster. The name *D. repens* has also been used for the California plants now called *D. donelliana*.

CORNACEAE. DOGWOOD FAMILY

Cornus

1. Flowers in a bractless cyme, appearing with or after the leaves
   2. Veins 3-4 on each side of midrib; leaf-blades 3-5 cm long . . . . *C. glabrata*
   2: Veins 4-7 on each side of midrib; leaf-blades mostly 5-10 cm long
      3. Style < 2.5 mm long; petals 2-3 mm long; cyme and lower surfaces of leaves variously pubescent but not hirsute; endocarp smooth, usually at least as long as broad . . . . *C. sericea* ssp. *sericea*
      3: Style 2.5 mm long or more; petals 3-4 mm long; cyme and lower surface of leaves densely hirsute; endocarp ridged, mostly broader than long . . . . *C. sericea* ssp. *occidentalis*

1: Flowers in a head subtended by 4-7 persistent, white, petaloid bracts 4-6 cm long, 3-6 cm broad, appearing at anthesis . . . . *C. nuttallii*

Cornus glabrata Bentham. BROWN DOGWOOD. Occasional, moist places: Pitkin Marsh, *Rubtzoff* (1952); Laguna de Santa Rosa near Trenton, *Rubtzoff* (1960); Franz Valley Road in Knights Valley, acc. de Nevers (1992); Alexander Valley; Skaggs Springs, *Hawver* (1915); Lake Sonoma area. < 5000 ft. May-June.

Cornus nuttallii Audubon. MOUNTAIN DOGWOOD. Occasional, mountain woods: Stewarts Point-Skaggs Springs Road 10 mi. e. of Hwy. 1, *Braden* (1970 NCC); Cazadero, *Graff* (1922); St. Helena Road 1-2 mi. from county line, acc. Vale (1990); Ida Clayton Road quad.; 2404 ft. elev. w. flank Mt. St. Helena, *Grummer* (1981 PUA). < 7200 ft. Apr.-July.

Cornus sericea Linnaeus ssp. **occidentalis** (Torrey & A. Gray) Fosberg. WESTERN DOGWOOD. Occasional, moist areas: Bohemian Hwy. 1 mi. n. of Freestone, *Best* (1981); Spring Lake; Pepperwood. < 8000 ft. May-July. —*C. o.* (Torrey & A. Gray) Coville.

Cornus sericea ssp. **sericea**. AMERICAN DOGWOOD. Occasional, moist places: Willow Creek; Pitkin Marsh, *Howell* (1947, 1945); Cunningham Marsh, *Howell* (1946); Fairfield Osborn (S); Bouverie; Pepperwood; ca 2000 ft. elev. Anderson Trail on Black Mtn. between Geyser Peak and upper end of Pine Flat Road, acc. J. & B. Guggolz (1992). < 9000 ft. May-July. —*C.* × *californica* C. Meyer. *C. stolonifera* Michaux.

CRASSULACEAE. STONECROP FAMILY
(Key by Reid Moran)

1. Plants slender annuals; lowermost leaves or all leaves opposite, not in rosettes, < 7 mm long
   2. Leaves all opposite, connate at base; flowers in leaf axils; petals < 2 mm long; stamens as many as sepals . . . . *Crassula*
   2: Leaves mostly alternate, only the lower opposite, all distinct; flowers in terminal cymes; petals 2.5-4.5 mm long; stamens twice as many as sepals . . . . *Parvisedum*

1: Plants perennial herbs; leaves alternate, often in rosettes, often larger; stamens twice as many as sepals
   3. Flowering stems terminal; petals spreading, separate . . . . *Sedum*
   3: Flowering stems lateral from leaf rosettes; petals erect forming a tube united at base . . . . *Dudleya*

Crassula

1. Sepals acute to acuminate, usually longer than petals; flowers usually 2 per leaf-pair; follicles 1-2-seeded; terrestrial
   2. Sepals generally 4, erect; flowers sessile or pedicelled . . . . *C. connata*
   2: Sepals 3 with tip often outcurved; flowers subsessile . . . . *C. tillaea*

1: Sepals rounded to obtuse, much shorter than petals; flowers 1 per leaf-pair; follicles 6-17-seeded; aquatic, usually becoming stranded in dry ponds
   3. Follicles oblong, subtruncate, suture erect, straight, abruptly outcurved in upper one-fourth . *C. aquatica*
   3: Follicles lanceolate, obliquely acute, suture outcurved from near middle . . . . *C. solieri*

Crassula aquatica (Linnaeus) Schoenlein. Occasional, open damp places: low flat area between Westside Road and cliffs of Bodega Head near Bodega Harbor, *Connors* (1992 BML); sites w. of Santa Rosa in Laguna de Santa Rosa drainage, *Harrison* (1978); Button Ranch on Walker Road (U. C. property) between Petaluma-Valley Ford Road and Pepper Road, acc. Patterson (1990); Laguna de Santa Rosa e. of Sebastopol, *Rubtzoff* (1958, 1966); Todd Road 1.4 mi. from Hwy. 116 and 0.6 mi. from Llano Road,

Rubtzoff (1974); 6½ mi. n. of Penngrove on Lichau Road, *Rubtzoff* (1969); Chileno Valley, *Rubtzoff* (1972); 4¼ mi. w. of Glen Ellen on Bennett Valley Road, *Rubtzoff & Froebe* (1972); Bennett Mtn. Lake, *Rubtzoff* (1970); Spring Lake; Pepperwood. < 10000 ft. Mar.-July. —*Tillaea aquatica* Linnaeus.

Crassula connata (Ruiz López & Pavón) A. Berger. SAND PYGMY-WEED. Locally common, open places: The Sea Ranch; Salt Point; Bodega Head, *Best* (1987); Button Ranch on Walker Road (U. C. property) between Petaluma-Valley Ford Road and Pepper Road, acc. Patterson (1990); ca 8 mi. from Petaluma toward Sonoma, *Koch* (1939 NCC); w. of Sears Point Raceway in Tolay Cr. drainage, acc. Patterson (1990); Bouverie; Sugarloaf Ridge; Spring Lake; near Mark West Springs, *Heller & Brown* (1902); Pepperwood; Lake Sonoma area. < 5000 ft. Feb.-May. —*Tillaea erecta* Hooker & Arnott.

Crassula solieri (C. Gay) F. Meigen. Rare, pools, vernal situations: Sonoma Valley Regional Park, *Moran & Quibell* (1991 NCC). < 1650 ft. Mar.-July.

★ **Crassula tillaea** Lester-Garland. Rare, open places: low flat area between Westside Road and cliffs of Bodega Head near Bodega Harbor, *Connors* (1992 BML); near e. gate of Spring Lake Park, *Wagnon* (1992 NCC). < 1650 ft. Mar.-Apr. Native of Europe.

Dudleya

1. Plants ± glaucous; peduncle 4-25 cm; pedicel usually 5-15 mm . . . . *D. cymosa*
1: Plants farinose or green; peduncle 10-35 cm; pedicel 1-3 mm . . . . *D. farinosa*

Dudleya cymosa (Lemaire) Britton & Rose. Occasional, rocky steep slopes: ca 1 mi. nw. of Camp Meeker, acc. J. & B. Guggolz (1991); Rio Nido between Guerneville and Hacienda, *Lieberman* (1978 NCC); Button Ranch on Walker Road (U. C. property) between Petaluma-Valley Ford Road and Pepper Road, acc. Patterson (1990); Bouverie; lower part of Cavedale Road near monument, acc. Quibell (1991); Pine Flat Road ca 5 mi. n. of Hwy 128 intersection, *J. & B. Guggolz* (1990); between Jimtown and Mercuryville on Geysers Road, *Utech & Knight* (1986 CM); ca 2000 ft. elev. Anderson Trail on Black Mtn. between Geyser Peak and upper end of Pine Flat Road, acc. J. & B. Guggolz (1992); Geyser Peak, acc. Knight & Edwards (1986); 3200 ft. elev. Socrates Mine, acc. Clifton (1984). 150-9000 ft. Apr.-June.

Dudleya farinosa (Lindley) Britton & Rose. Locally common, coastal: The Sea Ranch; Gualala Point State Park (green form), *Knight* (1985); Salt Point; Duncan's Landing, acc. Knight (1986); Bodega Head; ½ mi. at s. side of entrance of Bodega Harbour development, *Huntington* (1989 NCC). < 500 ft. May-Sept. —*D. eastwoodiae* Rose, type locality: Bodega Point, *Eastwood* (1901). *Cotyledon e.* (Rose) Fedde.

Parvisedum pumilum (Bentham) R. T. Clausen. Locally occasional, rocky areas, vernal pools: e. side of Hwy. 12 ca ¼ mi. n. of Agua Caliente on outcrop of tuff, *Baker* (1952 NCC); 0.9 mi. off Hwy. 12 near Kenwood on private road in volcanic rock, *M. Bess* (1980 NCC); w. side of Buzzard Peak near Mt. Hood, *J. & B. Guggolz* (1991 NCC). < 4200 ft. Mar.-May. —*Sedella pumila* (Bentham) Britton & Rose.

Sedum

1. Plants annual or biennial; leaves oblong to narrowly ovate, becoming thin, dry, membranaceous *S. radiatum*
1: Plants perennial; leaves spatulate; remaining fleshy, glaucous . . . . *S. spathulifolium*

Sedum radiatum S. Watson. Rare, rocky places: Pepperwood; Stewarts Point-Skaggs Springs Road 4 mi. w. of Dry Cr. Road, *Ersland* (1973 NCC). < 7000 ft. June-Aug. —*S. stenopetalum* Pursh ssp. *r.* (S. Watson) R. T. Clausen.

Sedum spathulifolium Hooker. STONECROP. Common, rocky ledges: Stewarts Point-Skaggs Springs Road, *Immel* (1970); Willow Creek; St. Helena Road 1½ mi. from county line, acc. Knight (1990); Pepperwood; 0.3 airmile from confluence of Rancheria Cr. and Warm Springs Cr., *Immel* (1970); 1800 ft. elev. unit 14, *Malloch* (1977 PG&E). < 8300 ft. May-July.

CUCURBITACEAE. GOURD FAMILY

1. Fruit fleshy, smooth; flowers yellow-orange; corolla campanulate, 5-lobed . . . . *Cucurbita*
1: Fruit not fleshy, ± spinose; flowers white or cream . . . . *Marah*

★ **Cucurbita pepo** Linnaeus var. **medullosa** Alefeld. ZUCCHINI. Rare, escape from cultivation: Copeland Cr. between Country Club Drive and Snyder Lane, Rohnert Park, *Self* (1977 NCC); Lake Sonoma area. The common, garden plant with elongated, green fruits. Native of Mexico.

Marah

1. Mature corolla rotate, cream . . . . *M. fabaceus*
1: Mature corollas campanulate
2. Plants with ♂ flowers < 8 mm in diameter; leaves 3-8 cm broad, orbicular, 5-cleft, the lobes mostly obtuse; corolla greenish; fruit globose . . . . *M. watsonii*
2: Plants with ♂ flowers > 8 mm in diameter; leaves 8-30 cm broad, with broad rounded sinuses between the 5-7 ± shallow lobes; corolla white; fruit ovoid, attenuate both ends . . . . *M. oreganus*

Marah fabaceus (Naudin) E. Greene. var. **fabaceus**. WILD CUCUMBER, CALIFORNIA MAN-ROOT. Common, slopes, embankments: Salt Point; Duncans Mills quad.; mouth of Salmon Cr., *Knowlton* (1937 NCC); Bodega Head; Bodega Point, *Eastwood* (1915); Bodega Bay, *Heller & Brown* (1902); Fairfield Osborn (S); Bouverie; Sugarloaf Ridge; Spring Lake; Santa Rosa Cr., Santa Rosa Cr., *Heller & Brown* (1902); Pepperwood; Lake Sonoma area. < 5300 ft. Feb.-Apr.

Marah oreganus (Torrey & A. Gray) T. J. Howell. COAST MAN-ROOT. Common, canyons, hilly areas: The Sea Ranch; Salt Point; 98 ft. elev. Fort Ross, *Clifton* (1981 PUA); Willow Creek; s. of W. College Ave. in Laguna de Santa Rosa drainage, *Harrison* (1978 NCC); 50 yds. from top of Sonoma Mts., *Lowry* (1983 SFSU); w. face of Mt. Hood, *G. Smith & Stocking* (1960 NCC); 6 mi. e. of Santa Rosa on e. side of Rincon Valley, *Mallory* (1920); Franz Valley, Geysers-Healdsburg Road, *Braslin* (1975 NCC). < 6000 ft. Mar.-May.

Marah watsonii (Cogniaux) E. Greene. Acc. *Munz* (1959) reported to be in Sonoma County. < 4000 ft. Mar.-Apr.

CUSCUTACEAE. DODDER FAMILY

Cuscuta

1. Capsule globose, ± depressed
  2. Corolla with scalelike appendages attached to the tube below the stamens, corolla-lobes recurved
    3. Flowers 5-merous; capsule having a withered corolla about the base; corolla-lobes acute, spreading but with tips inflexed . . . . *C. pentagona*
    3: Flowers 4-merous; capsule having withered corolla at the top; corolla-lobes with long attenuate recurved tips . . . . *C. howelliana*
  2: Corolla lacking scalelike appendages at base of stamens
    4. Anthers oval; corolla-lobes spreading, starlike in fruit; flowers sessile . . *C. californica* var. *breviflora*
    4: Anthers linear-oblong; corolla not star-shaped; flowers pedicelled . . . . *C. c.* var. *californica*
1: Capsule ovoid, conic or beaked, commonly longer than thick
  5. Calyx shorter than the corolla-tube; flowers ca 5 mm long; calyx-lobes overlapping . . . . . *C. subinclusa*
  5: Calyx = or > corolla-tube; flowers mostly smaller, calyx-lobes not markedly overlapping . . . *C. salina*
    6. Perianth papillate; flowers ± 3 mm . . . . *C. s.* var. *papillata*
    6: Perianth not papillate; flowers 2-4.5 mm
      7. Flowers 3-4.5 mm; coastal salt marshes . . . . *C. s.* var. *major*
      7: Flowers 2-3 mm; inland salt flats . . . . *C. s.* var. *salina*

Cuscuta californica Hooker & Arnott var. **breviflora** Engelmann. Uncommon: Bodega Dunes State Park on *Ericameria*, *Connors* (1993 BML); Petaluma River on *Grindelia*, *E. Adams* (1969 NCC). < 8300 ft. June-Aug. —*C. occidentalis* Millspaugh.

Cuscuta californica var. **californica**. Uncommon, open areas: ca 1 mi. nw. of Camp Meeker, *J. & B. Guggolz* (1991 NCC); Occidental Road near Facendini Lane, *Wagnon* (1989 NCC); Pepperwood. < 8300 ft. May-Aug.

Cuscuta howelliana Rubtzoff. BOGGS LAKE DODDER. Occasional, vernal situations, hosts=*Lasthenia, Eryngium* and *Downingia*: Annadel at False Lake Meadow, *Harrison* (1989 SRJC); Windsor sewer project near Sonoma Co. airport, host=*Lasthenia glaberrima, Harrison* (1981 NCC); Arata Lane, Windsor, host= *Lasthenia burkei* acc. Potter & Noonan (1986); Arata Lane, Windsor, hosts=*Lasthenia, Eryngium* and *Downingia*, *Harrison* (1986 SRJC). < 2700 ft. May-Aug.

Cuscuta pentagona Engelmann. Occasional, on many hosts, including *Grindelia:* Laguna de Santa Rosa ne. of Graton ca ½ mi. n. of Guerneville Road crossing, *Rubtzoff* (1959); Laguna de Santa Rosa at crossing of Occidental Road, *Baker* (1946 NCC); near Trenton s. of Woolsey Road, *Rubtzoff* (1962); Forestville Marsh, *Rubtzoff* (1955); Laguna de Santa Rosa n. of Sebastopol at Occidental Road crossing, *Howell* (1946); Willow Brook at Stony Point Road, *Rubtzoff* (1968); parasitic on *Grindelia* at Petaluma River salt marsh, *Adams* (1969). < 1600 ft. July-Nov. —*C. campestris* Yuncker.

Cuscuta salina Engelmann var. **major** Yuncker. Rare, coastal salt marshes: Bodega Bay area, *Clifton* (1980 PUA). Usually < ± 350 ft. May-Sept.

Cuscuta salina var. **papillata** Yuncker. Salt flats, saline ponds: Bodega Head, acc. Connors (1991). Usually < 350 ft. May-Sept.

Cuscuta salina var. **salina**. Uncommon, usually inland salt flats: lower Tubbs Island near San Pablo Bay, *Knight* (1985); on *Salicornia virginica* in Bodega Bay Salt Marsh, acc. J. & B. Guggolz (1992). Usually < ± 350 ft. May-Sept.

Cuscuta subinclusa Durand & Hilgard. Occasional: Laguna de Santa Rosa e. of Sebastopol on *Symphoricarpos, Rubus, Toxicodendron,* and *Polygonum*, *Rubtzoff* (1958); off Los Alamos Road near end of Corrick Road, *Knight* (1964); Preston, ca 3 mi. n. of Healdsburg on *Salix, Carlson* (1917). < 5300 ft. June-Oct. —*C. ceanothi* Behr.

DATISCACEAE. Datisca family

Datisca glomerata (C. Presl) Baillon. Durango root. Locally common, seeps, dry stream beds: Vacation Beach downstream from Guerneville, *Rubtzoff* (1959); e. Guernewood near Guerneville, *Rubtzoff* (1955); Fife Cr. Marsh at Guerneville, *Rubtzoff* (1962); 5½ mi. e. of Guerneville at Forest Hills, *Rubtzoff* (1968); Los Guilicos, *Bioletti* (1893 JEPS); Spring Lake; Pepperwood; Preston n. of Healdsburg, *J. I. Carlson* (1917); along Maacama Cr. near confluence of Russian River near Biggio Road, *Wagnon* (1988 NCC); Lake Sonoma area; dry stream bed near bridge on Pine Flat Road, *W., I. & T. Knight & Edwards* (1985); 2190 ft. elev. Squaw Cr., acc. Clifton (1984). < 6600 ft. May-July.

DIPSACACEAE. Teasel family

Dipsacus

1. Bracts of involucre curving upwards, the longest = or > the head; bracts of receptacle exceeding the flowers, ending in long straight slender spine . . . . *D. fullonum*

1: Bracts of involucre spreading ± horizontally, frequently not overtopping the head; those of receptacle almost = to flowers, ending in a stiff recurved spine . . . . *D. sativus*

★ **Dipsacus fullonum** Linnaeus. Wild teasel. Uncommon: Gualala Point State Park, *Knight* (1984). < 5700 ft. Apr.-Aug. Native of Eurasia. —*D. sylvestris* Hudson.

★ **Dipsacus sativus** (Linnaeus) Honckeny. Fuller's teasel. Common, weed, roadsides, disturbed areas:The Sea Ranch, *Wells* (1981); Salt Point; corner D St. Ext. and San Antonio Road en route to Point Reyes Station, *Knight* (1963); Sonoma Valley, *Krause* (1893); ¼ mile s. of jct. of Hwy. 12 and Glen Ellen Road, *Ruckert* (1960); Rincon Valley, *Wells* (1985); Bennett Valley, *Knight* (1963); Bennett Valley Road near Enterprise Road, *Stocking* (1963 NCC); Spring Lake; Pepperwood; Healdsburg, *Rattan* (1877). < 2700 ft. May-July. Naturalized from Europe. —*D. fullonum*, misapplied. Fruit head long used to raise nap on woolen cloth.

DROSERACEAE. Sundew family

Drosera rotundifolia Linnaeus. Round-leaved sundew. Rare, swamps, sphagnum bogs: Pitkin Marsh, *Howell* (1945). Probable farthest s. location in the Coast Ranges. < 7200 ft. July-Aug.

ELATINACEAE. WATERWORT FAMILY

Elatine

1. Flowers 4-merous; stamens 8 . . . . *E. californica*
1: Flowers 3-merous; stamens 3 or varying from 1 to 6
   2. Stamens 3, alternate with the carpels . . . . *E. brachysperma*
   2: Stamens 3-6, when 3, opposite the carpels . . . . *E. heterandra*

Elatine brachysperma A. Gray. Rare, ditches, vernal pools: Pepperwood. —*E. obovata* (Fassett) H. Mason. 100-1700 ft. Apr.-Sept.

Elatine californica A. Gray. WATERWORT. Rare, mud flats, water borders: Todd Road area of Laguna de Santa Rosa drainage, *Harrison* (1978 NCC); laguna in Chileno Valley, *Rubtzoff* (1972); 5 mi. n. of Sears Point, *Rubtzoff* (1970); Pepperwood. 50-6400 ft. Mar.-Aug.

Elatine heterandra H. Mason. Occasional, mud of vernal swales, water ca 1 ft. deep: Laguna de Santa Rosa e. of Graton, *Rubtzoff* (1966); 1 site sw. and 3 sites w. of Santa Rosa in Laguna de Santa Rosa drainage, *Harrison* (1978 NCC); Button Ranch on Walker Road (U.C. property) between Petaluma-Valley Ford Road and Pepper Road, acc. Connors (1992); laguna at Chileno Valley, *Rubtzoff* (1972); Bennett Mtn. Lake, *Rubtzoff* (1964). < 5000 ft.

ERICACEAE. HEATHER FAMILY

1. Plant not green; leaves scale-like . . . . *(Monotropoideae)*
   2. Ovary 1 loculed; fruit a berry
      3. Petals fused; anthers elongate, dehiscent by separate slits . . . . *Hemitones*
      3: Petals free, anthers horseshoe shaped, dehiscent by 1 unified slit . . . . *Pityopus*
   2: Ovary 4-5 loculed; fruit a capsule . . . . *Allotropa*
1: Plants green; leaves expanded (except for some Pyrola)
   4. Plants not woody; leaves basal or on short stems, alternate to whorled . . . . *(Pyroloideae)*
      5. Flowers 1-few in an umbel-like raceme . . . . *Chimaphila*
      5: Flowers few-many in an elongate raceme . . . . *Pyrola*
   4: Plants woody, leaves cauline, alternate
      6. Ovary inferior; fruit a berry . . . . *Vaccinium*
      6: Ovary superior, fruit a berry, drupe or capsule
         7. Fruit a capsule enclosed in a fleshy calyx . . . . *Gaultheria*
         7: Fruit not enclosed in a calyx
            8. Fruit a berry-like drupe
               9. Ovary and fruit smooth . . . . *Arctostaphylos*
               9: Ovary and fruit papilate . . . . *Arbutus*
            8: Fruit a capsule
               10. Fruit dehiscent from the base to the tip; petals free . . . . *Ledum*
               10: Fruit dehiscent from the tip to the base; corolla bell-to-funnel-shaped . . . . *Rhododendron*

Allotropa virgata A. Gray. SUGAR STICK. Occasional, thick humus: Alliance Redwood Camp, 6250 Bohemian Hwy., *Van Soelen* (1977 NCC); near Camp Meeker, *Baker* (1933 NCC); 1100 ft. elev. Devil's Kitchen, Pepperwood, *de Nevers* (1991). < 10000 ft. June-Aug.

Arbutus menziesii Pursh. PACIFIC MADRONE, MADROÑO. Common, forest situations: The Sea Ranch; Salt Point; near Occidental, acc. Knight (1991); Bouverie; Harrison Grade Road, acc. Knight (1991); Pocket Canyon Road between Forestville and Guerneville, acc. Knight (1990); Tresch Ranch, Walker Road near Two Rock, acc. J. & B. Guggolz (1992); Sugarloaf Ridge; Mt. Hood, *Knight* (1964); Spring Lake; Howarth Park in Santa Rosa, *Knight* (1985); Pepperwood; between Alpine Road and Petrified Forest Road on Calistoga Road, *Knight* (1963); chaparral area sw. of Bradford Mtn. 1½ mi. w. of Dry Cr., acc. J. & B. Guggolz (1992); Lake Sonoma area; 3080 ft. elev. Socrates Mine, acc. Clifton (1984); ca 2000 ft. elev. Anderson Trail on Black Mtn. between Geyser Peak and upper end of Pine Flat Road, acc. J. & B. Guggolz (1992); 2040 ft. elev. Squaw Cr., acc. Clifton (1984); 2440 ft. elev. Caldwell Ranch, acc. Clifton (1984); 3000 ft. elev. Caldwell Pines, acc. Clifton (1984); in all quads. except Bodega Head. < 5000 ft. Mar.-May.

Arctostaphylos

1. Plants with basal burl
  2. Plants totally without glands . . . . *A. glandulosa* fma. *cushingiana*
  2: Plants with glands . . . . *A. glandulosa* fma. *glandulosa*
1: Plants without basal burl
  3. Leaves pallid or whitish
    4. Leaves glabrous, bracts deltoid-acuminate . . . . *A. viscida* ssp. *pulchella*
    4: Leaves canescent, bracts foliaceous, lanceolate, campanulate
      5. Fruit usually pubescent, sometimes glabrate, not viscid . . . . *A. canescens* ssp. *canescens*
      5: Fruit viscid-pubescent . . . . *A. c.* ssp. *sonomensis*
  3: Leaves green
    6. Flowers 4-merous . . . . *A. nummularia*
    6: Flowers 5-merous
      7. Leaves lance-ovate to oblanceolate, ± elliptic, acute at both ends, smooth
        8. Inflorescence rachises thin
          9. Shrubs erect
            10. Leaves bright green, shiny; inflorescence axis glabrous or puberulent, not glandular . . . . *A. stanfordiana* ssp. *stanfordiana*
            10: Leaves slightly glaucous; inflorescence axis and bracts finely glandular-bristly . . . . *A. s.* ssp. *raichi*
          9: Shrubs procumbent . . . . *A. s.* ssp. *decumbens*
        8: Inflorescence rachises medium . . . . *A. hispidula*
      7: Leaves otherwise
        11. Branches prostrate; Stump Beach at Salt Point St. Pk. . . . . *A. uva-ursi* fma. *coactilis*
        11: Branches not prostrate, sprawling to erect
          12. Branchlets densely cinereous-pubescent with long bristly hairs many of which may be gland-tipped . . . . *A. columbiana*
          12: Branchlets otherwise
            13. Leaves ± variable in size and shape, larger ones 2.5-3 cm long, 1.5-2 cm broad, usually broadly ovate with truncate or rounded base
              14. Twigs, inflorescence axis, and petiole densely finely glandular-bristly; leaf-blade glandular, rough; petiole 3-6 mm; immature inflorescence axis 10-15 mm . . . . *A. bakeri* ssp. *bakeri*
              14: Twigs, inflorescence axis, and petiole usually puberulent to fine tomentose and sparsely glandular; leaf-blade not glandular, sparsely puberulent, smooth or slightly rough; petiole 4-8 mm, immature inflorescence axis 15-20 mm . . . . *A. b.* ssp. *sublaevis*
            13: Leaves generally ovate; eglandular throughout
              15. Plants subarborescent 3-6 m tall
                16. Fruit eglandular
                  17. Leaves bright green . . . . *A. manzanita* ssp. *manzanita*
                  17: Leaves glaucescent . . . . *A. m.* ssp. *glaucescens*
                16: Fruit evanescently glandular . . . . *A. m.* ssp. *elegans*
              15: Plants mostly erect-sprawling, infrequently erect but not tall or subarborescent . . . . *A. densiflora*

Arctostaphylos bakeri Eastwood ssp. **bakeri**. BAKER'S MANZANITA. Locally common, dry areas of serpentine: 2 mi. nw. of Camp Meeker, acc. to J. & B. Guggolz (1991); near Camp Meeker, *Eastwood & Howell* (1939), *Huntington* 1987 NCC); 2 mi. e. of Occidental at Facendini Lane and Occidental Road, *Howell* (type 1934); 492 ft. elev. Harrison Grade Road, *Clifton* (1981 PUA); w. end of Sweetwater Springs Road near old mine, acc. Knight (1986); Reservoir Hill, Healdsburg, *Raiche* (1986 NCC), actually not good Baker's Manzanita, acc. the late J. B. Roof. —*A. manzanita* Parry ssp. *b.* (Eastwood) P. V. Wells. 600-2000 ft. Feb.-Apr.

Arctostaphylos bakeri ssp. **sublaevis** P. V. Wells. THE CEDARS MANZANITA. Locally occasional, serpentine: The Cedars on divide above Big Austin Cr., *P. V. Wells* (type 1987 UC), acc. Raiche (1985). < 2500 ft. Feb.-Apr.

Arctostaphylos canescens Eastwood ssp. **canescens**. HOARY MANZANITA. Common, sandstone or shale ridges and slopes, chaparral: between Fort Ross and Stewarts Point, *Howell* (1934); both Trinity and Cavedale roads, acc. Knight (1989); Sugarloaf Ridge; Mt. Hood, *Knight* (1964); Rincon Ridge, *Knight* (1984); geysers Unit 18, *Quibell* (1980 NCC); 2400 ft. elev. Little Geysers, *Malloch* (1977 PG&E); ca

½ mi. ne. of Eureka Mine, *Crane* (1980 PG&E); w. margin of Little Geysers, *Crane* (1980 PG&E). Hybridizes with *A. viscida*: The Geysers Resort, *Knight* (1968); Los Alamos Road e. of Santa Rosa, *Knight* (1963). < 5000 ft. Jan.-May.

Arctostaphylos canescens Eastwood ssp. **sonomensis** (Eastwood) P. V. Wells. SONOMA MANZANITA. Rare, chaparral: Montecito Ave. near corner of Chanate Road, acc. Knight (1963); Fountain Grove II on Rincon Ridge, *J. & B. Guggolz* (1991 UC); Rincon Ridge near Santa Rosa, *Baker* (type 1933). < 4000 ft. Jan.-Apr. —*A. s.* Eastwood.

Arctostaphylos columbiana Piper. HAIRY MANZANITA. Locally common, usually coastal, dry rocky or clayey slopes, edges of woods: The Sea Ranch, *McKenzie* (1981); near Timber Cove, *Howell* (1934). < 2700 ft. Mar.-May. A cross with *A. uva-ursi* appears at Havens Neck, ca 4 mi. into Mendocino Co. and also it appears sporadically on road cuts n. of Anchor Bay along Hwy. 1 in Mendocino Co. during the last 2 decades, acc. Knight (1989). This hybrid was described by E. Greene as *A. media, Pittonia 2:171*, (1891). < 500 ft. Mar.-June.

Arctostaphylos densiflora M. S. Baker. VINE HILL MANZANITA. Locally occasional, sandy soil, shale outcrops; endagered: Vine Hill region n. of Sebastopol. A strange assortment of manzanitas of varying sizes, some with basal burls and some without, grow in the area (especially along Guerneville Road between Laguna and Vine Hill roads) as a result of hybridization. < ± 400 ft. Mar.-Apr.

Arctostaphylos glandulosa Eastwood forma **cushingiana** (Eastwood) P. V. Wells. CUSHING MANZANITA. Common, ridges and slopes in loose chaparral interspersed with forest: ca 1000 ft. elev. Cavedale Road, acc. Knight (1990); Trinity Grade, *Howell* (1933); between Alpine and Porter Cr. roads on Calistoga Road, *Knight* (1963); Pepperwood, acc. Knight (1988); chaparral area sw. of Bradford Mtn. 1½ mi. w. of Dry Cr., acc. J. & B. Guggolz (1992); 3300 and 3400 ft. elev. at Squaw Cr., acc. Clifton (1983). < 4300 ft. Jan.-Apr. —*A. c.* Eastwood. *A. g.* var. *c.* (Eastwood) Adams.

Arctostaphylos glandulosa ssp. **glandulosa**. EASTWOOD MANZANITA. Infrequent, slopes and ridges, usually at forest edges or in loose chaparral: Bouverie; Los Alamos Can., *Baker* (1928 NCC); Rincon Ridge, *Howell* (1931); Pepperwood; Lake Sonoma area; Kelly Road between Lake Sonoma and Annapolis, *Knight & Edwards* (1985); Mercuryville on Geysers Road, acc. Knight (1986); ¼ mi. ne. of Eureka Mine, *Crane* (1980 PG&E); substantial population in area around pipe gate at end of Pine Flat Road, acc. Knight (1990); 3080 and 3200 ft. elev. Socrates Mine, acc. Clifton (1984); ca 2000 ft. elev. Anderson Trail on Black Mtn. between Geyser Peak and upper end of Pine Flat Road, acc. J. & B. Guggolz (1992); 3300 ft. elev. crest of Mayacmas Mts. at Caldwell Pines, acc. Stone (1989). < 4000 ft. Jan.-Apr.

Arctostaphylos hispidula T. J. Howell. ROUGH MANZANITA. Locally occasional, barrens, loose chaparral: along Rockpile and Kelly roads, both collections by *Knight & Edwards* (1985). 750-2000 ft. Mar.-Apr. —*A. stanfordiana* Parry ssp. *h.* (T. J. Howell) Adams.

Arctostaphylos manzanita Parry ssp. **elegans** (Jepson) P. V. Wells. Sparsely occasional, soils derived from andesitic or quartz basaltic parent material: 4298 ft. elev. Mt. St. Helena lookout, *Grummer* (1978 PUA); 2440 ft. Caldwell Ranch, acc. Clifton (1984); 3300 ft. elev. Sonoma-Lake co. line, n. of Cobb Mtn. on crest of Mayacmas Mts. at Caldwell Pines, *Stone* (1989 JEPS). 1200-4900 ft. Mar.-May. —*A. e.* Jepson.

Arctostaphylos manzanita ssp. **glaucescens** P. V. Wells. GLAUCESCENT MANZANITA. Common, woodlands, grassland: Annadel; Sonoma Mtn. area, acc. Knight (1990); Russian River drainage from Windsor northward; Annapolis, Fort Ross, Cazadero, Sebastopol, Healdsburg, Lake Sonoma, Geyserville, Alexander Valley, Ida Clayton Road, Socrates Mine, Geothermal and Big Sulphur Cr. quads; chaparral area sw. of Bradford Mtn. 1½ mi. w. of Dry Cr., acc. J. & B. Guggolz (1992); prevalent along Pine Flat Road at ca 1600 ft. elev. Frequently it is scattered within populations of the species. < 2500 ft. Feb.-Apr.

Arctostaphylos manzanita ssp. **manzanita**. COMMON MANZANITA. Common, grassland, chaparral: Bouverie; Sonoma Mtn. Road near Enterprise Road, *Knight* (1964); Trinity Road, *Best* (1981); Mt. Hood, *Knight* (1964); 1200 ft. elev. Calistoga Road, *Clifton* (1981 PUA); between Spring Mtn. and Porter Cr. roads on Calistoga Road, *Knight* (1963); Pepperwood; Mill Cr. Road w. of Healdsburg, *Knight* (1964); Lake Sonoma area; ca 2000 ft. elev. Anderson Trail on Black Mtn. between Geyser Peak and upper end of Pine Flat Road, acc. J. & B. Guggolz (1992); Pine Mtn. Road, *Knight* (1963); 2040 ft. elev. Squaw Cr., acc. Clifton (1984); 2440 ft. elev. Caldwell Ranch, acc. Clifton (1984). < 4000 ft. Feb.-Apr.

Arctostaphylos manzanita × A. stanfordiana. Dense population of glossy-leaved, treelike, non-burl-forming manzanitas extending for ½ mi. along Annapolis Road near jct. of Old Quarry Road, near county refuse disposal site, 625 ft. elev., *Knight* (1990 NCC). The population needs more study. Maybe it needs a name! 2000 ft. ±. Feb.-Apr.

Arctostaphylos nummularia A. Gray. SHATTERBERRY, FIRE MANZANITA. Locally occasional, coastal terraces, pygmy forest conditions: ridge between Hwy. 1 and Gualala River at The Sea Ranch (population now almost extirpated due to recent logging), *Wells* (1984); prostrate and erect shrubs at Stump Beach in Salt Point State Park, *Knight* (1984). < 2000 ft. Mar.-May. —*A. sensitiva* (Jepson) Munz.

Arctostaphylos nummularia × **A. uva-ursi**. Stump Beach area at Salt Point, acc. Knight (1989). < 250 ft. Mar.-May.

Arctostaphylos stanfordiana C. Parry ssp. **decumbens** P. V. Wells. RINCON MANZANITA. Infrequent, chaparral ridges, oak woodland growing with the species: w. slope of Buzzard Peak in Mt. Hood Range, *J. & B. Guggolz* (1992 NCC); on Rincon Ridge between Fountaingrove Parkway and Montecito Drive, *Knight* (1990 NCC); ca 1050 ft. elev. Dry Cr. area at base of sw. shoulder of Bradford Mtn., *Raiche* (1984); chaparral area sw. of Bradford Mtn. 1½ mi. w. of Dry Cr., acc. J. & B. Guggolz with W. Knight (1992). ± 400 ft. Feb.-Apr. —*A. s.* var. *repens* Roof, Rincon Ridge, *Roof* (type 1972 CAS).

Arctostaphylos stanfordiana ssp. **raichei** W. Knight. Allegedly found in small populations from The Cedars southward to Austin Cr. Recreation Area on serpentine. 1000-3300 ft. Feb.-Apr.

Arctostaphylos stanfordiana ssp. **stanfordiana**. STANFORD MANZANITA. Common, chaparral, slopes, ridges: near Cazadero, *Raiche* (1982); Bouverie; Cavedale Road, *Knight* (1964); Sugarloaf Ridge; w. slope of Buzzard Peak in Mt. Hood Range, *J. & B. Guggolz* (1992 NCC UC); between Alpine and Porter Cr. roads on Calistoga Road, *Knight* (1963); 3 mi. ne. of central Santa Rosa at Rincon Ridge, *Howell* (1931); Pepperwood; Franz Valley School Road, acc. Lovell (1978); ca 1050 ft. elev. Dry Cr. area at base of sw. shoulder of Bradford Mtn., acc. J. & B. Guggolz & W. Knight (1992); St. Helena Road, acc. Quibell (1991); Ida Clayton Road, *Stein* (1981); Kortum Can. Road near county line, acc. Quibell (1991); 4298 ft. elev. Mt. St. Helena lookout, *Grummer* (1978 PUA); chaparral area sw. of Bradford Mtn. 1½ mi. w. of Dry Cr., acc. J. & B. Guggolz (1992); ca 2000 ft. elev. Anderson Trail on Black Mtn. between Geyser Peak and upper end of Pine Flat Road, acc. J. & B. Guggolz (1992); along road between geysers Units 14 and 18, *Crane* (1980 PG&E); PG&E geysers Unit 18, *Quibell* (1980 NCC); 2400 ft. elev. Little Geysers, *Malloch* (1977 PG&E). 300-4500 ft. Feb.-Apr.

Arctostaphylos uva-ursi (Linnaeus) Sprengel forma **coactilis** (Fernald & J. F. Macbride) P. V. Wells. BEARBERRY. Rare, littoral terraces: Stump Beach in Salt Point State Park, *Knight* (1984). < 10,000 ft. Mar.-June.

Arctostaphylos viscida Parry ssp. **pulchella** (T. J. Howell) P. V. Wells. WHITE-LEAVED MANZANITA. Common, serpentine, chaparral: in The Cedars, acc. Knight & Edwards (1985); Sonoma Cr. Can. near Mt. Hood, *Howell* (1933); Sugarloaf Ridge; Foote's Ravine at n. end of Knights Valley, *Knight* (1964); end of Pine Flat Road, acc. Knight & Edwards (1984); 3290 ft. elev. Socrates Mine, acc. Clifton (1984); PG&E geysers Unit 13, *Quibell* (1981 NCC); geysers Unit 17 at 3200 ft., *Malloch* (1977 PG&E). 500-3300 ft. Feb.-Apr. —*A. pulchella* T. J. Howell.

Chimaphila

1. Leaves lanceolate to elliptic; inflorescence 1-3-flowered; bracts persistent . . . . *C. menziesii*
1: Leaves oblanceolate; inflorescence 3-10-flowered; bracts deciduous . . . . *C. umbellata*

Chimaphila menziesii (D. Don) Sprengel. PIPSISSEWA, LITTLE PRINCE'S PINE. Rare, shady woods: Stewarts Point quad.; 2900 ft. elev. w. flank Mt. St. Helena, *Grummer* (1978 PUA). < 8300 ft. June-Aug.

Chimaphila umbellata (Linnaeus) Bartram. PRINCE'S PINE. Rare, dry, forest edges: 2000 ft. elev. near geysers Unit 18, *Malloch* (1977 PG&E). < 9700 ft. June-Aug. —*C. u.* ssp. *occidentalis* (Rydberg) S. F. Blake.

Gaultheria shallon Pursh. SALAL. Locally common, exposed maritime headlands, wooded and brushy slopes: The Sea Ranch; Rockpile Cr. and Annapolis quads.; 6 mi. n. of Stewarts Point, *Knight* (1964); Stump Beach at Salt Point State Park, *Utech* (1988 CM); Vine Hill region, acc. J. B. Roof (1980); Sugarloaf Ridge. < 2700 ft. Apr.-July.

Hemitomes congestum A. Gray. GNOME PLANT. Uncommon, in humus situations: The Sea Ranch, acc. *Mahaffey* (1990); Salt Point, acc. C. Tarp. No substantiating herbarium specimens. < 9000 ft. May-July. —*Newberrya congesta* Torrey.

Ledum glandulosum Nuttall. WESTERN LABRADOR TEA. Locally occasional, coastal, swampy areas: mouth of Gualala River quad.; The Sea Ranch; Rockpile Cr. quad.; Salt Point, *Utech* (1988 CM); 5 mi. n. of Fort Ross near Hwy. 1, *Cantelow* (1941); Pitkin Marsh, *Rubtzoff* (1951). < 2700 ft. Apr.-July. —*Rhododendron neoglandulosum* Harmaja.

Pityopus californicus (Eastwood) H. Copeland. CALIFORNIA PINEFOOT. Uncommon, deep shade: 8 mi. e. of Fort Ross at Murphy Mill (probably near Mill Gulch), *Bowerman* (1954 JEPS). < 6000 ft. May-July.

Pyrola picta Smith. SHINLEAF, WHITE VEINED WINTERGREEN. Occasional, wooded slopes: Annapolis quad.; near Cazadero, *Charles* (1925); Camp Meeker at Grub Cr., *Harrison* (1977 SRJC); bottom of Harrison Grade Road under bay trees, acc. Quibell (1991); Pocket Canyon between Forestville and Guerneville, *Gilbert* (1980 SRJC); Pepperwood; 2300 ft. elev. near geysers Unit 18, *Malloch* (1977 PG&E). < 8000 ft. June-Aug. —*P. aphylla* Smith.

Rhododendron

1. Leaves persistent, leathery; flowers rose; stamens 10 . . . . *R. macrophyllum*
1: Leaves deciduous, not leathery; flowers white to cream or with some pink or pale orange; stamens 5 . . . . *R. occidentale*

Rhododendron macrophyllum D. Don. CALIFORNIA ROSE-BAY. Locally common, coastal, dryish to damp hillslopes, forest shade: The Sea Ranch; Rockpile Cr. and Stewarts Point quads.; Salt Point, *Utech* (1988 CM); Kruse Rhododendron State Preserve. < 3700 ft. Apr.-July. —*R. californicum* Hooker.

Rhododendron occidentale (Torrey & A. Gray) A. Gray. WESTERN AZALEA. Occasional: stream banks, moist places: The Sea Ranch at end of Deer Trail Road, *Wells* (1984); Annapolis, *Sutliffe* (1920); 2 mi. n. of Camp Meeker along creek, acc. J. & B. Guggolz (1991); Graton Road 2.7 mi. e. of Occidental, *Rubtzoff* (1957); Russian River between Vacation Beach and e. Guernewood, *Rubtzoff* (1959); Forestville Marsh, *Rubtzoff* (1955); Pitkin Marsh, *Rubtzoff* (1951); Laguna de Santa Rosa e. of Graton, *Rubtzoff* (1960); creek 1 mi. sw. of Cunningham, *Quibell* (1982 NCC); Cunningham Marsh, *Rubtzoff & Harrison* (1978); 4¼ mi. ne. of Bloomfield, *Rubtzoff* (1970); end of Gehricke Road n. of Sonoma, acc. Knight & Edwards (1984); Bouverie; 1½ mi. up Cavedale Road from Hwy. 12, *Utech & Knight* (1986 CM); Sugarloaf Ridge; Pepperwood; Gates Road off of Calistoga Road, acc. Wells et al. (1984). Var. *sonomense* (E. Greene) Rehder, is based on shrub with smaller leaves and rose to whitish flowers with salmon spot in center (*R. sonomense* E. Greene, type locality: Mt. St. Helena, *E. Greene*): 1500 and 2400 ft. elev. w. flank Mt. St. Helena, acc. Grummer (1981), acc. Grummer & Hemphill (1983); 1800 ft. elev. along a tributary of Big Sulphur Cr. upstream from Unit 14, *Malloch* (1977 PG&E). < 7300 ft. Apr.-Aug. —*Azalea occidentalis* Torrey & A. Gray, type locality: Laguna de Santa Rosa, *Bigelow* (1854 U).

Vaccinium

1. Leaves thick and leathery, persistent; filaments hairy; flowers in racemes . . . . *V. ovatum*
1: Leaves thin, deciduous; filaments glabrous; flowers 1 to 2 or 4, in the axils
    2. Branchlets terete or nearly so; leaves gradually narrowed toward base . . . . *V. cespitosum*
    2: Branchlets angled; leaves abruptly narrowed or rounded toward base . . . . *V. parvifolium*

Vaccinium cespitosum Michaux. DWARF BILBERRY. Rare, damp areas: Perry Marsh, *Baker* (1940). < 10500 ft. May-June. —*V. arbuscula* (A. Gray) Merriam.

Vaccinium ovatum Pursh. CALIFORNIA HUCKLEBERRY. Common, mostly coastal, dry slopes, canyons: mouth of Gualala River quad.; The Sea Ranch on Timber Ridge, acc. Knight (1990); Annapolis, acc. Knight (1985); Stewarts Point quad.; Stump Beach at Salt Point State Park, acc. Knight (1990); Willow Creek; 2 mi. nw. of Camp Meeker, acc. J. & B. Guggolz (1991); Sugarloaf Ridge; Mill Cr. Road w. of Healdsburg, *Knight* (1964). < 2700 ft. Mar.-May. Var. *saporosum* Jepson, currently not recognized.

Vaccinium parvifolium Smith. RED HUCKLEBERRY. Occasional, shaded woods: The Sea Ranch, acc. Wells (1985); Salt Point; in redwoods 1 mi. n. of Fort Ross, *Howell* (1930); s. of Clipper Mill Bridge along Skaggs Springs Stewarts Point Road, acc. Lovell (1981). < 4700 ft. May-June.

EUPHORBIACEAE. SPURGE FAMILY

1. Flowers not subtended by involucre simulating calyx; perianth present in the ♂ flowers, absent in the ♀ flowers . . . . . *Eremocarpus*
1: Flowers ♀ and ♂ surrounded by one involucre simulating a calyx; perianth none or present as a single scale
   2. Glands of involucre without petaloid appendanges; leaves alternate (except *E. lathyris*) . . . . . *Euphorbia*
   2: Glands of involucre with petaloid appendages or, if these wanting, the leaves all opposite and in equilateral bases . . . . . *Chamaesyce*

Chamaesyce (Principally adapted from key of D. Koutnik.)

1. Involucre and fruit hairy; stem and leaves generally hairy or glabrate
   2. Fruit > 1.5 mm; seed > 1 mm, rugose; gland appendage narrower than gland *C. serpyllifolia* ssp. *hirtula*
   2: Fruit < 1.5 mm; seed < 1 mm, ± wrinkled; gland appendage = to gland width . . . . *C. maculata*
1: Involucre and fruit glabrous; stem and leaves glabrous (rarely hairy)
   3. Stipules fused into broad membranous scale . . . . . *C. serpens*
   3: Stipules separate
      4. Leaf margins entire . . . . . *C. ocellata*
      4: Leaf margins finely toothed, at least toward the tip . . . . . *C. serpyllifolia* ssp. *serpyllifolia*

★ **Chamaesyce maculata** (Linnaeus) Small. SPOTTED SPURGE. Very common, weed: 5½ mi. e. of Guerneville at Forest Hills; *Rubtzoff* (1965); Sonoma State Univ. campus, *Howell & Best* (1986); Kentucky Street, Petaluma, *Knight* (1963); Rincon Valley, *Wells* (1985); Spring Lake Village; 2 mi. e. of Santa Rosa, *Howell* (1926); Pepperwood. < 700 ft. Apr.-Oct. Native, Atlantic Coast to Mississippi Valley and S. America. —*Euphorbia m.* Linnaeus.

Chamaesyce ocellata (Durand & Hilgard) Millspaugh. Rare, dry sandy open places: Sonoma, *Behr* (1860s U). < 1650 ft. May-Oct. —*Euphorbia o.* Durand & Hilgard.

★ **Chamaesyce serpens** (Kunth) Small. Uncommon, flower beds: Royal Oaks Mobile Home Park, *Knight* (1987 NCC); Spring Lake Village. < 700 ft. Apr.-Oct. N. & S. America —*Euphorbia s.* Kunth.

Chamaesyce serpyllifolia (Persoon) Small var. **hirtula** (S. Watson) Koutnik. Uncommon, dry places: Spring Lake. < 8300 ft. Aug.-Oct. —*Euphorbia s.* Persoon var. *h.* (S. Watson) Wheeler.

Chamaesyce serpyllifolia var. **serpyllifolia**. THYME-LEAVED SPURGE. Common, dry disturbed places: Austin Cr. Marsh, *Rubtzoff* (1954); Vacation Beach near Guerneville, *Rubtzoff* (1958); Forest Hills near Hacienda, *Rubtzoff* (1961); Laguna de Santa Rosa e. of Graton, *Rubtzoff* (1965 NCC); Laguna de Santa Rosa n. of Sebastopol on Occidental Road, *Rubtzoff* (1958); Laguna de Santa Rosa n. of Cunningham, *Rubtzoff* (1958); s. of Piner Road in Laguna de Santa Rosa drainage, *Harrison* (1978 NCC); Todd Road Preserve; on road between Cotati and Santa Rosa, *Eastwood* (1921); vernal pool 6½ miles ne. of Penngrove, *Rubtzoff* (1964); Magnolia Ave. near cemetery w. of Petaluma, *Knight* (1963); Rincon Valley, *Wells* (1985); Spring Lake; Santa Rosa Cr. e. of Santa Rosa, *Heller* (1902); bed of Santa Rosa Cr., *Baker* (1931 NCC); Howarth Park, Santa Rosa, *Knight* (1964); Healdsburg, *Rattan* (1877). < 8300 ft. Aug.-Oct. —*Euphorbia s.* Persoon.

Eremocarpus setigerus (Hooker) Bentham. DOVE WEED, TURKEY MULLEIN. Common, open areas: The Sea Ranch, acc. Wells (1981); Salt Point; Fort Ross and Cazadero quads.; Willow Cr. Marsh, *Rubtzoff* (1959); ca 1 mi. nw. of Camp Meeker, acc. J. & B. Guggolz (1991); Guerneville Marsh, *Rubtzoff* (1954); Todd Road Preserve (H&K); corner Todd Road and US 101, *Knight* (1963); along banks of Copeland Cr., Rohnert Park, *Self* (1979 NCC); Fairfield Osborn (S); Sugarloaf Ridge; Spring Lake; Mark West Cr. n. of Santa Rosa, *Heller* (1902); Pepperwood; s. Fitch Mtn. Road, *Rubtzoff* (1978); chaparral area sw. of Bradford Mtn. 1½ mi. w. of Dry Cr., acc. J. & B. Guggolz (1992); Lake Sonoma area; ca 2000 ft. elev. Anderson Trail on Black Mtn. between Geyser Peak and upper end of Pine Flat Road, acc. J. & B. Guggolz (1992). < 3300 ft. May-Oct.

Euphorbia

1. Leaves opposite, 4-ranked; fruit > 8 mm . . . . . *E. lathyris*
1: Leaves alternate; fruit < 6 mm
   2. Leaf margin finely toothed; gland margin entire or truncate
      3. Plants perennial, seed surface smooth . . . . . *E. oblongata*
      3: Plants annual, seed surface sculptured, netlike . . . . . *E. spathulata*

2: Leaf margin entire; gland horned or scalloped
4. Leaves linear, sometimes oblanceolate; seed surface smooth . . . . *E. esula*
4: Leaves obovate to ovate or elliptic; seed surface dotted or sculptured, netlike
5. Seed dotted; fruit 2-keeled on lobes . . . . *E. peplus*
5: Seed sculpture low, netlike; fruit lobes smooth . . . . *E. crenulata*

Euphorbia crenulata Engelmann. CHINESE CAPS. Occasional, dry places: Sonoma Can. near Kenwood, *Jepson* (1923 JEPS); 6 mi. from Petrified Forest on road to Knights Valley, *Howell* (1929); Lake Sonoma area. < 5300 ft. Mar.-Aug.

★ **Euphorbia esula** Linnaeus. LEAFY SPURGE. Rare, waste places: Stewarts Point, acc. *Barbe* (1991); Franz Valley School Road, *de Nevers* (1992). < 4700 ft. June-July. Native of Europe.

★ **Euphorbia lathyris** Linnaeus. GOPHER PLANT, CAPER SPURGE. Common, weed, gardens, damp waste places: Salmon Cr. w. of Highway, *Best* (1986); Bodega Dunes, *Connors* (1991 BML); Bodega Ave. near Paula Lane, Petaluma, *Knight* (1963). < 700 ft. Most of year. Naturalized from Europe.

★ **Euphorbia oblongata** Grisebach. OBLONG SPURGE. Common, waste places: 14 locations throughout the county, acc. *Barbe* (1991). < 700 ft. Native of Eurasia.

★ **Euphorbia peplus** Linnaeus. PETTY SPURGE. Frequent, gardens, moist places: Timber Cove, *McKenzie* (1981); Stewarts Point, *Baker* (1932 NCC); Fife Cr. Marsh, *Rubtzoff* (1961); Gravenstein Hwy. near Sebastopol, *Stocking* (1966); Royal Oaks Mobile Home Park, acc. Knight (1990); Spring Lake; Pepperwood; Lake Sonoma area. < 1000 ft. Feb.-Aug. Naturalized from Europe.

Euphorbia spathulata Lamarck. WART SPURGE. Occasional, open disturbed areas: Bohemian Hwy. 3 mi. n. of Occidental, *Best* (1981); between Sonoma and Boyes Hot Springs, *Howell* (1929); Pepperwood. < 4300 ft. Mar.-June. —*E. dictyosperma* Fischer & C. Meyer, type locality: Bodega Bay.

FABACEAE [LEGUMINOSAE]. PEA FAMILY

Key ± acc. *Howell* (1949).

1. Corolla regular, inconspicuous; stamens numerous, conspicuous
2. Stamens distinct; flowers in globose heads < 1 cm in diameter . . . . *Acacia*
2: Stamens united near base; flowers in oblong spikes about 2.5 cm across . . . . *Albizia*
1: Corolla irregular or papilionaceous, usually conspicuous, stamens 10, usually concealed in the keel by the two lower petals at anthesis (in *Amorpha* only the banner is present)
3. Corolla irregular; leaves cordate to reniform, ± leathery (flowers present before leaves) . . . . . . *Cercis*
3: Corolla papilionaceous; leaves not cordate to reniform, nor ± leathery
4. Leaflets > 3 and palmate . . . . *Lupinus*
4: Leaflets none or 1 to many, pinnate if > 3
5. Plants shrubs and trees
6. Leaflets > 5
7. Leaves punctate-glandular, odorous; corolla reduced to purple banner; stipules not spiny . . . . *Amorpha*
7: Leaves not punctate-glandular or odorous; stipules frequently spiny (except *Sesbania*)
8. Corolla white; pod flat . . . . *Robinia*
8: Corolla entirely burnt orange; pod 4-winged . . . . *Sesbania*
6: Leaflets none or 1-5
9. Stems spinescent
10. Leaves with 1-3 leaflets; corolla bright purplish-rose . . . . *Pickeringia*
10: Leaves with mature plants reduced to a spiny petiole; corolla yellow . . . . *Ulex*
9: Stems not spinescent
11. Flowers 1 cm long or less, in sessile or shortly pedunculate umbels; legumes indehiscent . . . . *Lotus*
11: Flowers 1 cm long or more, not in umbels; legumes dehiscent
12. Branchlets strongly angled or pubescent; calyx 2-lipped
13. Style usually abruptly curved ± at middle; upper lip of calyx barely 2-lobed . . . . *Cytisus*
13: Style ± abruptly bent at tip; upper lip of calyx strongly 2-lobed *Genista*
12: Branchlets terete, subglabrous, calyx 1-lipped, split deeply above . . *Spartium*
5: Plants annual and perennial herbs
14. Flowers solitary
15. Stipules reduced to small or inconspicuous glands . . . . *Lotus*

15: Stipules not reduced to glands
16. Style tipped by a tuft or ring of hairs . . . . *Vicia*
16: Style hairy only on concave side . . . . *Lathyrus*
14: Flowers 2 to many (sometimes 1 in *Lathyrus aphaca*)
17. Flowers in sessile or pedunculate umbels; leaflets entire . . . . *Lotus*
17: Flowers in 2's, racemes, spikes, or heads, or if umbellate then the leaflets denticulate
18. Leaflets 0-3
19. Leaflets 0, stipules appearing leafletlike . . . . *Lathyrus aphaca*
19: Leaflets 3
20. Herbage and calyces glandular-punctate
21. Corolla at least partly blue to purple; calyx not swollen in fruit . . . . *Hoita*
21: Corolla white to yellow; calyx conspicuously swollen in fruit . . . . *Rupertia*
20: Herbage and calyces not glandular-punctate
22. Leaflets entire; stamens distinct; stipules large, foliaceous; corolla bright yellow . . . . *Thermopsis*
22: Leaflets usually denticulate; one stamen free, the rest united
23. Corolla withering, persistent . . . . *Trifolium*
23: Corolla deciduous
24. Inflorescence contracted or dense; style subulate . . . . *Medicago*
24: Inflorescences elongate, racemose; style filiform . . . . *Melilotus*
18: Leaflets > 3, or in *Lathyrus* and *Pisum* sometimes only 2, leaf then terminated by tendril
25. Leaves odd-pinnate
26. Fruit not prickly . . . . *Astragalus*
26: Fruit prickly, ± hooked . . . . *Glycyrrhiza*
25: Leaves pinnate, usually terminated by a tendril or the tendril rudimentary
27. Style tipped by a tuft or ring of hairs . . . . *Vicia*
27: Style hairy only on concave side
28. Stipules usually smaller than the leaflets (or large in *Lathyrus littoralis*); style not conspicuously flattened and folded *Lathyrus*
28: Stipules about = in size to leaflets; style conspicuously flattened and folded . . . . *Pisum*

Acacia

1. Leaves reduced to phyllodes, phyllodes with 2 or more main veins from base
2. Phyllodes lanceolate to oblanceolate, somewhat sickle-shaped; flowers in heads; phyllodes 12-25 mm wide . . . . *A. melanoxylon*
2: Phyllodes oblong-lanceolate, both margins curved outward; flowers in spikes; phyllodes 11-16 mm wide . . . . *A. longifolia*
1: Leaves bipinnately compound, leaves with 8-25 pairs of pinnae . . . . *A. dealbata*

★ **Acacia dealbata** Link. SILVER WATTLE. Common, adventive: near Analy High School in Sebastopol, *Koch* (1938 NCC); Llano Road, *Howell* (1986); 4.8 mi. up Cavedale Road, *Knight* (1984); Trinity Road, *Best* (1982), *Knight & Howell* (1984); Spring Lake; 3½ mi. from Santa Rosa courthouse toward Healdsburg, *Koch* (1938 NCC). < 1600 ft. Feb.-Mar. Native of se. Australia. —*A. decurrens* Willdenow var. *dealbata* (Link) F. Mueller.

★ **Acacia longifolia** (Andrews) Willdenow. SYDNEY GOLDEN WATTLE. Uncommon, escaped ornamental: Bodega Head. < 500 ft. May. Native of e. Australia.

★ **Acacia melanoxylon** R. Brown. BLACKWOOD ACACIA. Uncommon, spontaneous from seed of nearby cultivated plants: sprouts, Sonoma State Univ. campus, *Glasner* (1987 NCC); Alta Vista Lane on Rincon Ridge, Santa Rosa, *Knight* (1984). < 800 ft. May-June. Native of e. Australia.

★ **Albizia lophantha** (Willdenow) Bentham. PLUME ACACIA. Uncommon, escape: Lichau Road off Petaluma Hill Road, *Stocking* (1967 NCC). < 1000 ft. Early Spring. Native of sw. Australia. —*A. distachya* (Ventenat) J. F. Macbride.

Amorpha californica Nuttall var. **napensis** Jepson. FALSE INDIGO. Common, dry brushy or wooded slopes: Cazadero en route to Seaview, *Brandegee* (1893), *Bacigalupi* (1927); 1200 ft. elev. 4½ mi. w. of Cazadero on Fort Ross Road, *Atkinson* (1978 NCC), (1980 NCC); Duncans Mills, *Curran* (1886); Austin Cr., *Davy* (1894); hillside near redwood glade on Russian River, *Tryner* (1938 NCC); road from Guerneville to Monte Rio, *McMinn* (1922), *G. Johnson* (1931 NCC); Norrbom Road near Sonoma, acc. Knight & Edwards (1985); Bouverie; Gates Road off Calistoga Road, acc. Wells et al.(1984); Los Alamos Road, *Knight* (1963); Burke's Sanitarium near Mark West Springs, *Heller* (1902); Mark West Springs, *Mexia* (1929); Pepperwood; Knights Valley, *Breedlove* (1980); 8.2 mi. e. of Windsor, *Vestal* (1947). < 4000 ft. May-July.

Astragalus

1. Plants annual with small flowers, the keel not > 8 mm long; racemes capitate, subumbellate or, if racemose, then not > 10-flowered, nor > 3 cm long in fruit
    2. Pod < 5 mm long, broadly ovate or suborbicular in dorsal view, transversely wrinkled, 2-locular, mostly 2-seeded;flowers spreading or nodding at anthesis, at length racemose; pod deflexed, obcompressed, the valves spreading from an open dorsal groove and incurved at base toward the concavely arched ventral suture . . . . . *A. gambelianus*
    2: Pod larger, either strongly turgid or elongate, several-seeded; the flowers racemose or subumbellate; keel-petals surpassed by the wings or, if exserted, then the herbage green and glabrate; pod linear or narrowly oblong or, if body ovoid, then the beak subulate and nearly as long
        3. Keel-petals much shorter than the wings; pod variable in shape, sessile
            4. Pod fertile in the lower half, this ovoid-oblong, abruptly passing into a narrow subulate-acuminate beak nearly as long, silky-strigulose . . . . . *A. breweri*
            4: Pod fertile along the middle 2/3 or nearly the whole length, linear-oblong to narrowly lunate, sparsely strigulose or glabrate
                5. Fruit on erect peduncles; seeds pitted . . . . . *A. tener*
                5: Fruit on spreading or reflexed peduncles; seeds smooth . . . . . *A. rattanii*
        3: Keel-petals longer and much wider than the wings; pod linear, attenuate at both ends and elevated on a slender gynophore 1.5-2.5 mm long . . . . . *A. clarianus*

1: Plants perennial; leaflets of at least some developed leaves (especially those subtending the peduncles) > 23 in number; calyx pedicelled; flowers racemose, or if subsessile then both flowers and pods deflexed
    6. Stipules, at least the lowermost 1-3 (often leafless) nodes, ± united opposite the petiole
        7. Flowers large, the banner > 1 cm long; pod 2.5-5 cm long . . . . . *A. nuttallii* var. *virgatus*
        7: Flowers smaller, the banner 8-10 mm long; pod not > 1 cm long . . . . . *A. pycnostachyus*
    6: Stipules all free from each other opposite the petiole, or so early ruptured by the growing stem as to appear so
        8. Flowers very small and numerous, 20-95 to the raceme, reflexed, the banner 5-6 mm long; racemes very narrow and elongate, the axis 1-3 dm long . . . . . *A. clevelandii*
        8: Flowers fewer or, if nearly as numerous, then banner at least 9 mm long; racemes shorter . . . . . *A. nuttallii* var. *virgatus*

Astragalus breweri A. Gray. SONOMA LOCO. BREWER'S MILKVETCH. Occasional, meadows, grassy hillsides, often on serpentine: Bodega Bay; Cazadero en route to Seaview, *Brandegee* (1893), *Bacigalupi* (1927); 1200 ft. elev. 4½ mi. w. of Cazadero on Fort Ross Road, *Atkinson* (1978 NCC), (1980 NCC); Duncans Mills, *Curran* (1886); Austin Cr., *Davy* (1894); hillside near redwood glade on Russian River, *Tryner* (1938 NCC); road from Guerneville to Monte Rio, *McMinn* (1922), *G. Johnson* (1931 NCC); Norrbom Road near Sonoma, acc. Knight & Edwards (1985); Bouverie; Gates Road off Calistoga Road, acc. Wells et al.(1984); Los Alamos Road, *Knight* (1963); Burke's Sanitarium near Mark West Springs, *Heller* (1902); Mark West Springs, *Mexia* (1929); Pepperwood; Knights Valley, *Breedlove* (1980); 8.2 mi. e. of Windsor, *Vestal* (1947). < 2400 ft. Apr.-May.

Astragalus clarianus Jepson. CLARA HUNT'S MILKVETCH. Uncommon, grassy hillsides: field opposite Alpine School near Calistoga Road, *Grummer* (1986 PUA); volcanic clay soil 7 mi. ne. of Santa Rosa, *Barneby* (1954); St. Helena Road, acc. Lovell (1978). < 900 ft. Apr.-May.

Astragalus clevelandii E. Greene. CLEVELAND'S MILKVETCH. Mentioned in C.N.P.S. Inventory, acc. *Skinner & Pavlik* (1994) as being in some adjacent counties but there are no known available specimens. Would be expected in wet serpentine areas. 600-5000 ft. June-Sept.

Astragalus gambelianus E. Sheldon. Occasional, grassy hillsides, sometimes shady: hill back of Baker house in Adobe Can., *Baker* (1925 NCC); hills e. of Santa Rosa, *Heller & Brown* (1902); Pepperwood; Lake Sonoma area. 150-3000 ft. Mar.-June.

Astragalus nuttallii (Torrey & A. Gray) J. T. Howell var. **virgatus** (A. Gray) Barneby. Uncommon, coastal strand: Salt Point, *Wells* (1982). < 500 ft. Apr.-July.

Astragalus pycnostachyus A. Gray. Salt marshes and creeks near the ocean: acc. *Munz p. 877*, (1959), to be expected from San Mateo to Humboldt counties. < 100 ft. June-0ct.

Astragalus rattanii A. Gray. RATTAN'S MILKVETCH. Uncommon, sandbars along streams: Sulphur Cr., *Baker* (1930).
100-2700 ft. Apr.-July.

Astragalus tener A. Gray. Uncommon, grassy alkaline flats: Petaluma, *Congdon* (1880). < 200 ft, Mar.-June.

Cercis occidentalis Torrey. WESTERN REDBUD, JUDAS TREE. Occasional, foothills, dry slopes, canyons: a planted specimen with 6 trunks at 100-6th St., Petaluma, measures 40 ft. tall and 45 ft. wide; Geysers-Healdsburg Road, *Breslin* (1975 NCC); Geyser Peak, *Knight & Edwards* (1986); grade between Black Mtn. and Geyser Peak, *McMurphy* (1909); Little Sulphur Cr., *Geis* (1902). 300-5000 ft. Mar.-May.

★ **Cytisus scoparius** (Linnaeus) Link forma **scoparius**. SCOTCH BROOM. Common, roadsides: Annapolis quad.; Salt Point; Hwy. 1 near Timber Cove, *McKenzie* (1980); Moscow Road between Monte Rio and Duncans Mills, acc. Knight (1990); ca 1 mi. nw. of Camp Meeker, acc. J. & B. Guggolz (1991); Harrison Grade Road, acc. Quibell (1991); railroad crossing s. end of Graton, *L. L. Davis* (1975 NCC); (wing of inflorescence red) ca 150 yds. w. of jct. of Watertrough Road on n. side of Bodega Hwy., *W. I. & T. Knight* (1992 NCC); jct. Pepper Lane and Pepper Road in Stony Point area, *Stocking* (1963 NCC); Adobe Can., acc. Baker (1936); Pepperwood; Healdsburg, Geyserville, Alexander Valley and N. Fork Buckeye Cr. quads. < 3300 ft. Apr.-June. Forma **andreanus** (Puissant) Zabel has pink flowers instead of yellow: Adobe Can., acc. *Baker* (1954). < 500 ft. Apr.-June. Native of Europe.

★ **Genista monspessulana** (Linnaeus) L. Johnson. FRENCH BROOM. Common, roadsides: The Sea Ranch; Hwy. 1 near Salt Point, acc. Knight (1990); Annapolis Road between The Sea Ranch and Annapolis, acc. Knight (1990); Stewarts Point-Skaggs Springs Road, acc. Knight (1990); Timber Cove area on Hwy. 1, *McKenzie* (1981); Bodega Head; ca 1 mi. nw. of Camp Meeker, acc. J. & B. Guggolz (1991); Harrison Grade Road, acc. Quibell (1991); along US 101 between San Antonio Cr. and Petaluma, *Howell* (1981); Cavedale Road, acc. Knight (1990); Pythian Road near Los Guilicos School, *Wiley* (1965 NCC); Joost Ranch on Calistoga Road, *Knight* (1963); Gates Road off Calistoga Road, acc. Wells et al. (1984); Spring Lake; Pepperwood; chaparral area sw. of Bradford Mtn., acc. J. & B. Guggolz (1992); Pine Flat Road, acc. Knight (1990): ca 2000 ft.elev. Anderson Trail on Black Mtn. between Geyser Peak and upper end of Pine Flat Road, acc. J. & B. Guggolz (1992). < 1700 ft. Mar.-May. Native of Mediterrranean and Azores. —*Cytisus monspessulanus* Linnaeus.

Glycyrrhiza lepidota Pursh. WILD LICORICE. Occasional, moist waste places: Vacation Beach 1½ mi. down Russian River from Guerneville, *Rubtzoff* (1959 NCC); Russian River at Summerhome Park, *de Nevers* (1981); Forest Hills 5½ mi. upstream from Guerneville across from Hacienda, *Rubtzoff* (1959 NCC); Healdsburg, *Rattan* (1877); Lake Sonoma area (at dam site before construction of dam). < 8000 ft. May-July.

Hoita

1. Stem prostrate (including stolon) to decumbent, erect parts < 6.5 cm; leaflets obovate to round; flower 9-10 mm . . . . . *H. orbicularis*

1: Stem erect, < 2 m; leaflets lanceolate to round; flower 12-23 mm . . . . . *H. macrostachya*

Hoita macrostachya (de Candolle) Rydberg. Common, moist places: Occidental, *Howell* (1936); marsh w. of Graton toward Occidental, *Baker* (1931 NCC); Laguna de Santa Rosa, *Rubtzoff* (1962 NCC); 3 mi. up Cavedale Road from Hwy. 12, *Knight* (1984); Spring Lake; bridge over Santa Rosa Cr. at Santa Rosa Ave., *H. A. Thomas* (1934 NCC); Burke's Sanitarium (near Mark West Springs), *Heller* (1902); Mark West Springs Road 2 mi. from US 101, *Knight* (1963); Pepperwood; 10½ mi. w. of Skaggs Springs, *Raven* (1958); w. flank Mt. St. Helena, *Grummer* (1981 PUA); ca 2000 ft. elev. Anderson Trail on Black Mtn. between Geyser Peak and upper end of Pine Flat Road, acc. J. & B. Guggolz (1992); Geysers Road near Little Sulphur Cr. bridge, acc. Lovell (1981); Caldwell Ranch, acc. Clifton (1984). < 8300 ft. May-Aug. —*Psoralea m.* de Candolle var. *rhombifolia* Torrey.

Hoita orbicularis (Lindley) Rydberg. Common, moist places: Pitkin Marsh, *Rubtzoff* (1951); Perry Marsh, *Howell* (1945); Laguna de Santa Rosa, *Rubtzoff* (1962); Lichau Marsh (probably Fairfield Osborn), *Rubtzoff* (1968); Petaluma, *Congdon* (1880). < 7300 ft. May-July. —*Psoralea o.* Lindley.

Lathyrus

1. Leaflets 0; stipules leafletlike . . . . *L. aphaca*
1: Leaflets present; stipules not leafletlike
  2. Leaflets 2; plants introduced
    3. Plants annual
      4. Flowers < 1.5 cm long
        5. Peduncles 1-flowered; plants glabrous
          6. Stem 20-100 cm, winged
            7. Peduncles 10-30 mm; flowers reddish-purple . . . . *L. cicera*
            7: Peduncles 30-60 mm; flowers white, pink or blue . . . . *L. sativus*
          6: Stem 10-50 cm, not winged
            8. Flowers orange-red; pod 30-70 mm long x 4-7 mm wide . . . . *L. sphaericus*
            8: Flowers purple or pale blue; pod 25-50 mm long × 3-4 mm wide . . . . . *L. angulatus*
        5: Peduncles usually 2-flowered, sometimes 1-; plants hirsute at least about the nodes and each hair with bulbous base on fruit; stems winged . . . . *L. hirsutus*
      4: Flowers > 1.5 cm long
        9. Plants glabrous: leaflets linear to lance-linear; flowers deep purple . . . . *L. tingitanus*
        9: Plants rough-hairy; leaflets oval or oblong; flowers varicolored . . . . *L. odoratus*
    3: Plants perennial . . . . *L. latifolius*
  2: Leaflets at least 4 on some of the leaves; native perennial plants
    10. Tendrils absent or reduced and bristlelike, not prehensile
      11. Plants densely silky-villous; flowers white to pink or purple; on or near dunes . . *L. littoralis*
      11: Plants not densely silky-villous; flowers pale lilac to blue; open woods . . . . *L. torreyi*
    10: Tendrils well-developed, prehensile on at least some of the leaves
      12. Stems winged . . . . *L. jepsonii* ssp. *californicus*
      12: Stems not winged, at most sharply angled
        13. Flowers tan to yellowish or orange with purple or greenish pencilling 10-15 mm long; plants glabrous except for ciliate calyx-teeth, pods glabrous . . . . *L. sulphureus*
        13: Flowers white or cream with purple to pink lines, to pink or lavender, orchid or reddish-purple, usually fading yellowish, or drying yellowish or orange 15-20 mm long; pods initially pubescent . . . . *L. vestitus*

★ **Lathyrus angulatus** Linnaeus. Infrequent, open slopes: Bodega Head, *Best* (1985); Bouverie; 6½ mi. n. of Sonoma on Hwy. 12, *Crampton* (1966); Pepperwood; ½ mi. s. of Asti jct. on w. side of US 101, *J. & B. Guggolz* (1991 NCC). < 1000 ft. Native of Europe.

★ **Lathyrus aphaca** Linnaeus. Uncommon, escape: Calistoga Road (probably Mark West Springs Road) 3 mi. e. of Old Redwood Hwy., *J. Smith* (1974 NCC). < 1700 ft. Spring and summer. Native of Eurasia.

★ **Lathyrus cicera** Linnaeus. Occasional, escape: 1 site n. of Hwy. 12 in Laguna de Santa Rosa drainage, *Harrison* (1978 NCC); Sonoma Valley Regional Park, *Wagnon* (1989 NCC); Hwy. 12 and Madrone Road e. of Eldridge, *Best* (1990); Annadel, acc. Harrison (1978); St. Helena Road, *Wagnon* (1988). < 1700 ft. Native of Eurasia.

★ **Lathyrus hirsutus** Linnaeus. CALEY PEA. Uncommon, hay fields: around Hinebaugh Cr., Rohnert Park at s. end of Laguna de Santa Rosa drainage, *Harrison* (1978 NCC); Rohnert Park, *T. C. Fuller* (1969); Spring Lake. < 3300 ft. May-July. Native of Europe.

Lathyrus jepsonii E. Greene var. **californicus** (S. Watson) Hoover. Fairly common, along streams, sandy slopes: roadside near Plantation, *Baker* (1930 NCC); Guerneville, *Rattan* (1877); n. of Hwy. 12 in Laguna de Santa Rosa drainage, *Harrison* (1978 NCC); road to Cooper's Grove beyond the school on Sonoma Mtn. Road, *Baker* (1930 NCC); Santa Rosa Cr. e. of Santa Rosa, *Heller* (1902); near Mark West Springs, *Eastwood & Howell* (1938); 2 mi. n. of Windsor, *Eastwood & Howell* (1936); Lake Sonoma area. < 5000 ft. Apr.-June. —*L. californicus* S. Watson. *L. watsonii* White; type locality: Sonoma Co. *L. j.* ssp. *c.* C. Hitchcock.

★ **Lathyrus latifolius** Linnaeus. PERENNIAL SWEET PEA. Coommon, garden escape: Bohemian Hwy. 1 mi. n. of Occidental, *Best* (1981); Bodega Hwy. at Watson School e. of Bodega, *Best* (1981); Guerneville, *Lieberman* (1979 NCC); Roblar Road just w. of dump in Stony Point area, *Wells* (1981); Pepperwood; Pine Mtn. Road, *G. Smith* (1981). Usually < 6600 ft. May-Sept. Native of Europe.

Lathyrus littoralis Endlicher. Occasional, sandy areas along coastal strand: Wright's Beach, acc. Connors (1992); beach at Salmon Cr., *Baker* (1926 NCC); 2 mi. n. of Bodega Bay, *Bacigalupi* (1928). ± < 15 ft. Apr.-July.

★ **Lathyrus odoratus** Linnaeus. SWEET PEA. Occasional, garden escape: ¼ mi. s. on E. Cotati Ave. near railroad tracks, *Juul* (1973 NCC); Spring Lake; Pepperwood. Usually < 3300 ft. Apr.-June. Native of Europe.

★ **Lathyrus sativus** Linnaeus. GRASS PEA. Uncommon, waste places: along Llano Road s. of Todd Road near sewage plant, *Howell* (1986). < 300 ft. Native of Eurasia.

★ **Lathyrus sphaericus** A. J. Retzius. Uncommon, adventive: between Jimtown and Mercuryville on Geysers Road, *Knight & Utech* (1986 CM); Franz Valley, *Ordano* (1982). < 3300 ft. Native of Eurasia.

Lathyrus sulphureus A. Gray. Occasional, dry slopes: Geyserville and Alexander Valley quads. < 8300 ft. Apr.-July. The characters differentiating *L. vestitus* from *L. sulphureus* sometimes overlap considerably; pods pubescent on the former and pods glabrous on the latter may be the best character.

★ **Lathyrus tingitanus** Linnaeus. TANGIER PEA. Occasional, adventive: The Sea Ranch;. ca 1 mi. e. of Jenner, *Wells* (1985); Fife Cr. near Guerneville, *Rubtzoff* (1961); Cunningham Station, *Baker* (1947 NCC); property bordered by Chianti and Dutcher Cr. roads w. of US 101 and w. of Geyserville, acc. J. & B. Guggolz (1991). < 1700 ft. May-July. Native of Eurasia.

Lathyrus torreyi A. Gray. Occasional, open woods: near Cazadero, *Heller* (1903); Graton to Occidental, *Baker* (1926 NCC); Harrison Grade Road, *Baker* (1928 NCC). < 2700 ft. May-June.

Lathyrus vestitus Nuttall. HILLSIDE PEA. Common, brushy and wooded areas: The Sea Ranch; near Plantation, *Bailey* (1930 NCC); 249 ft. elev. Salt Point, *Clifton* (1981 PUA); 1 mi. n. of Fort Ross, *Baker* (1932 NCC); ca 1 mi. nw. of Camp Meeker, acc. J. & B. Guggolz (1991); near Bodega Bay, *Baker* (1932); Willow Cr.Marsh, *Utech* (1933 CM); Duncans Mills, *M. Jones* (1882); Guerneville, *Rattan* (1877); Sonoma Mtn. Road near Folger Ranch, *Knight* (1964); Crane Can. Road near Grange Road, *Wiley* (1965 NCC); Fairfield Osborn (*); Bouverie; 4 mi. from Bennett Valley, *Kindie* (1933 NCC); ½ mi. w. of Mark West Springs, *Eastwood & Howell* (1938); Pepperwood; Fitch Mtn., *Rubtzoff* (1952); Skaggs Springs, *Hawver* (1915); Sonoma Lake area; property bordered by Chianti and Dutcher Cr. roads w. of US 101 and w. of Geyserville, acc. J. & B. Guggolz (1991); ca 2000 ft. elev. Anderson Trail on Black Mtn. between Geyser Peak and upper end of Pine Flat Road, acc. J. & B. Guggolz (1992); Geyser Peak, *Knight & Edwards* (1986). < 5000 ft. Apr.-June. —*L. puberulus* White, type locality: Sonoma County. *L. v.* ssp. *puberulus* (White) C. L. Hitchcock. *L. bolanderi* S. Watson. *L. v.* ssp. *b.* (S. Watson) C. L. Hitchcock.

Lotus

GROUPS:

1. Leaflets 5, 3 near axis tip, lower pair large, stipular in position; stipules glandlike, often not apparent **Group 1**
1: Leaflets 3-15, lowest not stipular in position; stipules as preceding or not
  2. Stipules scarious or leafletlike, occasionally fragile and quickly lost; leaflets 3-15, pinnately arranged, often opposite or subopposite . . . . **Group 2**
  2: Stipules glandlike, bumplike, or conic, often not apparent; leaflets 3-9, pinnately or ± palmately arranged, usually irregularly arrayed on leaf-axis
    3. Fruit indehiscent, usually not flat, usually included or moderately exserted from calyx (occasionally conspicuously exserted and curved most of length), usually with curved, 2-3 mm beak; usually perennial . . . . **Group 3**
    3: Fruit dehiscent, usually flat, usually nearly entirely exserted and straight, usually with straight or curved 0.5-1.5 mm beak; annual or perennial
      4. Annual; fruit usually straight; wings < or ± = keel; stigma glabrous . . . . **Group 4**
      4: Perennial (except *L. strigosus*); fruit straight or ± curved; wings usually > keel; stigma almost glabrous or finely hairy . . . . **Group 5**

GROUP 1

1. Plants annual; hairy; inflorescence usually 1-3-flowered, 5-8 mm . . . . *L. angustissimus*
1: Plants perennial; glabrous or hairy; inflorescence 3-12-flowered, 8-14 mm
  2. Stems ± erect, ± strong; broad leaflets . . . . *L. corniculatus*
  2: Stems decumbent, weaker; narrower leaflets . . . . *L. tenuis*

GROUP 2

1. Peduncle bract 0 or just below umbel, usually simple or 2-3-parted; fruit 1.5-3 mm wide
  2. Petal-claws included to slightly exserted from calyx; plant often hairy . . . . *L. oblongifolius*
  2: Petal-claws clearly exserted from calyx; plant ± glabrous . . . . *L. formosissimus*

1: Peduncle bract usually well below umbel, usually 3-5-parted; occasionally fragile and lost; fruit 3-5 mm wide
   3. Hairs usually persistent, spreading; stipules often wide, leafletlike, persistent and sometimes becoming scarious; corolla pink to red-purple
      4. Plant resinous-glandular or viscid throughout; balsam-scented . . . . *L. balsamiferus*
      4: Plant not resinous-glandular or viscid throughout; not scented . . . . *L. stipularis*
   3: Hairs usually almost 0 or appressed; stipules scarious, fragile, not leafletlike, usually some lost; corolla initially white, pink or yellow-green
      5. Flowers usually 6-10; corolla 10-12 mm, white or pink, becoming striate, darker; pedicel 2-3 mm, slightly longer in fruit; leaflet length usually 2-3 × width . . . . *L. aboriginus*
      5: Flowers usually 12-20; corolla 12-17 mm, yellow green, becoming dark-blotched; pedicel 3-6 mm, > 3 mm at least in fruit; leaflet length usually 1-2.5 × width . . . . *L. crassifolius*

GROUP 3

1. Leaflets glabrous or finely strigose; hairs straight or slightly wavy
   2. Plants with stem usually ascending or erect, bushy-branched, often ± 1 m; stem slightly woody (but green), sparsely leafy; corolla yellow; calyx-lobes usually 1-2 mm, not curved outward . . . . *L. scoparius*
   2: Plants prostrate or low-ascending, < 1 m; stem not woody, leafy; corolla pink, white or yellow; calyx-lobes usually < 1 mm or curved outward
      3. Calyx-lobes slender, some or all curved outward or hooked; corolla usually white to pinkish (to brick-red when dry) . . . . *L. benthamii*
      3: Calyx-lobes wide, not curved outward, not hooked; corolla yellow . . . . *L. junceus* var. *bioletti*

1: Leaflets villous especially when young, later ± glabrate; stem hairs often stiff, straight; several prostrate stems . . . . *L. heermannii*

GROUP 4

1. Flowers usually 1 per leaf-axil; peduncle << 1 cm, bracts 0
   2. Corolla yellow, reddening with age; plant prostrate to low-ascending and calyx-lobes 0.8-2 × tube
      3. Calyx-lobes 1-2 × tube; fruit usually 3-4 mm wide; plants usually with soft, spreading hairs . . . . *L. humistratus*
      3: Calyx-lobes ± 0.8-1.2 x tube, fruit usually 2.3-3 mm wide; plants inconspicuously strigose or with soft, spreading hairs . . . . *L. wrangelianus*
   2: Corolla whitish-cream, with back of banner tinged purple; plant erect, branched from base; calyx-lobes ± 1.5 × tube . . . . *L. denticulatus*

1: Flowers 1-several per leaf-axil; peduncle usually > 1 cm, usually bracted
   4. Calyx-lobes >> tube; leaflets usually 3 (frequently 1 on upper leaves) . . . . *L. purshianus*
   4: Calyx-lobes ± = tube; leaflets 3-9 throughout
      5. Stem erect or ascending; leaf-axis not conspicuously flattened; corolla pink or salmon, quickly fading; fruit straight, margin often wavy . . . . *L. micranthus*
      5: Stem prostrate, often mat-forming; leaf-axis conspicuously flattened; corolla yellow; fruit curved near tip, margin not wavy . . . . *L. strigosus*

GROUP 5

1. Plants usually prostrate and mat-forming; inflorescence usually 1-2-flowered; corolla 5-10 mm; leaf-axis flat, ± bladelike; fruit usually curved only at or near tip . . . . *L. strigosus*

1: Plants with woody base, prostrate (mat-forming or not) to erect; inflorescence 1-9-flowered; corolla 15-24 mm; leaf-axis not flat or bladelike; fruit usually straight . . . . *L. grandiflorus*

Lotus aboriginus Jepson. Uncommon, borders of woods: near Kashia Indian School on Stewarts Point-Skaggs Springs Road, *Bailey* (1930 NCC); near Vine Hill schoolhouse, *Baker* (1929 NCC). < 2700 ft. May-July. —*Hosakia rosea* Eastwood.

★ **Lotus angustissimus** Linnaeus. Common, adventive: The Sea Ranch near recreation center, *J. H. Thomas* (1976); 0.9 mi. s. of Stewarts Point, *Crampton* (1957); Gerstle Cove at Salt Point State Park, *Wells* (1982); Ocean Cove between Stillwater Cove and Salt Point, *Hastings* (1940 NCC); Stillwater Cove between Timber Cove and Salt Point, *Eastwood & Howell* (1938); Fort Ross, *Crampton* (1955); Fort Ross area near school's reservoir, *Wagnon* (1989 NCC); Kolmer Gulch between Fort Ross and Timber Cove, *Hoover* (1964); 1½ mi. n. of Fort Ross on Hwy. 1, *Ferris* (1937); n. of Jenner, *Ripley & Barneby* (1945); 1 mi. s. of bridge across Russian River, *Keck* (1949); Hwy. 1 n. of Wright's Beach, *Baker* (1935); near Salmon Cr., *Connors* (1991 BML); Salmon Cr. Marsh, *Rubtzoff* (1962, 1967, 1968 NCC); Todd Road just e. of Llano Road along roadside, *Knight & Howell* (1987); Spring Lake. < 3300 ft. June-Aug. Native of Old World.

Lotus balsamiferus (Kellogg) E. Greene. Uncommon, dry wooded slopes: St. Helena Road near Napa Co. line, *Knight & Edwards* (1985); w. side of Dry Cr. Valley n. of Crane Cr. s. of Bradford Mtn., *Raiche* (1985). < 4000 ft. April-June. —*Hosackia balsamifera* Kellogg. *L. stipularis* (Bentham) E. Greene var. *b.* (Kellogg) Abrams.

Lotus benthamii E. Greene. Uncommon, rocky and sandy places: Rocky Point between Stewarts Point and Horseshoe Cove, *Knight* (1964); Fort Ross, *Heller* (1903). < 700 ft. Mar.-July.

★ **Lotus corniculatus** Linnaeus. (See *L. tenuis* for taxonomic clarification.) Uncommon, adventive: The Sea Ranch; s. of Bloomfield between Garrison and Carmody roads toward Two Rock, acc. Patterson (1990); Button Ranch on Walker Road (U. C. property) between Petaluma-Valley Ford Road and Pepper Road, acc. Patterson (1990); n. of jct. of Walker and Pepper roads e. of Two Rock, acc. Patterson (1990); Watson's pasture on Stony Point Road, *Knight* (1963). < 3300 ft. June-Sept. Native of Europe.

Lotus crassifolius (Bentham) E. Greene. Uncommon, dry flats and banks: Geyser Peak (Sulphur Peak), *Knight & Edwards* (1986); Cooley Ranch near Sonoma-Mendocino line ca 8 airmiles w. of Cloverdale, *Cooley* (1992). 1000-7000 ft. May-Aug.

Lotus denticulatus (Drew) E. Greene. Uncommon, gravelly areas and grassy slopes: ca 1 mi. n. of Mark West bridge, *Baker* (1934 NCC). < 4700 ft. May-July.

Lotus formosissimus E. Greene. Common, moist areas: The Sea Ranch; Salt Point; Hwy. 1 at entrance to Goat Rock State Park, *Best* (1981); above Doran Beach in Bodega Harbour development, *Huntington* (1989 NCC); Pitkin Marsh, *Rubtzoff* (1951); Cunningham Marsh, *Howell* (1947); Two Rock, *Congdon* (1880); Kenwood, *Michener & Bioletti* (1893); near Windsor, *Heller & Brown* (1902). < 2300 ft. Mar.-July.

Lotus grandiflorus (Bentham) E. Greene. Occasional, dry slopes: Monte Rio, acc. Jepson, *Fl. of Ca, Vol. 2* (1936); s. of Guerneville, *Rubtzoff* (1953); ca 5 mi. nw. of Occidental, *Wagnon* (1939 NCC). < 5000 ft. Apr.-July.

Lotus heermannii (Durand & Hilgard) E. Greene var. **orbicularis** (A. Gray) Isely. Occasional, coastal strand, dunes: 8.8 mi. s. of Stewarts Point, *Koch* (1940 NCC); Bodega Point, *Eastwood* (1915); Bodega Rock, *Baker* (1930 NCC); Bodega Head, *Ferris & Morey* (1958). < 850 ft. Apr.-July. —*Lotus h.* var. *eriophorus* (E. Greene) Ottley.

Lotus humistratus E. Greene. Very common, various habitats: sandbar in S. Fork of Gualala River near Hot Spot at The Sea Ranch, *Wells* (1984); ½ mi. n. of Stewarts Point, *Wells* (1985); Cazadero quad.; Fife Cr. near Guerneville, *Rubtzoff* (1961); jct. of Facendini Lane and Occidental Road, *Heblach* (1973 NCC); outcrop of serpentine along road between Sebastopol and Occidental, *Baker* (1956); Fairfield Osborn (S); Petaluma, *Eastwood* (1921); Sugarloaf Ridge; Spring Lake; Gates Road off Calistoga Road at Humbug Creek, acc. Wells et al. (1984); Santa Rosa Can., *Van Steenberge* (1932 NCC); near Mark West Springs, *Heller & Brown* (1902); Pepperwood; Healdsburg, *Rattan* (1877); 350-400 ft. elev. Old Skaggs Springs, *Utech, Sato & Ohara* (1990 CM); Lake Sonoma area; ca 2000 ft. elev. Anderson Trail on Black Mtn. between Geyser Peak and upper end of Pine Flat Road, acc. J. & B. Guggolz (1992); Geyser Peak, *Knight & Edwards* (1986); 2100 ft. elev. near Unit 18, *Malloch* (1978 PG&E); 2400 ft. elev. Little Geysers, *Malloch* (1977 PG&E); Cloverdale, *Geis* (1902). < 5600 ft. Mar.-June.

Lotus junceus (Bentham) E. Greene var. **biolettii** (E. Greene) Ottley. Occasional, coastal: Stewarts Point, *Abrams* (1920); Salt Point; Timber Cove, *Hoover* (1964); near Plantation, *Baker* (1930 NCC); hwy. n. of Fort Ross, *Baker* (1932 NCC); Kolmer Gulch near Fort Ross, *Hoover* (1964). < 1700 ft. Apr.-July. —*L. b.* E. Greene.

Lotus micranthus Bentham. Occasional, vernal pools, open slopes, plains: The Sea Ranch; Fort Ross, Cazadero and Duncans Mills quads.; Willow Cr. Marsh, *Rubtzoff* (1959); Bodega Dunes, *Connors* (1992 BML); hills near Guerneville, *Rubtzoff* (1953); near Todd Road in Laguna de Santa Rosa drainage, *Harrison* (1978 NCC); Lafferty Ranch on Sonoma Mtn., acc. Arnold et al. (1971); Tresch Ranch, Walker Road near Two Rock, acc. J. & B. Guggolz (1992); Bouverie; Sugarloaf Ridge; Spring Lake; Santa Rosa, *Eastwood* (1924); Pepperwood; Petrified Forest, *Eastwood* (1915); Fitch Mtn., *Rubtzoff* (1952); Lake Sonoma, Geyserville, Alexander Valley and Ida Clayton Road quads.; 3000 ft. elev. near Socrates Mine, *Malloch* (1977 PG&E). < 4300 ft. Mar.-May.

Lotus oblongifolius (Bentham) E. Greene. Uncommon, wet areas: Pitkin Marsh, *Rubtzoff* (1951), *Howell* (1951). < 8000 ft. May-Sept. —*Lotus o.* var. *nevadensis* (A. Gray) Munz.

Lotus purshianus (Bentham) Clements & E. G. Clements. SPANISH CLOVER. Very common, disturbed areas, fields: The Sea Ranch; Pitkin Marsh, *Rubtzoff* (1951); Bodega Road 3 mi. e. of Valley Ford, *Howell* (1930); Freestone, *Huntington* (1987 NCC); Vacation Beach near Guerneville, *Rubtzoff* (1958); Laguna de Santa Rosa, *Rubtzoff* (1958, 1962); Sebastopol, *Dows* (1907); e. of Cunningham near Llano Road, *Rubtzoff* (1958 NCC); Todd Road Preserve (H&K); Fairfield Osborn (S); Bennett Mtn. Lake, *Rubtzoff* (1955); Joost Ranch on Calistoga Road, *Knight* (1963); Button Ranch on Walker Road (U. C. property) between Petaluma-Valley Ford Road and Pepper Road, acc. Patterson (1990); w. of Sears Point Raceway in Tolay Cr. drainage, acc. Patterson (1990); Spring Lake; Friends House in Rincon Valley, *Wells* (1983, 1990); Howarth Park, Santa Rosa, *Knight* (1964); Blue Bell Lane in Santa Rosa, *Knight* (1963); Santa Rosa, *Rattan* (1877), *Heller* (1935); near Burke's Sanitarium near Mark West Springs, *Heller* (1902); Mark West Can., *Baker* (1925); Mark West Springs Road, *McLeod* (1930 NCC); Pepperwood; Petrified Forest, *Eastwood* (1915); Geyserville, *Heller* (1923); Asti, *Knight* (1963); Cloverdale, *Bufford* (1929 NCC). < 8000 ft. May-Oct. —*L. americanus* Bischoff. *Acmispon aestivalis* Heller, type locality, Santa Rosa, *Heller* (1902). *A. gracilis* Heller, type locality, Burke's Sanitarium near Mark West Springs, *Heller* (1902). *A. sparsiflorus* Heller, type locality: Russian River near Cloverdale, *Heller* (1902).

Lotus scoparius (Nuttall) Ottley. DEERWEED, CALIFORNIA BROOM. Common, dry slopes: Annapolis and Cazadero quads.; s. of Guerneville, *Rubtzoff* (1953); Bouverie; Sugarloaf Ridge; foot of Mt. Hood, *Heller* (1902); Joost Ranch on Calistoga Road, *Knight* (1963); Annadel, *Wells* (1984); Spring Lake; Pepperwood; Fitch Mtn., *Rubtzoff* (1952); chaparral area sw. of Bradford Mtn. 1½ mi. w. of Dry Cr., acc. J. & B. Guggolz (1992); ca 2000 ft. elev. Anderson Trail on Black Mtn. between Geyser Peak and upper end of Pine Flat Road, acc. J. & B. Guggolz (1992); Geyser Peak, *Knight & Edwards* (1986). < 5000 ft. Mar.-Aug.

Lotus stipularis (Bentham) E. Greene. Uncommon, chaparral: Fort Ross, *Baker* (1930 NCC); Vine Hill schoolhouse, *Baker* (1928 NCC); Mt. Hood, *Bioletti* (1893); 1000 ft. elev. unnamed ridge on w. side of Dry Cr. Valley n. of Crane Cr., *Raiche* (1985). < 3300 ft. Apr.-June.

Lotus strigosus (Nuttall) E. Greene. Uncommon, dry disturbed areas: Bodega Dunes, *Connors* (1991 BML). < 7700 ft. Mar.-June.

★ **Lotus tenuis** Waldstein-Wartemberg & Kitaibel. BIRDSFOOT TREFOIL. Common, especially at edge of asphalt road surfaces: acc. *Rubtzoff & Gankin, The Four Seasons, 8:2, p. 49,* (1988), "In California floristic literature in general, no distinction is made between *L. corniculatus* Linnaeus and *L. tenuis*, the latter being included in *L. corniculatus* in a broad sense. In *Flora Europaea 2:174* (1968), *L. tenuis* is treated as one of 12 species in the *L. corniculatus* group. It is stated there, however, that this group is sometimes treated as a single species with a number of subspecies and varieties. In agronomy and seed commerce the two plants are regarded as being distinct. In a pamphlet entitled *Birdsfoot Trefoil*, published by Volkman's Seed, San Francisco, probably in the mid or late 1950s, they treated as 'Broadleaf (erect)' *L. corniculatus* and as 'Narrowleaf (prostrate)' *L. tenuis*. According to this pamphlet, the two, although both are used as pasture components, differ ecologically as well — it being stated, for example, that *L. corniculatus* is tolerant of acid soils, *L. tenuis*, of saline. *Lotus corniculatus* is the plant with stronger, more erect stems and broad leaflets whereas the locally much commoner *L. tenuis* is the plant with weaker and decumbent stems and narrow leaflets. Most of the specimens in the county seem to be the latter type": The Sea Ranch, *J. H. Thomas* (1976); Willow Cr. Marsh, *Rubtzoff* (1959); town of Bodega, *Rose* (1962); Bodega dunes, *Connors* (1991); Austin Cr. Marsh, *Rubtzoff* (1960); Goat Rock State Beach, *Rubtzoff* (1962); Freestone, *Huntington* (1987 NCC); Austin Cr. Marsh, *Rubtzoff* (1960); Laguna de Santa Rosa ne. of Graton, *Rubtzoff* (1962); Rohnert Park, *Caruso* (1979 NCC); Spring Lake; Fairfield Osborn, *Quibell* (1972 NCC); Pepperwood. < 3300 ft. June-Sept. Native of Europe.

Lotus wrangelianus Fischer & C. Meyer. CALF LOTUS. Occasional, dry grassy areas: The Sea Ranch; Fort Ross, *Heller* (1903); Duncans Mills quad.; Willow Cr. Marsh, *Rubtzoff* (1959); 1½ mi. n. of Wright's Beach, *Cantelow* (1941); Salmon Cr., *Baker* (1930 NCC), *Howell* (1931); Bodega Head; n. of Bodega Bay, *Baker* (1938 NCC); Bay Flat Road near Spud Point Marina at Bodega Harbor, *Clifton* 1987 PUA); Sebastopol quad.; Pepper Road w. of Mecham Road in Stony Point district, *Lamb* (1970 NCC); Santa Rosa, *Mallory* (1920); Pepperwood; Lake Sonoma area; Alexander Valley and Ida Clayton Road quads. < 5000 ft. Mar.-June. —*L. subpinnatus* Lagasca, misapplied.

Lupinus

1. Plants perennial
    2. Plants shrubby or with definite woody stems of some length above the roots
        3. Banner ± pubescent on the back; plants mostly appressed-silky; keel narrow with long slender acumen
            4. Keel not ciliate on edges or with only an occasional hair; petioles 2-3.5 cm long . *L. chamissonis*
            4: Keel definitely ciliate along upper edge toward acumen, keel narrowed toward base; petioles 2-8 cm long

5. Stems not woody, mostly 2-4 dm long; petioles mostly 4-8 cm long . . . *L. albifrons* var. *collinus*
5: Stems woody, mostly 6-30 dm tall; petioles various
6. Bracts mostly 10-15 mm long, exceeding flower buds . . . . *L. a.* var. *douglasii*
6: Bracts mostly 4-8 mm long, not or scarcely exceeding buds . . . . . . *L. a.* var. *albifrons*
3: Banner glabrous on back; plants appressed-silky to tomentose or spreading-villous; keel mostly not markedly slender at acumen
7. Stems erect or nearly so
8. Herbage puberulent to pubescent to ± silky; racemes mostly 1-3 dm long, flowers predominantly yellow, (sometimes mixed with lavender or blue) . . . . *L. arboreus* var. *arboreus*
8: Herbage ± villous; racemes 0.5-1.0 dm long; banner mostly yellow, wings blue . . . . *L. a.* var. *eximius*
7: Stems decumbent or prostrate . . . . *L. variicolor*
2: Plants strictly herbaceous, i.e. stems not woody above the caudex
9. Leaves glabrous on upper surface or essentially so
10. Keel naked on upper edges; stems 5-15 dm tall, stout, mostly unbranched, fistulose, pubescence minute, spreading or appressed; corolla 12-14 mm long; wet places . . . . *L. polyphyllus*
10: Keel ciliate on upper edges; stems ± erect, branched, 3-12 dm tall, not fistulose, subglabrous to minutely strigose; corolla 10-14 mm long; dry areas of open woods and thickets . . *L. latifolius*
9: Leaves ± hairy above
11. Banner glabrous on back
12. Keel ± ciliate on upper edges at least toward tip, roots yellow . . . . *L. tidestromii*
12: Keel glabrous on upper edges or practically so, roots not yellow
13. Flowers 10-12 mm long, pale yellow or blue to lilac, banner obovate; pods strigose 2.5-3 cm long . . . . *L. adsurgens*
13: Flowers 12-18 mm long, violet or blue to lilac or white, banner roundish or oblong; pods silky-hairy ca 3-3.5 cm long
14. Stems decumbent or ascending 3-4 mm thick; flowers 12-14 mm long; banner roundish . . . . *L. formosus* var. *formosus*
14: Stems stout 5-7 mm thick; flowers 16-18 mm long; banner oblong . . . . *L. f.* var. *robustus*
11: Banner pubescent on back, the pubescence occasionally almost concealed by the upper lip of the calyx
15. Erect to decumbent, branched caudex, stems 1.5-3 (-5) dm long, mostly unbranched; leaflets 6-7, spatulate- obovate, rounded or retuse at apex, 3-4 cm long, 1-2 cm wide; flowers scattered or subverticillate, petals purplish-blue 14-16 mm long; keel ciliate on upper edges and on lower edges near claws; pods pubescent 2- 2.5 cm long . . . . *L. sericatus*
15: Plants matted, stems 2-4 dm long; leaflets 7-10, oblanceolate to spatulate, acute or rounded apically, 1-3 cm long, 4-10 mm wide; flowers mostly whorled, petals blue to red-purple or lavender 10-14 mm long; keel often nonciliate; pods villous-strigose 3-5 cm long . . . . *L. albifrons* var. *collinus*
1: Plants annual
16. Keel ciliate on upper margins near claws
17. Keel ciliate on lower edges also, near claws
18. Flowers yellowish; pods ca 1.5 cm long . . . . *L. luteolus*
18: Flowers deep purple-blue; pods 4-5 cm long; stems hollow . . . . *L. succulentus*
17: Keel not ciliate on lower edges
19. Pubescence of stems and peduncles very short ± appressed, barely 0.5 mm long; petals white, tinted or veined with rose or violet . . . . *L. microcarpus* var. *densiflorus*
19: Pubescence of stems and peduncles longer, spreading or retrorse 1-1.5 mm long; flowers white or cream to dark purple . . . . *L. m.* var. *microcarpus*
16: Keel ciliate on upper margins toward the apex or not ciliate at all
20. Pedicels slender 4-10 mm long; flowers 10-15 mm long
21. Keel entire, not toothed; spreading hairs generally many; pods 5.5-6.5 mm wide . *L. nanus*
21: Keel with a distinct tooth on the upper margin near the middle; spreading hairs generally few; pods 7.5-8.5 mm wide . . . . *L. affinis*
20: Pedicels 1-3 mm long; flowers 4-8 (-12) mm long
22. Banner 6-9 mm long, keel ciliate on upper edges toward apex; pods 1.5-3 cm long, 4-5.4 mm wide, strigose; seeds 5-8, 2-3 mm long . . . . *L. bicolor*
22: Banner 7-8.5 mm long, keel essentially glabrous; pods 2.5-3 cm long, 6-9 mm wide, shaggy-villous; seeds 4-6, 4-5 mm long . . . . *L. pachylobus*

Lupinus adsurgens E. Drew. Uncommon, forest floor: 2760 ft. elev. Caldwell Ranch, *Clifton* (1984 PUA). 2500-5000 ft. May-July.

Lupinus affinis J. G. Agardh. Occasional, grassy slopes, embankments: Larkspur Rock 1½ mi. e. of Bodega Bay, *Eastwood & Howell* (1939); Hessel Road near jct. of Gravenstein Hwy., acc. J. & B. Guggolz (1991); hills e. of Geyserville on road to The Geysers, *Howell* (1976). < 2700 ft. Mar.-May.

Lupinus albifrons Bentham var. **albifrons**. Common, grassy hills: The Sea Ranch, *Wells* (1984); Forestville, *C. P. Smith* (1908); Fairfield Osborn (S); hills e. of Agua Caliente on e. edge of Sonoma Valley, *Raven* (1957); Bouverie, *Knight* (1986); Trinity Road, *Weaver* (1975 NCC); w. base of Mt. Hood, *Heller & Brown* (1902); Sugarloaf Ridge; near Mark West Springs, *Eastwood* (1938); Pepperwood; Lake Sonoma area; n. of Cloverdale, *Eastwood* (1939). < 5000 ft. Mar.-June.

Lupinus albifrons var. **collinus** E. Greene. Occasional, rocky places: Bodega Bay, *Barbour* (1971 BML); 2 mi. s. of Bodega Bay, *Heller* (1940); near Guerneville, *Rubtzoff* (1953); w. base of Mt. Hood, *Heller & Brown* (1902); ca 2000 ft. elev. Anderson Trail on Black Mtn. between Geyser Peak and upper end of Pine Flat Road, acc. J. & B. Guggolz (1992). < 6600 ft. Mar.-June.

Lupinus albifrons var. **douglasii** (J. G. Agardh) C. P. Smith. Uncommon, open slopes: Petaluma, *Congdon* (1880). < 1700 ft. Mar.-June.

Lupinus arboreus Sims var. **arboreus**. YELLOW BUSH LUPINE. Common, sandy places: The Sea Ranch at Crow's Nest Drive and Black Point Reach, *Wells* (1981); 7 mi. s. of Gualala, *J. H. Thomas* (1976); Salt Point; n. of Fort Ross on Hwy. 1, *Cantelow* (1940); n. of bridge at Russian Gulch on Hwy. 1, *Wells* (1985); near Jenner, *Baker* (1935 NCC); s. end of Russian River Bridge near Bridgehaven, *Cantelow* (1940); Willow Creek; Goat Rock State Beach, *Wells* (1981); Coleman Valley Road, *Best* (1981); 2½ mi. n. of Bodega Bay, *Cantelow* (1939); 49 ft. elev. Bodega Head, *Clifton* (1977 PUA); Bodega Point, *Eastwood* (1915); Bodega Bay, *McIntyre* (1968 NCC); Doran Beach at Bodega Harbour development, *Huntington* (1989 NCC); road cut on Gravenstein Hwy. between Cotati and Stony Point Road (possibly planted by Caltrans), acc. Knight (1990). < 330 ft. Mar.-June. —*L. rivularis* Lindley.

Lupinus arboreus var. **eximius** (Burtt Davy) C. P. Smith. Uncommon, coastal scrub: Goat Rock Beach, *Best* (1984); Bodega Point, *Eastwood* (1915); Bodega Bay, *Baker* (1925 NCC). Plant generally much hairier than var. *arboreus*, and sometimes considered indistinct. < 2000 ft. Apr.-July.

Lupinus bicolor Lindley. MINIATURE LUPINE. Common, sandy places: Austin Cr. near Cazadero, *Bacigalupi* (1926); Goat Rock Beach, *Best* (1984); 6 mi. n. of Bodega Bay, *Eastwood & Howell* (1937); 2 mi. n. of Bodega Bay, *Bacigalupi* (1928); Bodega Bay, *Eastwood & Howell* (1937); Bodega Head; Willow Cr. Marsh, *Utech & Knight* (1988 CM); Todd Road Preserve, *Self* (1980 NCC); near Lichau Road e. of Petaluma Hill Road, *Wiley* (1965 NCC); Petaluma Hill Road, *Stocking* (1965 NCC); Fairfield Osborn (H); Button Ranch on Walker Road (U. C. property) between Petaluma-Valley Ford Road and Pepper Road, acc. Patterson (1990); 1½ mi. nw. of Petaluma, *Ott* (1934 NCC); Respini Ranch in Chileno Valley, *Knight* (1964); w. of Sears Point Raceway in Tolay Cr. drainage, acc. Patterson (1990); Sonoma Valley, *Jepson* (JEPS); Sonoma State Univ. campus, *McNally* (1970 NCC); Sonoma, *Wiesendanger* (1927); Bouverie; Spring Lake; Friends House at Rincon Valley, *Wells* (1985); ½ mile n. of Santa Rosa, *Hagg* (1934 NCC); Pepperwood; 4 mi. w. of Calistoga w. of county line, *Wiggins* (1949); 7 mi. from Calistoga in Knights Valley, *Howell* (1934); Independence School near Healdsburg, *Abrams* (1918); Lake Sonoma area; e. of Geyserville on road to The Geysers, *Howell* (1976); between Jimtown and Mercuryville on Geysers Road, *Utech & Knight* (1986 CM); Geyser Peak, *Knight & Edwards* (1986); 3 mi. from Jimtown en route to The Geysers, *Knight & Edwards* (1986); PG&E geysers Unit 13, *Quibell* (1981 NCC). < 5300 ft. Mar.-June. —*L. b.* ssp. *marginatus* D. Dunn. *L. b.* ssp. *microphyllus* (S. Watson) D. Dunn. *L. b.* ssp. *pipersmithii* (Heller) D. Dunn. *L. b.* var. *tridentatus* Eastwood, type locality, Santa Rosa, *Eastwood* (1921). *L. b.* ssp. *umbellatus* (E. Greene) D. Dunn. These subspecies and varieties are highly variable and indistinguishable geographically and morphologically. *L. micranthus* Lindley. *L. polycarpus* E. Greene.

Lupinus chamissonis Eschscholtz. Locally common, dunes, sandy beaches: Russian River near Monte Rio, *Eastwood & Howell*; Bodega Point, *Eastwood* (1915); Bodega Head; Larkspur Rock e. of Bodega Bay, *Eastwood & Howell* (1939). < ± 500 ft. Mar.-July.

Lupinus formosus E. Greene var. **formosus**. PALE SUMMER LUPINE. Common, dry sandy places, open fields: w. of Bodega Bay, *Baker* (1925 NCC); Todd Road Preserve (H&K); field at ne. corner Llano and Todd roads, acc. Knight (1986); Sonoma State Univ. campus, *Quibell* (1975 NCC); 2 mi. e. of Schellville, *Heller* (1940); ½ mi. n. of El Verano, *Cantelow* (1936); Bennett Valley, *Bufford* (1930 NCC), *Knight* (1963); Spring Lake; near se. boundary of Santa Rosa, *Howell* (1926); Santa Rosa Jr. Coll. campus, *Badger* (1939 NCC); near Mark West Springs, *Eastwood & Howell* (1938); Lake Sonoma area; 4200 ft.

elev. units 20 and 21, *Malloch* (1977 PG&E); 1 mi. s. of Cloverdale along US 101, *Wolfe* (1931 US). < 10000 ft. Apr.-Oct. —*L. f.* var. *bridgesii* (S. Watson) E. Greene. *L. lutosus* Heller, type locality: Alexander Valley, *Heller* (1940 UC).

Lupinus formosus var. **robustus** C. P. Smith. Uncommon, grassland: Lafferty Ranch on Sonoma Mtn., acc. Arnold et al. (1978). < 5000 ft. Apr.-Oct. —*L. sonomensis* Heller, type locality: 0.2 mi. e. of Schellville, *Heller* (1940).

Lupinus latifolius J. G. Agardh. Common, thickets, open woods: Harrison Grade Road, *Badger* (1939 NCC); 1 mi. w. of Forestville, *C. P. Smith* (1908); Trinity Road near Hwy. 12, *Cantelow* (1940); Hidden Oaks Road on w. side of w. end of Sonoma Mtn. Road, *Quibell* (1988 NCC); Button Ranch on Walker Road (U. C. property) between Petaluma-Valley Ford Road and Pepper Road, acc. Patterson (1990); w. of Sears Point Raceway in Tolay Cr. drainage, acc. Patterson (1990); Bennett Valley, *Newberry* (1931 NCC); Annadel, *Wells* (1984); Santa Rosa, *Eastwood* (1893); near Mark West Springs, *Eastwood & Howell* (1938); Pepperwood; Lake Sonoma area; Skaggs Springs, *Hawver* (1915). < 11500 ft. Apr.-July.

Lupinus luteolus Kellogg. BUTTER LUPINE. Uncommon, sandy places: Petrified Forest, *Brandegee* (1892); Spring Mtn. Road near summit, acc. Baker (1934). < 6300 ft. May-Aug.

Lupinus microcarpus Sims var. **densiflorus** (Bentham) Jepson. Common, open fields, hillsides: s. of Fort Ross on Hwy. 1, *C. P. Smith* (1903); Jenner, *Bacigalupi* (1922); town of Bodega, *Bacigalupi* (1922); 2 m. w. of Valley Ford, *Heller* (1940); hills e. of Petaluma, *Heller* (1938); Santa Rosa, *Heller* (1902); Knights Valley, *H. E. & S. T. Parks* (1930); n. of Cloverdale, *Cantelow* (1941). —*L. d.* Bentham. *L. d.* var. *persecundus* C. P. Smith, type locality: Duncans Mills, *M. Jones* (1882 UC). *L. d.* var. *aureus* (Kellogg) Munz — a cultivar name 'Ed Gedling' was assigned to an orange-colored form of this typically yellow lupine and it was planted by Caltrans on roadcuts: Petaluma-Point Reyes Road 1.4 mi. from Petaluma Blvd., S., *Riggins* (1982 OBI); on roadcut between Franz Valley Road and Petrified Forest Road on Porter Cr. Road, acc. Knight (1990). < 5300 ft. Apr.-June.

Lupinus microcarpus var. **microcarpus**. CHICK LUPINE. Common, grassland: 6 mi. n. of Jenner, *Eastwood & Howell* (1937); Hwy. 1 just n. of Jenner on cliffside above mouth of Russian River, *Huntington* (1989 NCC); edge of Hwy. 116 above Russian River near Hwy. 1, *Best* (1981); Goat Rock State Park, *Best* (1984); 2 mi. nw. of Camp Meeker, *J. & B. Guggolz* (1990 NCC, JEPS); Bodega Head; Bodega Point, *Eastwood* (1915); 1 mi. e. of Bodega Bay, *Howell* (1930), *Davidson* (1971 BML); Larkspur Rock near Bodega Harbor, *Brunk* (1933 NCC); hwy. between Bodega Bay and Bodega, *Baker* (1926 NCC); jct. Hwy. 1 and Valley Ford-Freestone Road, *Knight & Utech* (1986). —*L. densiflorus* var. *crinitus* Eastwood, type locality: Bodega Point, *Chandler* (U). *L. d.* var. *palustris* (Kellogg) C. P. Smith. < 5300 ft. Apr.-June.

Lupinus nanus Bentham. Very common, fields, grassy slopes: The Sea Ranch, *Wells* (1981); Bodega Point, *Eastwood* (1915); Bodega Head; Pitkin Marsh, *Rubtzoff* (1952); Sebastopol, *Dows* (1907); near Sebastopol, *Seager* (1927 NCC); Todd Road Preserve; Cotati, *McIntyre* (1968 NCC); Lafferty Ranch on Sonoma Mtn., acc. Arnold et al. (1971); Tresch Ranch, Walker Road near Two Rock, acc. J. & B. Guggolz (1992); Sonoma Valley, *Jepson* (1914); Bouverie; Sugarloaf Ridge; Bennett Valley Road, *Baker* (1930 NCC); jct. Melitta-Mark West-Petrified Forest roads (probably jct. of Hwy. 12 and Calistoga Road), *C. P. Smith* (1948); e. of Santa Rosa, *Heller* (1940); hills se. of Santa Rosa, *Heller & Brown* (1902); Spring Lake; Pepperwood; Old Redwood Hwy. nw. of Windsor exit, *de Nevers & Kaza* (1981); Felta Cr. area sw. of Healdsburg, *Bice* (1895 NCC); Lake Sonoma area; Geysers-Healdsburg Road ca 5 mi. n. of winery, *Breslin* (1975 NCC); between Jimtown and Mercuryville en route to The Geysers, *Knight & Edwards* (1986); Sulphur Cr. Road between Cloverdale and The Geysers, *Knight & Edwards* (1986). < 4300 ft. Apr-May. —*L. n.* ssp. *latifolius* (Bentham) D. Dunn. *L. vallicola* Heller ssp. *apricus* (E. Greene) D. Dunn.

Lupinus pachylobus E. Greene. BIG POD LUPINE. Grassy slopes, disturbed or open sites: campus Sonoma State University, *Stocking* (1965 NCC); Fairfield Osborn (*); Sugarloaf Ridge; Santa Rosa, *Mallory* (1920); Lake Sonoma area. Alleged to intergrade with *L. bicolor*. < 2000 ft. Mar.-Apr.

Lupinus polyphyllus Lindley. Occasional, moist areas: Pitkin Marsh, *Quibell* (1987); Two Rock, *Congdon* (1880); near Kenwood, *Bufford* (1930 NCC). < 10000 ft. May-Aug.

Lupinus sericatus Kellogg. COBB MOUNTAIN LUPINE. Occasional, open slopes, sometimes wooded: near summit of Triniti P. O. Road in Mt. Hood Range (probably Trinity Road near Napa Co. line), *Howell* (1935); ca 3600 ft. elev. Pine Mtn. ca half way between Mt. St. Helena and Cobb Mtn. on Sonoma-Lake co. line at USGS Benchmark 361, acc. Skee Hamann (1976); Socrates Mine Road just across Lake-Sonoma co. line in Sonoma Co. acc. Ed Deering (1991); Kelly Road ca 200 ft. nw. of its jct. with Rockpile Road, *J. & B. Guggolz* (1986 NCC). 1600-5000 ft. Apr.-June.

Lupinus succulentus Koch. SUCCULENT LUPINE. Occasional, grassy slopes, flats: Todd Road Preserve; near San Antonio Cr. on US 101, *Howell* (1982); vacant lot corner Petaluma Blvd. S. and B St., Petaluma, *Knight* (1965); hwy. e. of Santa Rosa, *Newberry* (1931 NCC); Geyser Road on Sulphur Cr., *Baker* (1930 NCC); Fitch Mtn., *Reiners* (1930 NCC); between Jimtown and Mercuryville on Geysers Road, *Utech & Knight* (1986 CM). < 2700 ft. Feb.-May.

Lupinus tidestromii E. Greene. TIDESTROM'S LUPINE. Uncommon, bluff bases, dunes: s. side of mouth of Russian River, *Best* (1981), *J. & B. Guggolz* (1991 JEPS); 59 ft. elev. Bodega Head, *Clifton* (1980 PUA). < 300 ft. May-June. —*L. t.* var. *layneae* (Eastwood) Munz, indistinct.

Lupinus variicolor Steudel. Common, grassy fields, slopes, dunes: visitors center at Gualala Point Regional Park, *Knight* (1954); The Sea Ranch; near Stewarts Point, *Abrams* (1920); Salt Point; n. of Fort Ross, *Cantelow* (1940); Russian Gulch n. of Jenner, *Howe* (1970 NCC); Goat Rock, *Best* (1984); 6 mi. n. of Bodega Bay, *Eastwood & Howell* (1937); Bodega Head; Larkspur Rock e. of Bodega Bay, *Eastwood & Howell* (1939); 2 mi. n. of Bodega (probably Bodega Bay), *Bacigalupi* (1928); ½ mi. e. of Bodega Bay, *Best* (1985); near Sebastopol, *Baker* (1929 NCC); Petaluma, *Congdon* (1880); Dry Cr. near Healdsburg, *Cantelow* (1934). < 1600 ft. Apr.July.

Medicago

1. Fruit 1-seeded, reniform, not prickly, black; flower 23 mm . . . . *M. lupulina*
1: Fruit several-seeded, spirally coiled, rarely sickle shaped, prickly or not, tan or dark; flower 311 mm
  2. Corolla 611 mm, violet or violetgreen to greenish yellow, rarely yellow; perennial, usually erect; fruit not prickly . . . . *M. sativa*
  2: Corolla 3-6 mm, yellow; annual; usually decumbent or low; fruit prickly (except some *M. polymorpha*)
    3. Leaflet with a dark central spot, usually obcordate, length ± = width, leaf sometimes subpalmate . . . . *M. arabica*
    3: Leaflet without a dark central spot, usually wedge-shaped or obovate, length 1-2 × width; leaf pinnate
      4. Leaflet 8-20 mm, usually glabrous; corolla 3.5-6 mm; common . . . . *M. polymorpha*
      4: Leaflet 3-10 mm, usually strigose; corolla 2.5-4mm; uncommon . . . . *M. praecox*

★ **Medicago arabica** (Linnaeus) Hudson. SPOTTED MEDIC, SPOTTED BURCLOVER. Common, grassy places: The Sea Ranch; Willow Creek; Bodega Head, *Knight* (1963); Duncans Mills, *Rubtzoff* (1961); Fife Cr. Marsh near Guerneville, *Rubtzoff* (1960); Laguna de Santa Rosa ne. of Graton, *Rubtzoff* (1962); e. of Agua Caliente in Sonoma Valley, *Howell* (1948); Bouverie; Lake Sonoma area. < 6600 ft. Apr.-June. Native of Eurasia.

★ **Medicago lupulina** Linnaeus. BLACK MEDIC. Common, waste places: 2.3 mi. s. of Stewarts Point on Hwy. 1, *Frenkel* (1964); Bodega Head Marsh, *Rubtzoff* (1954); Austin Cr. Marsh, *Rubtzoff* (1960); Fife Cr. Marsh near Guerneville, *Rubtzoff* (1960); Guerneville Marsh, *Rubtzoff* (1962); Tresch Ranch, Walker Road near Two Rock, acc. J. & B. Guggolz (1992); Spring Lake; Pepperwood; Dry Cr. Road 6 mi. w. of US 101, *Best* (1981); Lake Sonoma area; property bordered by Chianti and Dutcher Cr. roads w. of US 101 and w. of Geyserville, acc. J. & B. Guggolz (1991). < 8300 ft. Apr.-July. Native of Europe.

★ **Medicago polymorpha** Linnaeus. CALIFORNIA BURCLOVER. Common, waste places, grassland: The Sea Ranch; s. end of The Sea Ranch, *J. H. Thomas* (1976); Cazadero quad.; Willow Creek; Bodega Head; Bodega Head Marsh, *Rubtzoff* (1960); near Guerneville at Russian River, *Rubtzoff* (1956); Fife Cr. Marsh near Guerneville, *Rubtzoff* (1960); Laguna de Santa Rosa, *Rubtzoff* (1960); e. of Cunningham near Llano Road, *Rubtzoff* (1966); one of most common weeds of Laguna de Santa Rosa drainage, *Harrison* (1978 NCC); Todd Road Preserve; Fairfield Osborn (H); Tresch Ranch, Walker Road near Two Rock, acc. J. & B. Guggolz (1992); Petaluma, *Eastwood* (1921); w. of Sears Point Raceway in Tolay Cr. drainage, acc. Patterson (1990); Bouverie; Sugarloaf Ridge; Gates Road off Calistoga Road, acc. Wells et al. (1984); Spring Lake; Santa Rosa, *Heller* (1902); Pepperwood; Fitch Mtn., *Rubtzoff* (1962); Alexander Valley and Ida Clayton Road quads. < 5000 ft. Mar.-June. Native of Eurasia. —*M. p.* var. *brevispina* (Bentham) Heynhold. *M. p.* var. *vulgaris* (Bentham) Shinners. *M. hispida* Gaertner var. *h.*, *M. h.* var. *confinis* (Koch) Burnat.

★ **Medicago praecox** de Candolle. Uncommon, gravelly ground: Fitch Mtn., *Rubtzoff* (1964). < 5000 ft. Apr.-May. Native of Mediterranean region.

★ **Medicago sativa** Linnaeus. ALFALFA. Occasional, waste places: Hwy. 1 w. of turnoff to Bodega (probably near Valley Ford), *Wells* (1983); corner US 1 and Todd Road, *Knight* (1963); Santa Rosa, *Heller* (1902); Pepperwood; Lake Sonoma area. < 5000 ft. Apr.-Oct. Native of Old World.

Melilotus

1. Flowers yellow
    2. Flowers 2-3 mm long, on pedicels < 1 mm long . . . . *M. indicus*
    2: Flowers 5-7 mm long, on pedicels 1.5-2 mm long . . . . *M. officinalis*

1: Flowers white, 4-5 mm long . . . . *M. albus*

★ **Melilotus albus** Medikus. WHITE SWEETCLOVER. Common, damp waste places: Salt Point, acc. Wells (1984); Bodega Head; ½ mi. e. of Duncans Mills, *Kawahara* (1968); 2 mi. n. of Sebastopol, *Tryner* (1938 NCC); Todd Road Preserve; corner US 101 and Todd Road, *Knight* (1963); Copeland Cr. in Rohnert Park, *Self* (1979 NCC); Pepper Road nw. of Petaluma, *Stocking* (1965 NCC); Bouverie; Spring Lake; ne. end of Santa Rosa Jr. Coll. campus, *Hagg* (1932 NCC); Pepperwood; Lake Sonoma area. < 5000 ft. May-Sept. Native of Eurasia.

★ **Melilotus indicus** Allioni. SOURCLOVER. Common, waste places: The Sea Ranch; Salt Point; Goat Rock State Park, acc. Wells (1981); Bodega Head; w. of Santa Rosa in Laguna de Santa Rosa drainage, *Harrison* (1978 NCC); garden weed at Magnolia and Marshall aves. w. of Petaluma, *Knight* (1963); Sugarloaf Ridge; Spring Lake; ½ mi. e. on Carr Ave. in Santa Rosa, *Van Steenberge* (1932 NCC); s. end of Santa Rosa Jr. Coll. campus, *Hagg & Schmidt* (1932 NCC); between Jimtown and Mercuryville on Geysers Road, *Utech & Knight* (1986 CM); ca 2000 ft. elev. Anderson Trail on Black Mtn. between Geyser Peak and upper end of Pine Flat Road, acc. J. & B. Guggolz (1992). < 5000 ft. Apr.-Oct. Native of Eurasia.

★ **Melilotus officinalis** (Linnaeus) Pallas. YELLOW SWEETCLOVER. Uncommon, pastures, waste places: Fairfield Osborn (S). < 5000 ft. May-Aug. Native of Eurasia.

Pickeringia montana Nuttall. CHAPARRAL PEA. Common, dry slopes, ridges, chaparral; rarely fruits, spreads by underground roots after fire: The Sea Ranch Timber Preserve, *Wells* (1983); Annapolis and Cazadero quads.; ca 1 mi. nw. of Camp Meeker, acc. J. & B. Guggolz (1991); Harrison Grade Road, *Utech & Knight* (1986 CM); Bouverie; 4 mi. up Trinity Road, *Ruckert* (1960); Los Alamos Road, *Knight* (1963); Sugarloaf Ridge; chaparral between Santa Rosa and Mt. St. Helena, *Voznesenskii* (1840-1841 LE); Pepperwood; Petrified Forest, *Heller* (1902); Dry Cr. hills, *Hill* (1929 NCC); Ida Clayton Road quad.; summit of Pine Flat Road, *Knight* (1963); Geyser Peak, acc. Knight & Edwards (1986); Mt. St. Helena lookout, *Ground* (1968 PUA); chaparral area sw. of Bradford Mtn. 1½ mi. w. of Dry Cr., acc. J. & B. Guggolz (1992); ca 2000 ft. elev. Anderson Trail on Black Mtn. between Geyser Peak and upper end of Pine Flat Road, acc. J. & B. Guggolz (1992); Squaw Cr., acc. Clifton (1983); The Geysers, *Carruth* (1899); Mt. Tom quad. < 5000 ft. May-Aug.

★ **Pisum sativum** Linnaeus. GARDEN PEA, FIELD PEA. Uncommon, waif from cultivation: vernal flat between Cotati and Santa Rosa, *Raven* (1950). < 3300 ft. Spring. Native of Eurasia.

★ **Robinia pseudoacacia** Linnaeus. BLACK LOCUST. Occasional, escape: Gravenstein Hwy. near Sebastopol, *Wiley* (1965 NCC); near Crane Can., *C. Nelson* (1968 NCC); near Los Guilicos school, *Wiley* (1965 NCC). < 6300 ft. May-June. Native of e. U. S. but first cultivated in Europe.

Rupertia physodes (Hooker) Grimes. CALIFORNIA-TEA. Common, openings, brush, woods: Russian settlement, *Voznesenskii* (1841 LE); Cazadero, *Bacigalupi* (1926); Russian River, *Curran* (1886); Santa Rosa Road, *Congdon* (1880); Sonoma Mtn. Road between Cooper's Grove and Enterprise Road, acc. Quibell (1991); Fairfield Osborn (H); Bouverie; Cavedale Road just above monument, acc. Quibell (1991); Sugarloaf Ridge; Gates Road off Calistoga Road, acc. Wells et al. (1984); Los Alamos Road, *Knight* (1963); Spring Lake; Pepperwood; Petrified Forest, *Heller* (1902), *Eastwood* (1915); Fitch Mtn., *Rubtzoff* (1952); Skaggs Springs, *Hawver* (1915); Lake Sonoma area; Pine Flat Road, *Knight* (1985); 1½ mi. nw. of US 101 between Cloverdale and Boonville, *Breedlove* (1962). < 8300 ft. Apr.-June. —*Psoralea p.* Hooker.

★ **Sesbania tripletii** Host. SCARLET WISTERIA TREE. Occasional, escape from cultivation: willow thicket at Wohler Road bridge on Russian River (near Trenton), *Raiche* (1990 JEPS); Royal Oaks Mobile Home Park, Petaluma, acc. Knight (1991); flood channel near Spring Lake Park, Santa Rosa, *Wagnon* (1988 NCC). < 1700 ft. Apr.-Oct. Native of Argentina. —*S. punicea* Bentham. *Daubentonia punicea* de Candolle. *Piscidia p.* Cavanilles.

★ **Spartium junceum** Linnaeus. SPANISH BROOM. Occasional, waste places: 3.1 mi. n. of Fort Ross Road on Seaview Road, *Frenkel* (1964); Spring Lake; near Petrified Forest, *Ground* (1968 PUA); scattered occasionally along US 101 from Healdsburg to Cloverdale, especially at Lytton, acc. Knight (1993). < 2000 ft. Apr.-June. Native of Mediterranean region.

Thermopsis macrophylla Hooker & Arnott. FALSE-LUPINE. Common, open areas: The Sea Ranch; near Fort Ross, *McKenzie* (1977); 3.7 w. of Cazadero on Fort Ross Road, *Atkinson* (1978 NCC); ca 1 mi. nw. of Camp Meeker, acc. J. & B. Guggolz (1991); Pitkin Marsh, *Rubtzoff* (1951); Sebastopol, *Blankenship* (1920 NCC); jct. Sonoma Mtn. Road and Enterprise Road, acc. Quibell (1991); w. side of Sonoma Mtn. (probably upper part of Lichau Road), *Stocking* (1965 NCC); 1 mi. from summit of Sonoma Mtn., *Lowry* (1983 SFSU); 2¼ mi. s. of jct. of Dana St. and Wooddale aves., Petaluma, *Quibell & McCurry* (1980 NCC); 1 mi. s. of rock quarry on US 101 s. of Petaluma on 2nd roadcut past Kastania Road, acc. Knight (1990); 1½ mi. up Cavedale Road, *Utech & Knight* (1986 CM); 1 mi. up from Trinity Road jct. on Cavedale Road, acc. Knight (1986); Sugarloaf Ridge; Bennett Valley Road, *Stocking* (1964 NCC); Rincon Grade, *L. Roof* (1933 NCC); Pepperwood; Petrified Forest, *Eastwood* (1915); near Contact Mine in geothermal area, *Malloch* (1977 PG&E). < 7000 ft. May-June.

Trifolium: (Key by Catherine Best)

1. Heads on the same plant both involucrate and not involucrate . . . . *T. buckwestiorum*
1: Heads on the same plant either all involucrate or all not involucrate
  2. Heads involucrate, the involucre reduced to a ring in *T. depauperatum* var. *depauperatum*
    3. Corolla conspicuously inflated
      4. Corolla 10-20 mm long, generally yellowish . . . . *T. fucatum*
      4: Corolla 4-9 mm long, reddish, pink or cream
        5. Involucre reduced to a ring . . . . *T. depauperatum* var. *d.*
        5: Involucre lobed
          6. Involucral bracts wide, irregularly shaped, often ½ fused, margins wide, scarious . . . . *T. d.* var. *amplectens*
          6: Involucral bracts not as above
            7. Involucral bracts < 1.5 mm long, fused at the base, heads 1-1.5cm wide . . . . *T. d. var hydrophilum*
            7: Involucral bracts 2-2.5 mm long, often ± free at the base; heads 0.5-1 cm wide . . . . *T. d.* var. *truncatum*
    3: Corolla not inflated (*T. barbigerum* inflated in fruit)
      8. Involucre bowl-shaped
        9. Lobes of the involucre toothed
          10. Calyx-teeth plumose, the lower usually forked
            11. Corolla 5-6 mm long, < or = the calyx . . . . . . *T. barbigerum* var. *barbigerum*
            11: Corolla 10-12 mm long, > the calyx . . . . *T. b.* var. *andrewsii*
          10: Calyx-teeth ciliate, all simple . . . . *T. microdon*
        9: Lobes of the involucre entire . . . . *T. microcephalum*
      8: Involucre flat or of separate segments
        12. Plants annual
          13. Plants glandular-pubescent . . . . *T. obtusiflorum*
          13: Plants not glandular-pubescent
            14. Calyx-teeth dilated, abruptly narrowed to an awnlike tip, frequently tridentate . . . . *T. willdenovii*
            14: Calxy-teeth not dilated, gradually narrowed to an awnlike tip
              15. Corolla much longer than the 20-nerved calyx, flowers many *T. variegatum*
              15: Corolla ca as long as the 10-nerved calyx, flowers few . . . *T. oliganthum*
        12: Plants perennial
          16. Flowers subtended by narrow bracts, the outer connate and involucral; calyx inflated in age; flowers pink . . . . *T. fragiferum*
          16: Flowers not subtended by bracts, calyx not inflated in age, flowers reddish

. . . . *T. wormskioeldii*

2: Heads not involucrate
17. Plants annual
18. Calyces tomentose, becoming inflated in age; heads cottony balls in fruit, up to 1 cm. in diameter when pressed . . . . *T. tomentosum*
18: Calyces and heads not as above
19. Heads sessile above the uppermost leaves and stipules which may simulate an involucre
20. Plants glabrous . . . . *T. glomeratum*
20: Plants pubescent
21. Flowers showy, rose color . . . . *T hirtum*
21: Flowers not showy, pink or purple
22. Calyx conspicuously veined, inflated in age, heads rarely in pairs . . . . *T. striatum*
22: Calyx not conspicuously veined, not inflated in age, heads usually in pairs . . . . *T. macraei*
19: Heads not sessile above the uppermost leaves and stipules
23. Flowers sessile, not reflexed in age
24. Heads cylindrical, corolla pink or crimson
25. Corolla pink, calyx lobes becoming hard in age . . . *T. angustifolium*
25: Corolla crimson, calyx lobes not becoming hard in age . *T. incarnatum*
24: Heads globose to ovoid, corolla purple
26. Corolla > or ca = the calyx in length
27. Stems stout, corolla 12-15 mm long . . . . *T. amoenum*
27: Stems slender, corolla 5-10 mm long
28. Corolla exserted from the calyx . . . . *T. albopurpureum* var. *dichotomum*
28: Corolla ca = the calyx in length *T. a.* var. *albopurpureum*
26: Corolla < calyx in length . . . . *T. a.* var. *olivaceum*
23: Flowers pedicellate, reflexed in age
29. Flowers cream, pink or purple
30. Inflorescence of few fertile flowers with petals, later many sterile flowers without petals develop forming a bur, (petals cream, upper half of the calyx tube usually dark red) . . . . *T. subterraneum*
30: Inflorescence with many fertile flowers, no sterile flowers forming a bur
31. Peduncle < the subtending leaves in length . . . . . *T. cernuum*
31: Peduncle > the subtending leaves in length
32. Peduncle glabrous
33. Calyx-teeth ciliate with short flat appendages *T. ciliolatum*
33: Calyx-teeth not ciliate . . . . *T. gracilentum*
32: Peduncle villous near the top
34. Leaflets deeply bifid at the apex . *T. bifidum* var. *b.*
34: Leaflets entire or emarginate at the apex . . . . *T. b.* var. *decipiens*
29: Flowers yellow
35. Heads 5-7 mm broad, banner not striate . . . . *T. dubium*
35: Heads 8-10 mm broad, banner obviously striate . . . . . *T. campestre*
17: Plants perennial
36. Heads terminal or subterminal, usually sessile above uppermost leaves and stipules *T. pratense*
36: Heads axillary, not sessile above the uppermost leaves and stipules
37. Stems creeping, rooting at the nodes; calyx glabrous; flowers white . . . . *T. repens*
37: Stems erect or nearly so, not rooting at the nodes; calyx pubescent in the sinuses; flowers pink . . . . *T. hybridum*

Trifolium albopurpureum Torrey & A. Gray var. **albopurpureum**. RANCHERIA CLOVER. INDIAN CLOVER. Occasional, grassy slopes, brushy places: Willow Creek Road s. of Russian River, acc. Best (1981); Petaluma, *Knowlton* (1937 NCC); Spring Mtn. Road w. of St. Helena, *Howell* (1931). Usually < 3700 ft. Mar.-June.

Trifolium albopurpureum var. **dichotomum** (Hooker & Arnott) Isely. INDIAN CLOVER. Occasional, rocky slopes, ridges: Monte Rio, *Armstrong* (1930 NCC); Morelli Lane n. of Occidental on serpentine, *Best* (1981); Sebastopol quad.; near Kenwood, *Baker* (1928 NCC); Mt. Hood, acc. Baker; Mark West Springs, *Heller & Brown* (1923); Leslie Road near jct. with Mark West Springs Road, *Wagnon* (1989

NCC); Pepperwood; w. of Healdsburg, *Heller & Brown* (1923); Lake Sonoma area on Rockpile Road ridge on serpentine, *Raiche* (1986 NCC); Ida Clayton Road quad.; 2400 ft. elev. Little Geysers below cypresses, *Malloch* (1977 PG&E). —*T. petrophilum* Heller, type locality: Mt. St. Helena, *C. F. Baker* (1907 U). < 4300 ft. Mar.-May.

Trifolium albopurpureum var. **olivaceum** (E. Greene) Isely. VALLEY CLOVER. Uncommon, open grassy areas: Calistoga Road, *Novak* (1980 SRJC); Alexander Valley quad.; Squaw Cr., acc. Clifton (1984). < 3000 ft. Mar.-May. —*T. o.* var. *columbinum* (E. Greene) Jepson. *T. o.* E. Greene.

Trifolium amoenum E. Greene. SHOWY INDIAN CLOVER. At times, presumed to have been extinct, rich fields, swales: Fort Ross-Russian River-Bodega areas, *Voznesenskii* (1840-1841 LE); w. of Occidental, *Connors* (1993); Bodega, *Chandler* (1900 UC); Pitkin Marsh, acc. Baker (1954); 4 mi w. of Santa Rosa near Wright School, acc. Baker (1954); O'Farrell Hill region w. of Sebastopol, *Wagnon* (1940 NCC); w. of Valley Ford, *Heller* (1940); Stony Point, *Howell* (1938); vicinity of Petaluma, *Tracy* (1921); Petaluma-Marshall Road 2 mi. from Western Ave., *Hoover* (1969 SBBG); Santa Rosa, *Congdon* (1880); roadside 1 mi. n. of Santa Rosa High School, *McDonald* (1929 NCC). < 500 ft. Apr.-June.

★ **Trifolium angustifolium** Linnaeus. Uncommon, disturbed areas: Pepperwood, acc. Best (1995). < 1000 ft. Apr.-June. Native of Mediterranian region.

Trifolium barbigerum Torrey var. **andrewsii** A. Gray, SONOMA CLOVER, GRAY'S CLOVER. Wet meadows, foothill slopes: Skylane Blvd. at n. edge of county airport, *Albright* (1979 SRJC); Windsor, *Howell* (1943); off Shiloh Road near Windsor, *Harrison* (1982 SRJC); Fulton, *Crampton* (1955); Trenton near Russian River, *Crampton* (1955). < 2000 ft. Apr.-June. —*T. grayi* Lojacono-Pojero.

Trifolium barbigerum var. **barbigerum**. COLONY CLOVER. Common, coastal, inner hills: The Sea Ranch, *Wells* (1981); Salt Point State Park, *Wagnon* (1989 NCC); Fort Ross, *Heller* (1903); Wright's Beach, acc. Baker; Bodega Head; ca 1 mi. nw. of Camp Meeker, *J. & B. Guggolz* (1991 UC); Todd Road Preserve; Fairfield Osborn (H); Santa Rosa, *Eastwood* (1921); Pepperwood; Fitch Mtn., *Hill* (1930 NCC); Lake Sonoma area. Usually < 2300 ft. Apr.-June.

Trifolium bifidum A. Gray var. **bifidum.** PINOLE CLOVER. Uncommon, grassy places: The Sea Ranch; Duncans Mills quad.; Willow Cr. Road 6 mi. se. of the mouth of the Russian River, *Best* (1981); Vacation Beach near Guerneville, *Rubtzoff* (1956); Fairfield Osborn (H); Bouverie; Adobe Can., acc. Baker (1931); Lake Sonoma, Geyserville and Alexander Valley quads. < 2000 ft. Apr.-June.

Trifolium bifidum var. **decipiens** E. Greene. Occasional, grassy places: Cazadero, *Bacigalupi* (1926); Santa Rosa, *Heller & Brown* (1902); s. of Hwy. 12 w. of Santa Rosa in Laguna de Santa Rosa drainage, *Harrison* (1978 NCC); Independence School near Healdsburg, *Abrams* (1918); Preston n. of Healdsburg, *Carlson* (1917). < 3300 ft. Apr.-June.

Trifolium buckwestiorum Isely. Rare, known only from Sonoma and Santa Cruz counties, waste or grassy areas: n. of Cazadero, *Connnors* (1995 BML); Petaluma-Tomales Road, opposite the Coast Guard Station, acc. R. Morgan (1995). < 600 ft. Apr.-June.

★ **Trifolium campestre** Schreber. HOP CLOVER. Occasional, lawns, garden areas: near El Verano, *Best* (1980); trail to pond in Jack London State Park, *Wagnon* (1989 NCC); Shiloh Co. Park n. of Santa Rosa, *Wagnon* (1989 NCC); Pepperwood. Usually < 1000 ft. May-Sept. Introduced from Europe. —*T. procumbens* Linnaeus, misapplied.

★ **Trifolium cernuum** Brotero. MEDITERRANEAN CLOVER. Rare, possibly new in CA, grassy places, lake shores: Spring Lake Village; Spring Lake. Usually < 1000 ft. Apr.-June. Native of sw. Europe.

Trifolium ciliolatum Bentham. TREE CLOVER. Occasional, inner hills, grassy flats: Camp Meeker, *Armstrong* (1930); s. of Hwy. 12 w. of Santa Rosa in Laguna de Santa Rosa drainage, *Harrison* (1978 NCC); Fairfield Osborn, *Quibell* (1972 NCC); e. foothills of Sonoma Mtn. w. of Arnold Drive, *Rubtzoff* (1971); Bouverie; Bennett Valley, acc. Baker; along Leslie Road near jct. with Mark West Road, *Wagnon* (1989 NCC); Geyserville and Alexander Valley quads. < 5700 ft. Mar.-June.

Trifolium depauperatum Desvaux var. **amplectens** (Torrey & A. Gray) L. F. McDermott. PUFFY CLOVER, BLADDER CLOVER. Occasional, grassland: eastern foothills of Sonoma Mts. on Sobre Vista Road 2.9 mi. w. of Arnold Drive, *Rubtzoff* (1971); s. of Hwy. 12 w. of Santa Rosa in Laguna de Santa Rosa drainage, *Harrison* (1978 NCC); Fairfield Osborn (H); Bennett Valley, *Stocking* (1964 NCC); Sugarloaf Ridge; Spring Lake; Santa Rosa, *Stocking* (1969 NCC); near Healdsburg, *Hill* (1930); Lake Sonoma area. < 2600 ft. Apr.-June. —*T. a.* Torrey & A. Gray.

Trifolium depauperatum var. **depauperatum**. DWARF SAC CLOVER, ALKALI CLOVER. Common, inner hills, moist flats: Fort Ross and Cazadero quads.; s. of Hwy. 12 w. of Santa Rosa in Laguna de Santa Rosa drainage, *Harrison* (1978 NCC); Todd Road Preserve; Agua Caliente in Sonoma Valley, *Howell* (1948);

Pepperwood; near Healdsburg, *Abrams* (1918); Lake Sonoma area on Rockpile Road ridge on serpentine, *Raiche* (1986). < 3000 ft. Mar.-May. Native also in Chile. Acc. Jepson. *Fl. of CA, vol. 2 , p. 291.*, (1936) cites a form with narrow leaflets from Sonoma Valley as var. **angustatum** (E. Greene) Jepson.

Trifolium depauperatum var. **hydrophilum** (E. Grene) Isely. Uncommon, salt marshes, open areas in alkaline soils: 2 mi se. of jct. of Hwy 121 and Hwy 116, e. of Arnold Drive near Sonoma Creek, *Connors* (1995 BML). < 900 ft. Apr.-June.

Trifolium depauperatum var. **truncatum** (E. Greene) Isely. Occasional, interior hills, grassy places: near Sonoma, *Brewer* (1962); Bennett Valley, *Heller* (1902); Pepperwood; Lake Sonoma area. < 2700 ft. Apr.-June. —*T. amplectens* var. *t.* (E. Greene) Jepson.

★ **Trifolium dubium** Sibthorp. SHAMROCK. Common, fields, lawns, waste places: The Sea Ranch, *J. H. Thomas* (1976), *Wells* (1981); Stewarts Point, acc. Baker; Salt Point; Fort Ross, *Wagnon* (1989); Pitkin Marsh, *Rubtzoff* (1951); Laguna de Santa Rosa, *Rubtzoff* (1960); near Todd Road in Laguna de Santa Rosa drainage, *Harrison* (1978 NCC); Todd Road Preserve; Fairfield Osborn (H); Bouverie; Spring Lake; n. side of county airport on Skylane Blvd., *Wagnon* (1989 NCC); Pepperwood; Lake Sonoma area. Usually < 1600 ft. May-July. Introduced from the Old World.

★ **Trifolium fragiferum** Linnaeus. STRAWBERRY CLOVER. Common, waste places: The Sea Ranch, *Wells* (1981); Stoetz Lane 2 mi. ne. of Occidental, *Best* (1980); Todd Road near Cunningham, *Best* (1980); Spring Lake; Rincon Valley, acc. Wells (1984). < 5000 ft. May-Sept. Native of Europe.

Trifolium fucatum Lindley. BULL CLOVER. Common, widespread, wet hills, seepages: Bodega Point, *Eastwood* (1915); Bodega, *Clifton* (1974 PUA); n. of Occidental, acc. Best; Bennett Valley, acc. Baker; Mark West Springs, *Heller & Brown* (1902); Pepperwood; Lake Sonoma area; property bordered by Chianti and Dutcher Cr. roads w. of US 101 and w. of Geyserville, acc. J. & B. Guggolz (1991). This is a variable species and includes several forms that have been described as species. A speculative form is *T. virescens* E. Greene, moist flats: s. of Hwy. 12 w. of Santa Rosa in Laguna de Santa Rosa drainage, *Harrison* (1978 NCC). The type of *T. fucatum* was a plant cultivated in England from seed collected in California by David Douglas. < 3700 ft. Apr.-June. —*T. physopetalum* Fischer & C. Meyer, type collected by Russians in the 1830s.

★ **Trifolium glomeratum** Linnaeus. Becoming established: edge of lawn at Salt Point Lodge at Ocean Cove, *Wagnon* (1989 NCC); Fort Ross area near school's reservoir, *Wagnon* (1989 NCC); s. of Hwy. 12 w. of Santa Rosa in Laguna de Santa Rosa drainage, *Harrison* (1978 NCC); Sonoma, *Aden* (1982); Spring Lake; edge of Channel Drive near Santa Rosa Cr. bridge near Montgomery Drive, *Wagnon* (1989 NCC). < 4500 ft. Apr.-June. Native of Europe.

Trifolium gracilentum Torrey & A. Gray. PINPOINT CLOVER. Occasional, grassy slopes: Cazadero quad.; Bodega Point, *Eastwood* (1915); Bodega Head, *Connors* (1992 BML); Todd Road Preserve; Petaluma, *Congdon* (1880); Sonoma, *Bigelow,* (1857); Santa Rosa Cr., acc. Jepson, *Fl. of Ca. vol. 2, p. 309* (1935); Healdsburg, *Heller & Brown* (1902); Felta Cr. area sw. of Healdsburg, *Bice* (1895 NCC); Lake Sonoma area on Rockpile Road ridge on serpentine, *Raiche* (1986 NCC). < 6000 ft. Apr.-June.

★ **Trifolium hirtum** Allioni. ROSE CLOVER. Common, unpastured fields, weedy: Fairfield Osborn (H); Bouverie; Glen Ellen, acc. Best (1984); Annadel; along Santa Rosa Cr. in e. Santa Rosa, *Wagnon* (1989 NCC); Spring Lake; w. of Santa Rosa 0.1 mi. s. of Piner Road and ¼ mi. w. of Marlow Road, *Harrison* (1978 NCC); Pepperwood; Lake Sonoma area; property bordered by Chianti and Dutcher Cr. roads w. of US 101 and w. of Geyserville, acc. J. & B. Guggolz (1991); ca 2000 ft. elev. Anderson Trail on Black Mtn. between Geyser Peak and upper end of Pine Flat Road, acc. J. & B. Guggolz (1992); Cloverdale, *Crampton* (1957). < 6800 ft. Apr.-May. Introduced from Eurasia and n. Africa.

★ **Trifolium hybridum** Linnaeus. ALSIKE CLOVER. Uncommon, lawns, damp places, meadows: Guerneville Marsh, *Rubtzoff* (1954); Spring Lake; Spring Lake Village. < 5000 ft. May-Oct. Introduced from Europe.

★ **Trifolium incarnatum** Linnaeus. CRIMSON CLOVER. Occasional, pastures, roadsides: se. of mouth of Russian River, *Best* (1981); Gravenstein Hwy. near Sebastopol, *Best* (1981); s. of W. College Ave. in Laguna de Santa Rosa drainage, *Harrison* (1978 NCC); Spring Lake; Pepperwood; Healdsburg, acc. *Jepson, Fl. of CA, vol. 2, p. 306*, (1936); between Jimtown and Mercuryville on Geysers Road, *Utech & Knight* (1986 NCC). Usually < 1000 ft. May-Aug. Native of Europe.

Trifolium macraei Hooker & Arnott. CHILE CLOVER. Occasional, sandy well-drained soil, coastal: Stewarts Point, acc. Baker (1936); Salt Point, *Wagnon* (1989 NCC); Bodega Point, *Eastwood* (1915); Bodega Head; Felta Cr. area sw. of Healdsburg, *Bice* (1895 NCC). < 2000 ft. Apr.-June. Also native in Chile.

Trifolium microcephalum Pursh. MAIDEN CLOVER. Common, hills and valleys: The Sea Ranch acc. Wells (1981); Timber Cove, *Wagnon* (1989 NCC); Fort Ross quad.; Austin Cr. n. of Cazadero, *Rubtzoff* (1970); Duncans Mills quad.; near Todd Road in Laguna de Santa Rosa drainage, *Harrison* (1978 NCC); Petaluma, *Congdon* (1880); Fairfield Osborn (H); Sugarloaf Ridge; Franz Cr. on Chalk Hill Road, *Wagnon* (1989 NCC); Pepperwood; Healdsburg, *Heller & Brown* (1902); Lake Sonoma area; Geyserville, Alexander Valley and Ida Clayton Road quads.; 2400 ft. elev. Little Geysers, *Malloch* (1977 PG&E). < 9000 ft. Apr.-Aug.

Trifolium microdon Hooker & Arnott. SQUARE-HEAD CLOVER. Valparaiso Clover. Common, grasslands: Bodega Head, *Connors* (1992 BML); Bodega Road e. of Valley Ford, *Howell* (1930); Occidental Marsh, *Rubtzoff* (1957); 2 sites w. of Santa Rosa in Laguna de Santa Rosa drainage, *Harrison* (1978 NCC); Todd Road Preserve; Sonoma Mtn., *Rubtzoff* (1971); Sonoma, *Brewer* (1862); Sugarloaf Ridge; Healdsburg, *Hill* (1929); Lake Sonoma quad. < 3600 ft. Mar.-June. Also native in Chile.

Trifolium obtusiflorum Hooker. CREEK CLOVER. Uncommon, damp hillsides, gravelly stream beds: Cazadero, *Rubtzoff* (1970); Sonoma, *Heller* (1902); Fitch Mtn., *Hill* (1929 NCC); Preston n. of Cloverdale, *Carlson* (1917). < 6000 ft. Apr.-July.

Trifolium oliganthum Steudel. LANKY CLOVER, FEW-FLOWERED CLOVER. Occasional, on grassy slopes: Occidental, acc. Baker (1936); Bouverie; Santa Rosa Cr., *Heller & Brown* (1902); Pepperwood; Fitch Mtn., *Rubtzoff* (1952); Squaw Cr., acc. Clifton (1984). —*T. o.* var. *sonomense* E. Greene, type locality: Knights Valley (1894). < 3300 ft. Mar.-June.

★ **Trifolium pratense** Linnaeus. RED CLOVER. Widely cultivated, occasional, growing spontaneously: The Sea Ranch, acc. *Mahaffey* (1990); Guerneville Marsh, *Rubtzoff* (1970); Forestville, *Tryner* (1939 NCC); 2 sites w. of Santa Rosa in Laguna de Santa Rosa drainage, *Harrison* (1978 NCC); Spring Lake. Especially < 3300 ft. Apr.-Oct. Native of the Old World.

★ **Trifolium repens** Linnaeus. WHITE CLOVER. Common, frequently planted in lawn mixtures, waste places. Ladino and Salina clovers are cultivars which are widely grown in the county: 7 mi. s. of Gualala, *J. H. Thomas* (1976); Willow Creek; Bodega Head; Vacation Beach on Russian River, *Rubtzoff* (1959); Forest Hills across the river from Hacienda, *Rubtzoff* (1968); Pitkin Marsh, *Rubtzoff* (1951); Laguna de Santa Rosa ne. of Graton near Laguna Road, *Rubtzoff* (1960); s. of W. College Ave. in Laguna de Santa Rosa drainage, *Harrison* (1978 NCC); Todd Road Preserve; Spring Lake. Usually < 5000 ft. Apr.-Dec. Introduced from Eurasia.

★ **Trifolium striatum** Linnaeus. Uncommon, recently introduced, interior grassland: Pepperwood. Native of Europe. < 3000 ft. Apr.-June.

★ **Trifolium subterraneum** Linnaeus. SUBTERRANEAN CLOVER. Cultivated, widely naturalized: Pitkin Marsh, *Dempster* (1958), *Rubtzoff* (1970); Laguna de Santa Rosa, *Rubtzoff* (1966); Todd Road Preserve; Fairfield Osborn (H); Sonoma Valley, *Howell* (1955); Spring Lake; Pepperwood; Lake Sonoma area; property bordered by Chianti and Dutcher Cr. roads w. of US 101 and w. of Geyserville, acc. J. & B. Guggolz (1991); between Jimtown and Mercuryville on Geysers Road, *Utech & Knight* (1986 CM). < 3300 ft. Apr.June. Native of Old World.

★ **Trifolium tomentosum** Linneaus. Uncommon, grassy areas: 2 mi. se. of jct. of Hwy 121 & Hwy 116 e. of Arnold Dr. near Sonoma Cr., *Connors* (1996 BML). < 900 ft. April-June. Native of the Mediterranean region.

Trifolium variegatum Nuttall. WHITE TIP CLOVER. Common, pastures, damp meadows, moist hillsides: The Sea Ranch, *Wells* (1984); Timber Cove, *Wagnon* (1989 NCC); 1½ mi. n. of Jenner on Hwy. 1, *Huntington* (1989 NCC); Cazadero quad.; Willow Cr. Marsh, *Rubtzoff* (1959); Harrison Grade Road, *Wagnon* (1989 NCC); 3 sites w. of Santa Rosa in Laguna de Santa Rosa drainage, *Harrison* (1978 NCC); Todd Road Preserve; Grill Ranch w. of Cotati, *J. & B. Guggolz* (1989 NCC); Sonoma State Univ. campus, *Stocking* (1969 NCC); Fairfield Osborn; Bouverie; Santa Rosa, *Heller* (1902); near Healdsburg, *Hill* (1930 NCC); property bordered by Chianti and Dutcher Cr. roads w. of US 101 and w. of Geyserville, acc. J. & B. Guggolz (1991); w. flank Mt. St. Helena, *Grummer* (1979 PUA). < 3700 ft. Apr.-June. —*T. appendiculatum* Nuttall.

Trifolium willdenovii Sprengel. TOMCAT CLOVER. Common, widespread, coastal and interior hills, meadows: The Sea Ranch, *Wells* (1984); Bodega Head, *Connors* (1992 BML); 2 sites w. of Santa Rosa in Laguna de Santa Rosa drainage, *Harrison* (1978 NCC); Todd Road Preserve (S&T); Fairfield Osborn (H); Sonoma Valley, acc. Baker; Bouverie; Sugarloaf Ridge; Pepperwood; vernal pool 2 mi. nw. of Windsor by Old Redwood Hwy., *de Nevers* (1981); Lake Sonoma area; property bordered by Chianti and Dutcher Cr. roads w. of US 101 and w. of Geyserville, acc. J. & B. Guggolz (1991); Ida Clayton Road quad.; near Contact Mine on Pine Flat Road, *Malloch* (1977 PG&E). < 5700 ft. Mar.-June. —*T. tridentatum* Lindley.

Trifolium wormskioeldii Lehmann. COW CLOVER. Frequent, marshy places: The Sea Ranch, *Thomas* (1976); Salt Point; Arched Rock quad.; Bodega Head; Pitkin Marsh, *Rubtzoff* (1951); Cunningham Marsh, *Howell* (1947); Todd Road Preserve; Spring Lake; Lake Sonoma area. < 10500 ft. May-Oct. —*T. involucratum* of authors. *T. fimbriatum* Lindley.

★ **Ulex europaea** Linnaeus. GORSE, FURZE. Locally common, adventive: s. of town of Bodega, *Cantelow* (1937), *Knight* (1963), *J. H. Thomas* (1966); Valley Ford-Bodega Bay Hwy., *Knight* (1963), *Clifton* (1981 PUA); 1 mi. s. of Bodega Bay, *Juul* (1973 NCC); w. of Sebastopol, *Heller & Brown* (1902); large population in Blucher Cr. watershed sw. of Cunningham, acc. B. Guggolz (1991). < 1300 ft. Feb.-July. Native of Europe.

Vicia (Key ± acc. D. Isely, *Hickman* [1993]).

1. Plant erect; leaflet 5-12 cm; tendrils 0 . . . . *V. faba*
1: Plant sprawling or climbing; leaflet usually 1-4 cm; tendrils usually present
  2. Inflorescence a sessile or barely peduncled cluster of 1-4 pedicelled flowers
    3. Flower 5-6 mm . . . . *V. lathyroides*
    3: Flower 10-30 mm
      4. Corolla pink-purple to whitish, banner glabrous; fruit initially hairy, quickly glabrous
        5. Calyx 7-12 mm; corolla 1-1.8 cm, pink-purple to ± white; naturalized, not cultivated . . . . *V. sativa* ssp. *nigra*
        5: Calyx 10-15 mm; corolla 1.8-3 cm, usually pink-purple; naturalized and cultivated . . . . *V. s.* ssp. *sativa*
      4: Corolla yellow or yellow-white, sometimes purple-blotched, banner puberulent on back; fruit persistently hairy, or both
        6. Banner puberulent on back; calyx lobes ± =; fruit hairs not from conspicuous tubercles . . . . *V. pannonica*
        6: Banner glabrous; calyx lobes unequal; fruit hairs from tubercles obvious on lower margin of fruit . . . . *V. lutea*
  2: Inflorescence a peduncled raceme
    7. Flower 3-9 mm
      8. Fruit hairy, stalklike base 0 . . . . *V. hirsuta*
      8: Fruit glabrous or ± strigose, stalklike base ± 1-2 mm
        9. Fruit rounded at tip, with a short point on 1 side, 1-1.3 cm; leaflets 4-10, linear to elliptic; peduncle threadlike 1-3 cm, often curved, 1-3 flowered . . . . *V. tetrasperma*
        9: Fruit tapered at tip, 1-3.5 cm; leaflets or peduncle not as previous; flowers crowded or 1 . . . . *V. ludoviciana*
    7: Flower > 9 mm
      10. Leaf usually 1-1.5 dm; leaflets 16-24, larger ones 2.5-4 cm . . . . *V. gigantea*
      10: Leaf rarely > 1 dm; leaflets either fewer or smaller than previous
        11. Flowers not crowded, on 1 side of axis (except sometimes when pressed), > 1.5-2.5 cm, when pressed, 2.5-3.5× width; stalklike fruit base 2-5 mm . . . . *V. americana*
        11: Flowers crowded, usually on 1 side of axis, usually < 1.6 cm, length when pressed usually 4-6 × width; stalklike fruit base 1.5-3 mm
          12. Inflorescence ± = subtending leaf, 3-12-flowered; corolla dark or red-purple; fruit puberulent . . . . *V. benghalensis*
          12: Inflorescence often > subtending leaf, usually 10-25-flowered; corolla blue-purple or lavender to ± white; fruit glabrous
            13. Calyx generally obliquely attached but only slightly lopsided-swollen at base; flower length when pressed ± 3- 3.5 × width; fruit 6-7 mm wide; perennial . . . . *V. cracca*
            13: Calyx both obliquely attached and lopsided-swollen at base; flowers when pressed 4-6 × width; fruit 6-10 mm wide; annual
              14. Hairs on stems, leaves 0 or not obvious ± 1 mm; lower calyx lobe narrowly lanceolate, often 1-2 mm . . . . *V. villosa* ssp. *varia*
              14: Hairs on upper stems, leaves obvious 1-2 mm; lower calyx lobe needle- or threadlike 2-4 mm . . . . *V. v.* ssp. *villosa*

Vicia americana Willdenow var. **americana**. AMERICAN VETCH. Common, open places: The Sea Ranch, acc. *Mahaffey* (1990); Annapolis quad.; Salt Point; Bodega Head; Bodega Bay; Larkspur Rock 1 mi. e. of Bodega Bay, *Baker* (1932 NCC); Freestone Marsh, *Rubtzoff* (1959); fork of road past Green Valley School w. of Graton, *Kindig* (1933 NCC); Fairfield Osborn (H); near Cooper's Grove on Sonoma Mtn. Road, *Baker* (1930 NCC); Bouverie; Adobe Can. near Kenwood Springs, *L. Roof* (1932 NCC); Pepperwood; Lake Sonoma area. < 8000 ft. Apr.-June. —*V. a.* ssp. *oregana* (Nuttall) Abrams. *V. a.* var. *truncata* (Nuttall) Brewer. *V. a.* var. *linearis* S. Watson. *V. californica* E. Greene.

★ **Vicia benghalensis** Linnaeus. PURPLE VETCH. Common, waste places: 1 mi. e. of Sebastopol on Bodega Hwy., *Baker* (1935 NCC); widespread in Laguna de Santa Rosa drainage, *Harrison* (1978 NCC); Spring Lake; 2 mi. n. of Windsor, *Howell* (1933); vineyard 2 mi. s. of Asti, *Bacigalupi* (1948); ½ mi. s. of Asti, *Wiggins & Ferris* (1943). < 700 ft. May-June. Native of Mediterranean region.

★ **Vicia cracca** Linnaeus. COW VETCH, BLUE VETCH, BIRD VETCH. Uncommon, roadsides: between Valley Ford and Bodega Bay on n. side of Hwy. 1 on steep hillside, *Galey* (1991 DAV). < 5000 ft. June-Aug. Native of Eurasia.

★ **Vicia faba** Linnaeus. HORSE BEAN, BROAD BEAN. Uncommon, cultivated, escaping: pastures around Lakeville, acc. Knight (1990); Petaluma Hill Road, acc. Knight (1990). < 700 ft. Native of Eurasia.

Vicia gigantea Hooker. Occasional, moist places: The Sea Ranch; Salt Point; 2.3 mi. s. of Stewarts Point, *Cantelow* (1939); Bodega Bay, *Heller & Brown* (1902); Bodega Head. < 700 ft. Mar.-June.

★ **Vicia hirsuta** (Linnaeus) S. F. Gray. Occasional, adventive: Sonoma State Univ. campus, *Arnold* (1963 NCC); Rincon Valley, *Arnold* (1963 NCC); Lake Sonoma area. Usually < 700 ft. May-July. Native of Europe.

★ **Vicia lathyroides** Linnaeus. Rare, wet areas: Pepperwood, *de Nevers* (1992). < 3300 ft. Apr.-June. Native of most of Europe northward to sw. Finland.

Vicia ludoviciana Nuttall. Uncommon, brushy, wooded or grassy areas: 1 mi. w. of Petaluma in old quarry, *Knowlton* (1937 NCC); hillside opposite Maplewood (M. Baker's property in Adobe Can.), acc. Baker (1936). < 3300 ft. Apr.-June. —*V. exigua* Nuttall.

★ **Vicia lutea** Linnaeus. Uncommon, established in Bay Region ±: Pepperwood. < 1000 ft. Apr.-June. Native of s. Europe and n. Africa

★ **Vicia pannonica** Crantz. HUNGARIAN VETCH. Occasional, adventive: Petaluma Hill Road, *Knight* (1965); Sonoma Hwy. near Petaluma, *Baker* (1945 NCC); w. outskirts of Sonoma, *Baker* (1945 NCC); field near Rincon Valley School, *Baker* (1931 NCC). < 700 ft. Native of Europe.

★ **Vicia sativa** Linnaeus ssp. **nigra** (Linnaeus) Erhart. COMMON VETCH, NARROW-LEAVED VETCH. Common, waste places: Salt Point State Park at milepost 41 s. of Stump Beach, *Wells* (1981); mouth of Salmon Cr., *Baker* (1930 NCC); Larkspur Rock e. of Bodega Bay, *Baker* (1938 NCC); jct. Valley Ford-Freestone Road and Valley Ford Cutoff, *Utech & Knight* (1986 CM); Fife Cr. at Guerneville, *Rubtzoff* (1960); Pitkin Marsh, *Rubtzoff* (1951); Roblar Road in Stony Point district, *Knight* (1963); El Verano, *Baker* (1940 NCC); Bouverie; Spring Lake; Pepperwood; 680 ft. elev. ca 8 mi. ene. of Healdsburg, *J. & B. Guggolz* (1991 NCC); Geyserville & Alexander Valley quads. < 5000 ft. Apr.-June. Native of Old World. —*V. angustifolia* Reichard var. *a.*

★ **Vicia sativa** ssp. **sativa**. SPRING VETCH. Common, naturalized, waste places: The Sea Ranch, acc. Wells (1985); between Hwy. 1 and ocean s. of s. end of Myers Grade, *J. & B. Guggolz* (1991 NCC); Morelli Lane 2 mi. ne. of Occidental, *Best* (1980); Laguna de Santa Rosa e. of Cunningham, *Rubtzoff* (1966 NCC); Fairfield Osborn (S); Bouverie; Sugarloaf Ridge; between Santa Rosa and Agua Caliente on Hwy. 12, *Heller & Brown* (1902); Sugarloaf Ridge; 1 mi. e. of Santa Rosa, *Kindig* (1933 NCC); Spring Lake; 3 mi. w. of Santa Rosa on Hwy. 12, *Hagg* (1934 NCC); Pepperwood; Fitch Mtn., *Rubtzoff (1952).* < 5000 ft. Apr.-July. Native of Europe. —*V. angustifolia* var. *segetalis* (Thuillier) Koch.

★ **Vicia tetrasperma** (Linnaeus) Schreber. Occasional, ± naturalized: Salmon Cr. Marsh, *Rubtzoff* (1968); between Hwy. 1 and ocean s. of s. end of Myers Grade, *J. & B. Guggolz* (1991 NCC); Bohemian Hwy. between Freestone and Occidental on e. side of road, *J. & B. Guggolz* (1989 NCC); Freestone Marsh, *Rubtzoff* (1959 NCC); 1 mi. se. of Valley Ford, *T. C. Fuller* (1966); Spring Lake. Especially < 3300 ft. May-July. Native of Europe.

★ **Vicia villosa** Roth ssp. **varia** (Host) Corbiere. Occasional, adventive: The Sea Ranch, *Howell* (1981), *Wells* (1981); near Cunningham, *Howell* (1947); 10 mi. s. of Santa Rosa, *Eastwood & Howell* (1936); E. Cotati Ave. and Snyder Lane in Rohnert Park, *Waitman* (1985 NCC); Fairfield Osborn (H); Chileno Valley, *Howell* (1944); Bouverie; s. of Glen Ellen, *McClintock* (1969); Bennett Valley, *Knight* (1963); Spring Lake; embankment on St. Helena Road (white form), *Wagnon* (1988 NCC); 0.9 mi. up hwy. to

Boonville from US 101, *Breedlove* (1962). < 3300 ft. Spring. Native of Mediterranean region. —*V. v.* Host. *V. dasycarpa* Tenore.

★ **Vicia villosa** ssp. **villosa**. WINTER VETCH, HAIRY VETCH. Common, waste places: Fife Cr. Marsh in Guerneville, *Rubtzoff* (1961 NCC); near Guerneville, *Rubtzoff* (1956); near Cunningham, *Howell* (1947); Todd Road Preserve; Adobe Can. near base of Mt. Hood, *Howell* (1945); 6 mi. out Bennett Valley, *Kindig* (1933 NCC); US 101 n. of Santa Rosa, *Winblad* (1941); Lake Sonoma area. < 3300 ft. Apr.-July. Native of Europe.

FAGACEAE. OAK FAMILY

1. Fruit a spiny bur enclosing 1-3 nuts; leaves persistent, yellowish-gray or rusty-tomentose beneath *Chrysolepis*
1: Fruit an acorn with a cuplike involucre
    2. Flowers ♂ in erect dense catkins with persistent bracts; ♀ flowers at base of ♂ catkins; leaves with prominent parallel lateral veins . . . . *Lithocarpus*
    2: Flowers ♂ in ± drooping lax catkins with deciduous bracts; ♀ flowers in axillary clusters; leaves usually not so veined . . . . *Quercus*

Chrysolepis

1. Tree 15-45 m tall; leaves not folded . . . . *C. chrysophylla* var. *chrysophylla*
1: Shrub or small tree to ca 5 (10) m tall; leaves commonly folded upward on each side along midrib . . . . *C. c.* var. *minor*

Chrysolepis chrysophylla (Hooker) Hjelmquist var. **chrysophylla**. GIANT CHINQUAPIN. Occasional, tree, in nw. part of county in woods near the coast, inland through the hills n. of the lower Russian River: The Sea Ranch, ½ mi. s. of Gualala River near Hwy. 1, *Rhyne, Madison & Wells* (1985); Rockpile Cr., Annapolis and Stewarts Point quads. < 6600 ft. June-Sept. —*Castanopsis c.* (Douglas) A. deCandolle var. *c.*

Chrysolepis chrysophylla var. **minor** (Bentham) Munz. DWARF CHINQUAPIN. Occasional, pygmy forest near the coast, chaparral on dry rocky interior ridges: The Sea Ranch, acc. Wells (1983); Annapolis, Stewarts Point, Fort Ross and Arched Rock quads.; trail to pygmy forest in Salt Point State Park, *Huntington* (1989 NCC); near Kenwood, acc. Baker (1935); Geothermal quad. 6000 ft. June-Sept. —*Castanopsis c.* var. *m.* (Bentham) A. deCandolle.

Lithocarpus

1. Plant a tree 20-45 m tall; leaves with prominent parallel lateral veins ending in sharp teeth, 4-12 cm long; cups with scales narrow, spreading . . . . *L. densiflorus* var. *densiflorus*
1: Plant a shrub to ca 3 m high; leaves entire, 1.5-6 cm long, less conspicuously veined; cups with subulate recurving scales . . . . *L. d.* var. *echinoides*

Lithocarpus densiflorus (Hooker & Arnott) Rehder var. **densiflorus**. TAN OAK, TANBARK OAK. Common, borders of redwood forests, wooded canyons, ravines: The Sea Ranch ¾ mi. s. of Gualala River, *Rhyne, Madison & Wells* (1985); King Ridge Road between Plantation and Cazadero, *Wells* (1985); near Monte Rio, *McMinn* (1922); Bouverie; Pepperwood; chaparral area sw. of Bradford Mtn. 1½ mi. w. of Dry Cr., acc. J. & B. Guggolz (1992); Healdsburg, Geyserville, Alexander Valley, Cloverdale and Mt. Tom quads. < 5000 ft. June-Oct.

Lithocarpus densiflorus var. **echinoides** (R. Brown) Abrams. DWARF TANBARK. Occasional, dry slopes: Rockpile Cr., Salt Point, Fort Ross, Arched Rock, Ida Clayton Road and N. Fork Buckeye Cr. quads. < 6600 ft. June-Aug.

Quercus: (Principally adapted from key by S. J. Barnhart, Santa Rosa Junior College).

1. Plants evergreen, the leaves usually with spines or bristles along margin; margin not lobed
    2. Plants typically shrubs
        3. Leaves ± plane or undulate, or undulate, glossy above . . . . *Q. berberidifolia*
        3: Leaves convex, margins revolute, not glossy above; usually on serpentine . . . . *Q. durata*
    2: Plants typically trees
        4. Leaves bicolored, dark green above and pale below; young leaves often densely pubescent on lower surfaces, cup scales tuberculate . . . . *Q. chrysolepis*
        4: Leaves not bicolored, ± same color on both surfaces

5. Leaves plane, oblong to ovate-lanceolate; lower surface ± shiny, yellow green; twig galls common, cup scales flat . . . . *Q. wislizeni*
5: Leaves convex, broadly elliptical or roundish; tufts of hairs in axils of principal veins beneath; twig galls rare . . . . *Q. agrifolia*

1: Plants deciduous, the leaves typically with lobed margins
6. Leaves small (2.5-7.5 cm long), usually blue-green, unlobed to very shallow lobes < 1/3 of the way to midrib . . . . *Q. douglasii*
6: Leaves larger (6.5-15.3 cm long), not blue-green, distinctly lobed usually at least ½ of the way to midrib
7. Lobes of leaves bristle-tipped, acorn cup with overlapping scales . . . . *Q. kelloggii*
7: Lobes of leaves not bristle-tipped; acorn cups tuberculate
8. Leaves dull green and hairy above, lobe tips truncate; acorns 3-5 cm long, gradually tapering to apex; cups deep; terminal buds naked . . . . *Q. lobata*
8: Leaves shiny, glossy green above, lobe tips acuminate; acorns 2-3 cm long, rounded; cups shallow; terminal buds densely hairy . . . . *Q. garryana*

Quercus agrifolia Neé var. **agrifolia**. COAST LIVE OAK, ENCINA. Common, flatlands, slopes: Willow Creek; Pitkin Marsh, *Rubtzoff* (1951); 1700 ft. elev. 1 mi. from top of Sonoma Mtn. Road, *Lowry* (1983 SFSU); half way up Mt. Hood from Pythian Road, *Knight* (1964); Joost Ranch on Calistoga Road between Alpine and Porter Cr. roads, *Knight* (1963); Santa Rosa Cr., *Heller & Brown* (1902); Pepperwood; chaparral area sw. of Bradford Mtn. 1½ mi. w. of Dry Cr., acc. J. & B. Guggolz (1992); Healdsburg, Geyserville, Ida Clayton Road, Geothermal and Big Sulphur Cr. quads.; ca 2000 ft. elev. Anderson Trail on Black Mtn. between Geyser Peak and upper end of Pine Flat Road, acc. J. & B. Guggolz (1992); Cloverdale, *Kellogg & Harford* (1868-1869). < 5000 ft. Mar.-Apr.

Quercus agrifolia var. **frutescens** Engelmann. Uncommon, open rocky ground, chaparral: Alexander Valley quad.; w. flank Mt. St. Helena, acc. Grummer & Hemphill (1983); Squaw Cr., acc. Clifton (1984). A shrubby version, acc. Breedlove (1992), really not sufficiently distinct. Habitats of species.

Quercus berberidifolia Liebmann. CALIFORNIA SCRUB OAK. Common, rocky slopes, flats, open woods, chaparral: Cazadero quad.; hills e. of Sonoma, *Howell* (1946); Bouverie; Sugarloaf Ridge; St. Helena Road, *Baker* (1927); Annadel, *Wells* (1984); Spring Lake; Rincon Ridge, acc. Knight, (1990); Pepperwood; Petrified Forest, *Howell* (1926); Lake Sonoma, Healdsburg, Geyserville and Ida Clayton Road quads; 1000 ft. elev. Pine Flat Road, *Knight* (1963), *W., I. & T. Knight* (1985); 3200 ft. elev. Socrates Mine, acc. Clifton (1984); ca 2000 ft. elev. Anderson Trail on Black Mtn. between Geyser Peak and upper end of Pine Flat Road, acc. J. & B. Guggolz (1992); Geothermal quad.; 3400 ft. elev. Squaw Cr., acc. Clifton, (1983); Big Sulphur Cr. quad.; Pine Mtn. Road ne. of Cloverdale, *Knight* (1963). Segregated from *Q. dumosa* Nuttall. < 5000 ft. Mar.-May.

Quercus chrysolepis Liebmann. CANYON LIVE OAK, MAUL OAK. Common, canyons, ridges: Annapolis, Stewarts Point, Salt Point, Fort Ross, Cazadero and Arched Rock quads.; ca 1200 ft. elev. Cavedale Road, acc. Knight (1990); end of Pythian Road to Mt. Hood below Johnson's pond, *Knight* (1964); 350-400 ft. elev. Old Skaggs Springs Road, *Utech, Sato & Ohara* (1990 CM); ca 2000 ft. elev. Anderson Trail on Black Mtn. between Geyser Peak and upper end of Pine Flat Road, acc. J. & B. Guggolz (1992); Lake Sonoma, Geyserville, Ida Clayton Road, Socrates Mine, Geothermal, Big Sulphur Cr., Cloverdale and Mt. Tom quads. < 8600 ft. Apr.-May. -*Q. c.* var. *nana* (Jepson) Jepson.

Quercus douglasii Hooker & Arnott. BLUE OAK, DOUGLAS OAK. Common, dry hills, flats, away from coast: 3.2 mi. n. of Sears Point, *Raven* (1955); Alta Vista Ranch n. of Sonoma, *Rose* (1964); Bouverie; Sonoma Valley Reg. Park, acc. Barnhart (1992); Jack London Ranch in Sonoma Valley, *Sutliffe* (1921); Dry exposures in Annadel State Park, acc. Barnhart (1992); Spring Lake; Pepperwood; Tombs Cr., Geyserville, Alexander Valley and Ida Clayton Road quads.; 2600 ft. elev. Coldwater Cr. Can. ½ airmile sw. of county line at Caldwell Pines, acc. Stone (1989); 2440 ft. elev. Squaw Cr., acc. Clifton (1984); Cloverdale and Mt. Tom quads. < 4000 ft. Apr.-May.

Quercus durata Jepson. LEATHER OAK. Common, open areas, serpentine: Fort Ross and Cazadero quads.; Bouverie; ca 1000 ft. elev. Cavedale Road, acc. Knight (1990); Los Alamos Road e. of Santa Rosa, acc. Knight (1985); Pepperwood; 1½ mi. from Lake Sonoma park headquarters on Stewarts Point Skaggs Springs Road, *Knight* (1990 NCC); 900 ft. elev. Ida Clayton Road, acc. Knight (1989); Dry Cr. Road w. of Warm Springs Dam, *Breedlove* (1962); chaparral area sw. of Bradford Mtn. 1½ mi. w. of Dry Cr., acc. J. & B. Guggolz (1992); Geyserville, Alexander Valley and Ida Clayton Road quads.; 3200-3290 ft. elev. Socrates Mine, acc. Clifton (1984); 3060-3300 ft. elev. Squaw Cr., acc. Clifton (1983); Pine Mtn. Road ne. of Cloverdale, acc. Knight (1988); 950 ft. elev. 4 mi. nw. of US 101 on road to Boonville, *Breedlove* (1962), *Quibell* (1988 NCC). < 5000 ft. Apr.-May.

Quercus garryana Douglas var. **breweri** (Engelmann) Jepson. Occasional, chaparral: 3600 ft. elev. Geyser Rock in geothermal area, acc. Clifton (1983). 2000-6000 ft. Apr.-June.

Quercus garryana Douglas var. **garryana**. GARRY OAK, OREGON OAK. Common, open woods, hills, grassland: Annapolis and Salt Point quads.; Cazadero near Austin Cr. Can., *Bacigalupi* (1947); Duncans Mills and Sebastopol quads.; Fairfield Osborn; 6 mi. ne. of Penngrove, *Rose* (1964); Bouverie; Sugarloaf Ridge; n. slope exposures Annadel State Park, acc. Barnhart (1992); Spring Lake; Santa Rosa, *Eastwood* (1921); Mark West, *Abrams* (1916); Pepperwood; Healdsburg to Cloverdale, *Abrams* (1922); Healdsburg, Tombs Cr., Lake Sonoma, Geyserville and Ida Clayton Road quads.; 1998 ft. elev. Wildhorse Cr., *Clifton* (1984 PUA); Big Sulphur Cr., Cloverdale, Mt. Tom and N. Fork Buckeye Cr. quads. < 6000 ft. Apr.-June.

Quercus kelloggii Newberry. KELLOGG OAK, CALIFORNIA BLACK OAK. Common, wooded slopes, ± open areas: Llano Road ½ mi. from Gravenstein Hwy., *Howell & Knight* (1986); 1600 ft. elev. Sonoma Mtn. Road, *Lowry* (1983 SFSU); Jack London's Ranch near Glen Ellen, *Sutliffe* (1921); Healdsburg, Tombs Cr., Lake Sonoma, Geyserville, Ida Clayton Road, Big Sulphur Cr., Cloverdale, Mt. Tom and N. Fork Buckeye Cr. quads.; 3000 ft. elev. Caldwell Pines, acc. Clifton (1984). < 8000 ft. Apr.-May. —*Q. sonomensis* Bentham. This was based on a collection made by Karl Theodore Hartweg in 1846 near Sonoma. Acc. Jepson, *Erythea 5:35,* (1897) this was the first collection of the Kellogg Oak. Newberry's name was published in 1856, Bentham's in 1864.

Quercus lobata Neé. VALLEY OAK, ROBLE. Common, bottomlands, adjacent hills: Fort Ross, Cazadero, Tombs Cr., Lake Sonoma, Geyserville, Ida Clayton Road, Big Sulphur Cr., Cloverdale and N. Fork Buckeye Cr. quads.; between Sebastopol and Santa Rosa in Laguna de Santa Rosa drainage, *Harrison* (1978 NCC); Bouverie; Rincon Valley, *Wells* (1984); Pepperwood; ca 2000 ft. elev. Anderson Trail on Black Mtn. between Geyser Peak and upper end of Pine Flat Road, acc. J. & B. Guggolz (1992). Acc. Ruth Pratt, biologist, U. S. Fish & Wildlife Service, *The Press Democrat, Dec. 12, 1992,* Valley Oaks are reproducing between Lynch Creek and Corona Road and between Old Redwood Hwy. and Petaluma River. < 5700 ft. Mar.-Apr.

Quercus parvula E. Greene var. **shrevei** (C. H. Muller) K. Nixon. In Kevin C. Nixon's master's thesis, unpublished July 1980, he gives the distribution from Santa Barbara County to Mt. Tamalpais in Marin County, with a possiblility that it occurs as far north as Mendocino County. Specimens, which might fit the description, have been seen between Mercuryville and The Geysers, acc. Knight & Edwards (1986). Also Steve Barnhart reports specimens in the Salt Point area which resemble var. *shrevei*. The type locality is Palo Colorado Can., Monterey Co. Nixon's description: "Subgenus *Erythrobalanus*, small to large evergreen trees, shrubby in exposed sites, to 20 m tall, main trunk straight, crown narrow to ± spreading; twigs glabrous to lightly pubescent, greenish or brown, fluted; buds ovate to long fusiform, light brown to (usually) dark brown, with silky pubescence on acute bud scales, 4-12 mm long; leaf blades (25) 35-75 (105) mm long, oblong to lanceolate, leathery, usually entire, sometimes spinose (esp. in shrubby forms), planar to rolled (in shrubby forms), glabrous at maturity; petioles 2-25 mm long, lightly pubescent on adaxial surface only; ♂ catkins 20 to 80 mm long, at base of new growth and remotely spaced in leaf axils of elongating shoots; stamens 4-5 (-7), glabrous, slightly to completely exserted; ♀ flowers in pairs or solitary, subsessile or on a short peduncle, staminodes or stamens occasionally present, styles 3-4 (-5), dark brown to yellowish, flaring and reflexed; acorns maturing in two seasons, subglobose to cylindric, flat, blunt, rounded, or slightly tapered apically, light matte brown, with light pubescence, especially toward the tip, 20-45 mm long; cups usually deep, the rim often infolded so that the outer becomes an inner surface, scales acute, silky pubescent, deep brown with a silver sheen, blooming April to May." Within its distribution, *Q. parvula* var. *shrevei* is known to hybridize with *Q. wislizeni* var. *frutescens* and *Q. agrifolia*.

Quercus wislizeni A. deCandolle var. **frutescens** Engelmann. DWARF INTERIOR LIVE OAK. Occasional, shrubby: Rockpile Cr., Annapolis, Fort Ross, Cazadero, Arched Rock, Big Sulphur Cr., Cloverdale, Mt. Tom and N. Fork Buckeye Cr. quads.; Bouverie; ca 1100 ft. elev. Cavedale Road, *Knight* (1984); Rincon Ridge, *Knight* (1984); 3300 ft. elev. crest of Mayacmas Mts. at Caldwell Pines, acc. Stone (1989). Misapplied (in part) to *Q. parvula* E. Greene. Habitat of species.

Quercus wislizeni var. **wislizeni**. INTERIOR LIVE OAK. Common, forest woodland, chaparral, foothills, mountains: Rockpile Cr., Annapolis, Fort Ross, Cazadero, Arched Rock, Healdsburg, Geyserville, Alexander Valley and Ida Clayton Road quads.; Pepperwood; chaparral area sw. of Bradford Mtn. 1½ mi. w. of Dry Cr., acc. J. & B. Guggolz (1992); 2600 ft. elev. Coldwater Cr. Can. ca ½ airmiles sw. of Lake Co. line at Caldwell Pines, *Stone* (1989 JEPS); Pine Mtn. Road, *Knight* (1963). <5400 ft. Mar.-May.

HYBRIDS

Many species of *Quercus* are prone to hybridize and among the 9 species that are known to grow in Sonoma County, 9 (or possibly 10) different hybrids may be reported. Two notable areas where hybrids have been detected are the Pepperwood Preserve northeast of Mark West Springs, and along Cavedale Road north of Agua Caliente. At Pepperwood 8 or 9 species with 6 different hybrids have been collected, while along Cavedale Road in less than 3 miles from State Highway 12, 7 (or possibly 8) species together with 5 (or possibly 7) different hybrids have been found. All Sonoma County oaks except *Q. chrysolepis* have been reported in one or more crosses. The Goldcup Oak shows no propensity to hybridize; apparently it intergrades only with close relatives (e.g. *Q. dunnii* Kellogg and *Q. vaccinifolia* Kellogg). There is a small area n. of Cazadero where 6-7 species grow together at a Boy Scout facility at the Bergman Ranch. Since inter-subgeneric crosses in *Quercus* are not known to occur in the wild, the hybrids are listed here according to subgenus, first by the parent-species names followed by a specific name if one has been given or assigned to the hybrid.

SUBGENUS Erythrobalanus — Black Oaks

Quercus agrifolia × **Q. kelloggii**. = **Q.** × **chasei** McMinn, Babcock & Righter. Voight Ranch near Cunningham Sta., acc. Baker; Pepperwood, *de Nevers;* Sonoma Mtn. near Jack London State Historical Park, *O'Meara* (1984); between Mercuryville and Sulphur Cr. Road on Geysers Road, *Knight* (1986 NCC).

Quercus agrifolia × **Q. wislizeni** var. **frutescens**. Variable shrubby oaks in the vicinity of suspected parents on Cavedale Road may represent this hybrid, *Knight & Howell* (1984).

Quercus agrifolia × **Q. wislizeni** var. **wislizeni**. Perraco Road off of Grange Road, acc. Quibell (1991); Mt. Hood above Los Guilicos, acc. Barnhart (1992).

Quercus kelloggii × **Q. wislizeni**. ORACLE OAK. Occasional, probably the most frequent of the hybrid oaks, tends to be semideciduous. = **Q.** × **moreha** Kellogg emend. K. Brandegee. Has sometimes been spelled, but incorrectly, *Q. morehus*.

SUBGENUS Lepidobalanus — White Oaks

Quercus berberidifolia × **Q. durata**. Variation of forms on serpentine: Cavedale Road, *Knight & Howell* (1984).

Quercus berberidifolia × **Q. garryana**. = **Q.** × **howellii** Tucker: near Rocky Mtn. n. of Guerneville, acc. Tucker; Fairfield Osborn (H); Pepperwood; Shurtleff Ranch ca 4 mi. ne. of E. Windsor, acc. Tucker.

Quercus berberidifolia × **Q. lobata**. = **Q.** × **townei** E. J. Palmer; **Q. dumosa** var. **kinseliae** C. H. Muller: Cavedale Road, *Knight & Howell* (1984).

Quercus douglasii × **Q. berberidifolia**. Cavedale Road, *Howell & Knight* (1984).

Quercus douglasii × **Q. durata**. Near overlook at Lake Sonoma, acc. Gankin (1992).

Quercus douglasii × **Q. garryana**. = **Q.** × **eplingii** C. H. Muller: 3.2 mi. n. of Sears Point, *Raven* (1955); very common in Annadel State Park, acc. Barnhart (1992); Pepperwood; 1000 ft. elev. Pine Mtn. Road, *W., I. & T. Knight* (1985).

Quercus douglasii × **Q. lobata**. = **Q.** × **jolonensis** Sargent: Cavedale Road, *Knight & Howell* (1984).

Quercus durata × **Q. garryana**. = **Q.** × **subconvexa** Tucker: Pepperwood.

FRANKENIACEAE. FRANKENIA FAMILY

Frankenia salina (Molina) I. M. Johnston. Locally common, alkali flats: Bodega Head, *Clifton* (1980 PUA); Bodega Harbor Marsh, *Rubtzoff* (1955); Gilardi's Resort near Lakeville, *Knight* (1989). < 2600 ft. June-Oct. —*F. grandifolia* Chamisso & Schlechtendal.

GARRYACEAE. SILK-TASSEL FAMILY

Garrya

1. Lower surface of leaves glabrous or nearly so, leaves plane . . . . *G. fremontii*
1: Lower surface of leaves usually covered with curly or wavy hairs
  2. Tomentum of under surface of leaves of very short curled hairs forming a dense felt, margins strongly undulate . . . . *G. elliptica*
  2: Tomentum of under surface of leaves of long wavy hairs which are mostly ascending, margins plane . . . . *G. congdonii*

Garrya congdonii Eastwood. CONGDON'S SILK-TASSEL. Uncommon, dry canyons, ridges: near summit of Mt. Hood, *Baker* (1899 NCC, 1903 NCC); 3200 ft. elev. upper Squaw Cr. drainage 1.2 airmiles wnw. of Geyser Rock, acc. Stone (1983). 600-2750 ft. Feb.-Apr.

Garrya elliptica Lindley. COAST SILK-TASSEL. Occasional, dry slopes, ridges: The Sea Ranch; Stewarts Point, *Abrams* (1920); 1-2 mi. n. of Kruse Ranch Road on Coast Hwy., *McKenzie* (1980); 49 ft. elev. Gerstle Cove at Salt Point State Park, *Clifton* (1981 PUA); Willow Creek; within ½ mi. of mouth of Estero Americano on Marin Co. side, acc. Connors (1990); Fairfield Osborn; near Healdsburg, *McClisit* (1928). < 2700 ft. Jan.-Mar.

Garrya fremontii Torrey. FREMONT'S SILK-TASSEL. Locally common, away from coast: upper Hooker Can. on Cavedale Road, *Knight* (1964); near Kenwood, *Baker* (1946); Sugarloaf Ridge; Pepperwood; Bradford Mtn. w. of Dry Creek Valley, acc. Raiche (1985); Pine Flat Road, *Knight* (1963); between Black Mtn. and Geyser Peak, *McMurphy* (1909); Kortum Can. Road e. of Petrified Forest, *Knight & Edwards* (1985); chaparral area sw. of Bradford Mtn. 1½ mi. w. of Dry Cr., acc. J. & B. Guggolz (1992); ca 2000 ft. elev. Anderson Trail on Black Mtn. between Geyser Peak and upper end of Pine Flat Road, acc. J. & B. Guggolz (1992); Geyser Peak, *Knight & Edwards* (1986); Pine Mtn. Road ne. of Cloverdale, *Knight* (1963). 1000-7600 ft. Jan.-Apr.

GENTIANACEAE. GENTIAN FAMILY

1. Corolla rotate with conspicuous fringed glands on upper surface . . . . *Swertia*
1: Corolla campanulate to funnelform or salverform, without such glands
  2. Anthers coiled or spirally twisted after anthesis; flowers mostly red or pink . . . . *Centaurium*
  2: Anthers not coiled or twisted after anthesis
    3. Anthers cordate-ovate; corolla yellow, ca 6 mm long, short-salverform . . . . *Cicendia*
    3: Anthers oblong; corolla mostly blue or bluish, campanulate to funnelform . . . . *Gentiana*

Centaurium

1. Flowers sessile or subsessile, the pedicels usually not over 1 mm long and the cymules densely flowered . . . . *C. muehlenbergii*
1: Flowers pedicellate, the pedicels 1-many mm long and the cymules loosely flowered
  2. Corolla-lobes usually 3-4 mm, ca = tube . . . . *C. tenuiflorum*
  2: Corolla-lobes > 5 mm long, definitely longer or shorter than tube
    3. Corolla-lobes < ½ as long as tube, 5-5.5 mm long; anthers 1.5-2.5 mm long, oblong . . . . *C. davyi*
    3: Corolla-lobes > ½ as long as tube, 8-10 mm long; anthers 3.5 mm long, linear . . . *C. trichanthum*

Centaurium davyi (Jepson) Abrams. Uncommon, low damp places: Newanga Road, Santa Rosa, *Knight* (1965); Pepperwood, *Breedlove* (1980); Cherry Cr. Road w. of Cloverdale, *Knight & Edwards* (1986). < 3300 ft. May-Aug.

Centaurium muehlenbergii (Grisebach) Piper. CANCHALAGUA. Uncommon, mostly damp places, stream-sides, also clay soil of open slopes: Bodega Head, *Connors* (1992 BML); ca 1 mi. nw. of Camp Meeker, acc. J. & B. Guggolz (1991); Stoetz Lane ne. of Occidental, *Best* (1981); Pitkin Marsh, *Rubtzoff* (1951, 1976); 7 sites w. and sw. of Santa Rosa in Laguna de Santa Rosa drainage, acc. Harrison (1978); Button Ranch on Walker Road (U. C. property) between Petaluma-Valley Ford Road and Pepper Road, acc. Connors (1992); Bouverie; Coney Lake near Kenwood, *Baker* (1954 NCC); Bennett Mtn. Lake,

Rubtzoff, (1961); Annadel; e. foothills of Sonoma Mtn., *Rubtzoff* (1971); Spring Lake, *Wells* (1987); Stewarts Point-Skaggs Springs Road near Lake Sonoma overlook, *Best* (1981); ca 2000 ft. elev. Anderson Trail on Black Mtn. between Geyser Peak and upper end of Pine Flat Road, acc. J. & B. Guggolz (1992); 2840 ft. elev. Bear Can. between Bacon Flat and Big Basin in upper geothermal area, acc. Clifton (1984); 2420-2800 ft. elev. Caldwell Ranch, acc. Clifton (1984). < 1700 ft. June-Aug. —*C. floribundum* (Bentham) B. L. Robinson.

★ **Centaurium tenuiflorum** Hoffmannsegg & Link. Uncommon, hydrothermally altered soil, w.-facing slopes: 2300 ft. elev. in upper Big Sulphur Cr. drainage at Little Geysers, *Stone* (1983 JEPS). Native of Europe.

Centaurium trichanthum (Grisebach) Robinson. Occasional, moist places: Sebastopol Road 4 mi. w. of Santa Rosa, *Howell* (1947), *Rubtzoff* (1966); flats midway between Santa Rosa and Cotati, *Eastwood & Howell* (1938); w. side of US 101 n. of Wilfred Ave., *Baker* (1940); Annadel, *Wells* (1984); 3160-3200 ft. elev. Socrates Mine, acc. Clifton (1984); 2300 ft. elev. Big Sulphur Cr. drainage at Little Geysers, acc. Stone (1983). < 2700 ft. May-Aug.

Cicendia quadrangularis (Lamarck) Grisebach. Occasional, grassy places, woodlands: Salt Point; Plantation, *Koch* (1940 NCC); 4 sites w. and sw. of Santa Rosa in Laguna de Santa Rosa drainage, acc. Harrison (1978); Todd Road Preserve; 6½ mi. ne. of Penngrove (now Fairfield Osborn), *Rubtzoff* (1969); Sobre Vista Road 0.9 mi. w. of Arnold Dr., El Verano, *Rubtzoff* (1971); Bouverie; 4¼ mi. w. of Glen Ellen on Bennett Valley Road, *Rubtzoff & Froebe* (1971); Bennett Mtn. Lake, *Rubtzoff* (1963); Santa Rosa, *Eastwood* (1921); Pepperwood; Lake Sonoma area. < 9000 ft. Mar.-May. —*Microcala quadrangularis* (Lamarck) Grisebach.

Gentiana

1. Sinus-plaits of corolla entire; at least some flowers without subtending ovate bracts, on pedicels to 7 cm long; calyx-lobes ± unequal 6-12 mm long . . . . *G. sceptrum*
1: Sinus-plaits of corolla usually divided into 2 acuminate teeth, each flower subtended by 2 linear bracts, bracts lance-ovate to ovate; calyx-lobes subequal 8-12 mm long . . . . *G. affinis*

Gentiana affinis Grisebach var. **ovata** A. Gray. Occasional, grassy and brushy places, w. part of county: The Sea Ranch; Stewarts Point, *Abrams* (1920), *Parks* (1925); Hwy. 1 between Stewarts Point and Fort Ross, *Baker* (1931 NCC); Salt Point; 3.3 mi. e. of Bodega Bay, *Carter & Crum* (1941); w. bank of Estero Lane ca 1 mi. w. of jct. with Franklin School Road between Valley Ford and Bodega Bay, *Quibell* (1975 NCC); Bohemian Grove near Monte Rio, *Rixford* (1916). < 7600 ft. June-Aug. —*G. oregana* Engelmann, misapplied.

Gentiana sceptrum Grisebach. Uncommon, wet meadows, bogs, w. part of county: The Sea Ranch, *Wells* (1978); Pitkin Marsh, *Rubtzoff* (1951); Perry Marsh, *Baker & Howell* (1945), still extant in upper marsh, acc. J. & B. Guggolz (1992). < 4000 ft. July-Sept.

Swertia albicaulis (Grisebach) Kunze var. **nitida** (Bentham) Jepson. Uncommon, dry slopes, yellow pine forest, foothill woodland, chaparral: Franz Valley Road (collected for several years in the '80s for the Jr. Coll. wildflower show), *Barnhart* (1981 SRJC); Pine Mtn. fire road in Mayacmas Mts., *Raiche* (1984). 500-6300 ft. May-July. —*S. n.* Jepson. *Frasera a.* Grisebach ssp. *n.* (Bentham) Post. *F. n.* Bentham.

GERANIACEAE. Geranium family

1. Petals unequal (occasionally upper 2 different from lower 3 in shape) . . . . *Pelargonium*
1: Petals equal
    2. Leaves pinnate, lobed to dissected; fertile stamens 5 . . . . *Erodium*
    2: Leaves palmate; fertile stamens usually 10 . . . . *Geranium*

Erodium

1. The leaves simple, often lobed or divided; style-column 5-12 cm long
    2. Style-column 5-9 cm long; concavities at top of fruit subtended by a single fold, the upper part of the carpel pubescent;sepals with a short green tip . . . . *E. brachycarpum*

2: Style-column 9-12 cm long; concavities at top of fruit subtended by 2 folds; the upper part of the carpel glabrous; sepals with a prominent reddish mucro . . . . *E. botrys*
1: Leaflets pinnate; style-column 2.5-4 cm long
3. Leaflets broad, coarsely toothed or serrate; sepal-tips not setose; claws of petals glabrous . *E. moschatum*
3: Leaflets pinnately lobed or divided; sepal-tips setose; claws of petals ciliate . . . . *E. cicutarium*

★ **Erodium botrys** (Cavanilles) Bertoloni. BROADLEAF FILAREE. Common, grassy places: The Sea Ranch; Fort Ross, Cazadero and Duncans Mills quads.; Pitkin Marsh, *Rubtzoff* (1952); Todd Road Preserve (H&K); Crane Cr. Reg. Park off Pressley Road, acc. Quibell (1991); Tresch Ranch, Walker Road near Two Rock, acc. J. & B. Guggolz (1992); Chileno Valley, *Knight* (1964); Bouverie; Sugarloaf Ridge; Spring Lake; Pepperwood; 3 mi. s. of Healdsburg, *Heller & Brown* (1902); chaparral area sw. of Bradford Mtn. 1½ mi. w. of Dry Cr., acc. J. & B. Guggolz (1992); Lake Sonoma area; Alexander Valley and Ida Clayton Road quads. < 3300 ft. Mar.-May. Native of Mediterranean region.

★ **Erodium brachycarpum** (Godron) Thellung. FOOTHILL FILAREE. Common, grassy places: Cazadero quad.; Todd Road Preserve; Fairfield Osborn (S); Button Ranch on Walker Road (U. C. property) between Petaluma-Valley Ford Road and Pepper Road, acc. Patterson (1990); 4 mi. sw. of Sonoma, *Howell* (1957); 1 mi. s. of Sonoma, *Howell* (1957); hills e. of Agua Caliente in Sonoma Valley, *Howell* (1948); Bouverie; Sugarloaf Ridge; Spring Lake Village; Pepperwood; Healdsburg quad.; 350-400 ft. elev. Old Skaggs Springs Road, *Utech, Sato & Ohara* (1990 CM); Asti, *J. H. Thomas* (1966); Ida Clayton Road quad. < 3300 ft. Apr.-Aug. Naturalized from N. Africa. —*E. obtusiplicatum* (Maire, Weiller & Wilczek) J. T. Howell.

★ **Erodium cicutarium** (Linnaeus) L'Héritier. RED-STEMMED FILAREE. Very common, open cultivated dry places: The Sea Ranch; Fort Ross-Russian River-Bodega areas, *Voznesenskii* (1840-41 LE); Bodega Head Marsh, *Rubtzoff* (1960); Willow Creek; Bodega Head, *Knight* (1963); Fairfield Osborn (H); ½ mi. from Lichau Road on Pressley Road (e. of Petaluma Hill Road), *McNally* (1970 NCC); Tresch Ranch, Walker Road near Two Rock, acc. J. & B. Guggolz (1992); ½ mi. nw. of Petaluma in cemetery, *Ott* (1934 NCC); w. of Sears Point Raceway in Tolay Cr. drainage, acc. Patterson (1990); Bouverie; Sugarloaf Ridge; Pepperwood; Ida Clayton Road quad.; geysers Unit 13, *Quibell* (1981); 3200 ft. elev. Socrates Mine, acc. Clifton (1984); 3600 ft. Geyser Rock, acc. Clifton (1983). < 6600 ft. Feb.-May. Naturalized from Mediterranean region.

★ **Erodium moschatum** (Linnaeus) L'Héritier. WHITE-STEMMED FILAREE. Common, loamy soils: The Sea Ranch; 7 mi. s. of Gualala, *J. H. Thomas* (1976); Duncans Mills quad.; Fort Ross-Russian River-Bodega area, *Voznesenskii* (1840-41 LE); Bodega Head; corner Magnolia and Marshall aves. w. of Petaluma, acc. Knight (1964); Bouverie; Rincon Valley, *Wells* (1984); Santa Rosa Cr., *Heller & Brown* (1902); Spring Lake Village. < 5000 ft. Feb.-May. Naturalized from Mediterranean region.

Geranium

1. Stems 3-10 dm long; flowers to 13 mm across . . . . *G. robertianum*
1: Stems 1-6 dm long; flowers smaller
2. Sepals prominently awned or subulate-tipped; seeds reticulate
3. Plants perennial from carrotlike root; no hairs with gland-tips
4. Stems with retrorse-spreading stiff shining hairs; petals scarcely longer than sepals; seeds coarsely reticulate . . . . *G. potentilloides*
4: Stems with retrorse-appressed dull hairs; petals ca twice as long as sepals; seeds finely reticulate . . . . *G. retrorsum*
3: Plants annual
5. Fruiting pedicels much longer than calyx; beak of mature fruit 3-6 mm long . . . *G. bicknellii*
5: Fruiting pedicels scarcely or not longer than calyx; beak of mature fruit 1-2 mm long
6. Carpels with short stiff subequal spreading hairs; lobes of upper leaves acute; pits of seed square or rounded . . . . *G. dissectum*
6: Carpel with long ± unequal ascending hairs; lobes of leaves obtuse; pits of seed ± elongate . . . . *G. carolinianum*
2: Sepals awnless, at most with callous tips; seeds smooth or minutely granular . . . . *G. molle*

Geranium bicknellii Britton. BICKNELL'S CRANESBILL. Uncommon, open disturbed places: Petaluma Hill Road, *Knight* (1965); Pepperwood; 2420 ft. elev. Caldwell Ranch, *Clifton* (1984 PUA); 2190 ft. elev. Squaw Cr., acc. Clifton (1984). Var. *longipes* (S. Watson) Fernald, currently not recognized. 200-5000 ft. July-Aug.

Geranium carolinianum Linnaeus. Occasional, shady places: Fort Ross-Russian River-Bodega area, *Voznesenskii* (1840-41 LE); Willow Creek; Larkspur Rock ca 1½ mi. e. of Bodega Bay, *Ott* (1934 NCC); between Strawberry School and Cooper's Grove on Sonoma Mtn. Road, *Baker* (1931 NCC); Sugarloaf Ridge; Santa Rosa Cr., *Heller* (1902); Pepperwood; Cloverdale, *Bolander* (1865 UC). < 5600 ft. Apr.-July.

★ **Geranium dissectum** Linnaeus. Common, open waste places: The Sea Ranch; Salt Point; Fort Ross, Duncans Mills and Arched Rock quads.; Cunningham Marsh, *Howell* (1947); 3 sites w. of Santa Rosa in Laguna de Santa Rosa drainage, acc. Harrison (1978); Todd Road Preserve; between Strawberry School and Cooper's Grove on Sonoma Mtn. Road, *Baker* (1931 NCC); Fairfield Osborn (*); Tresch Ranch, Walker Road near Two Rock, acc. J. & B. Guggolz (1992); Royal Oaks Mobile Home Park on N. McDowell Blvd., Petaluma, *Knight* (1984); Bouverie; Sugarloaf Ridge; Gates Road off Calistoga Road, acc. Wells et al. (1984); Spring Lake; Santa Rosa Jr. Coll. campus, *Hagg & Schmidt* (1932 NCC); Lake Sonoma area; Geyserville, Alexander Valley and Ida Clayton Road quads. < 4000 ft. Mar.-June. Naturalized from Europe.

★ **Geranium molle** Linnaeus. Common, moist sites, usually about garden areas: The Sea Ranch; Cazadero, *Bacigalupi* (1926); Duncans Mills Marsh, *Rubtzoff* (1960); Bodega Head Marsh, *Rubtzoff* (1960); Bodega Point, *Eastwood* (1915); Bodega Head, *Ferris & Morey* (1958); 1 mi. e. of Bodega Bay, *Baker* (1935 NCC); Coleman Valley, *Baker* (1939 NCC); Fairfield Osborn (*); Tresch Ranch, Walker Road near Two Rock, acc. J. & B. Guggolz (1992); Bouverie; Annadel, *Wells* (1984); Spring Lake; Spring Lake Village; Pepperwood; Lake Sonoma area; Geyserville quad.; 2040 ft. elev. Squaw Cr., acc. Clifton (1984). < 3300 ft. May-July. Native of Europe.

★ **Geranium potentilloides** deCandolle. Uncommon, moist, shaded sites, possibly poorly collected: Larkspur Rock 1 mi. e. of Bodega Bay, *Baker* (1936 NCC). < 1700 ft. Native of Australasia. —*G. pilosum* Forster f.

★ **Geranium retrorsum** L'Héritier. Common, grassy, waste places: 1 mi. n. of Fort Ross, *Baker* (1932); Duncans Mills quad.; Willow Cr. Marsh, *Rubtzoff* (1959 NCC); Bodega Head, *Ferris & Morey* (1958); s. of Bodega Bay, *Baker* (1930 NCC); on Bodega Hwy. at Watson School ca 3 mi. w. of Freestone, *Best* (1981); Freestone, *Huntington* (1987); Montgomery Drive near Spring Lake Park, Santa Rosa, *Wagnon* (1988 NCC). < 1700 ft. May-Sept. Native of Australasia.

★ **Geranium robertianum** Linnaeus. RED ROBIN. Uncommon, garden weed: Royal Oaks Mobile Home Park on N. McDowell Blvd., Petaluma, *Knight* (1986). < 400 ft. Native of Eurasia.

Pelargonium

1. Inflorescence dense, pedicels < 6 mm . . . . *P. vitifolium*

1: Inflorescence open, pedicels usually > 6 mm . . . . *P. grossularioides*

★ **Pelargonium grossularioides** (Linnaeus) L'Héritier. Uncommon, waste ground: Mrs. Pitkin's garden, Sebastopol, *Howell* (1951); Jack London State Park near London's old residence near Glen Ellen, *Wagnon* (1988 NCC). < 1000 ft. Apr.-July. Native of S. Africa.

★ **Pelargonium vitifolium** (Linnaeus) L'Héritier. Uncommon, garden conditions, escape from cultivation: Hwy. 1 n. of Fort Ross, *Best* (1984); Bodega Bay, *Howell* (1955). < 1000 ft. Spring. Native of S. Africa.

GROSSULARIACEAE. GOOSEBERRY FAMILY
[Segregated from SAXIFRAGACEAE]

Ribes

1. Plants currants; nodal spines lacking; berry spineless; flowers usually several to many in a raceme . . . . *R. sanguineum var. glutinosum*

1: Plants gooseberries; nodal spines present; berry often spiny; flowers mostly 1-few

  2. Ovary and berry without spines or prickles . . . . *R. divaricatum*

  2: Ovary and berry with spines or bristles

    3. Free part of flower-tube ca as broad as long

      4. Young twigs usually bristly on internodes; leaves glandular beneath

        5. Flowers purplish or reddish; ovary with some hairs or bristles that are not gland-tipped . . . . *R. menziesii*

        5: Flowers greenish-white; ovary with all bristles gland-tipped . . . . *R. victoris*

      4: Young twigs not bristly on internodes; leaves not glandular . . . . *R. californicum*

    3: Free part of flower-tube much longer than broad . . . . *R. roezlii* var. *cruentum*

Ribes californicum Hooker & Arnott. HILLSIDE GOOSEBERRY. Common, rocky canyons, open slopes: Plantation Road ca 3 mi. from Hwy. 1, *Koch* (1940); woods near Cazadero, *Heller* (1903); Fairfield Osborn (S); Button Ranch on Walker Road (U. C. property) between Petaluma-Valley Ford Road and Pepper Road, acc. Connors (1992); San Antonio Cr. near bridge ca 6 mi. s. of Petaluma, *Knight* (1964); Cooper's Grove, *Baker* (1928 NCC); Bennett Valley, *Heller & Brown* (1902); hills e. of Santa Rosa, *Heller & Brown* (1902); Pepperwood; Lake Sonoma area; w. of Cloverdale, *Baker* (1924 NCC), *Applegate* (1934). < 3300 ft. Feb.-Mar. —*Grossularia californica* Coville & Britton.

Ribes divaricatum Douglas var. **pubiflorum** Koehne. STRAGGLY GOOSEBERRY. Occasional, stream banks, shady canyons: Russian River near Montesano, *Sargent* (1982); Salmon Cr. near Bodega, acc. Quibell (1991); Pitkin Masrsh, *Rubtzoff* (1951); Sonoma Cr. at e. base of Mt. Hood, *Heller & Brown* (1902). < 2300 ft. Mar.-May. —*Grossularia divaricata* Coville & Britton.

Ribes menziesii Pursh. CANYON GOOSEBERRY. Common, wooded canyons: coastal canyon 8 mi. s. of Fort Ross, *Howell* (1934); 2 mi. s. of Seaview, *A. Menzies* (1963); Willow Creek; Bodega Bay, *McMinn* (1922); hillside forest in vicinity of Coleman Valley Road, *Rubtzoff* (1969); Salmon Cr. Road 3¾ mi. w. of Bodega, *Rubtzoff* (1969); Hoods Peak, *Raven* (1954). Mostly < 2000 ft. Mar.-Apr. —*R. menziesii* var. *leptosmum* (Coville) Jepson, indistinct. *Grossularia menziesii* Coville & Britton. *G. leptosma* Coville.

Ribes roezlii Regel var. **cruentum** (E. Greene) Rehder. SIERRA GOOSEBERRY. Occasional, open wooded slopes: Cazadero, *Heller* (1902 UC); Agua Caliente, Sonoma Valley, *Heggie* (1943); e. base of Mt. Hood, *Heller & Brown* (1902); Mt. Hood, *Bioletti* (1893 UC); Lake Sonoma quad.; slopes of Mayacmas Range on road from Geysers to Geyserville, *Bacigalupi* (1948); Little Sulphur Cr. 500-8700 ft. Apr.-June. —*R. cruentum E.* Greene, type locality: Sonoma County (most probably Hoods Peak Range). *Grossularia cruenta* Coville & Britton.

Ribes sanguineum Pursh var. **glutinosum** (Bentham) Loudon. PINK-FLOWERING CURRANT. Common, open places, among trees and brush: The Sea Ranch; Kolmer Gulch s. of The Sea Ranch, *McKenzie* (1981); hills e. of Stewarts Point, *Baker* (1920 NCC); Sea View Road, *Baker* (JEPS); Willow Creek; Bodega Bay, *Best* (1988); Coleman Valley Road 4 mi. w. of Occidental, *McNally* (1970 NCC); Lichau Road off Petaluma Hill Road, *Stocking* (1964 NCC); Fairfield Osborn (S); Crane Can., *Stocking* (1964 NCC); hills se. of Santa Rosa, *Heller & Brown* (1902). < 3300 ft. Mar.-Apr. —*R. glutinosum* Bentham.

Ribes victoris E. Greene. VICTOR'S GOOSEBERRY. Occasional, shady canyons, wooded slopes: woods near Cazadero, *Heller* (1903); near stream on Lichau Road, *Stocking* (1964 NCC); along Sonoma Cr. at foot of Mt. Hood, *Heller* (1902); Adobe Can., *Michner & Bioletti* (1893 UC); Los Guilicos, *Michener & Bioletti* (1892 UC). < 1000 ft. Mar.-Apr. —*Grossularia v.* Coville & Britton.

HALORAGACEAE. WATER-MILFOIL FAMILY
[Including HIPPURIDACEAE]

1. Stems erect; leaves simple and entire . . . . . *Hippuris*
1: Stems lax; submersed leaves finely dissected . . . . . *Myriophyllum*

Hippuris vulgaris Linnaeus. MARE'S TAIL. Uncommon, permanent shallow quiet water: Duncans Mills Marsh, *Rubtzoff* (1970); Bodega Head Marsh, *Rubtzoff* (1955). < 9000 ft. July-Sept.

Myriophyllum

1. Flowers and fruits in axils of ordinary emersed foliage leaves; flowers dioecious . . . . . *M. aquaticum*
1: Flowers and fruits in terminal bracteate spikes, occasionally the spike proliferating vegetatively at apex; flowers unisexual or sometimes some perfect
   2. Bracts shorter than flowers or fruits . . . . . *M. sibiricum*
   2: Bracts longer than flowers or fruits
      3. Stamens 4; bracts serrate; spikes not proliferating terminally . . . . . *M. hippuroides*
      3: Stamens 8; bracts pectinately lobed; spikes proliferating into a vegetative shoot at apex, acc. *Mason* (1957) . . . . . *M. verticillatum*

★ **Myriophyllum aquaticum** (Vellozo) Verdcourt. PARROT'S FEATHER. Uncommon, escape from cultivation, artificial ponds, wet creek beds: Cunningham Marsh, *Rubtzoff* (1961); Laguna de Santa Rosa area e. of Cunningham, *Rubtzoff* (1963); Laguna de Santa Rosa floodplain w. of Llano Road, acc. Lovell (1983). < 1700 ft. Summer. Native of S. America. —*M. brasiliense* Cambessèdes. *Enydria aquatica* Vellozo.

Myriophyllum hippuroides Torrey & A. Gray. Uncommon, margins of ponds: Atascadero Cr. Marsh w. of Sebastopol, *Rubtzoff* (1962); Bennett Mtn. Lake, *Rubtzoff* (1963); Lake Ilsanjo, Annadel, *Wagnon* (1991 NCC). < 3700 ft. May-July.
★ **Myriophyllum sibiricum** V. Komarov. Uncommon, ponds, quiet streams: Spring Lake; Pepperwood. < 8600 ft. June-Sept. Native of British Columbia, e. U.S. and Eurasia. —*M. exalbescens* Fernald.
★ **Myriophyllum verticillatum** Linnaeus. Rare, ponds: Duncans Mills Marsh, *Rubtzoff* (1955). < 5000 ft. June-July. Native of Eurasia.

HIPPOCASTANACEAE. Horse-chestnut, Buckeye family

Aesculus californica (Spach) Nuttall. CALIFORNIA BUCKEYE. Common, widespread, canyons, dry slopes, edge of flatlands: Salt Point; Fort Ross, Cazadero and Arched Rock quads.; Fife Cr. Marsh, *Rubtzoff* (1960); Crane Can., *Stocking* (1963 NCC); Fairfield Osborn (S); Railroad Ave. at crossing with Petaluma Hill Road, *Juul* (1973 NCC); 1 mi. from top of Sonoma Mtn. Road, *Lowry* (1983 SFSU); Tresch Ranch, Walker Road near Two Rock, acc. J. & B. Guggolz (1992); US 101 between San Antonio Cr. and Petaluma, *Howell* (1981); Bouverie; Adobe Can., *Bately* (1932 NCC); Pepperwood; Healdsburg and Tombs Cr. quads.; Lake Sonoma area; Geyserville, Alexander Valley, Ida Clayton Road, Geothermal, Big Sulphur Cr., Cloverdale, Mt. Tom and N. Fork Buckeye Cr. quads. < 5700 ft. May-June.

HYDROPHYLLACEAE. Waterleaf family

1. Styles ± united
    2. Perennial herbs
        3. Leaves entire or toothed, but not deeply divided or lobed
            4. Herbs with bulblike or tuberous base; leaves reniform . . . . *Romanzoffia*
            4: Herbs not bulblike at base; leaves longer than wide . . . . *Phacelia*
        3: Leaves deeply divided or lobed . . . . *Phacelia*
    2: Annual herbs
        5. Ovary 1-celled, with broad placentae; inflorescence not usually a clearly marked cyme; calyx usually with ± evident auricles between the lobes
            6. Stems recurved-prickly; capsule prickly or bristly; seeds without a cucullus . . . . *Pholistoma*
            6: Stems usually not recurved-prickly; capsule pubescent; seeds with a ± terminal cucullus *Nemophila*
        5: Ovary falsely 2-celled by intrusion of narrow placentae; inflorescence usually an evident coiled cyme; calyx without auricles between the lobes
            7. Corolla mostly deciduous, blue to purplish or white, if persistent the flowers erect . . . . *Phacelia*
            7: Corolla persistent, yellowish or reddish, pendulous . . . . *Emmenanthe*
1: Styles separate to base . . . . *Eriodictyon*

Emmenanthe penduliflora Bentham. WHISPERING BELLS. Occasional, dry places, particularly after burns or disturbances: ca 1200 ft. elev. sw. slope of Mt. Hood near base of Sugarloaf Ridge, *Baker* (1947 NCC); Fitch Mtn., *Rubtzoff* (1964); near The Geysers on road to Dunlevy property on a burn, *Harrison* (1979 SRJC); ca 2000 ft. elev. Anderson Trail on Black Mtn. between Geyser Peak and upper end of Pine Flat Road, acc. J. & B. Guggolz (1992); Cloverdale grade (possibly old road between Cloverdale and Hopland), *Hoak* (1931). < 7300 ft. Apr.-July.

Eriodictyon californicum (Hooker & Arnott) Torrey. YERBA SANTA. Common, dry ridges, rocky slopes: Fort Ross and Cazadero quads.; ca 1 mi. nw. of Camp Meeker, acc. J. & B. Guggolz (1991); Bohemian Hwy. 3 mi. n. of Occidental, *Best* (1981); 4 mi. ne. of Occidental, *Brundage* (1933 NCC); Harrison Grade Road summit on serpentine, acc. Quibell (1991); Bouverie; Mt. Hood, *Knight* (1964); Sugarloaf Ridge; Gates Road off Calistoga Road, acc. Wells et al. (1984); Spring Mtn. Road, *Knight* (1963); Pepperwood; Petrified Forest, *Hemphill* (1970 PUA); chaparral area sw. of Bradford Mtn. 1½ mi. w. of Dry Cr., acc. J. & B. Guggolz (1992); probably on trip to Mt. St. Helena from Fort Ross, *Voznesenskii* (1840-1841 LE); Tombs Cr. quad.; Lake Sonoma area; summit of Pine Flat Road, *Knight* (1963); ca 2000 ft. elev. Anderson Trail on Black Mtn. between Geyser Peak and upper end of Pine Flat Road, acc. J. & B. Guggolz (1992); Geyser Peak, *Knight & Edwards* (1986); Ida Clayton Road quad.; 4298 ft. elev. Mt. St. Helena lookout, *Grummer* (1978 PUA); Little Geysers, acc. Quibell (1980); Cloverdale and Mt. Tom quads. < 6400 ft. May-July.

Nemophila

1. Corollas 1.5-4 cm broad
   2. Corollas with blue periphery or conspicuously dark-veined, often white with black spots in center; stems ± pubescent . . . . . *N. menziesii* ssp. *menziesii*
   2: Corollas white with obvious dark spots nearly to margin; stems subglabrous . . . . *N. m.* ssp. *atomaria*
1: Corollas mostly 0.1-1 cm broad
   3. Leaves all opposite
      4. Auricles 1/3 as long as sepals in fruit; corolla ± shallowly campanulate; leaves deeply divided . . . . . *N. pedunculata*
      4: Auricles minute or obsolete; corolla bowl-shaped; leaves not deeply divided . . . . *N. parviflora*
   3: Leaves opposite on lower stems, alternate above
      5. Corolla prominently exserted from calyx . . . . . *N. heterophylla*
      5: Corolla scarcely exserted from calyx . . . . . *N. parviflora*

Nemophila heterophylla Fischer & C. Meyer. Common, light shade, canyons, slopes: mts. in w. part of county, *Krauss* (1894); several mi. e. of Gualala River on Stewarts Point-Skaggs Springs Road, *Heller* (1940); woods near Cazadero, *Heller* (1903); Willow Creek; near Camp Meeker, *Hagg* (1934 NCC); Salmon Cr. 2 mi. ne. of Bodega, *Constance* (1938); on road to Bodega Point, *Eastwood* (1915); Bodega Bay, *Wrangel* (type U); Bodega Head; e. entrance to Coleman Valley Road, *Koch* (1940 NCC); Neeley Hills s. of Guerneville, *Rubtzoff* (1953); Russian River near Trenton, *Heller & Brown* (1902); 2½ mi. w. of Graton, *L. L. Davis* (1975 NCC); Fairfield Osborn (S); Bouverie; Trinity Road, *Cantelow* (1940); Sonoma Cr. Can. (probably Adobe Can.), *Eastwood & Howell* (1940); Sugarloaf Ridge; Los Guilicos, *McNally* (1970 NCC); Spring Lake; Pepperwood; 6 mi. from Petrified Forest on road to Kellogg and Knights Valley, *Howell* (1930); Petrified Forest, *Eastwood* (1915); Windsor, *Swett* (1901); Lake Sonoma area; Skaggs Springs, *Hawver* (1915); Pine Flat, *Baker* (1934 NCC); Geyserville, *Rodda* (1929); 6.8 mi. s. of The Geysers, *Cantelow* (1935); geysers Unit 13, *Quibell* (1981 NCC); 2200 ft. elev. Unit 15, *Malloch* (1977 PG&E); 2100 ft. elev. n. of Unit 16, *Cooney* (1977 PG&E); Cloverdale grade (probably old highway between Cloverdale and Hopland), *Eastwood & Howell* (1934). < 5700 ft. Mar.-July. —*N. inaequalis* Eastwood, type locality: Russian River, *T. Brandegee* [U].

Nemophila menziesii Hooker & Arnott ssp. **atomaria** (Fischer & C. Meyer) Brand. BABY WHITE-EYES. Occasional, moist slopes, open areas: Fort Ross, *Wrangel* (1842 LE); ½ mi. s. of Fort Ross, *Constance* (1938); 30 ft. elev. Wright's Beach, *Brantley* (1970 PUA); Bodega Head; Bouverie; Kenwood, *Jepson* (JEPS); Sugarloaf Ridge; Pepperwood; Lake Sonoma area; 3 mi. e. of Geyserville, *Baker* (1932 NCC); Geyser Peak, acc. Knight & Edwards (1986). < 5000 ft. Mar.-June. —*N. atomaria,* type locality: Fort Ross, *Wrangel* [U].

Nemophila menziesii ssp. **menziesii**. BABY BLUE-EYES. Common, moist slopes, flats: The Sea Ranch; Fort Ross, *Wrangel* (1842 LE); hills at Camp Meeker, *Hagg* (1934 NCC); Bodega Head; Pressley Road 4 mi. from Petaluma Hill Road, *McNally* (1970 NCC); Fairfield Osborn (S); Tresch Ranch, Walker Road near Two Rock, acc. J. & B. Guggolz (1992); near Sonoma, *Baker* (1952 NCC); El Verano, *Griffin* (1922); along highway between El Verano and Glen Ellen, *Baker* (1944 NCC); Bouverie; Kenwood, *Baker* (1902); Rincon Valley, *Mallory* (1920); field ne. of Santa Rosa, *Copeland* (1933); Pepperwood; Lake Sonoma area. < 5300 ft. Feb.-June. —*N. liniflora* Fischer & C. Meyer, type locality: Fort Ross, *Wrangel* [U]. *N. menziesii* var. *venosa* (Jepson) Brand.

Nemophila parviflora Bentham. Fairly common, moist shade: Salt Point; Fort Ross, *Heller* (1903); Russian Gulch n. of Jenner, *Constance* (1938); Monte Rio, *Sutliffe* (1927); e. entrance to Coleman Valley, *Koch* (1940 NCC); Bittner Road between Joy and Occidental roads (Occidental Road is probably Bohemian Hwy.), *Ferris & Lorraine* (1961); Neeley Hills s. of Guerneville, *Rubtzoff* (1953); Sonoma Cr. Can. (probably Adobe Canyon), *Eastwood & Howell* (1940); between Cloverdale and The Geysers at jct. with Geysers Road above Sulphur Cr., *Knight & Edwards* (1986). < 5400 ft. Apr.-June.

Nemophila pedunculata Bentham. Rare, moist open or shaded areas: The Geysers geothermal area, *Quibell* (1980 NCC). < 8000 ft. Apr.-Aug.

Phacelia

1. Plants perennial or biennial from ± woody taproot or branched caudex
   2. Ovules and seeds usually many (4+) . . . . . *P. bolanderi*
   2: Ovules 4; seeds 1-4
      3. Inflorescence mostly elongate, narrow, dense, consisting of many short lateral cymes along a central axis; stems mostly simple

4. Lower leaves mostly with 2-several pairs of lateral leaflets; biennial . *P. heterophylla* ssp. *virgata*
4: Lower leaves mostly entire, or with 1 pair of basal pinnae; perennial . . . . *P. nemoralis*
3: Inflorescence mostly shorter, more open, widely branched
5. Calyx-lobes broadly lanceolate to ovate, overlapping in fruit . . . . *P. imbricata*
5: Calyx-lobes linear-lanceolate to linear, not overlapping in fruit
6. Plants not obviously cespitose; stems not very glandular
7. Corolla blue to purplish or lavender, 5-6 mm long . . . . *P. californica*
7: Corolla white, 7-9 mm long . . . . *P. egena*
6: Plants obviously cespitose from much-branched caudices and making patches; stems glandular . . . . *P. corymbosa*
1: Plants annual from slender taproot
8. Leaves pinnately toothed to compound, their divisions further toothed or divided
9. Leaves not deeply divided, the blades ovate to rounded, shallowly lobed; stems prickly-hispid . . . . *P. malvifolia*
9: Leaves mostly deeply divided nearly or quite to midrib; stems mostly less hispid
10. Calyx much enlarged and ± chartaceous and veiny in fruit, lobes ovate to lance-oblong 4-6 mm long in flower; corolla blue with paler center, broadly campanulate 8-10 mm long; stamens 9-13 mm long . . . . *P. ciliata*
10: Calyx ± enlarged in fruit, lobes mostly oblanceolate to obovate, 4-5 mm long and closely investing capsule, lobes not dimorphic, always entire, sometimes ± unequal; corolla promptly deciduous, blue or bluish, or light tan; stamens mostly not long-exserted . . . . *P. distans*
8: Leaves entire to pinnate, but then with entire divisions
11. Corolla-scales present, attached to tube at base of each filament, deciduous
12. Corolla rotate to campanulate, lavender to violet ca 10-12 (-18) mm long; stamens sometimes hairy; style 6-10mm long . . . . *P. divaricata*
12: Corolla tubular-campanulate, lavender to purple with yellow tube 7-11 mm long; stamens glabrous; style 3-4 mm long . . . . *P. suaveolens*
11: Corolla-scales lacking, the filament-bases sometimes dilated or winged . . . . *P. campanularia*

Phacelia bolanderi A. Gray. Occasional, dry to moist places, ± coastal influence: Stewarts Point, *M. Parks* (1925); 4 mi. e. of Stewarts Point, *Okada* (1975); bridge near Gualala River between Stewarts Point and Skaggs Springs, *Heller* (1940); Cazadero, *Graff* (1926); Stewarts Point-Skaggs Springs Road 9 mi. w. of US 101, *Best* (1981). < 4700 ft. May-July.

Phacelia californica Chamisso. Common, canyons, rocky bluffs: The Sea Ranch; Salt Point; 89 ft. elev. Roemer Gulch n. of Fort Ross, *Clifton* (1981 PUA); 1 mi. n. of Fort Ross, *Howell* (1930); Russian Gulch n. of Jenner, *Thomsen* (1973 NCC); Kruze Ranch on coast at Salt Point State Park campsites, *F. W. Hoffman* (1950 NCC); 23 ft. elev. Jenner, *Aguillon* (1975 PUA); Bodega Head; 1 mi. e. of Bodega Bay, *Howell* (1930); Larkspur Rock e. of Bodega Bay, *Ott* (1934 NCC); Bodega Road 8 mi. e. of Valley Ford (probably near Two Rock), *Howell* (1930); Russian River near Monte Rio, *Eastwood & Howell* (1939); roadside e. of Vine Hill School, *Baker* (1940 NCC); 2 mi. n. of Sebastopol, *Howell* (1937); along W. College Ave. 8 mi. w. of Santa Rosa, *Howell* (1931); Sonoma Valley; Pepperwood; 2 mi. e. of Petrified Forest, *Howell* (1937); 4 mi. from Calistoga en route to Petrified Forest, *Howell* (1929); Geyser Peak, *Knight & Edwards* (1986). < 3300 ft. Apr.-July.

★ **Phacelia campanularia** A. Gray. Probably planted by Caltrans: roadcut on Mark West Springs Road, *Harrison & Leveque* (1977 SRJC). < 5300 ft. Feb.-Apr. Native of w. Colorado Desert.

Phacelia ciliata Bentham. Uncommon, vernal flats: between Cotati and Santa Rosa, *Raven* (1950). < 5600 ft. Mar.-May.

Phacelia corymbosa Jepson. Uncommon, dry open places, serpentine: Layton chrome mine on Austin Cr., *F. W. Hoffman* (1950); n. of Gray Cr. e. of its jct. with E. Austin Cr.; ridge above Pine Flat ne. of Healdsburg; the latter two sites acc. Heckard, *U. C. Publications in Botany, 32:1 p. 51,* (1960). < 9600 ft. May-Aug.

Phacelia distans Bentham. WILD-HELIOTROPE. Common, fields, slopes: Bodega (possibly Bodega Bay), *Hinds* (1844 type [U]); Bodega Point, *Eastwood* (1915, 1917); Bodega Head; sand hills at n. end of Bodega Bay, *Howell* (1930); 1 mi. e. of Bodega Bay, *Howell* (1930); Fairfield Osborn (H); Petaluma, *Eastwood* (1921); 1 mi. n. of Sonoma, *Rose* (1965); Button Ranch on Walker Road (U. C. property) between Petaluma-Valley Ford Road and Pepper Road, acc. Patterson (1990); w. of Sears Point Raceway in Tolay Cr. drainage, acc. Patterson (1990); Bouverie, *McClintock* (1969); mouth of Sonoma Cr. Can. (probably Adobe Can.), *Howell* (1935); Kenwood, *A. & R. A. Nelson* (1933); Sugarloaf Ridge; 9 mi. n. of Kenwood, *Cantelow* (1938); Bennett Valley Road, *Chester* (1929 NCC); Santa Rosa Cr. Can., *Baker* (JEPS); Franz

Valley Road, *Quibell* (1979 NCC); road to Petrified Forest 7½ mi. from Santa Rosa, *Howell* (1929); near Mark West Springs, *Heller & Brown* (1902); Pepperwood; Ida Clayton Road, *McNally* (1970 NCC), *Berser* (1979 NCC); Knights Valley, *Heller & Brown* (1902), *Heller* (1941); Lake Sonoma area. < 7000 ft. Mar.-June.

Phacelia divaricata (Bentham) A. Gray. Occasional, open slopes, flats: Fort Ross; mouth of Salmon Cr., *Howell* (1931); s. of Bodega Bay, *Baker* (1930); Bodega, *G. & E. A. Jacobsen* (1929 NCC); Skaggs Springs, *H. Edwards* (1877 U); Lake Sonoma area. < 5000 ft. Mar.-May. —*Eutoca wrangeliana* Fischer & C. Meyer, type locality: Fort Ross, *Douglas* (U).

Phacelia egena (Brand) J. T. Howell. Uncommon, dry areas: Layton mine, headwaters of Big Austin Cr., *F. W. Hoffman* (1947); Bennett Valley, *Leitholt* (1893); 3698 ft. elev. Cobb Mtn., *Jokerst* (1985 PUA); 3500 ft. elev. on graywacke sandstone in upper Cobb Cr. drainage in ne. part of county, acc. Stone (1985). < 8300 ft. May-June.

Phacelia heterophylla Pursh ssp. **virgata** (E. Greene) Heckard. Uncommon, hillsides: 3080 ft. elev. Socrates Mine, *Clifton* (1984 PUA). < 9700 ft. May-July.

Phacelia imbricata E. Greene. Common, dry rocky places: 2 mi. n. of Sonoma between Sonoma and Boyes Hot Springs, *Howell* (1929); Bouverie; Bennett Valley, *Cass* (1929 NCC); Spring Lake; 7½ mi. from Santa Rosa en route to Petrified Forest, *Howell* (1929); Pepperwood, *Breedlove* (1983); 4 mi. from Calistoga en route to Petrified Forest, *Howell* (1929); Healdsburg, *O'Toole* (1979 NCC); property bordered by Chianti and Dutcher Cr. roads w. of US 101 and w. of Geyserville, acc. J. & B. Guggolz (1991). < 9600 ft. Apr.-June.

Phacelia malvifolia Chamisso. STINGING PHACELIA. Uncommon, gravelly and sandy places: Salt Point; Hwy. 1 Timber Cove, *Althouse* (1979 SRJC); mouth of Russian River at Jenner, *Harrison* (1981 SRJC). < 4700 ft. Apr.-July.

Phacelia nemoralis E. Greene. Uncommon, damp slopes: Petaluma, *Tidestrom* (U); Bouverie; 2½ mi. ne. of Alexander Valley, *Heckard* (1952). < 2300 ft. Apr.-July.

Phacelia suaveolens E. Greene. Rare, usually in burned area: Petrified Forest, *E. Greene* (probably 1890s, type ND); Franz Valley School Road near old school on s. roadside, acc. Knight (1965). < 5600 ft. May-Aug.

Pholistoma auritum (Lindley) Lilja. Rare, shaded slopes: Bouverie; Trinity-Cavedale area, *Tunison* (1975 SRJC). < 6300 ft. Mar.-May.

Romanzoffia californica E. Greene. Occasional, shaded or open, damp, rocky areas: n.-facing rocky roadcuts on Stewarts Point-Skaggs Springs Road just e. of Annapolis bridge, acc. Quibell (1991); Bodega Point, *Eastwood* (1917); Larkspur Rock e. of Bodega Bay, *Howell* (1931); e.-facing cliff to e. of 13808 Green Valley Road nw. of Graton, *Rockwood* (1991 NCC); Hwy. 128 nw. of Cloverdale, acc. Quibell (1991). < 2700 ft. Mar.-May. —*R. suksdorfii* E. Greene.

JUGLANDACEAE. WALNUT FAMILY

Juglans

1. Leaflets 7-9 (13), entire or nearly so . . . . *J. regia*

1: Leaflets 11-19, serrate

  2. Leaflets bright green and glabrous above, paler and villose-pubescent on main veins beneath; fruit ± flattened at ends; shell of nuts almost smooth . . . . *J. californica* var. *hindsii*

  2: Leaflets glabrous and ± glossy above, soft-pubescent beneath; fruit globular; shell of nuts irregularly ridged . . . . *J. nigra*

Juglans californica. S. Watson var. **hindsii** Jepson. CALIFORNIA BLACK WALNUT. Scarce, except for 2 groves along Russian River: ¼ mi. below bridge at Hacienda, *Knight* (1988 NCC)--the population actually extends from Mirabel to Korbel along River Road; probable offspring of planted material 1/10 mi. s. of Petaluma Ave. on Arnold Dr. near El Verano, *Howell & Knight,* (1984); Pepperwood (possibly planted), *de Nevers* (1981); along Maacama Cr. near confluence of Russian River near Biggio Road, *Wagnon* (1988 NCC). Debatable as to whether or not the species is native as the nuts may be a result of trade between Indian tribes in Napa, e. Contra Costa or Sacramento County areas. < 800 ft. Apr.-May. —*J. hindsii* (Jepson) R. E. Smith.

★ **Juglans nigra** Linnaeus. BLACK WALNUT. Uncommon, possibly planted: Vacation Beach near Guerneville, *Rubtzoff* (1953). Native in e. and central U. S.

Juglans 'paradox'. It is believed that Luther Burbank was the first to use the name for a hybrid walnut that he produced by crossing the California black walnut, *Juglans californica var. hindsii*, and the English Walnut, *Juglans regia*. There used to be a large mature paradox walnut tree growing in a vacant lot near the south end of Old Redwood Highway, Santa Rosa. Paradox trees produce fertile seed. Luther Burbank introduced about a dozen species and varieties including *J. nigra* Linnaeus and *J. cinerea* Linnaeus. Along with the native *J. californica* var. *hindsii*, they became involved in varying degrees of hybridization and therefore it may be difficult to determine some of the walnut trees in the Santa Rosa area.

★ **Juglans regia** Linnaeus. ENGLISH WALNUT. Waif along roadside: between E. Cotati Ave. and Rohnert Park Expressway on e. side of Petaluma Hill Road, acc. Knight (1990). < 700 ft. Native in se. Europe, the Himalayas and China.

LAMIACEAE [LABIATAE]. MINT FAMILY

1. Ovary 4-lobed, style not basal; nutlets almost completely united, laterally attached
    2. Corolla very irregular, apparently 1-lipped; stamens moderately exserted . . . . *Teucrium*
    2: Corolla nearly equally 5-lobed; stamens long-exserted . . . . *Trichostema*

1: Ovary deeply 4-parted, style basal; nutlets almost distinct, basally attached
    3. Calyx 2-lipped, lips entire . . . . *Scutellaria*
    3: Calyx regularly 5-toothed (10-toothed in *Marrubium*) or 2-lipped with 3 teeth on upper lip and 2 teeth on lower lip
        4. Stamens and style included in corolla-tube; calyx-teeth 10, spinescent, recurved and hooked in fruit . . . . *Marrubium*
        4: Stamens and style exserted beyond the corolla-tube; calyx-teeth 5, not recurved and hooked in fruit
            5. Corolla strongly 2-lipped, the lips unequal
                6. Upper pair of stamens > lower
                    7. Anther-sacs parallel or nearly so . . . . *Agastache*
                    7: Anther-sacs divergent . . . . *Nepeta*
                6: Upper pair of stamens < lower or =
                    8. Calyx closed in fruit, its upper lip truncate with 3 cusps on the margin, lower lip 2-cleft . . . . *Prunella*
                    8: Calyx not closed in fruit, regularly 5-toothed, or if 2-lobed the upper lip not truncate
                        9. Upper lip of corolla concave
                            10. Fertile stamens 4
                                11. Calyx-teeth not spine-tipped or mucronate; corolla without hairy ring within . . . . *Lamium*
                                11: Calyx-teeth spine-tipped or mucronate; corolla-tube with a hairy ring within . . . . *Stachys*
                            10: Fertile stamens 2 . . . . *Salvia*
                        9: Upper lip of corolla plane
                            12. Flowers solitary or few in the axils of the leaves
                                13. Corolla with a hairy ring near the base within . . . . *Lepechinia*
                                13: Corolla naked within
                                    14. Plants upright; corolla-tube curved upward . . . . *Melissa*
                                    14: Plants trailing; corolla-tube straight . . . . *Satureja*
                            12: Flowers capitate, spicate or capitate-verticillate in the axils of leaves or bracts . . . . *Pogogyne*
            5: Corolla regular or nearly so, lobes nearly or quite equal
                15. Flowers in dense terminal heads, stamens 4; plants usually of dry habitats . . . *Monardella*
                15: Flowers in axillary whorls; plants of wet habitats
                    16. Stamens 2 . . . . *Lycopus*
                    16: Stamens 4 . . . . *Mentha*

Agastache urticifolia (Bentham) Kuntze. HORSEMINT. Uncommon, moist places: Guerneville, near river, acc. Jepson, *Fl. of Ca. 3; Vol. II,* (1943) *Davy*; Dry Cr. near Healdsburg, *Carruth* (1900); roadside, Franz Valley, *Hemphill* (1936 PUA). < 10000 ft. June-Aug. —*A. glaucifolia* Heller, type locality: Knights Valley, *Heller* (1902).

Lamium

1. Upper leaves sessile or clasping; corolla purple-red . . . . *L. amplexicaule*
1: Upper leaves petioled; corolla lilac . . . . *L. purpureum*

★ **Lamium amplexicaule** Linnaeus. GIRAFFE HEADS. Occasional, weed, waste places: Camp Meeker, *Waitman* (1985 NCC); 1 mi. n. of Sebastopol, *Trynor* (1939 NCC); Blank Road near Gravenstein Hwy., acc. J. & B. Guggolz (1991); w. of Santa Rosa in Laguna de Santa Rosa drainage, *Harrison* (1978 NCC); Adobe Can., *L. Roof* (1953 NCC); Spring Lake; Ridgway Ave., Santa Rosa, *Cossin* (NCC); Pepperwood. < 2700 ft. Mar.-Aug. Native of Eurasia.
★ **Lamium purpureum** Linnaeus. RED HENBIT. Occasional, weed, waste places: The Sea Ranch, *Wells* (1985); road parallel to Bohemian Hwy. ca ¼ mi. n. of Occidental, *Best* (1981); w. of Santa Rosa in Laguna de Santa Rosa drainage, *Harrison* (1978 NCC); Adobe Can., Baker's property, *Howell* (1937); Brookside Garden in Adobe Can., *Baker* (1938 NCC); Spring Lake Village; Lake Ralphine e. of parking area, Santa Rosa, *Dudik* (1975 NCC); Pepperwood. < 1000 ft. Apr.-Sept. Native of Europe.

Lepechinia calycina (Bentham) Epling. PITCHER SAGE. Occasional, open slopes: Cazadero quad.; upper Hooker Can. on Cavedale Road, *Knight* (1964); Lake Sonoma area; between Jimtown and Mercuryville on Geysers Road, *Utech & Knight* (1986 CM); ca 2000 ft. elev. Anderson Trail on Black Mtn. between Geyser Peak and upper end of Pine Flat Road, acc. J. & B. Guggolz (1992); se. trail to Geyser Peak above Little Sulphur Cr., *Stocking* (1967 NCC). < 3000 ft. Apr.-June. —*Sphacele c.* Bentham.

Lycopus americanus W. C. Barton. BUGLEWEED, WATER-HOREHOUND. Occasional, wet areas: in willows at Wohler Road bridge on Russian River, *Raiche* (1990 JEPS); Laguna de Santa Rosa ne. of Graton, *Rubtzoff* (1961); 2 mi. e. of Graton, *Rubtzoff* (1960); Russian River at Fitch Mtn. grove 2 mi. e. of Healdsburg, *Rubtzoff* (1962). < 3300 ft. Aug.-Sept.

★ **Marrubium vulgare** Linnaeus. HOREHOUND. Common, waste places, former residential sites: Willow Creek; Bodega Head; Todd Road Preserve; Fairfield Osborn (H); Magnolia Ave., Petaluma, *Knight* (1963); Mt. Hood, *Knight* (1964); Bennett Valley, *Knight* (1963); Spring Lake; around Santa Rosa, *Brunk* (1933 NCC); Pepperwood; ca 2000 ft. elev. Anderson Trail on Black Mtn. between Geyser Peak and upper end of Pine Flat Road, acc. J. & B. Guggolz (1992). < 2000 ft. Spring and summer. Native of Europe.

★ **Melissa officinalis** Linnaeus. LEMON BALM, BEE BALM. Occasional, open woods, waste places: ca 1 mi. nw. of Camp Meeker, acc. J. & B. Guggolz (1991); Guerneville, Jepson, *Fl. of Ca, 3; Vol. II, p. 414*; Copeland Cr. between Snyder Lane and Sonoma State Univ., *Self* (1979 NCC); Kenwood, *Baker* (1932 NCC); bed of slough draining Kenwood Marsh, *Howell* (1932); Spring Lake; Mill Cr. Road near end, *Knight & Edwards* (1986). < 2700 ft. June-Sept. Native of s. Europe.

Mentha

1. Flower-whorls forming terminal spikes, or some of the lower in the leaf-axils and ± distant
   2. Plants glabrous or almost so
      3. Leaves sessile or almost so; spikes slender, ± interrupted . . . . *M. spicata*
      3: Leaves all distinctly petioled; spikes thick and mostly dense . . . . *M. × piperita*
   2: Plants tomentose or villous-tomentose . . . . *M. suaveolens*
1: Flower-whorls all distant and in leaf-axils; leaves exceeding the whorls, or those of the upper whorls often reduced and shorter than the whorls in *M. pulegium*
   4. Leaves rounded or obtuse at apex, those subtending the upper flower-whorls shorter than the whorls and turned downward . . . . *M. pulegium*
   4: Leaves acute at apex, those of the upper flower-whorls exceeding whorls and not turned downward . . . . *M. arvensis*

Mentha arvensis Linnaeus. FIELD MINT. Occasional, moist places: Perry Marsh, *Howell* (1945); 2 mi. n. of Sebastopol, *Baker* (1945 NCC); Pitkin Marsh, *Rubtzoff* (1951 CAS, 1957 NCC); Laguna de Santa Rosa at crossing of Occidental Road, *Baker* (1946 NCC); Spring Lake; w. of Windsor near Russian River, *Heller* (1902); Alexander Valley, *Jepson* (1921 JEPS); 2720 ft. elev. Caldwell Ranch, acc. Clifton (1984). Plant with angles and the sides of stem ca equally pubescent with hairs spreading are forma **lanata** (Piper) S. R. Stewart. *M. a.* var. *villosa*, currently not recognized. Also on Atlantic Coast and Eurasia. < 8000 ft. July-Oct. —*M. a.* ssp. *haplocalyx* Briquet.

★ **Mentha × piperita** Linnaeus. PEPPERMINT. BERGAMOT. Occasional, wet areas: Pitkin Marsh, *Rubtzoff* (1951); margin of creek at Hwy. 12 se. of Forestville, *Rubtzoff* (1951); Kenwood, *Baker* (1931 NCC); edge of Lake Ralphine, Santa Rosa, *Harrison* (1978 SRJC); bank of Porter Cr. opposite Mark West Springs, *Baker* (1945 NCC); Pepperwood. < 8300 ft. July-Oct. Native of Europe. —*M. citrata* Ehrhart.

Mentha pulegium Linnaeus. PENNYROYAL. Common, low damp places: The Sea Ranch; Meyers Grade Road n. of Jenner, *Wells* (1981); Bodega Head; Freestone, *Huntington* (1987 NCC); Estero Americano on side road off Hwy. 1, 3 mi. w. of Valley Ford, *Knight* (1963); Pitkin Marsh, *Baker* (1928 NCC), *Rubtzoff* (1951); Todd Road Preserve; ¼ mi. e. of Cotati, *Stocking* (1963); Copeland Cr. at Sonoma State Univ. campus, *Self* (1979 NCC); Fairfield Osborn (S); Tresch Ranch, Walker Road near Two Rock, acc. J. & B. Guggolz (1992); Theodore King Ranch on Old Redwood Hwy. 2½ mi. n. of Petaluma, *Knight* (1963); Kenwood, *Baker* (1932 NCC); Bennett Valley, *Knight* (1963); Annadel, *Wells* (1984); Spring Lake, Santa Rosa, *Knight* (1984); Burbank garden, Santa Rosa, *Mrs. Luther Burbank* (1926); Pepperwood; Knights Valley, *Tracy* (1935); Windsor, *Heller* (1923); Lake Sonoma area. < 3300 ft. June-Sept. Native of Europe.

★ **Mentha spicata** Linnaeus. SPEARMINT. Common, marshes, moist fields: Salt Point; near creek at n. edge of Occidental, *Best* (1983); near community center at Occidental, *Best & Wells* (1983); between Sebastopol and Cunningham, *Baker* (1946 NCC); Railroad Ave. in ditch near Cypress Ave. near Cotati, *Knight* (1963); jct. Sonoma Mtn. Road and Pressley Road, acc. Quibell (1991); ditch near Penngrove, *Knight* (1963); Ducker Cr. in Rincon Valley, *Wells* (1985); Spring Lake. < 5700 ft. July-Oct. Native of Europe.

★ **Mentha suaveolens** Ehrhart. APPLE MINT. Uncommon, moist places: Green Valley Marsh nw. of Graton, *Harrison* (1974 SRJC). < 4000 ft. July-Oct. Native of Europe. —*M. rotundifolia* (Linnaeus) Hudson, misapplied.

Monardella

1. Plants annual . . . . *M. undulata*
1: Plants perennial
    2. Outer bracts erect, sheathing, not leaflike . . . . *M. viridis*
    2: Outer bracts reflexed, leaflike in texture and shape
        3. Leaves thickish . . . . *M. villosa* ssp. *franciscana*
        3: Leaves not thick
            4. Upper parts of plant and lower surface of leaves pubescent to villous . . . . *M. v.* ssp. *villosa*
            4: Upper parts of plant and lower leaf-surfaces puberulent at most
                5. Bracts leaflike, not markedly ciliate . . . . *M. sheltonii*
                5: Bracts membranous, purple, ciliate . . . . *M. purpurea*

Monardella purpurea T. J. Howell. SISKIYOU MONARDELLA. Occasional, serpentine: 1024 ft. elev. Coleman Valley Road, *Clifton* (1984 PUA); Stoetz Lane ne. of Occidental, *Hoover* (1965); 2 mi. ne. of Occidental along Harrison Grade Road, *Baker* (1946 NCC). < 4700 ft. June-Aug. —*M. odoratissima* ssp. *neglecta* (E. Greene) Epling.

Monardella sheltonii T. J. Howell. Occasional, dry areas, sometimes on serpentine: Duncans Mills, *M. E. Jones* (1882); 3200 ft. elev. Socrates Mine, acc. Clifton (1984); 3300 ft. elev. Squaw Cr., acc. Clifton (1983); 2440 ft. elev. Caldwell Ranch, acc. Clifton (1984). < 6600 ft. June-Aug. —*M. coriacea* Heller, type locality: Russian River at Healdsburg, *Heller* (1902). *M. villosa* ssp. *s.* (Torrey) Epling.

Monardella undulata Bentham. CURLY-LEAVED MONARDELLA. Habitat has been partially destroyed: Larkspur Rock e. of Bodega Bay, acc. Baker (1954). < 1000 ft. May-July.

Monardella villosa Bentham ssp. **franciscana** (Elmer) Jokerst. Occasional, dry gravelly or rocky places: near Cazadero, *Graff* (1928); knob near Doran Park entrance at Bodega Bay, *Knight* (1965); Stewarts Point-Skaggs Springs Road 18 mi. e. of Stewarts Point, *Best* (1981); Geyser Peak, *Knight & Edwards* (1986). < 1400 ft. June-Aug. —*M. v.* var. *f.* (Elmer) Epling.

Monardella villosa ssp. **villosa**. COYOTE MINT. Common, dry rocky or gravelly areas: The Sea Ranch; Stewarts Point, *Abrams* (1920); 18 mi. e. of Stewarts Point on Stewarts Point-Skaggs Springs Road, *Best* (1981); 49 ft. elev. Gerstle Cove at Salt Point State Park, *Clifton* (1981 PUA); 1 mi. n. of Fort Ross, *Howell* (1930); Jenner, *Mason* (1928); 3.2 mi. n. of Jenner, *Cantelow* (1939); n. of Jenner, *Ripley & Barneby* (1945); coastal bluffs s. of mouth of Russian River, *J. H. Thomas* (1969); 8½ mi. n. of Bodega Bay, *Howell* (1930); cliffs below Westside Regional Park on Bodega Bay, *Hitchcock & Martin* (1939); near Cazadero, *Graff* (1928); Fairfield Osborn (S); Petaluma, *Congdon* (1880); Bouverie; 9 mi. n. of Sonoma,

Rose (1965); Bennett Valley, *Leitholt* (1893); Spring Lake; between Mark West Springs (probably Franz Valley Road), *Heller* (1902); Pepperwood; between Knights Valley and Mark West Springs (probably Franz Valley Road), *Heller* (1902); 6 mi. from Petrified Forest en route to Knights Valley, *Howell* (1929); w. of Calistoga, *Krauss* (1894); Fitch Mtn., *Rubtzoff* (1952); chaparral area sw. of Bradford Mtn. 1½ mi. w. of Dry Cr., acc. J. & B. Guggolz (1992); Lake Sonoma area; ca 2000 ft. elev. Anderson Trail on Black Mtn. between Geyser Peak and upper end of Pine Flat Road, acc. J. & B. Guggolz (1992); The Geysers, *Baker* (1898 UC); 3600 ft. elev. Geyser Rock, acc. Clifton (1983); 3620 ft. elev. Anderson Cr. 1 mi. sw. of Cobb Mtn. summit, acc. Clifton (1984); 3700 ft. elev. Cobb Mtn., *Jokerst* (1985); 2440 ft. elev. Caldwell Ranch, acc. Clifton (1984); 3000 ft. elev. Caldwell Pines, acc. Clifton (1984). < 4300 ft. June-Aug. —*M. mollis* Heller. *M. subserrata* E. Greene, type locality: Sonoma County (probably in the ne. part), *Dunn* (1890 UC). *M. villosa* ssp. *subserrata* (E. Greene) Epling.

Monardella viridis Jepson. GREEN MONARDELLA. Occasional, dry slopes, ridges: Lafferty Ranch on Sonoma Mtn., acc. Arnold et al. (1971); 2½ mi. e. of Glen Ellen, *Baker* (1934 NCC); Sugarloaf Ridge; Fountaingrove property on Rincon Ridge, *Knight* (1990 NCC); Pepperwood, *Knight & Edwards* (1986 NCC); Mountain Home area ne. of Petrified Forest, *Stocking* (1968 NCC); Sonoma Co. (probably Knights Valley), *Krauss* (1894). < 3300 ft. July-Sept.

Nepeta cataria Linnaeus. CATNIP. Occasional, weed, commonly near dwellings: Russian River near Guerneville, *Davy* (1894 JEPS); Petaluma, *Chesnut* (1887 JEPS); near Baker's house in Adobe Can., *Baker* (1927 NCC); Spring Lake Village. < 4300 ft. July-Sept. Native of Eurasia.

Pogogyne

1. Stamens all 4 with fertile anthers; style obviously bearded below the subequal stigma lobes; corolla 9-20 mm long, plainly exserted . . . . *P. douglasii*

1: Stamens upper sterile; style sparsely hairy immediately below the unequal stigma lobes; corolla 3-12 mm long, ± hidden in heads
   2. Stems mostly prostrate or ascending, slender; inflorescence 8-12 mm broad; corolla 3-5 mm long . . . . *P. serpylloides*
   2: Stems erect or nearly so, robust; inflorescence mostly 12-20 mm broad; corolla 4-8 mm long . . . . *P. zizyphoroides*

Pogogyne douglasii Bentham. Occasional, low places, beds of winter pools; threatened by continuing development in the sw. Santa Rosa Plain and Windsor; many locations have been extirpated: Laguna de Santa Rosa n. of Occidental Road, *Baker* (1946 NCC); 4 mi. w. of Santa Rosa on road to Sebastopol, *Howell* (1947); 3 sites w. and sw. of Santa Rosa in Laguna de Santa Rosa drainage, *Harrison* (1978 NCC); e. of Wright Road, *Harrison* (1979 NCC); Stony Point Road in Todd Dist. *Focht* (1932 NCC); vernal pool on Bennett Valley Road, *Harrison* (1974); 4¼ mi. w. of Glen Ellen on Bennett Valley Road, *Rubtzoff & Froebe* (1971); Bennett Valley Road just w. of Enterprise Road on n. side, acc. Quibell (1991); some vernal pools in Laguna de Santa Rosa and Windsor areas, acc. Marco E. Waaland (1989); Annadel in False Lake meadow, *Herrick* (1982 NCC); jct. Old Redwood Hwy. and Starr Road, *J. Smith* (1974 NCC); 2 mi. n. of Windsor, *Eastwood & Howell* (1936); s. of Shiloh Road e. of Hwy. 101, *J. & B. Guggolz* (1988 SRJC); n. of Arata Lane e. of Hwy. 101, *Harrison* (1979 SRJC), *J. & B. Guggolz* (1988 SRJC). < 3000 ft. May-July. —*Pogogyne d.* ssp. *minor* J. T. Howell.

Pogogyne serpylloides (Torrey) A. Gray. Rare, various habitats: The Sea Ranch; dry pool 5 mi. e. of Bennett Mtn. Lake, *Rubtzoff* (1964). < 4000 ft. Apr.-June. —*P. s.* ssp. *intermedia* J. T. Howell, currently not recognized.

Pogogyne zizyphoroides Bentham. Uncommon, open areas: Salt Point; ca half way between the ne. corner of Sonoma County airport and Windsor on Sanders Lane, *J. & B. Guggolz* (1986 NCC). < 1300 ft. July-Aug.

Prunella

1. Leaves ovate to oblong-ovate, rounded at base, ½ as broad as long . . . . *P. vulgaris* var. *vulgaris*

1: Leaves lanceolate, gradually narrowed at base, about 1/3 as broad as long . . . . *P. v.* var. *lanceolata*

Prunella vulgaris Linnaeus var. **lanceolata** (Barton) Fernald. SELF-HEAL. Common, moist areas: The Sea Ranch; 0.7 mi. s. of Gualala on coastal bluff, *J. H. Thomas* (1976); Annapolis, *Sutcliffe* (1920); Stewarts Point quad.; Salt Point (Stump Beach); Fort Ross-Russian River, Bodega areas, *Voznesenskii* (1840-1841 LE); Duncan Mills Marsh, *Rubtzoff* (1957); Hwy. 1 se. of Salmon Cr. crossing, *Rubtzoff*

(1971); 5½ mi. e. of Bodega Bay (near Valley Ford), *Cantelow* (1941); Freestone Marsh, *Rubtzoff* (1959); Guerneville, *Rattan* (1877); between Guerneville and Vacation Beach, *Rubtzoff* (1959); Guerneville Marsh, *Rubtzoff* (1954); Forestville, *Marasco* (1970 NCC); upper Green Valley Marsh, *Rubtzoff* (1974); Pitkin Marsh, *Rubtzoff* (1951), *Quibell* (1987 NCC); Sebastopol, *Dows* (1907), *Bately* (1932 NCC); Laguna de Santa Rosa e. of Graton, *Rubtzoff* (1966); Vine Hill School, *Howell* (1945); Cunningham Marsh, *Howell* (1947); Button Ranch on Walker Road (U. C. property) between Petaluma-Valley Ford Road and Pepper Road, acc. Connors (1992); Stony Point Road in Todd Dist., *Focht* (1932); Bennett Mtn. Lake, *Rubtzoff* (1961); near Windsor, *Carruth* (1901); ne. slopes of Fitch Mtn., *Rubtzoff* (1952); near Lake Sonoma on Stewarts Point-Skaggs Springs Road, *Best* (1981); Lake Sonoma area. < 8000 ft. May-Sept. —*P. v.* var. *atropurpurea* Fernald, type—Santa Rosa, *Heller* (1902 GH).

Prunella vulgaris var. **vulgaris**. SELF-HEAL. Occasional, moist areas: Kenwood Marsh, *Rubtzoff, Bunyan, Harrison & Weigner* (1974); Pepperwood. < 700 ft. Summer.

Salvia

1. Plants annual; stems mostly simple 1-5 dm tall, mostly retrorse hairs; leaves mostly basal; calyx purplish . . . . . *S. columbariae*
1: Plants perennial; numerous branches or stems, no retrorse hairs; ± leafy throughout; calyx not purplish
  2. Height 1-1.5 m . . . . . *S. microphylla*
  2: Height considerably less
    3. Plants erect to 6 dm tall, leaves simple, petiolate, oblong, ± narrowed at base, rugose . *S. officinalis*
    3: Plants creeping with leafy ± matted stems, quite leafless flowering stems, leaves elliptic-obovate, not rugose . . . . . *S. sonomensis*

Salvia columbariae Bentham. CHIA. Occasional, dry open disturbed areas: Pepperwood; Lake Sonoma area; ca 2000 ft. elev. Anderson Trail on Black Mtn. between Geyser Peak and upper end of Pine Flat Road, acc. J. & B. Guggolz (1992); Geyser Peak, *Knight & Edwards* (1986); 2000 ft. elev. Big Sulphur Cr. drainage, *Malloch* (1977 PG&E). < 4000 ft. Mar.-June.

★ **Salvia microphylla** Kunth. Escaped exotic: along road near Cunningham Marsh, *Howell* (1947). Native of Mexico. —*S. grahamii* Bentham.

★ **Salvia officinalis** Linnaeus. SAGE. Rare, damp areas: ¼ mi. n. of Graton Road, *Ferrar* (1979 NCC). Native of Mediterranean area.

Salvia sonomensis E. Greene. SONOMA SAGE. Common, dry slopes: Bouverie; up Cavedale Road from Hwy. 12, *Utech & Knight* (1986 CM); upper Hooker Can. on Cavedale Road, *Knight* (1964); 750 ft. elev. near ridge top w. of Stuart Can. at Bouverie, acc. Stone (1985); Sugarloaf Ridge; w. of Buzzard Peak in Mt. Hood Range, acc. J. & B. Guggolz (1992); Spring Mtn. Road, *Knight* (1963); Calistoga Road at Joost Ranch, *Knight* (1963); Gates Road off Calistoga Road, acc. Wells et al. (1984); Pepperwood; Franz Valley Road 8 mi. nw. of Calistoga, *K. J. Murray* (1980 SFSU); Franz Valley Road, *Clifton* (1973 PUA); along ridge n. of Dorman Can. Road near Sonoma-Napa co. line, acc. Knight & Edwards (1986). Type locality: maybe in Sonoma Co., "Hab. in California septentrionale *Douglas*." < 6600 ft. May-June.

Satureja douglasii (Bentham) Briquet. YERBA BUENA. Common, shady woods: The Sea Ranch; Stewarts Point quad.; Willow Creek; Salmon Cr. Road w. of town of Bodega, *Knight* (1963); w. of Occidental on Coleman Valley Road, *Stocking* (1969 NCC); Hidden Oaks Road on w. side of w. end of Sonoma Mtn. Road, *Quibell* (1988 NCC); Button Ranch on Walker Road (U. C. property) between Petaluma-Valley Ford Road and Pepper Road, acc. Connors (1992); Bouverie; Lake Sonoma area; Pepperwood. < 3000 ft. Apr.-Sept. —*Micromeria douglasii* Bentham. *M. chamissonis* E. Greene.

Scutellaria

1. Corolla white to light yellow, occasionally tinged pink or blue, lower lip not mottled . . . . . . *S. californica*
1: Corolla blue to violet-blue, lower lip usually white-patched
  2. Stem and leaf hairs 1-3 mm, spreading; upper cauline leaf blades usually crenate; plants with tubers . . . . . *S. tuberosa*
  2: Stem and leaf hairs << 0.5 mm, curled or appressed; upper cauline leaf blades entire; rhizomes slender, tubers ± 0
    3. Flower 12-22 mm; leaf length 2-3 × width; corolla hairs not gland-tipped . . . . . . *S. antirrhinoides*
    3: Flower 25-35 mm; leaf length 3-7 × width; corolla hairs gland-tipped . . . . . . *S. siphocampyloides*

Scutellaria antirrhinoides Bentham. SKULLCAP. Occasional, moist or drying areas: near Cazadero, *Graff* (1928); Piner Road ca 1½ mi. w. of Piner School near Santa Rosa, *Comalli* (1939); near Russian River w. of Windsor, *Heller* (1902); Fitch Mtn., *Rubtzoff* (1952); Lake Sonoma area; s. slope of Black Mtn. between Geyser Peak and upper end of Pine Flat Road, acc. J. & B. Guggolz (1992). < 6600 ft. June-Sept. —*S. viarum* Heller, type locality: Windsor, *Heller* (UC).

Scutellaria californica A. Gray. CALIFORNIA SKULLCAP. Occasional, ± gravelly areas: Bouverie; Sonoma Can., Kenwood (probably Adobe Canyon); Santa Rosa; Pepperwood; near Petrified Forest, *Howell* (1937); Lake Sonoma area; property bordered by Chianti and Dutcher Cr. roads w. of US 101 and w. of Geyserville, acc. J. & B. Guggolz (1991); 3300 ft. elev. upper Squaw Cr. drainage 1.2 airmiles nw. of Geyser Rock, acc. Stone (1983). < 6600 ft. June-July.

Scutellaria siphocampyloides Vatke. AUSTIN'S SKULLCAP. Uncommon, rocky or gravelly areas: near Mirabel Mine (probably at The Cedars), *Baker* (1936). 1000-7700 ft. May-July. —*S. austinae* Eastwood.

Scutellaria tuberosa Bentham. DANNIE'S SKULLCAP. Occasional, shady woods, open brush: Cavedale Road in low chaparral, acc. Quibell (1991); Schultz Can. in Annadel on s. facing slopes above creek, acc. Quibell (1991); Trinity Mtn. e. of Glen Ellen, *Howell* (1937); Lake Sonoma area. 500-3300 ft. Mar.-July.

Stachys

1. Annual; rhizomes none . . . . . . *S. arvensis*
1: Perennial; rhizomes slender
   2. Ring of hairs on inner surface of corolla-tube not indicated on the outer surface by a constriction; ring of hairs within corolla-tube oblique or horizontal and near the middle of tube; corolla-tube > 15 mm . *S. chamissonis*
   2: Ring of hairs on the inner surface of corolla-tube indicated on the outside by an oblique constriction of tube, this more pronounced and often slightly saccate on the anterior side; corolla-tube < 15 mm
      3. Upper lip of corolla 2-2.5 mm long, corolla white . . . . . . *S. stricta*
      3: Upper lip of corolla 3-6 mm long; corolla pale rose and mottled or whitish
         4. Plants ± pubescent with straight hairs . . . . . . *S. ajugoides*
         4: Plants ± densely arachnoid-pubescent with whitish cobwebby hairs . . . . . . *S. albens*

Stachys ajugoides Bentham. HEDGE NETTLE. Very common, springy ravines, canyons: s. part of The Sea Ranch, *J. H. Thomas* (1976); ca ½ mi. n. of Stewarts Point, *Wells* (1985); Salt Point; 5 mi. n. of Jenner, *N. J. Murray* (1984 SFSU); Russian Gulch n. of Jenner, *Epling & Robison* (1935 NCC); Willow Creek; Goat Rock State Beach, *Wells* (1981); 10 ft. elev. Salmon Cr., *Reeve* (1980 PUA); Bodega Point, *Eastwood* (1915); Bodega Head, *Ferris & Morey* (1958); Larkspur Rock e. of Bodega Bay, *Ott* (1934 NCC); Bodega, *Rose* (1962); 4½ mi. w. of Cazadero on Fort Ross Road, *Quibell & Atkinson* (1980 NCC); Cunningham Marsh, *Howell* (1947); 4 mi. e. of Valley Ford, *Howell* (1930); Pitkin Marsh, *Rubtzoff* (1951); Rohnert Park from Sonoma St. to Snyder Lane, *Caruso* (1979 NCC); Fairfield Osborn (*); Petaluma, *Congdon* (1880); near Sears Point jct., *Baker* (1943 NCC); along hwy. n. of Sears Point, *Baker* (1946); Schellville, *Epling & Robison* (1935 NCC); Bouverie; Sugarloaf Ridge; between Santa Rosa and Agua Caliente, *Heller & Brown* (1902); Annadel, *Wells* (1984); end of Fern Glenn Drive, Santa Rosa, *Dudik* (1975 NCC); Pepperwood; corner Calistoga and Porter Cr. roads, *Knight* (1964); Lake Sonoma area; ca 2000 ft. elev. Anderson Trail on Black Mtn. between Geyser Peak and upper end of Pine Flat Road, acc. J. & B. Guggolz (1992); Geyser Peak, acc. Knight & Edwards (1986); 3800 ft. w. flank Mt. St. Helena, *J. C. Edwards* (1970 PUA); 2100 ft. elev. below geysers Unit 16, *Malloch* (1977 PG&E). *Stachys ajugoides* var. *a.* and *S. a.* var. *rigida* are accompanied by descriptions covering plants that are widely variable and in many cases form intermediates. Therefore we consider the epithet *S. ajugoides* best fits this odd assortment. < 8300 ft. May-Sept. —*S. ajugoides* var. *rigida* (Nuttall) Jepson & Hoover. *S. rigida* Nuttall ssp. *rigida*. *S. rigida* ssp. *quercetorum* (Heller) Epling. *S. emersonii* Piper. *S. mexicana* Bentham.

Stachys albens A. Gray. WHITE STACHYS. Occasional, very wet areas: ca 1 mi. nw. of Camp Meeker, acc. J. & B. Guggolz (1991); Adobe Can. at Golden Bear Lodge at creekside, acc. Knight (1965); Sugarloaf Ridge; Lake Sonoma area. < 10000 ft. May-Oct.

★ **Stachys arvensis** Linnaeus. Locally common, vineyards: Alexander Valley, acc. Doreen Smith. < 500 ft. May-Sept. Native of Europe.

Stachys chamissonis Bentham. COAST HEDGE NETTLE. Common, wet places, swamps: The Sea Ranch; Gualala River near Mendocino-Sonoma line, *Abrams* (1920); near Stewarts Point, *Carruth* (1901); 249 ft. elev. Salt Point, *Clifton* (1981 PUA); Bodega Bay Marsh, *Rubtzoff* (1962); Willow Cr. Marsh, *Rubtzoff* (1959); Duncans Mills, *M. Jones* (1882); Duncans Mills Marsh, *Rubtzoff* (1955); Forestville Marsh, *Rubtzoff* (1955); marshy area s. of Hwy. 116 near Bloomfield Road, *J. & B. Guggolz* (1992 JEPS); Laguna de Santa Rosa e. of Graton, *Rubtzoff* (1960); Laguna de Santa Rosa e. of Cunningham, *Rubtzoff* (1958); Cunningham Marsh, *Rubtzoff* (1961). < 500 ft. June-Oct.

Stachys stricta E. Greene. Uncommon, edges of marshy ground: Bouverie; Pepperwood; Knights Valley, *E. Greene* (type, U), *Davy* (1894 JEPS), *Harrison* (1979 NCC); 2420 ft. elev. Caldwell Ranch, *Clifton* (1984 PUA). < 2500 ft. June-Sept

★ **Teucrium fruticans** Linnaeus. GERMANDER. Uncommon, persisting from cultivation at former residential site: Fairfield Osborn (H). Native of Europe.

Trichostema

1. Lower petioles 5-15 mm long; leaves rather remote, not apiculate . . . . *T. laxum*
1: Lower petioles not > 4 mm long; leaves crowded, apiculate . . . . *T. lanceolatum*

Trichostema lanceolatum Bentham. VINEGAR WEED. Common, open places: The Sea Ranch, acc. *Mahaffey* (1990); Fort Ross quad.; Little Cat Mtn. near Cazadero, *Winter* (1934 NCC); Russian River, *Carruth* (1900); London's Place, Glen Ellen, *Sutliffe* (1921); Sonoma Cr. w. of Agua Caliente, *Rubtzoff* (1973); Bouverie; Los Alamos Road, *Knight* (1963); Spring Lake; w. of Santa Rosa in Laguna de Santa Rosa drainage, *Harrison* (1978 NCC); n. of Santa Rosa, *Wells* (1986); Pepperwood; Lake Sonoma area; near lowest elev. at start of Pine Mtn. Road, *Knight* (1963). < 3300 ft. Aug.-Oct.

Trichostema laxum A. Gray. TURPENTINE WEED. Common, low fields, gravelly sands of stream bottoms, ± on serpentine: 2 mi. from Monte Rio on Camp Meeker Road, *Howell* (1931); n. of Duvoul Cr. n. of Camp Meeker, *Edwards* (1984), *J. & B. Guggolz* (1990 JEPS); Stoetz Lane on serpentine, *Best* (1981); near lake on Stoetz Lane ca 2 mi. ne. of Occidental, *Best* (1983); Harrison Grade Road, *Baker* (1946 NCC), *Stocking* (1966 NCC), *Hoover* (1965); Vacation Beach near Guerneville, *Raven* (1962); Russian River, *Carruth* (1900); Russian River at Forest Hills, *Rubtzoff* (1966); Preston (n. of Healdsburg), *Carlson* (1919); chaparral area sw. of Bradford Mtn. 1½ mi. w. of Dry Cr., acc. J. & B. Guggolz (1992); Geyser Can. in ne. Sonoma Co., *Brewer* (1862 JEPS); The Geysers, *Carruth* (1899). < 6600 ft. July-Sept.

LAURACEAE. LAUREL FAMILY

Umbellularia californica (Hooker & Arnott) Nuttall. CALIFORNIA BAY, CALIFORNIA LAUREL, PEPPERWOOD. Common, canyons, valleys: The Sea Ranch; Salt Point; Russian California, *Voznesenskii* (1840); Willow Creek; The Cedars, acc. Raiche, (1985); Monte Rio, *H. Mason* (1925); Graton-Occidental Road, *Hoover* (1968); Roberts Road (off Petaluma Hill Road), *Stocking* (1963 NCC); Lichau Road (off Roberts Road), *Stocking* (1963 NCC); Fairfield Osborn (*); Tresch Ranch, Walker Road near Two Rock, acc. J. & B. Guggolz (1992); Bouverie; Kenwood, *A. & R. Nelson* (1933); Sugarloaf Ridge; Mt. Hood, *Knight* (1964); Gates Road off Calistoga Road, acc. Wells et al. (1984); Los Alamos Road, *Knight* (1963); Spring Lake; hills e. of Santa Rosa, *Heller & Brown* (1902); Pepperwood; chaparral area sw. of Bradford Mtn. 1½ mi. w. of Dry Cr., acc. J. & B. Guggolz (1992); Lake Sonoma area; summit of Pine Flat Road, *Knight* (1963); ca 2000 ft. elev. Anderson Trail on Black Mtn. between Geyser Peak and upper end of Pine Flat Road, acc. J. & B. Guggolz (1992); Geyser Peak, *Knight & Edwards* (1986); 2600 ft. elev. Coldwater Cr. Can. sw. of county line at Caldwell Pines, acc. Stone (1989); Asti, *Knight* (1963). < 5300 ft. Dec.-May.

LENTIBULARIACEAE. BLADDERWORT FAMILY

Utricularia vulgaris Linnaeus. COMMON BLADDERWORT. Uncommon, quiet water: big pond on ridge of The Sea Ranch, acc. C. Tarp (1986). < 1700 ft. July-Sept.

LIMNANTHACEAE. FALSE MERMAID FAMILY

Limnanthes

1. Plants with 1st few foliage leaves on seedlings entire, lobes of leaves on adult plants narrowly obovate with entire margins . . . . *L. vinculans*

1: Plants with leaves otherwise
  2. Sepals with long white fine ± curly hairs at least within, occasionally also without; leaves ± pilose; petals scarcely if at all emarginate . . . . *L. alba*
  2: Sepals mostly glabrous, sometimes slightly coarse-ciliate; leaves glabrous; petals often emarginate; stamens 5-8 mm long
    3. Petals yellow with white tips . . . . *L. douglasii* ssp. *douglasii*
    3: Petals white . . . . *L. d.* ssp. *nivea*

Limnanthes alba Bentham. WHITE MEADOWFOAM. Occasional, low moist open areas: near Guerneville Hwy. in Laguna de Santa Rosa drainage, *Harrison* (1978 NCC); e. foothills of Sonoma Mts. on Sobre Vista Road 0.9 mi. w. of Arnold Drive near El Verano, *Rubtzoff* (1971); 300 ft. elev. hills w. of Agua Caliente bordering Sonoma Valley, *Rose* (1951); Bouverie; 1 mi. ne. of Glen Ellen, ne. of Warfield, *Baker* (1923 NCC); near Santa Rosa, *Alexander* (1944); Lake Sonoma area. < 4700 ft. Apr.-June.

Limnanthes douglasii R. Brown ssp. **douglasii**. COMMON MEADOWFOAM. Common, moist places: Salt Point; Willow Creek; 6 mi. n. of Salmon Cr. on Hwy. 1, *Rubtzoff* (1971); Bodega Bay, *Rose* (1933); Bodega Point, *Eastwood* (1915); s. of Hwy. 116 near Blank Road, acc. J. & B. Guggolz (1990); base of hills e. of Petaluma, *C. P. Smith* (1921); Petaluma, *Eastwood* (1921); adjacent to jct. of Adobe and Sonoma Mtn. roads, *Lowry* (1983 SFSU); foothills of Sonoma Mts. on Sobre Vista Road 0.9 mi. w. of Arnold Drive near El Verano, *Rubtzoff* (1971); El Verano, *Griffin* (1922); Sonoma, *Rose* (1965); Sugarloaf Ridge; Los Guilicos Valley, *Hoover* (1939); fields e. of Santa Rosa, *Heller & Brown* (1902); Spring Lake; Santa Rosa, *Eastwood* (1921); n. side of Hwy. 116, ½ mi. nw. of Stony Point Road crossing near Santa Rosa, *Rubtzoff* (1966); Pepperwood; w. of Knights Valley, *Breedlove* (1983); 4 mi. w. of Calistoga, *Wiggins* (1949); E. Windsor and W. Windsor, *Heller* (1940). < 2300 ft. Mar.-May.

Limnanthes douglasii ssp. **nivea** (C. Mason) C. Mason. SNOWY MEADOWFOAM. Common, vernally moist places: Willow Cr. Road 1 mi. from Bridgehaven, *Best* (1981); Occidental Marsh, *Rubtzoff* (1957); Freestone Marsh, *Rubtzoff* (1959); near Freestone, *Rose* (1962); Graton Road about 2 mi. w. of Graton, *Knight* (1967); Atascadero Cr. Marsh, *Rubtzoff* (1965); Laguna de Santa Rosa e. of Sebastopol, *Rubtzoff* (1969); fields near Sebastopol, *Mason* (1928); Todd Road Preserve; Todd Road 1.4 mi. from Hwy. 116 and 0.6 mi. from Llano Road, *Rubtzoff* (1966); Stony Point Marsh, *Rubtzoff* (1955); Santa Rosa, *Abrams* (1918); property bordered by Chianti and Dutcher Cr. roads w. of US 101 and w. of Geyserville, acc. J. & B. Guggolz (1991). < 3300 ft. Mar.-May.

Limnanthes vinculans Ornduff. SEBASTOPOL MEADOWFOAM. Threatened by urbanization and agriculture: in 25 vernal pools in Laguna de Santa Rosa and Windsor areas, acc. Marco E. Waaland (1989). Specific locations: acc. Ornduff, *Brittonia 21:11,* (1969), Santa Rosa, Sebastopol, Camp Meeker, Two Rock; Atascadero Cr. Marsh nw. of Sebastopol n. of Green Valley Road, *Rubtzoff* (1958); Cunningham Marsh, *Howell* (1947); n. side of Hessel Road at jct. of hwy. in wet pasture se. of Cunningham, acc. Lovell (1980); 5 locations in Laguna de Santa Rosa drainage, *Harrison* (1978 NCC); Todd Road Preserve 1.4 mi. from Hwy 116 and 0.6 mi. from Llano Road, *Rubtzoff* (type 1966 UC), *Self* (1980 NCC); Horn Ave. (may be extinct) ½ mi. e. of Santa Rosa Ave., n. of Rohnert Park, *Quibell* (1983 NCC); nw. corner of jct. of Scenic and Whistler aves., Santa Rosa, acc. Lovell (1980). It was believed to be endemic to Sonoma County until it was discovered at Yountville in 1992, Napa County. < 1000 ft.

LINACEAE. FLAX FAMILY

1. Styles 5, ovary chambers 10, fruit 10-seeded . . . . *Linum*

1: Styles 2-3, ovary chambers 4 or 6, fruit 4 or 6 seeded . . . . *Hesperolinon*

Hesperolinon

1. Carpels and styles 2 . . . . *H. bicarpellatum*
1: Carpels and styles 3
  2. Petals white to pinkish
    3. Petals 1.5-3.5 mm long, not or little spreading; filaments included . . . . *H. micranthum*
    3: Petals 4-6 mm long, widely spreading; filaments exserted
      4. Pedicels 1-5 mm long in flower, not strongly reflexed, the buds not pendent . . . . *H. disjunctum*
      4: Pedicels 5-15 mm long in flower, early deflexed from branch, the buds pendent . *H. spergulinum*
  2: Petals yellow . . . . *H. clevelandii*

Hesperolinon bicarpellatum (H. K. Sharsmith) H. K. Sharsmith. TWO-CARPELLATE DWARF FLAX. Uncommon, creek beds: 2700 ft. elev. w. flank Mt. St. Helena, *Grummer* (1979 PUA). 200-3300 ft. June. —*Linum b.* H. K. Sharsmith.

Hesperolinon clevelandii Small. Uncommon, serpentine: sw. slope Mt. Hood, *Baker* (1947). 500-4700 ft. May-July. —*Linum c.* E. Greene.

Hesperolinon disjunctum H. K. Sharsmith. Uncommon, openings in brush: 1198 ft. elev. Calistoga Road, *Clifton* (1981 PUA). 330-3300 ft.

Hesperolinon micranthum (A. Gray) Small. Occasional, often on serpentine: Cazadero quad.; 1024 ft. elev. Coleman Valley Road, *Clifton* (1984 PUA); Harrison Grade Road, *Baker* (1946 NCC), *H. Sharsmith* (1946), *Ferris & Lorraine* (1961); Trinity Road, *Craig* (1941); chaparral slope of Mt. Hood, *Baker* (1947 NCC); Lake Sonoma quad.; 2000 ft. elev. near geysers Unit 18, *Malloch* (1977 PG&E). 160-6600 ft. May-July. —*Linum m.* A. Gray.

Hesperolinon spergulinum (A. Gray) Small. Occasional, dry slopes, gravelly barrens of serpentine: Bohemian Hwy. 3.8 mi. nw. of Occidental, *Huntington* (1986 NCC); ca 2 mi. ne. of Occidental along Harrison Grade Road, *Baker* (1947); Pepperwood; near Petrified Forest, *Howell* (1937); Lake Sonoma area; 3200 ft. elev. upper Squaw Cr. drainage 0.7 airmile nw. of Geyser Rock, acc. Stone (1983). 330-3300 ft. June-July. —*Linum s.* A. Gray, type locality: Cloverdale, *Bolander* (1865 UC).

Linum

1. Petals yellow . . . . *L. trigynum*
1: Petals blue (rarely white in *L. usitatissimum*)
  2. Stigmas elongate
    3. Petals ca 1.5 cm long; fruit 7-10 mm long; leaves narrowly lanceolate; seeds 4 mm long, brown, shiny but ± roughened . . . . *L. usitatissimum*
    3: Petals ca 1 cm long; fruit 5 mm long; leaves linear; seeds 3 mm long, brownish-green, shining and smooth . . . . *L. bienne*
  2: Stigmas nearly as broad as long . . . . *L. lewisii*

★ **Linum bienne** Miller. Common, grassy places: The Sea Ranch; Stewarts Point quad.; ½ mi. n. of Stewarts Point along Hwy. 1, *Wells* (1985); Gerstle Cove at Salt Point State Park, *Wells* (1982); Wright's Beach, *Hastings* (1940 NCC); Bodega Head, acc. Connors (1991); Bodega Bay across highway from Tides Restaurant, *Knight* (1963); Larkspur Rock e. of Bodega Bay, *Eastwood & Howell* (1939); ca 1 mi. nw. of Camp Meeker, acc. J. & B. Guggolz (1991); Sweetwater Springs Road between Guerneville and Healdsburg, *Knight & Edwards* (1986 NCC); Harrison Grade Road, *G. L. Smith* (1958); Todd Road area of Laguna de Santa Rosa drainage, *Harrison* (1978 NCC); Tresch Ranch, Walker Road near Two Rock, acc. J. & B. Guggolz (1992); Chileno Valley, *Howell* (1944); Pepperwood; between Jimtown and Mercuryville on Geysers Road, *Knight* (1986 CM). < 3300 ft. Apr.-Aug. Introduced from Mediterranean region. —*L. angustifolium* Hudson.

Linum lewisii Pursh. Uncommon, dry ridges, slopes: Graton, *Baker* (1928); Annadel. 400-11400 ft. May-Sept.

★ **Linum trigynum** Linneaus. Rare, rocky slopes: Ft. Ross acc. Mahaffey (1996). New to California. Native of S. Europe. -*L. gallicum* Linneaus.

Linum usitatissimum Linnaeus. COMMON FLAX. Uncommon, casual weed, waste places: vacant lot, Sebastopol, *Howell* (1957 NCC); 2 mi. e. of Sonoma on Hwy. 37, *McNally* (1970 NCC); Spring Lake Village. < 400 ft. Feb.-May. Introduced from Europe.

LOASACEAE. Loasa family

Mentzelia

1. Plants perennial; sepals 15-46 mm; petals 30-80 mm; seeds lenticular, winged . . . . *M. laevicaulis*
1: Plants annual; sepals less; petals less; seeds not winged
   2. Sepals 1-3.5 mm; petals 2-6 mm; seeds prismatic, angles usually sharp, grooved, sides smooth, shiny . . . . *M. dispersa*
   2: Sepals 9-19 mm; petals 20-40 mm; seeds grainlike, irregularly angled, minutely papilate . . . . *M. lindleyi*

Mentzelia dispersa S. Watson. Uncommon, dry, disturbed, sandy to rocky areas: ca 2000 ft. elev. Anderson Trail on Black Mtn. between Geyser Peak and upper end of Pine Flat Road, *J. & B. Guggolz* (1992 NCC); 3300 ft. elev. upper Squaw Cr. drainage 1.2 airmiles nw. of Geyser Rock, acc. Stone (1983). < 8500 ft. May-Aug.

Mentzelia laevicaulis (Hooker) Torrey & A. Gray. Blazing star. Uncommon, sand and gravel bars in river bottoms or on dry hillsides: Vacation Beach near Guerneville, *Rubtzoff* (1961); Russian River near Cloverdale, *Rattan* (1877); Russian River bed, *Geis* (1902); Sulphur Cr. ½ mi. from US 101, *Knight* (1963). < 9000 ft. June-Oct.

★ **Mentzelia lindleyi** Torrey & A. Gray. Possible garden escape: near El Verano, *Alford* (1972 SRJC). < 4000 ft. Mar.-May. Native from w. Fresno Co. northward to Alameda Co.

LYTHRACEAE. Loosestrife family

1. Flower-tube short, campanulate to globular; leaves opposite; petals 4 . . . . *Ammannia*
1: Flower-tube cylindric; leaves mainly alternate, petals usually 6 . . . . *Lythrum*

Ammannia coccinea Rottboell. Uncommon, wet places: Russian River at Vacation Beach 1½ mi. downstream from Guerneville, *Rubtzoff* (1956); Laguna de Santa Rosa e. of Graton, *Rubtzoff* (1963); 0.3 mi. n. of laguna crossing at Stony Point Road in Laguna de Santa Rosa drainage, *Harrison* (1978 NCC); Todd Road Preserve; Copeland Cr. ¼ mi. e. of hwy. in Rohnert Park, *Self* (1979 NCC); Spring Lake. < 1000 ft. May-Oct. —*A. auriculata* Willdenow, misapplied.

Lythrum

1. Flowers sessile or nearly so
   2. Plants annual; not stoloniferous . . . . *L. hyssopifolium*
   2: Plants perennial; ± stoloniferous . . . . *L. adsurgens*
1: Flowers short-pedicelled
   3. Plants annual . . . . *L. tribracteatum*
   3: Plants perennial
      4. Flowers numerous in showy spike; petals 8-14 mm; stamens twice as many as petals . . . *L. salicaria*
      4: Flowers single in axillary leaf bases; petals 4-8 mm; stamens same number as petals . *L. californicum*

Lythrum adsurgens E. Greene. Uncommon, moist places: 1 mi. s. of the Gualala River on Hwy. 1, *Wolf* (1927); Pitkin Marsh, *Rubtzoff* (1951); Bloomfield between Two Rock and Valley Ford, *Congdon* (1880). < 1000 ft. Apr.-Oct. Native of Europe.

Lythrum californicum Torrey & A. Gray. California loostrife. Uncommon, moist places: Kenwood, *Baker* (1932 NCC). < 7300 ft. Apr.-Oct.

★ **Lythrum hyssopifolium** Linnaeus. Loosestrife. Common, wet and sometimes dry places: The Sea Ranch; Horseshoe Cove, *Knight* (1963); Goat Rock State Beach, *Wells* (1981); n. of Bodega Dunes Co. Campground, *Stocking* (1963 NCC), *Connors* (1991 BML); Pitkin Marsh, *Rubtzoff* (1951); Perry Marsh, *Howell* (1945); 2 mi. n. of Sebastopol, *Baker* (1936 NCC); 4 mi. e. of Valley Ford, *Howell* (1930); 8 mi. w. of Petaluma on Bodega Hwy., *Howell* (1933); Copeland Cr. ¼ mi. e. of US 101, Rohnert Park, *Self* (1979 NCC); Railroad Ave. near Cotati, *Knight* (1963); slough draining Kenwood Marsh, *Howell* (1945); Macartney's Woods near Kenwood, *Baker* (1927); Bennett Valley, *Heller* (1902); Spring Lake; near Santa Rosa, *Eastwood* (1921); Green Hill Drive, Santa Rosa, *Stocking* (1963 NCC); Ducker Cr., Rincon Valley,

Wells (1985); Pepperwood; 2 mi. n. of Windsor, *Eastwood & Howell* (1936); Lake Sonoma area. < 5300 ft. Apr.-Oct. Native of Europe.
★ **Lythrum salicaria** Linnaeus. PURPLE LOOSESTRIFE. Noxious weed: edge of hop field w. of Santa Rosa on W. College Ave. ca 1 mi. w. of College Station, *Baker* (1948), acc. *Barbe* (1991). < 3300 ft. Native of Europe.
★ **Lythrum tribracteatum** C. P. Sprengel. Occasional, moist places: 5 mi. n. of Sears Point Road, *Rubtzoff* (1970). < 5000 ft. May-June. Native of Europe.

MALVACEAE. MALLOW FAMILY

1. Style-branches terminating in capitate or truncate stigmas
   2. Involucel of bractlets below calyx wanting . . . . . *Abutilon*
   2: Involucel of 1-several bractlets below calyx present
      3. Ovules 2 or more per carpel; carpels hirsute . . . . . *Modiola*
      3: Ovules solitary; carpels not hirsute
         4. Plants not beset with small scurfy scales; flowers purple . . . . . *Malvastrum*
         4: Plants beset with small scurfy scales; flowers yellowish . . . . . *Malvella*
1: Style-branches filiform, longitudinally stigmatic on the inner side
   5. Involucel of 3-9 connate bractlets
      6. Axis of fruit surpassing the carpels, forming a cone or projection in the center . . . . . *Lavatera*
      6: Axis of fruit not extending above the ring of carpels . . . . . *Alcea*
   5: Involucel of 1-3 distinct bractlets, or lacking
      7. Bractlets of involucel 3; weedy introduced annuals . . . . . *Malva*
      7: Bractlets 0 or 1; native annuals or perennials . . . . . *Sidalcea*

★ **Abutilon theophrasti** Medikus. VELVET-LEAF. Uncommon, escape: Harrison Grade Road, *Allen* (1987 SRJC); laguna in Chileno Valley, *Rubtzoff* (1971); Basch Ranch on Bodega Avenue west of Petaluma, *Knight* (1963); flower bed in Royal Oak Mobile Home Park in Petaluma, *Knight* (1985); Pepperwood. < 350 ft. July-Aug. Native of s. Asia.

★ **Alcea rosea** Linnaeus. HOLLYHOCK. Uncommon, garden escape: Fort Ross-Russian River-Bodega areas, *Voznesenskii* (1840-1841 LE); flower bed in Royal Oak Mobile Home Park in Petaluma, *Knight* (1985). < 300 ft. July-Aug. Native of China. —*Althaea r.* (Linnaeus) Cavanilles.

Lavatera

1. Involucel surpassing calyx; leaves densely soft-downy . . . . . *L. arborea*
1: Involucel shorter than calyx; leaves sparsely pubescent . . . . . *L. cretica*

★ **Lavatera arborea** Linnaeus. TREE MALLOW. Uncommon, naturalized, mostly coastal: Salmon Cr. w. of Hwy. 1, *Best* (1986); n. shore of Bodega Bay, *Best* (1981); Lucas Wharf at Bodega Harbor, *Wells* (1982). < 350 ft. June-July. Native of Europe.
★ **Lavatera cretica** Linnaeus. Uncommon, naturalized, probably not usually collected: waste ground along Copeland Cr. at Sonoma State Univ. in Rohnert Park, *Howell* (1987). < 400 ft. May-July. Native of Mediterranean region.

Malva

1. Bractlets of involucel ovate to oblong
   2. Petals 2-4 × length of calyx; corolla 2-3 cm wide . . . . . *M. sylvestris*
   2: Petals 1-2 × length of calyx; corolla 1-1.5 cm wide . . . . . *M. nicaeensis*
1: Bractlets linear to nearly so
   3. Claws of petals glabrous; calyx enlarged in fruit and widely spreading, veiny-reticulate . . . *M. parviflora*
   3: Claws of petals bearded; calyx barely enlarged and mostly closed over fruit, scarcely reticulate *M. neglecta*

★ **Malva nicaeensis** Allioni. BULL MALLOW. Common, weed, waste places: s. end of The Sea Ranch, *J. H. Thomas* (1976); Fife Cr. Marsh, *Rubtzoff* (1960); Guerneville, *Davy* (1895 UC); Pitkin Marsh, *Rubtzoff* (1951); Laguna de Santa Rosa, *Rubtzoff* (1958); Felta Cr. area sw. of Healdsburg, *Bice* (1895 NCC). < 1300 ft. Mar.-Sept. Native of Eurasia.

★ **Malva parviflora** Linnaeus. CHEESEWEED, LITTLE MALLOW. Common, poorly collected, waste places: Tresch Ranch, Walker Road near Two Rock, acc. J. & B. Guggolz (1992); corner Mission Drive and Middlefield Road in Petaluma, *Knight* (1963); jct. Magnolia and Marshall avenues west of Petaluma, acc. Knight (1964); bed of Santa Rosa Creek, *Baker* (1930 NCC); Burbank Ave. near Hearn Ave. in Santa Rosa, *Alimonti* (1988 NCC); Pepperwood. < 5000 ft. Nearly all year. Native of Eurasia.
★ **Malva neglecta** Wallroth. COMMON MALLOW, CHEESES. Rare, disturbed ground: near W. College Ave. in Laguna de Santa Rosa drainage, *Harrison* (1978 NCC). < 10000 ft. May-Oct. Native of Eurasia.
★ **Malva sylvestris** Linnaeus. HIGH MALLOW. Uncommon, garden escape: vacant lot in Healdsburg, *Baker* (1943 NCC), *Howell* (1943 NCC). < 1000 ft. Native of Eurasia.

★ **Malvastrum capense** Gray & Harvey. Uncommon, garden escape: corner McConnell and Mendocino aves., Santa Rosa, *Hastings* (1939 NCC). Native of S. Africa.

Malvella leprosa (Ortega) Kaprovickas. ALKALI-MALLOW, WHITE WEED. Uncommon, vernal pools, waste places: 200 ft. elev. Willow Cr. Road, *Knight* (1986 NCC); Stony Point Road 0.3 mi. n. of its crossing of the Laguna de Santa Rosa between Cotati and Santa Rosa, *Harrison* (1978 NCC); large population on both sides of Corona Road 200 yds. w. of jct. with N. McDowell Blvd., *Knight* (1984); w. of Sears Point Raceway in Tolay Cr. drainage, acc. Patterson (1990). < 3300 ft. May-Oct. —*Sida l.* (Ortega) K. Schumann var. *hederacea* (Hooker) K. Schumann.

★ **Modiola caroliniana** (Linnaeus) G. Don. Occasional, naturalized: Fort Ross, *Baker* (1932 NCC); Wright's Beach, *Ripley & Barneby* (1945), *Frankel* (1964); Salmon Cr., *Bacigalupi* (1928); near Kenwood, *Baker* (1932 NCC). Usually < 1350 ft. Apr.-Sept. Native of tropical America

Sidalcea: (Key ± acc. S. R. Hill, *Hickman*, [1993]).

1. Plants annual (except *S. calycosa* ssp. *rhizomata* sometimes a weak perennial from rhizomes); filament series unequal; usual flowers in spring
  2. Upper stipules and bracts compounded into linear segments; bracts ± = calyx; leaf-lobes not linear . . . . *S. diploscypha*
  2: Upper stipules usually simple to 2-lobed; bracts usually < calyx; leaf-lobes usually linear
    3. Bracts 2-5 mm. not silky-hairy, not hiding calyx; petals 12-20 mm . . . . . . *S. calycosa* ssp. *calycosa*
    3: Bracts 8-12 mm, silky-hairy, obscuring calyx; petals 20-25 mm . . . . . . *S. c.* ssp. *rhizomata*

1: Plants perennial; filament series ± =; usually flowers in late spring or summer
  4. Leaves usually evenly arrayed on stem, upper and lower usually similar and lobed < halfway to base; fruit segment beakless . . . . . . *S. hickmanii* ssp. *viridis*
  4: Leaves mostly from near stem-base, lower usually ± crenate, upper usually very deeply lobed (nearly compound); fruit segment usually with a short beak
    5. Inflorescence densely head-or spikelike, not 1-sided; calyces in flower usually overlapping; petals < 15 mm . . . . . . *S. oregana* ssp. *valida*
    5: Inflorescence usually not densely head- or spikelike, occasionally 1-sided; calyces in flower usually not overlapping; if inflorescence ± congested, petals usually > 15 mm
      6. Leaves (except lowest) nearly compound, lobes ternately divided into linear segments; stem-base decumbent, rooting . . . . . . *S. malvaeflora* ssp. *laciniata*
      6: Leaves (especially mid-stem leaves) unlobed or at least lobes not ternately divided into linear segments; stem base rooting or not
        7. Basal leaf-blade usually < 2 cm; calyx and stipules usually purple; pedicel hairlike . . . . . . *S. m.* ssp. *purpurea*
        7: Basal leaf-blade often > 2 cm; calyx and stipules not purple; pedicel stouter . . . . . . *S. m.* ssp. *malvaeflora*

Sidalcea calycosa M. E. Jones ssp. **calycosa**. SWAMP CHECKER. Occasional, wet places: Duncans Mills, *M. Jones* (type 1882 UC); Pitkin Marsh, *Rubtzoff* (1951); Perry Marsh, *Howell* (1945); Valley Ford, *K. Brandegee* (1886 UC); Pyle-Madera vernal pools off Wright Road w. of Santa Rosa, before they were "trashed;" *Burt* (1980 SRJC); Petaluma, *Congdon* (1880); Santa Rosa, *K. Brandegee* (1886 UC). < 4000 ft. Apr.-July.

Sidalcea calycosa ssp. **rhizomata** (Jepson) Munz. Uncommon, among tussocks of rush and sedge: The Sea Ranch, *G. E. Snyder* (1991); Stewarts Point, *Abrams* (1920). < 100 ft. May-July. —*S. r.* Jepson.

Sidalcea diploscypha (Torrey & A. Gray) Bentham. Occasional, open hillsides, plains: Cazadero quad.; Hessel Road near Gravenstein Hwy., acc. J. & B. Guggolz (1991); jct. of Occidental Road and

Facendini Lane near Occidental, *Ferris & Lorraine* (1961); Todd Road Preserve; Petaluma, *Congdon* (1880); abundant in foothills e. of Santa Rosa, *Harrison* (1978 NCC); Spring Lake; w. of Calistoga, *Krauss* (1894); Healdsburg, *Cantelow* (1934); Felta Cr. area sw. of Healdsburg, *Bice* (1895 NCC); Lake Sonoma area; property bordered by Chianti and Dutcher Cr. roads w. of US 101 and w. of Geyserville, acc. J. & B. Guggolz (1991); between Jimtown & Mercuryville on Geysers Road, *Utech & Knight* (1986 CM); White Rock Ranch near Geysers Road, *J. & B. Guggolz* (1992 NCC). < 2700 ft. Apr.-June.

Sidalcea hickmanii E. Greene ssp. **viridis** C. L. Hitchcock. MARIN CHECKERBLOOM. Infrequent, coastal brush: Arched Rock, acc. Smith & Pavlik *C.N.P.S. Inv.* (1994); 1 mi. n. of Salmon Cr., *Baker* (1938), *R. Evans* (1986). < 1350 ft. June.

Sidalcea malvaeflora (de Candolle) Bentham ssp. **laciniata** C. L. Hitchcock. Uncommon, open grassy slopes: Doran Beach near Bodega Harbour development, *Huntington* (1989 NCC); Todd Road Preserve; Crane Can., *Juul* (1973 NCC); Grange Road 0.4 mi. n. of Cuenza Road, *Lehner* (1975 NCC); Hidden Acres Road near jct. with Hidden Oaks Road e. of Grange Road, acc. Quibell (1991); Petaluma, *Eastwood* (1952); Sugarloaf Ridge; Pepperwood; 397 ft. elev. Calistoga Road, *Fallon* (1970 PUA); property bordered by Chianti and Dutcher Cr. roads w. of US 101 and w. of Geyserville, acc. J. & B. Guggolz (1991). Usually < 1700 ft. Spring-summer.

Sidalcea malvaeflora ssp. **malvaeflora**. Common, open grassy slopes: The Sea Ranch; Sebastopol, *Bately* (1932 NCC); Hwy. 12 to Sebastopol near Wright's School, *Baker* (1929); Button Ranch on Walker Road (U. C. property) between Petaluma-Valley Ford Road and Pepper Road, acc. Connors (1992). Usually < 1700 ft. Spring-summer.

Sidalcea malvaeflora ssp. **purpurea** C. L. Hitchcock. Uncommon, coastal: Gerstle Cove at Salt Point State Park, *Gazinsky* (1987 NCC); 80 ft. elev. ocean side of Hwy. 1, ½ mi. n. of Moon Rocks at Salt Point State Park, *Huntington* (1989 NCC); 197 ft. elev. The Butcherknife in Cazadero quad., *Rue* (1973 PUA). Usually < 200 ft. Spring-summer.

Sidalcea oregana ssp. **valida** (E. Greene) C. L. Hitchcock. KENWOOD MARSH CHECKERBLOOM. Occasional, wet places: w. of Kenwood, *Baker* (1932 NCC), acc. J. & B. Guggolz (1987), acc. E. Parsons (1991); Mark West Springs, acc. Skinner & Pavlik, C.N.P.S. Inventory (1988); Knights Valley, *E. Greene* (1894 ND); Knights Valley marshes ca 1½ mi. w. of Franz Valley Road and 1 mi. sw. of Hwy. 128, *Harrison* (1979 NCC); Bavarian Lion Ranch, Knights Valley s. of Redwood Cr., acc. Rubtzoff (1991). —*S. v.* E. Greene, type locality: Knights Valley, *E. Greene* (1894 U). Only 2 extant locations remain of < 250 plants. ± 500 ft. June-July.

MARTYNIACEAE. MARTYNIA FAMILY

Probobiscidea

1. Calyx composed of 5 free sepals; body of fruit prickly . . . . *P. lutea*
1: Calyx ± spathaceous, dentate or lobed above, cleft to base below; body of fruit roughly sculptured . . . . *P. louisianica*

★ **Proboscidea louisianica** (Miller) Thellung. COMMON UNICORN PLANT. Uncommon, weed, low moist ground: orchards along US 101 2.1 mi. n. of jct. to Skaggs Springs (probably Dry Cr. exit), *Raven* (1956). < 1700 ft. Summer. Native of s. United States.

★ **Probiscidea lutea** (Lindley) Stapf. YELLOW UNICORN PLANT. Uncommon, weed, disturbed areas; vicinity of Santa Rosa, collector unknown (1975 NCC). < 1700 ft. Late summer. Native of Texas, Mexico and S. America. —*Ibicella* (Lindley) Van Eseltine.

MYOPORACEAE. MYOPORUM FAMILY

★ **Myoporum laetum** Forster f. Uncommon, seedlings: Salmon Cr., acc. Connors (1991); Bodega Head, acc. Connors (1991). < 700 ft. Native of New Zealand.

MYRICACEAE. WAX MYRTLE FAMILY

Myrica californica Chamisso & Schlechtendal. WAX MYRTLE. Locally common, canyons, moist slopes, coastal shelves: The Sea Ranch; Stewarts Point, *Baker* (1898 NCC); e. of Stewarts Point near Stewarts Point Skaggs Springs Hwy., *Baker* (1941 NCC); Salt Point; Old Plantation Road near Salt Point area, *Ott* (1933 NCC); Seaview (inland from Fort Ross), *Baker* (1898 NCC); Bodega Head; Duncans Mills Marsh, *Rubtzoff* (1957); Camp Meeker, *Howell* (1924); Pitkin Marsh, *Howell & Stacey* (1936), *Rubtzoff* (1951); Laguna de Santa Rosa e. of Cunningham near Llano Road, *Rubtzoff* (1962); Button Ranch on Walker Road (U. C. property) between Petaluma-Valley Ford Road and Pepper Road, acc. Connors (1992). < 500 ft. Mar.-Apr.

MYRTACEAE. MYRTLE FAMILY

Eucalyptus

1. Tree ultimately 35-38 m tall; trunk typically curved; leaves lance-shaped, medium green; bark sometimes persistent towards base; capsules in long clusters; flower bud cap relatively smooth, often beaked . . . *E. camaldulensis*

1: Tree ultimately 45-60 m tall; trunks straight; young leaves opposite, oval, silvery; mature leaves, heavy masses,sickle-shaped, dark green; bark exfoliates in strips; capsules mostly solitary in axils; flower bud cap warty, ribbed, blue-gray . . . . *E. globulus*

★ **Eucalyptus camaldulensis** Dehnhardt. RED GUM. Commonly planted, seedlings: Napa St., Sonoma, *Best* (1984); corner Petaluma Ave. and Arnold Drive near El Verano, *Knight & Howell* (1984). < 1000 ft. Native of Australia.

★ **Eucalyptus globulus** Labillardière. BLUE GUM. Commonly planted, seedlings: The Sea Ranch; Willow Creek; Bodega Head; US 101 just n. of San Antonio Cr., *Howell* (1981); Spring Lake. < 1000 ft. Naturalized from Tasmania and Victoria, Australia.

NYCTAGINACEAE. FOUR O'CLOCK FAMILY

Abronia

1. Flowers yellow; leaf-blades round to oval 1-4 (-6) cm long . . . . *A. latifolia*

1: Flowers rose; leaf-blades oval to oval-rhombic to lance-oblong . . . . *A. umbellata*

Abronia latifolia Eschscholtz. SAND VERBENA. Locally common, coastal strand: The Sea Ranch; Goat Rock beach, acc. Knight (1987); Bodega Head. < 400 ft. May-Oct.

Abronia umbellata Lamarck ssp. **umbellata.** Rare, coastal strand: mouth of Salmon Cr., *Bacigalupi* (1928); s. spit of Bodega Bay at Doran Beach, acc. Duebendorfer (1987); Bodega Dunes, *Connors* (1992 BML); mouth of Estero Americano on Marin Co. side, acc. Connors (1990). < 500 ft. Most of year. Ssp. **brevifolia** (Standley) Munz is listed in *Skinner & Pavlik* (1994) for Sonoma County, but positive information is lacking.

NYMPHAEACEAE. WATER-LILY FAMILY

Nuphar lutea (Linnaeus) Sibthorp & Smith ssp. **polysepala** (Engelmann) E. Beal. COW LILY, YELLOW POND-LILY. Occasional, slow streams, ponds: near airfield at The Sea Ranch, *Wells* (1984); ca 1 mi. se. of Jenner near jct. of hwys. 1 and 116, *Rubtzoff* (1964); w. of Duncans Mills ca 500 ft. sw. of jct. Hwy. 116 and Hwy. 1, *Van Kirk* (1981 NCC); Duncans Mills Marsh; *Rubtzoff* (1954); Pitkin Marsh, *Rubtzoff* (1951), *Quibell* (1981 NCC); Laguna de Santa Rosa e. of Graton, *Rubtzoff* (1960); Sebastopol, *McMurphy* (1908); Santa Rosa, Jepson, *Fl. Calif. 1:512* (1922). < 8300 ft. Apr.-Sept. —*N. p.* Engelmann.

OLEACEAE. Olive family

1. Leaves simple, densely silvery scaly below; fruit a drupe . . . . *Olea*
1: Leaves compound, not silvery scaly; fruit a samara . . . . *Fraxinus*

Fraxinus

1. Plant an arborescent shrub 2-7 m high; corolla present, petals 2, white; style obscurely lobed . . . *F. dipetala*
1: Plant a tree 10-25 m high; corolla absent; style obviously 2-lobed . . . . *F. latifolia*

Fraxinus dipetala Hooker & Arnott. California ash. Rare, dry slopes: 350-400 ft. elev. Old Skaggs Springs Road, *Utech, Sato & Ohara* (1990 CM); Lake Sonoma area. 300-4300 ft. Apr.-May.

Fraxinus latifolia Bentham. Oregon ash. Common, near streams, canyons: Fort Ross area, *Voznesenskii* (1841); Duncans Mills Marsh, *Rubtzoff* (1957); ne. of Camp Meeker, *Stocking* (1966 NCC); Pitkin Marsh, *Rubtzoff* (1957); Laguna e. of Sebastopol, *Howell* (1951); Laguna de Santa Rosa se. of Sebastopol near Cooper Road, *Rubtzoff* (1964); Bouverie; Brush Cr. in Rincon Valley, *Rose* (1968); Spring Lake; Petrified Forest, *McMinn* (1922); Spring Mtn. Road, *Knight* (1963); Porter Cr. sw. of Healdsburg, *Stocking* (1963 NCC); Lake Sonoma area; 3150 ft. elev. at upper Squaw Cr. drainage 1.2 airmiles nw. of Geyser Peak, acc. Stone (1983). < 5700 ft. Mar.-May. —*F. oregona* Nuttall.

★ **Olea europea** Linnaeus. Olive. Occasional seedlings: sw. of Sonoma, *Howell* (1964); Mt. Hood, *Knight* (1967); Rincon Ridge on Montecito Ave. opposite Manzanita Road, *Wells* (1989); Alta Vista Lane on Rincon Ridge, *Knight* (1984); Spring Lake; along Drucker Creek, Rincon Valley, *Wells* (1985). Native of Mediterranean region.

ONAGRACEAE. Evening-primrose family

1. Sepals persistent . . . . *Ludwigia*
1: Sepals deciduous after anthesis
  2. Flowers 2-merous; fruit bristly with hooked hairs . . . . *Circaea*
  2: Flowers normally 4-merous; fruit with no hooked hairs
    3. Seeds with tuft of hairs (coma) at one end . . . . *Epilobium*
    3: Seeds without a coma
      4. Fruit a dehiscent capsule, although occasionally tardily so
        5. Stigma capitate, entire . . . . *Camissonia*
        5: Stigma 4-lobed
          6. Petals 2-lobed, small . . . . *[Boisduvalia] Epilobium*
          6: Petals entire, usually larger
            7. Anthers erect, attached near base . . . . *Clarkia*
            7: Anthers versatile, attached near their middle . . . . *Oenothera*
      4: Fruit indehiscent, nutlike and hard . . . . *Gaura*

Camissonia

1. Plants mostly acaulescent and cespitose; ovary fertile in lower portion only, the upper portion sterile, tubular, elongate, subfiliform and simulating a floral-tube
  2. Fruit winged; annual . . . . *C. graciliflora*
  2: Fruit not winged; perennial . . . . *C. ovata*
1: Plants caulescent; ovary fertile to near summit, not prolonged into a long sterile portion, though sometimes beaked
  3. Capsule linear, cylindrical, often torulose, curved to straight; leaves < 2 mm wide
    4. Plants low, usually < 1.5 dm tall; capsule 2.5-3.5 cm long, ending in a well-defined beak *C. contorta*
    4: Plants 1.5-3 dm tall; capsule shorter, not beaked . . . . *C. strigulosa*
  3: Capsule quadrate, when mature curved or contorted; leaves 5-20 mm wide
    5. Plants perennial, often ± woody near base . . . . *C. cheiranthifolia*
    5: Plants annual
      6. Stems semi-prostrate; cauline leaves lance-oblong, obtuse, sessile but not clasping . *C. micrantha*
      6: Stems erect or ascending; cauline leaves oblong-ovate to broadly ovate, acute, with subcordate clasping base . . . . *C. hirtella*

Camissonia cheiranthifolia (Sprengel) Raimann. BEACH EVENING-PRIMROSE. Occasional, mostly coastal strand: The Sea Ranch; Horseshoe Cove, *Knight* (1963); s. side of mouth of Russian River in dunes, *Best* (1981); Bodega Head; n. end of Bodega Bay, *Howell* (1930); Bodega Bay, *Baker* (1925 NCC); Bodega Bay dunes, acc. Connors (1991). < 350 ft. Apr.-Aug. —*Oenothera c.* Sprengel.

Camissonia contorta (Douglas) Raven. Uncommon, dry loose soil: 3 mi. s. of Sebastopol, *Breedlove* (1962); Kenwood, *Harrison* (1978 SRJC). < 7700 ft. May-June. —*Oenothera c.* Douglas.

Camissonia graciliflora (Hooker & Arnott) Raven. Uncommon, grassy places: Annadel Heights, *Harrison* (1976 SRJC); Pepperwood; Lake Sonoma area. < 2700 ft. Mar.-May. —*Oenothera g.* Hooker & Arnott.

Camissonia hirtella (E. Greene) Raven. Uncommon, chaparral: The Geysers, *Harrison* (1979 SRJC), 3300 ft. elev. Squaw Cr., acc. Clifton (1983). < 7700 ft. Mar.-Aug. —*Oenothera micrantha* Hornemann var. *jonesii* (Leveille) Munz.

Camissonia micrantha (Sprengel) Raven. Occasional, coastal: near mouth of Salmon Cr., *Baker* (1930 NCC); Bodega Point, *Eastwood* (1916); Bodega Marine Reserve, *Best* (1980); Bodega (probably Bodega Bay), *Rose* (1962); dunes between Bodega Bay and Bodega Head, *Raven* (1963). < 1000 ft. Mar.-May. —*Oenothera m.* Sprengel.

Camissonia ovata (Torrey & A. Gray) Raven. SUN CUP. Common, grassy areas: The Sea Ranch; Tin Barn Road near Buddhist Temple, acc. Knight (1991); Bodega Head quad.; Hessel Road near Gravenstein Hwy., acc. J. & B. Guggolz (1991); 2 sites w. of Santa Rosa in Laguna de Santa Rosa drainage, *Harrison* (1978 NCC); Todd Road Preserve; Carinelli property, 2800 Llano Road, acc. Lovell (1985); off Grange Road near Hidden Oaks Road, acc. Quibell (1991); Fairfield Osborn; Button Ranch on Walker Road (U. C. property) between Petaluma-Valley Ford Road and Pepper Road, acc. Connors (1992); Tresch Ranch, Walker Road near Two Rock, acc. J. & B. Guggolz (1992); Chileno Valley, *Knight* (1964); cemetery in Petaluma, *Ott* (1934 NCC); sw. of Sonoma, *Howell* (1949); Bouverie; Sonoma Valley Regional Park, acc. Quibell (1991); Spring Lake; Pepperwood; Felta Cr. area sw. of Healdsburg, *Bice* (1895 NCC); property bordered by Chianti and Dutcher Cr. roads w. of US 101 and w. of Geyserville, acc. J. & B. Guggolz (1991). < 1700 ft. Mar.-June. —*Oenothera o.* Torrey & A. Gray.

Camissonia strigulosa (Fischer & C. Meyer) Raven. Occasional, sandy places, coastal: Bodega Marine Reserve, *Best* (1980); between Bodega Bay and Bodega Head, *Raven* (1963); Bodega Harbor, *Heller & Brown* (1902). < 7000 ft. —*Oenothera contorta* Douglas var. *s.* (Fischer & C. Meyer) Munz.

Circaea alpina Linnaeus ssp. **pacifica** (Ascherson & Magnus) Raven. ENCHANTER'S NIGHTSHADE. Rare, deep woods: Guerneville, *Rattan* (1877). < 9000 ft. June-Aug. —*C. p.* Ascherson & Magnus.

Clarkia: (Key ± acc. H. Lewis, *Hickman* [1993]).

1. Petal-blade lobed . . . . . *C. concinna*
1: Petal-blade not or not obviously lobed
  2. Petal-claw 2-lobed or -toothed, not entire
    3. Petals 6-14 mm; stigma not beyond anthers . . . . . *C. rhomboidea*
    3: Petals > 12 mm; stigma beyond anthers . . . . . *C. virgata*
  2: Petal-claw ± 0 or entire
    4. Inflorescence axis in bud recurved at tip, in flower ± straight; buds pendent
      5. Petals 6-23 mm, spot 0, base darker or not; stigma not beyond anthers . . *C. gracilis* ssp. *gracilis*
      5: Petals 20-40 mm, with a red spot near middle; stigma beyond anthers . . . *C. g.* ssp. *sonomensis*
    4: Inflorescence axis in bud usually ± straight; buds erect or reflexed
      6. Flower buds reflexed . . . . . *C. unguiculata*
      6: Flower buds erect
        7. Ovary, immature fruit 4-grooved
          8. Stem decumbent; inflorescence internode usually < subtending flower . . . . . *C. amoena* ssp. *amoena*
          8: Stem erect; inflorescence internode usually > subtending flower . . . *C. a.* ssp. *huntiana*
        7: Ovary, immature fruit 8-grooved
          9. Stigma not beyond anthers; petals usually 5-15 mm
            10. Stem decumbent to prostrate; leaf oblanceolate, elliptic, or obovate, tip usually obtuse . . . . . *C. davyi*
            10: Stem erect; leaf linear to ovate, tip usually acute . . *C. purpurea* ssp. *quadrivulnera*

9: Stigma usually beyond anthers; petals usually 10-60 mm
  11. Leaves definitely imbricated, lanceolate to elliptic or ovate, length usually < 5 × width; inflorescence dense . . . . *C. imbricata*
  11: Leaves not imbricated, linear to narrow-lanceolate, length > 5 × width; inflorescence open . . . . *C. purpurea* ssp. *viminea*

Clarkia amoena (Lehmann) Nelson & J. F. Macbride ssp. **amoena**. FAREWELL-TO-SPRING, SUMMER'S DARLING. Common, grassy flats and slopes: The Sea Ranch; bluffs 15 miles n. of Fort Ross, *Covel* (1938); 8 mi. s. of Fort Ross, *Howell* (1930); 5 mi. n. of Jenner, *Munz* (1936); Bodega Point, *Eastwood* (1915); Bodega Bay, *Eastwood & Howell* (1938); Bodega Bay on road to Doran Park, *H. & M. Lewis* (1947); Monte Rio, *Hoak* (1912); Button Ranch on Walker Road (U. C. property) between Petaluma-Valley Ford Road and Pepper Road, acc. Connors (1992); 4 mi. w. of Petaluma, *Rubtzoff* (1962); Sugarloaf Ridge; Bennett Valley, *Leithvet* (1893); Skaggs Springs, *Hawver* (1915); Pine Flat Road 10 mi. sw. of geysers Unit 18, *Malloch* (1977 PG&E); 0.9 mi. nw. of Cloverdale to Boonville on Hwy. 128, *C. & P. McMillan* (1951 NCC), *Breedlove* (1962); Alder Glen Springs (probably on Old Redwood Hwy. n. of Cloverdale), *Eastwood & Howell* (1938). < 2000 ft. June-Aug. —*Godetia amoena* G. Don.

Clarkia amoena ssp. **huntiana** (Jepson) Harlan Lewis & M. Lewis. Occasional, openings of woods: ca 1 mi. nw. of Camp Meeker, acc. J. & B. Guggolz (1991); between e. Guernewood and Vacation Beach w. of Guerneville, *Rubtzoff* (1951); Sweetwater Springs Road 2 mi. n. of Guerneville, *Standish* (1961 NCC); corner Roblar and Steinbeck roads w. of Stony Point, *Knight* (1963); Ida Clayton Road on serpentine, *Howe* (1970 NCC). < 2300 ft. June-Aug.

Clarkia concinna (Fischer & C. Meyer) E. Greene. RED RIBBONS. Common, grassy or brushy embankments: Larkspur Rock e. of Bodega Bay, *Hastings* (1940 NCC); ca 1 mi. nw. of Camp Meeker, acc. J. & B. Guggolz (1991); near Russian River Terrace, *Bately* (1932 NCC); Fairfield Osborn (S); Bouverie; Gates Road off Calistoga Road, acc. Wells et al. (1984); Pepperwood; 350-400 ft. elev. Old Skaggs Springs Road, *Utech, Sato & Ohara* (1990 CM); Lake Sonoma area; 2400 ft. elev. w. flank Mt. St. Helena, *Grummer* (1981 PUA). < 5000 ft. May-July. —*Eucharidium concinnum* Fischer & C. Meyer, type locality: Fort Ross.

Clarkia davyi (Jepson) Harlan Lewis & M. Lewis. Uncommon, mostly sandy areas: The Sea Ranch 7 mi. s. of Gualala, *Thomas* (1976); Salt Point. < 350 ft. June-July. —*Godetia quadrivulnera* Spach var. *davyi* Jepson.

Clarkia gracilis (Piper) Nelson & J. F. Macbride ssp. **gracilis**. Uncommon, shady situations: Fairfield Osborn (H); Pepperwood. < 5000 ft. Apr.-July. —*Godetia g.* Piper.

Clarkia gracilis ssp. **sonomensis** (C. L. Hitchcock) Harlan Lewis & M. Lewis. Occasional, shady situations: ca 1 mi. nw. of Camp Meeker, acc. J. & B. Guggolz (1991); ca ½ mi. s. of Eldridge in Sonoma Valley, *Baker* (1937 NCC); ½ mi. e. of Glen Ellen, *C. L. Hitchcock* (1928); Bennett Valley Road, *Van Steenberge* (1932 NCC); Pepperwood; 350-400 ft. elev. Old Skaggs Springs Road, *Utech, Sato & Ohara* (1990 CM); Lake Sonoma area; Alder Springs on grade n. of Cloverdale, *Eastwood & Howell* (1933). < 1700 ft. Apr.-July. —*Godetia amoena* G. Don var. *sonomensis* C. L. Hitchcock, type locality: ½ mi. e. of Glen Ellen, *C. L. Hitchcock* (1928).

Clarkia imbricata Harlan Lewis & M. Lewis. VINEHILL CLARKIA. Almost extinct, growing among low grasses, roadsides, fields, orchards: 5 mi. n. of Sebastopol on road to Forestville, *Howell* (1936); Pitkin Ranch, *Howell* (1945); Vine Hill region *Harlan Lewis & M. Lewis* (type 1957), acc. Knight (1990), acc. Rae (1991). < 350 ft. June-July.

Clarkia purpurea (Curtis) Nelson & J. F. Macbride ssp. **quadrivulnera** (Douglas) Harlan Lewis & M. Lewis. Common, open areas: ca 1 mi. nw. of Camp Meeker, acc. J. & B. Guggolz (1991); near Vine Hill school, *Howell* (1955); Sebastopol, *Dows* (1907); Fairfield Osborn, *Harlan Lewis & M. Lewis* (1981); near cemetery w. of Petaluma, *Knight,* (1963); between Sonoma and Boyes Hot Springs, *Howell* (1929); vicinity of jct. of Sonoma Mtn. Road and Warm Springs Road w. of Glen Ellen, *Baker* (1930 NCC); Pepperwood; 2 mi. e. of Petrified Forest, *Howell* (1937); Knights Valley, *H. E & S. T.* Parks (1930); Lake Sonoma area; Ida Clayton Road quad.; ca 2000 ft. elev. Anderson Trail on Black Mtn. between Geyser Peak and upper end of Pine Flat Road, acc. J. & B. Guggolz (1992). < 2700 ft. Apr.-July. —*Godetia quadrivulnera* Spach. *Oenothera q.* Douglas var. *hirsuta* Kellogg, type locality: Petaluma, Harford.

Clarkia purpurea ssp. **viminea** (Douglas) Harlan Lewis & M. Lewis. Common, open areas: Monte Rio, *Hoak* (1912); near cemetery w. of Petaluma, *Knight* (1963); Bouverie; Cooper's Grove on Sonoma Mtn. Road, *Baker* (1930 NCC); vicinity of jct. of Sonoma Mtn. Road and Warm Springs roads w. of Glen

Ellen, *Baker* (1930 NCC); Sugarloaf Ridge; Santa Rosa, *Heller* (1902); Skaggs Springs site, *Best* (1981). < 3000 ft. May-July. —*Godetia v.* Spach.

Clarkia rhomboidea Douglas. DIAMOND CLARKIA. Uncommon, grassy fields, hillsides: 2900 and 3150 ft. elev. w. flank Mt. St. Helena, *Grummer* (1979 PUA); Cates property on Pine Flat Road, *Harrison* (1979 SRJC). < 8300 ft. May-July. —*C. virgata* E. Greene, type locality: Sonoma County (possibly in Mt. Hood area), *Bioletti* (1893 U). *Godetia latifolia* Nelson & Kennedy.

Clarkia unguiculata Lindley. Occasional, dry shady slopes: Crane Can., *Stocking* (1965 NCC); Sonoma Mtn. Road near Enterprise Road, *Utech & Knight* (1986 CM); Bouverie; Mark West Can., *Baker* (1925); Pepperwood; s. of Hwy. 128 above Redwood Cr. near Knights Valley, *J. & B. Guggolz* (1990); Lake Sonoma area; between Jimtown & Mercuryville on Geysers Road, *Utech & Knight* (1986 CM). < 5000 ft. May-June. —*C. elegans* Douglas.

Clarkia virgata E. Greene. SIERRA CLARKIA. Acc. *Erythea 3:123*, (1895) Bioletti reported the type of this species from Sonoma and Amador cos. No specimen has ever been filed from Sonoma Co. but Bioletti did most of his collecting in the Mt. Hood area during 1893. < 3700 ft. May-July.

Epilobium

1. Leaves alternate; hypanthium 0; petals entire; stamens subequal, in 1 whorl; pollen bluish gray, grains shed singly . . . . . *E. angustifolium* ssp. *circumvagum*

1: Leaves opposite at least at base; hypanthium 0.3-34 mm;; petals notched; stamens in 2 unequal whorls; pollen cream, grains usually shed in 4's
- 2. Plants annual; leaves opposite only near base; lower stem peeling; seed hair-tuft 0
  - 3. Fruit wall tough, ± indehiscent below; seeds in 2 rows per chamber
    - 4. Fruit 8-12 mm, sharply 4-angled . . . . . *E. cleistogamum*
    - 4: Fruit 5-8 mm, cylindric . . . . . *E. pygmaeum*
  - 3: Fruit wall flexible, dehiscent to base; seeds in 1 row per chamber
    - 5. Fruit not beaked, central axis remaining intact . . . . . *E. densiflorum*
    - 5: Fruit beaked, central axis disintegrating . . . . . *E. torreyi*
- 2: Plants annual to subshrub; leaves usually opposite up to inflorescence; lower stem usually not peeling; seeds hair-tufted
  - 6. Hypanthium 17-34 mm; corolla red-orange . . . . . *E. canum*
  - 6: Hypanthium usually 0.4-8 (-16) mm; corolla white to purple
    - 7. Plants annual, stem peeling below; leaves ± narrowly lanceolate, upper generally clustered
      - 8. Plant 2-20 dm, ± glandular; hypanthium 1.5-16 mm; sepals 2-8 mm, petals 2-15 mm; seed 1.4-2.7 mm . . . . . *E. brachycarpum*
      - 8: Plants 0.5-4 dm, nonglandular; hypanthium 0.4-1 mm, sepals 1.5-4 mm, petals 1.8-5 mm; seed 0.6-1.2 mm
        - 9. Leaves usually folded along midrib, upper clustered; inflorescence dense; petals 1.8-2.5; seeds 0.6-0.9 mm, papillate . . . . . *E. foliosum*
        - 9: Leaves flat, not clustered; inflorescence open; petals 2-5 mm; seeds 0.8-1.2 mm, netted . . . . . *E. minutum*
    - 7: Plants perennial, stem not peeling, leaves narrowly lanceolate to ovate, not clustered
      - 10. Leaf-veins ± obscure; stem with fleshy shoot; seed papillate . . . . . *E. halleanum*
      - 10: Leaf-veins obvious; stem with rosette or fleshy shoot; seed ridged
        - 11. Petals 2-6 (-9) mm, white to pink; leaf lanceolate, reduced in open inflorescence; stem with basal rosette . . . . . *E. ciliatum* ssp. *ciliatum*
        - 11: Petals 4-15 mm, pink to rose-purple; leaf lanceolate to ovate, little reduced on crowded inflorescence
          - 12. Stem with fleshy shoot; inflorescence long . . . . . *E. c.* ssp. *glandulosum*
          - 12: Stem with leafy rosette; inflorescence ± flat-topped . . . . . *E. c.* ssp. *watsonii*

Epilobium angustifolium Linnaeus ssp. **circumvagum** Mosquin. FIREWEED. Uncommon, fairly moist places: Salt Point; near Russian River Sta. on N. Pacific Coast Railroad, *Bolton* (1896). < 11400 ft. July-Aug. Also native in Eurasia.

Epilobium brachycarpum C. Presl. Common, open, usually dry, sometimes disturbed areas: near community center in Occidental, *Best & Wells* (1983); Todd Road Preserve; Oakmont between Rincon Valley and Los Guilicos, *Best* (1988); Spring Lake; Pepperwood; 2000 ft. elev. e. of geysers Unit 18, *Malloch* (1977 PG&E). < 11200 ft. June-Sept. —*E. paniculatum* Torrey & A. Gray.

Epilobium canum (E. Greene) Raven. CALIFORNIA FUCHSIA, ZAUSCHNERIA. Common, dry mostly gravelly or stony areas: Fairfield Osborn (S); Sugarloaf Ridge; Pepperwood; Lake Sonoma area; 2400 ft. elev. w. flank Mt. St. Helena, *Grummer* (1981 PUA); Russian River n. of Cloverdale, *Heller* (1902). < 5000 ft. Aug.-Oct. —*Zauschneria californica* C. Presl. *Z. velutina* Eastwood, type locality: Russian River n. of Cloverdale, *Heller* (1902 UC).

Epilobium ciliatum Rafinesque ssp. **ciliatum**. NORTHERN WILLOW HERB. Common, moist places, disturbed areas: Stewarts Point-Skaggs Springs Road 6 mi. e. of Stewarts Point, *Best* (1981); 6 mi. n. of Bodega Bay, *Munz* (1936); Guerneville, *Rattan* (1877); slough draining Kenwood Marsh, *Howell* (1932); Santa Rosa Cr. e. of Santa Rosa, *Heller* (1902); Pepperwood. < 13000 ft. July-Sept. —*E. adenocaulon* Haussknecht. *E. a.* var. *parishii* (Trelease) Munz. *E. a.* var. *holosericeum* (Trelease) Munz. *E. a.* var. *occidentale* Trelease. *E. californicum* Haussknecht, type locality: Fort Ross, *Wrangel*.

Epilobium ciliatum ssp. **glandulosum** (Lehmann) P. Hoch & Raven. Uncommon, wet places: Cunningham Station, *Baker* (1947 NCC); slough draining Kenwood Marsh, *Howell* (1932). < 11500 ft. June-Aug. —*E. brevistylum* Barbey var. *b.*

Epilobium ciliatum ssp. **watsonii** (Barbey) P. Hoch & Raven. WESTERN WILLOW HERB. Occasional, moist places: The Sea Ranch; 3.7 mi. s. of Jenner, *Clausen* (1935); Roblar Road near county dump (near Stony Point), *Knight* (1963); e. of Bennett Valley, *Baker* (1931 NCC). < 1250 ft. May-Aug. —*E. watsonii* Barbey, type locality: near Fort Ross. *E. franciscanum* Barbey. *E. w.* Barbey var. *f.* (Barbey) Jepson.

Epilobium cleistogamum (Curran) P. Hoch & Raven. Rare, dried vernal beds: Spring Lake. < 1000 ft. Apr.-June. —*Boisduvalia cleistogama* Curran.

Epilobium densiflorum (Lindley) P. Hoch & Raven. Common, moist areas: Stoetz Lane ne. of Occidental, *Best* (1984); Todd Road Preserve, *Howell* (1987); Spring Lake; Pepperwood; 2000 ft. elev. e. of geysers Unit 18, *Malloch* (1977 PG&E). < 8800 ft. May-Aug. —*Boisduvalia densiflora* (Lindley) S. Watson.

Epilobium foliosum (Torrey & A. Gray) Suksdorf. Occasional, dry disturbed open places: Camp Meeker, *Sutliffe* (1927); 1024 ft. elev. Coleman Valley Road, *Clifton* (1984 PUA); Cavedale Road, *Knight, Buford, Wood & Bartholomew* (1985); Pepperwood, *Breedlove* (1983); Geysers Road between Jimtown and Mercuryville, *Knight & Edwards* (1986). < 8000 ft. Apr.-Aug. —*E. minutum* Lindley var. *biolettii* E. Greene. *E. m.* Lehmann var. *f.* Torrey & A. Gray.

Epilobium halleanum Haussknecht. Streambanks, moist meadows: Cunningham Marsh, *Howell* (1947). < 11500 ft. June-Aug. —*E. brevistylum* var. *ursinum* (Trelease) Jepson.

Epilobium minutum Lehmann. SLENDER ANNUAL FIREWEED. Uncommon, dry disturbed open places, often on serpentine: Bohemian Hwy. 3.6 mi. nw. of Occidental, *Huntington* (1987 NCC); Harrison Grade Road, *Utech & Knight* (1986 CM); Tombs Cr. quad., 3500 ft. elev. on serpentine, 200 yds. n. of Geyser Rock, *Malloch, Burton & Crane* (1978 PG&E). 330-5300 ft. Apr.-Aug.

Epilobium pygmaeum (Spegazzini) P. Hoch & Raven. Infrequent, vernal pools, dry mud flats: 2 sites w. of Santa Rosa in Laguna de Santa Rosa drainage, *Harrison* (1978 NCC); laguna in Chileno Valley, *Rubtzoff* (1959); Spring Lake; Pepperwood. < 8300 ft. May-Aug. —*Boisduvalia glabella* (Nuttall) Walpers, including var. *campestris* (Jepson) Jepson.

Epilobium torreyi (S. Watson) P. Hoch & Raven. Uncommon, moist areas: 4 sites w. and sw. of Santa Rosa in Laguna de Santa Rosa, *Harrison* (1978 NCC); Fairfield Osborn (H); Jack London State Park near Glen Ellen, *Wagnon* (1989 NCC); Pepperwood; Healdsburg, *Rattan* (1877). < 8600 ft. May-July. —*Gayophytum strictum* A. Gray, type locality: Cloverdale, *Bolander* (1867 UC). *Boisduvalia stricta* (A. Gray) E. Greene.

★ **Gaura drummondii** (Spach) Torrey & A. Gray. Rare, waste places; noxious weed: county line e. of Schellville, acc. *Barbe* (1991). Native of Texas to central Mexico. < 1000 ft. May-Sept. —*G. odorata* Lagasca, misapplied.

Ludwigia

1. Stamens in 2 series, mostly 8 or 10 in number; petals 10-20 mm long
   2. Flowering stems usually floating or creeping; leaves oblong, 1-10 cm long; bracteoles at base of ovary deltoid; capsule mostly 2-3 mm thick . . . . . *L. peploides*
   2: Flowering stems usually erect; leaves lanceolate, mostly 5-10 cm long; bracteoles lanceolate; capsule 3-4 mm thick . . . . . *L. hexapetala*

1: Stamens in 1 series, 4-5 in number; petals none or small and quickly shed . . . . . *L. palustris*

★ **Ludwigia hexapetala** (Hooker & Arnott) Zardini, Gu & Raven. WATER PRIMROSE. Locally common, sluggish water, stagnant pools: both sides of the summer crossing at confluence of Hulbert Cr. and Russian River 2 mi. downstream from Guerneville, *Knight* (1990); Pepperwood; just s. of the summer crossing below Asti, *Guggolz* (1990). < 1000 ft. June-Aug. Native of S. America and se. United States. —*L. uruguayensis* Hara. *Jussiaea u.* (Cambessèdes) Hara.

Ludwigia palustris (Linnaeus) Elliott. Occasional, wet places: Pitkin Marsh, *Howell* (1945), *Baker* (1945 NCC), *Rubtzoff* (1951 NCC); Perry Marsh, *Howell* (1945); Laguna de Santa Rosa e. of Graton, *Rubtzoff* (1966); lake on Cannon's place (probably near present Fairfield Osborn), *Baugh* (1967 NCC); Spring Lake; Mark West Cr. near Mark West Resort, acc. de Nevers (1992); along Franz Cr. at Franz Valley Road, acc. de Nevers (1992); Yellowjacket Ranch in Knights Valley, acc. de Nevers (1992). < 3300 ft. June-Sept. —*L. p.* var. *pacifica* Fernald & Griscom.

Ludwigia peploides (Kunth) Raven. Occasional, pools, sluggish streams: Pitkin Marsh, *Rubtzoff* (1951 NCC, 1952 NCC); Todd Road Preserve; Copeland Cr., Rohnert Park, *Self* (1979 NCC); McDermott Lake on Lichau Road (probably now it is Fairfield Osborn), *Stocking* (1966 NCC); Lake Ilsanjo at Annadel, *Stocking* (1969 NCC); Spring Lake; Newanga Road in Santa Rosa, *Knight* (1965). < 3000 ft. May-Oct. —*Jussiaea repens* Linnaeus var. *p.* (Humboldt, Bonpland & Kunth) Grisebach.

Oenothera elata Kunth ssp. **hookeri** (Torrey & A. Gray) W. Dietrich & W. L. Wagner. EVENING-PRIMROSE. Uncommon, damp or seepage areas: The Sea Ranch; Hwy. 1 n. of Salmon Cr.; acc. Connors (1991); Pocket Canyon Road w. of Forestville near quarry, acc. Knight (1990); Graton Road between Dyer Road and Atascadero Cr., acc. W. I. & T. Knight (1992). < 700 ft. June-Sept. —*O. hookeri* Torrey & A. Gray.

OROBANCHACEAE. BROOM-RAPE FAMILY

1. Capsule 2-valved with 2 placentae on each valve; filaments not with hairy tuft at base . . . . . *Orobanche*

1: Capsule 4-valved with 1 placenta on each valve; filaments with tuft of hairs at base . . . . . *Boschniakia*

Boschniakia strobilacea A. Gray. CALIFORNIA GROUND-CONE. Occasional: Salt Point pygmy forest under *Arctostaphylos columbiana*, acc. Kjeldsen (1991); w. side Mt. St. Helena, *de Nevers* (1986); Mill Cr. w. of Healdsburg, *Mullin* (1933 NCC); Fitch Mtn., *Cannon* (1991 NCC). < 10000 ft. May-July.

Orobanche

1. Flowers on elongate scapelike pedicels, the pedicels without bractlets
   2. Pedicels 1-3 × as long as stem; calyx-lobes longer than tube . . . . . *O. uniflora*
   2: Pedicels many, not exceeding length of stem; calyx-lobes not longer than tube . . . . . *O. fasciculata*

1: Flowers on shorter pedicels, these subtended by bractlets
   3. Stems with a thickened tuberlike base and branched above . . . . . *O. bulbosa*
   3: Stems not having a thickened tuberlike base
      4. Flowers, especially the lower, on pedicels up to 2 or 3 cm long; inflorescence often corymbose
         5. Lower lip of corolla spreading 7-15 mm long; anthers woolly
            6. Plants 10-35 cm tall, stem below the inflorescence generally > 6 cm; inflorescence elongate 5-20 cm; dry places . . . . . *O. californica* ssp. *jepsonii*
            6: Plants 4-10 cm tall, stem below the inflorescence < 5 cm; inflorescence ± flat or convex-topped cluster; moist or vernally wet places . . . . . *O. c.* ssp. *grayana*
         5: Lower lip of corolla erect 4-6 (-8) mm long; anthers subglabrous . . . . . *O. c.* ssp. *californica*
      4: Flowers sessile or nearly so; inflorescence spicate . . . . . *O. valida* ssp. *howellii*

Orobanche bulbosa G. Beck. Occasional, on *Adenostoma* and associates: Duncan's Landing between Goat Rock and Salmon Cr., *de Nevers* (1981); s. slope of Sugarloaf Peak e. of Kenwood, *Baker* (1947 NCC); Jensen Ranch in Dry Cr. area w. of Healdsburg, *Baker* (1936 NCC); 2800 ft. elev. w. flank Mt. St. Helena, *Grummer* (1980 PUA); 2980 ft. elev. on Rattlesnake Fire Road ese. of Mercuryville in geothermal area, *Raiche* (1991 JEPS); 3200 ft. elev. Socrates Mine, *Clifton* (1984 PUA). < 5700 ft. Mostly Apr.-July.

Orobanche californica Chamisso & Schlechtendal ssp. **californica** Occasional, on members of Asteraceae: Plantation, *Zanelli* (1925); Bodega Marine Reserve, *Connors* (1991 BML); with *Grindelia* in a dry bed at a rill n. of Duvoul Cr. n. of Camp Meeker on serpentine, *Edwards* (1985); Sebastopol, *Sturgeon* (1928); Russian River near Cloverdale, *Lili* (1933 JEPS). < 500 ft. May-Aug. —*O. grayana* G. Beck var. *violacea* (Eastwood) Munz, misapplied. *O. g.* var. *nelsonii* Munz, misapplied.

Orobanche californica ssp. **grayana** (G. Beck) Heckard. Occasional: near the laguna in Chileno Valley, *Rubtzoff* (1964); Pepperwood; w. side of Russian River on Healdsburg-Guerneville Road, *Baker* (1937 NCC). < 6000 ft. June-Sept.

Orobanche californica ssp. **jepsonii** (Munz) Heckard. Uncommon: Pepperwood. < 7300 ft. June-Oct.

Orobanche fasciculata Nuttall. CLUSTERED BROOM-RAPE. Occasional: ca 1 mi. nw. of Camp Meeker, acc. J. & B. Guggolz (1991); Mark West Road, *Baker* (1937 NCC); Lake Sonoma area; Geyser Peak, *Knight & Edwards* (1986); Socrates Mine Road 200 yds. downhill from crest of hills at Socrates Mine, *Malloch* (1977 PG&E); 2400 ft. elev. at Little Geysers below cypresses, *Malloch* (1977 PG&E). < 11000 ft. Apr.-July. —*O. f.* var. *franciscana* Achey.

Orobanche uniflora Linnaeus. NAKED BROOM-RAPE. Occasional, rocky hillsides: Westminster Woods at Camp Meeker on *Sedum* on rocky outcrops, acc. Harrison (1992); Armstrong Grove on Saxifragaceae, acc. Harrison (1992); Petaluma, *Congdon* (1877); Annadel, acc. Harrison (1992); Pepperwood, *Almeda* (1982); Sausal Cr.in Pine Flat region, *Applegate* (1934); 4000 ft. elev. w. flank Mt. St. Helena, *Grummer* (1980 PUA). < 10500 ft. May-Aug. —*O. u.* var. *minuta* (Suksdorf) Achey. *O. u.* var. *sedi* (Suksdorf) Achey. *O. u.* var. *purpurea* (Heller) Achey.

Orobanche valida Jepson ssp. **howellii** Heckard & L. T. Collins. HOWELL'S BROOM-RAPE. Occasional: head of Jacket Cr. on Mt. St. Helena, *Hemphill* (1963). 600-5700 ft. June-July.

OXALIDACEAE. WOOD SORREL FAMILY

Oxalis

1. Petals yellow
    2. Plants acaulescent, with underground bulblets; petals ca 2 cm long . . . . *O. pes-caprae*
    2: Plants caulescent, without bulblets, petals 8-16 mm long
        3. Plants annual; leaf-bearing stems 3-7 cm high . . . . *O. laxa*
        3: Plants perennial; leaf-bearing stems 10-30 cm long
            4. Stems creeping from a slender taproot and rooting at nodes . . . . *O. corniculata*
            4: Stems erect or decumbent, not rooting at nodes
                5. Petals 12-16 mm long; seeds 2-2.4 mm long; leaflets 13-26 mm broad . . . . . . *O. suksdorfii*
                5: Petals 8-10 mm long; seeds 1.2-1.6 mm long; leaflets 7-16 mm broad . . . . *O. albicans*

1: Petals pinkish, purple or violet; annual or perennial
    6. Perennial; acaulescent; petals pinkish, often veined purple; flowers solitary; capsule round-ovoid, subglabrous, 7-8 mm long . . . . *O. oregana*
    6: Annual; erect; petals purple or violet; inflorescence in cyme, umbelliform with 4-10 flowers; capsule lanceolate,glabrous, 7-12 mm long; with bulbs . . . . *O. rubra*

Oxalis albicans Humboldt, Bonpland & Kunth ssp. **pilosa** (Nuttall) Eiten. Common, brushy hillsides, open hills: Stewarts Point, *Abrams* (1920); Salt Point; bluff above beach at mouth of Salmon Cr., *Ferris* (1937); coast n. of Bodega Bay, *Eastwood & Howell* (1939); Bodega Head, *Connors* (1992 BML); Larkspur Rock e. of Bodega Bay, *Eastwood & Howell* (1939); border of Cunningham Marsh, *Howell* (1947); Farrell Hill w. of Sebastopol, *Baker* (1936 NCC); 3 mi. s. of Healdsburg, *Heller & Brown* (1902). < 700 ft. Most of year. —*O. p.* Nuttall.

★ **Oxalis corniculata** Linnaeus. Common, garden and lawn weed: The Sea Ranch ca 7 mi. s. of Gualala, *J. H. Thomas* (1976); Fort Ross-Russian River-Bodega areas, *Voznesenskii* (1840-1841 LE); Bodega Head;

1 mi. n. of Sebastopol, *Tryner* (1938 NCC); Vallejo's home, Sonoma, *Eastwood* (1933); Bouverie; Spring Lake Village; Pepperwood; N. McDowell Blvd., Royal Oak Mobile Home Park, Petaluma, acc. Knight (1990); Southwest Blvd., Rohnert Park, *Caruso* (1977 NCC); corner Ridgway and Ripley, Santa Rosa, *Baker* (1938 NCC); 5½ mi. from Hwy. 12 (probably US 101) on Dry Cr. Road, *Ruckert* (1960). Usually < 8300 ft. ± all year. Native of Europe.

★ **Oxalis laxa** Hooker & Arnott. Rare, disturbed places: Pepperwood. < 3300 ft. Apr.-June. Native of Chile.

Oxalis oregana Nuttall. REDWOOD SORREL. Locally common, shady woods: The Sea Ranch, *Wells* (1981); Salt Point; 8 mi. s. of Fort Ross, *Howell* (1934); across Russian River from Duncans Mills, *Meagher* (1947 NCC); Duncans Mills, *Rattan* (1880); Duncans Mills Marsh, *Rubtzoff* (1957); Camp Meeker, *Gurevitch* (1968 NCC); Neely Hill area s. of Guerneville, *Rubtzoff* (1953); Guerneville, *Congdon* (1880), *Fallon* (1970 PUA); 2 mi. w. of Forestville, *McNally* (1970 NCC); among redwood trees w. of Sebastopol, *Heller & Brown* (1902); Cooper's Grove on Sonoma Mtn. Road, acc. Quibell (1991). < 3300 ft. Apr.-Sept.

★ **Oxalis pes-caprae** Linnaeus. BERMUDA BUTTERCUP. Abundant, fields, orchards, waste places: ½ mi. n. of Stewarts Point on Hwy. 1, *D. M. & S. H. Porter* (1967); 49 ft. elev. Duncan's Landing s. of Wright's Beach, *Brantley* (1970 PUA); 20 ft. elev. Salmon Cr., *Brantley* (1970 PUA); Bodega Head; Bouverie; Pepperwood. < 1700 ft. Nov.-Mar. Native of S. Africa.

★ **Oxalis rubra** Saint-Hilaire. Occasional, lawns, waste places: Salmon Cr. w. of Hwy. 1, *Best* (1985); ca 1 mi. w. of Sebastopol on Bodega Hwy., *Best* (1985); Railroad Ave. near Cotati, *Juul* (1973 NCC). Native of S. America.

Oxalis suksdorfii Trelease. Uncommon, partly shady: Larkspur Rock e. of Bodega Bay, *Eastwood & Howell* (1939). < 2300 ft. May-July.

PAPAVERACEAE. POPPY FAMILY
[FUMARIACEAE included]

1. Flowers biradial; outer 2 petals alike, both pouched at base; leaves dissected . . . . . . *Dicentra*
1: Flowers radial; petals 4, 6, or more, not pouched at base; leaves entire to dissected
  2. Plant a subshrub to small tree
    3. Petals 4, yellow; leaf entire or minutely toothed . . . . . . *Dendromecon*
    3: Petals 6, white; leaf deeply lobed . . . . . . *Romneya*
  2: Plant an herb, sometimes woody at base
    4. Leaves entire to minutely toothed, cauline opposite; petals usually 6
      5. Stamens many, sometimes 8-16 in 2 series; fruit a capsule, not composed of separating segments . . . . . . *Meconella*
      5: Stamens > 12; longitudinal segments (carpels) of fruit 9-18, separating . . . . . . *Platystemon*
    4: Leaves dissected into ± linear segments, cauline alternate; petals 4 . . . . . . *Eschscholzia*

Dendromecon rigida Bentham. BUSH POPPY. Rather common, dry slopes, stony washes: ca 1 mi. nw. of Camp Meeker, acc. J. & B. Guggolz (1991); Bohemian Hwy. 3.6 mi. nw. of Occidental, *Huntington* (1986 NCC); Stoetz Lane ne. of Occidental, *Best* (1981); Green Valley Road 0.8 mi. e. of jct. with Harrison Grade Road w. of Graton, *Ferris & Lorraine* (1961); Bouverie; 1½ mi. up Cavedale Road, *Utech & Knight* (1986 CM); Cavedale Road, *Quibell* (1979 NCC); Trinity Road 2 mi. from jct. of Hwy. 12, *Reickert* (1960); Trinity Road, *Clausen* (1933), *Cantelow* (1940); Sugarloaf Ridge; Mt. Hood, *Baker* (1947 NCC); 4300 ft. elev. w. flank Mt. St. Helena, *Grummer* (1978 PUA); Socrates Mine Road just e. of the mine, *Malloch* (1977 PG&E). < 6000 ft. Apr.-Aug.

Dicentra

1. Stems 5-16 dm; taprooted; corolla deciduous; flowers erect, golden yellow . . . . . . *D. chrysantha*
1: Stems 0.5-4 dm high; rhizomatous; corolla ± persistent; flowers nodding, rose-purple to whitish *D. formosa*

Dicentra chrysantha (Hooker & Arnott) Walpers. GOLDEN EAR-DROPS. Uncommon, burned or disturbed areas: Little Cr. bridge on Healdsburg-Geysers Road, *Harrison* (1982 SRJC); ca 2000 ft. elev.

near county road marker 17.76 Little Sulphur Cr. in geothermal area, *Raiche* (1991 JEPS). < 6000 ft. Apr.-Sept.

Dicentra formosa (Haworth) Walpers. BLEEDING HEART. Occasional, damp ± shaded places: Forest Hills 5½ mi. e. of Guerneville, *Rubtzoff* (1975). < 7300 ft. Mar.-July.

Eschscholzia

1\. Torus with only an erect hyaline rim, outer rim absent; annuals with entire cotyledons . . . . . . *E. cespitosa*

1: Torus with 2 rims, the inner erect and hyaline, outer spreading; biennials with 2-cleft cotyledons *E. californica*

Eschscholzia cespitosa Bentham. TUFTED POPPY. Uncommon, dry flats, brushy slopes: Bohemian Hwy. 3 mi. n. of Occidental, *Best* (1981); near Camp Meeker, *Giampaoli* (1968 NCC); 1 mi. se. of Rio Nido toward Hacienda, *Ernst* (1959); Lake Sonoma area; Skaggs Springs, *Hastings* 1940 NCC). < 5000 ft. Mar.-June.

Eschscholzia californica Chamisso. CALIFORNIA POPPY. Common, grassy open places: The Sea Ranch; Horseshoe Cove, *Knight* (1963); Salt Point; Willow Creek; near Duncan's Landing s. of Wright's Beach, *Stuckrath* (1977 PUA); coast n. of Bodega Bay, *Eastwood & Howell* (1939); Bodega Point, *Eastwood* (1899 UC); Todd Road Preserve; Fairfield Osborn (S); Tresch Ranch, Walker Road near Two Rock, acc. J. & B. Guggolz (1992); between San Antonio Cr. and Petaluma on US 101, *Howell* (1981); riverbank at foot of Jefferson St. in Petaluma, *Knight* (1963); w. of Sears Point Raceway in Tolay Cr. drainage, acc. Patterson (1990); Bouverie; Sugarloaf Ridge; Spring Lake; Pepperwood; chaparral area sw. of Bradford Mtn. 1½ mi. w. of Dry Cr., acc. J. & B. Guggolz (1992); Lake Sonoma area. < 6600 ft. Feb.-Sept. —*E. debilis* E. Greene, type locality: Russian River Sta., *Newell* (1900 U).

Meconella

1\. Leaves all basal; flowers scapose, yellowish . . . . *M. linearis*

1: Leaves basal and cauline; flowers on branches as well as main stems, white . . . . *M. californica*

Meconella californica Torrey. Uncommon, open rocky slopes: Larkspur Rock e. of Bodega Bay, *Baker et al.* (1926 NCC), *Howell* (1931), *Ott* (1934 NCC), acc. Quibell (1975); Bodega Point, *Eastwood* (1917); Tresch Ranch, Walker Road near Two Rock, acc. J. & B. Guggolz (1992); Button Ranch on Walker Road (U.C. property) between Petaluma-Valley Ford Road and Pepper Road, acc. Connors (1992). < 3300 ft. Mar.-May.

Meconella linearis (Bentham) Nelson & J. F. Macbride: uncommon, dunes: just n. of Bodega Marine Reserve, *Connors* (1992 BML); Bodega Marine Reserve, *Connors* (1990 BML, 1992); Tresch Ranch near Two Rock, acc. J. & B. Guggolz (1990). < 3300 ft. Mar.-June. —*Hesperomecon pulchellum* E. Greene, type locality, Fort Ross. *H. l.* (Bentham) E. Greene.

Platystemon californicus Bentham. CREAM CUPS. Common, open grassy clayey or sandy places: The Sea Ranch; Horseshoe Cove, *Knight* (1963); Salt Point; Fort Ross-Russian River-Bodega areas, *Voznesenskii* (1840- 1841 LE); Willow Creek; 30 ft. elev. Wright's Beach, *Brantley* (1970 PUA); 30 ft. elev. Duncan's Landing, *Brantley* (1970 PUA); 6 mi. n. of Bodega Bay, *Eastwood* & Howell (1937); Respini Ranch in Chileno Valley, *Knight* (1964); cultivated field ¼ mi. s. of Sonoma State Univ., *Juul* (1973 NCC); Petaluma, *Thorne* (1971 NCC); corner Hearn and Moorland aves., Santa Rosa, *Smith* (1975 NCC); Lake Sonoma area. < 3300 ft. Mar.-May. —*P. greeneanus* Fedde, type locality: Mendocino, *H. E. Brown* (1902 UC) and Bodega Point and Point Reyes, *Eastwood* (1902 UC).

★ **Romneya coulteri** Harvey. MATILIJA POPPY. Largest flowers of any plant native to Ca. Rare, suckering from planted material: Bennett Valley Road 4 mi. e. of jct. with Grange Road in pasture, *Marasco* (1970 NCC). < 4000 ft. May-July. Native of Orange and San Diego counties.

PHILADELPHACEAE. Mock-orange family
[Segregated from SAXIFRAGACEAE]

Whipplea modesta Torrey. Yerba de selva, Modesty. Common, shady banks: The Sea Ranch, *Wells* (1983); Annapolis and Stewarts Point quads.; Salt Point; Seaview Road, *Knight* (1964), *McKenzie* (1977 NCC); Duncans Mills quad.; Willow Creek; Camp Meeker, *Howell* (1924); Coleman Valley Road, *L. L. Davis* (1975 NCC); Sonoma Mtn. Road, *Baker* (1924 NCC); Bouveric; Hooker Can. on Cavedale Road, *Knight* (1964); Gates Road off Calistoga Road, acc. Wells et al. (1984); along St. Helena Road, *Thomas* (1978); Pepperwood; chaparral area sw. of Bradford Mtn. 1½ mi. w. of Dry Cr., acc. J. & B. Guggolz (1992); Lake Sonoma area; property bordered by Chianti and Dutcher Cr. roads w. of US 101 and w. of Geyserville, acc. J. & B. Guggolz (1991). < 4700 ft. Apr.-June.

PHRYMACEAE. Loopseed family

★ **Phryma leptostachya** Linnaeus. A waif in the garden at Friend's House in Rincon Valley, *Wells* (1985). A monotypic family. Native from New Brunswick to Manitoba, s. to Florida and Kansas; also e. Asia.

PLANTAGINACEAE. Plantain family

Plantago

1. Perennial
    2. Plant with clustered fibrous roots
        3. Leaves widely elliptic to ± cordate, narrowed abruptly to petiole; corolla-lobes spreading . . *P. major*
        3: Leaves elliptic-oblanceolate, tapered gradually to petiole; corolla-lobes usually enclosing fruit, together beaklike . . . . *P. subnuda*
    2: Plant taprooted
        4. Leaves deeply pinnately lobed . . . . *P. coronopus*
        4: Leaves entire or lobes few and shallow
            5. Spike 2-8 cm, ± ovoid, dense, becoming cylindric in fruit . . . . *P. lanceolata*
            5: Spike 8-18 cm, narrowly cylindric, usually ± loose near base . . . . . *P. maritima* var. *maritima*
                6. Plants tall, narrow-leaved from wet places . . . . *P. m.* var. *juncoides*
                6: Plants low with spreading or ascending peduncles from dry places or coastal bluffs . . . . *P. m.* var. *californica*

1: Annual, sometimes biennial
    7. Leaves deeply cut, lobes ascending, sharp; spike nodding, becoming erect or wavy in fruit *P. coronopus*
    7: Leaves ± entire; spikes erect
        8. Corolla-lobes lanceolate, acute
            9. Leaves narrowly oblanceolate; inflorescence dense; corolla-lobes ± 2 mm . *P. truncata* ssp. *firma*
            9: Leaves threadlike to linear; inflorescence ± loose; corolla-lobes ± 0.5 mm . . . . . . *P. elongata*
        8: Corolla-lobes round-ovate, usually obtuse . . . . *P. erecta*

★ **Plantago coronopus** Linnaeus. Locally common, coastal strand and inland: Salt Point parking area, *Vale & Wagnon* (1988 NCC); Bodega Head, acc. Connors (1991); Pepperwood. < 1000 ft. Apr.-July. Native of Europe.

Plantago elongata Pursh. Uncommon, salt marshes, pond margins: coast n. of Salmon Cr., *Baker* (1930 NCC); Bodega Point, *Eastwood* (1915); Petaluma, *Congdon* (1880); 5 mi. n. of Sears Point, *Rubtzoff* (1970). < 700 ft. Apr.-June. —*P. e.* ssp. *pentasperma* I. J. Bassett, indistinct. *P. bigelovii* A. Gray var. *b.*, indistinct.

Plantago erecta E. Morris. Common, open dry areas: The Sea Ranch, *J. H. Thomas* (1976); Salt Point; 1 mi. n. of Fort Ross, *Howell* (1930); Duncans Mills and Arched Rock quads.; Bodega Head; 2 sites w. of Santa Rosa in Laguna de Santa Rosa, *Harrison* (1978 NCC); Stony Point, *Congdon* (1880); cemetery in Petaluma, *Ott* (1934 NCC); n. slope Sonoma Mtn., *Baker* (1926 NCC); Tresch Ranch, Walker Road near Two Rock, acc. J. & B. Guggolz (1992); Petaluma, *Eastwood* (1921); Bouverie; Warfield Station at jct. of Hwy. 12 and Trinity Road, *Koch* (1940 NCC); Sugarloaf Ridge; w. base of Mt. Hood, *Heller* (1902); end of Newanga Road, Santa Rosa, *Eusland* (1973 NCC); Spring Lake; Lake Ralphine in Howarth Park, Santa Rosa, *Stocking* (1968 NCC); near Mark West Springs, *Heller* (1902); Pepperwood; Healdsburg quad.; Lake Sonoma area; 2400 ft. elev. at Little Geysers, *Malloch* (1977 PG&E); near Mendocino Co. line on Hwy. 128 nw. of Cloverdale, *Quibell* (1988 NCC). < 2300 ft. Mar.-May. —*P. hookeriana* Fischer & C. Meyer var. *californica* (E. Greene) I. Poe.

★ **Plantago lanceolata** Linnaeus. RIBGRASS, ENGLISH PLANTAIN. Common, weed, lawns, moist waste places: The Sea Ranch, *Wells* (1981); Salt Point; Duncans Mills and Arched Rock quads.; Bodega Head; Bodega Road near Bodega, *Baker* (1929 NCC); near Freestone, *Koch* (1940 NCC); Pitkin Marsh, *Rubtzoff* (1951); Todd Road Preserve; Fairfield Osborn (S); Willow Brook n. of Petaluma, *Knight* (1987); Bouverie; Sugarloaf Ridge; Gates Road off Calistoga Road, acc. Wells et al. (1984); Spring Lake; field near 4408 Langner Ave. in Santa Rosa; *R. F. Smith* (1974 NCC); Pepperwood; Preston (n. of Cloverdale), *Carlson* (1917); chaparral area sw. of Bradford Mtn. 1½ mi. w. of Dry Cr., acc. J. & B. Guggolz (1992); Lake Sonoma area. < 5300 ft. Apr.-Aug. Native of Europe.

★ **Plantago major** Linnaeus. COMMON PLANTAIN. Weed, damp waste places, septic tank leachfields: Salt Point; Willow Creek; w. side of Bodega Head, *Best* (1986); 98 ft. elev. Atascadero Cr. w. of Sebastopol, *Clifton* (1981 PUA); Fairfield Osborn (S); ne. corner Marshall and Magnolia aves. in Petaluma, acc. Knight (1965); (S); Spring Lake. < 7300 ft. Apr.-Sept. Native of Europe.

Plantago maritima Linnaeus var. **maritima**. Occasional, saline areas: The Sea Ranch, *J. H. Thomas* (1976), *Wells* (1981); cliffs at Salt Point, *Best* (1984); Bodega Marine Reserve Marsh, *Connors* (1990 BML); Petaluma, *Eastwood* (1921). < 500 ft. June-Aug. —*P. m.* var. *californica* (Fernald) Pilger and *P. m.* var. *juncoides* (Lamarck) A. Gray, indistinct. Referring to the previous plant, acc. Peter Connors, in the Bodega Head area these varieties are very distinct. Var. *juncoides* occurs in Bodega Harbor salt marshes, whereas var. *californica* is in the lower edge of coastal bluffs with salt spray influence.

Plantago subnuda Pilger. Common, moist banks: The Sea Ranch, *Wells* (1985); Horseshoe Cove, *Knight* (1963); Goat Rock State Beach, *Wells* (1981), *Best* (1983); 1 mi. n. of Salmon Cr., *Keck & Clausen* (1935); n. end of Bodega Harbor, *Howell* (1930); Pitkin Marsh, *Rubtzoff* (1951), *Quibell* (1987 NCC); Cunningham Marsh, *Howell* (1946); Button Ranch on Walker Road (U. C. property) between Petaluma-Valley Ford Road and Pepper Road, acc. Connors (1992). < 1000 ft. May-Sept. —*P. hirtella* Humboldt, Bonpland & Kunth var. *galeottiana* (Decaisne) Pilger.

★ **Plantago truncata** Chamisso ssp. **firma** (Kunze) Pilger. Uncommon, open grassy slopes, meadows that are wet in the spring: vernal pool 4½ mi. w. of Glen Ellen on Bennett Valley Road, *Rubtzoff & Froebe* (1971); summit of Bennett Valley Road, *Eastwood & Howell* (1940). < 1000 ft. Apr.-May. Native of Chile. —*P. f.* Kunze.

PLUMBAGINACEAE. LEADWORT FAMILY

1. Inflorescence paniculate on small secund spikes; leaves broad . . . . *Limonium*
1: Inflorescence capitate; leaves linear . . . . *Armeria*

Armeria maritima (Miller) Willdenow ssp. **californica** (Boissier) Porsild. THRIFT, SEA PINK. Common, sandy places, coastal bluffs: The Sea Ranch; Horseshoe Cove, *Knight* (1963); ½ mi. n. of Moon Rocks Campground at Salt Point State Park, *Huntington* (1989 NCC); 25 ft. elev. Goat Rock, *Woodhouse* (1975 PUA); 49 ft. elev. Duncan's Landing, *Brantley* (1970 PUA); Salmon Cr., *Cummings* (1970), *Reeve* (1980 PUA); Bodega Head, *Marasco* (1970 NCC); Bodega Point, *Eastwood* (1915); 98 ft. elev. Bodega Bay, *Rue* (1973 PUA), *Clifton* (1974 PUA). < 700 ft. Apr.-Aug. —*A. m.* var. *c.* (Boissier) G. Lawrence.

Limonium californicum (Boissier) A. A. Heller. SEA-LAVENDER, WESTERN MARSH-ROSEMARY. Occasional, coastal salt marshes: Salt Point; Sonoma coast n. of Bodega Bay, *Eastwood & Howell* (1939); Bodega Marine Reserve salt marsh, *Connors* (1991 BML); Bodega Bay, *Clifton* (1980 PUA). < 200 ft. July-Dec.

POLEMONIACEAE. PHLOX FAMILY

1. Calyx green-herbaceous, becoming translucent in age, growing with the capsule, not ruptured by it
    2. Calyx-sinuses not distended; leaves pinnately compound; perennial . . . . *Polemonium*
    2: Calyx-sinuses distended in a lip; leaves simple, entire to bipinnately dissected; annual . . . . . . *Collomia*
1: Calyx-membrane between green ribs, and either growing with the capsule or splitting in sinuses
    3. Leaves opposite, especially near base of plant, the upper sometimes alternate, mostly cauline
        4. Leaves entire; stamens unequally inserted . . . . *Phlox*
        4: Leaves palmately cleft, rarely entire; stamens equally inserted . . . . *Linanthus*
    3: Leaves alternate, often with a basal rosette
        5. Calyx-lobes subequal; flowers solitary, cymose, or in heads; bracts and calyx without rigid spinose lobes
            6. Upper cauline leaves usually much reduced; leaf-lobes with fine-pointed tips; seeds usually light brown, ends angular . . . . *Gilia*
            6: Upper cauline leaves quite well-developed; leaf-lobes with blunt tips; seeds mostly dark brown to black, ends rounded . . . . *Allophyllum*
        5: Calyx-lobes unequal; flowers in dense bracteate heads; bracts and calyx with rigid spinose lobes . . . . *Navarretia*

Allophyllum divaricatum (Nuttall) A. D. Grant & V. Grant. STRAGGLING VOLCANIC GILIA. Occasional, disturbed open places: s. slope of Mt. St. Helena near Sonoma-Napa county line, *Baker* (1926); Pepperwood; ca 2000 ft. elev. Anderson Trail on Black Mtn. between Geyser Peak and upper end of Pine Flat Road, acc. J. & B. Guggolz (1992). 1000-6000 ft. Apr.-June. —*Gilia gilioides* (Bentham) E. Greene ssp. *volcanica* (Brand) H. Mason & A. D. Grant. *G. divaricata* Nuttall.

Collomia

1. Stems usually simple, erect, 1-10 dm high; inflorescence headlike, terminal; flowers yellow to orange . . . . *C. grandiflora*
1: Stems diffuse, branched from base, 0.5-2 dm high; inflorescence with flowers in axillary or terminal clusters; flowers white to pink . . . . *C. heterophylla*

Collomia grandiflora Lindley. LARGE-FLOWERED COLLOMIA. Occasional, dry, wooded or open slopes: Pitkin Marsh, *Stocking* (1965 NCC); near Skaggs Springs, *Hawver* (1915); Lake Sonoma area. < 8300 ft. Apr.-July.

Collomia heterophylla Hooker. VARIED-LEAVED COLLOMIA. Common, disturbed areas: The Sea Ranch, *Wells* (1976); Willow Cr. Marsh, *Rubtzoff* (1959); Bouverie, *de Nevers* (1981); Spring Lake; Pepperwood. < 4300 ft. Apr.-June.

Gilia

1. Inflorescence stems generally glandular, corolla throat yellow with purple spots
    2. Corolla 10-19 mm, throat well exserted from calyx . . . . *G. tricolor*
    2: Corolla 7-11 mm, throat partly included in calyx
        3. Calyx in fruit 8-13 mm, generally much longer than capsule . . . . *G. millefoliata*
        3: Calyx in fruit 5-8 mm, generally equal to, or shorter than the capsule . . . . *G. clivorum*
1: Inflorescence stems generally glabrous or tomentose, corolla throat neither yellow nor spotted
    4. Flowers in loose, few-flowered clusters . . . . *G. achilleifolia* ssp. *multicaulis*
    4: Flowers in dense spheric heads
        5. Corolla 5-8 mm, lobes linear, < 1-2 mm wide; calyx-lobes acute, erect
            6. Heads glabrous or sparsely hairy to glandular at base . . . . *G. capitata* ssp. *capitata*
            6: Heads tomentose at base . . . . *G. c.* ssp. *tomentosa*
        5: Corolla 7-13 mm, lobes oblong, ± 3 mm wide; calyx-lobes acuminate, tips recurved . . . . *G. c.* ssp. *chamissonis*

Gilia achilleifolia Bentham ssp. **multicaulis** (Bentham) A. D. & V. Grant. MANY-STEMMED GILIA. Infrequent, shaded woods: Rand Can. (e. of Crane Can.), *Baker* (1925); Cloverdale, *Brandegee* (1890). < 4000 ft. May-June. —*G. m.* Bentham.

Gilia capitata Sims ssp. **capitata**. BLUE FIELD GILIA. Common, well-drained areas: Occidental, *Baker* (1947 NCC); Guerneville, *Rattan* (1877); end of Stoetz Road between Occidental and Graton, *Utech & Knight* (1986 CM); Sebastopol, *Dows* (1907); near Wake Robin Lodge (ca 1½ mi. w. of Glen Ellen), *Baker* (1938 NCC); 1½ mi. up Cavedale Road from Hwy. 12, *Utech & Knight* (1986 CM); Sugarloaf Ridge; between Mark West Springs and Petrified Forest, *Abrams* (1916); Pepperwood; Petrified Forest, *Heller* (1902), *Eastwood* (1915); s. slopes of Fitch Mtn., *Rubtzoff* (1952); Skaggs Springs, *Hawver* (1915); Lake Sonoma area; ca 2000 ft. elev. Anderson Trail on Black Mtn. between Geyser Peak and upper end of Pine Flat Road, acc. J. & B. Guggolz (1992). < 6300 ft. May-July. —*G. pallida* Heller, type locality: Petrified Forest, *Heller* (1902 UC).

Gilia capitata ssp. **chamissonis** (E. Greene) V. Grant. DUNE GILIA. Occasional, sand hills near the coast: 11 mi. n. of Bodega Bay, *V. & A. Grant* (1948); near Salmon Cr. bridge, *Winter* (1932 NCC); Bodega Head Marsh, *Rubtzoff* (1960); Bodega Harbor Marsh, *Rubtzoff* (1955). < 200 ft. May-July. —*G. achilleifolia* ssp. *chamissonis* (E. Greene) Brand. *G. chamissonis* E. Greene.

Gilia capitata ssp. **tomentosa** (Brand) V. Grant. Occasional, rocky seabluffs: Stewarts Point, *Abrams* (1920); Miller Cr. Can. in Salt Point State Park, *Wells* (1982); 3 mi. n. of Jenner, *Cantelow* (1941); e. of Bodega Bay, *Howell* (1930). < 100 ft. May-July. —*G. achilleifolia* Bentham ssp. *chamissonis* (E. Greene) V. Grant var. *t.* Eastwood, type locality: Bodega Point, *Eastwood* (1915).

Gilia clivorum (Jepson) V. Grant. Rare, interior, grassy areas away from coast: Knights Valley, *Leithold* (1895). < 3700 ft. Mar.-May.

Gilia millefoliata Fischer & C. Meyer. Occasional, established dunes, coastal strand: Bodega Bay, *Eastwood* (type 1901); w. side of Bodega Head, *Baker* (1930 NCC); 10 ft. elev. Bay Flat Road near Spud Point Marina at Bodega Harbor, *Clifton* (1980 PUA); dunes along Westside Road, *Connors* (1992 BML). < 35 ft. Apr-June. —*G. multicaulis* Bentham ssp. *m.* (Fischer & C. Meyer) H. Mason & A. D. Grant.

Gilia tricolor Bentham. BIRD'S EYES. Locally common, open grassy plains and slopes: Lafferty Ranch at Sonoma Mtn., acc. Arnold (1971); Willow Brook n. of Petaluma (appears to have been planted), *Knight & Wagnon* (1987 NCC); El Verano, *Griffin* (1922); Mt. Hood, *Eastwood* (1893); hills e. of Santa Rosa, *Heller & Brown* (1902); Santa Rosa, *Eastwood* (1921); Pepperwood; Knights Valley, *Edwards* (1877); Lake Sonoma area. < 4000 ft. Usually Mar.-Apr.

Linanthus

1. Corolla salverform; tube 1½-4 x calyx
   2. Calyx membrane usually as wide as ribs . . . . . *L. ciliatus*
   2: Calyx membrane much less wide than ribs or obscure
      3. Corolla-tube and -lobes bright yellow; leaf lobes needlelike, usually ascending . . . . . *L. acicularis*
      3: Corolla-tube or -lobes (not both) yellow; leaf lobes linear-lanceolate or oblanceolate, usually spreading
         4. Stigma less than 1 mm; corolla-lobes 2-3 mm long . . . . . *L. bicolor*
         4: Stigma more than 1 mm; corolla-lobes 5-8 mm long
            5. Calyx usually hairy only on lobe margins . . . . . *L. androsaceus*
            5: Calyx usually hairy throughout . . . . . *L. parviflorus*

1: Corolla funnel-shaped, tube included or to 2 x calyx
   6. Corolla open in daytime, closing in evening; corolla-lobes white to lavender . . . . . *L. grandiflorus*
   6: Corolla usually open in evening or during overcast day, usually closed in daytime; corolla-lobes white with purple edge . . . . . *L. dichotomus*

Linanthus acicularis E. Greene. BRISTLY LINANTHUS. Somewhat common, dry grassy places: Fort Ross, *Rodda* (1929); Fairfield Osborn; Sonoma Cr. Can., *Howell* (1933); hills e. of Agua Caliente at edge of Sonoma Valley, *Howell* (1948); Bouverie; Lake Sonoma area; Geyserville, *Bailey* (1925); Geysers geothermal area, *Quibell* (1980 NCC). < 2300 ft. Apr.-June.

Linanthus androsaceus (Bentham) E. Greene. SHOWER GILIA. Plentiful, grassy slopes, open places: Stewarts Point, *Abrams* (1920); near Bodega Head, *Howell* (1930); Lafferty Ranch at Sonoma Mtn., acc. Arnold et al. (1971); Tresch Ranch, Walker Road near Two Rock, acc. J. & B. Guggolz (1992); Button Ranch on Walker Road (U. C. property) between Petaluma-Valley Ford Road and Pepper Road, acc. Connors (1992); Fairfield Osborn (H); Bouverie; Santa Rosa Cr. Can., *Howell & Keck* (1931); Pepperwood; near Hwy. 128 n. of Cloverdale, *Grimes* (1981); Skaggs Springs, *Hawver* (1915); on a slope of serpentine ca 3 mi. w. of Warm Springs Dam, *Breedlove* (1976); between Jimtown and Mercuryville on Geysers Road, *Utech & Knight* (1986 CM); 2400 ft. elev. Little Geysers, *Malloch* (1977 PG&E). < 4000 ft. Apr.-June.

Linanthus bicolor (Nuttall) E. Greene. BABY STARS. Common, grassy and open woodland areas, chaparral: Salt Point; Fort Ross quad.; Cazadero, *Bacigalupi* (1926); Duncans Mills quad.; 0.3 mi. off end of Alta Monte Road e. of Rohnert Park, *Lehner* (1975 NCC); Fairfield Osborn (H); Tresch Ranch, Walker Road near Two Rock, acc. J. & B. Guggolz (1992); Sugarloaf Ridge; ca 11 mi. n. of Sonoma, *Mason* (1943); Bennett Valley Road just e. of Enterprise Road, acc. Quibell (1991); Bennett Valley Road w. of Grange Road, acc. Quibell (1991); hills at e. end of Sonoma Ave., Santa Rosa, *Mallory* (1920); 5.4 mi. w. of Mark West Springs, *Eastwood & Howell* (1938); Ida Clayton Road quad.; 3 mi. e. of Warm Springs Dam, *Breedlove* (1976); Lake Sonoma area; ca 2000 ft. elev. Anderson Trail on Black Mtn. between Geyser Peak and upper end of Pine Flat Road, acc. J. & B. Guggolz (1992). < 5700 ft. Mar.-June.

Linanthus ciliatus (Bentham) E. Greene. WHISKER BRUSH. Bristly-leaved Linanthus. Occasional, dry open places: Kenwood, *Baker* (1925 NCC); Pepperwood; Skaggs Springs, *Hawver* (1915). < 10000 ft. Apr.-July.

Linanthus dichotomus Bentham. EVENING SNOW. Occasional, dry hillslopes, sandy or gravelly flats: Petaluma Hill Road, *Chiriboga* (1974 SRJC); se. slope of Mt. Hood, *Baker* (1927 NCC); Rincon Valley, *Mallory* (1920); near Mark West Springs, *Heller & Brown* (1902); Saddle Mtn. e. of Calistoga Road, *Harrison* (1978 SRJC). < 5700 ft. Apr.-June.

Linanthus grandiflorus (Bentham) E. Greene. LARGE-FLOWER LINANTHUS. Acc. *Munz* (1959, p. 513) and *Abrams* (Vol. III, 1951, p. 426) but no specimens are at CAS/DS, NCC or PUA. < 4000 ft. Apr.-July.

Linanthus parviflorus (Bentham) E. Greene. Common, open places, woods, chaparral: cliffs above Wrights Beach, *Ripley & Barneby* (1945); near Bodega Bay, *Litton* (1927 NCC); bluffs n. of Estero Americano at coastal county line, *Breedlove* (1976); Bouverie, *McClintock* (1969); along road by Santa Rosa Cr. Can., *Baker* (1931); Santa Rosa, *Eastwood* (1921); Monan's Rill e. of Calistoga Road and n. of St. Helena Road, *Wells* (1985); Pepperwood; Petrified Forest, *Eastwood* (1915). < 4000 ft. Apr.-June. Segregated from *L. androsaceus* (Bentham) E. Greene.

Navarretia

1. Calyx-ribs strap-shaped, wider at the base than membrane; lobes of bract-tip 7-many; fruit dehiscing from the base upward, not to tip
    2. Corolla bright blue with purple throat . . . . . *N. pubescens*
    2: Corolla yellow or off-white . . . . . *N. cotulifolia*

1: Calyx-ribs tapered, narrower at base than membrane; lobes of bract-tip usually 0-few; fruit dehiscing from tip downward, or if dehiscing upward, then only when wet
    3. Plant generally not glandular; seeds stuck together until wet; fruit translucent, appressed to seeds, opening at base when wet
        4. Axis of inner bracts concave-expanded at base, corolla-lobes ovate
            5. Bract tips with 3 spreading spines; corolla-lobes with 1 vein entering base, corolla pale blue to white; stigmas 2 . . . . . *N. intertexta*
            5: Bract tips with multiple fine spines; corolla-lobes with 3 veins entering base, corolla pale blue; stigmas 3
                6. Calyx and bracts glabrous; bract-tips entire; calyx-lobes entire . . . . . *N. subuligera*
                6: Calyx and bracts appressed-hairy; bract-tips many-lobed; calyx-lobes toothed . . . *N. tagetina*
        4: Axis of inner bracts broadly membranous-winged at base, not concave-expanded; corolla-lobes linear or narrowly ovate
            7. Corolla tube exserted, throat 2-3 mm wide, lobes narrowly ovate, corolla white . . . . . *N. leucocephala* ssp. *leucocephala*
            7: Corolla tube included, throat 1-2 mm wide, lobes linear; corolla blue or white
                8. Stems low, branches spreading; corolla blue, rare . . . . . *N. l.* ssp. *plieantha*
                8: Stems usually erect, branches ascending; corolla white, not rare . . . . . *N. l.* ssp. *bakeri*
    3: Plants usually glandular; seeds not stuck together; fruit opaque, not appressed to seeds, dehiscing downward from tip to base
        9. Branches in pairs or whorls below heads; bracts palmately lobed . . . . . *N. divaricata* ssp. *vividior*
        9: Branching not as previous; bracts palmately or pinnately lobed
            10. Bract-base not concave, lobes many, usually forked on back, stamens attached in tube

11. Corolla dark blue, 9-12 mm; outer bracts with middle lobe lanceolate, long tapered; foliage with skunky odor . . . . *N. squarrosa*
11: Corolla light blue, 5-7 mm; outer bracts with middle lobe ovate-acuminate, without skunky odor . . . . *N. mellita*
10: Bracts or their bases concave, lobes few, not forked on back; stamens attached in tube
12. Outer bracts tipped with 3 spines; stamens short, slightly exserted . . . . . *N. atractyloides*
12: Outer bracts not tipped with 3 spines; stamens long, well-exserted
13. Outer bracts pinnate, much longer than wide, usually much exceeding the head . . . . *N. viscidula*
13: Outer bracts palmate, little longer than wide and not exceeding the head . *N. heterodoxa*

Navarretia atractyloides (Bentham) Hooker & Arnott. Uncommon, dry areas: banks of Russian River n. of Cloverdale, *Heller* (1902). 300-3000 ft. May-July.

Navarretia cotulifolia (Bentham) Hooker & Arnott. COTULA NAVARRETIA. Occasional, moist soil: Laguna de Santa Rosa near Graton, *Rubtzoff* (1966); ne. of Graton at end of Piner Road, *Rubtzoff* (1969); Todd Road near Llano Road, *Rubtzoff, Harrison & Jones* (1974); Santa Rosa, *Call* (1885), *Heller* (1902); on volcanic soil, vernally wet ground of Parktrail Drive off Summerfield Road before development, Santa Rosa. *Allen* (1983 SRJC). < 400 ft. May-June.

Navarretia divaricata (A. Gray) E. Greene ssp. **vividior** (Jepson & V. L. Bailey) H. Mason. Uncommon, bare soil, slopes, hillsides: 2400 and 3100 ft. elev. w. flank Mt. St. Helena, *Grummer* (1979, 1982 PUA). 1000-8600 ft. June-Aug.

Navarretia heterodoxa (E. Greene) E. Greene. CALISTOGA NAVARRETIA. Fairly common, dry open rocky places: 2 mi. e. of Duncans Mills on sand bar, *Roos* (1950); Sonoma, *Burlingame* (1921); Mt. Hood area, *Baker* (1950 NCC); Santa Rosa, *Rattan* (1877); Pepperwood; ca 2 mi. e. of Petrified Forest, *Howell* (1937); Lake Sonoma area. < 3000 ft. June-July. —*Gilia h. E.* Greene, type locality: hills w. of Calistoga (possibly in Sonoma Co.), *E. Greene* (U).

Navarretia intertexta (Bentham) Hooker. NEEDLE-LEAVED NAVARRETIA. Common, vernal pools, moist places: Duncans Mills, *Roos* (1950); Laguna de Santa Rosa 1 mi. se. of Trenton, *Rubtzoff* (1961); 4 sites w. of Santa Rosa in Laguna de Santa Rosa drainage, *Harrison* (1978 NCC); Fairfield Osborn (H); e. foothills of Sonoma Mts. on Sobre Vista Road 0.9 mi. w. of Arnold Drive near El Verano, *Rubtzoff & Froebe* (1971); Bouverie; 4¼ mi. w. of Glen Ellen on Bennett Valley Road, *Rubtzoff & Froebe* (1971); Kenwood, *Bioletti* (1893); Bennett Mtn. Lake, *Rubtzoff* (1961); Annadel, *Wells* (1984); Spring Lake; 850 ft. elev. Shiloh Ranch in valley of small tributary of Mark West Cr. drainage, 2.3 airmiles ne. of jct. of Shiloh and Faught roads, 0.6 airline miles wsw. of Mark West Springs, acc. Stone (1985); Pepperwood, *Breedlove* (1980); Petrified Forest, *Howell* (1937); near Healdsburg, *Carruth* (1900); Healdsburg, *Rattan* (1877); Fitch Mtn., *Rubtzoff* (1962); head of Maacama Cr. near Knights Valley, *H. Mason* (1930); Lake Sonoma area. < 7600 ft. May-July.

Navarretia leucocephala Bentham ssp. **bakeri** (H. Mason) A. G. Day. Common, vernal pools: Laguna de Santa Rosa e. of Graton, *Rubtzoff* (1963); ne. of Graton at end of Piner Road, *Rubtzoff* (1966); Wright School near Santa Rosa, *Ewan* (1935); 3 sites w. of Santa Rosa in Laguna de Santa Rosa drainage, *Harrison* (1978 NCC); Todd Road ca 0.6 mi. w. of Llano Road, *Rubtzoff* (1974); Bennett Valley Road 4½ mi. w. of Glen Ellen, *Rubtzoff & Froebe* (1974); Annadel; Santa Rosa, *Eastwood* (1921); Shiloh Road area n. of Santa Rosa, *McCarten* (1985), *J. & B. Guggolz* (1989 JEPS); ca 2 mi. n. of Windsor, *Eastwood & Howell* (1938). < 5700 ft. May-July. —*N. b.* H. Mason.

Navarretia leucocephala ssp. **leucocephala**. WHITE-FLOWERED NAVARRETIA. Common, vernal pools, valley grassland, foothill woodland, chaparral: midway between Santa Rosa and Cotati, *Eastwood & Howell* (1938); Adobe Can. between Mt. Hood and Sugarloaf Ridge, *Baker* (1927 NCC); on slope of Mt. Hood in burned chaparral, *Baker* (1947 NCC). < 5500 ft. Apr.-May.

Navarretia leucocephala ssp. **plieantha** (H. Mason) A. G. Day. MANY-FLOWERED NAVARRETIA. Threatened by grazing and off-road vehicle use: Bennett Mtn. Lake, *Rubtzoff* (1961); 1 vernal pool near ne. corner of Sonoma County Airport, acc. Marco E. Waaland (1989); Shiloh Ranch near Santa Rosa, *Stone* (1985); 2 mi. sw. of Windsor s. of Saunders Road, *J. & B. Guggolz* (1986 JEPS). < 3100 ft. May-June. —*N. p.* H. Mason.

Navarretia mellita E. Greene. HONEY-SCENTED NAVARRETIA. Fairly common, dry slopes, flats: Stump Beach in Salt Point State Park, *Wells* (1983); Petaluma-Valley Ford Hwy. e. of Valley Ford, *Howell* (1930); ca 1 mi. up Adobe Can., *Baker* (1927 NCC); near Petrified Forest, *Howell* (1937); Dry Cr. area,

Harrison (1981 SRJC); Stewarts Point Skaggs Springs Road 6.7 mi. w. of Skaggs Springs, *Raven & Snow* (1958); The Geysers, *Harrison* (1978, 1982 SRJC). < 4000 ft. May-July.

Navarretia pubescens (Bentham) Hooker & Arnott. DOWNY NAVARRETIA. Occasional, dried depressions, dry slopes: Fort Ross quad.; Sonoma Valley Reg. Park, *Harrison* (1979 SRJC), *Russo* (1984 SRJC); Bennett Valley Road near Sonoma Cr. bridge, *Baker* (1927); Pepperwood; near Maacama Cr. in Knights Valley, *Mason* (1940); Healdsburg dump site, *Harrison* (1986 SRJC); canyon near Geyserville, *McMurphy* (1909). < 3700 ft. May-July.

Navarretia squarrosa (Eschscholtz) Hooker & Arnott. SKUNKWEED. Abundant, dry flats, fields; The Sea Ranch; 49 ft. elev. Gerstle Cove at Salt Point State Park, *Clifton* (1981 PUA); 1 mi. n. of Fort Ross, *Howell* (1930); Bodega Head dunes, Connors (1991); Willow Cr. Marsh, *Rubtzoff* (1959); 2 mi. e. of Duncans Mills, *Roos* (1950); Guerneville Marsh, *Rubtzoff* (1954); Vacation Beach near Guerneville, *Rubtzoff* (1956); Forest Hills across Russian River from Hacienda, *Rubtzoff* (1965); Pitkin Marsh, *Rubtzoff* (1952); Sebastopol, *Dows* (1907); 6 mi. n. of Bodega, *Munz* (1936); Bodega, *Rose* (1962); 4 mi. e. of Valley Ford en route to Two Rock, *Howell* (1930); 5 mi. ne. of Graton, *Rose* (1969); 3 sites w. of Santa Rosa in Laguna de Santa Rosa drainage, *Harrison* (1978 NCC); Todd Road 1.4 mi. from Hwy. 116, *Rubtzoff* (1966); Stony Point, *Howell* (1938); Sugarloaf Ridge; Spring Lake; between Petrified Forest and Mark West Springs, *Abrams* (1916); Lake Sonoma area. < 2700 ft. June-Aug. —*N. pterosperma* Eastwood, type locality: Bodega Point, *Eastwood* (1904).

Navarretia subuligera E. Greene. AWL-LEAVED NAVARRETIA. Acc. *Smith & Pavlik* (1994) the plant occurs in Sonoma Co. No known record has been found. < 3700ft. May-Aug.

Navarretia tagetina E. Greene. MARIGOLD NAVARRETIA. Occasional, moist and grassy places: 4 sites w. and sw. of Santa Rosa in Laguna de Santa Rosa drainage, *Harrison* (1978 NCC); Todd Road Preserve, *Rubtzoff* (1966); Bennett Valley Road 4¼ mi. w. of Glen Ellen, *Rubtzoff & Froebe* (1971); False Lake Meadow in Annadel, *Rubtzoff & Tunison* (1976); Pepperwood; ca 2 mi. n. of Windsor, *Eastwood & Howell* (1936). 100-2000 ft. Apr.-June.

Navarretia viscidula Bentham. STICKY NAVARRETIA. Common, clayey to rocky places: ca 1 mi. nw. of Camp Meeker, *J. & B. Guggolz* (1991 NCC); Pitkin Ranch, *Baker* (1928 NCC); Pitkin Marsh, *Rubtzoff* (1951); Laguna de Santa Rosa e. of Graton, *Rubtzoff* (1966); Cunningham Marsh, *Howell* (1947); 5 mi. s. of Sebastopol, *Rubtzoff* (1957); Green Valley Road in Graton area, *Rubtzoff & Harrison* (1974), *J. & B. Guggolz* (1991 JEPS); w. of Santa Rosa in Laguna de Santa Rosa drainage, *Harrison* (1978 NCC); Fulton, *McMurphy* (1906); Sonoma, *Burlingame* (1921); between Healdsburg and Geyserville, *Heller* (1902); along Stewarts Point Skaggs Springs Road more than 12 mi. w. of site of the former town of Skaggs Springs, *Sharsmith* (1952). 300-2600 ft. June-July.

Phlox

1. Plants annual; flowers inconspicuous . . . . *P. gracilis*
1: Plants perennial; flowers showy . . . . *P. speciosa* ssp. *nitida*

Phlox gracilis E. Greene. SLENDER PHLOX. Common, moist grassy slopes, bottomlands, forests: near Hwy. 1 n. of Fort Ross, *Cantelow* (1940); Fairfield Osborn (H); Bouverie; Cavedale Road, *Raven* (1954); near Santa Rosa, *Eastwood* (1921); hills e. of Santa Rosa, *Heller & Brown* (1902); Spring Lake; Pepperwood; Lake Sonoma area; PG&E geysers Unit 13, *Quibell* (1981 NCC). < 11000 ft. Mar.-Aug. —*Microsteris g.* (Hooker) E. Greene.

Phlox speciosa Pursh ssp. **nitida** (Suksdorf) Wherry. SHOWY PHLOX. Uncommon, rocky hillsides, wooded slopes, chaparral: nw. of Healdsburg on w. side of Dry Cr. Valley, *Raiche* (1985); ca 3¼ mi. s. of Skaggs Springs, Morrell (1925); Skaggs Springs, *Purdy* (1931); near road to marina at Lake Sonoma, acc. P. Gilbert (1989). < 8000 ft. Apr.-June. —*P. s.* ssp. *occidentalis* (Durand) Wherry var. *n.* Suksdorf.

Polemonium carneum A. Gray. GREAT POLEMONIUM. Uncommon, brushy and grassy places: ca 1¼ mi. e. of Bodega Bay, *Baker* (1935); near Bodega, *Robinson* (1925 NCC). < 6000 ft. Apr.-Aug.

POLYGALACEAE. Milkwort family

Polygala

1. Flowers of 2 kinds, those in racemes at base of plant lacking petals, others with petals; sepals glabrous except for the ciliate margins; capsule thin-walled, reticulate . . . . . *P. californica*
1: Flowers all having petals; sepals often pubescent; capsule thick-walled, scarcely reticulate . . . . . *P. cornuta*

Polygala californica Nuttall. California milkwort. Common, openings, brush, woods: The Sea Ranch; Salt Point; 1 mi. n. of Fort Ross, *Ott* (1933 NCC); Cazadero, *Bacigalupi* (1926); hills s. of Guerneville, *Rubtzoff* (1952); 1.2 mi. w. of Occidental on Coleman Valley Road, *Wendt* (1975); Joy Road grade 1.8 mi. from jct. of Bodega Hwy., *Ferris & Lorraine* (1961); Guerneville, *Rattan* (1877); Green Valley Road 0.8 mi. e. of jct. of Harrison Grade Road, *Ferris & Lorraine* (1961); Fairfield Osborn (S); Bouverie; Trinity Road ¾ mi. from jct. of Hwy. 12, *Craig* (1941), *Reickert* (1960); Glen Ellen, *Rattan* (1883); Sugarloaf Ridge; Los Guilicos, *Stocking* (1963 NCC); Mt. Hood, *Michener & Bioletti* (1893); Gates Road off Calistoga Road along Humbug Creek, acc. Wells et al. (1984); Annadel, *Wells* (1984); Rincon Ridge on Montecito Drive near jct. of Chanate Road, acc. Knight (1964); Burke's Sanitarium near Mark West Springs, *Heller* (1902); Pepperwood; Fitch Mtn., *Rubtzoff* (1952); chaparral area sw. of Bradford Mtn. 1½ mi. w. of Dry Cr., acc. J. & B. Guggolz (1992); 350-400 ft. elev. Old Skaggs Springs Road, *Utech, Sato & Ohara* (1990 CM); Geyserville, Alexander Valley and Ida Clayton Road quads.; 2800 ft. elev. w. flank Mt. St. Helena, *Grummer* (1981 PUA); ca 2000 ft. elev. Anderson Trail on Black Mtn. between Geyser Peak and upper end of Pine Flat Road, acc. J. & B. Guggolz (1992); Hwy. 128 1½ mi. nw. of US 101 near Cloverdale, *Breedlove* (1981); near Mendocino Co. line on Hwy. 128 nw. of Cloverdale, *Quibell* (1988 NCC). < 4700 ft. Mar.-July.

Polygala cornuta Kellogg. Horned milkwort. Rare, canyons: 3600 ft. elev. w. flank Mt. St. Helena, *Grummer* (1980 PUA). 300-7000 ft. June-Aug.

POLYGONACEAE. Buckwheat family

(Principally adapted from key of J. C. Hickman.)

1. Leaves basal or opposite, stipules not obvious or 0; nodes mostly not swollen; flowers usually in involucres
   2. Flowers subtended by bracts or not, but not enclosed in an involucre
      3. Bracts solitary, 2-lobed, enlarged, reticulate and saccate in fruit, (often reddish); leaves rounded . . . . . *Pterostegia*
      3: Bracts whorled, hooked at tip; leaves linear . . . . . *Lastarriaea*
   2: Flowers enclosed in a tubular to campanulate involucre
      4. Involucre with 4-5-spine- or bristle-tipped hooked teeth . . . . . *Chorizanthe*
      4: Involucre with 3-8 teeth or lobes, not bristle or spine-tipped . . . . . *Eriogonum*
1: Leaves alternate or ± basal; stipules obvious sheathing the stem; nodes ± swollen; flowers not in involucres
   5. Calyx-lobes usually 4 or 6 (flower parts in 3s) . . . . . *Rumex*
   5: Calyx-lobes 5 (rarely 4 or 6)
      6. Leaf-blade linear to lanceolate; inflorescence 1-8-flowered axillary clusters; achene little if at all exceeding perianth . . . . . *Polygonum*
      6: Leaf-blade triangular-cordate, inflorescence > 10 flowered, the top ± flat; achene much exceeding the perianth . . . . . *Fagopyrum*

Chorizanthe

1. Involucres not both 6-tooth and 6-ribbed . . . . . *C. polygonoides*
1: Involucres 6-toothed and 6-ribbed
   2. Teeth of involucres without a scarious membrane along their margin
      3. Outer and inner calyx-lobes alike or nearly so
         4. Involucral teeth hooked at the tip . . . . . *C. cuspidata* var. *cuspidata*
         4: Involucral teeth straight . . . . . *C. c.* var. *villosa*
      3: Outer and inner calyx-lobes definitely unlike, the inner much smaller . . . . . *C. valida*
   2: Teeth of involucres with a scarious membrane along their margin . . . . . *C. membranacea*

Chorizanthe cuspidata S. Watson var. **cuspidata**. SPINEFLOWER. Occasional, sandy places: n. end of Bodega Harbor, *Howell* (1930); 16 ft. elev. Bay Flat Road near Spud Point Marina at Bodega Harbor, *Clifton* (1980 PUA). < 850 ft. May-July.

Chorizanthe cuspidata var. **villosa** (Eastwood) Munz. Occasional, coastal, sandy areas: Horseshoe Cove, *Knight* (1963); Bodega Bay, *K. Brandegee* (1905 UC); n. end of Bodega Bay, *Howell* (1930); Bodega Head. < 750 ft. May-July. —*C. v.* Eastwood, type locality: Bodega Point, *Eastwood* (1900).

Chorizanthe membranacea Bentham. PINK SPINEFLOWER. Occasional, dry rocky slopes: side canyon above Kenwood Springs near Adobe Can., *Baker* (1926 NCC); Lake Sonoma area; Geyserville, *Rodda* (1929); Pine Flat Road, *J. & B. Guggolz* (1990 JEPS); along road up Sulphur Cr., *Baker* (1930 NCC); The Geysers, *Bolander* (1866). < 5400 ft. Apr.-July.

Chorizanthe polygonoides Torrey & A. Gray. KNOTWEED SPINEFLOWER. Rare, dry gravelly or sandy places: near Baker Ranch in Adobe Can., *Bately* (1932 NCC); near Spring Mtn. Road ca 10 mi. out of Santa Rosa, *Baker* (1930 NCC); Petrified Forest, *Eastwood* (1915). < 4700 ft. Apr.-July.

Chorizanthe valida S. Watson. Rare, sandy places: Russian colony, *Vosnesenskii* (isotype 1841 LE); Sonoma, Petaluma, Russian River, acc. Jepson, *Fl. of CA, 1:394* (1922); Sebastopol, *Dows* (1907). < 1000 ft. June-Aug.

Eriogonum

1. Calyx stipelike at the attenuated base; bracts leafy, indefinite in number

  2. Involucres with lobes at least 0.5 × as long as tube and usually reflexed or spreading . . *E. compositum*

  2: Involucres with lobes much shorter than tube, toothlike and suberect

    3. Calyx ochroleucous; leaves broadly ovate, rounded or subcordate at base . . . . *E. nervulosum*

    3: Calyx yellow; leaves obovate to spatulate, attenuated at base . . . . *E. ternatum*

1: Calyx not stipelike at the base; bracts not leafy, regularly 3 in number

  4. Involucres campanulate to turbinate, not angled or ribbed, 4 or 5-toothed or -lobed . . . . . *E. hirtiflorum*

  4: Involucres cylindric or cylindric-turbinate to turbinate or prismatic

    5. Plants annual; leaves mostly in basal rosettes . . . . *E. luteolum*

    5: Plants perennial; leaves often cauline as well as basal

      6. Involucres solitary at the nodes, the lateral ones appressed to the branches; Outer Coast Range form . . . . *E. nudum*

      6: Involucres mostly clustered or in heads

        7. Plants essentially herbaceous, only the base woody

          8. Involucres and flowering stems tomentose; heads 1-few . . . . *E. latifolium*

          8: Involucres and flowering stems glabrous, or if tomentose then plants with many heads; (interior form) . . . . *E. nudum*

        7: Plants definitely shrubby . . . . *E. fasciculatum* var. *foliolosum*

Eriogonum compositum Bentham. Occasional, dry rocky banks, sometimes on serpentine, hilly/montane areas: Rocky Mtn. on Austin Cr., serpentine, *Hoffman* (1948); Duvoul Cr. ca 2 mi. n. of Camp Meeker, *Raiche* (1984); 3.9 mi. on Sweetwater Springs Road from Armstrong Woods Road, *Raiche* (1983); 1700 ft. elev. Prichett Peaks e. of Lake Sonoma and sw. of Cloverdale, *Raiche* (1984); Geyser Peak, *Knight & Edwards* (1986); 3500 ft. elev. on graywacke sandstone in upper Cobb Cr. drainage, acc. Stone (1989). < 8300 ft. May-July.

★ **Eriogonum fasciculatum** Bentham var. (Nuttall) Abrams. CALIFORNIA BUCKWHEAT. Rare, dry slopes: near jct. Rockpile and Stewarts Point-Skaggs Springs roads, probably planted by Caltrans, *Guggolz* (1975); *Best* (1983), *Knight* (1984). < 3000 ft. Mar.-Oct. Introduced from s. California.

Eriogonum hirtiflorum S. Watson. Locally common, gravelly openings in chaparral: 3470 ft. elev. 1.0 airmiles nw. of Geyser Rock, acc. Stone (with Clifton and Davilla), (1983). 1600-6600 ft. June-Oct.

Eriogonum latifolium Smith. COAST BUCKWHEAT. Common, littoral, dry, often stony areas: 0.7 mi. s. of Gualala on The Sea Ranch, *J. H. Thomas* (1976); Horseshoe Cove, *Knight* (1963); Salt Point; 1 mi. n. of Fort Ross, *Howell* (1930); Jenner, *Mason* (1928); near mouth of Russian River, *Dudley* (1893); Ocean View between Duncan's Landing and Wright's Beach, *Holmgren & Reveal* (1966); parking lot s. of Duncan's Landing, *Wells* (1983); 8½ mi. n. of Bodega Bay, *Howell* (1930); 4 mi. n. of Bodega Bay, *Howell* (1930); ½ mi. n. of Salmon Cr., *Balls & Everett* (1951); 2 mi. n. of Bodega Bay, *Bracelin* (1930), *Munz* (1936); Bodega Point, *Eastwood* (1915); 1 mi. e. of Bodega Bay, *Howell* (1930); Bodega Dunes Campground, *Hitchcock & Martin* (1939); 59 ft. elev. Bodega Head, *Clifton* (1980 PUA); Larkspur Rock

e. of Bodega Bay, *Baker* (1935 NCC); cliff along Hwy. 1 near Bodega Bay, *Baker* (1931 NCC). < 500 ft. June-Sept.

Eriogonum luteolum E. Greene. Locally common, shallow gravelly, sandy, dry rocky places, sometimes on serpentine: 2 mi. from Monte Rio on road to Camp Meeker, *Howell* (1931); near Occidental, *M. Williams & Roderick* (1978); 2 mi. ne. of Occidental on Stoetz Lane on serpentine, *Best* (1981); probably between Santa Rosa and Mt. St. Helena, *Voznesenskii* (1840-1841 LE); 5 mi. w. of Petaluma, *Morgan* (1955); Sonoma Mtn., *Leitholt* (1893); Sonoma Cr. in Sonoma Valley, *Howell* (1932); Sugarloaf Ridge State Park on trail to Mt. Hood, *W. Ross* (1984); Los Alamos Road on serpentine, *Knight* (1963); 4 mi. from Petrified Forest on road to Knights Valley, *Howell* (1929); Lake Sonoma area; Red Hill off Pine Flat Road, *W., I. & T. Knight & Edwards* (1985); ca 2000 ft. elev. Anderson Trail on Black Mtn. between Geyser Peak and upper end of Pine Flat Road, acc. J. & B. Guggolz (1992); 3200 ft. elev. upper Squaw Cr. drainage, acc. Stone (1983); Russian River near Cloverdale, *Geis* (1902); 2700 ft. elev. The Geysers, *G. Smith* (1961). 350-8000 ft. June-Sept. —*E. vimineum* Douglas var. *californicum* Gandoger, type locality: Petaluma, *Tidestrom* (UC).

Eriogonum nervulosum (S. Stokes) Reveal. SNOW MTN. BUCKWHEAT. Occasional, outcrops of serpentine: The Cedars at Layton Mine on Austin Cr., *F. W. Hoffman* (1947); 1220 ft. elev. The Cedars near Layton Mine, *Raiche* (1985); serpentine barrens along Sonoma/Lake county boundary in Mayacmas Mts., *Specht* (1977 HSC); ca 2400 ft. elev. on fire road near jct. of Ida Clayton and Western Mine roads, acc. *Raiche* (1984); Union Oil geothermal development, acc. Heckard & Hickman (1984); ne. of Big Sulphur Cr. along county line, *York, Jensen & Crane* (1982 PG&E); localities reported within 2 mi. along the Lake-Sonoma county line near crest of Mayacmas Mts., acc. *Kruckeberg* (1984). 1000-7000 ft. May-Aug. —*E. ursinum* S. Watson var. *n.* S. Stokes.

Eriogonum nudum Bentham var. **nudum**. NUDE BUCKWHEAT. Common, dry flats, slopes: King's Ridge Road n. of Cazadero, *Wells* (1985); Morelli Lane 1 mi. ne. of Occidental, *Best* (1981); Fairfield Osborn (S); 5 mi. w. of Petaluma, *Morgan* (1955); Hwy. 12, 1¼ mi. n. of Kenwood, *Baker* (1931 NCC); Mt. Hood, *Knight* (1964); Bennett Valley, *Leitholt* (1893); Los Alamos Road, *Knight* (1963); Pepperwood; 2 mi. w. of Petrified Forest, *Rose* (1933); Skaggs Springs, *Geis* (1902); Lake Sonoma area; between Healdsburg and Geyserville, *Heller* (1902); ca 2000 ft. elev. Anderson Trail on Black Mtn. between Geyser Peak and upper end of Pine Flat Road, acc. J. & B. Guggolz (1992); Sulphur Cr. Road about 1 mi. from US 101, *Knight* (1963). < 8000 ft. June-Nov. Var. **auriculatum** Bentham designates stout flowering stems not strongly fistulose; leaf margins strongly wavy; flowers white to pink, subglabrous. 200-5400 ft. July-Sept. Var. **oblongifolium** S. Watson designates stems spreading, often with stout or fistulose branches; leaf margins flat to slightly wavy; flowers white or yellow, often pubescent. 500-7000 ft. May-Aug. Var. **pubiflorum** Bentham designates slender stems barely or not fistulose; leaf margins flat; flowers white or pale yellow, obviously hairy. 150-6600 ft. June-Sept. Some of the var. *nudum* collections may apply to the last three varieties.

Eriogonum ternatum T. J. Howell. TERNATE BUCKWHEAT. Locally common, serpentine: The Cedars, 5 mi. n. of Cazadero on King Ridge Road, *Raiche* (1984); 1200 ft. elev, The Cedars, headwaters of Big Austin Cr. near Layton Mine, *Raiche* (1985); 1800 ft. elev. The Cedars, n. of Cazadero, *Raiche* (1984); near Red Slide on E. Austin Cr., The Cedars *Raiche* (1984). 1000-7300 ft. June-Aug.

★ **Fagopyrum esculentum** Moench. BUCKWHEAT. Escape, flower bed: Spring Lake Village. Native of Eurasia. < 1700 ft. —*F. sagittatum* Gilibert. Toxic to livestock.

Lastarriaea coriacea (Goodman) Hoover. Infrequent, sandy to gravelly soils along coast: Sonoma County, acc. Reveal, *Phytologia 66 (4) p. 413,* (1989). < 3000 ft. Apr.-June. —*Chorizanthe chilensis* Remy var. *californica* (H. Gross) Goodman. *C. coriacea* Goodman.

Polygonum

1. Flowers in axillary fascicles, or in spikes with leafy bracts; ochrea hyaline and finally mostly 2-lobed and lacerate
    2. Leaves not jointed with the ochrea, 3-nerved; flowers solitary in the axils, sessile
        3. Plants perennial, 2-6 dm high . . . . *P. bolanderi*
        3: Plants annual, 0.5-2 dm high . . . . *P. californicum*
    2: Leaves jointed with the ochrea, 1-nerved; flowers mostly 2 or more in axils of full-sized leaves, short-pedicelled
        4. Plants perennial . . . . *P. paronychia*

4: Plants annual
  5. Fruit ± 3.5 mm, shiny
    6. Flower, 1 erect; fruit dark brown . . . . *P. marinense*
    6: Flowers, 1-3 drooping; fruit black . . . . *P. douglasii*
  5: Fruit < 2.5 mm (except some late in season), dark brown, ± dull
    7. Calyx ± 2.5 mm exserted from stipule . . . . *P. arenastrum*
    7: Calyx ± 2 mm ± enclosed in stipule . . . . *P. prolificum*

1: Flowers fascicled in terminal dense to open spikelike racemes, or in panicles without leafy bracts; ochreae rarely hyaline or lacerate
  8. Inflorescence of open panicles and the leaves broad or the inflorescence of small axillary clusters and terminal spike with the leaves cordate-sagittate
    9. Stems twining; annual; leaves ovate-sagittate . . . . *P. convolvulus*
    9: Stems erect; stout perennials; leaves ovate
      10. Leaves rounded at base, abruptly acuminate at apex, 5-15 cm long . . . . *P. cuspidatum*
      10: Leaves cordate at base, gradually narrowed at apex, 15-30 cm long . . . . *P. sachalinense*
  8: Inflorescence elongate, spiciform, not in ample panicles or leaves not cordate-sagittate
    11. Stems simple; leaves basal and long-petioled or cauline and short-petioled or sessile; inflorescence a single terminal spike . . . . *P. bistortoides*
    11: Stems branched; leaves all cauline and similar; spikes terminating main stems and branches
      12. The ochrea eciliate or nearly so, at least in maturity
        13. Plants in and about water
          14. Peduncles glabrous; spikes ovoid or oblong, mostly < 3 cm . . . . *P. amphibium* var. *stipulaceum*
          14: Peduncles pubescent; spikes slender-cylindric, mostly 3-10 cm *P. a.* var. *emersum*
        13: Plants not aquatic
          15. Peduncles with or without sessile glands; spikes ± arched, slender-cylindric . . . . *P. lapathifolium*
          15: Peduncles with stalked glands; spikes erect, oblong-cylindric . . . *P. pensylvanicum*
      12: The ochrea fringed with bristly cilia
        16. Plants annual
          17. Calyx punctate with glands; spikes slender, arching . . . . *P. hydropiper*
          17: Calyx not glandular-punctate; spikes dense, erect . . . . *P. persicaria*
        16: Plants perennial
          18. Calyx greenish, obviously glandular-punctate . . . . *P. punctatum*
          18: Calyx white to rose, not glandular or punctate . . . . *P. hydropiperoides*

Polygonum amphibium Linnaeus var. **emersum** Michaux. KELP, WATER SMARTWEED. Occasional, in and about ponds: Bodega Head, *Rubtzoff* (1954); Laguna de Santa Rosa, *Rubtzoff* (1958); laguna in Chileno Valley, *Rubtzoff* (1962); bed of slough draining Kenwood Marsh; *Howell* (1932); entrance to Stevens Ranch at Kenwood, *Baker* (1932 NCC); Spring Lake. < 5000 ft. June-Oct. —*Persicaria alismaefolia* E. Greene, type locality: Cloverdale, *Heller* (1902 U). *Polygonum coccinium* Muhlenberg. Noxious weed.

Polygonum amphibium var. **stipulaceum** Coleman. WATER SMARTWEED. Occasional, ponds, lakes: 0.7 se. of Salmon Cr. crossing, *Rubtzoff* (1971 NCC); w. side of Hwy. 1 se. of Salmon Cr., *Rubtzoff* (1971); 69 ft. elev. Salmon Cr., *Clifton* (1984 PUA); Laguna de Santa Rosa, *Rubtzoff* (1960); Howarth Park, Santa Rosa, *Knight* (1964); Spring Lake Reservoir, Santa Rosa, *Rubtzoff* (1971). < 10000 ft. July-Sept. —*P. amphibium* var. *natans* Michaux. *Persicaria amphibia* Linnaeus var. *stipulacea* (Coleman) Hara.

★ **Polygonum arenastrum** Boreau. DOORWEED, COMMON KNOTWEED. Common, weed, hard-packed soil, waste places: Stump Beach at Salt Point State Park; Todd Road Preserve; Copeland Cr. near Rohnert Park, *Self* (1979); Fairfield Osborn (H); Petaluma, *Eastwood* (1921); w. of Sears Point Raceway in Tolay Cr. drainage, acc. Patterson (1990); Kenwood Marsh, *Howell* (1934); Sugarloaf Ridge; Spring Lake Village; Lake Sonoma area. acc. *Mertens & Raven* (1965). < 8300 ft. May-Nov. Naturalized from Eurasia. —*P. aviculare* Linnaeus, misapplied.

Polygonum bistortoides Pursh. MARSH BUCKWHEAT, WESTERN BISTORT. Uncommon, wet soil of marshes: Pitkin Marsh, *Rubt-zoff* (1952); marshy area s. of Hwy. 116 near Bloomfield Road, acc. J. & B. Guggolz (1992). < 10000 ft. June-Aug. —*Bistorta b.* (Pursh) Small.

Polygonum bolanderi Brewer. Occasional, dry gravelly and rocky places: K Ranch in Sonoma Valley, *Krauss* (1893); Mt. Hood, *Brewer* (1862 UC), *Bolander* (1864 UC); 2 mi. from Petrified Forest en route to Calistoga, *Howell* (1926); 3½ mi. w. of Calistoga, *G. L. Stebbins* (1966); Mountain Home Road near Petrified Forest, *Stocking* (1966 NCC). 750-5000 ft. June-Nov.

Polygonum californicum Meissner. Rare, dry sandy flats: ca 3/4 mi. n. of Camp Meeker, *J. & B. Guggolz* (1991 NCC); Napa-Sonoma county line on Kortum Can. Road, *Knight & Edwards* (1985). < 4000 ft. May-Oct.

★ **Polygonum convolvulus** Linnaeus. BLACK BINDWEED. Adventive: Wild Rose Drive, Santa Rosa, *Harrison* (1974 SRJC); Penny Island, mouth of Russian River, growing around abandoned farm buildings, acc. Harrison (1992); Pepperwood. < 3400 ft. May-Sept. Naturalized from Europe.

★ **Polygonum cuspidatum** Siebold & Zuccarini. JAPANESE KNOTWEED. Occasional, escape from cultivation: Gates Road off Calistoga Road, acc. Wells et al. (1984); 7 locations with specimens on deposit at CDA from central to southern part of county, acc. *Barbe* (1991). < 3400 ft. Native of Japan.

Polygonum douglasii E. Greene. Rare, fairly dry areas: off Bohemian Hwy. between Camp Meeker and Bohemian Grove, *J. & B. Guggolz* (JEPS 1991). < 10000 ft. June-Sept.

★ **Polygonum hydropiper** Linnaeus. MARSHPEPPER. WATERPEPPER. Common, moist places: Russian River bottom at Forest Hills across river from Hacienda, *Rubtzoff* (1968); Laguna de Santa Rosa near Occidental Road, *Howell* (1946), *Rubtzoff* (1970 NCC); slough draining Kenwood Marsh, *Howell* (1932); Newanga Road near Spring Lake Reservoir in Santa Rosa, *Knight* (1965). Probably naturalized from Europe. < 5000 ft. July-Oct. —*Persicaria hydropiper* (Linnaeus) Spach.

Polygonum hydropiperoides Michaux. WATERPEPPER. Common, moist places: 1 mi. e. of Jenner, *Rubtzoff* (1964); Bodega Head Marsh, *Rubtzoff* (1954); Stoetz Lane near pond 2 mi. ne. of Occidental, *Best* (1981); Pitkin Marsh, *Rubtzoff* (1951); laguna n. of Sebastopol, *Trynor* (1938 NCC), *Howell* (1946); Atascadero Cr. Marsh w. of Sebastopol, *Rubtzoff* (1958); Laguna de Santa Rosa, *Rubtzoff* (1958); entrance to Stevens Ranch at Kenwood, *Baker* (1930). < 5000 ft. June-Oct. Often confused with *P. persicaria*. Var. *asperifolium* E. E. Stanford, currently not recognized. —*Persicaria h.* (Michaux) Small.

Polygonum lapathifolium Linnaeus. WILLOW WEED. Common, moist places: e. of Guerneville, *Rubtzoff* (1955); Hacienda, *Wagnon & Vale* (1988 NCC); Laguna de Santa Rosa ne. of Graton, *Baker* (1946 NCC), *Rubtzoff* (1958); ditch on Rainsville Road near Liberty Road nw. of Petaluma, *Knight* (1963); Willow Brook near Corona Road n. of Petaluma, *Knight* (1987 NCC); ditch on Bodega Ave. near Cleveland Lane w. of Petaluma, *Knight* (1963); slough draining Kenwood Marsh, *Howell* (1932); Fitch Mtn., *Rubtzoff* (1964). < 5000 ft. June-Oct. —*P. pensylvanicum* (Linnaeus) Gomes, misapplied. *Persicaria lapathifolia* (Linnaeus) A. Gray.

Polygonum marinense T. Mertens & Raven. MARIN KNOTWEED. Uncommon, salt marsh: Bodega Marine Reserve marsh near B.M.L. housing, *Connors* (BML 1991) < 35 ft. May-Oct.

Polygonum paronychia Chamisso & Schlechtendal. Occasional, coastal strand: Horseshoe Cove, *Knight* (1963); s. side of mouth of Russian River, *Best* (1981); sand hills at mouth of Salmon Cr., *Baker* (1930); 10 ft. elev. Salmon Cr. beach, *Clifton* (1980 PUA); near Salmon Cr., *Cantelow* (1940); Bodega Point, *Eastwood* (1915); Bodega dunes, *Connors* (1990 BML). < 500 ft. Mar.-Sept.

★ **Polygonum pensylvanicum** Linnaeus. PINKWEED. Uncommon, marshes: Hacienda, *Knight* (1988 NCC); slough draining Kenwood Marsh, *Howell* (1932). ± 100 ft. July-Sept. Introduced from e. U. S.

★ **Polygonum persicaria** Linnaeus. LADY'S THUMB. Common, moist, waste places: Willow Cr. Marsh, *Rubtzoff* (1959); Salmon Cr., *Rubtzoff* (1969); Stoetz Lane near the pond 2 mi. ne. of Occidental, *Best* (1981); Guerneville Marsh, *Rubtzoff* (1954); Laguna de Santa Rosa at crossing of Occidental Road, *Baker* (1946), *Rubtzoff* (1966); Atascadero Cr. Marsh w. of Sebastopol, *Rubtzoff* (1962); 5 mi. n. of Sears Point, *Rubtzoff* (1970); along Santa Rosa Cr. in e. Santa Rosa, acc. Wagnon (1989); Spring Lake; Pepperwood; Fitch Mtn., *Rubtzoff* (1964); Lake Sonoma area. < 5000 ft. June-Nov. Native of Europe. —*Persicaria maculata* (Rafinesque) A. Löve & D. Löve.

★ **Polygonum prolificum** (Small) Robinson. Uncommon, waste areas: Schollenberger Park, Petaluma, *W., I. & T. Knight* (1986 NCC); flower bed at Royal Oaks Mobile Home Community, Petaluma, *Knight* (1991 NCC). < 6700 ft. Native of e. U. S.

Polygonum punctatum Elliott. WATER SMARTWEED. Common, low moist places: Guerneville Marsh, *Rubtzoff* (1954); Pitkin Marsh, *Howell & Stacey* (1936), *Rubtzoff* (1952); Twin Pine Marsh 2 mi. n. of Sebastopol, *Baker* (1945); lagoon at Sebastopol, *Heller* (1902); marshy area s. of Hwy. 116 near Bloomfield Road, acc. J. & B. Guggolz (1992); Laguna de Santa Rosa at Occidental Road crossing, *Howell* (1946), *Rubtzoff* (1966); Llano Road n. of Gravenstein Hwy., *Howell & Knight* (1986 NCC); slough

draining Kenwood Marsh, *Howell* (1932); Ducker Cr. in Rincon Valley, *Wells* (1985); Spring Lake; Newanga Road, Santa Rosa, *Knight* (1965). < 5000 ft. July-Oct. —*Persicaria punctata* (Elliott) Small.
★ **Polygonum sachalinense** Maxim. GIANT KNOTWEED, SACALINE. Rare, waste places: Bloomfield-Sebastopol area, acc. *Barbe* (1991). < 1700 ft. Native of Japan.

Pterostegia drymarioides Fischer & C. Meyer. Common, shady places: dunes at s. side of mouth of Russian River, *Best* (1981); Bodega Head; Bodega Point, *Eastwood* (1900); Bodega Bay, *Heller & Brown* (1902); Pepperwood; near old Skaggs Springs site, *Quibell* (1980). Type locality: Bodega (probably in vicinity of Bodega Bay).

Rumex: (Partially adapted from key by J. C. Hickman).

1. Flowers dioecious (rarely some bisexual); stem erect, slender, < 4 dm; flowers ± red; tubercles 0 . . . *R. acetosella*
1: Flowers usually bisexual; stem decumbent to erect, generally stout, usually > 4 dm; flowers usually green; tubercles various or 0
  2. Stem prostrate to ± erect, with late-flowering axillary shoots; basal leaves 0
    3. Tubercles inconspicuous or 0; inner perianth-lobes triangular, teeth at base short-triangular . . . *R. salicifolius* var. *denticulatus*
    3: Tubercles 1-3, usually obvious; inner perianth-lobes lanceolate to round, teeth usually 0
      4. Tubercle 1 obvious, width > 1/3 width of inner perianth-lobe
        5. Inflorescence ± open; leaves thin, linear to lanceolate . . . *R. s.* var. *salicifolius*
        5: Inflorescence dense; leaves ± thick, some oblong to elliptic . . . *R. s.* var. *crassus*
      4: Tubercles 3 obvious ± =, width > or < 1/3 width of inner perianth-lobe . . . *R. s.* var. *transitorius*
  2: Stem erect, axillary shoots 0; leaves basal during early season
    6. Tubercles inconspicuous or 0 . . . *R. occidentalis*
    6: Tubercles 1-3, obvious
      7. Margin of inner perianth lobe entire
        8. Inner perianth lobe 2-3 mm, tubercle 1/3 width of lobe or wider; inflorescence open, interrupted . . . *R. conglomeratus*
        8: Inner perianth lobe 5-6 mm, tubercle << 1/3 width of lobe; inflorescence dense, continuous . . . *R. crispus*
      7: Margin of inner perianth lobe with 0.5-4 mm teeth
        9. Plants perennial
          10. Largest leaf 20-75 cm; blade lance-ovate, cordate; inner perianth-lobe ovate-triangular . . . *R. obtusifolius*
          10: Largest leaf < 20 cm, if larger, blade narrowly lanceolate, tapered; inner perianth-lobe ovate to round-cordate . . . *R. pulcher*
        9: Plants annual . . . *R. maritimus*

★ **Rumex acetosella** Linnaeus. SHEEP SORREL. Common, weed, open and brushy areas, grassland: The Sea Ranch; Horseshoe Cove, *Knight* (1963); Salt Point; Willow Creek; Bodega Head; Kings Ridge Road ca 4 mi. n. of Cazadero, *Wells* (1985); Duncans Mills and Arched Rock quads.; Freestone, *Huntington* (1987 NCC); Pitkin Marsh, *Rubtzoff* (1951); Sebastopol, *Randall* (1907); 2 sites w. of Santa Rosa in Laguna de Santa Rosa drainage, *Harrison* (1978 NCC); Todd Road Preserve; Fairfield Osborn (S); Bouverie; Sugarloaf Ridge; Bennett Valley, *Heller* (1902); Annadel, *Wells* (1984); Santa Rosa, *Heller* (1902); campus of Santa Rosa Jr. Coll., *Hagg & Schmidt* (1932 NCC); Pepperwood; Ida Clayton Road quad.; between Jimtown and Mercuryville on Geysers Road, *Utech & Knight* (1986 CM); ca 2000 ft. elev. Anderson Trail on Black Mtn. between Geyser Peak and upper end of Pine Flat Road, acc. J. & B. Guggolz (1992). < 10000 ft. Mar.-Aug. Native of Eurasia.
★ **Rumex conglomeratus** Murray. Common, low moist areas: The Sea Ranch; Horseshoe Cove, *Knight* (1963); 1024 ft. elev. Coleman Valley Road, *Clifton* (1984 PUA); Bodega Head; Pitkin Marsh, *Howell* (1938), *Rubtzoff* (1951 NCC); foot of Mt. Hood, *Heller* (1902); Bennett Valley, *Heller* (1902); Ducker Cr. in Rincon Valley, *Wells* (1985); 4.6 mi. w. of Warm Springs Dam in Lake Sonoma area, *Thomas* (1976); 2420 ft. elev. Caldwell Ranch, *Clifton* (1984 PUA). < 5000 ft. Apr.-Oct. Naturalized from Europe.

Rumex crispus Linnaeus. CURLY DOCK. Common, weed, low places: The Sea Ranch; Horseshoe Cove, *Knight* (1963); Bodega Bay, *Barbour* (1967 BML); Freestone, *Huntington* (1987 NCC); Pitkin Marsh, *Rubtzoff* (1951); Petaluma-Valley Ford Road just e. of jct. with Roblar Road, acc. Quibell (1991); 1 mi. se. of Bloomfield on road to Two Rock, *Howell* (1955); several sites w. of Santa Rosa in Laguna de Santa Rosa, *Harrison* (1978 NCC); Todd Road Preserve; Fairfield Osborn (S); vicinity of Petaluma, *Howell* (1981); Schollenberger Park, Petaluma, *W., I. & T. Knight* (1986 NCC); Bouverie; Sugarloaf Ridge; Spring Lake; Blue Bell Lane, Santa Rosa, *Knight* (1963); Pepperwood; n. of Windsor, *Baker* (1943); Lake Sonoma area; 2420 ft. elev. Caldwell Ranch, *Clifton* (1984 PUA). < 8300 ft. ± year around. Native of Eurasia.

Rumex maritimus Linnaeus. GOLDEN DOCK. Uncommon, wet ground: found abundantly in late 1978 on effluent-sprayed land e. of the Sebastopol Campground (Hwy. 12 at the Laguna crossing). Rubtzoff (personal communication with Harrison) has collected it in the Laguna de Santa Rosa area but has never found it in abundance at other sites. It grows in brackish wetland and may be another indicator of the saline nature of the effluent, *Harrison* (1978 SRJC). < 6700 ft. May-Sept. —*R. fueginus* Philippi.

★ **Rumex obtusifolius** Linnaeus. BITTER DOCK. Uncommon, low moist areas: along Russian River, *Rubtzoff* (1987). Ssp. *agrestis* (Fries) Danser, currently not recognized. < 5000 ft. June-Dec. Native of Europe.

Rumex occidentalis S. Watson. WESTERN DOCK. Occasional, brackish marshy areas: The Sea Ranch, *Wells* (1987); Bodega Head, *Connors* (1991 BML); Petaluma, *Congdon* (1890), *Eastwood* (1921). < 5000 ft. Aug.-Sept. —*R. fenestratus* E. Greene.

★ **Rumex pulcher** Linnaeus. FIDDLE DOCK. Common, weed, waste places: The Sea Ranch; 7 mi. s. of Gualala, *J. H. Thomas* (1976); Willow Creek; Bodega Head; w. side of Bodega Harbor, *Best* (1986); Freestone, *Huntington* (1987); Pitkin Marsh, *Rubtzoff* (1951); 2 sites w. and sw. of Santa Rosa in Laguna de Santa Rosa drainage, *Harrison* (1978 NCC); Llano Road n. of Gravenstein Hwy., *Howell & Knight* (1986); Bouverie; jct. Hwy. 37 and Hwy. 121 (near Sears Point Raceway), *Baker* (1946 NCC); Santa Rosa, *Winters* (1935 NCC); Pepperwood; Lake Sonoma area; Alexander Valley quad.; between Jimtown and Mercuryville on Geysers Road, *Utech & Knight* (1986 CM); 1 mi. s. of Cloverdale, *Baker* (1943 NCC). Usually < 5000 ft. May-Sept. Native of Mediterranean area.

Rumex salicifolius J. A. Weinmann var. **crassus** (Rechinger f.) J. T. Howell. Occasional, rocky ocean bluffs, dunes: The Sea Ranch; Horseshoe Cove, *Knight* (1963); Salt Point; 2.6 mi. n. of Jenner, *Raven & Snow* (1958); Salmon Cr. beach parking lot, *Wells* (1982); Bodega dunes, *Connors* (1990 BML). < 350 ft. May-Sept. —*R. c.* Rechinger f.

Rumex salicifolius var. **denticulatus** Torrey. Occasional, moist places: Sonoma State Univ. campus, *Howell* (1986); Sonoma-Napa county line on Hwy. 12, *Raven* (1952); Alder Glen Springs (old road between Cloverdale and Hopland), *Eastwood & Howell* (1938). < 12000 ft. May-Sept. —*R. s.* forma *escallosus* J. T. Howell. *R. californicus* Rechinger f.

Rumex salicifolius var. **salicifolius**. WILLOW DOCK. Occasional, seepages, slopes, moist places, low elevations: 10 ft. elev. Bodega Head, *Clifton* (1981 PUA); w. of Santa Rosa in Laguna de Santa Rosa drainage, *Harrison* (1978 NCC); 1 mi. sw. of Sonoma, *Howell* (1957). < 10000 ft. May-Sept. Coastal forms intergrade with var. *crassus*.

Rumex salicifolius var. **transitorius** (Rechinger f.) J. Hickman. Occasional, moist ground, marshes: Goat Rock Beach off Hwy. 1, *Best* (1982); ca 1 mi. nw. of Camp Meeker, acc. J. & B. Guggolz (1991); Santa Rosa Jr. Coll. campus, *Baker* (1945 NCC); Alder Glen Springs (on old highway between Cloverdale and Hopland), *Eastwood & Howell* (1938). < 6600 ft. May-Sept.

PORTULACACEAE. PURSLANE FAMILY

1. Fruit 2-3-valved, dehiscing from tip
  - 2. Inflorescence an umbel or panicle of headlike clusters, dense; sepals in fruit usually round to reniform, ± scarious . . . . *Calyptridium quadripetalum*
  - 2: Inflorescence a cyme, raceme, panicle, or flowers solitary, usually open; sepals in fruit usually not round or reniform, usually not scarious
    - 3. Style 2-branched (or stigmas 2, sessile); petals fused at base, tube expanding in fruit, caplike; fruit 2-valved . . . . *Calyptridium umbellatum*
    - 3: Styles 3, occasionally fused at base; petals ± free, tube not expanding in fruit, not caplike, fruit usually 3-valved
      - 4. Cauline leaves alternate
        - 5. Inflorescence not secund; petals usually red; ovules many; seeds 6-40 per fruit . . *Calandrinia*
        - 5: Inflorescence usually secund; petals white to pink; ovules usually 3; seeds 1-3 per fruit

. . . . *Montia linearis*

4: Cauline leaves opposite

6. Leaves mostly basal; cauline leaves usually 2, free or fused . . . . *Claytonia*

6: Leaves ± gradually reduced upward; cauline leaves usually 4+, not fused . . *Montia fontana*

1: Fruit circumscissile

7. Sepals free, 2-8; ovary superior; fruit circumscissile near base . . . . *Lewisia*

7: Sepals fused at base, 2-lobed; ovary partly inferior; fruit circumscissile near middle . . . . *Portulaca*

Calandrinia

1. Branches mostly spreading-trailing; capsule enveloped by the fruiting calyx . . . . *C. ciliata*

1: Branches ascending; capsule nearly 2 × as long as the fruiting calyx . . . . *C. breweri*

Calandrinia breweri S. Watson. Uncommon, gravelly slopes, disturbed areas, burns: Ward Ranch in Kenwood, acc. Baker (1932); w. slope of Mt. Hood, *Baker* (1947 NCC). < 4000 ft. Mar.-June.

Calandrinia ciliata (Ruiz López & Pavón) de Candolle. RED MAIDS. Common, cultivated fields, open grassy places: The Sea Ranch; Horseshoe Cove, *Knight* (1963); Willow Creek; Bodega Head; Pitkin Marsh, *Rubtzoff* (1952); Todd Road Preserve; Fairfield Osborn (S); cemetery, Petaluma, *Ott* (1934 NCC); 997 ft. elev. hills near Sonoma, *Haney* (1972 PUA); Pickleberry Vineyard off Enterprise Road between Bennett Valley and Sonoma Mtn. Roads, *Coster* (1985 NCC); Sonoma Ave. n. Santa Rosa, *Mallory* (1920); Sugarloaf Ridge; Spring Lake; n. of Santa Rosa near Jr. Coll., *Eastwood & Howell* (1936); Pepperwood; 3 mi. s. of Healdsburg, *Heller & Brown* (1902); Lake Sonoma area; Little Geysers geothermal area, *Quibell* (1980 NCC). < 7300 ft. Feb.-May. —*C. c. var. menziesii* (Hooker) J. F. Macbride, currently not recognized.

Calyptridium

1. Plants annual, spreading or decumbent base 5-10 cm high; basal leaves oblong-spatulate; sepals round-reniform, white scarious and rose-tinged with greenish center; capsule oblong-oval, 10-20-seeded . . . *C. quadripetalum*

1: Plants biennial or perennial, erect or ascending, scapelike 8-36 cm high; leaves in dense rosette,spatulate, cauline leaves few; sepals orbicular, emarginate at base and apex, equal, wholly scarious or with mere greenish center, dull white or often pink; capsule globose-ovate, 1-2- seeded . . . . *C. umbellatum*

Calyptridium quadripetalum S. Watson. FOUR-PETALED PUSSYPAWS. Uncommon, open areas: Mayacmas Mts. at nw. end of Pine Flat Road, *Raiche* (1984); 3290 ft. elev. Socrates Mine, acc. Clifton (1984); The Geysers, *E. Greene* (1892), acc. E. Greene, *Flora Franciscana, p. 182* (1894). 1400-6700 ft. Apr.-June.

Calyptridium umbellatum (Torrey) E. Greene. PUSSYPAWS. Uncommon, loose sandy or gravelly areas: Sulphur Cr. en route to The Geysers, *Baker* (1932 NCC). < 14,000 ft. May-Aug.

Claytonia

1. Plants annual

2. Seeds dull black, ± tubercled at 20×, appendage brownish, fleshy or shrunken

3. Petals 5-8 mm; styles maturing well after anthers (cross-pollinated); inflorescence many-flowered, long-stalked above cauline leaf-pair . . . . *C. gypsophiloides*

3: Petals 2-5 mm; styles maturing ± with anthers (self-pollinated); inflorescence few-many flowered, stalk 0 to long

4. Cauline leaves free or fused and appearing 2-lobed, lobes 2-30 mm; petals 2-5 mm . . . . *C. exigua* ssp. *exigua*

4: Cauline leaves fused around stem into a round or squarish disk 3-15 mm wide; petals 2 mm . . . . *C. e.* ssp. *glauca*

2: Seeds shiny black, smooth at 20×, appendage white, fleshy

5. Basal leaves linear or blade > 3 × longer than wide, tapered gradually to petiole, cauline leaf pair ± disklike; petals 1-6 mm; (often weedy) . . . . *C. parviflora*

5: Basal leaf-blades < 3 × longer than wide (except seedlings), base abruptly tapered to cordate; petals < 5 mm

6. Basal leaves usually many, prostrate to spreading, smaller inward, blades widely elliptic to obovate, base wedge-shaped; cauline leaf-pair generally unequally fused; inflorescence ± sessile; only petioles often red; petals minute . . . . *C. rubra* ssp. *depressa*
6: Basal leaves few-many, spreading to erect, ± unequal, blades obtuse to acute, usually longer than wide, base wedge-shaped; cauline leaves generally equally fused, cauline leaf disk teeth usually obtuse or acute; petals minute to > sepals; (often weedy) . . . . *C. perfoliata*

1: Plants perennial . . . . *C. sibirica*

Claytonia exigua Torrey & A. Gray ssp. **exigua**. Uncommon, dry rocky or gravelly areas, sometimes on serpentine: Stoetz Lane near Occidental, *Best* (1984); 3060 ft. elev. Squaw Cr., acc. Clifton (1983). < 6700 ft. Apr.-June. —*C. tenuifolia* Torrey & A. Gray. *Montia e.* (Torrey & A. Gray) Jepson.

Claytonia exigua ssp. **glauca** (Torrey & A. Gray) John M. Miller & Chambers. Uncommon, roadcuts, various exposures: 2200 ft. elev. near well pad below geysers Unit 15, *Malloch* (1977 PG&E). < 6700 ft. Mar.-June. —*C. perfoliata* Willdenow var. *g.* Nuttall.

Claytonia gypsophiloides Fischer & C. Meyer. Common, moist loose soil, sometimes on serpentine: type locality: Fort Ross, acc. *Erythea 1:38*, (1893); Bohemian Hwy. 3.6 mi. nw. of Occidental, *Huntington* (1987 NCC); Graton Road 1 mi. e. of Occidental, *Best* (1982); hills near Sonoma, acc. Baker; region around Santa Rosa, *Baker* (1926 NCC); Pepperwood; Porter Cr. Road, *Stocking* (1963 NCC); 14 mi. sw. of Skaggs Springs, *Hastings* (1926 NCC); Lake Sonoma area; 780 ft. elev. Knights Valley, *Clifton* (1974 PUA); Geysers-Healdsburg Road, *Breslin* (1975 NCC); Geysers geothermal area, *Quibell* (1980 NCC); Geyser Peak, *Knight & Edwards* (1986); < 4300 ft. Mar.-May. —*Montia g.* (Fischer & C. Meyer) T. J. Howell.

Claytonia parviflora Hooker. Uncommon, rather dry shaded slopes: region of Monte Rio, acc. Baker (1932); 2 mi. up Adobe Can., acc. Baker (1932). Sugarloaf Ridge State Park, acc. Bowcutt (1993) < 7700 ft. Feb.-May. —*Montia perfoliata* Willdenow forma *parviflora* (Hooker) J. T. Howell.

Claytonia perfoliata Willdenow. MINER'S LETTUCE. Common, shaded vernally moist areas: The Sea Ranch; Salt Point; type locality, Fort Ross; 3.7 mi. w. of Cazadero on Fort Ross Road, *Quibell & Atkinson* (1980 NCC); Willow Creek; Bodega Head: Sebastopol quad.; 3 sites w. of Santa Rosa in Laguna de Santa Rosa drainage, *Harrison* (1978 NCC); Todd Road Preserve; Fairfield Osborn (*); outskirts of Sonoma, *Howell* (1946); Bouverie; Sugarloaf Ridge; Spring Lake; Pepperwood; Lake Sonoma area. 6700 ft. Feb.-May. —*Montia p.* (Willdenow) T. J. Howell var. *p.*

Claytonia rubra (T. J. Howell) Tidestrom ssp. **depressa** (A. Gray) John M. Miller & Chambers. Uncommon, with the species: Geysers geothermal area, *Quibell* (1980 NCC). < 8300 ft. Apr.-June. —*Montia perfoliata* var. *d.* (A. Gray) Jepson.

Claytonia sibirica Linnaeus. CANDY FLOWER. Common, moist shaded areas: The Sea Ranch; Gualala River near Mendocino-Sonoma co. line, *Abrams* (1900); ca 1 mi. e. of Hwy. 1 from Stewarts Point, *Stocking* (1969 NCC); Bodega Bay near Inn at the Tides, acc. Connors (1991); Spring Mtn. Road, *Okada* (1975 NCC). < 4300 ft. Mar.-Aug. —*Montia s.* (Linnaeus) T. J. Howell.

Lewisia rediviva Pursh. BITTERROOT. Uncommon, rocky areas, loose gravelly slopes, sometimes on serpentine: Arnold Drive (Hwy. 121) ca 3 mi. n. of Sears Point, acc. G. Smith (1984); near Sonoma, *Brewer* (JEPS 1862); Annadel Park, acc. Betty Guggolz (1984); Hammedin Ranch in Dry Cr. Valley, acc. Baker (1932); 3500 ft. elev. on serpentine 0.4 mi. n. of The Geysers, *Malloch* (1978 PG&E). 200-10000 ft. Mar.-June.

Montia

1. Plants prostrate to erect, sometimes floating, often matted; leaves opposite; petals 1-2 mm, white . *M. fontana*
1: Plants erect, often much branched, never floating; leaves all alternate; petals 4-6.5 mm, white or pale pink . . . . *M. linearis*

Montia fontana Linnaeus. WATER CHICKWEED, BLINKS. Occasional, pools, slopes: jct. Stewarts Point-Skaggs Springs and Annapolis roads, *Bacigalupi* (1948); Stoetz Lane 2 mi. ne. of Occidental, *Best* (1982); s. of Piner Road in Laguna de Santa Rosa drainage, *Harrison* (1978 NCC); Todd Road Preserve; Santa Rosa, *Eastwood* (1921); Pepperwood; geysers Unit 13, *Quibell* (1980 NCC); 2190 ft. elev. Squaw Cr., acc. Clifton (1984). < 10700 ft. Mar.-June. —*M. hallii* (A. Gray) E. Greene. *M. minor* Gmelin. *Claytonia h.* A. Gray.

Montia linearis (Hooker) E. Greene. Uncommon, moist meadows, banks: Santa Rosa, *Eastwood* (1921). < 7700 ft. Apr.-June. —*Claytonia l.* Douglas.

★ **Portulaca oleracea** Linnaeus. COMMON PURSLANE. Common, weed, gardens, waste places: ne. corner Magnolia and Marshall aves. w. of Petaluma, acc. Knight (1939); Spring Lake; Spencer Ave., Santa Rosa, *H. Thomas* (1934 NCC); Bambi Lane, Santa Rosa, *Wells* (1989); Pepperwood. < 5700 ft. May-Sept. Naturalized from Europe.

PRIMULACEAE. PRIMROSE FAMILY

1. Plants scapose, leaves all basal; flowers in bracteate umbels . . . . . . *Dodecatheon*
1: Plants with leafy stems; flowers not in umbels
  2. Leaves opposite or whorled; corolla evident if present
    3. Main leaves in single whorl at summit of stem . . . . . . *Trientalis*
    3: Main leaves distributed along the stems
      4. Corolla lacking, calyx petaloid; flowers solitary, axillary, sessile . . . . . . *Glaux*
      4: Corolla present, calyx not petaloid; flowers not solitary and sessile in axils . . . . . . *Anagallis*
  2. Leaves alternate (except the lowest); corolla shorter than calyx . . . . . . *Centunculus*

★ **Anagallis arvensis** Linnaeus. SCARLET PIMPERNEL, POOR MAN'S WEATHERGLASS. Common, weed: The Sea Ranch; Horseshoe Cove, *Knight* (1963); Salt Point; Fort Ross-Russian River-Bodega areas, *Voznesenskii* (1840-1841 LE); Willow Creek Marsh, *Rubtzoff* (1959); Bodega Head; bluff above Bodega Bay, *Cantelow* (1936); Russian River at Vacation Beach, *Rubtzoff* (1956); Guerneville Marsh, *Rubtzoff* (1954); Russian River at Forest Hills, *Rubtzoff* (1974); Forestville Marsh, *Rubtzoff* (1955); Pitkin Marsh, *Rubtzoff* (1952); Laguna de Santa Rosa e. of Cunningham, *Rubtzoff* (1958); Todd Road Preserve; Copeland Cr., Rohnert Park, *Self* (1979 NCC); Stony Point Marsh, *Rubtzoff* (1955); ne. of Penngrove on Lichau Road, *Rubtzoff* (1969); Spring Lake; sw. end of Santa Rosa Jr. Coll. campus, *Hagg & Schmidt* (1932 NCC); Fairfield Osborn (*); Tresch Ranch, Walker Road near Two Rock, acc. J. & B. Guggolz (1992); hills e. of Agua Caliente, *Howell* (1948); Bouverie; Pepperwood; chaparral area sw. of Bradford Mtn. 1½ mi. w. of Dry Cr., acc. J. & B. Guggolz (1992); Lake Sonoma area; Ida Clayton Road quad.; ca 2000 ft. elev. Anderson Trail on Black Mtn. between Geyser Peak and upper end of Pine Flat Road, acc. J. & B. Guggolz (1992); between Jimtown and Mercuryville on Geysers Road, *Utech & Knight* (1986 CM). < 3300 ft. Mar.-July. Naturalized from Europe.

Centunculus minimus Linnaeus. CHAFFWEED. Occasional, moist places: Salt Point; Pitkin Marsh, *Rubtzoff* (1952); Laguna de Santa Rosa ne. of Graton, *Rubtzoff* (1960); Atascadero Cr. Marsh w. of Sebastopol, *Rubtzoff* (1972); Todd Road 1.4 mi. from Hwy 116, *Rubtzoff* (1974); 6½ mi. ne. of Penngrove on Lichau Road, *Rubtzoff* (1971); Macartney Woods in Adobe Canyon, *Baker* (1927 NCC); 4¼ mi. w. of Glen Ellen on Bennett Valley Road, *Rubtzoff* (1971); Bennett Mtn. Lake, *Rubtzoff* (1963); Pepperwood. < 3100 ft. Apr.-July. Cosmopolitan. —*Anagallis minimus* (Linnaeus) Krause.

Dodecatheon hendersonii A. Gray ssp. **hendersonii**. SHOOTING STAR. Common, grassland; shady areas: Bodega Bay; Camp Meeker, *Howell* (1924); Sonoma Mtn. Road, *Knight* (1964); Fairfield Osborn (S); Tresch Ranch, Walker Road near Two Rock, acc. J. & B. Guggolz (1992); Petaluma cemetery, *Ott* (1934 NCC); Bouverie; jct. of Trinity and Cavedale roads, *Chisani* (1957 NCC); Sugarloaf Ridge; Los Guilicos, *Stocking* (1960 NCC); Bennett Valley Road ½ mi. e. of Farmers Lane, *Parsons* (1983 SFSU); Spring Lake; Pepperwood; Lake Sonoma area; Pine Flat Road, *Baker* (1934 NCC); Geyser Peak, acc. Knight & Edwards (1985); geysers Unit 13, *Quibell* (1981 NCC). Ssp. **cruciatum** (E. Greene) H. J. Thompson, a 4-merous form, usually currently not recognized: Hidden Acres Road w. of Sonoma Mtn. Road, acc. Quibell (1991). < 6700 ft. Feb.-May.

Glaux maritima Linnaeus. SEA-MILKWORT. Rare, salt marshes: Petaluma, *Davy* (1893 UC). < 100 ft. May-July.

Trientalis latifolia Hooker. STARFLOWER. Common, shady places: The Sea Ranch, *Wells* (1981); Salt Point; Bodega Head; s. of Guerneville, *Rubtzoff* (1953); Roberts Road 3.8 mi. from Petaluma Hill Road, *D. Carlson* (1978 NCC); Bouverie; Sugarloaf Ridge; Pepperwood; 997 ft. elev. Franz Valley School Road, *Klusman* (1970 PUA); W. Dry Cr. Road 5 mi. nw. of Healdsburg, *Reickert* (1960 NCC); Lake Sonoma area; 350-400 ft. elev. Old Skaggs Springs Road, *Utech, Sato & Ohara* (1990 GM); between Jimtown and Mercuryville on Geysers Road, *Utech & Knight* (1986 CM); Geyser Peak, acc. Knight & Edwards (1986). —*T. borealis* Rafinesque ssp. *l.* (Hooker) Hultén.

RANUNCULACEAE. CROWFOOT FAMILY

1. Flowers not subtended by a finely dissected involucre
  2. Flowers irregular; upper sepal spurred . . . . *Delphinium*
  2: Flowers regular
    3. Sepal spurs 1-3 mm, plants scapose with simple narrow basal leaves . . . . *Myosurus*
    3: Sepal spurs none; leaves usually dissected or compound (*Caltha* and some *Ranunculus* simple)
      4. Pistil 1; fruit a berry, leaves 1-3 ternate . . . . *Actaea*
      4: Pistils 2-many; fruit an achene or follicle
        5. Ovary with 2 or more ovules; fruit a follicle
          6. Petals spurred; leaves deeply 3 lobed . . . . *Aquilegia*
          6: Petals none; sepals petal-like; leaves simple . . . . *Caltha*
        5: Ovary with 1 ovule; fruit an achene
          7. Cauline leaves opposite or whorled
            8. Leaves opposite; plant a woody vine . . . . *Clematis*
            8: Leaves whorled; plant herbaceous . . . . *Anemone*
          7: Cauline leaves alternate or none
            9. Flowers perfect, petals yellow or white; anthers yellow, leaves entire to compound . . . . *Ranunculus*
            9: Flowers unisexual; plants dioecious, petals none; sepals greenish white to purple, leaves ternately decompound . . . . *Thalistrum*
1: Flowers subtended by a finely dissected involucre . . . . *Nigella*

Actaea rubra (Aiton) Willdenow. BANEBERRY. Occasional, moist woods: Annapolis quad.; near Cazadero, *Charles* (1925); Camp Meeker at Grub Cr., *Harrison* (1977 SRJC); bottom of Harrison Grade Road under bay trees, acc. Quibell (1991); Pocket Canyon between Forestville and Guerneville, *Gilbert* (1980 SRJC); Pepperwood; 2300 ft. elev. near geysers Unit 18, *Malloch* (1977 PG&E). < 9300 ft. May-June. —*A. spicata* Linnaeus subsp *arguta* (Torrey & A. Gray) E. Murray. *A. r.* ssp. *a.* (Nuttall) Hultén, currently not recognized.

Anemone oregana A. Gray. WINDFLOWER. Occasional, moist wooded areas: Fort Ross, *Davy* (1890s JEPS); Willow Creek; Cooper's Grove on Sonoma Mtn. Road, *Baker* (1936 NCC); ca half-way up Trinity Road past the quarry, acc. Knight (1964); Annadel; Lake Sonoma area. < 2000 ft. Apr.-May. —*A. quinquefolia* Linnaeus ssp. *grayi* (Behr & Kellogg) E. Murray.

Aquileqia

1. Petal spur ± straight to ± incurved, mouth ± 90° to floral axis, ± round; petal blade 1-8 mm . *A. formosa*
1: Petal spur ± outcurved, mouth > 90° to floral axis, elliptic to triangular; petal blade 0 . . . . . . *A. eximia*

Aquilegia eximia Planchon. Uncommon, very damp areas: Adobe Can., *Baker* (1921 NCC); 3160 ft. elev. Socrates Mine, acc. Clifton (1984). < 6000 ft. May-Aug. —*A. tracyi* Jepson. *A. fontinalis* J. T. Howell.

Aquilegia formosa Fischer. COLUMBINE. Common, damp areas: The Sea Ranch, *Wells* (1981); Willow Creek; 2 mi. w. of Bodega, *Chenosorth* (1964 NCC); 5 mi. w. of Joy Road on Coleman Valley Road, *Hannan* (1981 NCC); Freestone, *Dudik* (1975 NCC); near Cooper's Grove on Sonoma Mtn. Road, *Utech & Knight* (1986 CM); Pressley Road just w. of jct. with Sonoma Mtn. Road, acc. Quibell (1991); Fairfield Osborn (S); Tresch Ranch, Walker Road near Two Rock, acc. J. & B. Guggolz (1992); 6 mi. w. of

Petaluma near Rainsville Road, *Ott* (1934 NCC); Bouverie; Pepperwood; 997 ft. elev. Franz Valley School Road, *Klusman* (1970 PUA); Felta Cr. area sw. of Healdsburg, *Bice* (1895 NCC); 350-400 ft. elev. Old Skaggs Springs Road, *Utech, Sato & Ohara* (1990 CM); Lake Sonoma area. < 11000 ft. June-Aug. —*A. truncata* Fischer & C. Meyer, type locality: Fort Ross. *A. f.* var. *t.* (Fischer & C. Meyer) Baker, currently not recognized.

★ **Caltha palustris** Linnaeus. MARSH MARIGOLD. Infrequent, marshy areas: Forestville Marsh, *Rubtzoff* (1955). < 500 ft. Native in n. and e. N. America and arctic and temperate Eurasia.

Clematis

1. Leaflets 5-7; inflorescence cymose-paniculate, usually many-flowered . . . . *C. ligusticifolia*
1: Leaflets 3; inflorescence 1-3-flowered . . . . *C. lasiantha*

Clematis lasiantha Nuttall. PIPESTEMS. Common, scandent, near streams, canyons: Bouverie; lower Cavedale Road at monument, acc. Quibell (1991); Santa Rosa Cr. Can., *Gossin* (1933 NCC); end of Fern Glenn Drive, Santa Rosa, *Dudik* (1975 NCC); Pepperwood; Lake Sonoma area; 2700 ft. elev. w. flank Mt. St. Helena, *Grummer* (1979 PUA); Geyser Peak, *Knight & Edwards* (1986); 2500 ft. elev. canyonside near Cloverdale, *Acorne* (1968 NCC). < 6600 ft. Mar.-June.

Clematis ligusticifolia Nuttall. VIRGIN'S BOWER. Occasional, scandent upon adjacent woody vegetation, moist places: Russian River at Vacation Beach, *Rubtzoff* (1962); Forest Hills across Russian River from Hacienda, *Rubtzoff* (1962); along Santa Rosa Cr., e. Santa Rosa, *Wagnon* (1989 NCC); Russian River, *Meyer* (1976 SRJC), *Harrison* (1989 SRJC); Lake Sonoma area; ca 2000 ft. elev. Anderson Trail on Black Mtn. between Geyser Peak and upper end of Pine Flat Road, acc. J. & B. Guggolz (1992). < 8000 ft. Mar.-Aug.

Delphinium

1. Sepals ± red or yellow; angle between claw and blade of lower petals 140-180°, blade hairs 0 or obscure to naked eye
   2. Sepals bright yellow, not red . . . . *D. luteum*
   2: Sepals scarlet to orange-red, rarely yellow but not brightly so . . . . *D. nudicaule*
1: Sepals not red or yellow (sometimes maroon or pink); blade of lower petal angled 60-130° to claw, its hairs apparent to naked eye
   3. Plant with all the following: inflorescence ± secund; flowers usually on and after mid-June; stem usually < 40 cm;pedicels ascending; seed winged, without inflated collar at widest end; root < 5 cm . . *D. depauperatum*
   3: Plant with 2 or more exceptions to the characters in 3
      4. Stem-base usually much narrower than weakly attached root; seed with inflated collar at widest end, unwinged; inflorescence in mid-flower usually with < 6 flowers on main axis or oldest open-flowered pedicels usually > 25 mm and > 2 × youngest; fruit curved
         5. Sepals ± maroon; plants within populations often dissimilar, often ± infertile . . . . *D. nudicaule × decorum*
         5: Sepals not maroon; plants within populations usually similar, usually ± fertile
            6. Most leaves on upper 1/3 of stem; stem usually > 50 cm; flowers on main inflorescence axis usually > 15 . . . . *D. bakeri*
            6: Most leaves on lower 1/3 of stem; stem usually < 50 cm; flowers on main inflorescence axis often < 20
               7. Sepals dark blue-purple (often faded, mottled on herbarium specimens), puberulent outside, usually not reflexed; lower stem hairy; lower petal blade often > 6 mm . . . *D. decorum*
               7: Sepals usually bright blue to white or pink (usually not faded, not mottled on herbarium specimens), mostly glabrous, reflexed (if sepals purplish and puberulent outside, then lower stem ± glabrous or lower petal blade< 6 mm) . . . . *D. patens*
      4: Stem-base usually ± as wide as adjacent root, attachment firm; seed without inflated collar at widest end, winged or not; inflorescence in mid-flower with usually > 6 flowers on main axis or oldest open- flowered pedicel usually < 25 mm or < 2 × youngest; fruit ± straight
         8. Root > 15 cm, usually with enlarged, whitish (brown on herbarium specimens) buds; stems usually > 1 m, often 2 or more per root system . . . . *D. californicum*
         8: Root < 15 cm (if larger, then enlarged buds 0); stems usually < 1 m, usually 1 per root system

9. Plants with some hairs of lower petioles > 0.5 mm, white, straight, spreading; seeds unwinged . . . . *D. variegatum*
9: Plants with hairs of lower petioles 0-0.5 mm, straight, curved, or both; seeds winged . . . . *D. hesperium*

Delphinium bakeri Ewan. BAKER'S LARKSPUR. Rare, low brush: Coleman Valley, *Baker* (type 1939 COLO), *Lohman* (1946); Hedrin Ranch in Coleman Valley, acc. Etienne & Niehaus (1977). 300-1000 ft. Apr.-May.

Delphinium californicum Torrey & A. Gray. Uncommon, moist, partly shaded places, near thickets: Salmon Cr. Road, *J. & B. Guggolz* (1993 NCC). < 3300 ft. May-June.

Delphinium decorum Fischer & C. Meyer. YELLOW-TINGE LARKSPUR. Common, open grassy areas: The Sea Ranch, *Wells* (1981); Stewarts Point, *Baker* (1932 NCC); Bodega Head; Larkspur Rock e. of Bodega Bay, *Howell* (1931), *Ott* (1934 NCC); Sugarloaf Ridge. Type locality: Fort Ross, Port Bodega. Apparent hybrid: *D. decorum* × *D. nudicaule*; see master's thesis of Ed Guerrant at Sonoma State Univ. < 7600 ft. Mar.-May.

Delphinium depauperatum Torrey & A. Gray. Uncommon, edge of woods, damp thickets: Ida Clayton Road quad. < 8300 ft. May-July.

Delphinium hesperium A. Gray. WESTERN LARKSPUR, COAST LARKSPUR. Common, grassy areas: n. of Jenner, *Ripley & Barneby* (1945); Cazadero, *Graff* (1925); 1 mi. on Joy Road from jct. of Bodega Hwy., *Ferris & Lorraine* (1961); Hwy. 116 2 mi. w. of Forestville, *Rubtzoff* (1974); (1940); Calistoga Road, *Baker* (1935 NCC); Santa Rosa, *Heller* (1902); near Petrified Forest, *Eastwood* (1915); Felta Cr. area sw. of Healdsburg, *Bice* (1895 NCC); Healdsburg, *G. Hofmann* (1981 NCC); 5 mi. e. of Geyserville, *Cantelow* (1938); Lake Sonoma area. < 3700 ft. May-June.

Delphinium luteum A. A. Heller. YELLOW LARKSPUR, GOLDEN LARKSPUR. Rare, openings near brushy slopes; only 2 extant populations left, < 60 plants remaining: near Bodega Bay on road to Bodega, *Heller* (type 1902 UC), *Eastwood* (1917); *Baker* (1926), *Heller* (1940); Larkspur Rock e. of Bodega Bay, *Baker* (1930 NCC), *Howell* (1931); Bodega, *L. Burbank* (1917); site between Valley Ford and Bodega Bay on Hwy. 1, *J. & B. Guggolz* (1987 NCC); 2 mi. w. of Petaluma on D St. Ext. toward Point Reyes Sta., *Good* (1962 HSC). Near Walker Cr. in Marin Co. there are hybrids involving this species with *D. nudicaule* and possibly *D. decorum* showing a variety of colors, sometimes yellow, acc. Quibell (1991). < 200 ft. Mar.-May.

Delphinium nudicaule Torrey & A. Gray. RED LARKSPUR. Common, dry areas, woods, among shrubs: The Sea Ranch, acc. Wells (1985); Salt Point; Bohemian Hwy. 3 mi. n. of Occidental, *Best* (1981); Bohemian Hwy. 2 mi. nw. of Dutch Bill Cr., *Stocking* (1964 NCC); 0.8 mi. from jct. of Armstrong Woods Road and Sweetwater Springs Road, *Cannon* (1980 NCC); Crane Can., Stocking (1963 NCC); 1 mi. from top of Sonoma Mtn., *Lowry* (1983 SFSU); Fairfield Osborn (S); Bouverie; Sugarloaf Ridge; Enterprise Road between Bennett Valley and Sonoma Mtn. roads, acc. Quibell (1991); Gates Road off Calistoga Road, acc. Wells et al. (1984); Felta Cr. area sw. of Healdsburg, *Bice* (1895 NCC); 4 mi. w. on Dry Cr. Road, *Juul* (1973 NCC); Pepperwood; 350-400 ft. elev. Old Skaggs Springs Road, *Utech, Sato & Ohara* (1990 CM); Lake Sonoma area;Geyser Peak, acc. Knight & Edwards (1986). < 8500 ft. Mar.-June.

Delphinium patens Bentham. SPREADING LARKSPUR. Occasional, thicket and woods borders: Tresch Ranch, Walker Road near Two Rock, *J. & B. Guggolz* (1992 JEPS); e. of Spring Lake Park, *Lovell* (1982 JEPS); Mark West Springs Road, *Baker* (1927 NCC), *Christensen* (1929 NCC); near Mark West Springs, *Dudley* (1902); Pepperwood. <3700 ft. Mar.-May. —*D. decorum* Fischer & C. Meyer var. *sonomensis* Eastwood, type locality: *Eastwood*, Altruria — a communal area near Mark West Springs in the 1890s, (1900).

Delphinium variegatum Torrey & A. Gray ssp. **variegatum**. ROYAL LARKSPUR. Common, open grassy hills, woods: Todd Road Preserve; Fairfield Osborn (H); s. of Schellville, *Cantelow* (1940); 1 mi. s. of El Verano, *Baker* (NCC), *Howell* (1941); 1½ mi. up Cavedale Road from Hwy. 12, *Utech & Knight* (1986 CM); foot of Mt. Hood, *Beals* (1930 NCC); Bennett Valley, *Baker* (1928), *Beals* (1930); e. of Santa Rosa, *Heller & Brown* (1902); Santa Rosa, *K. Brandegee* (UC); white flowers, e. of Spring Lake Park, *Lovell* (1982 JEPS); Wilfred Ave., Santa Rosa, *Broderson* (1985 NCC); 4 mi. w. of Mark West Springs, *Eastwood & Howell* (1938); Pepperwood. Within the morphological range of variation, some of these might apply: ssp. **apiculatum** (E. Greene) Ewan forma **apiculatum** designates hairier plants. Ssp. **apiculatum** forma **emiliae** (E. Greene) Ewan has few leaves and sepals 8-10 mm long incurved at the tips: Los

Guilicos, *Ewan* (1935 UC). Ssp. **apiculatum** forma **superbum** Ewan has large flowers. < 2700 ft. Apr.-May. —*D. emiliae* E. Greene, type locality: Emily Gibbons Booth Ranch in Knights Valley, *E. Greene* (1894 U).

Myosurus minimus Linnaeus. MOUSE-TAIL. Rare, moist places such as vernal pools: Occidental Road Marsh near jct. of High School Rd., *Harrison & Quibell* (1978 NCC). < 5000 ft. Mar.-June.

★ **Nigella damascena** Linnaeus. LOVE IN A MIST. Uncommon, garden plant: spreading along Calder Ave. parking strip in Sebastopol, *Ruddick* (1989 NCC). Native of s. Europe.

Ranunculus

- 1. Petals white, aquatics with submerged leaves dissected; achenes transversely ridged
  - 2. Receptacle glabrous; style 2-3 × longer than ovary at anthesis . . . . . *R. lobbii*
  - 2: Receptacle short-haired; style not longer than ovary at anthesis . . . . . *R. aquatilis*
- 1: Petals yellow; plants mostly terrestrial; achenes not transversely ridged
  - 3. Plants annual
    - 4. Leaves shallowly toothed or entire; stems reclining, rooting at lower nodes . . . . . *R. pusillus*
    - 4: Leaves 3 (or 3-5) -foliolate or -parted; stems ascending, usually not rooting at nodes
      - 5. Achenes covered with spines, hooked hairs or papillae
        - 6. Petals 5-8 mm. long; achenes 5-6 mm long
          - 7. Achenes 10-20, not spiny on margins; stems glabrous . . . . . *R. muricatus*
          - 7: Achenes 5, with long spines on margins; stems thinly hirsute . . . . . *R. arvensis*
        - 6: Petals 1-3 mm long (or none); achenes 1.5-2 mm long
          - 8. Petals generally 0, 1 or 2; achene-papillae with hooked bristles . . . . . *R. hebecarpus*
          - 8: Petals 5; achene-papillae terminating in hooked spines . . . . . *R. parviflorus*
      - 5: Achenes smooth, not spiny or papillate . . . . . *R. uncinatus*
  - 3: Plants perennial
    - 9. Plants, at least the stems, obviously pubescent or hirsute
      - 10. Plants with runners, rooting at nodes . . . . . *R. repens*
      - 10: Plants without runners
        - 11. Receptacle pubescent or hispid
          - 12. Achenes smooth, beaks straight, ca 4 mm long
            - 13. Leaves hairy, leaflets ovate to cuneate . . *R. orthorhynchus* var. *orthorhynchus*
            - 13: Leaves glabrous, leaflets ovate to ± round . . . . . *R. o.* var. *bloomeri*
          - 12: Achenes generally papillate, beaks curved at tip, 0.3 mm long . . . . . *R. sardous*
        - 11: Receptacle glabrous
          - 14. Petals mostly 5-6, mostly 1-2 × as long as broad . . . . . *R. occidentalis*
          - 14: Petals mostly 8-15, mostly 2-3 × as long as broad . . . . . *R. californicus*
    - 9: Plants glabrous or nearly so . . . . . *R. flammula*

Ranunculus aquatilis Linnaeus. WATER BUTTERCUP. Uncommon, slow streams, ponds, ditches: Duncans Mills Marsh, *Rubtzoff* (1964); Bodega Bay Marsh, *Rubtzoff* (1962); 2 mi. n. on Adobe Road from jct. of Sonoma Mtn. and Adobe roads, *Lowry* (1983); Button Ranch on Walker Road (U. C. property) between Petaluma-Valley Ford Road and Pepper Road, acc. Connors (1992); Boyes Hot Springs, *Rattan* (1884); Bennett Mtn. Lake, *Rubtzoff* (1962); Spring Lake; Pepperwood; near Windsor, *Heller & Brown* (1902). < 9600 ft. Apr.-July.

★ **Ranunculus arvensis** Linnaeus. HUNGER WEED. Uncommon, dry bed of vernal pool: n. of Hwy. 12 in Laguna de Santa Rosa drainage, *Harrison* (1978 NCC); Petaluma Hill Road at first creek crossing n. of Rohnert Park Expressway e. side of road, acc. Quibell (1985); Bennett Valley Road 4¼ mi. w. of Glen Ellen, *Rubtzoff & Froebe* (1971). < 3300 ft. May-June. Native of Europe.

Ranunculus californicus Bentham. CALIFORNIA BUTTERCUP. Common, vernally moist meadows, slopes: The Sea Ranch; s. part of The Sea Ranch, *J. H. Thomas* (1976); Rocky Point a little s. of Gualala, *Knight* (1964); Hwy. 1 near Plantation turnoff, *Knight* (1964); Cazadero, *Bacigalupi* (1926); near Trenton on River Road, *Heller* (1902); Freestone Marsh, *Rubtzoff* (1962); Duncan's Landing, *Knight & Edwards* (1986); Bodega Bay, *Heller* (1902); 2.3 mi. n. of Bodega Bay on Hwy. 1, *Cantelow* (1936); Bodega Head; Laguna de Santa Rosa e. of Sebastopol, *Rubtzoff* (1966); se. of Sebastopol, *Rubtzoff* (1966); e. of Cunningham, *Rubtzoff* (1966); Todd Road Preserve; Pressley Road off Roberts Road which is off Petaluma Hill Road, *Knight* (1964); Roberts Road, *Stocking* (1963 NCC); ¼ mile s. of Sonoma State Univ., *Juul* (1973 NCC); Fairfield Osborn (*); 5 mi. e. of Valley Ford on Valley Ford-Petaluma Hwy., *Quibell* (1970 NCC); Button Ranch on Walker Road (U. C. property) between Petaluma-Valley Ford Road and Pepper Road, acc. Patterson (1990); Stony Point Marsh, *Rubtzoff* (1955); Upper Stemple Cr. Marsh, *Peñalosa* (1961); cemetery w. of Petaluma, *Ott* (1934 NCC); corner La Cresta and Summit drives in Petaluma, *Knight* (1964); Petaluma, *M. Jones* (1924); I St. Ext. ca 1½ mi. from San Antonio Road s. of Petaluma, *Knight* (1964); near Boyes Hot Springs 2 mi. n. of Sonoma, *Howell* (1931); Bouverie; Trinity Road, *Kawahara* (1968); Sugarloaf Ridge; Bennett Mtn. Lake, *Rubtzoff* (1962); e. end of Sonoma Ave., Santa Rosa, *Mallory* (1920); near Santa Rosa, *Heller & Brown* (1902); Santa Rosa, *Eastwood* (1921); Tombs Cr. quad.; near Windsor, *Heller & Brown* (1902); Independence School near Healdsburg, *Abrams* (1918). (Some populations from which Rubtzoff collected specimens on Lichau Road in the 1960s almost resemble *R. canus* Bentham as body of achene approaches 5 mm. and there are sometimes less than 8 petals.) < 8000 ft. Feb.-May. —*R. c.* var. *cuneatus* E. Greene and *R. c.* var. *gratus* Jepson, currently not recognized.

Ranunculus flammula Linnaeus. CROWFOOT. CREEPING SPEARWORT. Uncommon, marshy muddy places: The Sea Ranch. < 7700 ft. July-Aug. —*R. f.* var. *ovalis* (Bigelow) L. Benson, currently not recognized.

Ranunculus hebecarpus Hooker & Arnott. DOWNY BUTTERCUP. Uncommon, shady places: The Sea Ranch, acc. *Mahaffey* (1990); Bouverie. < 4300 ft. Mar.-May.

Ranunculus lobbii (Hiern) A. Gray. LOBB'S AQUATIC BUTTERCUP. Common, floating in shallow vernal pools: Cazadero, *Davy* (1890s JEPS); Upper Stemple Cr. Marsh, *Peñalosa* (1961); pond in Americana Cr. area near Petaluma-Valley Ford Road s. of Bloomfield, *Rubtzoff* (1962); 3 sites w. and sw. of Santa Rosa in Laguna de Santa Rosa drainage, *Harrison* (1978 NCC); near Todd Road between Hwy. 116 and Llano Road, *Rubtzoff* (1972); Todd Road Preserve; Sonoma State Univ. campus, *McNally* (1970 NCC); along sewer line at Cotati, *Harrison* (1989 SRJC); Game Refuge Road off Petaluma Hill Road, *Koch* (1938 NCC); Fairfield Osborn (*); 2 mi. n. on Adobe Road from jct. of Sonoma Mtn. Road, *Lowry* (1983 SFSU); Sonoma Valley Regional Park vernal pool, *Barnhart* (1989 SRJC); ponds on e. side of Taylor Mtn. s. of Santa Rosa, *Quibell* (1981 NCC); 850 ft. elev. Shiloh Ranch 2.3 airmiles ne. of intersection of Shiloh and Faught roads in valley of small tributary in Mark West Cr. drainage, acc. Stone (1985); near Windsor, *Heller & Brown* (1902). < 1000 ft. Feb.-Apr.

★ **Ranunculus muricatus** Linnaeus. PRICKLE-FRUITED BUTTERCUP. Common, moist places: Cazadero, *Bacigalupi* (1926); Duncans Mills & Occidental marshes, *Rubtzoff* (1957); Guerneville Marsh, *Rubtzoff* (1955); Pitkin Marsh, *Rubtzoff* (1952); Forestville Marsh, *Rubtzoff* (1955); Bodega Head Marsh, *Rubtzoff* (1960); upper Stemple Cr. Marsh, *Rubtzoff* (1961); Forest Hills across Russian River from Hacienda, *Rubtzoff & Harrison* (1970); Hessel Road w. of Hwy. 116, *Rubtzoff* (1966); 6½ mi. ne. of Penngrove, *Rubtzoff* (1969); Todd Road Preserve; Crane Cr. Reg. Park e. of Petaluma Hill Road on Pressley Road, acc. Quibell (1991); Lichau Road, *Stocking* (1964 NCC); Fairfield Osborn (*); 2 mi. n. of jct. of Sonoma Mtn. and Adobe roads on Adobe Road, *Lowry* (1983 SFSU); Tresch Ranch, Walker Road near Two Rock, acc. J. & B. Guggolz (1992); Stony Point Marsh, *Rubtzoff* (1955); Petaluma, *Eastwood* (1921); D St. Ext. 2 mi. ne. of jct. with San Antonio Road s. of Petaluma, *Quibell* (1972 NCC); Petaluma, *Eastwood* (1921), *Johnson* (1967); D St. s. of Petaluma, *Lamb* (1970 NCC); between Jess Ave. and US 101, Petaluma, *M. P. Johnson* (1967); w. of Sears Point Raceway in Tolay Cr. drainage, acc. Patterson (1990); near Sonoma, *Heller & Brown* (1902); Bouverie; Sonoma Valley; Sugarloaf Ridge; Spring Lake; Pepperwood; Lake Sonoma area. < 2300 ft. May-June. Native of Old World.

Ranunculus occidentalis Nuttall. WESTERN BUTTERCUP. Common, meadows, flats, open woodland: upper Green Valley Marsh nw. of Graton, *Rubtzoff* (1974); Pepperwood; Lake Sonoma area; Skaggs Springs, *Hawver* (1915); Alexander Valley quad.; between Jimtown and Mercuryville on Geysers Road, *Utech & Knight* (1986 CM); ca 2000 ft. elev. Anderson Trail on Black Mtn. between Geyser Peak and upper end of Pine Flat Road, acc. J. & B. Guggolz (1992). < 7300 ft. Mar.-May.

Ranunculus orthorhynchus Hooker var. **bloomeri** (S. Watson) L. Benson. BLOOMER'S BUTTERCUP. Common, meadows: a specimen labelled "Sonoma County," *Congdon* (1880); The Sea Ranch, acc. *Mahaffey* 1990; Bodega Point, Eastwood (1917); 4 mi. se. of Bodega Bay, *Hagg* (1934 NCC); Pitkin Marsh, *Rubtzoff* (1951); Griffith Woods on Laguna Road near Vine Hill Dist., *McDonnell* (1929 NCC); 2½ mi. from Trenton in Russian River area, *C. Clark* (1930 NCC); Occidental Marsh, *Rubtzoff* (1957); Freestone Marsh, *Rubtzoff* (1958); Atascadero Cr. Marsh, *Rubtzoff* (1958); upper Stemple Cr. Marsh, *Rubtzoff* (1961); Cunningham Marsh, *Howell* (1947); Hwy. 116 ½ mi. nw. of Stony Point Road, *Rubtzoff*

(1966); Hessel Road w. of Hwy. 116, *Rubtzoff* (1966); Todd Road Preserve; Lakeville, *Howell* (1934); Sugarloaf Ridge; Pepperwood; Knights Valley, *Wiggins* (1932); near Windsor, *Heller & Brown* (1902); Alexander Valley, *Cantelow* (1937); 7½ mi. w. of Calistoga, *Cantelow* (1937). < 1500 ft. Mar.-May.

Ranunculus orthorhynchus var. **orthorhynchus** A. Gray. Uncommon, meadows: Bodega Bay, *Heller & Brown* (1902); Freestone Marsh, *Rose* (1962); between Calistoga and Knights Valley, *Wiggins* (1932). < 7700 ft. May-July. —*R. o.* var. *platyphyllus* A. Gray.

★ **Ranunculus parviflorus** Linnaeus. Uncommon, flower beds, waste places: Willow Cr. Road 6 mi. w. of Occidental, *Best* (1981); Spring Lake Village. < 4000 ft. June-Aug. Native of Old World.

Ranunculus pusillus Poiret. Common, marshy areas, shallow water: Guerneville Marsh, *Rubtzoff* (1962); Pitkin Marsh, *Howell* (1945, 1951); Atascadero Cr. Marsh, *Rubtzoff* (1972); all over the Laguna de Santa Rosa in vernal pools in early spring, often weedily profuse, *Harrison* (1979 NCC); Fairfield Osborn (H); Sonoma Mtn., *Baker & Griffin* (1925); 6½ mi. ne. of Penngrove on Lichau Road, *Rubtzoff* (1969); Chileno Valley near county line, *Howell* (1947); near Wright School between Santa Rosa and Sebastopol, *Baker* (1934 NCC); Bennett Mtn. Lake, *Rubtzoff* (1963); Spring Lake; Old Redwood Hwy. 2 mi. nw. of Windsor exit from Hwy. 101, *de Nevers* (1981). < 1700 ft. Apr.-May. —*R. biolettii* E. Greene, type locality: Mt. Hood, *Bioletti* (1889 UC).

★ **Ranunculus repens** Linnaeus. CREEPING BUTTERCUP. Occasional, lawns, pastures, low places: The Sea Ranch; Willow Creek; Guerneville Marsh, *Rubtzoff* (1954); Pitkin Marsh, *Rubtzoff* (1951); Griffith Woods on Laguna Road near Vine Hill Dist., *C. Clark* (1930 NCC). < 700 ft. May-Aug. Native of Old World.

★ **Ranunculus sardous** Crantz. HAIRY BUTTERCUP. Uncommon, waste places: sw. corner Occidental Road at High School Road near Sebastopol, *Harrison* (1985 SRJC). < 1000 ft. May-July. Native of Europe. —*R. parvulus* Linnaeus.

Ranunculus uncinatus D. Don. Infrequent, moist shady places: The Sea Ranch, *Wells* (1981); Duncans Mills Marsh, *Rubtzoff* (1955); Austin Cr. Marsh between Hwy. 116 and Cazadero, *Rubtzoff* (1955); Lake Sonoma quad.; Geysers geothermal area, *Quibell* (1980 NCC). < 8000 ft. May-July. Plants with fruiting body stiff-hairy, especially on margin, have been called var. **parviflorus** (Torrey) L. Benson: Duncans Mills Marsh, *Rubtzoff* (1955); Austin Cr. Marsh between Hwy. 116 and Cazadero, *Rubtzoff* (1955); Lake Sonoma quad. < 2000 ft. May-July.

Thalictrum fendleri A. Gray var. **polycarpum** Torrey. MEADOW-RUE. Common, canyons, shady slopes: Willow Creek Marsh, *Rubtzoff* (1959); Austin Cr. Marsh between Hwy. 116 and Cazadero, *Rubtzoff* (1960); Sonoma Mtn. Road just s. of Hidden Acres Road and Peracca Road at creek crossing e. of Grange Road, acc. Quibell (1991); 6 mi. w. of Petaluma near Rainsville Road, *Ott* (1934 NCC); Enterprise Road between Bennett Valley Road and Sonoma Mtn. Road, acc. Quibell (1985); Spring Mtn. Road, *L. Roof* (1933 NCC); Pepperwood; near Windsor, *Heller & Brown* (1902); Dry Cr. w. of Healdsburg, *Cantelow* (1934); Lake Sonoma area. < 5000 ft. Mar.-June. —*T. p.* (Torrey) S. Watson.

RESEDACEAE. MIGNONETTE FAMILY

Reseda luteola Linnaeus. DYER'S ROCKET. Occasional, adventive, waste places: Russian Gulch drainage n. of Jenner, *Harrison* (1988 SRJC); Willow Cr. Road ca 3 mi. from Coleman Valley Road, *W., I. & T. Knight* (1985); Pepperwood; ca 2100 ft. elev. on Ida Clayton Road, *Barnhart* (1987 SRJC). < 2200 ft. May-Oct. Native of Europe.

RHAMNACEAE. BUCKTHORN FAMILY

1. Fruit fleshy with black or red berries; leaves with several pairs of lateral veins; flowers not showy, greenish-yellow . . . . *Rhamnus*

1: Fruit dry; leaves with 1-3 pairs of lateral veins; flowers showy, white, creamy or blue . . . . *Ceanothus*

Ceanothus

There are many known hybrids between the species of the same section. This tendency for *Ceanothus* to hybridize often makes identification difficult. Frequently these crosses have been found wherever the parental species overlap. Examples are: *C. confusus* × *C. jepsonii*, *C. divergens* × *C. ramulosus*, *C. foliosus* × *C. parryi*, *C. incanus* × *C. thyrsiflorus* and *C. integerrimus* × *C. thyrsiflorus*. In a few areas other than in Sonoma County, hybrids have been found involving the species of both sections, *Cerastes* and *Ceanothus*, acc. *Van Rensselaer & McMinn* (1942).

Early in 1930 M. S. Baker discovered a remarkable thicket of hybrids involving species of section *Cerastes* on the lower western slope of the Hood Mountain Range about 6 miles from Santa Rosa on the road to Sonoma. On two occasions, in May, 1935 and April 1937, Baker and Howell visited the perplexing population and altogether more than 20 collections were made. Howell, *Leaflets of Western Botany 2:259-262* (1940) analyzed the variations and concluded that 4 species were responsible, either directly or through other hybrids: *C. cuneatus*, *C. gloriosus*, *C. prostratus* and *C. purpureus*. acc. M. A. Nobs, *Carnegie Institution of Washington Publication 623, p. 72 (1963)* agrees with this analysis that the population originated "as a quadruple hybrid."

1. Leaves soft, not brittle
  2. Branchlets spinose
    3. Branchlets stout, thornlike, ca 3 cm ± . . . . *C. incanus* forma *incanus*
    3: Branchlets stouter, thornlike structures much larger . . . . *C. i.* forma *spinosissimus*
  2: Branchlets not spinose
    4. Leaves with 3 main veins from base
      5. Leaf margins denticulate, black glands on denticulations . . . . *C. oliganthus* var. *sorediatus*
      5: Leaf margins not denticulate, no black glands
        6. Leaves glandular-varnished above, with balsamlike odor . . . . *C. velutinus* var. *hookeri*
        6: Leaves not so, branchlets angled or ridged
          7. Leaves usually grayish-silky beneath, margins revolute between teeth . . . . . . *C. griseus*
          7: Leaves not silky beneath, margins not revolute . . . . *C. thyrsiflorus*
    4: Leaves without 3 main veins from base
      8. Leaves small, ± wavy-margined
        9. Leaves oblong-elliptic, margins glandular-denticulate . . . . *C. foliosus* var. *foliosus*
        9: Leaves broadly elliptic to obovate, margins ± plane, often with few glands *C. f.* var. *vineatus*
      8: Leaves not small or wavy-margined
        10. Leaves 4-5 × longer than wide, not thin, not semideciduous . . . . *C. parryi*
        10: Leaves oblong to oblong-elliptical, thinnish, semideciduous
          11. Leaves pinna-veined or with the basal pair of lateral veins a little more prominent than the other lateral veins . . . . *C. integerrimus* var. *integerrimus*
          11: Leaves 3-veined (ribbed) from base . . . . *C. i.* var. *californicus*
1: Leaves hard, will snap when pressurized between thumb and fingers . . . . [Sect. *Cerastes*]
  12. Plants prostrate to sprawling
    13. Leaves typically 20-50 mm long, margins with numerous teeth, oval to elliptic or obovate . . . . *C. gloriosus* var. *gloriosus*
    13: Leaves smaller, narrowly oblanceolate, cuneate-spatulate, or narrow-elliptic to oval
      14. Leaves 6-32 mm long, ovate to oval, often few coarse teeth near apex, or more finely toothed with 4-7 on each side, rarely almost to quite entire, often undulate and slightly troughed above; plants low, ± spreading, occasionally prostrate . . . . *C. prostratus* var. *occidentalis*
      14: Leaves 4-20 mm long; usually basally cuneate, mostly 3-toothed at apex only, occasionally with teeth on sides, margins slightly revolute; plants sometimes forming mats via rooting nodes, sometimes decumbent, low-spreading
        15. Leaves 6-12 mm wide, 10-22 mm long; mostly not on serpentine . . . . . *C. confusus*
        15: Leaves 3-6 mm wide, 6-16 mm long; on serpentine . . . . *C. pumilus*
  12: Plants erect
    16. Leaves deflexed; on serpentine
      17. Flowers violet-blue . . . . *C. jepsonii* var. *jepsonii*
      17: Flowers white . . . . *C. j.* var. *albiflorus*
    16: Leaves not deflexed, toothed nearly all around

18. Leaves mostly < 12 mm long; infrequent on serpentine . . . . *C. sonomensis*
18: Leaves larger
  19. Plants with divergent, arching stems; leaves ± dentate
    20. Leaves rounded, truncate or retuse at apex; dentate or ± spinosely toothed 15-30 . . . . *C. gloriosus* var. *exaltatus*
    20: Leaves toothed at apex, 5-8 coarse, spinescent teeth or almost lobed *C. divergens*
  19: Plants with erect branches; leaves on mature branches typically entire . . *C. cuneatus*

Ceanothus confusus J. T. Howell. RINCON RIDGE CEANOTHUS. Formerly locally common, but now endangered: Mt. Hood, *Howell* (1933); Rincon Ridge (ne. sector of Santa Rosa), *Howell* (1937, type); ca ½ mi. past summit on Dorman Can. Road, acc. J. & B Guggolz (1986); in appressed groundcover mats at sw. base of Bradford Mtn., acc. J. & B. Guggolz & W. Knight (1992); n. slope Red Mtn. between Hot Springs Road and Cherry Cr. Road on serpentine (w. of Cloverdale), *Raiche* (1985). < 1700 ft. Feb.-Apr. —*C. divergens* Parry ssp. *c.* (J. T. Howell) Abrams.

Ceanothus cuneatus (Hooker) Nuttall. BUCKBRUSH. Common, chaparral: Cazadero quad.; Sonoma Mtn. Road to Cooper's Grove, *Baker* (1924); ne. of Sonoma, acc. Barbour; Mt. Hood Range, *Howell* (1935); Sugarloaf Ridge; Rincon Ridge, *Howell* (1931); Calistoga Road between Alpine Road and Porter Cr. Road, *Knight* (1963); 1200 ft. elev. Calistoga Road, *Clifton* (1981 PUA); Petrified Forest Road, *Callizo* (1978 PUA); Pepperwood, *Breedlove* (1980); Petrified Forest, *Howell* (1937); Tombs Cr., Lake Sonoma area, Geyserville, Alexander Valley and Ida Clayton Road quads.; ca 2000 ft. elev. Anderson Trail on Black Mtn. between Geyser Peak and upper end of Pine Flat Road, acc. J. & B. Guggolz (1992); summit Pine Flat Road, *Knight* (1963); Big Sulphur Cr. quad. < 6000 ft. Mar.-May. —*C. ramulosus* (E. Greene) McMinn.

Ceanothus divergens C. Parry. CALISTOGA CEANOTHUS. Rare, chaparral: Schultz Road, Kenwood, *Harrison* (1978 SRJC); Franz Valley School Road (near old schoolhouse), *Callizo* (1979 PUA), *Knight & Edwards* (1985); Mt. Hood, *Howell* (1935); S. Burma Trail, Annadel, acc. Lovell (1976); Pepperwood. A considerable population exists on Kortum Can. just over the county line in Napa Co. < 1700 ft. Feb-Mar. —*C. prostratus* Bentham var. *d.* (C. Parry) K. Brandegee.

Ceanothus foliosus C. Parry var. **foliosus**. WAVYLEAF CEANOTHUS. Fairly common, rocky slopes, dry ridges: The Sea Ranch, *Wells* (1981); Annapolis and Stewarts Point quads.; Salt Point; Fort Ross, Cazadero and Arched Rock quads.; ca 1 mi. nw. of Camp Meeker, acc. J. & B. Guggolz (1991); 750 ft. elev. near ridge top w. of Stuart Can. at Bouverie, acc. Stone (1985); Hooker Can. on Cavedale Road, *Knight* (1964); Sugarloaf Ridge; Mt. Hood, *Michener & Bioletti* (1893); Calistoga Road between Alpine and Porter Cr. roads, *Knight* (1963); Pepperwood; base of sw. shoulder of Bradford Mtn., acc. J. & B. Guggolz & W. Knight (1992); Ida Clayton Road quad.; geysers Unit 18, *Malloch* (1977 PG&E), acc. Quibell (1980); Pine Mtn. Road, *Knight* (1963). < 5000 ft. Mar.-May.

Ceanothus foliosus var. **vineatus** McMinn. VINE HILL CEANOTHUS. Uncommon, rolling hills: type locality: near Vine Hill schoolhouse ca 10 mi. w. of Santa Rosa, *McMinn*; Vine Hill School Road, *Knight* (1963). < 1000 ft. Mar.-May.

Ceanothus gloriosus J. T. Howell var. **exaltatus** J. T. Howell. NAVARRO CEANOTHUS. Occasional, openings along roads, chaparral: type locality: Vine Hill region, *Howell* (1931); The Sea Ranch Timber Preserve of pygmy forest, *Wells* (1980); along Guerneville Road w. of Laguna Road, acc. Lovell (1980); Los Guilicos Valley, acc. Baker (1932). < 1700 ft. Mar.-May.

Ceanothus gloriosus var. **gloriosus**. GLORY MAT. Uncommon, littoral bluffs and terraces, brushy slopes: The Sea Ranch; Stewarts Point, *Howell* (1934); Salt Point; near Bodega Bay, *Best* (1983). 1700 ft. Mar.-May.

Ceanothus griseus (Trelease) McMinn. CARMEL CEANOTHUS. Occasional, coastal ravines, mountains: 6 mi. n. of Stewarts Point, *Knight* (1964); Salt Point; Fort Ross, *Heller* (1903). Hybridizes with *C. foliosus* var. *foliosus* at Salt Point. < 700 ft. Mar.- May.

Ceanothus incanus Torrey & A. Gray forma **incanus**. COAST WHITETHORN. Occasional, moist slopes, valley flats: summit Pine Flat Road, acc. Knight (1989); crest between Geyser Mtn. (probably Geyser Peak) and The Geysers, *McMurphy* (1909); ca 2000 ft. elev. Anderson Trail on Black Mtn. between Geyser Peak and upper end of Pine Flat Road, acc. J. & B. Guggolz (1992). < 3000 ft. Apr.-May.

Ceanothus incanus forma **spinosissimus** Klein. Uncommon: Kelly Road in flatland few mi. e.of Annapolis, acc. Knight & Edwards (1985); 3 mi. nnw. of Mt. St. Helena, Sonoma County (probably Ida Clayton Road), *Klein* (type 1969 UC).

Ceanothus integerrimus Hooker & Arnott var. **californicus** (Kellogg) G. T. Benson. Occasional, mountain slopes: 3300 ft. elev. crest of Mayacmas Mts. at Caldwell Pines, acc. Stone (1989); 2000 ft. elev. near Cloverdale, *Muth* (1968 PUA). 2000-7000 ft. May-June.

Ceanothus integerrimus Hooker & Arnott var. **integerrimus**. DEER BRUSH. Occasional, mountain slopes and flats: Annapolis, Fort Ross, Cazadero, Healdsburg, Lake Sonoma, Geyserville, Alexander Valley and Ida Clayton Road quads.; Pine Flat Road near summit, *Knight* (1963). 500-5000 ft. May-June.

Ceanothus jepsonii E. Greene var. **albiflorus** J. T. Howell. Occasional, serpentine: 1400 ft. elev. w. slope of Mt. St. Helena on Ida Clayton Road, *Stein* (1981); Dorman Can. in Dry Cr. area w. of Healdsburg, acc. Knight & Edwards (1985). < 2000 ft. Mar.-Apr.

Ceanothus jepsonii E. Greene var. **jepsonii**. MUSK BRUSH. Common, usually areas of serpentine: Bohemian Hwy. ca 3 mi. toward Monte Rio from Camp Meeker, *Eastwood & Howell* (1939), *Huntington* (1987 NCC); Morelli Lane near county dump near Occidental, *Knight* (1963); Stoetz Lane between Occidental and Graton, *Utech & Knight* (1986 CM); outcrop of serpentine on Joy Road (sw. of Occidental), acc. Knight (1988); near top of Harrison Grade Road, acc. Quibell (1991); Los Alamos Road near PG&E tower, *Harrison* (1982 SRJC); Ida Clayton Road, *Winter* (1978 SRJC); Pine Flat Road, *Knight* (1963); outcrop of serpentine 2.6 mi. from Lake Sonoma park headquarters on Stewarts Point Skaggs Springs Road, *Knight* (1990 NCC); Los Alamos Road at PG&E tower site, *Harrison* (1982); Lake Sonoma area; Geyserville, Alexander Valley and Socrates Mine quads.; geysers Unit 13, acc. Quibell (1980); dirt road leading to site I and F, etc. ca ¼ mi. above jct. with Geysers Rock Road, *Cooney* (1977 PG&E); near Mendocino Co. line on Hwy. 128 nw. of Cloverdale, *Quibell* (1987 NCC); Mt. Tom quad. < 3000 ft. Mar.-Apr.

Ceanothus × mendocinensis McMinn. A hybrid between *C. thyrsiflorus* and *C. velutinus* var. *hookeri*: on road from Seaview to Plantation (on ridge e. of Fort Ross), *Wagnon* (1939 NCC); half way between Seaview and Fort Ross, *Roderick* (1961). < 2000 ft. Apr.-May.

Ceanothus oliganthus Nuttall var. **sorediatus** (Hooker & Arnott) Hoover. JIM BRUSH. Rare, woodland or brushy areas of hillsides: Occidental Ridge, 5 mi. w. of Sebastopol, *Trigeiro* (1929 NCC); Griffith Ranch, Kenwood Valley, 8 mi. e. of Santa Rosa, *Whitney* (1928 NCC); 550 ft. elev. across from entrance of Sky Hi Ranch, 7060 Sonoma Mtn. Road, 0.2 mi w. of jct. with Enterprise Road, *W., I. & T. Knight* (1994 NCC); 0.3 mi. n. of Petrified Forest entrance, *Clausen* (1935). < 3300 ft. Feb.-May. —*C. s.* Hooker & Arnott.

Ceanothus parryi Trelease. PARRY CEANOTHUS. Common, wooded canyon slopes: Fort Ross quad.; near Cazadero, *Congdon* (1892), *McMinn* (1922 UC), *Bacigalupi* (1926); Camp Meeker, *McMinn* (1926 UC); Occidental, *Rowntree* (1933), acc. Knight (1988); s. of Guerneville in Neeley Hill area, *Rubtzoff* (1953); Bouverie; 1½ mi. up Cavedale Road from Hwy. 12, *Utech & Knight* (1986 CM); Trinity Road, acc. Knight (1965); Mt. Hood, *Heller & Brown* (1902); Rincon Ridge, *Howell* (1941); Pepperwood, *Breedlove* (1980); chaparral area sw. of Bradford Mtn. 1½ mi. w. of Dry Cr., acc. J. & B. Guggolz (1992); Lake Sonoma area. < 2500 ft. Apr.-May.

Ceanothus prostratus Bentham. WESTERN SQUAW CARPET. Occasional, chaparral, Douglas-fir forest: Rincon Ridge (ne. sector of Santa Rosa), *Clausen* (1933); 4298 ft. elev. Mt. St. Helena lookout, *Grummer* (1978 PUA); 2700 ft. elev. 0.8 mi. in Napa Co. e. of county line on Western Mine Road, *Stein* (1981); geysers Unit 18, acc. Quibell (1980); 3300 ft. elev. crest of Mayacmas Mts. at Caldwell Pines, acc. Stone (1989). As of 1985, acc. Knight, the beautiful matlike population at Parker Hill in n. Santa Rosa appears to have been extirpated by grading in preparation for a commercial and housing development. This same latter population has been annotated *C. confusus* on some herbarium sheets. < 7300 ft. Apr.-June. C. p. var. *occidentalis* McMinn.

Ceanothus pumilus E. Greene. SISKIYOU MAT. Uncommon, serpentine outcrops: Fred McMurray ranch sw. of Healdsburg, acc. Nobs (1963); ca half way down road to marina in Lake Sonoma area, acc. Knight (1986). 600-6000 ft. Apr.-May. —*C. prostratus* Bentham var. *profugus* Jepson.

Ceanothus sonomensis J. T. Howell. SONOMA CEANOTHUS. Occasional, chaparral: near Macnab Cypress population on Cavedale Road, *Knight* (1984); ca 1000 ft. elev. Trinity Road ca 2 mi. e. of Glen Ellen, *Howell* (type 1933); 1200 ft. elev. along unpaved road s. of Trinity Road and w. of cul-de-sac on Manzanita Lane, acc. Stone (1984); sw. exposure Mt. Hood, *Gaffke* (1976 SRJC); s. slope of Sugarloaf Peak (in Sugarloaf State Park), *Baker* (1947); Annadel, acc. Lovell (1976); Circle C Ranch n. of Gates Road, Franz Valley, acc. Lovell (1978). < 2700 ft. Feb.-Apr.

Ceanothus thyrsiflorus Eschscholtz. BLUE BLOSSOM. Occasional, canyon slopes of wooded hills: The Sea Ranch, *Wells* (1983); Annapolis and Stewarts Point quads.; Salt Point; Cazadero and Duncans Mills quads.; Willow Creek; Sebastopol, *Yates* (UC); US 101 at Railroad Ave. interchange between Petaluma and Cotati, *Knight* (1963); Button Ranch on Walker Road (U. C. property) between Petaluma-Valley Ford Road and Pepper Road, acc. Patterson (1990); Healdsburg quad. < 2000 ft. Mar.-June. Hybridizes with *C. griseus* at Salt Point.

Ceanothus velutinus Douglas var. **hookeri** M. Johnston. TOBACCO BRUSH. Fairly common, wooded slopes: Salt Point; Seaview (along ridge road e. of Fort Ross), *Abrams* (1920); between Monte Rio and Guerneville, *McMinn* (1920s); summit of Pine Flat Road, *Knight* (1963). < 3000 ft. Apr.-July. —*C. v.* var. *laevigatus* (Hooker) Torrey & A. Gray.

Rhamnus

1. Petals lacking; winter-buds with bud-scales
   2. Plants mostly < 1 m high; branchlets ± divaricate, ending in weak spines; leaf-blades 10-17 mm, petioles 1-4 mm long . . . . *R. crocea*
   2: Plants 1.5-4 m high; branchlets not divaricate, seldom ending in weak spines; leaf-blades 18-60 mm long, petioles 2-8 mm long . . . . *R. ilicifolia*

1: Petals present; winter-buds without bud-scales
   3. Leaves persistent, ± thick and coriaceous, mostly 3-8 cm long; seeds mostly 2
      4. Leaves glabrous or subglabrous beneath; blades oblong to elliptic . . . . . . *R. californica*
      4: Leaves pubescent or tomentose below; blades narrowly elliptical . . . . . . *R. tomentella*
   3: Leaves deciduous, ± thin and not coriaceous, 8-20 cm long; seeds mostly 3 . . . . . *R. purshiana*

Rhamnus californica Eschscholtz. CALIFORNIA COFFEEBERRY. Common, coastal sandy and rocky places, inland hillsides and ravines: Gualala Point Regional Park, *Knight* (1984); The Sea Ranch, *Wells* (1983); Annapolis and Stewarts Point quads.; Salt Point; Cazadero quad.; Duncans Mills, *Rubtzoff* (1951); Arched Rock quad.; Bodega Dunes, 3-3½ ft. tall, smaller ovate elliptic-leaved form, Bodega, *Connors* (1991 BML); Bodega Head; 2 mi. n. of Guerneville, *Usinger* (1924); Wiggins Hill w. of Petaluma, *Knight* (1963); Fairfield Osborn (*); Bouverie; Sugarloaf Ridge; Spring Lake; 3 mi. n. of Santa Rosa (possibly Rincon Ridge, unusual form), *Howell* (1938); Pepperwood; chaparral area sw. of Bradford Mtn. 1½ mi. w. of Dry Cr., acc. J. & B. Guggolz (1992); Lake Sonoma area; Geyserville and Ida Clayton Road quads.; ca 2000 ft. elev. Anderson Trail on Black Mtn. between Geyser Peak and upper end of Pine Flat Road, acc. J. & B. Guggolz (1992); ¼ mi. nw. of Eureka Mine in geysers area, *Crane* (1980 PG&E); geysers Unit 13 near Lincoln Rock, acc. Quibell (1980); geysers Unit 18, acc. Quibell (1980); Cloverdale quad. < 6600 ft. May-July.

Rhamnus crocea Nuttall. SPINY REDBERRY. Occasional, canyons, dry washes: Sonoma Cr. at foot of Mt. Hood, *Heller* (1902); Annadel on Steve's N. Trail near *Paxistima myrsinites*, acc. Quibell (1991); Lake Sonoma area; property bordered by Chianti and Dutcher Cr. roads w. of US 101 and w. of Geyserville, acc. J. & B. Guggolz (1991); ca 2000 ft. elev. Anderson Trail on Black Mtn. between Geyser Peak and upper end of Pine Flat Road, acc. J. & B. Guggolz (1992); Pine Mtn. Road, *Knight* (1963); near Mendocino Co. line on Hwy. 128 nw. of Cloverdale, *Quibell* (1988 NCC). < 3300 ft. Mar.-Apr.

Rhamnus ilicifolia Kellogg. HOLLY-LEAF REDBERRY. Occasional, dry slopes: 2-3 mi. w. of Skaggs Springs (on Stewarts Point-Skaggs Springs Road), *Baker* (1927); Button Ranch on Walker Road (U. C. property) between Petaluma-Valley Ford Road and Pepper Road, acc. Patterson (1990). < 6600 ft. Mar.-June. —*R. crocea* ssp. *i.* (Kellogg) C. B. Wolf.

Rhamnus purshiana de Candolle. CASCARA SAGRADA. Uncommon, moist places: The Sea Ranch on ridge e. of Hwy. 1, *Wells* (1984); Salt Point; Willow Creek; Bodega Head (to 6 m tall), *Barbour* (1968 BML). < 6600 ft. May-July.

Rhamnus tomentella Bentham. HOARY COFFEEBERRY. Uncommon, mostly on serpentine: Reservoir Hill ne. of Healdsburg, *Raiche* (1985); near summit of Pine Flat Road, acc. Knight (1963); Big Sulphur Cr. quad. < 3300 ft. May-June. —*R. californica* Eschscholtz ssp. *t.* (Bentham) C. B. Wolf.

ROSACEAE. Rose family

1. Ovary inferior; fruit a pome
  2. Margins of leaves entire . . . . . *Cotoneaster*
  2: Margins of leaves toothed
    3. Leaves persistent
      4. Flowers in corymbs; pistils 5 . . . . . *Pyracantha*
      4: Flowers in panicles; pistil 1 . . . . . *Heteromeles*
    3: Leaves deciduous
      5. Flowers in cymose corymbs, clusters
        6. Leaves serrulate to serrate; fruit 4-10 cm across; carpels with leathery or papery walls . *Pyrus*
        6: Leaves crenate, ± doubly serrate or lobed; fruit 10-12 mm long, with bony central nutlets . . . . . *Crataegus*
      5: Flowers solitary, racemose or in cymes
        7. Inflorescence a raceme; ovary-cells 6-10, each 1-ovuled . . . . . *Amelanchier*
        7: Inflorescence not racemose; ovary cells mostly 5, each 2-ovuled
          8. Styles usually separate; fruit with grit-cells forming a ring beneath epidermis; leaves involute in bud . . . . . *Pyrus*
          8: Styles usually connate at base; fruit without grit-cells; leaves revolute or plicate in bud . *Malus*

1: Ovary or ovaries superior; fruit not a pome
  9. Fruit of 1-5 dehiscent follicles . . . . . *Physocarpus*
  9: Fruit of indehiscent achenes or of drupelets, or of large drupes
    10. Fruit of dry achenes, which may be imbedded in the surface of fleshy receptacle (e.g., strawberry) or enclosed in a fleshy cup-shaped structure (e.g., rose); petals various in number or 0
      11. Leaves simple, sometimes 3-lobed; woody
        12. Petals present
          13. Leaves sublinear; ovule and seed 1 . . . . . *Adenostoma*
          13: Leaves obovate to ovate; ovules 2; seed mostly 1 . . . . . *Holodiscus*
        12: Petals absent . . . . . *Cercocarpus*
      11: Leaves compound or pinnately dissected
        14. Plants shrubby . . . . . *Rosa*
        14: Plants herbaceous
          15. Styles jointed with the ovary and deciduous
            16. Filaments dilated, 10 in number; floral-tube usually deep . . . . . *Horkelia*
            16: Filaments filiform or narrow, mostly 20 or more; floral-tube shallow or none
              17. Receptacle not enlarged in fruit . . . . . *Potentilla*
              17: Receptacle enlarged in fruit
                18. Receptacle spongy; flowers yellow . . . . . *Duchesnea*
                18: Receptacle fleshy; flowers white . . . . . *Fragaria*
          15: Styles not jointed with the ovary, persistent
            19. Pistils many; petals 4-8 mm long . . . . . *Geum*
            19: Pistil 1; flowers smaller
              20. Floral-tube not prickly; petals 0
                21. Leaves palmately lobed; plants not > 1 dm high . . . . *Aphanes*
                21: Leaves pinnate; plants 1-6 dm high . . . . . *Sanguisorba*
              20: Floral-tube armed with prickles
                22. Prickles on floral-tube retrorsely barbed; petals 0; sepals mostly 4 . . . . . *Acaena*
                22: Prickles hooked; petals 5, yellow; sepals 5 . . . . . *Agrimonia*
    10: Fruit of drupes or drupelets; petals 5
      23. Carpels becoming drupelets which are ± coherent and form a fleshy "berry"; plants mostly ± woody, often prickly; leaves mostly pinnate . . . . . *Rubus*
      23: Carpels becoming larger solitary drupes; trees or shrubs with simple leaves
        24. Pistil 1; flowers bisexual; leaves mostly serrate . . . . . *Prunus*
        24: Pistils usually 5; flowers unisexual or bisexual; leaves entire . . . . . *Oemleria*

Acaena

1. Inflorescence a head; prickles 4, on fruiting top, 10-15 mm; plant mat-forming from stolons *A. novae-zelandiae*
1: Inflorescence ± a spike, interrupted below; prickles > 4, scattered on fruit, usually 1-3 mm; plant from ± woody, branched caudex . . . . *A. pinnatifida* var. *californica*

★ **Acaena novae-zelandiae** T. Kirk. BIDDY-BIDDY. Locally abundant: Horseshoe Cove in Salt Point State Park, *Raiche* (1985); vicinity of Jenner and Occidental, acc. *Barbe* (1989). < 700 ft. Native of New Zealand. Noxious weed.

Acaena pinnatifida Ruiz López & Pavón var. **californica** (Bitter) Jepson. Occasional, sandy and rocky areas: The Sea Ranch; Horseshoe Cove, *Knight* (1963); Bodega Head and Dunes, *Connors* (1990 BML); near Occidental, *Marac* (1925 NCC); Estero Americano near Valley Ford, *Holve* (1991 NCC); Button Ranch on Walker Road (U.C. property) between Petaluma-Valley Ford Road and Pepper Road, acc. Connors (1992); Debbie Hill Road (US 101 Frontage Road) off of Railroad Ave., acc. Quibell (1991); s. of Coast Guard Training Center near Two Rock, *J. & B. Guggolz* (1990 JEPS). < 1500 ft. Apr.-June. —*A. c.* Bitter.

Adenostoma fasciculatum Hooker & Arnott. CHAMISE. Common, dry ridges and slopes: Annapolis, Fort Ross, Cazadero, Duncans Mills and Arched Rock quads.; Fairfield Osborn (S); Bouverie; 1½ mi. up Cavedale Road from Hwy. 12, *Utech & Knight* (1986 CM); Sugarloaf Ridge; Calistoga Road between St. Helena and Porter Cr. roads on Joost Ranch, *Knight* (1963); Gates Road off Calistoga Road, acc. Wells et al. (1984); Annadel near top of Steve's N. Trail, acc. Quibell (1991); w. face of Taylor Mtn. s. of Santa Rosa, acc. Quibell (1991); Pepperwood; Ida Clayton Road quad.; probably between Santa Rosa and Mt. St. Helena, *Voznesenskii* (1840-1841 LE); chaparral area sw. of Bradford Mtn. 1½ mi. w. of Dry Cr., acc. J. & B. Guggolz (1992); summit of Pine Flat Road, *Knight* (1963); ca 2000 ft. elev. Anderson Trail on Black Mtn. between Geyser Peak and upper end of Pine Flat Road, acc. J. & B. Guggolz (1992); Geyser Peak, *Knight & Edwards* (1986); geysers Unit 13, acc. Quibell (1981); Big Sulphur Cr., Cloverdale, Mt. Tom and N. Fork Buckeye Cr. quads. < 5400 ft. May-June.

Agrimonia gryposepala Wallroth. COMMON AGRIMONY. Uncommon, borders of woods: Ida Clayton Road in marsh, *Harrison* (1979 SRJC). < 5700 ft. July-Aug.

Amelanchier alnifolia (Nuttall) Nuttall. SERVICE-BERRY. Common, dry rocky and gravelly slopes, flats: Salt Point; 8½ mi. n. of Bodega Bay, *Howell* (1930); near Camp Meeker on road to Occidental, *Eastwood & Howell* (1939), *Barnhart* (1977 SRJC); Sonoma Mtn. Road near jct. Hidden Acres Road near jct. of Hidden Oaks Road, acc. Quibell (1991); Annadel, *Harrison* (1976 SRJC); Lake Sonoma area. < 8600 ft. Apr.-June. —*A. a.* Nuttall ssp. *pallida* (E. Greene) E. Murray.

Aphanes occidentalis (Nuttall) Rydberg. WESTERN LADY'S MANTLE. Occasional, rocky and brushy slopes: Bodega Dunes, *Connors* (1992 BML); Fairfield Osborn (H); Bouverie; Spring Lake; Pepperwood; Lake Sonoma area; near Mendocino Co. line on Hwy. 128 nw. of Cloverdale, *Quibell* (1988 NCC). < 4000 ft. Mar.-June. —*Alchemilla arvensis* (Linnaeus) Scopoli. *Alchemilla occidentalis* Nuttall.

Cercocarpus betuloides Torrey & A. Gray. MOUNTAIN-MAHOGANY. Common, dry slopes, washes: Fairfield Osborn (S); Bouverie; Cavedale Road off Trinity Road, *Stocking* (1963 NCC); Adobe Can., *Hatch* (1933 NCC); Gates Road off Calistoga Road, acc. Wells et al. (1984); Spring Lake; Pepperwood; Petrified Forest, *Howell* (1937); chaparral area sw. of Bradford Mtn. 1½ mi. w. of Dry Cr., acc. J. & B. Guggolz (1992); Lake Sonoma area; Pine Flat Road near summit, *Knight* (1963); ca 2000 ft. elev. Anderson Trail on Black Mtn. between Geyser Peak and upper end of Pine Flat Road, acc. J. & B. Guggolz (1992); Geysers and Big Sulphur Cr. quads.; Pine Mtn. Road ne. of Cloverdale at Harold Smith Ranch, *Knight* (1963); Cloverdale quad. < 8300 ft. Mar.-May.

★ **Cotoneaster pannosa** Franchet. Uncommon, escape from cultivation: Salt Point, *Wells* (1982); Rincon Ridge by path n. of Montecito Ave. opposite Manzanita Ave., *Wells* (1989). Usually < 3300 ft. Native of sw. China.

Crataegus douglasii Lindley. HAWTHORN. Occasional, near streams, meadows: hwy. between Bodega Bay and town of Bodega, *Baker* (1925 NCC); Bohemian Hwy. 1 mi. n. of Freestone, *Best* (1981); Forestville Marsh, *Rubtzoff* (1958); Pitkin Marsh, *L. Roof* (1933 NCC), *Howell* (1945), *Rubtzoff* (1951); Atascadero Cr. Marsh, *Rubtzoff* (1958); Elphick Road near Sebastopol, *R. Thompson* (1975 NCC); Ragle Ranch Reg. Park w. of Sebastopol, *Wagnon* (1990 NCC); Sebastopol, *McMurphy* (1908); marshy area s. of Hwy. 116 near Lone Pine Road, acc. Lovell (1977); Petaluma-Valley Ford Road e. of Valley Ford, acc. Quibell (1991); tributary of American Cr. 2 mi. nw. of Bloomfield, *Rubtzoff* (1973). < 5700 ft. May-June. —*C. gaylussacia* Heller, type locality: along lagoon at Sebastopol, *Heller* (1902).

★ **Duchesnea indica** Linnaeus. MOCK-STRAWBERRY. Uncommon, used as groundcover, escaping: 3 mi. n. of Sonoma, *Farnum* (1963); Gates Road off Calistoga Road, acc. Wells et al. (1984). Usually < 1700 ft. May-Aug. Native of India.

Fragaria

1. Leaves thick, coriaceous, strongly reticulate with ridged veinlets; flowers 20-35 mm broad . . . . *F. chiloensis*
1: Leaves thin, scarcely reticulate; flowers mostly 10-20 mm broad . . . . *F. vesca*

Fragaria chiloensis (Linnaeus) Duchesne. BEACH STRAWBERRY. Common, coastal strand: The Sea Ranch; s. end of The Sea Ranch, *Thomas* (1976); Salt Point; Fort Ross, *Davy* (1890s UC); Russian colony, *Voznesenskii* (1840-1841 LE); Salmon Cr., *Hastings* (1940 NCC); 2 mi. n. of Bodega Bay, *McNally* (1970 NCC); Bodega Bay, *Heller & Brown* (1902), *Eastwood* (1917); Bodega Head; Bodega Harbour development ½ mi. s. of entrance, *Huntington* (1989 NCC). < 700 ft. Mar.-Aug.

Fragaria vesca Linnaeus ssp. **vesca**. WOOD STRAWBERRY. Common, shaded, fairly damp places: The Sea Ranch; Annapolis and Stewarts Point quads.; Rocky Point between Stewarts Point and Salt Point, *Knight* (1963); Salt Point; Duncans Mills Marsh, *Rubtzoff* (1963); Willow Creek; Arched Rock quad.; Bodega Bay, *McNally* (1970 NCC); Larkspur Rock e. of Bodega Bay, *Ott* (1934 NCC); w. of Austin Cr., *Warfield* (1964 NCC); near Guerneville, *Rubtzoff* (1953); Russian River near Trenton, *Heller & Brown* (1902); Hidden Acres Road between Sonoma Mtn. and Grange roads, acc. Quibell (1991); Fairfield Osborn (S); Trinity Road, *Stocking* (1966 NCC); Gates Road off Calistoga Road, acc. Wells et al. (1984); Pepperwood; Petrified Forest, *Eastwood* (1910); chaparral area sw. of Bradford Mtn. 1½ mi. w. of Dry Cr., acc. J. & B. Guggolz (1992); Lake Sonoma area. < 6600 ft. Mar.-June. —*F. californica* Chamisso & Schlechtendal. *F. v.* ssp. *c.* (Chamisso & Schlechtendal) Staudt.

Geum macrophyllum Willdenow. BIGLEAF AVENS. Rare, moist places: Valley Ford, *Baker* (1940 NCC). < 11000 ft. May-Aug.

Heteromeles arbutifolia (Lindley) M. Roemer. TOYON, CHRISTMAS BERRY. Common, semi-dry brushy slopes, canyons: The Sea Ranch, *Wells* (1981); Annapolis quad.; Salt Point; Fort Ross, Cazadero and Sebastopol quads.; Hidden Acres Road between Sonoma Mtn. and Grange roads, acc. Quibell (1991); Fairfield Osborn (S); Tresch Ranch, Walker Road near Two Rock, acc. J. & B. Guggolz (1992); Bouverie; Mt. Hood, *Knight* (1964); Sugarloaf Ridge; Spring Lake; Pepperwood; Healdsburg, Tombs Cr., Lake Sonoma, Geyserville and Alexander Valley quads.; Ida Clayton Road, *Milburn* (1969 NCC); summit of Pine Flat Road, *Knight* (1963); ca 2000 ft. elev. Anderson Trail on Black Mtn. between Geyser Peak and upper end of Pine Flat Road, acc. J. & B. Guggolz (1992); near Lincoln Rock geysers Unit 13, acc. Quibell (1981); Pine Mtn. Road, *Knight* (1963). < 4300 ft. June-July. —*Photinia a.* Lindley.

Holodiscus discolor (Pursh) Maximowicz. OCEANSPRAY, CREAM BUSH. Common, brushy areas: The Sea Ranch, acc. Wells (1985); Duncans Mills quad.; Willow Creek; 10 mi. e. of Stewarts Point, *Best* 1981); Camp Meeker, *Howell* (1924); Sweetwater Springs Road 2 mi. e. of Armstrong Redwoods State Preserve, *Stocking* (1963 NCC); Fairfield Osborn (S); Tresch Ranch near Two Rock, acc. J. & B. Guggolz (1992); Bouverie; Sugarloaf Ridge; Joost Ranch, Calistoga Road between St. Helena and Porter Cr. roads, *Knight* (1963); St. Helena Road, *Knight* (1963); Pepperwood; Lake Sonoma area; 3500 ft. elev. upper Cobb Cr. drainage, acc. Stone (1985). Var. *franciscanus* (Rydberg) Jepson, currently not recognized. < 4300 ft. May-Aug. —*H. boursieri* (Carriére) Rehder, in part.

Horkelia

1. Pistils > 50; bractlets ovate, sepaloid; antisepalous anthers
   2. Flower-tube glabrous within; sepals greenish within; antisepalous anthers 1-1.4 mm long . . . . . *H. californica* ssp. *dissita*
   2: Flower-tube hairy within; sepals purple-flecked within; antisepalous anthers 1.3-1.8 mm long . . . . . *H. c.* ssp. *californica*
1: Pistils < 30 (occasionally more in *H. marinensis*); bractlets lanceolate to narrowly linear, much smaller than sepals; anthers rarely up to 1 mm long
   3. Leaflets serrate to shallowly lobed 7-12 pairs; (herbage densely shaggy-villous) . . . . . *H. marinensis*
   3: Leaflets deeply palmatifid, 10-20 pairs (villous) . . . . . *H. tenuiloba*

Horkelia californica Chamisso & Schlechtendal ssp. **californica** Occasional, coastal influence, grassy slopes: The Sea Ranch; Russian Gulch n. of Jenner, *Harrison* (1989 SRJC); Duncan's Landing between mouth of Russian River and Bodega Bay, *Harrison* (1981 SRJC); 8 mi. n. of Bodega Bay, *Howell* (1936); Bodega Head, *Harrison* (1976 SRJC). < 1300 ft. May-Aug.

Horkelia californica ssp. **dissita** (Crum) B. Ertter. Uncommon, moist, often shady areas: Coleman Valley, *Baker* (1929 NCC); Russian River, *Carruth* (1900); w. of Windsor near Russian River, *Heller* (1902); Healdsburg, *Rattan* (1877). < 5300 ft. May-Aug. —*H. elata* (E. Greene) Rydberg.

Horkelia marinensis (Elmer) Crum. POINT REYES HORKELIA. Sandy coastal flats: Although listed by the California Natural Diversity Data Base as occurring in Sonoma County, no documented location information is available for this taxon. < 1200 ft. May-Sept.

Horkelia tenuiloba (Torrey) A. Gray. THIN-LOBED HORKELIA. Common, silty or sandy meadows: Plantation, *Bailey* (1925 NCC); Pitkin Marsh, *Baker* (1928 NCC); 5 mi. n. of Sebastopol on road to Forestville, *Howell* (1936); Vine Hill Region, *J. & B. Guggolz* (1990 JEPS); w. of Healdsburg, acc. Raiche (1985); chaparral area sw. of Bradford Mtn. 1½ mi. w. of Dry Cr., acc. J. & B. Guggolz (1992). < 1700 ft. Apr.-July. —*H. fusca* Lindley var. *t.* Torrey, type locality: laguna of Santa Rosa Cr., *Bigelow* (1854 UC). *Potentilla t.* E. Greene. *P. micheneri* E. Greene.

Malus fusca (Rafinesque) C. K. Schneider. OREGON CRAB APPLE. Rare, moist open woods: Sebastopol, Agua Caliente in Sonoma Valley; Santa Rosa Cr., acc. Parke, *Explorations and Surveys for a Railroad Route from the Mississippi River to the Pacific Ocean, 4:85* (1856); along creek on road to Petrified Forest, acc. Kuhn, *Madroño 1:27,* (1916); Lake Sonoma area. < 2700 ft. Apr.-June. —*Pyrus rivularis* Douglas.

Oemleria cerasiformis (Hooker & Arnott) J. W. Landon. OSO BERRY. Common, brushland, moist wooded ravines: Fort Ross near wide place at San Andreas Fault, *McKenzie* (1981); Duncans Mills, *Davy* (1890s JEPS); Willow Creek; slope of serpentine n. of Camp Meeker, *Stocking* (1964 NCC); near Camp Meeker, *Howell* (1924), *Warfield* (1964 NCC); Bouverie; Gates Road off Calistoga Road, acc. Wells et al. (1984); Channel Drive next to sewer pond on the way to Annadel, acc. Harrison (1992); Mark West Cr. Road, *Giampaoli* (1968 NCC); Pepperwood; Lake Sonoma area. < 5700 ft. Feb.-Apr. —*Osmaronia c.* (Torrey & A. Gray) E. Greene.

Physocarpus capitatus (Pursh) Kuntze. NINEBARK. Common, north slopes, moist banks: The Sea Ranch; Cazadero, *Congdon* (1892); lower Salmon Cr. Marsh, *Rubtzoff* (1957); Larkspur Rock e. of Bodega Bay, *Ott* (1934 NCC); Occidental along Bohemian Hwy., *Stocking* (1968 NCC); Coleman Valley Road, *Stocking* (1969 NCC); Bodega, *Rowntree* (1932); 1 mi. n. of Freestone on Bohemian Hwy., *Best* (1981); Vacation Beach near Guerneville, *Rubtzoff* (1953); Fife Cr. Marsh near Guerneville, *Rubtzoff* (1960); Hacienda, acc. Knight (1988); Guerneville, *Rattan* (1877); Harrison Grade Road in creekbed between Graton and Occidental, *Knight* (1963); Pitkin Marsh, *Rubtzoff* (1951); Forestville Marsh, *Rubtzoff* (1958); Bouverie; Matanzas Cr. w. of Grange Road; *Stocking* (1966 NCC); Sonoma Mtn. Road w. of jct. at Enterprise Road, acc. Quibell (1991); Spring Cr. trail into Annadel, above the dam, along the creek, acc. Harrison (1992); Santa Rosa, *Heller* (1902); near Mark West Springs at Burke's Sanitarium, *Heller* (1902); Pepperwood; 350-400 ft. elev. Old Skaggs Springs Road, *Utech, Sato & Ohara* (1990 CM); Lake Sonoma area. < 4700 ft. May-July.

Potentilla

1. Flowers solitary, axillary, long-pedicelled; stems slender, prostrate, often rooting at nodes . . . . . *P. anserina* ssp. *pacifica*
1: Flowers cymose, short-pedicelled; stems not stoloniferous
  2. Styles terminal or nearly so
    3. Basal leaves definitely pinnate, sparingly strigose, leaflets 9-13, palmately cleft to ca middle into 3-6 lanceolate divisions; petals 5-7 mm long . . . . . *P. hickmanii*
    3: Basal leaves digitate, pilose, hirsute to tomentulose to glabrate, leaflets 5-7, oblanceolate, divided ca halfway into coarse lanceolate teeth; petals 8-11 mm long . . . . . *P. gracilis* var. *fastigiata*
  2: Styles lateral or nearly basal
    4. Cymes usually leafy, petals and calyx-lobes not reflexed in anthesis . . *P. glandulosa* ssp. *glandulosa*
    4: Cymes not leafy, petals and calyx-lobes usually reflexed in anthesis . . . . *P. glandulosa* ssp. *reflexa*

Potentilla anserina Linnaeus ssp. **pacifica** (T. J. Howell) Rousi. Common, wet places, seepages, coastal influence:The Sea Ranch; Salt Point; Horseshoe Cove, *Knight* (1963); Willow Creek Marsh, *Rubtzoff* (1959); Coleman Valley Marsh, *Rubtzoff* (1954); Bodega Head; Bodega Bay, *Eastwood* (1917), *McNally* (1970 NCC); Bodega Head Marsh, *Rubtzoff* (1960); just s. of Bodega Bay e. of Hwy. 1 across from entrance to Doran Park, acc. Quibell (1991); 2½ mi. nw. of Bodega (probably Salmon Cr. Road), *Cantelow* (1938); American Cr. ½ mi. e. of Valley Ford, *Rubtzoff* (1961). < 500 ft. Apr.-Aug. —*P. egedii* Wormskiöld var. *grandis* Torrey & A. Gray. *P. a.* Linnaeus var. *g.* Torrey & A. Gray.

Potentilla glandulosa Lindley ssp. **glandulosa**. STICKY CINQUEFOIL. Common, moist to dryish open areas: Salt Point; Fort Ross, *Wrangel* (1840 LE); Arched Rock quad.; 8½ mi. n. of Bodega Bay, *Howell* (1930); 1 mi. e. of Bodega Bay, *Howell* (1930); Bodega Harbour development on Pinnacle Gulch Trail, *Huntington* (1989 NCC); ca 3 mi. w. of Occidental on Coleman Valley Road, *Stocking* (1969 NCC); Tresch Ranch, Walker Road near Two Rock, acc. J. & B. Guggolz (1992); Rainsville Road w. of Petaluma, *Ott* (1934 NCC); Bouverie; Bennett Valley Road s. of summit, *Baker* (1931 NCC); Gates Road off Calistoga Road, acc. Wells et al. (1984); upper road in Santa Rosa Cr. Can., *H. Thomas* (1933 NCC); Pepperwood; chaparral area sw. of Bradford Mtn. 1½ mi. w. of Dry Cr., acc. J. & B. Guggolz (1992); Lake Sonoma area; 2400 ft. near geysers Unit 18, *Malloch* (1977 PG&E). < 4000 ft. May-July. —*P. wrangeliana* Fischer & Avé-Lallemant, type locality: Fort Ross-Russian River-Bodega area cultivated from seed in St. Petersburg Bot. Gar. *P. g.* var. *w.* (Fischer & Avé-Lallemant) Wolf. *Drymocallis w.* (Fischer & Avé-Lallemant) Rydberg.

Potentilla glandulosa ssp. **reflexa** (E. Greene) Keck. Occasional, mostly dryish slopes: mountains near Calistoga (probably Knights Valley area), *Krauss* (1894). < 8600 ft. May-July.

Potentilla gracilis Hooker var. **fastigiata** (Nuttall) S. Watson. Possibly extinct in the county. < 11800 ft. June-Aug. —*P. blaschkeana* Turczaninow: type locality: Russian settlement (maybe Fort Ross), *Blaschke*. *P. g.* ssp. *nuttallii* (Lehmann) Keck.

Potentilla hickmanii Eastwood. Presumed extinct in the county, marshy places: 1 to 1½ mi. w. of Cunningham, *Howell* (1947). < 400 ft. Apr.-Aug.

Prunus

1. Plants shrubby to subarborescent, < 6 m tall
  2. Many of branchlets spinose . . . . . *P. subcordata*
  2: Branchlets not spinose
    3. Twigs bitter in taste and odor when broken . . . . . *P. emarginata*
    3: Twigs not bitter; fruit ± edible
      4. Flowers several on slender pedicels; fruit yellow to red, pulp sweet to ± bitter, > 1 cm in diameter . . . . . *P. avium*
      4: Flowers 12 to many in elongate racemes, pedicels relatively short; fruit dark red, slightly astringent or bitter, 5-6 mm in diameter . . . . . *P. virginiana* var. *demissa*
1: Plants arborescent, 8 m. or more tall
  5. Flowers mostly solitary, appearing with or before leaves . . . . . *P. cerasifera*
  5: Flowers in clusters
    6. Flowers appearing with leaves . . . . . *P. domestica*
    6: Flowers appearing before leaves . . . . . *P. angustifolia*

★ **Prunus angustifolia** Marsh var. **varians** Wight & Hedrick. BIG CHICKASAW PLUM. Rare, escape from cultivation: Pitkin Marsh, *Rubtzoff* (1952 NCC). A horticultural introduction. Native of Oklahoma and Texas.
★ **Prunus avium** Linnaeus. SWEET CHERRY. Rare, escape from cultivation: seedlings and young trees growing spontaneously along Sonoma Mtn. Road between Enterprise Road and Cooper's Grove, *Knight & Wagnon* (1984). Native of Eurasia.
★ **Prunus cerasifera** Ehrhart. CHERRY PLUM, MYROBALAN. Occasional, escape from cultivation: jct. of Sonoma Mtn. Road and Bennett Valley Road ca 1 mi. w. of Glen Ellen, *Knight & Edwards* (1986); Channel Drive at Annadel, acc. Wagnon (1988); Spring Lake; along Santa Rosa Cr. in e. Santa Rosa, acc. Wagnon (1989). < 700 ft. Probably native to the Caucasus Mts. and sw. Asia.
★ **Prunus domestica** Linnaeus. COMMON GARDEN PLUM. Escape from cultivation: jct. of Santa Rosa Cr. and Channel Dr. at Annadel, acc. Wagnon (1990). Exact origin unknown but of European or the Eurasian region.

Prunus emarginata (Hooker) Walpers. BITTER CHERRY. Uncommon, brushy slopes: Fort Ross, *Rawlins* (1942); hills se. of Santa Rosa, *Heller & Brown* (1902); Pepperwood. < 9400 ft. Apr.-May.

Prunus subcordata Bentham. SIERRA PLUM. Occasional, dry rocky areas: Occidental Road just w. of Coffee Lane, Santa Rosa, acc. Isaac (1991); Calistoga Road near jct. of Alpine Road on e. side of road, acc. Knight (1966); Hope Valley n. of Barber's (probably near present day Doyle Park in Santa Rosa), *Baker* (1928 NCC); corner Mill Station Road and Cherry Ridge Road, acc. Isaac (1990). <6300 ft. Mar.-May.

Prunus virginiana Linnaeus var. **demissa** (Nuttall) Torrey. WESTERN CHOKE-CHERRY. Occasional, dampish areas, woods, brushy slopes, flats: Pocket Can. along Pocket Cr., *Gilbert* (1980 SRJC); 1 mi. from Old Adobe e. of Petaluma, *Koch* (1939 NCC); Adobe Can. at Kenwood Springs, *Brunk* (1933 NCC); Pepperwood; Porter Cr. Road, *Stocking* (1967 NCC). < 9700 ft. May-June. —*P. d.* (Nuttall) Walpers.

Pyracantha

1. Calyx and underside of leaves grayish-tomentose . . . . *P. angustifolia*
1: Calyx and underside of leaves glabrous or lightly pubescent
  2. Corymbs pubescent; leaves mostly narrow-elliptic, acute . . . . *P. coccinea*
  2: Corymbs glabrous (except in some varieties); leaves oblong to oblong-lanceolate, acute . . . . *P. crenulata*

★ **Pyracantha angustifolia** (Franchet) C. Schneider. Escape from cultivation: along Santa Rosa Cr. in e. Santa Rosa, acc. Wagnon (1989). < 700 ft. Native of sw. China.

Pyracantha coccinea Roemer. FIRETHORN. Escape from cultivation: Spring Lake. Native of Italy to w. Asia.
★ **Pyracantha crenulata** Roemer. Escape from cultivation: across from Temelec subdivision on Arnold Drive near El Verano, *Knight & Howell* (1984); Pepperwood. Native of the Himalaya Mts. of Asia.

Pyrus communus Linnaeus. COMMON PEAR. Escape from cultivation: Ragle Ranch Regional Park near Sebastopol, acc. Wagnon (1990). Native of Europe and w. Asia.

Rosa

1. Sepals pectinate-glandular on margins to pinnatifid, especially on the attenuate tips
  2. Leaflets resinous-pubescent or glandular beneath, rounded to acute at apex, usually doubly serrate
    3. Prickles strongly hooked or curved; sepals reflexed or spreading in fruit, finally deciduous . *R. eglanteria*
    3: Prickles straight; sepals persistent and erect in fruit . . . . *R. nutkana*
  2: Leaflets glabrate beneath, acuminate, sharply and closely serrate . . . . *R. canina*
1: Sepals mostly entire
  4. Hips clothed with gland-tipped bristles; plants 1-3 dm tall . . . . *R. spithamea* var. *sonomensis*
  4: Hips lacking gland-tipped bristles; plants larger
    5. Sepals and styles deciduous in fruit; pedicels 1-3 cm long; prickles not recurved-flattened
      6. Pedicels ± reflexed in fruit; hips 6-8 mm thick at maturity; plants 1 (-3) m high . *R. gymnocarpa*
      6: Pedicels not reflexed in fruit; hips 4-6 mm thick at maturity; branches ± climbing to 3 m (flowers often double) . . . . *R. multiflora*
    5: Sepals and styles persistent in fruit; pedicels generally < 2 cm long; prickles recurved-flattened . . . . *R. californica*

Rosa californica Chamisso & Schlechtendal. Common, rather moist places: Willow Creek; Vacation Beach near Guerneville, *Rubtzoff* (1953); Pitkin Marsh, *Rubtzoff* (1951); 2 sites w. of Santa Rosa in Laguna de Santa Rosa drainage, *Harrison* (1978 NCC); 4 mi. e. of Valley Ford on road to Petaluma, *Howell* (1930); Pressley Road near Summit View Ranch at gateway, *Knight* (1964); Petaluma Valley, *Krauss* (1893); hills e. of Petaluma, *McMurphy* (1910); bank of slough draining Kenwood Marsh, *Howell* (1932); Bennett Valley Road near Enterprise Road, *Stocking* (1968 NCC); ne. of Spring Lake across Montgomery Drive, Santa Rosa, *B. A. Wood* (1973 NCC); Matanzas Cr. at Doyle Park, Santa Rosa, *Howell* (1926); road between Petrified Forest and Mark West Springs, *Abrams* (1916); Pepperwood; 2 mi. n. of Windsor, *Eastwood & Howell* (1936); Skaggs Springs, *Hawver* (1915). < 5400 ft. May-Aug.

★ **Rosa canina** Linnaeus. DOG ROSE. Naturalized about old settlements: Wildwood Farm 1 mi. s. of Kenwood on Hwy. 12, *Knight* (1991 NCC). < 3000 ft. May-June. Native of Eurasia.

★ **Rosa eglanteria** Linnaeus. SWEET-BRIER. EGLANTINE. Common, pastures, waste places: The Sea Ranch, acc. Wells (1981); Salt Point; Bodega Head; ca 1 mi. nw. of Camp Meeker, acc. J. & B. Guggolz (1991); Occidental and Freestone marshes, *Rubtzoff* (1959); Ragle Ranch Reg. Park, Sebastopol, *Knight & Wagnon* (1991 NCC); Cunningham Marsh, *Howell* (1946); w. of Santa Rosa in Laguna de Santa Rosa drainage, *Harrison* (1978 NCC); Hessel Road near Gravenstein Hwy., acc. J. & B. Guggolz (1991); Blank Road near Gravenstein Hwy., acc. J. & B. Guggolz (1991); s. of Hwy. 1 between Valley Ford and Bodega Bay, acc. Knight (1990); Franc Ave., Sebastopol, *Bately* (1932 NCC); 4 mi. e. of Valley Ford, *Howell* (1930); Tresch Ranch, Walker Road near Two Rock, acc. J. & B. Guggolz (1992); jct. Stony Point and Meacham roads, *Knight* (1963); Stony Point Marsh, *Rubtzoff* (1955); Two Rock, *Rose* (1955); Lake Sonoma area. Native of Europe. —*R. rubiginosa* Linnaeus.

Rosa gymnocarpa Nuttall. WOOD ROSE. Common, usually shady woods: The Sea Ranch, acc. Wells (1985); Annapolis quad.; Salt Point; n. of Fort Ross, *Baker* (1932 NCC); Duncans Mills Marsh, *Rubtzoff* (1957); w. of Austin Cr. along Russian River, *Warfield* (1964 NCC); knob at Doran Park entrance on Bodega Bay, *Knight* (1964); Guerneville Marsh, *Rubtzoff* (1954); Fairfield Osborn (*); Tresch Ranch, Walker Road near Two Rock, acc. J. & B. Guggolz (1992); Cooper's Grove on Sonoma Mtn. Road, *Knight* (1964); Mt. Hood, *Knight* (1964); Pepperwood; 350-400 ft. elev. Old Skaggs Springs Road, *Utech, Sato & Ohara* (1990 CM); Lake Sonoma area; Geyserville quad.; upper Squaw Cr. drainage 0.7 airmiles nw. of Geyser Rock, acc. Stone (1983). < 6600 ft. May-July.

★ **Rosa multiflora** Thunberg. Occasional, introduction which has seeded in grassy areas: Ragle Ranch Reg. Park, Sebastopol, *Knight & Wagnon* (1991 NCC); Wildwood Farm, 1 mi. s. of Kenwood on Hwy. 12, *Knight* (1991 NCC). Native of Japan and China.

Rosa nutkana J. S. Presl. NOOTKA ROSE. Uncommon, damp slopes, flats: The Sea Ranch; Salt Point; 2 mi. s. of Fort Ross, *Howell* (1937). < 2300 ft. May-July.

Rosa spithamea S. Watson var. **sonomensis** (E. Greene) Jepson. SONOMA ROSE. Common, dry slopes, chaparral: Mt. Hood, *Baker* (1947 NCC); Pepperwood; Petrified Forest, *E. Greene* (1890s UC); Skaggs Springs, *Hawver* (1915); Ida Clayton Road under chaparral, acc. Quibell (1991); chaparral area sw. of Bradford Mtn. 1½ mi. w. of Dry Cr., acc. J. & B. Guggolz (1992); ca 2000 ft. elev. Anderson Trail on Black Mtn. between Geyser Peak and upper end of Pine Flat Road, acc. J. & B. Guggolz (1992); Geyser Peak, *Knight & Edwards* (1986). < 4000 ft. May-Aug. —*R. sonomensis* E. Greene, type locality: Petrified Forest, *E. Greene* (U).

Rubus

1. Leaves simple, palmately 5-lobed; stems without prickles
    2. Leaves not whitish-velutinous beneath . . . . *R. parviflorus* var. *parviflorus*
    2: Leaves whitish-velutinous beneath . . . . *R. p.* var. *velutinus*

1: Leaves generally 3-5-foliolate; stems mostly prickly
    3. Berries joined to fleshy receptacle and falling with it or separately, fruit oblong
        4. Flowers borne in small cymose or racemose clusters; prickles slender; leaves mostly 3-foliolate, hairs few beneath; fruit to 2 cm long . . . . *R. ursinus* var. *ursinus*
            5. Plants with leaves felted-pubescent beneath . . . . *R. u.* var. *sirbenus*
            5: Plants with fruit to 6 cm long . . . . *R. u.* var. *titanus*
        4: Flowers borne in large terminal panicles; prickles broad-based and stout; leaves usually 5-foliolate
            6. Canes and inflorescence generally without prickles; leaf margins finely serrate . . . . *R. ulmifolius* var. *inermis*
            6: Canes and inflorescence with prickles; leaf-margins unequally and coarsely toothed or serrate . . . . *R. discolor*

3: Berries forming a hollow cap which separates from dry receptacle as single complete fruit, berry width = length
   7. Flowers white, petals mostly 1 cm or less; stems very glaucous . . . . *R. leucodermis*
   7: Flowers red-purple, petals 1.5-2 cm; stems not glaucous
      8. Leaves sparingly pubescent beneath . . . . *R. spectabilis* var. *spectabilis*
      8: Leaves densely pilose beneath . . . . *R. s.* var. *franciscanus*

★ **Rubus discolor** Weihe & Nees. HIMALAYA BERRY. Common, thickets, widely naturalized: Salt Point; Bodega Head; throughout the Laguna de Santa Rosa drainage, *Harrison* (1978 NCC); Todd Road Preserve; Fairfield Osborn (S); Stony Point along Roblar Road, acc. Knight (1990); Hidden Acres Road between Sonoma Mtn. and Grange roads, acc. Quibell (1991); Button Ranch on Walker Road (U. C. property) between Petaluma-Valley Ford Road and Pepper Road, acc. Patterson (1990); along US 101 between San Antonio Cr. and Petaluma, *Howell* (1981); Bouverie; Ducker Cr., Rincon Valley, *Wells* (1985); Spring Lake; Pepperwood; Lake Sonoma area. < 5400 ft. May-July. Native of Eurasia. —*R. procerus* P. J. Mueller.

Rubus leucodermis Torrey & A. Gray. WESTERN RASPBERRY, BLACKCAP RASPBERRY. Occasional, slopes, canyons: Salt Point; ca 1 mi. nw. of Camp Meeker, acc. J. & B. Guggolz (1991); between Duncans Mills and Guerneville, *Davy* (1896 JEPS); Pepperwood; Lake Sonoma area; Pine Flat Road at Circle 8 Ranch, *Knight* (1963); ca 2000 ft. elev. Anderson Trail on Black Mtn. between Geyser Peak and upper end of Pine Flat Road, acc. J. & B. Guggolz (1992). < 8000 ft. Apr.-July.

Rubus parviflorus Nuttall var. **parviflorus.** THIMBLEBERRY. Common, open woods, canyons: Salt Point; Fort Ross Gulch on Hwy. 1, *McKenzie* (1981); Duncans Mills and Arched Rock quads.; Willow Creek; Bodega Bay-Valley Ford Hwy., *Knight* (1963); Fairfield Osborn (S); Bouverie; 4.9 mi. up Cavedale Road from Hwy. 12, *Knight* (1984); Sugarloaf Ridge; Pepperwood; Lake Sonoma area. < 8300 ft. Mar.-Aug.

Rubus parviflorus var. **velutinus** (Hooker & Arnott) E. Greene. Uncommon, usually coastal: The Sea Ranch. < 500 ft. Mar.-Aug.

Rubus spectabilis Pursh var. **franciscanus** (Rydberg) J. T. Howell. Uncommon, coastal: Bodega Head. < 500 ft. Mar.-June.

Rubus spectabilis var. **spectabilis.** SALMONBERRY. Occasional, moist areas, thickets: The Sea Ranch; Salt Point; Willow Creek; Bodega Head; Bodega Bay, *Heller & Brown* (1902); ca 3 mi. w. of Bodega near coast (possibly on Salmon Cr. Road), *Ott* (1934 NCC). < 1000 ft. Mar.-June.

★ **Rubus ulmifolius** Schott var. **inermis** (Willdenow) W. O. Focke. Infrequent, garden escape: across the entrance of Shell Beach between Goat Rock and Wright's beaches, *Wagnon* (1991 NCC); between Enterprise Road and Cooper's Grove on Sonoma Mtn. Road, low-growing and mounding, *Knight & Wagnon* (1984); bank of Sonoma Cr. near Kenwood, *Baker* (1944); Ducker Cr., Rincon Valley, *Wells* (1985); bank of Santa Rosa Cr. by Brookwood Ave. bridge, Santa Rosa, *Wagnon* (1990 NCC). < 1700 ft. European garden plant.

Rubus ursinus Chamisso & Schlechtendal var. **sirbenus** (L. Bailey) J. T. Howell. Infrequent, wooded slopes: 4.9 mi. up Cavedale Road from Hwy. 12, *Knight* (1984). < 5000 ft. Mar.-July.

Rubus ursinus var. **titanus** L. Bailey. MAMMOTH BERRY. Remnants of once prolific agricultural product: Ragle Ranch Park w. of Sebastopol, acc. Wagnon (1990); occasional along water courses, Laguna de Santa Rosa, *Harrison* (1978 NCC); Lake Sonoma area. < 500 ft. Mar.-July.

Rubus ursinus var. **ursinus**. CALIFORNIA BLACKBERRY. Common, woods, rather damp places: The Sea Ranch; Kolmer Gulch between Timber Cove and Fort Ross, *McKenzie* (1981); Salt Point; Duncans Mills, *M. Jones* (1882 UC); Willow Creek; Bodega Head; Fairfield Osborn (S); Sonoma Mtn. Road, *Warfield* (1964 NCC); Sugarloaf Ridge; Spring Lake; Woolsey Road near Mark West Cr., *Comalli* (1939 NCC); Mill Creek Road, *Knight* (1964); Lake Sonoma area. < 5000 ft. Mar.-July. —*R. vitifolius* Chamisso & Schlechtendal. *R. lemurum* W. S. Brown, type locality: in ravine 15 mi. n. of Fort Ross, *Brown* (UC).

★ **Sanguisorba minor** Scopoli ssp. **muricata** Briquet. GARDEN BURNET. Infrequent, escape from cultivation: The Sea Ranch; toward s. end Chalk Hill Road, acc. Knight (1964). < 5400 ft. May-July. Native of Eurasia.

RUBIACEAE. MADDER FAMILY

(Key ± acc. L. Dempster, *Hickman* [1979]).

1. Calyx of 6 teeth; flowers sessile in few-flowered heads, subtended by deeply divided involucre . . *Sherardia*
1: Calyx not evident; flowers pedicellate, solitary or in 3-flowered or imperfect cymules or panicles of cymules; involucre none . . . . *Galium*

Galium:

1. Annuals
   2. Mericarps sausage-shaped; fruits with uncinate hairs not evenly distributed on the two mericarps; leaves 4-6 at a node . . . . *G. murale*
   2: Mericarps ± spheroidal; leaves in whorls of 5 to 8
      3. Plants very slender, erect; leaves 6 mm long or less
         4. Fruits with evenly distributed uncinate hairs (rarely glabrous); leaves eventually reflexed . . . . *G. parisiense*
         4: Fruits glabrous; leaves seldom or little reflexed . . . . *G. divaricatum*
      3: Plants coarse, usually climbing or decumbent; leaves 13-32 mm long
         5. Fruits tuberculate, over 4 mm wide, on stout recurved pedicels; basal part of stems not appearing dead at anthesis . . . . *G. tricornutum*
         5: Fruits evenly set with uncinate hairs, 4 mm wide or less, on slender spreading pedicels; basal part of stems appearing dead at anthesis . . . . *G. aparine*

1: Perennials
   6. Corollas usually 3-lobed; mericarps smooth, hard and spherical . . . . *G. trifidum* var. *pacificum*
   6: Corollas 4-lobed; mericarps variously rough, hairy, or fleshy, never as preceding
      7. Fruits densely covered with soft white or brownish uncinate hairs which nearly = or exceed the body in length; leaves in whorls of 6 . . . . *G. triflorum*
      7: Fruits glabrous or short-haired, tuberculate; leaves 4 at a node
         8. Plants low, matted, tufted or reclining, not woody above ground
            9. Leaves subulate or lanceolate, gradually narrowed to a pungent apex; upper leaves often triangular or rhombic in cross section . . . . *G. andrewsii*
            9: Leaves linear to elliptical, abrupt at apex and not pungent
               10. Leaves notably shiny, commonly 6-10 mm long, sparsely and inconspicuously pubescent, elliptical or often obovate, narrowed gradually to a short petiole, obtuse or rounded (rarely acute) at apex, ± coriaceous, hairs apically directed; stems ± glabrous . *G. muricatum*
               10: Leaves seldom shiny, usually noticeably pubescent or hispid, occasionally small, but commonly larger 6-20 mm long, ovate or elliptical (rarely lanceolate), usually thin, hairs mostly divaricate; stems generally pubescent; Coast Range form . . . . . *G. californicum*
         8: Plants taller, ± woody above ground, occasionally merely tufted from a woody base, but often climbing or sprawling
            11. Stems woody, often very long, slender, climbing or sprawling; stems and margins and ribs of leaves, scabrous with retrorse aculeate hairs; (leaves usually rounded or apiculate, the terminal hair ephemeral; growth usually hanging in tangled masses from supporting shrubs) *G. porrigens*
            11: Stems woody generally only at base of plants, or rarely climbing very high
               12. Leaf surfaces glabrous or nearly so . . . . *G. bolanderi*
               12: Leaves and stems hispid or canescent throughout; (woody stems wiry and very slender); coastal form . . . . *G. californicum*

Galium andrewsii A. Gray. PHLOX-LEAVED BEDSTRAW. Rare, dry ridges, benches: geysers Units 13 & 18, *Quibell* (1980 NCC). < 8600 ft. Apr.-June.

★ **Galium aparine** Linnaeus. GOOSE-GRASS. Common, shady banks: The Sea Ranch; Salt Point; Duncans Mills Marsh, *Rubtzoff* (1953); Freestone Marsh, *Rubtzoff* (1959); 0.7 mi. sw. of Salmon Cr. crossing of Hwy. 1, *Rubtzoff* (1971); Bodega Head; Forestville Marsh, *Rubtzoff* (1955); near Sebastopol, *Trynor* (1938 NCC); n. of Hwy 12 w. of Santa Rosa in Laguna de Santa Rosa drainage, *Harrison* (1978 NCC); Fairfield Osborn (*); Button Ranch on Walker Road (U. C. property) between Petaluma-Valley Ford Road and

Pepper Road, acc. Patterson (1990); Bouverie; Gates Road off Calistoga Road, acc. Wells et al. (1984); Rincon Valley, *Wells* (1990); Pepperwood; Lake Sonoma area; Geyserville quad. < 5000 ft. Mar.-July. —*G. spurium* Linnaeus var. *echinospermum* (Wallroth) Hayek. Perhaps a very early introduction from Europe.

Galium bolanderi A. Gray. BOLANDER'S BEDSTRAW. Uncommon, with pines and oaks or lush chaparral: Mt. Hood, *Bioletti* (1893 UC). < 8400 ft. May-Aug.

Galium californicum Hooker & Arnott. CALIFORNIA BEDSTRAW, CLEAVERS. Common, woods and open hills: Annapolis quad.; Stewarts Point, *Dempster & Stebbins* (UC); Salt Point; 1 mi. n. of Fort Ross, *Howell* (1930); Cazadero and Duncans Mills quads.; 1 mi. e. of Bodega Bay, *Howell* (1930); Graton Road, *Baker* (1926 NCC); near Sebastopol, *Trynor* (1938 NCC); Fairfield Osborn (H); Button Ranch on Walker Road (U. C. property) between Petaluma-Valley Ford Road and Pepper Road, acc. Patterson (1990); Bouverie; Pepperwood; Lake Sonoma area; ca 2000 ft. elev. Anderson Trail on Black Mtn. between Geyser Peak and upper end of Pine Flat Road, acc. J. & B. Guggolz (1992). < 5700 ft. Mar.-July.

★ **Galium divaricatum** Lamarck. LAMARCK'S BEDSTRAW. Occasional, grassy areas: Stewarts Point quad.; Freestone Marsh, *Rubtzoff* (1962); Forest Hills 5½ mi. e. of Guerneville, *Rubtzoff* (1974); 3 sites w. and sw. of Santa Rosa in Laguna de Santa Rosa drainage, *Harrison* (1978 NCC); jct. Todd and Llano roads, *Best* (1987); 6½ mi. ne. of Penngrove on Lichau Road, *Rubtzoff* (1959); Bouverie; mouth of Adobe Cr. Can., *Howell* (1934); Macartney Woods s. of Maplewood (Baker's home), *Baker* (1927 NCC); Mt. Hood, *Michener & Bioletti* (1893 UC); sw. slope of Mt. Hood, *Baker* (1947 NCC); 4½ mi. w. of Glen Ellen on Bennett Valley Road, *Rubtzoff & Froebe* (1971); Bennett Mtn. Lake, *Rubtzoff* (1963); Lake Sonoma area. < 2300 ft. May-July. Native of Europe.

★ **Galium murale** (Linnaeus) Allioni. TINY BEDSTRAW. Uncommon, grassy places: Fairfield Osborn (H); Spring Lake; Lake Sonoma area. < 2300 ft. Apr.-May. Native of Europe.

Galium muricatum Wight. HUMBOLDT BEDSTRAW. Uncommon, dense or light shade of damp forests: e. of Stewarts Point, *Dempster & Stebbins* (1968 UC). < 3700 ft. Mar.-July.

★ **Galium parisiense** Linnaeus. WALL BEDSTRAW. Occasional, waste open areas: Austin Cr. Beach ½ mi. n. of Cazadero, *Rubtzoff* (1970); ca 1 mi. nw. of Camp Meeker, acc. J. & B. Guggolz (1991); Vacation Beach near Guerneville, *Rubtzoff* (1956); upper Green Valley Marsh nw. of Graton, *Rubtzoff & Harrison* (1974); Fairfield Osborn (H); 6½ mi. ne of Penngrove on Lichau Road, *Rubtzoff* (1969); Bouverie; Jack London State Park, *Wagnon* (1989); Spring Lake; Pepperwood; Lake Sonoma area; Geyserville, Ida Clayton Road quad.; ca 2000 ft. elev. Anderson Trail on Black Mtn. between Geyser Peak and upper end of Pine Flat Road, acc. J. & B. Guggolz (1992). < 7300 ft. Apr.-Aug. Introduced from Mediterranean regions.

Galium porrigens Dempster. CLIMBING BEDSTRAW. Common, dry slopes, scandent: Salt Point; Fort Ross and Cazadero quads.; 8½ mi. n. of Bodega Bay, *Howell* (1930); Bodega Bay, *Howell* (1934); Bodega Head, *Connors* (1992 BML); Bodega Bay-Valley Ford Hwy., *Knight* (1964); Coleman Valley Road w. of Occidental, *Stocking* (1969 NCC); Stoetz Road ne. of Occidental, *Ferris & Lorraine* (1961); Harrison Grade Road from jct. of Green Valley Road, *Ferris & Lorraine* (1961); upper Green Valley Marsh area, *Rubtzoff* (1974); 1 mi. n. of Sebastopol on Molino Road, *Trynor* (1958 NCC); Crane Can., *Stocking* (1963 NCC); Fairfield Osborn (*); 2 mi. n. of Sonoma between Sonoma and Boyes Hot Springs, *Howell* (1929); Bouverie; Sugarloaf Ridge; Los Guilicos, *Stocking* (1966 NCC); Monan's Rill e. of Calistoga Road and n. of St. Helena Road, *Wells* (1985); Gates Road off Calistoga Road on Humbug Creek, acc. Wells et al. (1984); Pepperwood; Mill Cr. Road, *Knight* (1964); Healdsburg, *Hoemann* (1981 NCC); Fitch Mtn., *Rubtzoff* (1952); Preston (2 mi. n. of Cloverdale on US 101), *J. I. Carlson* (1917); Lake Sonoma area; Ida Clayton Road quad.; ca 2000 ft. elev. Anderson Trail on Black Mtn. between Geyser Peak and upper end of Pine Flat Road, acc. J. & B. Guggolz (1992); Geyser Peak, *Knight & Edwards* (1986); Geysers geothermal area, *Quibell* (1980 NCC); geysers Unit 13, *Quibell* (1981 NCC). < 3700 ft. Mar.-June. —Separated from *G. nuttallii* A. Gray.

★ **Galium tricornutum** Dandy. ROUGH CORN BEDSTRAW. Uncommon, introduction: Sebastopol, *K. Brandegee* (1905 UC). < 6700 ft. Apr.-May. Native of Europe.

Galium trifidum Linnaeus var. **pacificum** Wiegand. Common, wet places: The Sea Ranch; 2 mi. e. of Stewarts Point, *Nobs & G. Smith* (1949); Austin Cr. Marsh between Hwy. 116 and Cazadero, *Rubtzoff* (1954); Duncans Mills Marsh, *Rubtzoff* (1954); marshy spot near Salmon Cr., *Baker* (1930 NCC); lower Salmon Cr. Marsh, *Rubtzoff* (1962); Bodega Head Marsh, *Rubtzoff* (1954); Bodega Harbor Marsh, *Rubtzoff* (1955); Willow Creek, *Rubtzoff* (1959); Freestone Marsh, *Rubtzoff* (1959); Duncans Mills Marsh, *Rubtzoff* (1957); Forestville Marsh, *Rubtzoff* (1955); Atascadero Cr. Marsh, *Rubtzoff* (1958); Pitkin Marsh, *Howell*

(1937), *Rubtzoff* (1951 NCC); Cunningham Marsh ca 1 mi. sw. of Cunningham, *Howell* (1946, 1947); Chileno Valley, *Rubtzoff* (1971); marsh 3 mi. w. of Fulton on River Road, *Rubtzoff* (1970). < 8000 ft. June-Aug. —Has been confused with var. *subbiflorum* Wiegand.

Galium triflorum Michaux. SWEET-SCENTED BEDSTRAW. Uncommon, wooded places, moist shaded canyons: Annapolis, Stewarts Point and Salt Point quads.; redwood forest 1 mi. n. of Fort Ross, *Howell* (1930); Duncans Mills quad.; 8 mi. n. of Bodega Bay, *Howell* (1931); sw. of Occidental, *Hoover* (UC); Guerneville Marsh, *Rubtzoff* (1954); Bohemian Hwy. between Occidental and Freestone, *Baker* (1930 NCC, 1931 NCC); upper Green Valley Marsh nw. of Graton, *Rubtzoff* (1974); Button Ranch on Walker Road (U. C. property) between Petaluma-Valley Ford Road and Pepper Road, acc. Connors (1992); Pepperwood. < 10000 ft. May-July. Also Europe and Asia.

★ **Sherardia arvensis** Linnaeus. FIELD MADDER. Occasional, meadows, lawns, orchards, fields: The Sea Ranch; se. of lower Salmon Cr. Marsh, *Rubtzoff* (1967); 8 mi. n. of Bodega Bay, *Howell* (1931); Duncans Mills Marsh, *Rubtzoff* (1960); Todd Road Preserve; Hidden Oaks Road between Sonoma Mtn. and Grange roads, acc. Quibell (1991); Fairfield Osborn (H); common at Annadel, *Barnhart* (1989 SRJC); Spring Lake; Pepperwood; Healdsburg quad.; Lake Sonoma area. < 2000 ft. Jan.-July. Native of s. Europe.

RUTACEAE. RUE FAMILY

★ **Ruta graveolens** Linnaeus. RUE. Rare, garden origin: Russian Colony (perhaps Fort Ross), *Voznesenskii* (1840-1841 LE). Native of the Mediterranean region.

SALICACEAE. WILLOW FAMILY

1. Stamens 2-5, rarely to 9; buds with a single scale; bracts entire or merely toothed, persistent or tardily dehiscent; disk reduced to a gland . . . . *Salix*
1: Stamens numerous; buds with several imbricated scales; bracts fimbriate, caducous; flowers on a cup-shaped disk . . . . *Populus*

Populus: Identification of the numerous forms of poplars is difficult because of the variation which occurs in the size and shape of leaves among closely related species and upon different parts of the same tree. Much of this variation has originated from hybridization among our native species and those of Europe and Asia. In selecting leaves for identification, material should be taken from mature lateral branches and never from suckers or fast-growing new shoots.

1. Leaves white-tomentose beneath, palmately 3- or 5-lobed or toothed . . . . *P. alba*
1: Leaves glabrous on both surfaces or slightly pubescent on veins beneath, pinnately veined, not lobed
  2. Petioles round in cross-section, leaves distinctly glaucescent, whitish or rusty beneath, broadly ovate, 6.2-13 cm long . . . . *P. balsamifera* ssp. *trichocarpa*
  2: Petioles flattened; leaves often paler beneath but not distinctly glaucous or whitish
    3. Leaves 7.5-12.5 cm long . . . . *P.* × *canadensis*
    3: Leaves 4-8.7 long
      4. Trees very slender, ascending or appressed branches forming a narrow columnar crown, 13-30 m tall; leaves broadly cuneate at base, with small coarse teeth, 5-8 cm long . . . . . *P. nigra* var. *italica*
      4: Trees pyramidal or spreading, 12-30 m tall; leaves slightly cordate to truncate, coarsely crenate-dentate, 4-8 cm long . . . . *P. fremontii*

★ **Populus alba** Linnaeus. WHITE OR SILVER POPLAR, sometimes called SILVER MAPLE. Sometimes produce large colonies by root-suckering: Occidental Road half way between Jonive Road and Facendini Lane, acc. Knight (1990); Forestville Marsh, *Rubtzoff* (1972); Sonoma Mtn. Road at jct. with Pressley Road, acc. Quibell (1991); along Ducker Cr., Santa Rosa, *Wells* (1985); around Petaluma acc. Knight. < 6000 ft. Native of central Europe and Asia.

Populus balsamifera Linnaeus ssp. **trichocarpa** (Torrey & A. Gray) Brayshaw. BLACK COTTONWOOD. Rare, in the 1930's there were two large colonies, one above and one below the westward bend of the

Russian River, acc. *Griffin & Critchfield* (1972): opposite Hacienda Beach, *Rubtzoff* (1987); Healdsburg and Geyserville quads. < 9300 ft. Feb.-Apr. —*P. t.* Torrey & A. Gray.

★ **Populus × canadensis** Moench (*P. deltoides* Bartram × *P. nigra* Linnaeus). CAROLINA POPLAR. Although mostly grown horticulturally from a single ♂ clone, it suckers vigorously. Grown throughout the United States and Canada but uncommon in this county: "The old Griswold place," *Baker* (1920 NCC); Fife Cr. Marsh near Guerneville, *Rubtzoff* (1960).

Populus fremontii S. Watson. FREMONT COTTONWOOD, ALAMO. Occasional, especially along waterways, also used horticulturally: Vacation Beach near Guerneville, *Rubtzoff* (1959); Fife Cr. Marsh near Guerneville, *Rubtzoff* (1960); Odd Fellows Park on Russian River, *Knight* (1964); Forest Hills, *Rubtzoff* (1965); Forestville Marsh, *Rubtzoff* (1973); Brush Cr. dike between Mission Blvd. and Brush Cr. Road, Santa Rosa, *Wells* 1987); tributary of Mark West Cr. 3 mi. w. of Fulton, *Rubtzoff* (1970); Pepperwood, Lake Sonoma area; Big Sulphur Cr. and Cloverdale quads. < 6700 ft. Mar.-Apr.

★ **Populus nigra** Linnaeus var. **italica** Du Roi. LOMBARDY POPLAR. Commonly cultivated fastigiate tree that can grow to 100 ft. high and produce a colony of suckers: two trees with sucker-colonies in Santa Rosa, acc. Wells (1984). Native in s., central and e. Europe; only the male of the species has been imported into the United States. —*P. i.* Moench.

Salix: Various species are often difficult to identify because of the variations in foliar characters. There are exotic, weeping, semiweeping, corkscrew and yellow-twigged willows which have been variously named: **Salix blanda** Andersson, **S. babylonica** Linnaeus, **S. matsudana** Koidz, **S. tortuosa** Linnaeus and **S. vitellina** Linnaeus. These are all horticultural species.

1. Leaves grayish-white appearance on both sides, strap-shaped . . . . . *S. exigua*
1: Leaves green on top
  2. Leaf-blades entire or nearly so
    3. Shrub to ca 1 m tall, serpentine endemic . . . . . *S. breweri*
    3: Shrub > 1 m tall, off serpentine
      4. Leaves subrevolute, nearly entire
        5. Leaves glabrous beneath . . . . . *S. lasiolepis* var. *lasiolepis*
        5: Leaves ± hairy beneath, sometimes rusty-hairy . . . . . *S. l.* var. *bigelovii*
      4: Leaves not subrevolute or nearly entire
        6. Leaves densely silvery-white-hairy beneath, growing in thickets; mostly coastal orientation . . . . . *S. sitchensis*
        6: Leaves densely to thinly pubescent, usually individual small trees . . . . . *S. scouleriana*
  2: Leaf-blades distinctly serrulate
    7. Leaf-blades tending to be elliptic, strap-shaped, acute at both ends . . . . . *S. melanopsis*
    7: Leaf-blades lanceolate, wider
      8. Glands near junction of petiole and blade . . . . . *S. lucida* ssp. *lasiandra*
      8: Glands not near junction of petiole and blade . . . . . *S. laevigata*

Salix breweri Bebb. BREWER'S WILLOW. Rare, serpentine endemic: Layton Mine on Austin Cr., *Hoffman* (1947); base of Mt. Hood, *Baker* (1931 NCC), *Howell* (1933). < 4500 ft. Mar.-Apr.

Salix exigua Nuttall. NARROW-LEAVED WILLOW, SANDBAR WILLOW. Common, along streams: Austin Cr. Marsh between Hwy. 116 and Cazadero, *Rubtzoff* (1954); Willow Creek; along Russian River near Guerneville, acc. Knight (1990); Laguna de Santa Rosa near Trenton, *Rubtzoff* (1960); Todd Road 1.4 mi. from Hwy. 116 and 0.6 mi. from Llano Road, *Rubtzoff* (1966); Stony Point Road ½ mi. from Railroad Ave., *Knight* (1963); Petaluma Hill Road near jct. of Rohnert Park Expressway, acc. Knight (1990); Lynch Cr. near N. McDowell Blvd. in Petaluma, acc. Knight (1990); Bennett Valley, *Heller* (1902); Spring Lake; along Leslie Cr. 110 yds. above confluence of Mark West Cr., *de Nevers* (1981); St. Helena Road 1 mi. from Napa-Sonoma co. line, acc. Knight (1990); pond ca 3 mi. w. of Fulton on River Road, *Rubtzoff* (1969); near Windsor on the Russian River, *Heller* (1902); Russian River at Fitch Mtn. grove, *Rubtzoff* (1964); Maacama Creek 15 mi. nw. of Calistoga, *Ball, Bracelin, Parks & Parks* (1931); Lake Sonoma area. < 8000 ft. Mar.-May. —*S. hindsiana* Bentham. *S. sessilifolia* var. *h.* (Bentham) Andersson.

Salix laevigata Bebb. RED WILLOW. Common, streams: Vacation Beach near Guerneville, *Rubtzoff* (1957); Fife Cr. Marsh at Guerneville, *Rubtzoff* (1961); Laguna de Santa Rosa near Trenton, *Rubtzoff* (1960); Atascadero Cr. Marsh w. of Sebastopol, *Rubtzoff* (1966); San Antonio Cr. on e. side of Hwy. 101, *Rubtzoff* (1972); near Sonoma, *Heller* (1902); Kenwood Marsh, *Rubtzoff, Harrison, Hass & Quibell* (1974);

Sugarloaf Ridge; banks of Santa Rosa Cr., *Baker* (1930); 1 mi. s. of Santa Rosa on Santa Rosa Cr., *Ball & Parks* (1931); Spring Lake; River Road 3 mi. w. of Fulton, *Rubtzoff* (1969); Pepperwood; near Windsor on Russian River, *Heller* (1902); Russian River at Fitch Mtn. grove, *Rubtzoff* (1952); Lake Sonoma area; Big Sulphur Cr. quad. <5700 ft. Mar.-May.

Salix lasiolepis Bentham var. **bigelovii** Bentham. Occasional, near streams: Bodega Bay, *Rubtzoff & Crovello* (1964); 1.2 mi. e. of Bodega Bay, *Baker, Ball, Bracelin & Crum* (1931); Pitkin Marsh, *Rubtzoff* (1952); Laguna de Santa Rosa e. of Cunningham, *Rubtzoff* (1962); Willow Brook n. of Petaluma, acc. Knight & Wagnon (1989).

Salix lasiolepis var. **lasiolepis**. ARROYO WILLOW. Very common, along streams: Willow Cr. Marsh, *Rubtzoff* (1959); Bodega Head, acc. Connors (1991); Pitkin Marsh, *Rubtzoff* (1952); 5 mi. ne. of Penngrove on Lichau Road, *Rubtzoff* (1964); Willow Brook n. of Petaluma, acc. Knight & Wagnon (1988); Lynch Cr. near N. McDowell Blvd., Petaluma, acc. Knight (1990); hills 2 mi. se. of Santa Rosa, *Heller* (1902); Santa Rosa Cr. e. of Santa Rosa, *Heller* (1902); Spring Lake; Pepperwood; Maacama Cr. ca 15 mi. nw. of Calistoga, *Ball, Bracelin, Parks & Parks* (1931); Russian River at Fitch Mtn. grove, *Rubtzoff* (1952); Lake Sonoma area. < 9300 ft. Feb.-Apr.

Salix lucida Muhlenberg ssp. **lasiandra** (Bentham) E. Murray. SHINING WILLOW. Common, along streams: 1.2 mi. e. of Bodega Bay, *Baker, Ball, Bracelin & Crum* (1931); Bodega Point, *Eastwood* (1915); Vacation Beach near Guerneville, *Rubtzoff* (1953); Fife Cr. Marsh at Guerneville, *Rubtzoff* (1962); Pitkin Marsh, *Rubtzoff* (1951); Stony Point Marsh, *Rubtzoff* (1955); Norrbom Road near Sonoma, *Knight & Edwards* (1986); bank of slough draining Kenwood Marsh, *Howell* (1932); Santa Rosa Cr. e. of Santa Rosa, *Heller* (1902); Spring Lake; near Windsor on Russian River, *Heller* (1902); Big Sulphur Cr. quad. < 10500 ft. Mar.-May. —*S. lasiandra* Bentham.

Salix melanopsis Nuttall. DUSKY WILLOW. Rare, creek and river bottoms: Guerneville, *McMinn* (1922); Bohemian Grove near Monte Rio, *Setchell* (1931); Russian River flood plain at Hacienda, *Wagnon* (1988 NCC); Geyserville, *Tracy* (1932); along Maacama Cr. near confluence of Russian River near Biggio Road, *Wagnon* (1988 NCC). < 9000 ft. Mar.-May. —*S. exigua* ssp. *m.* Cronquist.

Salix scouleriana Hooker. SCOULER'S WILLOW. Occasional, dry habitats, usually slopes: The Sea Ranch, *Wells* (1981); Stewarts Point quad.; Salt Point; near Fort Ross, *Setchell* (1934); Duncans Mills quad.; 3 mi. w. of Sebastopol, *Baker* (1933); Bohemian Grove near Monte Rio, *Setchell* (1933); ca 1 mi. nw. of Camp Meeker, acc. J. & B. Guggolz (1991); 4 mi. s. of Camp Meeker, *Mexia* (1933); 2 mi. e. of Occidental; *Baker, Ball, Bracelin & Crum* (1931); Joy Road sw. of Occidental, *Schneebeli* (1963); Freestone Marsh, *Rubtzoff* (1962); 0.75 mi. nw. of Freestone, *Rubtzoff & Eckenwalder* (1973); Pitkin Marsh, *Rubtzoff* (1951) . < 11300 ft. Apr.-June. —*S. flavescens* Nuttall. *S. stagnalis* Nuttall. *S. brachystachys* Bentham. *S. capreoides* Andersson. *S. nuttallii* Sargent.

Salix sitchensis Bongard. SITKA WILLOW. Occasionally in interior of county, but especially along coast: The Sea Ranch on Hwy. 1 just n. of firehouse, *Quibell* (1988 NCC); Kruse Ranch, *Setchell* (1934); Salt Point, *Wells* (1982); Bodega Head, *Connors* (1992 BML); Austin Cr. ½ mi. n. of Cazadero, *Rubtzoff* (1970); ca 1 mi. e. of Bodega Bay, *Baker, Ball, Bracelin & Crum* (1931); Willow Creek Marsh, *Rubtzoff* (1959); between Monte Rio and Camp Meeker, *Setchell* (1933); Russian River between Vacation Beach and e. Guernewood, *Rubtzoff* (1959); Hacienda, *Wagnon* (1988 NCC); end of Nuns Can. Road off of Hwy. 12, acc. Wagnon (1991); Mt. Hood, *Heller* (1902); at confluence with Redwood Cr. and Maacama Cr., *J. & B. Guggolz* (1990 JEPS); Maacama Cr. 15 mi. nw. of Calistoga, *Ball, Bracelin, Parks & Parks* (1931). 1400 ft., higher in Siskiyou and Humboldt cos. Feb-Apr. —*S. coulteri* Andersson, now included.

SAXIFRAGACEAE. SAXIFRAGE FAMILY

[See PHILADELPHIACEAE & GROSSULARIACEAE]

(Key ± acc. P. E. Elvander, *Hickman* [1993])

1. Flowers solitary, showy; modified sterile stamens 5, toothed to fringed; styles minute or inconspicuous . . . . *Parnassia*

1: Flowers usually 2 or more, or not showy; modified sterile stamens 0; styles usually apparent; placentas 2-3, parietal or axile

  2. Stamens 3, rarely 2; petals 4 (rarely 5); hypanthium bilateral, open on 1 side . . . . *Tolmiea*

  2: Stamens 5 or 10; petals usually 5; hypanthium usually radial, closed

    3. Stamens 5

4. Leaves mostly basal; ovary chamber 1, placentas 2, parietal; inflorescence a panicle . . *Heuchera*
4: Leaves both basal and cauline; ovary chambers 2, placentas 2, axile; inflorescence ± round-topped . . . . . *Boykinia*

3: Stamens 10
5. Ovary chamber 1, placentas parietal
6. Styles and placentas 3 . . . . . *Lithophragma*
6: Styles and placentas 2
7. Inflorescence a panicle; fruit valves unequal, wide; petal lobes 0 . . . . . *Tiarella*
7: Inflorescence a raceme; fruit valves equal, narrow; petals usually 5 . . . . . *Tellima*
5: Ovary chambers 2, sometimes 3; placentas axile . . . . . *Saxifraga*

Boykinia occidentalis Torrey & A. Gray. Fairly common, damp shady places: The Sea Ranch, acc. *Mahaffey* (1990); along cr. s. of Mt. Hood among redwoods, *Stocking* (1968 NCC); along Sonoma Cr. at the foot of Mt. Hood, *Heller* (1902); Los Alamos Cr. Can., *Baker* (1927 NCC); Deer Cr. off Pine Flat Road, *Baker* (1928 NCC); The Geysers, *Carruth* (1899). < 5000 ft. June-July. —*B. elata* (Nuttall) E. Greene.

Heuchera

1. Styles 2-4 mm, exserted; hypanthium broadly obconic, base tapered; inflorescence usually open *H. micrantha*
1: Styles usually < 1.8 mm, barely exserted; hypanthium bell-shaped, base truncate; inflorescence usually dense . . . . . *H. pilosissima*

Heuchera micrantha Lindley. ALUMROOT. Common, moist banks of rocks and humus: The Sea Ranch; Salt Point; Cazadero, *Bacigalupi* (1926); Willow Creek; ca 1 mi. nw. of Camp Meeker, acc. J. & B. Guggolz (1991); Guerneville, *Rattan* (1877); Russian River Terrace, *Bately* (1932 NCC); Neeley Hills s. of Guerneville, *Rubtzoff* (1954); Sonoma Mtn. Road near Cooper's Grove, *Baker* (1934); Tresch Ranch, Walker Road near Two Rock, acc. J. & B. Guggolz (1992); Button Ranch on Walker Road (U. C. property) between Petaluma-Valley Ford Road and Pepper Road, acc. Connors (1992); Bouverie; Gates Road off Calistoga Road, acc. Wells et al. (1984); Burke's Sanitarium on Mark West Cr. Road, *Heller* (1902); Stewarts Point-Skaggs Springs Road 9 mi. e. of Stewarts Point, *Best* (1981); Skaggs Springs, *Hawver* (1915); Lake Sonoma area; Geysers-Healdsburg Road, *Bresslin* (1975 NCC). < 8300 ft. May-July. —*H. m.* var. *erubescens* (A. Braun & Bouché) Rosendahl.

Heuchera pilosissima Fischer & C. Meyer. COAST ALUMROOT. Common, wooded slopes, predominately coastal: Gualala Point State Park, *Knight* (1984); The Sea Ranch; Salt Point; Wright's Beach ocean bluffs, *Ripley & Barneby* (1945 NCC); Coleman Valley Road ca 5 mi. w. of Occidental, *Best* (1985); Bodega Point, *Eastwood* (1901, 1917); cliff near Larkspur Rock e. of Bodega Bay, *Baker* (1926 NCC); 1 mi. e. of Bodega Bay, *Howell* (1930); can. near Bodega Bay, *Heller* (1938). Type locality: Fort Ross area. < 1700 ft Apr.-June.

Lithophragma

1. Calyx-tube turbinate; petals mostly 3-toothed at apex, the central tooth rather larger; hypanthium ± long-obconic; ovary usually > half inferior . . . . . *L. affine*
1: Calyx-tube short campanulate, broad; petals not toothed; hypanthium spheric to bell-shaped; ovary < half inferior . . . . . *L. heterophyllum*

Lithophragma affine A. Gray. WOODLAND STAR. Common, moist grassy banks: Larkspur Rock 1 mi. e. of Bodega Bay, *Ott* (1934 NCC); 2 sites w. and sw. of Santa Rosa in Laguna de Santa Rosa drainage, *Harrison* (1978 NCC); Todd Road Preserve; Lafferty Ranch on Sonoma Mtn., acc. Arnold et al. (1971); Fairfield Osborn (S); Tresch Ranch, Walker Road near Two Rock, acc. J. & B. Guggolz (1992); Bouverie; Sugarloaf Ridge; Lake Sonoma area; meadow nw. of geysers Unit 13, *Self, Thompson & Quibell* (1980 NCC). < 6700 ft. Mar.-May.

Lithophragma heterophyllum (Hooker & Arnott) Torrey & A. Gray. HILLSIDE STAR. Common, shady slopes: Bodega Head; Vine Hill region, *Baker* (1932 NCC); Fairfield Osborn (S); Petaluma, *Congdon* (1883); Bouverie; Sugarloaf Ridge; Gates Road off Calistoga Road, acc. Wells et al. (1984); hills e. of Santa Rosa, *Heller* (1902); Santa Rosa, *Congdon* (1883); e. of Spring Lake, acc. Lovell (1982); Pepperwood; inland end Stewarts Point-Skaggs Springs Road, *R. J. Thompson* (1975 NCC); Lake Sonoma area; Sausal Cr. off Pine Flat Road, Applegate (1929); between Jimtown and Mercuryville on Geysers Road, *Utech & Knight* (1986 CM). < 5000 ft. Mar.-June.

Parnassia parviflora deCandolle. GRASS-OF-PARNASSUS. Occasional, wet meadows, bogs: Cazadero, *Graff* (1920); The Cedars, *Raiche* (1984); Mt. Hood, *Baker* (1925 NCC). < 10000 ft. July-Oct. —*P. palustris* Linnaeus var. *californica* A. Gray.

Saxifraga

1. Petals unequal, 2 of them 3-4 times longer than the other 3; plant extensively stoloniferous . . . *S. stolonifera*
1: Petals equal; rhizomes or bulblets
  2. Leaves ± round, base usually cordate . . . *S. mertensiana*
  2: Leaves not round, nor cordate
    3. Blade ± ovate; sepals usually spreading to rarely reflexed, ovate; petals 2-4 mm, > sepals, obovate . . . *S. integrifolia*
    3: Blade ovate; sepals usually reflexed, elliptic; petals 2.5-4.5 mm, > > sepals, elliptic to round
      4. Leaf-teeth shallow, sepals reflexed . . . *S. californica*
      4: Leaf-teeth coarse, sepals tend to ± spread . . . *S. fallax*

Saxifraga californica E. Greene. SAXIFRAGE. Common, shady often grassy banks: grade nw. of Cazadero, *Cantelow* (1939); Larkspur Rock 1 mi. e. of Bodega Bay, *Hagg* (1934 NCC); Todd Road Preserve; Sonoma Mtn. Road w. of Enterprise Road and e. of Cooper's Grove, acc. Quibell (1991); Bouverie; Hooker Cr. near Mt. Hood, *Raven* (1957); Sonoma Cr. at base of Mt. Hood, *Heller & Brown* (1902); Rincon Valley, *Mallory* (1920); Pepperwood; Lake Sonoma area. < 4000 ft. Feb.-June

Saxifraga fallax E. Greene. Uncommon, mountainous rocky areas: n. slope of Sonoma Mtn., *Baker* (1948 NCC). < 7500 ft. Mar.-May.

Saxifraga integrifolia Hooker. Occasional, rocky wet areas: Lovell Valley ca 4 mi. e. of Sonoma, *Bacigalupi* (1946); golf club links (possibly El Verano or Santa Rosa), *Baker* (1948 NCC); Santa Rosa, *Meyer* (1928); Pepperwood. < 3300 ft. Apr.-July. —*S. fragosa* Suksdorf. *S. i.* var. *claytoniifolia* (Canby) Rosendahl.

Saxifraga mertensiana Bongard. Common, damp rocky areas: Cazadero, *Heller* (1902 JEPS); Willow Cr. Road 1 mi. from Bridgehaven, *Utech* (1987 CM); common around Camp Meeker on wet rocky cliffs, *Travaglini* (1973 SRJC), *Symes* (1979 SRJC); Bodega Point, *Eastwood* (1915 JEPS); Larkspur Rock e. of Bodega Bay, *Ott* (1934 NCC); off Green Valley Road w. of Graton, acc. Rockwell (1991); Austin Cr. Road, *Rupp* (1970 NCC); grassy slope above Dry Cr. w. of Warm Springs, *Breedlove* (1981); 1 mi. w. of Skaggs Springs, *Bacigalupi* (1926); Lake Sonoma area; w. end Knights Valley, *R. J. Thompson* (1975 NCC). < 8300 ft.- Feb-July.

★ **Saxifraga stolonifera** Meerburgh. STRAWBERRY-GERANIUM. Naturalized, garden plant: Guernewood Park, *L. Roof* (1933 NCC); Sonoma Co., acc. *Munz* (1959). Native of Asia. —*S. sarmentosa* Linnaeus.

Tellima grandiflora (Pursh) Lindley. FRINGE CUPS. Common, rocky areas, damp woods: edge of forest near S. Fork Gualala River at The Sea Ranch, *Wells* (1981); Fort Ross, *Heller* (1903); Bodega, *Heller & Brown* (1902); Willow Cr. Road, *Utech* (1988 CM); Coleman Valley Road, *Stocking* (1969 NCC); 1 mi. nw. of Freestone, *Rose* (1962); Franklin School Road, Valley Ford, acc. Quibell (1991); Guerneville, *Congdon* (1880); Davis Cr. which flows into Green Valley Cr. ½ mi. e. of Bones Road nw. of Graton, *L. L. Davis* (1975 NCC); Button Ranch on Walker Road (U. C. property) between Petaluma-Valley Ford Road and Pepper Road, acc. Patterson (1990); Tresch Ranch, Walker Road near Two Rock, acc. J. & B. Guggolz (1992). < 6700 ft. Apr.-June.

Tiarella trifoliata Linnaeus var. **unifoliata** (Hooker) Kurtz. SUGAR-SCOOP, LACEFLOWER. Occasional, shady places :Stewarts Point-Skaggs Springs Road 3 mi. e. of Stewarts Point, *Best* (1981); under redwoods 1 mi. n. of Fort Ross, *Howell* (1930); Willow Cr. Road, *Utech* (1988 CM); under redwoods, Monte Rio, *Stocking* (1967 NCC); Camp Meeker, *Travaglini* (1973 SRJC); bank of Dutch Bill Cr. 1 mi. n. of Occidental, *Best* (1980); common in Armstrong Grove, acc. Harrison (1992). ± < 5000 ft. May-July. —*T. u.* Hooker.

Tolmiea menziesii (Pursh) Torrey & A. Gray. PIG-A-BACK PLANT. Rare, damp cool areas: Salt Point. < 6000 ft. May-June.

SCROPHULARIACEAE. FIGWORT FAMILY

1. Fertile stamens 5; leaves alternate; corolla almost regular, rotate . . . . *Verbascum*
1: Fertile stamens 4 or 2; leaves opposite or alternate; corolla generally ± labiate
  2. Plants acaulescent; corolla almost rotate, 1.5 mm long; anther cells wholly confluent . . . . . . *Limosella*
  2: Plants mostly caulescent; corolla mostly not rotate, larger; anther-cells distinct
    3. Stigmas distinct, flattened or platelike
      4. Corolla campanulate, not distinctly 2-lipped . . . . *Digitalis*
      4: Corolla almost 2-lipped
        5. Connective of stamens wider than the parallel anther-cells; corolla whitish with yellow tube . . . . *Gratiola*
        5: Connective of stamens not so expanded, the anther-cells not parallel
          6. Anther-bearing stamens 2 . . . . *Lindernia*
          6: Anther-bearing stamens 4
            7. Plants shrubby, corolla yellow . . . . [*Diplacus*] *Mimulus*
            7: Plants annual or biennial herbs; corolla purplish, violet, red to yellow or buff *Mimulus*
    3: Stigmas wholly united, capitate or pointed
      8. Corolla spurred or saccate on lower side of base, generally also with prominent palate; capsule opening by pores, or irregularly
        9. Corolla gibbous or saccate at base . . . . *Antirrhinum*
        9: Corolla with narrow spur at base
          10. Flowers in terminal racemes or spikes; stems ± erect at least in upper part; leaves linear to lanceolate or ovate . . . . *Linaria*
          10: Flowers in axils of broad leaves; stems trailing or twining
            11. Leaves glabrous, palmately veined and lobed; capsule opening by 2 pores; flowers violet . . . . *Cymbalaria*
            11: Leaves pubescent, pinnately veined, rounded to hastate; capsule ± circumscissile; flowers white or yellowish- white . . . . *Kickxia*
      8: Corolla not spurred or saccate on lower side at base, 2-lipped to nearly regular, without prominent palate; capsule valvate
        12. Corolla with upper lip flattened or widely arched, not forming a galea
          13. Upper lip of corolla 2-lobed
            14. Plants annual; corolla gibbous on upper side of base
              15. Lower lip of corolla appearing 2-lobed, the middle lobe keel-shaped and concealed by the lateral lobes; stamens included in keel . . . . . . *Collinsia*
              15: Lower lip of corolla 3-lobed; stamens exserted . . . . *Tonella*
            14: Plants perennial; corolla not gibbous on upper side
              16. Corolla ± brownish, mostly < 10 mm long, inflated, with 4 erect lobes and 1 reflexed; sterile stamen reduced to a scale or wanting . . . . *Scrophularia*
              16: Corollas mostly blue, lavender, rose, white, yellow, rarely brownish, 10-40 mm. long, tubular or funnnelform, ± bilabiate; sterile stamen usually ca = fertile ones
                17. Plant a shrub or subshrub 3-20 dm high . . . . *Keckiella*
                17: Plant suffrutescent, matted or forming clumps 1.5-5 dm high . . . . *Penstemon*
          13: Upper lip of corolla appearing 1-lobed by fusion of 2
            18. Stems elongating and leafy . . . . *Veronica*
            18: Stems scapose, leaves basal . . . . *Synthyris*
        12: Upper lip of corolla narrowly arched, forming a galea which encloses the anthers
          19. Anther-cells = in size and position; seed-coat not loose and not reticulate
            20. Leaves opposite; capsule symmetrical; plants annual or biennial
              21. Corolla white and pink; calyx-lobes not =, obtuse; flowers crowded in dense inflorescence with spreading bracts . . . . *Bellardia*
              21: Corolla yellow, (reddish-purple in *P. latifolia*); calyx-lobes similar, lanceolate; flowers scattered with ascending bracts . . . . *Parentucellia*

20: Leaves alternate or basal; capsule asymmetrical, generally decurved; plants perennial . . . . *Pedicularis*

19: Anther-cells unequally placed, the upper attached by its middle, the lower normally attached or abortive; seed-coat usually loose and reticulate

22. Corolla beak-tip closed (opening directed downward); stigma unexpanded (dotlike); seed attachment lateral . . . . *Cordylanthus*

22: Corolla beak-tip open; stigma expanded (headlike or 2-lobed); seed attachment terminal

23. Stamens each with 1 anther sac; lower corolla lip either obviously 3-pouched or corolla very small, 4-6 mm . . . . *Triphysaria*

23: Stamens each with 2 anther sacs; lower corolla lip 3-pouched or not, usually > 8 mm . . . . *Castilleja*

Antirrhinum (Acc. ± to key by D. M. Thompson).

1. Capsule not oblique, dehiscing by irregular bursting; stems twining and supported by the long capillary pedicels . . . . *A. kelloggii*

1: Capsule ± oblique, dehiscing by fairly definite terminal or subterminal pores

2. Stems self-supporting, lacking filiform twisted branchlets

3. Plants stout; flowers reddish, 16-19 mm long; corolla-tube merely saccate at base; plant glabrous throughout its vegetative parts; leaves linear . . . . *A. virga*

3: Plants slender; flowers 10-12 mm long, bluish with yellow palate; corolla-tube with spurlike base; calyx-hairs to 2/3 as long as calyx-segments . . . . *A. cornutum*

2: Stems in mature plants largely supported by twisted branchlets, or at least possessing them

4. Flowers 12-15 mm long, palate pubescent; dorsal calyx-segments 8-12 mm long . . . . *A. vexillo-calyculatum* ssp. *vexillo-calyculatum*

4: Flowers 10-12 mm long, palate hirsute-villose; dorsal calyx-segments 4-7 mm long . . . . *A. v.-c.* ssp. *breweri*

Antirrhinum cornutum Bentham. Uncommon, recently disturbed places: 3.3 mi. up Sonoma Mtn. Road, *Englehart* (1973 SRJC). < 4100 ft. May-July.

Antirrhinum kelloggii E. Greene. Uncommon, dry slopes, especially burns: near Kenwood, *Baker* (1947 NCC); Nuns Can. off Hwy. 12 near Kenwood, *Baker* (1925 NCC); Mt. Hood, *Baker* (1947 NCC). < 4300 ft. Mar.-May. —*A. hookerianum* Pennell.

Antirrhinum vexillo-calyculatum Kellogg ssp. **breweri** (A. Gray) D. Thompson. Uncommon, open rocky areas: 1¼ mi. nw. of Camp Meeker, *J. & B. Guggolz* (1991 NCC); Pepperwood. < 6700 ft. June-Aug. —*A. b.* A. Gray.

Antirrhinum vexillo-calyculatum ssp. **vexillo-calyculatum**. SNAPDRAGON. Common, rocky slopes, often on burns, sometimes on serpentine: Fort Ross-Russian River-Bodega areas, *Voznesenskii* (1840-41 LE); The Cedars, acc. Raiche (1984); Bohemian Hwy. 3.6 mi. nw. of Occidental, *Huntington* (1986 NCC); Pepperwood; Fitch Mtn., *Rubtzoff* (1952); Stewarts Point Road opposite entrance to View Point Road adjacent to Lake Sonoma, *Guggolz* (1977 NCC); ca 2000 ft. elev. Anderson Trail on Black Mtn. between Geyser Peak and upper end of Pine Flat Road, acc. J. & B. Guggolz (1992). Acc. D. M. Thompson, 3 collections of intermediates nearest to *A. vexillo-calyculatum* ssp. *v.-c.* x *A. v-c.* ssp. *breweri* are represented by: Russian Gulch Road near bridge on old road (probably close to Hwy. 1 above Jenner), *Baker* (1947 NCC); sides of Little Oak Mtn. near Cazadero, *V. Winter* (1932 NCC); Sulphur Cr. Hwy., *Baker* (1934 NCC). < 6600 ft. May-July. —*A. vagans* A. Gray.

Antirrhinum virga A. Gray. TALL SNAPDRAGON. Occasional, rocky and gravelly places: Fort Ross-Russian River-Bodega areas, *Voznesenskii* (1840-41 LE); Boyes Hot Springs, *Ackerman* (1985 SRJC); ca 300 ft. elev. Mountain Ave e. of Boyes Hot Springs, acc. J. & B. Guggolz (1992); Kenwood Springs, *Howell* (1932); Adobe Can,, *Howell* (1932); foot of Mt. Hood, *Heller* (1902); w. side of Dry Cr. Valley, *Raiche* (1985); 2700 ft. elev. upper Big Sulphur Cr. drainage 3 mi. due s. of Cobb Mtn. at summit above intersection of Kincaid (Burma) and Big Sulphur Cr. roads, acc. Stone (1983). < 6600 ft. June-July.

★ **Bellardia trixago** (Linnaeus) Allioni. Common, old fields, grassy areas: roadside 600 yds. e. of Two Rock, *V. Winter* (1935 NCC); Button Ranch on Walker Road (U. C. property) between Petaluma-Valley Ford Road and Pepper Road, acc. Patterson (1990); between Pepper Road and Two Rock Grange Hall on

Bodega Hwy., acc. Knight (1985); corner Payran and Jefferson sts., Petaluma, *Hilliard* (1979 NCC); w. of Sears Point Raceway in Tolay Cr. drainage, acc. Patterson (1990); e. boundary of Sonoma-Napa co. (probably Hwy. 12), *Kuhn*; Annadel; Pepperwood. < 1000 ft. Apr.-May. Native of S. Europe.

Castilleja (Key ± acc. T. I. Chuang & L. R. Heckard. *Hickman* (1993)].

1. Inflorescence not very reddish (usually green, white, yellow or purplish); flower ± short (usually bee-pollinated); lower lip 3-pouched, teeth usually ± erect, white or green to lavender, upper lip << 3 × lower [Owl's-clover]:
  2. Bracts green throughout
    3. Leaves and bracts entire, corolla pale yellow to orange . . . . *C. campestris*
    3: Leaves entire or lobed; bracts lobed; corolla yellow . . . . *C. rubicunda* ssp. *lithospermoides*
  2: Bracts tipped purplish, yellowish or whitish
    4. Corolla beak hooked, densely shaggy-hairy; filaments puberulent
      5. Upper bracts usually 5-7 mm wide, lobes < 2 mm, tipped pale lavender; inflorescence appearing alternately light- and dark-banded . . . . *C. exserta* ssp. *latifolia*
      5: Upper bracts < 5 mm wide, lobes > 2 mm; inflorescence not banded . . . . . . *C. e.* ssp. *exserta*
    4: Corolla beak straight, puberulent; filaments glabrous
      6. Inflorescence 10-20 mm wide; corolla linear, pouches ± 2 mm wide; stigma well-included . . . . *C. attenuata*
      6: Inflorescence 20-40 mm wide; corolla wider distally, pouches usually 3-5 mm wide; stigma usually ± exserted
        7. Branches usually many; leaf oblong > 3 mm wide; seed coat tight-fitting . . . . . *C. ambigua*
        7: Branches 0 or few from mid-stem; leaf linear-lanceolate < 3 mm wide; seed coat loose-fitting . . . . *C. densiflora*

1: Inflorescence ± red (yellow-orange); flower long (usually hummingbird-pollinated); lower lip of small, incurved,usually dark green (red) teeth, upper lip > 3 × lower [Indian paintbrush]:
  8. Plant white-woolly . . . . *C. foliolosa*
  8: Plant not white-woolly
    9. Calyx divided 2/3 in front, < 1/3 in back, lobes usually curved upward; corolla (including lower lip) usually curved out through front calyx sinus . . . . *C. subinclusa* ssp. *franciscana*
    9: Calyx divided ± 1/3-½ in front and back, lobes not curved upward; corolla usually curved out through front calyx sinus (lower lip included)
      10. Plant glandular-puberulent to -sticky below inflorescence
        11. Branches usually crowded below spike; leaves usually crowded; seed coat deeply netted; coastal . . . . *C. wightii*
        11: Branches few below spike; leaves usually ± well-spaced; seed coat shallowly netted; inland . . . . *C. applegatei* ssp. *martinii*
      10: Plant usually not glandular below spike (except barely in some *C. affinis*)
        12. Plant usually ± bristly-puberulent; leaves occasionally 5-lobed; dry places . . . . *C. affinis*
        12: Plant glabrous to sparsely puberulent; leaves entire; moist places . . . . *C. uliginosa*

Castilleja affinis Hooker & Arnott ssp. **affinis**. Occasional, dry wooded or brushy slopes: 8½ mi. n. of Bodega Bay, *Howell* (1930); Larkspur Rock 1 mi. e. of Bodega Bay, *Howell* (1931); Adobe Can., *Baker* (1927); hills ne. of Santa Rosa, *Heller & Brown* (1902); near Santa Rosa, *Alexander* (1944); Pepperwood; near Healdsburg, *Baker* (1929 NCC). < 4000 ft. Mar.-May.

Castilleja ambigua Hooker & Arnott ssp. **ambigua**. JOHNNY-NIP. Common, low saline areas: The Sea Ranch; Salt Point; 0.7 mi. se. of Salmon Cr., *Rubtzoff* (1971); 4 mi. n. of Bodega Bay, *Howell* (1930); Freestone Marsh, *Rubtzoff* (1959); Pitkin Marsh, *Howell* (1938, 1945); Atascadero Cr. Marsh w. of Sebastopol, *Rubtzoff* (1972); Cunningham Marsh, *Howell* (1947); 3 sites w. and sw. of Santa Rosa in Laguna de Santa Rosa drainage, *Harrison* (1978 NCC); 4 mi. e. of Valley Ford, *Howell* (1930); Bennett Mtn. Lake, *Rubtzoff* (1974); Pepperwood; vernal pool 2 mi. nw. of US 101 Windsor exit, *de Nevers* (1981). < 1000 ft. May-Aug.—*Orthocarpus sonomensis* Eastwood, type locality: 2 mi. n. of Windsor, *Eastwood & Howell* (1936). *O. castillejoides* Bentham.

Castilleja applegatei Fernald ssp. **martinii** (Abrams) Chuang & Heckard. Infrequent, dry slopes: Springs Road (probably St. Helena Road), *Okada* (1975); Pine Flat Road, *Barnhart* (1977 SRJC); Skaggs Springs, *Hawver* (1915); ca 0.5 mi. e. of U.S. 101 on Sulphur Cr. Road, *Andreae* (1992 NCC).> 9300 ft. May-June.

Castilleja attenuata (A. Gray) Chuang & Heckard. VALLEY TASSELS. Common, grassy areas: 5 sites w. and sw. of Santa Rosa in Laguna de Santa Rosa drainage, *Harrison* (1978 NCC); Pressley Road near Crane Creek Reg. Park, acc. Quibell (1991); Fairfield Osborn (S); e. foothills of Sonoma Mts. on Sobre Vista Road 0.9 mi. w. of Arnold Drive, *Rubtzoff* (1971 NCC; Bouverie; 4¼ mi. w. of Glen Ellen on Bennett Valley Road, *Rubtzoff & Froebe* (1971 NCC); Spring Lake; s. end of Santa Rosa fairgrounds, *Wagnon* (1938); Gates Road off Calistoga Road, acc. Wells et al. (1984); Spring Lake; Old Redwood Hwy. and Starr Road, *J. Smith* (1974 NCC); Pepperwood; Lake Sonoma area; property bordered by Chianti and Dutcher Cr. roads w. of US 101 and w. of Geyserville, acc. J. & B. Guggolz (1991); 2400 ft. elev. Little Geysers, *Malloch* (1977 PG&E). Usually < 4000 ft. Mar.-May. —*Orthocarpus attenuatus* A. Gray.

Castilleja campestris (Bentham) Chuang & Heckard. Occasional, moist areas, hog wallows, winter pools: near Wright School half way between Sebastopol and Santa Rosa, *Baker* (1935); 3 sites w. and sw. of Santa Rosa in Laguna de Santa Rosa drainage; *Harrison* (1978 NCC); Todd Road Preserve; False Lake Meadow in Annadel State Park, *Tunnison & Harrison* (1976); e. of Santa Rosa fairgrounds, *Wagnon* (1938). < 7700 ft. Apr.-July. —*Orthocarpus c.* Bentham.

Castilleja densiflora (Bentham) Chuang & Heckard. OWL'S-CLOVER. Common, grassy areas: 5 sites w. and sw. of Santa Rosa in Laguna de Santa Rosa drainage, *Harrison* (1978 NCC); Pressley Road near Crane Creek Reg. Park, acc. Quibell (1991); Fairfield Osborn (S); e. foothills of Sonoma Mts. on Sobre Vista Road 0.9 mi. w. of Arnold Drive, *Rubtzoff* (1971 NCC); Bouverie; 4¼ mi. w. of Glen Ellen on Bennett Valley Road, *Rubtzoff & Froebe* (1971 NCC); Spring Lake; s. end of Santa Rosa fairgrounds, *Wagnon* (1938); Gates Road off Calistoga Road, acc. Wells et al. (1984); Spring Lake; Old Redwood Hwy. and Starr Road, *J. Smith* (1974 NCC); Pepperwood; Lake Sonoma area; property bordered by Chianti and Dutcher Cr. roads w. of US 101 and w. of Geyserville, acc. J. & B. Guggolz (1991); 2400 ft. elev. Little Geysers, *Malloch* (1977 PG&E). < 4700 ft. Mar.-May. —*Orthocarpus densiflorus* Bentham.

Castilleja exserta (A. A. Heller) Chuang & Heckard ssp. **exserta**. PURPLE OWL'S-CLOVER. Common, grassy slopes, open areas: Bodega Point, *Eastwood* (1915); 4 mi. w. of Santa Rosa on Sebastopol Hwy., *Wagnon* (1938 NCC); El Verano, *Griffin* (1922); Sugarloaf Ridge; Santa Rosa, *Heller & Brown* (1902); 2 mi. nw. of Windsor exit from US 101, *de Nevers & Kaza* (1981). < 5400 ft. Mar.-May. —*Orthocarpus purpurascens* Bentham var. *p.*

Castilleja exserta ssp. **latifolia** (S. Watson) Chuang & Heckard. Infrequent, mostly on old dunes: Bodega Point, *Eastwood* (1915). < 700 ft. Mar.-May. —*Orthocarpus purpurascens* Bentham var. *latifolius* S. Watson.

Castilleja foliolosa Hooker & Arnott. CHAPARRAL PAINTBRUSH. Common, dry open and rocky places: Fairfield Osborn (S); Bouverie; 1½ mi. up Cavedale Road, *Utech & Knight* (1986 CM); Mt. Hood, *Lamb* (1970 NCC); Santa Rosa Cr. Can., *Mullin* (1933 NCC); Agua Caliente, Sonoma Valley, *Lehman* (1975 NCC); Sugarloaf Ridge; Spring Mtn. Road, *Knight* (1963); Pepperwood; chaparral area sw. of Bradford Mtn. 1½ mi. w. of Dry Cr., acc. J. & B. Guggolz (1992); Lake Sonoma area; ca 2000 ft. elev. Anderson Trail on Black Mtn. between Geyser Peak and upper end of Pine Flat Road, acc. J. & B. Guggolz (1992); Geyser Peak, acc. Knight & Edwards (1986); 2400 ft. elev. Little Geysers, *Malloch* (1977 PG&E). < 6000 ft. Mar.-June.

Castilleja rubicundula (Jepson) Chuang & Heckard ssp. **lithospermoides** (Bentham) Chuang & Heckard. CREAM SACS. Common, grassy open areas: The Sea Ranch, *Wells* (1982); Bodega Bay area, *Eastwood* (1915); jct. of Occidental Road and Facendini Lane, *Ferris & Lorraine* (1961); 2 sites w. of Santa Rosa in Laguna de Santa Rosa drainage, *Harrison* (1978 NCC); 1.4 mi. w. of Napa-Sonoma Co. line on Hwy. 12, *Thomas* (1966); Bennett Valley Road, *Baker* (1927 NCC); Pepperwood; Knights Valley, *E. Greene* (1894); Lake Sonoma area; property bordered by Chianti and Dutcher Cr. roads w. of US 101 and w. of Geyserville, acc. J. & B. Guggolz (1991); n. of Cloverdale, *Baker* (1924 NCC). < 3000 ft. Apr.-June. —*Orthocarpus l.* Bentham.

Castilleja subinclusa E. Greene ssp. **franciscana** (Pennell) Chuang & Heckard. Common, mostly coastal, sometimes inland, dry slopes: Miller Cr. Can. at Salt Point State Park, *Wells* (1982); Goat Rock Beach, *Best* (1984); 5 mi. n. of Bodega Bay, *Baker* (1931 NCC); Bodega Bay, *Heller & Brown* (1902); Larkspur Rock e. of Bodega Bay, *Howell* (1931), *Ott* (1934 NCC), *Eastwood & Howell* (1939); Bodega Harbour development on Pinnacle Gulch Trail, *Huntington* (1989 NCC); ca 1 mi. nw. of Camp Meeker, *J. & B. Guggolz* (1991 NCC); Hidden Acres Road above Sonoma Mtn. Road, *Quibell* (1988 NCC); Button Ranch on Walker Road (U. C. property) between Petaluma-Valley Ford Road and Pepper Road, acc. Connors (1992); 9 mi. from Petaluma en route to Marshall, *Koch* (1939 NCC); Dry Cr., *Cantelow* (1934), *Raiche* (1985). < 500 ft. Mar.-June. —*C. f.* Pennell.

Castilleja uliginosa Eastwood. Rare, marshy areas: Pitkin Marsh, *Baker* (1946 NCC), *Howell* (type 1937), *Rubtzoff & Heckard* (1971); Trembley's Marsh adjoining Pitkin Marsh, *Rubtzoff* (1953); s. end of n. part of Pitkin Marsh, acc. Lovell (1978). < 800 ft. June-July. Unclearly differentiated from *C. miniata* Hooker.

Castilleja wightii Elmer. Common, mostly coastal strand: The Sea Ranch; Miller Cr. at Salt Point State Park, *Wells* (1982); between Fort Ross and Jenner, *Beane & James* (1949), *Wells* (1985); Hwy. 1 just n. of Jenner above mouth of Russian River, *Huntington* (1989 NCC); Goat Rock Beach, *Wells* (1981); 1.2 mi. s. of bridge across Russian River at Bridgehaven; *Holmgren & Reveal* (1966); near Hwy. 1 on Coleman Valley Road, *J. H. Thomas* (1976); ca 1.6 mi. s. of Russian River bridge (probably near Wright's Beach), *Cantelow* (1939); Bodega Point, *Eastwood* (1915); Bodega Head; mouth of Estero Americano, acc. Quibell (1982). < 1000 ft. Mar.-July. —*C. inornata* Eastwood, type locality: Sonoma Coast n. of Bodega Bay, *Eastwood & Howell* (1939).

Collinsia (Key ± acc. E. C. Neese, *Hickman* (1993).

1. Flowers crowded in whorls; lower pedicels < calyx
   2. Inflorescence with corolla ± uniformly purple; upper lip with winged appendages > 1 mm near base . . . . . *C. greenei*
   2: Inflorescence with upper lip or entire corolla paler; upper lip wingless
      3. Inflorescence with upper filaments with curved basal spur > 1 mm projecting into pouch; corolla-throat as wide as long; calyx-lobes usually sharp-tipped . . . . . *C. heterophylla*
      3: Inflorescence with filaments ± unspurred; corolla throat longer than wide; calyx-lobes usually ± blunt-tipped . . . . . *C. tinctoria*

1: Flowers solitary or whorled, not very crowded; lower pedicels > calyx
   4. Inflorescence with largest corollas > 13 mm . . . . . *C. sparsiflora* var. *arvensis*
   4: Inflorescence with corollas < 13 mm
      5. Corolla 5-8 mm, throat barely angled to tube, pouch hidden by calyx . . . . . *C. s.* var. *collina*
      5: Corolla 7-13 mm, throat strongly angled to tube, pouch evident . . . . . *C. s.* var. *sparsiflora*

Collinsia greenei A. Gray. Uncommon, rocky and stony areas, sometimes on serpentine: Pine Flat and Middletown roads in Mayacmas Mts., acc. *Kruckeberg* (1984); 0.8 mi. w. of jct. of Socrates Mine Road with Pine Flat Road, *H. K. Sharsmith* (1952 NCC). < 8300 ft. May-July.

Collinsia heterophylla Buist. CHINESE HOUSES. Common, shady areas: Larkspur Rock 1 mi. e. of Bodega Bay, *Howell* (1931); 3 mi. w. of Occidental on Coleman Valley Road, *Stocking* (1969 NCC); jct. Hidden Oaks and Hidden Acres roads, and at Hidden Acres Road at w. end of Sonoma Mtn. Road, acc. Quibell (1991); Crane Can., *Stocking* (1963 NCC); Lichau Road e. of Petaluma Hill Road, *Stocking* (1964 NCC); Button Ranch on Walker Road (U. C. property) between Petaluma-Valley Ford Road and Pepper Road, acc. Patterson (1990); Bouverie; Sonoma Mtn. Road near Enterprise Road, *Utech & Knight* (1986 CM); Kenwood, *A. & R. Nelson* (1933); Sugarloaf Ridge; Pepperwood; Petrified Forest, *Heller* (1902); mts. w. of Calistoga, *Krauss* (1894); Felta Cr. area sw. of Healdsburg, *Bice* (1895 NCC); 350-400 ft. elev. Old Skaggs Springs Road, *Utech, Sato & Ohara* (1990 CM); Geysers-Healdsburg Road 10.1 mi. n. of red winery, *Breslin* (1975 NCC); between Jimtown and Mercuryville on Geysers Road, *Utech & Knight* (1986 CM); Skaggs Springs, *Heller* (1940); Lake Sonoma area. < 3400 ft. Mar.-June. —*C. bicolor* Bentham.

Collinsia sparsiflora Fischer & C. Meyer var. **arvensis** (E. Greene) Jepson. BLUE-EYED MARY. Common, grassy places: n. of Hwy. 12 w. of Santa Rosa in Laguna de Santa Rosa drainage, *Harrison* (1978 NCC); Petaluma Hill Road, *Focht* (1932 NCC); near sw. entrance of Sonoma State Univ. campus, acc. Quibell (1991); Boyes Hot Springs, *Keck & Howell* (1931); Kenwood, *Michener & Bioletti* (1893), *Ramp* (1925); Bennett Valley, *Baker* (1923 NCC); hills se. of Santa Rosa, *Heller & Brown* (1902); Spring Lake; Pepperwood, *Breedlove* (1983); Healdsburg, *Rattan* (1880); Lake Sonoma area; Pine Flat area, *Applegate* (1934). < 2500 ft. Mar.-May. —*C. a.* E. Greene; type locality: Knights Valley, *Greene* (1888 UC).

Collinsia sparsiflora var. **collina** (Jepson) V. Newsom. Uncommon, grassy areas: Rincon Valley, *Baker* (1928 NCC); Pepperwood. < 4000 ft. Mar.-June.

Collinsia sparsiflora var. **sparsiflora**. SLENDER BLUE-EYED MARY. Common, grassy areas: Fort Ross, *Heller* (1903); Coleman Valley, *Baker* (1929 NCC); Coleman Valley Road 5 mi. w. of Occidental, *Best* (1983); ca 1 mi. nw. of Camp Meeker, acc. J. & B. Guggolz (1991); Larkspur Rock 1½ mi. e. of Bodega

Bay, *Ott* (1934 NCC); n. of E. Railroad Ave. off Petaluma Hill Road, *Juul* (1973 NCC); Tresch Ranch, Walker Road near Two Rock, acc. J. & B. Guggolz (1992); Bouverie; Annadel; MacDonald Reservoir (now Lake Ralphine, Santa Rosa), *Baker* (1929 NCC); Pepperwood; Felta Cr. area sw. of Healdsburg, *Bice* (1895 NCC); Lake Sonoma area; geysers Unit 18, *Quibell* (1980 NCC); type locality: Fort Ross. < 3300 ft. Mar.-May.

Collinsia tinctoria Bentham. Occasional, mostly stony areas: near Cazadero, *Carruth* (1907); Duvoul Cr. serpentine near Camp Meeker, *Tokubo* (1977 SRJC); Camp Meeker, *Herran* (1982 SRJC); Bodega Head; Bouverie; ne. of Kenwood Springs, *Pennell* (1940); Pepperwood; ca 2000 ft. elev. Anderson Trail on Black Mtn. between Geyser Peak and upper end of Pine Flat Road, acc. J. & B. Guggolz (1992). < 8300 ft. May-Aug.

Cordylanthus

1. Plants erect 2-4 dm high, (*C. maritimus* ssp. *palustris* 2-4 dm high but decumbent); herbage very pale green; stamens 2 . . . . *C. mollis*

1: Plants taller; herbage green (sometimes tinted purplish); stamens 4
  - 2. Leaves oblong to lanceolate, at least the lower entire; flowers in an elongate spike; calyx spathelike, enclosing proximal part of corolla . . . . *C. maritimus* ssp. *palustris*
  - 2: Leaves or their segments linear to filiform; flowers in heads or scattered; calyx proper narrow, enclosing corolla at base only
    - 3. Outer bracts fan-shaped, only shallowly 3-7-lobed; floral bract and calyx roughened-papillate distally on outer surface; corolla < 1 cm long with yellow lateral pouches . . . . *C. pringlei*
    - 3: Outer bracts entire or 3-parted up to 2/3 their length; floral bract scabrous to puberulent; corolla 1-2 cm long with white or ivory lateral pouches
      - 4. Main stem glabrescent to densely glandular-puberulent with scattered stiff trichomes; outer bract (or its central segment if trifid) linear with narrow or expanded apex < 1.5 mm wide, hirsutulous to hirsute with longest trichomes < 2 mm; corolla with posterior (upper) surface of pouch and throat heavily blotched with maroon-purple
        - 5. Outer bracts entire; stems minutely and sparsely glandular-puberulent, especially below flowers . . . . *C. tenuis* ssp. *brunneus*
        - 5: Outer bracts 3-parted; stems largely glabrous, not glandular below flowers *C. t.* ssp. *capillaris*
      - 4: Main stems glandular-puberulent with dense, long weak trichomes; outer bract (or its central segment if trifid) narrowly oblong to linear with expanded apex > 1.5 mm wide, pilose with longest trichomes 2-3 mm; corolla with upper surface of pouch and throat lightly marked, if at all, with maroon-purple stripes . . . . *C. pilosus*

Cordylanthus maritimus Bentham ssp. **palustris** (Behr) Chuang & Heckard. PT. REYES BIRD'S-BEAK. Rare, coastal salt marshes: salt marsh on w. side of Bodega Bay, *Howell* (1955); Bodega Head; n. of Doran Co. Park, acc. Lovell (1982); Bodega Head Salt Marsh, acc. Lovell (1982). < 6400 ft. May-Oct. —*C. m.* Nuttall.

Cordylanthus mollis A. Gray ssp. **mollis**. SOFT BIRD'S-BEAK. Rare, estuarine conditions: Petaluma Marsh between San Antonio and Mudhen Slough, observed by J. Michaels (1978), reported to California Native Plant Society data base. < 40 ft. July-Nov.

Cordylanthus pilosus A. Gray. Occasional, dry openings in oak woodland, chaparral, areas of serpentine: Camp Meeker, *Hoffman* (JEPS, UC); 4.1 mi. up Cavedale Road from Hwy. 12 near the McNab Cypress, *Knight & Howell* (1984); 1½ mi. sw. of Kenwood on lower road to Glen Ellen, *Howell* (1932); Kenwood, *Baker* (1932 NCC); Howarth Park, Santa Rosa, *Knight* (1984); 2 mi. n. of Healdsburg, *Heckard et al.* (1986 JEPS). < 4200 ft. July-Sept.

Cordylanthus pringlei A. Gray. Uncommon, dry slopes and flats of serpentine, chaparral, mixed evergreen forests: headwaters of E. Austin Cr., *Hoffman* (1948 UC). < 5700 ft. Aug.-Sept.

Cordylanthus tenuis A. Gray ssp. **brunneus** (Jepson) Munz. SERPENTINE BIRD'S-BEAK. Uncommon, often on serpentine: Mt. Hood, *Pennell* (1940 UC); Pepperwood; near Bradford Mtn. w. of Healdsburg, *Raiche* (1985); chaparral area sw. of Bradford Mtn. 1½ mi. w. of Dry Cr., *J. & B. Guggolz* (1992 NCC). Usually < 3000 ft. July-Aug.

Cordylanthus tenuis ssp. **capillaris** (Pennell) Chuang & Heckard. PENNELL'S BIRD'S-BEAK. Locally common, areas of serpentine, chaparral: Camp Meeker, acc. Lovell (1978), *Edwards* (1984); 2 mi. nw. of Camp Meeker, *J. & B. Guggolz* (1991 JEPS); Harrison Grade Road between Occidental and Graton,

Knight (1963); 2 mi. ne. of Occidental along Harrison Grade Road, *Baker* (1947 NCC); along Stoetz Lane ca ½-1 mi. nw. of jct. with Harrison Grade Road, *Bacigalupi & Heckard* (1977 JEPS); base of sw. shoulder of Bradford Mtn., acc. J. & B. Guggolz & W. Knight (1992). < 1000 ft. July-Sept. —*C. c.* Pennell, type locality: serpentine area 2 mi. e. of Occidental, *Mason* (1950 UC).

★ **Cymbalaria muralis** Gaertner, Meyer & Scherbius. KENILWORTH-IVY. Uncommon, escaping in cool moist areas: Pepperwood; Healdsburg, *Albright* (1979 SRJC). Introduced from Europe. < 3300 ft. May-Sept. —*Linaria c.* (Linnaeus) P. Miller.

★ **Digitalis purpurea** Linnaeus. FOXGLOVE. Occasional, naturalized in shady places: Gualala River near mouth, acc. Quibell (1991); The Sea Ranch, acc. *Mahaffey* (1990); Salt Point; Bohemian Ave., Monte Rio, *Morse* (1981 NCC); ca 1 mi. nw. of Camp Meeker, acc. J. & B. Guggolz (1991); 3/4 mi. w. of Occidental, *Lu Watters* (1964); reseeds in flower beds at Royal Oaks Mobile Home Park, Petaluma, acc. Knight (1991); Bouverie; Pepperwood; Lake Sonoma area. < 3300 ft. May-Sept. Native of Eurasia.

Gratiola ebracteata Bentham. HEDGE-HYSSOP. Occasional, muddy areas: near Wright School w. of Santa Rosa, *Baker* (1930 NCC); 2 sites w. and sw. of Santa Rosa in Laguna de Santa Rosa drainage, *Harrison* (1978 NCC); Todd Road Preserve; 3 mi. e. of Valley Ford on road to Petaluma, *Howell* (1930); Bennett Valley Road near summit, *Baker* (1927 NCC). < 6600 ft. Apr.-Aug.

Keckiella

1. Corolla not red, not distinctly tubular
    2. Staminode glabrous; corolla white, tinged with pink, long-hairy externally . . . . *K. breviflora*
    2: Staminode densely bearded; corolla purplish-brown, lower lip pale yellow, brown-purple-lined, short-pubescent externally . . . . *K. lemmonii*

1: Corolla brick-red, distinctly tubular . . . . *K. corymbosa*

Keckiella breviflora (Lindley) Straw. GAPING BEARDTONGUE. Uncommon, dry rocky areas: Geyser Peak, acc. Knight & Edwards (1986). < 9000 ft. May-July. —*Penstemon breviflorus* Lindley.

Keckiella corymbosa (Bentham) Straw. REDWOOD PENSTEMON. Uncommon, rocky slopes, cliffs: e. of Cazadero, *Bacigalupi* (1947); Lake Sonoma area; Geyser Peak, *Knight & Edwards* (1986). < 5300 ft. June-Oct. —*Penstemon corymbosus* Bentham.

Keckiella lemmonii (A. Gray) Straw. LEMMON'S BEARDTONGUE. Rare, brushy or wooded slopes: Lake Sonoma area; Geysers Road near Sulphur Cr. bridge, acc. Lovell (1982). < 6400 ft. June-Aug. —*Penstemon l.* A. Gray.

Kickxia

1. Sepals ovate; leaves rounded or cordate at base; corolla (without the spur) 6-8 mm long . . . . *K. spuria*

1: Sepals lanceolate; leaves hastate-lobed at base; corolla 5 mm long . . . . *K. elatine*

★ **Kickxia elatine** (Linnaeus) Dumortier. FLUELLIN. Common, naturalized: Vacation Beach 2 mi. down stream from Guerneville, *Rubtzoff* (1962); Guerneville on road to Armstrong Redwoods State Park, *Bravo* (1982); 2 sites w. of Santa Rosa in Laguna de Santa Rosa drainage, *Harrison* (1978 NCC); mouth of Crane Can. along roadside in flatland, *Knight* (1963); Willow Brook, *Knight & Edwards* (1984); Spring Lake; Santa Rosa Jr. Coll. campus, *Baker* (1942); horse paddock e. of Parker Hill Road, Santa Rosa, *Wells* (1985); Pepperwood. < 3300 ft. June-Sept. Introduced from Europe.

★ **Kickxia spuria** (Linnaeus) Dumortier. Uncommon, alkaline soil: Schollenberger Park, Petaluma, *W., I, & T. Knight* (1987 NCC). Introduced from Europe.

Limosella

1. Style 0.2-0.4 mm long, usually sharply decurved at base; corolla-lobes acute, dull white or pinkish; leaves 2-8 mm wide . . . . *L. aquatica*

1: Style 0.5-1 mm long, straight or arcuately curved; corolla-lobes rounded, white or with violet tinge; leaves 0.5-2 mm wide . . . . *L. acaulis*

Limosella acaulis Sessé & Moçiño. Uncommon, muddy shores: Button Ranch on Walker Road (U. C. property) between Petaluma-Valley Ford Road and Pepper Road, *Connors* (1992 BML); Spring Lake. < 11000 ft. May-Oct.

Limosella aquatica Linnaeus. MUDWORT. Rare, drying muddy shores of ponds and streams: near Occidental Road near Sebastopol in Laguna de Santa Rosa drainage, *Harrison* (1978 NCC). < 10600 ft. June-Sept.

Linaria

1. Flowers bright yellow; corolla including the spur 2-3 cm long . . . . . *L. vulgaris*
1: Flowers not yellow
    2. Spur strongly curved, placed transversely or obliquely, corolla including spur 15-20 mm long, violet with whitish ridges instead of well-formed palate . . . . . *L. canadensis*
    2: Spur nearly straight, vertical; corolla 35-38 mm long including spur, violet-purple with small yellowish patch on palate . . . . . *L. maroccana*

Linaria canadensis (Linnaeus) Dumont de Courset. BLUE TOADFLAX. Occasional, waste places: Pitkin Marsh, *Rubtzoff* (1952 NCC); Petaluma, *Ott* (1933 NCC); Spring Lake Village; Stewarts Point-Skaggs Springs Road 8.2 mi. e. of Berkeley Youth Camp, *Quibell* (1980 NCC); 3000 ft. elev. Little Geysers, *Malloch* (1978 PG&E). < 6000 ft. Mar.-May. —*L. c.* var. *texana* (Scheele) Pennell, indistinct.

★ **Linaria maroccana** Hooker f. Uncommon, escape from cultivation: planted by Caltrans along US 101 at Piner Road, Santa Rosa, *de Nevers* (1982). < 1700 ft. Native of Mediterranean.

★ **Linaria vulgaris** Hill. BUTTER-AND-EGGS. Occasional, weed, escape: Bay Hill Road between Bodega Bay Hwy. and Hwy. 1, *Knight* (1965); w. of Bodega 2 mi. s. of Hwy. 1, along Estero Lane, *G. T. Robbins* (1958); Bodega, *Eastwood* (1915). < 3300 ft. Summer. Native of Mediterranean.

Lindernia dubia (Linnaeus) Pennell var. **anagallidea** (Michaux) Cooperrider. FALSE PIMPERNEL. Uncommon, damp places: jct. Starr Road and Old Redwood Hwy., *J. & B. Guggolz* (1995 UC); Ledson Marsh, *Harrison* (1979 SRJC). < 5300 ft. July-Sept. —*L. a.* (Michaux) Pennell.

Mimulus

1. Corolla yellow
    2. Plants shrubby . . . . . *M. aurantiacus*
    2: Plants herbaceous
        3. Corolla 2-lipped
            4. Uppermost calyx-lobe rarely > 2 × the length of the others . . . . . *M. guttatus*
            4: Uppermost calyx-lobe almost 3 × as long as the others . . . . . *M. nasutus*
        3: Corolla nearly regular, scarcely 2-lipped
            5. Plant perennial, ± slimy . . . . . *M. moschatus*
            5: Plant annual, not slimy
                6. Calyx not ridged or not low-winged along the septal midribs . . . . . *M. pilosus*
                6: Calyx ridged or low-winged along the septal midribs . . . . . *M. floribundus*
1: Corolla reddish or purplish
    7. Fruit symmetric, the walls papery
        8. Plant perennial . . . . . *M. cardinalis*
        8: Plant annual
            9. Corolla 6-10 mm long, mature calyx 3-8 mm. long . . . . . .*M. rattanii*
            9: Corolla 10-40 mm long, mature calyx 5-25 mm. long
                10. Corolla-lobes subequal, stigma-lobes subequal . . . . . *M. layneae*
                10: Corolla-lobes unequal, stigma-lobes unequal . . . . . *M. bolanderi*
    7: Fruit asymmetric, the walls firm and coriaceous
        11. Corolla slightly 2-lipped . . . . . *M. tricolor*
        11: Corolla strongly 2-lipped
            12. Lower lip of corolla < 1/3 as long as upper lip . . . . . *M. douglasii*
            12: Lower lip of corolla > 1/3 as long as upper lip
                13. Corolla 3-4.5 cm. long; mature calyx 12-15 mm. long . . . . . *M. kelloggii*
                13: Corolla 1.5-2.5 cm. long; mature calyx 9-12 mm. long . . . . . *M. congdonii*

Mimulus aurantiacus Curtis. NORTHERN STICKY MONKEYFLOWER. Very common, rocky areas: The Sea Ranch; Annapolis quad.; Salt Point; Fort Ross near San Andreas fault, *McKenzie* (1981); Cazadero, Duncans Mills and Arched Rock quads.; Willow Creek; Bodega Dunes, *Connors* (1991 BML); Bodega Head; Pinnacle Gulch Trail in Bodega Harbour development, *Huntington* (1989 NCC); 4 mi. ne. of Occidental, *Brundage* (1933 NCC); 2 mi. w. of Olivet Road on River Road between Fulton and Mirabel Park, *J. Smith* (1974 NCC); Fairfield Osborn (S); Button Ranch on Walker Road (U. C. property) between Petaluma-Valley Ford Road and Pepper Road, acc. Patterson (1990); Bouverie; 1½ mi. up Cavedale Road from Hwy. 12, *Utech & Knight* (1986 CM); Mt. Hood, *Lamb* (1970 NCC); Bennett Valley Road, acc. Quibell (1990); Spring Mtn. Road, *Knight* (1963); Annadel, acc. Quibell (1991); Pepperwood; Felta Cr. area sw. of Healdsburg, *Bice* (1895 NCC); chaparral area sw. of Bradford Mtn. 1½ mi. w. of Dry Cr., acc. J. & B. Guggolz (1992); 350-400 ft. elev. Old Skaggs Springs Road, *Utech, Sato & Ohara* (1990 CM); Lake Sonoma area; Geyserville and Ida Clayton Road quads.; ca 2000 ft. elev. Anderson Trail on Black Mtn. between Geyser Peak and upper end of Pine Flat Road, acc. J. & B. Guggolz (1992); Big Sulphur Cr., Cloverdale, Mt. Tom and N. Fork Buckeye Cr. quads. < 5300 ft. Mar.-Aug. —*Diplacus a.* (Curtis) Jepson.

Mimulus bolanderi A. Gray. Occasional, dry open areas such as burns: Mt. Hood, *Michener & Bioletti* (1892 MIN). Usually < 5700 ft. May-July.

Mimulus cardinalis Bentham. SCARLET MONKEYFLOWER. Fairly common, seeps, stream banks: Austin Cr. 3½ mi. s. and ½ mi. n. of Cazadero, *Rubtzoff* (1970); ½ mi. e. of Duncans Mills, *Kawahara* (1968); Willow Cr. Marsh, *Rubtzoff* (1959); Monte Rio, *Hoak* (1912); ca 1 mi. nw. of Camp Meeker, acc. J. & B. Guggolz (1991); Vacation Beach near Guerveville, *Rubtzoff* (1958); Fife Cr. Marsh at Guerneville, *Rubtzoff* (1959); Forest Hills across river from Hacienda, *Rubtzoff* (1965); Russian River at Mirabel Park, *Rubtzoff* (1970); Sebastopol, *Dows* (1907); Lake Sonoma area; Pine Flat Road, *Knight* (1985); ca 2000 ft. elev. Anderson Trail on Black Mtn. between Geyser Peak and upper end of Pine Flat Road, acc. J. & B. Guggolz (1992). < 8000 ft. Apr.-Oct.

Mimulus congdonii B. L. Robinson. Locally common, often on or around rocky outcrops, disturbed places: ca 2 mi. from Sonoma on road to Napa in chaparral burn, *Howell* (1954); 1200-1400 ft. elev. 3.6 mi. ne. of Hwy. 12 on Cavedale Road in burned area, *Ammirati* (1965 SFSU); Trinity Mtn. e. of Glen Ellen, *Howell* (1937); Hope Valley near Kenwood, *Baker* (1950 NCC); Adobe Can. opposite Bush's place, *Baker* (1927 NCC); near Mt. Hood, *Bioletti* (1894 UC); slopes of Mt. Hood, *Heller* (1902 NY); 1100 ft. elev. se. slope of Mt. Hood, *Baker* (1927); Los Alamos Can. (commonly known as Santa Rosa Cr. Can., 4-5 mi. e. of Santa Rosa), *Baker* (1927 JEPS); ca 4 mi. up St. Helena Road from Rincon summit, *Baker* (1928 NCC); burned flats near Franz Valley School, *Raven* (1965 MO); Franz Valley, *Howell* (1941); 400 ft. elev. 1¼ mi. from Calistoga Road on Franz Valley School Road in chaparral burn, *Ammirati* (1965 SFSU). < 3700 ft. Mar.-May. —*M. modestus* Eastwood.

Mimulus douglasii (Bentham) A. Gray. PURPLE MOUSE EARS. Locally common, moist meadows, rocky places, often on serpentine: near foot of Trinity Grade, *Baker* (1938 NCC); Macartney Woods near Baker's house in Adobe Can., *Baker* (1927 NCC); 1300 ft. elev. se. slope of Mt. Hood, *Baker* (1927 NCC); Sonoma Mtn. Road near Bruner's, not far from Warm Springs Road, *M. S. & A. G. Baker* (1925 NCC), *Baker & Griffin* (1925); Hope Valley near Kenwood, *Baker* (1950 NCC); on serpentine at Pepperwood, *Breedlove* (1983); 1.1 mi. n. of Porter Cr. Road along Franz Valley Road near roadcut of serpentine, *D. M. Thompson* (1989 RSA); Upper Dry Cr. on Jensen Ranch, *Baker* (1936 NCC); on serpentine at Lake Sonoma area Liberty Glen Campground entrance kiosk, *D. M. Thompson* (1989 RSA); 2400 ft. elev. Little Geysers, *Malloch* (1977 PG&E). < 4000 ft. Mar.-May.

Mimulus floribundus Lindley. Uncommon, moist areas: Russian River at Vacation Beach, *Rubtzoff* (1956); bed of Sonoma Cr., *Rattan* (1885); at spring on Cavedale Road, *Knight* (1984); Mill Cr. 4½ mi. from Healdsburg, *Thomas* (1934 NCC). < 8300 ft. Apr.-Aug.

Mimulus guttatus de Candolle. LARGE MONKEYFLOWER. Common, moist banks, open areas: The Sea Ranch; Stewarts Point-Skaggs Springs Road 15.6 mi. e. of Hwy. 1, *Alimonti* (1988 NCC); Horseshoe Cove, *Knight* (1963); Salt Point; 1 mi. n. of Meyers Grade jct. on Hwy. 1, *Huntington* (1989 NCC); Goat Rock Beach, *Wells* (1981); Bodega Head, *Knight* (1963); Stoetz Lane e. of Occidental, *Best* (1980); *Knight* (1985); Cunningham Marsh, *Howell* (1947); Todd Road Preserve; Copeland Cr. near Sonoma State Univ. campus, *Self* (1979 NCC); 1½ mi. from top of Sonoma Mtn. Road, *Lowry* (1983 SFSU); Fairfield Osborn (*); Button Ranch on Walker Road (U. C. property) between Petaluma-Valley Ford Road and Pepper Road, acc. Patterson (1990); hills e. of Agua Caliente bordering Sonoma Valley, *Howell* (1948); Bouverie; 1400 ft. elev. Cavedale Road, acc. Knight (1985); Sugarloaf Ridge; Pepperwood; 350-400 ft. elev. Old

Skaggs Springs Road, *Utech, Sato & Ohara* (1990 CM); ca 2000 ft. elev. Anderson Trail on Black Mtn. between Geyser Peak and upper end of Pine Flat Road, acc. J. & B. Guggolz (1992); Little Geysers, *Quibell* (1980 NCC). < 8300 ft. Mar.-Aug.

Acc. D. M. Thompson: "The taxa of the *M. guttatus* complex can appear distinct in small areas, but they are not when variation throughout their ranges is taken into consideration. I do not recognize ssp. *arvensis* (E. Greene) Munz, ssp. *litoralis* Pennell, *M. glareosus* E. Greene and *M. nasutus* E. Greene. This is a very complex species—best treated broadly until we know more about it."

Mimulus kelloggii (E. Greene) A. Gray. Common, damp disturbed areas: se. slope of Mt. Hood, *Baker* (1927); Santa Rosa-St. Helena Road ca 4 mi. e. of Rincon summit, *Baker* (1928 NCC); Spring Mtn. Road, *Chandler* (1907 UC), *Howell* (1931); Mark West Can., *Baker* (1927 NCC), *Learned* (1927 NCC); near Mark West Springs, *Heller & Brown* (1902 GH); Pepperwood; 1½ mi. w. of Petrified Forest, *Gould* (1938 UC); Petrified Forest, *Curran* (1884), *Eastwood* (1915 UC); 350 ft. elev. road from Mark West Springs to Kellogg, *Balls & Lenz* (1952 RSA); Geysers geothermal area, *Quibell* (1980 NCC); headwaters of Sulphur Cr., *Applegate* (1934); The Geysers near Pine Mtn. (probably Pine Flat Road), *Bolander* (1864 UC). < 5300 ft. Mar.-June.

Mimulus layneae (E. Greene) Jepson. LAYNE'S MONKEYFLOWER. Locally common, dry sandy disturbed areas: Sonoma, *Burlingame* (1921); Hope Valley near Kenwood, *Baker* (1950 NCC); Petrified Forest, *Eastwood* (1915); Franz Valley Road 1.25 mi. from the road to Calistoga in chaparral burn, *Ammirati* (1965 SFSU); 400 ft. elev. Franz Valley School Road 1 mi. w. of the Sonoma-Napa co. line in a chaparral burn, *Ammirati* (1965 SFSU); 720 ft. elev. along Franz Valley School Road 1.0-1¼ mi. w. of the Napa Co. line (¼ mi. w. of the school) on the 1964 burn, *Bacigalupi* (1965 JEPS); w. flank of Mt. St. Helena, *Grummer* (1983 PUA); 2400 ft. elev. Little Geysers, *Malloch* (1977 PG&E). < 6700 ft. May-Aug.

Mimulus moschatus Lindley. MUSK MONKEYFLOWER. Common, wet places: The Sea Ranch; 1 mi. s. of Gualala River, *Wolf* (1927); Stewarts Point, *Griffin* (1925); Duncans Mills Marsh, *Rubtzoff* (1955); Willow Cr. Marsh, *Rubtzoff* (1959); 1 mi. n. of Occidental on road parallel to Bohemian Hwy., *Best* (1980); Freestone Marsh, *Rubtzoff* (1959); Salmon Cr. 5 mi. w. of town of Bodega, *Rubtzoff* (1969); Bohemian Grove near Monte Rio, *Bacigalupi* (1921); Guerneville, *Rattan* (1877); Guerneville Marsh, *Rubtzoff* (1954); Forestville Marsh, *Rubtzoff* (1955); Pitkin Marsh, *Baker* (1928 NCC), *Howell* (1947); Laguna de Santa Rosa ne of Graton, *Rubtzoff* (1960). < 9700 ft. June-Aug. —*M. m.* var. *sessilifolius* A. Gray.

Mimulus nasutus E. Greene. Infrequent, wet sandy or gravelly places: hills e. of Agua Caliente bordering Sonoma Valley, *Howell* (1948); Bouverie; 1400 ft. elev. Cavedale Road, acc. Knight (1985). Type locality: Knights Valley and Skaggs Springs. < 1500 ft. Mar.-Aug. Sonoma County botanists prefer to have this remain as a species. See Thompson's analysis under *M. guttatus*.

Mimulus pilosus (Bentham) S. Watson. Common, moist sandy or gravelly areas: The Sea Ranch at sandbar in S. Fork of Gualala River, *Wells* (1984); Russian River, *Carruth* (1899); Vacation Beach near Guerneville, *Rubtzoff* (1958); Guerneville, *Rattan* (1877); Forest Hills across Russian River from Hacienda, *Rubtzoff* (1968); Santa Rosa, *Bellue* (1938); Pepperwood; Healdsburg, *Rattan* (1877); Fitch Mtn. grove 2 mi. e. of Healdsburg, *Rubtzoff* (1962); 2960 ft. elev. ephemeral creek bed ca 3/4 airline mi. s. of Mercuryville, *Raiche* (1991 JEPS); Preston, n. of Cloverdale, *Bellue* (1938); Pine Mtn. Road near Cloverdale, *Knight* (1963); Lake Sonoma area. < 8700 ft. Apr.-Sept. —*Mimetanthe pilosa* E. Greene.

Mimulus rattanii A. Gray. Infrequent, mostly on chaparral burns: 1400 ft. elev. 3.0 mi. ne. of Hwy. 12 in burned area (probably Cavedale Road), *Ammirati* (1965 SFSU). < 4200 ft. May-July.

Mimulus tricolor Lindley. Locally common, drying vernal pools, low moist meadow sites: 500 ft. elev. near Occidental, *Armstrong* (1934 UC); upper Green Valley Marsh nw. of Graton, *Rubtzoff* (1974); Laguna de Santa Rosa ca 1 mi. ese. of Sebastopol, *Rubtzoff, Harrison & Lennon* (1974); w. and sw. of Santa Rosa in Laguna de Santa Rosa drainage, *Harrison* (1978 NCC); Stony Point Road near the lagunas sw. of Santa Rosa, *Robbins* (1938 JEPS); vernal pool along Bennett Valley Road e. of summit, *Baker* (1952 JEPS); low ground near Santa Rosa, *Abrams* (1918); in field sw. of Santa Rosa, *Baker* (1899 UC); along electric railroad between Sebastopol and Santa Rosa, *K. Brandegee* (1905 UC); Stony Point Road s. of Wilfred Ave. e. side in vernal pool, acc. Quibell (1991); w. of Glen Ellen on Bennett Valley Road, *Rubtzoff & Froebe* (1971); 70 ft. elev. near "village store" 4 mi. nw. of Santa Rosa, *Darby* (1940 SACT); Fulton, *McMurphy* (1906); Old Redwood Hwy. just n. of jct. with Starr Road in Windsor area, *J. & B. Guggolz* (1988 NCC). < 4600 ft. Apr.-June.

Parentucellia

1. To 5 dm tall; leaves 2-4 cm long, corolla yellow with calyx ca 1/3 length of corolla . . . . *P. viscosa*
1: To 2.5 dm tall; leaves to 1 cm long; corolla reddish-purple, slender, mostly enclosed in completely fused calyx except for short, purplish-tipped calyx-lobes . . . . *P. latifolia*

Parentucellia latifolia (Linnaeus) Caruel. Uncommon, open areas: Walker Road between Petaluma-Valley Ford and Pepper roads on Tresch Ranch, *J. & B. Guggolz* (1989 JEPS). < 500 ft. Native of Mediterranean region.

★ **Parentucellia viscosa** (Linnaeus) Caruel. Common, roadsides, fields: The Sea Ranch; Willow Cr. Marsh, *Utech* (1988 CM); 3 mi. w. of Graton near creek which runs into Green Valley Cr., *L. L. Davis* (1975 NCC); 4 sites w. of Santa Rosa in Laguna de Santa Rosa drainage, *Harrison* (1978 NCC); Todd Road Preserve; near jct. Stony Point Road and W. Hearn Ave., *Quibell* (1982 NCC); Stony Point Road near Santa Rosa, *Dudik* (1975 NCC); w. end of Crane Can. Road on s. side, acc. Quibell (1991); Spring Lake; a northern Santa Rosa exit from US 101, *Wells* (1984); 2 mi. nw. of Windsor exit from US 101 on Old Redwood Hwy., *de Nevers & Kaza* (1981); Pepperwood; Lake Sonoma area; Alexander Valley quad.; Pine Flat Road ca 11 mi. sw. of geysers Unit 8 via Socrates Mine Road, *Malloch* (1977 PG&E); between Jimtown and Mercuryville on Geysers Road, *Utech & Knight* (1986 CM); ca 2000 ft. elev. Anderson Trail on Black Mtn. between Geyser Peak and upper end of Pine Flat Road, acc. J. & B. Guggolz (1992); 4 mi. n. of Asti, *Wiggins & Ferris* (1943); 1 mi. s. of Cloverdale, *Howell* (1943). < 2100 ft. Apr.-June. Native of British Isles, w. and s. Europe, N. Africa, Azores and Canary Ids.

Pedicularis densiflora Hooker. INDIAN WARRIOR. Common, dry brushy slopes: The Sea Ranch, *Wells* (1982); Camp Meeker, *Howell* (1924); Guerneville, *Rattan* (1877); Sonoma, *A. Menzies* (1962); 2 mi. from Hwy. 12 on Trinity Road, *Koch* (1938), *Ruckert* (1960); Bouverie; Gates Road off Calistoga Road, acc. Wells et al. (1984); Sugarloaf Ridge; Pepperwood; Ida Clayton Road, *J. Winter* (1979 NCC); chaparral area sw. of Bradford Mtn. 1½ mi. w. of Dry Cr., acc. J. & B. Guggolz (1992); Lake Sonoma area; property bordered by Chianti and Dutcher Cr. roads w. of US 101 and w. of Geyserville, acc. J. & B. Guggolz (1991); Geyser Peak, *Knight & Edwards* (1986); Little Geysers, *Quibell* (1980 NCC). < 7000 ft. Jan.-June.

Penstemon

1. Anthers glabrous or sparsely hairy, not comose; corolla rounded dorsally, corolla rose-violet with blue or lilac lobes, glabrous, gaping, 25-35 mm long; staminode glabrous
   2. Leaves mostly linear (some narrowly oblanceolate), 0.5-5 mm wide, usually with axillary leaf clusters . . . . *P. heterophyllus* ssp. *heterophyllus*
   2: Leaves narrowly oblanceolate, 2-7.5 mm wide, rarely with axillary leaf clusters . . . . . *P. h.* ssp. *purdyi*
1: Anthers densely comose; corolla rose-red 27-33 mm long, 8-12 mm wide when pressed, with 2 ventrally bearded ridges on floor of throat and prominently keeled dorsally, corolla-throat only slightly dilated; anthers included; staminode yellow-bearded . . . . *P. newberryi* ssp. *sonomensis*

Penstemon heterophyllus Lindley ssp. **heterophyllus**. Common, dry hillsides, rocky cliffs: Fairfield Osborn (S); near Agua Caliente at edge of Sonoma Valley, *Baker* (1899); Sugarloaf Ridge; Pepperwood; between Knights Valley and Mark West Springs, *Heller* (1902); Petrified Forest, *Eastwood* (1915); w. of Calistoga (probably Knights Valley), *Krauss* (1894); geothermal area, *Malloch* (1977 PG&E). < 5300 ft. Apr.-July.

Penstemon heterophyllus var. **purdyi** (Keck) McMinn. Occasional, dry hillsides, rocky cliffs: Cazadero, *Graff* (1925); ca 1 mi. nw. of Camp Meeker, acc. J. & B. Guggolz (1991); Mt. Hood, *Heller & Brown* (1902); near Mark West Springs, *Roderick* (1967); 1 mi. w. of Petrified Forest, *Haas* (1941); chaparral area sw. of Bradford Mtn. 1½ mi. w. of Dry Cr., acc. J. & B. Guggolz (1992); Geyser Peak, acc. Knight & Edwards (1986); Lake Sonoma area; ca 2000 ft. elev. Anderson Trail on Black Mtn. between Geyser Peak and upper end of Pine Flat Road, acc. J. & B. Guggolz (1992). < 5300 ft. May-June. —*P. h.* ssp. *p.* Keck.

Penstemon newberryi A. Gray var. **sonomensis** (E. Greene) Jepson. SONOMA PENSTEMON. Uncommon, rocky summit: Mt. Hood, *Michener & Bioletti* (1893), acc. Knight (1967); lookout on Mt. St. Helena, *J. & B. Guggolz* (1986 NCC). > 4500 ft. June-Aug. —*P. s.* E. Greene, type locality: Mt. Hood, *Atterbury* (1891 U). *P. n.* ssp. *s.* (E. Greene) Keck.

Scrophularia californica Chamisso & Schlechtendal. BEE PLANT, CALIFORNIA FIGWORT. Common, weedy, ± damp places: The Sea Ranch; Salt Point; redwood canyon 1 mi. n. of Fort Ross, *Howell* (1930); Willow Creek; Pinnacle Gulch Trail in Bodega Harbour development, *Huntington* (1989 NCC); Estero Americano, *Knight* (1963); Willow Cr. Road, *Utech* (1988 CM); Pitkin Marsh, *Rubtzoff* (1951); Sonoma Mtn. Road e. of jct. with Pressley Road, acc. Quibell (1991); Fairfield Osborn (S); Tresch Ranch, Walker Road near Two Rock, acc. J. & B. Guggolz (1992); 6 mi. w. of Petaluma near Rainsville Road, *Ott* (1934 NCC); I St. Ext. near San Antonio Road, *Lamb* (1970 NCC); Bennett Valley, *Knight* (1963); Spring Lake; Pepperwood; Fitch Mtn., *Rubtzoff* (1952); Lake Sonoma area; ca 2000 ft. elev. Anderson Trail on Black Mtn. between Geyser Peak and upper end of Pine Flat Road, acc. J. & B. Guggolz (1992). < 5000 ft. Feb.-July.

Synthyris reniformis (Douglas) Bentham var. **reniformis.** SNOW QUEEN. Uncommon, moist shaded embankments or forests: Fort Ross quad.; n. of Cazadero on roadcut, acc. Roderick & Knight, (1965); ½ mi. s. of Cazadero, *H. Thomas* (1933 NCC); 2¾ mi. from end of Mill Cr. Road, *Knight & Edwards* (1986); Lake Sonoma area, acc. Bunyan (1975). < 5000 ft. Feb.-Apr. —*S. r.* var. *cordata* A. Gray, indistinct.

Tonella tenella (Bentham) A. A. Heller. Occasional, mostly shady slopes: Graton-Camp Meeker Road, *Baker* (1926); Adobe Can., *Baker* (1926); borrow pit on Trinity Road, *Hardison* (1981 NCC); summit of Trinity Road ca ½ mi. past jct. of Cavedale Road, *Quibell & Hardison* (1981 NCC); hills e. of Santa Rosa, *Heller & Brown* (1902); Pepperwood; Lake Sonoma area. < 5300 ft. Mar.-May.

Triphysaria

1. Branches divergent from central erect axis; flowers showy, usually obviously exserted; galea straight or gradually curved; bracts pinnatifid; herbage not tinged purplish
   2. Galea purple; herbage pubescent
      3. Lower lip of corolla purplish at base, sacs light yellow, 3-4 mm deep . . . . *T. eriantha* ssp. *eriantha*
      3: Lower lip of corolla rose-pink or white turning rose, 4-5 mm deep . . . . *T. e.* ssp. *rosea*
   2: Galea yellowish; herbage glabrous, or puberulent in inflorescence; lower lip of corolla abruptly inflated 2-4 mm deep
      4. Lower lip of corolla white, fading rose, with greenish-yellow spot at base of each tooth . . . . *T. versicolor* ssp. *versicolor*
      4: Lower lip of corolla pale sulphur-yellow with purple dots on margins . . . . . *T. v.* ssp. *faucibarbata*
1: Branches many, weak, ascending from the base, the central stem indistinct; flowers minute, inconspicuous; galea sharply curved; bracts often bipinnatifid; herbage tinged purplish . . . . *T. pusilla*

Triphysaria eriantha (Bentham) Chuang & Heckard ssp. **eriantha**. JOHNNY-TUCK, BUTTER-AND-EGGS. Common, open places, flatlands: The Sea Ranch coastal bluffs, *J. H. Thomas* (1976); ca 3½ mi. e. of Occidental, *J. H. Thomas* (1955); Tresch Ranch, Walker Road near Two Rock, acc. J. & B. Guggolz (1992); s. of Schellville, *Ripley & Barneby* (1954); *Bioletti* (1892); 2 mi. n. of Sonoma, *Howell* (1931); Boyes Hot Springs, Sonoma Valley, *Keck* (1931); Bouverie; corner Hwy. 12 and Adobe Can. Road near Kenwood, *Knight* (1964); 4¼ mi. w. of Glen Ellen on Bennett Valley Road, *Rubtzoff & Froebe* (1971); Sugarloaf Ridge; Spring Lake; between Windsor and W. Windsor, *Heller* (1940); property bordered by Chianti and Dutcher Cr. roads w. of US 101 and w. of Geyserville, acc. J. & B. Guggolz (1991). < 4400 ft. Mar.-May. —*Orthocarpus erianthus* Bentham var. *e.*

Triphysaria eriantha ssp. **rosea** (A. Gray) Chuang & Heckard. Common, open grassy hills, coastal: The Sea Ranch, *Wells* (1981); Salt Point; Plantation Road 2 mi. from Hwy. 1, *Koch* (1940 NCC); 5.7 mi. n. of Fort Ross, *Cantelow* (1941); 1½ mi. n. of Jenner on Hwy. 1, *Huntington* (1989 NCC); Bodega Head, *Best* (1981). < 700 ft. Mar.-May. —*Orthocarpus erianthus* Bentham var. *roseus* A. Gray.

Triphysaria pusilla (Bentham) Chuang & Heckard. Common, grassy fields, partially bare areas: The Sea Ranch; Salt Point; Fort Ross-Russian River-Bodega areas, *Voznesenskii* (1840-1841 LE); Duncans Mills quad.; Bodega Head; 5 sites w. and sw. of Santa Rosa in Laguna de Santa Rosa drainage, *Harrison* (1978 NCC); Todd Dist., *Rubtzoff* (1974), *Self* (1980 NCC); Crane Can. Reg. Park, acc. Quibell (1991); Fairfield Osborn (*); Bouverie; Sugarloaf Ridge; Bennett Valley, *Heller & Brown* (1902), *Koch* (1939 NCC); Spring Lake; Santa Rosa, *Eastwood* (1921); Pepperwood; Lake Sonoma area; property bordered by Chianti and Dutcher Cr. roads w. of US 101 and w. of Geyserville, acc. J. & B. Guggolz (1991). < 1400 ft. Apr.-May. —*Orthocarpus pusillus* Bentham.

Triphysaria versicolor Fischer & C. Meyer ssp. **faucibarbata** (A. Gray) Chuang & Heckard. Common, open areas: 8 mi. n. of Bodega Bay, *Howell* (1930); Occidental Marsh, *Rubtzoff* (1957); Pitkin Marsh, *Rubtzoff* (1952); Laguna de Santa Rosa e. of Graton, *Rubtzoff* (1966 NCC); Laguna de Santa Rosa ca 1 mi. ese. of Sebastopol, *Rubtzoff* (1972); 4 mi. w. of Santa Rosa on Sebastopol Hwy., *Wagnon* (1938 NCC); Todd Road Preserve; Wilfred Ave. w. of Langer Ave. between US 101 and Stony Point Road, acc. Quibell (1990); Hessel Road near Gravenstein Hwy., acc. J. & B. Guggolz (1991); 8 mi. n. of Petaluma, *Cantelow* (1938); Button Ranch on Walker Road (U. C. property) between Petaluma-Valley Ford Road and Pepper Road, acc. Connors (1992); ca 1½ mi. nw. of Petaluma, *Ott* (1934 NCC); laguna in Chileno Valley, *Rubtzoff* (1972); w. outskirts of Sonoma along 5th St., *Baker* (1952 NCC); 9 mi. sw. of Sonoma on road to San Rafael, *Howell* (1931); 4¼ mi. w. of Glen Ellen on Bennett Valley Road, *Rubtzoff & Froebe* (1974); Bennett Mtn. Lake, *Rubtzoff* (1963); Annadel, *Wells* (1984); Spring Lake; Pepperwood; between Windsor and W. Windsor, *Heller* (1940). < 1700 ft. Apr.-June. —*Orthocarpus faucibarbatus* Bentham ssp. *f.*

Triphysaria versicolor ssp. **versicolor**. Occasional, near coast; few inland: Stewarts Point, *Abrams* (1917); Salt Point; Fort Ross, *Heller* (1902); 3 mi. s. of mouth of Russian River, *Keck & Clausen* (type 1939); Arched Rock quad.; Bodega Bay, *Heller* (1938); 4 mi. w. of Bodega, *Cantelow* (1938); quarry at Bodega Harbor, *Harrison* (1976 SRJC); w. of Llano Road and n. of Todd Road, acc. Lovell (1985). < 400 ft. Apr.-June. —*T. versicolor* Fischer & C. Meyer, type locality: Port Bodega. *Orthocarpus faucibarbatus* var. *albidus* Keck.

Verbascum

1. Plants very woolly; leaves entire, cauline leaves decurrent; stem stout . . . . *V. thapsus*
1: Plants with green herbage, with simple gland-tipped hairs; leaves dentate, cauline leaves not decurrent; stem not stout
  2. Pedicels 10-15 mm long; leaves sinuate-dentate, glabrous . . . . *V. blattaria*
  2: Pedicels 3-5 mm long; leaves sinuately denticulate, pubescent . . . . *V. virgatum*

★ **Verbascum blattaria** Linnaeus. MOTH MULLEIN. Occasional, weed, waste places: ½ mi. e. of Duncans Mills, *Kawahara* (1968); Bohemian Hwy. 3½ mi. n. of Occidental, *Best* (1981); Russian River near Guerneville, *Ferguson* (1919 JEPS); Willow Brook, *Knight & Edwards* (1984); Pressley Road just e. of jct. with Roberts and Lichau roads next to Copeland Cr. bridge, acc. Quibell (1991); Spring Lake; Lake Sonoma area; between Jimtown and Mercuryville on Geysers Road, *Utech & Knight* (1986 CM). < 5300 ft. May-Sept. Naturalized from Eurasia.

★ **Verbascum thapsus** Linnaeus. WOOLY MULLEIN. Common, weed, creek beds, waste places: The Sea Ranch; Mt. Hood, *Knight* (1964); Spring Lake; Pepperwood; Porter Cr. Road, *Stocking* (1963 NCC); 3½ mi. w. of Healdsburg, *Stocking* (1963 NCC); Lake Sonoma area; Geyser Peak, *Knight & Edwards* (1986); Russian River near Cloverdale, *Gels* (1903); ca 2000 ft. elev. Anderson Trail on Black Mtn. between Geyser Peak and upper end of Pine Flat Road, acc. J. & B. Guggolz (1992); Pine Mtn. Road ne. of Cloverdale, *Knight* (1963). < 7300 ft. June-Sept. Naturalized from Eurasia.

★ **Verbascum virgatum** Stokes. WAND MULLEIN. Uncommon, weed: Gravenstein Hwy. near US 101 jct., *Knight & P. Watters* (1963); Copeland Cr. near Sonoma State Univ. campus, *Self* (1979 NCC); Franz Cr. along Franz Valley Road, acc. de Nevers (1992). < 1000 ft. May-Sept. Naturalized from Eurasia.

Veronica

1. Main stem ending in a single racemelike inflorescence
  2. Plants perennial from underground rootstocks; flowers in upper axils only . *V. serpyllifolia* ssp. *humifusa*
  2: Plants annual, flowers from most axils
    3. Corolla 2-2.5 mm wide; pedicels 1-2 mm long
      4. Leaves linear-oblong to spatulate; corolla white . . . . *V. peregrina* ssp. *xalapensis*
      4: Leaves rounded to oval; corolla bright blue . . . . *V. arvensis*
    3: Corolla 6-11 mm wide; pedicels 10-25 mm long, leaves crenate or dentate; corolla deep blue . . . . *V. persica*
1: Main stem with lateral racemes below the tip
  5. Leaves of main stems definitely short-petioled; corolla 7-10 mm wide . . . . *V. americana*
  5: Leaves of main stems sessile, ± cordate-clasping; corolla 3-7 mm wide

6. Capsule ovoid to roundish, scarcely notched; leaves lanceolate to broader; corolla pale lavender with violet lines . . . . *V. anagallis-aquatica*
6: Capsule deeply obcordate, much broader than long; leaves linear to linear-lanceolate; corolla lilac or blue-lavender . . . . *V. scutellata*

Veronica americana (Rafinesque) Schweinitz. AMERICAN BROOKLIME. Common, near streams, wet places: n. of Crow's Nest and e. of Spyglass Lane at The Sea Ranch, *Wells* (1981); Stewarts Point, *Parks* (1925); Stewarts Point- Skaggs Springs Road 3 mi. e. of Stewarts Point, *Best* (1981); Annapolis, *Sutliffe* (1920); n. end of Bodega Bay, *Rose* (1964); Bodega Bay, *Baker* (1923 NCC); creek bed at Occidental, *Best* (1981); Pitkin Marsh, *Rubtzoff* (1951); 2 mi. s. of Graton, *Hastings* (1940 NCC); Analy township (Sebastopol), *Congdon* (1880); Fairfield Osborn (S); Button Ranch on Walker Road (U. C. property) between Petaluma-Valley Ford Road and Pepper Road, acc. Connors (1992); Bennett Valley, *Bately* (1932 NCC); Felta Cr. area sw. of Healdsburg, *Bice* (1895 NCC). < 11000 ft. May-Aug.

★ **Veronica anagallis-aquatica** Linnaeus. WATER SPEEDWELL. Common, wet places: Vacation Beach on Russian River, *Rubtzoff* (1968); Forest Hills near Hacienda on Russian River, *Rubtzoff* (1956, 1968); Laguna de Santa Rosa ne. of Graton n. of Guerneville Road crossing, *Rubtzoff* (1972); w. of Santa Rosa in Laguna de Santa Rosa drainage, *Harrison* (1978 NCC); Hinebaugh Cr., Rohnert Park, *Rubtzoff & Harrison* (1974); Copeland Cr., Rohnert Park, *Self* (1979 NCC); N. McDowell Blvd. in Petaluma, *Hilliard* (1979 NCC); Ducker Cr., Rincon Valley, *Wells* (1985); Spring Lake; tributary to Mark West Cr. 3 mi. w. of Fulton, *Rubtzoff* (1970, 1971); Franz Cr. along Franz Valley Road, acc. de Nevers (1992); Russian River near Fitch Mtn. grove 2 mi. e. of Healdsburg, *Rubtzoff* (1964). < 8300 ft. May-Sept. Native of Europe.

★ **Veronica arvensis** Linnaeus. CORN SPEEDWELL. Uncommon, weed in lawns, gardens, waste places: Spring Lake; Pepperwood, *Breedlove* (1983). < 5000 ft. Apr.-July. Native of Eurasia.

Veronica peregrina Linnaeus ssp. **xalapensis** (Kunth) Pennell. PURSLANE SPEEDWELL. Occasional, moist places: 3 sites w. of Santa Rosa in Laguna de Santa Rosa drainage, *Harrison* (1978 NCC); Todd Road Preserve; w. of Llano Road and n. of Todd Road, acc. Lovell (1986); N. McDowell Blvd., Petaluma, *Hilliard* (1979 NCC); Spring Lake; Pepperwood; Felta Cr. area sw. of Healdsburg, *Bice* (1895 NCC); Big Sulphur Cr., *J. & B. Guggolz* (1987 NCC). < 10300 ft. Apr.-Aug.

★ **Veronica persica** Poiret. PERSIAN SPEEDWELL. Occasional, weed, lawns, waste places: Pitkin Ranch, *Baker* (1928 NCC); Sonoma State Univ. campus w. of field house, *Stocking* (1969 NCC); Lafferty Ranch on Sonoma Mtn., acc. Arnold et al. (1971); Bouverie; Sugarloaf Ridge; Flamingo Plaza, Santa Rosa, *Wells* (1985); Lake Sonoma area. < 3700 ft. Feb.-May. Naturalized from Eurasia.

Veronica scutellata Linnaeus. MARSH SPEEDWELL. Occasional, wet places, swales: pond e. of Timber Ridge Road, The Sea Ranch, *Wells* (1984); 2 mi. e. of Stewarts Point, *Nobs & G. Smith* (1949); Pitkin Marsh, *Rubtzoff* (1951); Cunningham Marsh, *Howell* (1947). < 7300 ft. May-Aug.

★ **Veronica serpyllifolia** Linnaeus ssp. **humifusa** (Dickson) Syme. THYME-LEAVED SPEEDWELL. Uncommon, weed, lawns, moist places: Forest Hills across Russian River from Hacienda, *Rubtzoff* (1991); Pepperwood. < 10000 ft. Apr.-Aug. Native of Eurasia.

SIMAROUBACEAE. QUASSIA FAMILY

★ **Ailanthus altissima** (Miller) Swingle. TREE OF HEAVEN. Suckering along fences: across from Cinnabar Theater on Old Redwood Hwy., n. of Petaluma, acc. Knight (1990); scattered along Petaluma Hill Road between Roberts Road and Rohnert Park Expressway, acc. Knight (1991); corner Brigham Ave. and Hwy. 12, Santa Rosa, *Knight* (1989 NCC). < 4200 ft. June. Native of China.

SOLANACEAE. NIGHTSHADE FAMILY

1. Corolla rotate or nearly so
   2. Anthers connivent, longer than filaments; calyx remaining small
      3. Anthers dehiscing by terminal pores; plants prickly if with yellow flowers . . . . *Solanum*
      3: Anthers dehiscing lengthwise; plants not prickly, but with yellow flowers . . . . *Lycopersicon*
   2: Anthers not connivent, mostly shorter than filaments; calyx bladdery-inflated in fruit, early closed at top . . . . *Physalis*

1: Corolla not rotate
   4. Fruit a berry; corolla urceolate . . . . *Salpichroa*
   4: Fruit a capsule; corolla tubular to funnelform
      5. Capsule prickly, 25-50 mm long; flowers solitary, axillary . . . . *Datura*
      5: Capsule not prickly, 5-13 mm long; flowers in terminal racemes or panicles . . . . *Nicotiana*

Datura

1. Corolla 15-20 cm long; calyx tubular, not prismatic; capsule nodding . . . . *D. wrightii*
1: Corolla 6-12 cm long; calyx prismatic; capsule erect . . . . *D. stramonium*

★ **Datura stramonium** Linnaeus. JIMSON WEED. Locally common, weed, cultivated ground, waste places: Fuori Lane near Jonive Road, Occidental area, *Best* (1983); "sandy river deposits," Guerneville, acc. Davey (*Erythea 4:152.* [1896]); Todd Road Preserve; Copeland Cr. near Sonoma State Univ. campus, *Self* (1979 NCC); ca 1 mi. n. of Penngrove on Old Redwood Hwy, *Donovan* (1989 NCC); field near Petaluma, *Knight* (1963); pasture along Willow Brook near Petaluma, *Knight & Edwards* (1984); sand bar in Franz Cr. along Franz Valley Road, acc. de Nevers (1992). < 5000 ft. Summer. Native of Mexico. —*D. tatula* Linnaeus. *D. s.* var. *t.* (Linnaeus) Torrey.
★ **Datura wrightii** Regel. TOLGUACHA. Uncommon, gravelly or sandy dry open places: Lake Sonoma area. < 7300 ft. Apr.-Oct. Native of Mexico. —*D. meteloides* A. de Candolle.

★ **Lycopersicon esculentum** Miller. TOMATO. Garden escape: Fort Ross-Russian River-Bodega area, *Voznesenskii* (1840-1841 LE). < 400 ft. Summer-fall. Native of S. America.

Nicotiana

1. Plant a shrub; flowers yellow . . . . *N. glauca*
1: Plant an herb; flowers whitish or cream, sometimes tinged violet
   2. Leaves spatulate . . . . *N. sylvestris*
   2: Leaves not spatulate
      3. Cauline leaves sessile or nearly so, except near base of stem . . . . *N. quadrivalvis*
      3: Cauline leaves petioled . . . . *N. acuminata* var. *multiflora*

★ **Nicotiana acuminata** Hooker var. **multiflora** (Philippi). Reiche. Occasional, pastures, roadsides, along creeks: Copeland Cr. near Sonoma State Univ. campus, *Self* (1979 NCC); Willow Brook near Petaluma, *Knight & Edwards* (1984); near US 101 on San Antonio Cr., *Knight* (1963); Agua Caliente, Sonoma Valley, acc. Jepson, *Fl. Ca 3:453,* (1943); bed of Franz Cr. at bridge on Chalk Hill Road, *Wagnon* (1991 NCC). < 5300 ft. Summer. Native of S. America.

Nicotiana glauca Graham. TREE TOBACCO. Infrequent, waste places: Button Ranch on Walker Road (U. C. property) between Petaluma-Valley Ford Road and Pepper Road, acc. Patterson (1990). < 3700 ft. Spring-summer. Native of S. America.

Nicotiana quadrivalvis Pursh var. **quadrivalvis**. INDIAN TOBACCO. Infrequent, disturbed ground, pastures: Russian River Valley, *Carruth* (1899); near Cotati, *Knight* (1963); Buena Vista near Sonoma, *Behr* (1860s U); Preston, n. of Cloverdale, *Carlson* (1917); Lake Sonoma area. < 5000 ft. May-Oct. The plant of the N. Coast Ranges is taller than the typical form in the Sierra Nevada foothills and has been called var. **exaltata** Setchell (cf. Jepson, *Fl. Ca 3:454,* [1943]). —*N. bigelovii* (Torrey) S. Watson.

★ **Nicotiana sylvestris** Spegazzini & O. Comes. Escaping from cultivation and becoming established around Petaluma: ditch on Corona Road, *Knight* (1990); Live Oak Road, *Knight* (1963). < 700 ft. Native of S. America.

Physalis

1. Pedicels > fruiting calyx; corolla without dark center 3-8 mm broad . . . . *P. lancifolia*
1: Pedicels < fruiting calyx; corolla with dark center 10-15 mm broad . . . . *P. philadelphica*

★ **Physalis lancifolia** Nees. Rare, weed, roadsides: Stoetz Lane near Occidental, *Best* (1983); Skillman Lane near Petaluma, *Knight* (1963). < 700 ft. June-Sept. Native of S. America. —*P. angulata* Linnaeus var. *l.* (Nees) Waterfall.

★ **Physalis philadelphica** Lamarck. TOMATILLO. Rare, orchard weed, escape from gardens: Fort Ross-Russian River-Bodega areas, *Voznesenskii* (1840-1841 LE); gravel bar in Franz Cr. along Franz Valley Road, acc. de Nevers (1992). < 2300 ft. Summer. Native of Mexico. —*P. ixocarpa* Brotero.

★ **Salpichroa origanifolia** (Lamarck) Thellung. Rare, waste places: Pepperwood, *de Nevers* (1991). < 1000 ft. July-Oct. Native of S. America. —*S. rhomboidea* (Gillies & Hooker) Miers.

Solanum

1. Plants not prickly
  2. Corolla deeply 5-cleft, whitish; peduncles mostly longer than pedicels; leaves entire or shallowly toothed
    3. Stems obviously and persistently villous or hirsute; ripe berries greenish, yellow or reddish . . . . *S. sarrachoides*
    3: Stems sparsely pubescent or glabrous, at least in maturity; ripe berries black
      4. Larger inflorescences usually bifurcate, the peduncle deflexed at maturity; corolla 6-11 mm long; filaments not = in length . . . . *S. furcatum*
      4: Larger inflorescences rarely bifurcate, the peduncle erect at maturity; corolla 3-7 mm long; filaments subequal
        5. Inflorescence umbelliform; calyx-lobes distinct, reflexed at maturity; berry surface glossy; anthers 1-1.2 mm long . . . . *S. americanum*
        5: Inflorescence subracemose; calyx-lobes partly fused, not reflexed at maturity; berry surface dull; anthers 1.8-2.4 mm long . . . . *S. nigrum*
  2: Corolla angulately shallowly lobed, often purple; peduncles mostly shorter than pedicels; leaves subentire, sometimes lobed at base
    6. Pubescence of upper stems with some forked hairs or branched hairs . . . . *S. umbelliferum*
    6: Pubescence of upper stems lacking or of simple hairs . . . . *S. xantii*
1: Plants prickly with stiff spines
  7. Plants annual; leaves bipinnatifid with rounded lobes; berry partly or wholly invested by spiny calyx; corolla yellow . . . . *S. rostratum*
  7: Plants perennial; leaves entire, sinuate or lobed, but not pinnatifid; berry not enclosed by spiny or unarmed calyx
    8. Leaves entire to repand-dentate, permanently gray-canescent on both surfaces; corolla violet or blue; fruit ca 8-15 mm in diameter, orange . . . . *S. elaeagnifolium*
    8: Leaves lobed or deeply dentate, not permanently gray-canescent at least on upper surface; corolla violet; fruit 15-25 mm in diameter, yellow . . . . *S. carolinense*

★ **Solanum americanum** P. Miller. SMALL-FLOWERED NIGHTSHADE. Fairly common, weed, moist woods, thickets, cultivated ground, waste places: The Sea Ranch Lodge, *Wells* (1981); Salt Point; Timber Cove between Salt Point and Fort Ross, *McKenzie* (1981); Fort Ross-Russian River-Bodega areas, *Voznesenskii* (1840-1841 LE); s. side of mouth of Russian River, *Breedlove* (1961); Copeland Cr. near Sonoma State Univ. campus, *Self* (1979 NCC); Mark West Cr. at confluence of Leslie Cr., *de Nevers* (1981); Pepperwood; Lake Sonoma 2400 ft. elev. Little Geysers, *Malloch* (1977 PG&E). < 3300 ft. Apr.-Nov. Introduced from tropical America, not from Europe, acc. *Abrams, 3:676,* (1951), or the Old World, acc. *Munz, p. 597,* (1959). —*S. nodiflorum* Jacquin.

★ **Solanum carolinense** Linnaeus. CAROLINA HORSE-NETTLE. Locally established, weed: Sebastopol apple orchard on Laguna Road, *Fuller* (1967); w. end of Laguna de Santa Rosa, acc. *Barbe* (1991). < 700 ft. May-Aug. Even in e. United States, where it is native, it is often a troublesome weed.
★ **Solanum elaeagnifolium** Cavanilles. WHITE HORSE-NETTLE. Rare, weed, roadsides, pastures: mouth of Adobe Can., *Baker* (1945); 9 locations in e. part of county, acc. *Barbe* (1991). Quite widely a weed in s. California and Central Valley. < 4000 ft. May-Sept. Native of central U.S. to Mexico. Noxious weed.
★ **Solanum furcatum** Dunal. FORKED NIGHTSHADE. Occasional, waste places, fields, near coast: The Sea Ranch; Salt Point area, on bluffs at Stump Beach, acc. Wells (1983); Salt Point area, ½ mi n. of Moon Rocks campground, *Huntington* (1989 NCC). < 700 ft. May-Oct. Native of S. America.
★ **Solanum nigrum** Linnaeus. BLACK NIGHTSHADE. Occasional, waste places, fields: Pepperwood. < 700 ft. Mar.-Oct. Native of Eurasia.
★ **Solanum rostratum** Dunal. BUFFALO BERRY. Occasional, widespread weed in California: adventive at Maplewood, home of M. S. Baker on Sonoma Cr. near the mouth of Adobe Can., *Howell* (1936); Elnoka Lane off Hwy. 12 near Oakmont, Santa Rosa, many plants in vacant lot, acc. Wagnon (1990); behind Long's Drug Store in Flamingo Plaza, Santa Rosa, *Wagnon* (1990 NCC). < 1700 ft. May-Sept. Native of great plains e. of the Rocky Mts.
★ **Solanum sarrachoides** Sendtner. HAIRY NIGHTSHADE. Weed, gardens, waste ground, generally overlooked: Burbank Farm near Sebastopol, acc. Baker (1932); field bordering Kenwood Marsh, *Howell* (1932); Spring Lake. < 3300 ft. May-Oct. Native of S. America. —*S. villosum* of Jepson, not of P. Miller.

Solanum umbelliferum Eschscholtz. BLUE WITCH. Reported repeatedly "from Sonoma Co. Mendocino Co. southward" acc. *Abrams* and *Munz* floras; Mill Cr., acc. Baker (1932); Harrison & Barnhart checklist, but no specimen of this species has been seen in herbaria. < 5300 ft. Most of the year. Purported records are probably based on the following species:

Solanum xanti A. Gray var. **intermedium** Parish. XANTUS NIGHTSHADE. Somewhat common, brush of woods of inland hills and canyons: Willow Creek; Bouverie; Pepperwood; Fitch Mtn. near Healdsburg, *Rubtzoff* (1952); Pine Flat Road ne. of Healdsburg, *Applegate* (1934); Lake Sonoma area. < 9000 ft. Feb.-July. Acc. E. Greene, (*Erythea 3:72,* [1895]) described the plant that grows from Marin and Napa counties northward as a new species, *S. cupuliferum*. It may well be another recognizable geographic variety in the *S. xanti* complex and distinct from var. *intermedium*.

STERCULIACEAE. CACAO FAMILY

Fremontodendron californicum (Torrey) Coville. FLANNELBUSH, FREMONTIA. Uncommon, brushy slopes, serpentine: Austin Cr. State Recreation Area, acc. Raiche (1986). < 6700 ft. May. Used horticulturally. —*F. napense* Lloyd (in part); indistinct.

THYMELAEACEAE. DAPHNE FAMILY

Dirca occidentalis A. Gray. WESTERN LEATHERWOOD. Very rare, coastal brush area: 0.35 mi w. of Fitzpatrick Lane on e. side on Salmon Cr. Road w. of Bodega, *Knight & G. L. Stebbins* (1969). < 1000 ft. Jan.-Mar.

TROPAEOLACEAE. NASTURTIUM OR TROPAEOLUM FAMILY

★ **Tropaeolum majus** Linnaeus. GARDEN NASTURTIUM. Reseeding from cultivation: Royal Oak Mobile Home Park, N. McDowell Blvd., Petaluma, acc. Knight (1991). < 1000 ft. Mar.-Nov. Native of S. America.

ULMACEAE. ELM FAMILY

Ulmus

1. Leaf blade sometimes doubly serrate, very unequally oblique at base, lower surface or veins short-hairy *U. minor*
1: Leaf blade usually singly serrate, nearly equal or barely oblique at base, lower surface usually glabrous or vein axils puberulent . . . . *U. pumila*

★ **Ulmus minor** Miller. ENGLISH ELM, SMOOTH-LEAVED ELM. Uncommon, weed-tree suckers: Sonoma State Univ. site along Petaluma Hill Road, *Knight* (1963). < 2000 ft. Native of Europe. —*U. campestris* Linnaeus. *U. procera* Salisbury.

★ **Ulmus pumila** Linnaeus. SIBERIAN ELM. Uncommon, reproducing root-sprouts: Bodega Bay, *Connors* (1993). <5400 ft. Native of c. Asia.

URTICACEAE. NETTLE FAMILY

1. Plants creeping with tiny round leaves . . . . *Soleirolia*
1: Plants not creeping with comparatively much larger leaves
    2. Plants with stinging hairs; opposite leaves
        3. Pistillate calyx 4-parted, the segments almost distinct, the inner much the longer and enclosing the achene . . . . *Urtica*
        3: Pistillate calyx saccate, 2-4-toothed at the apex . . . . *Hesperocnide*
    2: Plants without stinging hairs; alternate leaves . . . . *Parietaria*

Hesperocnide tenella Torrey. WESTERN NETTLE. Uncommon, moist shaded areas: Bodega Dunes, *Connors* (1993 BML). < 3300 ft. Apr.-June.

Parietaria pensylvanica Willdenow. PELLITORY. Rare, moist shady places: The Geysers near footbridge, *Baker* (1932 NCC). < 8300 ft. May-July.

★ **Soleirolia soleirolii** (Requien) Dandy. BABY'S TEARS. Infrequent, garden plants perhaps established as weed; sometimes used as small-leaved groundcover around stepping stones: Sonoma State Univ. campus greenhouse, acc. Quibell (1990). Usually < 400 ft. Native of Corsica and Sardinia. —*Heloxine soleirolii* Requien.

Urtica

1. Plants perennial; ♀ and ♂ flowers in separate clusters
    2. Stems whitish-pubescent; leaf bases ± rounded, leaves > twice as long as wide; flower-clusters mostly exceeding upper leaves . . . . *U. dioica* ssp. *holosericea*
    2: Stems glabrous except sometimes for scattered bristles; leaf bases deeply cordate; leaves < twice as long as wide; flower-clusters mostly shorter than upper leaves . . . . *U. d.* ssp. *gracilis*
1: Plants annual; ♀ and ♂ flowers mixed in same cluster . . . . *U. urens*

Urtica dioica Linnaeus ssp. **gracilis** (Aiten) Selander. AMERICAN STINGING NETTLE. CALIFORNIA NETTLE. Occasional, wet brushy thickets: The Sea Ranch near S. Fork of Gualala River, *Wells* (1984); 0.2 mi. s. of Gualala on Hwy. 1, *Knight* (1964); The Sea Ranch; Salt Point; jct. of Hwys. 116 and 1 e. of Jenner, *Woodland* (1969); Willow Creek; n. end of Bodega Bay, *Rose* (1964); Bodega Head Marsh, *Rubtzoff* (1960); Freestone Marsh, *Rubtzoff* (1962); Neeley Hills s. of Guerneville, *Rubtzoff* (1951); Button Ranch on Walker Road (U. C. property) between Petaluma-Valley Ford Road and Pepper Road, acc. Connors (1992). < 10000 ft. Apr.-June. —*U. lyallii* S. Watson var. *californica* (E. Greene) Jepson. *U. californica* E. Greene.

Urtica dioica ssp. **holosericea** (Nuttall) Thorne. HOARY NETTLE. Common, damp areas: Bodega Harbor Marsh, *Rubtzoff* (1955); Bodega Head Marsh, *Rubtzoff* (1954); Duncans Mills, *M. Jones* (1892); Duncans Mills Marsh, *Rubtzoff* (1957); Coleman Valley Marsh, *Rubtzoff* (1964); Forest Hills across river from Hacienda, *Rubtzoff* (1965); Laguna de Santa Rosa ne. of Graton, *Rubtzoff* (1958); Copeland Cr., Rohnert Park, *Self* (1979); Lichau Road e. of Petaluma Hill Road, *C. Nelson* (1968 NCC); marsh 5 mi. ne. of Penngrove on Lichau Road, *Rubtzoff* (1964); Fairfield Osborn (*); Bouverie; Sugarloaf Ridge; top of Taylor Mtn. under redwoods, s. of Santa Rosa, acc. Quibell (1991); Pepperwood; 4 mi. w. of Calistoga on road to Santa Rosa, *Woodland* (1968); near Burke Sanitarium (near Mark West Springs), *Heller* (1902); Lake Sonoma area; ca 2000 ft. elev. Anderson Trail on Black Mtn. between Geyser Peak and upper end of Pine Flat Road, acc. J. & B. Guggolz (1992). < 10000 ft. June-Sept. —*U. h.* Nuttall.

★ **Urtica urens** Linnaeus. DWARF NETTLE. Common, garden, orchard or pasture weed: Royal Oaks Mobile Home Park on N. McDowell Blvd., Petaluma, *Knight* (1985); Petaluma Hill Road, *Koch* (1939 NCC); Middle Two Rock Road w. of Petaluma, *Anast* (1980 NCC); Lake Sonoma area. < 10000 ft. Jan.-Apr. Naturalized from Europe.

VALERIANACEAE. VALERIAN FAMILY

1. Plant with calyx-segments; biennial, sometimes suffruticose; corolla with tubular spur; stamen 1; leaves smooth, entire . . . . . *Centranthus*
1: Plant without calyx-segments; annual herb; corolla with tube usually spurred at base; stamens 3; leaves glabrous, entire or with a few teeth; need fruit for specific determinations . . . . . *Plectritis*

★ **Centranthus ruber** (Linnaeus) de Candolle. GARDEN VALERIAN, RED VALERIAN. Locally common, waste places: growing in bulkhead along embankment on n. side of Hwy. 1, Jenner, *W. I. & T. Knight* (1992 NCC). < 700 ft. Apr.-Aug. Native of Europe to sw. Asia.

Plectritis

1. Corolla rose to bright pink, 5-8 mm long
   2. Spur short and thick, < < tube and throat; keel of the convex side of the fruit acute, not grooved
      3. Fruits obviously winged and incurved . . . . . *P. congesta*
      3: Fruits apparently wingless . . . . . *P. brachystemon*
   2: Spur long and slender, = the tube and throat; keel on the convex side of the fruit broad with a groove down the center . . . . . *P. ciliosa*
1: Corolla white or pale pink, 2-4 mm long . . . . . *P. macrocera*

Plectritis brachystemon Fischer & C. Meyer. Occasional, grassy areas, open slopes: Fort Ross, *Fischer & C. Meyer* (type 1830s U). < 6000 ft. Mar.-June. —*P. congesta* ssp. *b.* (Fischer & C. Meyer) Moley. *P. magna* (E. Greene) Suksdorf. *P. samolifolia* (de Candolle) Hoeck, misapplied. *Betckea major* Fischer & C. Meyer, type locality: Fort Ross.

Plectritis ciliosa (E. Greene) Jepson. Occasional, grassy areas: hills e. of Santa Rosa, *Heller & Brown* (1902); Spring Lake; end of Fernglen Drive, Santa Rosa, *Dudik* (1975 NCC); Pepperwood. < 5700 ft. Mar.-May.

Plectritis congesta (Lindley) de Candolle ssp. **congesta**. Occasional, grassy areas, open slopes: Bodega Head, *Ferris & Morey* (1958); Santa Rosa Cr., *Heller & Brown* (1902); Pepperwood; The Geysers, *Quibell* (1980 NCC). < 3000 ft. Apr.-June. —*Valerianella magna* E. Greene, type locality: Knights Valley, *E. Greene* (1894 U).

Plectritis macrocera Torrey & A. Gray var. **macrocera**. Uncommon, moist woody or grassy areas: Calistoga Road, *Wiggins* (1949). < 6700 ft. Apr.-May. —*P. m.* var. *grayi* (Suksdorf) Dyal, currently not recognized.

VERBENACEAE. VERBENA OR VERVAIN FAMILY

1. Plants low, matted; white to pink or rose-colored flowers in dense, short axillary heads . . . . . *Phyla*
1: Plants erect; blue to purplish flowers in elongate, terminal spikes . . . . . *Verbena*

★ **Phyla nodiflora** (Linnaeus) E. Greene. LIPPIA. Common, covering the ground in moist places: Vacation Beach near Guerneville, *Rubtzoff* (1954); Laguna de Santa Rosa ne. of Graton near Guerneville Road, *Rubtzoff* (1958); w. of Santa Rosa in Laguna de Santa Rosa drainage, *Harrison* (1978 NCC); Gilardi's Resort, Lakeville, *Knight* (1984); Spring Lake Park, Santa Rosa, *Wells* (1987); Pepperwood. Naturalized from S. America < 1400 ft. May-Oct. *Lippia n.* (Linnaeus) Michaux var. *rosea.* (D. Don) Munz.

Verbena

1. Plants upright, stiff; stems ± 4-angled; simple below; spikes short, dense and in terminal cymose clusters with stems almost leafless just beneath; leaves sessile or partially clasping
   2. Bracts >> the calyx . . . . *V. rigida*
   2: Bracts ca = the calyx . . . . *V. bonariensis* var. *conglomerata*
1: Plants ascending, not stiff; stems not 4-angled; much branched, ultimately diffuse and ± procumbent, spikes in 3's at the end of the branches
   3. Spikes 1-3 dm long in fruit; leaves canescent, not scabrous above; inner face of nutlets without whitish papillae . . . . *V. lasiostachys* var. *lasiostachys*
   3: Spikes 0.3-1 dm long in fruit; leaves bright green, scabrous above; inner face of nutlets with whitish papillae . . . . *V. l.* var. *scabrida*

★ **Verbena bonariensis** Linnaeus var. **conglomerata** Briquet. Uncommon, waste places, gardens: Sebastopol, *Noldeke* (1961); Bambi Lane just w. of Middle Rincon Road, Santa Rosa, *Wells* (1992); Pepperwood. < 1100 ft. June-Oct. Native of S. America

Verbena lasiostachys Link var. **lasiostachys**. Uncommon, dry to moist areas: Todd Road Preserve; Spring Lake; Lake Sonoma area; Alexander Valley. < 8300 ft. May-Sept.

Verbena lasiostachys var. **scabrida** Moldenke. Uncommon, moist places: Lichau Road 5 mi. ne. of Penngrove, *Rubtzoff* (1964); Kenwood; Pepperwood; chaparral area sw. of Bradford Mtn. 1½ mi. w. of Dry Cr., acc. J. & B. Guggolz (1992). < 5000 ft. May-Oct. —*V. robusta* E. Greene.

★ **Verbena rigida** Sprengel. Uncommon, adventive: Santa Rosa Jr. Coll. campus, *Baker* (1933). Native of s. Brazil and Argentina.

VIOLACEAE. Violet family

Viola

1. Flowers blue or violet (8-12 mm long); lateral petals with a tuft of very slender elongated hairs at base . . . . *V. adunca*
1: Flowers yellow, or if some other color, at least with yellow centers; lateral petals with a tuft of very short clavate hairs at base
   2. Leaves not deeply cut or lobed, but with margins entire or serrate-crenate
      3. Corolla yellow on face except for dark veining
         4. Plants strictly erect with cauline leaves and flowers crowded at ends of stems which are naked below; capsule acute
            5. Upper petals not backed with brown; stipules ± entire . . . . *V. glabella*
            5: Upper petals backed with brown, stipules deeply toothed
               6. Leaves palmately 3-7 (-9) -cleft or -divided into lance-oblong lobes, ovate to wedge-shaped or almost reniform in outline . . . . *V. lobata* var. *lobata*
               6: Leaves not lobed, cordate-ovate, irregularly crenate . . . . *V. l.* var. *integrifolia*
         4: Plants erect to prostrate, but with leaves and flowers scattered along the stems; capsule obtuse
            7. Stems prostrate, stolonlike; leaves round-cordate . . . . *V. sempervirens*
            7: Stems ascending to erect; at least upper leaves narrowed
               8. Plants with crown of rootstock near the surface; flowers 2 cm or less across; cleistogamous flowers present in upper axils
                  9. Blades of basal leaves as long as or longer than wide
                     10. Petals 6-10 mm long; leaves ± spreading, greenish to grayish, usually tinted purple; upper leaves subentire or irregularly or few (2-4) -toothed
                     10: Petals 10-12 mm long; leaves erect, grayish, without purple tinting; upper leaves ± regularly 5-6-dentate on each edge . . . . *V. purpurea* var. *quercetorum*
                  9: Blades of basal leaves commonly wider than long . . . . *V. p.* var. *purpurea*
               8. Plants with rootstocks short (1-2 cm long), soft; flowers > 2 cm across; cleistogamous flowers lacking . . . . *V. pedunculata*
      3: Corolla white with purple veins or blotches; upper petals red-violet on back . . . . *V. ocellata*
   2: Leaves deeply cut or lobed, 2-3 times pinnatifid . . . . *V. douglasii*

Viola adunca Smith. WESTERN DOG VIOLET. Common, meadows: The Sea Ranch; Rocky Point between Salt Point and Stewarts Point, *Knight* (1964); 2 mi. s. of Stewarts Point, *Cantelow* (1939); Salt Point; Willow Creek; ditch near Goat Rock Beach, *Mayer* (1968 NCC); Bodega Bay, *Heller & Brown* (1902); Estero Road ½ mi. w. of Franklin School off Hwy. 1 w. of town of Bodega, *Frank* (1987 NCC); Cunningham, *Bauder* (1929 NCC); pond near geysers Unit 18, *Malloch* (1977 PG&E). <11700 ft. Mar.-July.

Viola douglasii Steudel. Occasional, grassy flats, slopes: El Verano, *Erafin* (1922); Melita Road e. of Santa Rosa, *Baker* (1906); near Windsor, *Heller & Brown* (1902); Arata Lane vernal pool, Windsor, acc. Harrison (1992); Windsor, *Ortens* (1978 SRJC). < 7700 ft. Mar.-May.

Viola glabella Nuttall. STREAM VIOLET. Occasional, wet places, stream banks, woods: The Sea Ranch, *Wells* (1984); Stewarts Point, *M. Parks* (1925); Cazadero, *Davy* (1896 UC); Duncans Mills, *Rattan* (1880), *Rubtzoff* (1958); Willow Creek; Villa Grande near Monte Rio, *Gorter* (1920 JEPS); Guerneville Marsh, *Rubtzoff* (1954); Pocket Can. between Forestville and Guerneville, acc. Baker (1954). < 8700 ft. Mar.-July.

Viola lobata Bentham ssp. **integrifolia** (S. Watson) R. J. Little. Occasional, dry slopes, open woods: Ida Clayton Road 1.4 mi. from county line, *Bacigalupi* (1956 JEPS); geysers Unit 13, *Quibell* (1981 NCC). < 5700 ft. Apr.-July.

Viola lobata var. **lobata**. PINE VIOLET. Occasional, dry slopes, open woods: Sturgeon's Mill on Furlong Road between Sebastopol and Occidental, *G. L. Smith* (1971 JEPS); summit of ridge 3 mi. se. of Sonoma, *Brewer* (1862 UC); Youngman Road near Warfield (probably near jct. of Hwy. 12 and Trinity Road), *Baker* (1936 NCC); Gates Road off Calistoga Road near Humbug Creek, acc. Wells et al. (1984); Spring Lake; Pepperwood; Ida Clayton Road 7.1 mi. from Hwy. 128, *Berger* (1979 NCC); w. side of Dry Cr. Valley n. of Crane Cr., *Raiche* (1985); chaparral area sw. of Bradford Mtn. 1½ mi. w. of Dry Cr., acc. J. & B. Guggolz (1992); Lake Sonoma area; Pine Mtn. near The Geysers, *Brewer* (1864 UC); geothermal area, *D. J. Hall* (1977 PG&E). < 7700 ft. Apr.-July.

Viola ocellata Torrey & A. Gray. WESTERN HEART'S EASE. Common, often on serpentine, thickets, rocky and grassy banks: near Plantation, *Koch* (1939 UC); Cazadero, *Congdon* (1880), *Randall* (1930 UC); Graton-Occidental Road e. of summit, acc. Baker; Bodega, *Krager* (1924 JEPS); Harrison Grade, *Baker* (1939 UC); near Graton, *Baker* (1939 NCC); Bouverie; Stuart Cr. Can. e. of Hwy. 12 in redwood forest, *J. Petersen* (1985 NCC); Adobe Can. at bridge, *Baker* (1906); w. of Healdsburg in Dry Cr. area, *Raiche* (1985); Lake Sonoma area; Cloverdale grade (probably old Redwood Hwy. n. of Cloverdale en route to Hopland), *Eastwood & Howell* (1934); Felta Cr. area nw. of Healdsburg, *Bice* (1895 NCC). < 3700 ft. Mar.-June.

Viola pedunculata Torrey & A. Gray. JOHNNY-JUMP-UP, WILD PANSY. Common, grassy slopes: The Sea Ranch, *Wells* (1984); Russian River below Guerneville, *Davy* (1894); fields se. of Valley Ford, acc. Quibell (1991); jct. of Cotati and Hill roads, *Baker* (1936); Stony Point Road ½ mi. s. of jct. with Meacham Road, *Lamb* (1970 NCC), *Quibell* (1970 NCC); e. side of Petaluma Valley, *C. P. Smith* (1906); 5 mi. from refuse disposal site on Stage Gulch Road between Adobe Road and Lakeville Hwy., *Lehman* (1975 NCC); 0.4 mi. n. of jct. of Snyder Lane and Petaluma Hill Road, *Stocking* (1964 NCC); near Sonoma, *Heller & Brown* (1902). < 3300 ft. Feb.-Apr.

Viola purpurea Kellogg ssp. **purpurea**. MOUNTAIN VIOLET. Uncommon, dry slopes: Geysers geothermal area, *Quibell* (1980 NCC). < 8300 ft. Apr.-June.

Viola purpurea ssp. **quercetorum** (M. Baker & J. Clausen) R. J. Little. Uncommon, dry brushy or grassy slopes: Pine Flat Road, *Baker* (1934), *Clausen* (1934). < 6000 ft. Mar.-June. —*V. q.* Baker & Clausen.

Viola sempervirens E. Greene. REDWOOD VIOLET, EVERGREEN VIOLET. Occasional, mostly shady woods: The Sea Ranch, *Wells* (1984); Gualala River, *Abrams* (1920); Annapolis, *Sutliffe* (1920); Stewarts Point quad.; Kruse Rhododendron State Park, *Wiggins* (1948); Salt Point; Cazadero, *Heller* (1903); Duncans Mills quad.; Valley Ford, *Lobenstein* (1906 UC); Felta Cr. area sw. of Healdsburg, *Bice* (1895 NCC). < 4700 ft. Feb.-Apr.

VISCACEAE. MISTLETOE FAMILY
[formerly part of LORANTHACEAE]

Acc. Job Kuijt, *Loranthaceae* has been split into three families. There are no true *Loranthaceae* here; it is a large-flowered tropical family. *Eremolepidaceae* is a S. American and s. Mexican family. All are parasitic.

1. Berry 6-10 mm, globose . . . . *Viscum*
1: Berry 4-5 mm
    2. Berry on a recurved pedicel, compressed; ♀ sepals 2; anthers 1-celled . . . . *Arceuthobium*
    2: Berry sessile, globose, ♀ sepals 3; anthers 2-celled . . . . *Phoradendron*

Arceuthobium campylopodum Engelmann. WESTERN DWARF MISTLETOE. Common, on foothill, ponderosa and shore pines: Salt Point; 2 mi. s. of Anchor Cove (probably Anchor Bay), *Way* (1927); 4000 ft. elev. w. flank Mt. St. Helena, *Grummer* (1978 PUA); Geysers geothermal area, *Quibell* (1979 NCC). < 8000 ft. Aug.-Nov.

Phoradendron

1. ♀ spikes with 2 flowers at each joint; leaves oblanceolate . . . . *P. densum*
1: ♀ spikes with 6 or more flowers at each joint; leaves obovate-elliptic
    2. Internodes and leaves densely short-villous; leaves narrow-obovate to roundish, 1.5-3 (-4) cm long . . . . *P. villosum*
    2: Internodes minutely canescent to canescent-tomentose, later ± glabrate; leaves thickish, elliptic obovate to oblong-oblanceolate obtuse, 1.5 cm long . . . . *P. macrophyllum*

Phoradendron densum Trelease. DENSE MISTLETOE. Occasional, on sargent cypress: Camp Meeker, *Stocking* (1968); Morelli Lane near county refuse area e. of Occidental, *Knight* (1963); Harrison Grade Road, *Best* (1981 NCC). < 7700 ft. June-Aug. —*P. bolleanum* (Seemann) Eichler ssp. *d.* (Torrey) Weins.

Phoradendron macrophyllum (Engelmann) Cockerell. BIG LEAF MISTLETOE. Occasional, on willows and woody dicots other than *Quercus*: Laguna de Santa Rosa ne. of Graton near Guerneville Road crossing on *Salix laevigata, Rubtzoff* (1965); on the cultivar *Fraxinus velutina* Pacific St. just e. of Humboldt St., Santa Rosa, *Wells* (1985). < 4000 ft. Dec.-Mar. —*P. flavescens* (Pursh) Nuttall var. *m.* Engelmann.

Phoradendron villosum (Nuttall) Nuttall. OAK MISTLETOE. Common, on oaks, sometimes on other trees: Hidden Acres Road between Sonoma Mtn. and Grange roads, acc. Quibell (1991); Sugarloaf Ridge; Spring Lake; Pepperwood; 4298 ft. elev. Mt. St. Helena lookout, *Grummer* (1978 PUA). < 7000 ft. July-Sept. —*P. v.* var. *rotundifolium* Trelease. *P. flavescens* var. *v.* (Nuttall) Engelmann.

Viscum album Linnaeus. EUROPEAN MISTLETOE. Acc. Job Kuijt, this county is the only place in the U.S. where this European genus is found: seen at over 50 sites in 1971 with heavy concentration w., s. and n. of Sebastopol and outliers in Graton; over 50 more in 1991 as far away as Occidental, Forestville, Fulton, Brush Creek Road and Cotati. Spreading rapidly on *Acer, Alnus, Betula, Crataegus, Malus, Populus, Robinia, Salix* and *Ulmus*. < 400 ft. Feb.-Mar.

VITACEAE. GRAPE FAMILY

Vitis

1. Plants compact bush, tendrils few or none . . . . *V. rupestris*
1: Plants vines, with tendrils
    2. Skin of fruit loose and easily separating from pulp; young growth ± tomentose . . . . *V. californica*
    2: Skin of fruit adhering and not easily separating from the pulp; young growth often glabrous *V. vinifera*

Vitis californica Bentham. CALIFORNIA WILD GRAPE. Common, scandent on trees and large shrubs — a weed, with the potential to kill the plant on which it climbs: Laguna de Santa Rosa near Trenton s. of Woolsey Road, *Rubtzoff* (1960); Bouverie; Trinity Road, *M. Schneider* (1986 NCC); Sugarloaf Ridge; Adobe Can. adjacent to Mt. Hood, *Heller & Brown* (1902); Mt. Hood below Johnson Lake, *Knight* (1964);

Spring Lake; along Santa Rosa Cr., e. Santa Rosa, acc. Wagnon (1989); Pepperwood; Fitch Mtn., *Rubtzoff* (1952); 4.6 mi. w. of Warm Springs Cr. between Skaggs Springs and Geyserville, *Thomas* (1981); Lake Sonoma area; ca 2000 ft. elev. Anderson Trail on Black Mtn. between Geyser Peak and upper end of Pine Flat Road, acc. J. & B. Guggolz (1992); along 1st stream beyond county dump on Pine Mtn. Road, *Knight* (1963). < 3300 ft. May-June.
★ **Vitis rupestris** Scheele. Uncommon, escaped rootstock for wine grapes: Dunbar Road near Glen Ellen, *Knight & Wagnon* (1984); along Santa Rosa Cr. flood channel adjacent to Montgomery Drive, Santa Rosa, acc. Wagnon (1988). < 3300 ft. Native in e. U. S.
★ **Vitis vinifera** Linnaeus. WINE GRAPE. Uncommon, escape from cultivation: Spring Lake Village. < 3300 ft. Probably native in se. Europe to w. India.

ZYGOPHYLLACEAE. CALTROP FAMILY

★ **Tribulus terrestris** Linnaeus. PUNCTURE VINE. Common, along roadsides, waste places: flea market between Cunningham and Sebastopol, acc. Knight (1990); Sonoma State Univ. campus near football field, acc. Knight (1990); river side of Weller St. near curb in mats, Petaluma, acc. Knight (1992); Spring Lake Park, Santa Rosa, *Wells* (1987); along Franz Valley Road and in Knights Valley, acc. de Nevers (1992); along Russian River near Cloverdale, *Filippi* (1964 NCC). < 3300 ft. Apr.-Oct. Native of Mediterranean.

MONOCOTS

1. Plants small, stranded on mud or floating on water, lacking delineation between stem and leaf, occasionally with 1 to few simple roots . . . . *(p 266) Lemnaceae*
1: Plants with stem and leaves, the leaves occasionally scalelike
  2. Perianth lacking or reduced, its parts frequently mere scales or bristles, not petal-like in texture and color
    3. Flowers in axils or husklike or chaffy scales and ± concealed by them, or styles and stamens exserted; flowers in heads, spikes or spikelets
      4. Leaf-sheaths usually split lengthwise on the side opposite the blade; leaves frequently 2-ranked; stems mostly terete and hollow; anthers versatile . . . . *(p 282) Poaceae*
      4: Leaf-sheaths continuous around stem or fractured when aged; leaves mostly 3-ranked; stems often triangular across, usually with a pith; anthers basifixed . . . . *(p 250) Cyperaceae*
    3: Flowers not in axils or chaffy bracts or, if subtended by bracts, exceeding or equalling them and not concealed
      5. Plants submerged or floating; flowers submersed or floating, or barely raised above the surface of the water
        6. Plants marine; flowers on 1 side of a flattened axis . . . . *(p 316) Zosteraceae*
        6: Plants of fresh or brackish water
          7. Flowers in heads or spikes . . . . *(p 313) Potamogetonaceae*
          7: Flowers axillary and solitary or very few in a cluster
            8. Leaves alternate . . . . *(p 313) Ruppiaceae*
            8: Leaves opposite
              9. Carpel 1; leaves mostly 1-3 cm long . . . . *(p 279) Najadaceae*
              9: Carpels 2 or more; leaves mostly 3-10 cm long . . . . . . *(p 316) Zannichelliaceae*
      5: Plants of shallow water or land; often flowers and leaves well-emersed
        10. Flowers on a spadix enclosed by a white or yellow spathe . . . . *(p 250) Araceae*
        10: Flower not on a spadix enclosed by a white or yellow spathe
          11. Inflorescence a dense elongate spike
            12. Plants 1-4 dm; stamens 3 or 6, ♂ and ♀ flowers mixed . . . . [*Triglochin*] *(p 265) Juncaginaceae*
            12: Plants 10-20 dm; stamens often 3, ♂ flowers in separate upper portion of spike . . . . *(p 315) Typhaceae*
          11: Inflorescence of subglobose heads, open clusters or racemes
            13. Flowers unisexual, the lower heads ♀ . . . . *(p 314) Sparganiaceae*
            13: Flowers bisexual, perianth segments distinguishable
              14. Ovaries 3 or 6 dividing at maturity [*Lilaea*] (*Lilaeaceae*) *(p 265)* Juncaginaceae
              14: Ovary 1, forming a capsule . . . . *(p 262) Juncaceae*
  2: Perianth well-developed, at least the inner segments petaloid
    15. Carpels ± free, 1-loculed, mostly 1-ovuled, maturing into whorl or lot of achenes *(p 249) Alismataceae*
    15: Carpels united into a mostly 3-12-loculed ovary maturing into a capsule or berry
      16. Leaves large; plant arborescent, unbranched . . . . *(p 250) Arecaceae*
      16: Leaves and plants otherwise
        17. Ovary superior
          18. Petals colored, sepals green; flowers in umbels . . . . *(p 250) Commelinaceae*
          18: Petals and sepals the same color, or if unlike, flowers not umbellate; (ovary inferior on *Amaryllis*) . . . . *(p 267) Liliaceae*
        17: Ovary inferior
          19. Plants aquatic with mostly submersed leaves . . . . *(p 260) Hydrocharitaceae*
          19: Plants terrestrial
            20. Flowers regular, ovary 3-loculed; stamens 3; leaves equitant *(p 260) Iridaceae*
            20: Flowers irregular, ovary 1-loculed; stamens 1-2; leaves not equitant . . . . *(p 279) Orchidaceae*

ALISMATACEAE. WATER-PLANTAIN FAMILY

1. Leaf-blades sagittate . . . . . *Sagittaria*
1: Leaf-blades not sagittate
   2. Petals ca 0.5 cm long . . . . . *Alisma*
   2: Petals ca 1 cm long . . . . . *Damasonium*

Alisma

1. Petals pink; leaf blades usually lanceolate, base tapered; lateral walls of the fruit thin and translucent . . . . . *A. lanceolatum*
1: Petals usually white, sometimes pink; leaf blade lanceolate to ovate, bases truncate to slightly lobed; lateral walls of the fruit thick and opaque . . . . . *A. plantago-aquatica*

★ **Alisma lanceolatum** Withering. LANCE-LEAVED WATER-PLANTAIN. Common, marshy shores, swales and pond margins: Willow Cr. Marsh, *Rubtzoff* (1959); Austin Cr. Marsh, *Rubtzoff* (1954); Laguna de Santa Rosa ne. of Graton, *Rubtzoff* (1958); w. of Santa Rosa in Laguna de Santa Rosa drainage, *Harrison* (1978 NCC); jct. of Meacham and Pepper roads, acc. Roderick (1970); along Stony Point Road ca ¼ mi. n. of jct. with Rainsville Road, acc. Roderick (1965); Bennett Lake Marsh, *Rubtzoff* (1954); Fitch Mtn., *Rubtzoff* (1962). < 1700 ft. Naturalized from Eurasia.

Alisma plantago-aquatica Linnaeus. Common, pond margins, swales, marshy shores: Willow Creek; Guerneville Marsh, *Rubtzoff* (1954); Vacation Beach near Guerneville, *Rubtzoff* (1956); Laguna de Santa Rosa e. of Graton, *Rubtzoff* (1963); Green Valley Cr. near Ross Station, Graton, *Rubtzoff* (1962); Pitkin Marsh, *Rubtzoff* (1951); Atascadero Cr. Marsh, *Rubtzoff* (1956); Cunningham Marsh, *Howell* (1947); 2 sites w. of Santa Rosa in Laguna de Santa Rosa drainage, *Harrison* (1978 NCC); Roseland Cr. between Sebastopol Road and Hearn Ave. on Stony Point Road, *Harrison* (1978 NCC); laguna in Chileno Valley, *Rubtzoff* (1972); San Antonio Road between D and I st. extensions, *Knight* (1963); 4.7 mi. s. of El Verano on Arnold Drive, *Raven* (1963); Bennett Mtn. Lake, *Rubtzoff* (1961); Spring Lake; Newanga Road, Santa Rosa, *Knight* (1965); Healdsburg, *Rattan* (1877, 1879). < 5400 ft. June-July. —*A. p.-a.* var. *americana* Schultes & Schultes, misapplied. *A. p.-a. brevipes* (E. Greene) Samuelsson. *A. trivale* Pursh.

Damasonium californicum Torrey. FRINGED WATER-PLANTAIN. Occasional, mud or shallow water, vernal marshes: Stemple Cr., *Morgan* (1961); 2 sites w. and sw. of Santa Rosa in Laguna de Santa Rosa drainage, *Harrison* (1978 NCC); Sebastopol, *Howell* (1937); Laguna de Santa Rosa near Todd Road, *Rubtzoff* (1958); Todd Road Preserve; Willow Brook 4 mi. n. of Petaluma on Stony Point Road, *Rubtzoff* (1965); Petaluma, *Congdon* (1880). < 5700 ft. Apr.-Aug. —*Machaerocarpus californicus* (Torrey) Small.

Sagittaria

1. Fruiting heads maturing on stout, recurved pedicels; flowering scapes 1 to many *S. montevidensis ssp. calycina*
1: Fruiting heads maturing on slender, spreading or ascending pedicels; flowering scapes usually 1 or 2
   2. Terminal leaf-lobe much longer than basal ones, often 2 × as long or longer . . . . . *S. cuneata*
   2: Terminal leaf-lobe subequal or slightly longer than the basal ones . . . . . *S. latifolia*

Sagittaria cuneata E. Sheldon. ARUM-LEAVED ARROWHEAD. Common, shallow ponds, streams, swampy areas: e. of Guernewood in small pool along Russian River, *Rubtzoff* (1955); Laguna de Santa Rosa e. of Graton, *Rubtzoff* (1963); lagoon at Sebastopol, *Heller* (1902); laguna in Chileno Valley, *Rubtzoff* (1972); 5 mi. n. of Sears Point in marshy pond margin, *Rubtzoff* (1971); Bennett Mtn. Lake, *Rubtzoff* (1963); lagoon near Fulton, *Carruth* (1901); Healdsburg, *Rattan* (1877). < 8300 ft. June-Aug.

Sagittaria latifolia Willdenow. BROAD-LEAVED ARROWHEAD, TULE POTATO, WAPATO. Common, slow streams or at edge of ponds: Forestville Marsh, *Rubtzoff* (1955); Pitkin Marsh, *Howell* (1945); Laguna de Santa Rosa, *Rubtzoff* (1958); slough n. of Cunningham Marsh along hwy., acc. Baker (1932); lagoon e. of Sebastopol, *Howell* (1946); Sonoma Valley, *Krauss* (1893). < 5000 ft. July-Aug.

Sagittaria montevidensis Chamisso & Schlechtendal ssp. **calycina** (Engelmann) C. Bogin. MONTEVIDEO ARROWHEAD. Uncommon, mud or shallow water at edge of ponds: 5 mi. n. of Sears Point, *Rubtzoff* (1970). < 1000 ft. July-Aug. —*S. calycina* Engelmann.

ARACEAE. ARUM FAMILY

Leaves ±whorled/ blade ± flat, petioled; inflorescence bract wide, ± stalked.

1. Inflorescence bract lemon-yellow; petiole < leaf-blade . . . . . *Lysichiton*

1: Inflorescence bract green or white (or light yellow in *Arum*); petiole ± = or > blade
   2. Peduncle usually = or > leaf; inflorescence bract white . . . . . *Zantedeschia*
   2: Peduncle usually < leaf; inflorescence bract greenish-white to yellow . . . . . *Arum*

★ **Arum italicum** Miller. ITALIAN ARUM. Possibly growing from garden discards: bank of Santa Rosa Cr. near Spring Lake, acc. Wagnon (1989); Ducker Cr., Rincon Valley, *Wells* (1985). < 1700 ft. Native of Eurasia.

Lysichiton americanum Hultén & St. John. YELLOW SKUNK CABBAGE. Possibly extinct, wet ground near all-year streams or perennial springs: Russian River, acc. *Jepson* (1911). < 4400 ft. Apr.-June. —*L. kamts-chatcense* Schott, misapplied.

★ **Zantedeschia aethiopica** (Linnaeus) Sprengel. CALLA, CALLA LILY. Possibly growing from garden discards: The Sea Ranch; Salt Point; just s. and n. of Harbour Way (e. of Bodega Bay) and Hwy. 1 on inland side of road in bog surrounded by willows, acc. Roderick (1975); Coleman Valley Road ca ¼ mi. from Hwy. 1 at old abandoned house (n. side of road), acc. Roderick (1989); Forestville Marsh, *Rubtzoff* (1974). < 1000 ft. Mar.-May. Native of South Africa.

ARECACEAE [PALMAE]. PALM FAMILY

★ **Phoenix canariensis** Chabaud. CANARY ISLAND DATE PALM. Ornamental: Scores of seedlings from under a tree at Magnolia and Marshall aves. w. of Petaluma, acc. Knight (1932). < 3400 ft. Native of Canary Islands.

COMMELINACEAE. SPIDERWORT FAMILY

★ **Tradescantia fluminensis** Velloso. WANDERING JEW, SPIDERWORT. Probably a product of garden discards: forming a dense population on edge of Channel Drive opposite a residence on bank of Santa Rosa Cr., *Wagnon* (1990 NCC). < 1700 ft. Native of S. America.

CYPERACEAE. SEDGE FAMILY

1. Flowers perfect, or perfect and ♂; achene naked
   2. Plants with all scales of spikelet bearing achenes, or empty basal scales not > 2; flowers all perfect
      3. Scales of spikelet spirally imbricated
         4. Bristles much exerted beyond the scale . . . . . *Eriophorum*
         4: Bristles included within the scales
            5. Style-base deciduous from the summit of the achene, tubercle none; 1 or more involucral bracts present . . . . . *Scirpus*
            5: Style-base persistent as a tubercle on the summit of achene; involucral bracts none *Eleocharis*
      3: Scales of spikelet 2-ranked . . . . . *Cyperus*
   2: Plants with basal 3-several scales of the spikelet empty; spikelet consisting of both ♂ and perfect flowers . . . . . *Rhynchospora*

1: Flowers all unisexual; achenes surrounded by a perigynium . . . . . *Carex*

Carex: Cap. italics = sections, lower case italics = species

1. Spikelet 1 on each culm, androgynous . . . . . *FIRMICULMES: C. multicaulis*

1: Spikelets several to many, occasionally 2 on each culm
   2. Styles 2, achenes lenticular

3. Spikelets sessile, short, with relatively few perigynia
  4. Spikelets androgynous or occasionally spikelets unisexual and the plants ± dioecious
    5. Inflorescence simple, the spikelets in spikes or heads
      6. Plants loosely cespitose or open, with stolons or long-creeping rootstocks
        7. Spikelets generally unisexual, the plants usually dioecious; perigynia usually with a prominent serrulate beak . . . . **DIVISAE:**
          a. Beak of perigynium from ½ as long to nearly as long as body . . . . *C. praegracilis*
          a: Beak of perigynium a third as long as body or less . . . . *C. simulata*
        7: Spikelets androgynous . . . . ***BRACTEOSAE:*** *C. tumulicola*
      6: Plants cespitose, rootstocks short or short-prolonged, not long-creeping . . . . ***VULPINAE:*** *C. stipata*
    5: Inflorescence ± compound, some of the spikelets on short-condensed or elongate branches
      8. Perigynia plano-convex . . . . ***MULTIFLORAE:***
        b. ♀ scales acute or cuspidate, not prominently awned, shorter than the perigynia *C. densa*
        b: ♀ scales conspicuously awned, the awn usually longer than the perigynium . *C. dudleyi*
      8: Perigynia thick plano-convex or unequally biconvex; inflorescence ± paniculate; ♀ scales brownish . . . . ***PANICULATAE:*** *C. cusickii*
  4: Spikelets gynaecandrous
    9. Margin of perigynia winged, the wing broad or narrow; perigynia not spongy-thickened at base . . . . ***OVALES:***
      c. Perigynia visable in the spikelets, the subtending scales shorter and usually narrower
        d. Inflorescence not obviously leafy-bracteate; perigynia plano-convex or flat (except where distended by achenes)
          e. Leaf sheaths white-hyaline ventrally
            f. Perigynia small 2.25-3.5 mm long . . . . *C. subfusca*
            f: Perigynia larger 3.5-8 mm long
              g. Perigynia strongly nerved on ventral side . . . . *C. harfordii*
              g: Perigynia with ventral side smooth or very finely nerved or nerved only near the base
                h. Inflorescence generally densely capitate, oblong to roundish
                  i. Perigynium-beak not hyaline at tip or only slightly so . . . . . . *C. pachystachya*
                  i: Perigynium-beak obviously hyaline at tip . . . . *C. subbracteata*
                h: Inflorescence usually elongate, capitate above, spicate-interrupted below *C. gracilior*
          e: Leaf-sheaths green-striate to the top ventrally . . . . *C. feta*
        d: Inflorescence usually densely capitate, obviously leafy-bracteate, the bracts often longer than head; perigynia flattened . . . . *C. athrostachya*
      c: Perigynia covered and nearly or quite concealed by the scales . . . . *C. ovalis*
    9: Margin of perigynia rounded or sharp-edged but not winged; perigynia ± spongy-thickened at base
      10. Perigynia spreading or ascending in spikelets ***STELLULATAE:*** *C. echinata* spp. *phyllomanica*
      10: Perigynia appressed . . . . ***DEWEYANAE:***
        j. Perigynia 3.5-4 mm. long, scarcely nerved, the beak bidentulate *C. deweyana* spp. *leptopoda*
        j: Perigynia 4-4.5 mm. long, strongly nerved dorsally, the beak deeply bidentate *C. bolanderi*

3: Spikelets stalked, the stalks usually elongate and obvious, or if lateral spikelets are sessile, then the spikelets elongate, linear or oblong with many perigynia
  11. Lowest bract long-sheathing, the blade leaflike; perigynia beakless or very, very short beak . . . . ***BICOLORES:***
    k. Perigynia short-tapering and sometimes a bit constricted at the apex . . . . *C. saliniformis*
    k: Perigynia blunt at apex . . . . *C. hassei*
  11: Lowest bract without sheaths or nearly so, the blade bractlike or leafless; perigynia beaked but the beak usually very short
    12. Achenes not constricted . . . . ***ACUTAE:***
      l. Leaf-sheaths not becoming filamentose when breaking . . . . *C. nebrascensis*
      l: Leaf-sheaths breaking and becoming filamentose
        m. Lowest bract of inflorescence not leaflike, generally small and much shorter than inflorescence . . . . *C. nudata*
        m: Lowest bract of inflorescence leaflike, shorter to longer than the inflorescence *C. barbarae*
    12: Achenes centrally constricted . . . . ***CRYPTOCARPAE:*** *C. obnupta*

2: Styles 3 (or 4); achenes triangular (or quadrangular)
 13. Perigynia pubescent except *C. mendocinensis*
  14. Achenes with sides convex above . . . . ***MONTANAE:***
   n. Perigynia many-nerved as well as 2-keeled; ♀ scales generally as long as perigynia or longer . . . . *C. globosa*
   n: Perigynia 2-keeled, otherwise nerveless or nearly so; ♀ scales mostly shorter than the perigynia . . . . *C. brevicaulis*
  14: Achenes with sides flat or concave
   15. Leaves glabrous . . . . ***HIRTAE:*** *C. lanuginosa*
   15: Leaves pubescent . . . . ***SILVATICAE:***
    o. Plants with ♀ spikelets oblong-cylindric, 5-9 mm wide, with closely packed perigynia, erect, not flexuous or nodding . . . . *C. gynodynama*
    o: Plants with ♀ spikelets linear, 3.5-6 mm wide, with loosely placed but approximate perigynia, flexuous-erector nodding . . . . *C. mendocinensis*
 13: Perigynia glabrous
  16. Style and achene not continuous, the style withering and deciduous
   17. Lowest bract of inflorescence long-sheathing
    18. Plants with obvious long-creeping stolons or rootstocks ***PANICEAE:*** *C. californica*
    18: Plants densely cespitose, the rootstocks very short or short-prolonged
     19. Rootstocks and base of plants obviously fibrous-coated . . . . ***FERRUGINEAE:*** *C. albida*
     19: Rootstocks and base of plants ± scaly, not fibrous coated
      20. Base of plant brownish, not reddish; leaves not rough or slender tipped . . . . ***LAXIFLORAE:*** *C. hendersonii*
      20: Base of plant strongly reddish-tinged; leaves and bracts distended to a rough slender tip . . . . ***SYLVATICAE:*** *C. mendocinensis*
   17: Lowest bract of inflorescence short-sheathed or sheathless
    21. Plants stoloniferous, blades 1-2 cm broad . . . . ***ANOMALAE:*** *C. amplifolia*
    21: Plants loosely to densely cespitose stoloniferous in c. buxbaumii, blades < 1 cm. broad . . . . ***ATRATAE:***
     p. Perigynia puncticulate, the beak roughened, bidentate . . . . *C. serratodens*
     p: Perigynia granular-roughened or papillose, the beak bidentulate . . . *C. buxbaumii*
  16: Style and achene continuous, the style indurate and permanent
   22. Plants with ♀ scales with awn much longer than the broadened base; perigynia closely and finely many-ribbed . . . . ***PSEUDO-CYPEREAE:*** *C. comosa*
   22: Plants with ♀ scales acute or awned, the scale-body broader and longer; perigynia coarsely fewer ribbed . . . . ***VESICARIAE:***
    q. Culms arising oblique to the ground; perigynia appressed-ascending even in fruit, rather loosely arranged in the spikelets in a few rows . . . . *C. vesicaria* var. *major*
    q: Culms absolutely erect; perigynia appressed-ascending when young, squarrose-spreading in fruit, more closely arranged in the spikelets in more numerous rows . . . . *C. utriculata*

Carex albida L. Bailey. WHITE SEDGE. Uncommon, open, marshy ground: Perry Marsh, *Hoover* (1939); 4 mi. nw. of Sebastopol 0.4 mi. n. of jct. of Hwy. 116 and Guerneville Road, acc. Horenstein (1988); Pitkin Marsh, *Rubtzoff* (1953), *J. & B. Guggolz* (1988 UC & NCC); Santa Rosa Cr. (perhaps Laguna de Santa Rosa), *Thurber, Bigelow* (type 1854 U). < 300 ft. —*C. sonomensis* Stacey, type locality: Pitkin Marsh, *Howell & Stacey* (1937).

Carex amplifolia Boott. Wet soil along streams amd in meadows: Sugarloaf Ridge State Park, acc. Bowcutt (1993). < 7200 ft.

Carex athrostachya Olney. SLENDER-BEAKED SEDGE. Infrequent, streamsides, moist or wet soil: 98 ft. elev. Atascadero Cr. w. of Sebastopol, *Clifton* (1981 PUA); laguna near Sebastopol, *Howell & Stacey* (1937); 3150 ft. elev. upper Squaw Cr. drainage 1.2 airmiles nw. of Geyser Rock, acc. Stone (1983). < 10,500 ft. Early July.

Carex barbarae Dewey. SANTA BARBARA SEDGE. Common, hills, seepages and slopes that may be dry in the summer: Cheney Gulch on Hwy. 1 e. of Bodega Bay, *Barbour* (1970 BML); Vacation Beach near Guerneville, *Rubtzoff* (1953); Forest Hill Marsh across Russian River from Hacienda, *Rubtzoff* (1955);

Sebastopol, *Congdon* (1880), *Heller* (1902), *Howell & Stacey* (1937); w. of Santa Rosa in Laguna de Santa Rosa drainage, *Rubtzoff* (1958), *Harrison* (1978 NCC); Button Ranch on Walker Road (U. C. property) between Petaluma-Valley Ford Road and Pepper Road, acc. Connors (1992); Willow Brook, *Knight & Wagnon* (1987); Los Guilicos Valley near Kenwood, *Bioletti* (1893); 2 mi. n. of Windsor, *Eastwood & Howell* (1936); along Leslie Cr. w. of Petrified Forest, *de Nevers* (1981); Pepperwood; e. of Petrified Forest, *Ripley & Barneby* (1945); between Healdsburg and Stewarts Point, *J. Kipping* (1973). Cited in *Erythea 8:77* (1922): Russian River, *Bolander* (1865); Mark West Creek, *Bolander* (1865); Skaggs Springs, *Geis* (1902). < 3000 ft. Late Mar. —*C. lacunarum* Holm, type locality: Sebastopol, *Heller* (1902).

Carex bolanderi Olney. BOLANDER'S SEDGE. Infrequent, moist or wet ground of springs, stream banks and marshes: 3 mi. e. of Stewarts Point, *Best* (1981); 1 mi. n. of Fort Ross, *Howell* (1930); Camp Meeker, acc. Stacey (1917); Adobe Can., *Bioletti* (1893); s. Guilicos Valley near Kenwood, *Bioletti* (1893); Santa Rosa, *Congdon* (1880); 2800 ft. elev. at upper Squaw Cr. 1.1 airmiles wnw. of Geyser Rock, acc. Stone (1983). < 8300 ft. Mid May.

Carex brevicaulis Mackenzie. Common, dry coastal hills and flats, grassland and brush, often forming low turflike patches: 6.5 mi. s. of Gualala River (now part of The Sea Ranch), *Eastwood, Howell & Stacey* (1937); Bodega Head, acc. Connors (1991); border of Pitkin Marsh, *Howell & Stacey* (1937); 3 mi. n. of Sebastopol, *Howell* (1938); Harrison Grade Road, *Thorne* (1965); Stony Point, *Howell* (1938). < 8300 ft.

Carex buxbaumii Wahlenberg. RED-BASED SEDGE. Infrequent, moist or wet ground: Pitkin Marsh, *Baker* (1945 NCC). < 11000 ft. Common and widespread in N. Hemisphere, rare in California.

Carex californica L. Bailey. CALIFORNIA SEDGE. Rare, threatened or endangered in California but common elsewhere: Bodega Head and Valley Ford quadrangles acc. *Skinner & Pavlik* (1994). Have not been able to find any substantiating herbarium specimens. < 300 ft.

Carex comosa Boott. Widely distributed but rarely seen in California, swamps and marshes: Guerneville, *Davy*, acc. *Mackenzie, p. 91*, (1922); Clark Marsh near Westside Park at Bodega Bay, *Connors* (1991 BML). < 1300 ft.

Carex cusickii Mackenzie. Rare in California marshes: Pitkin Marsh, *Howell & Stacey* (1936), *Rubtzoff* (1951 NCC). < 6700 ft. Early June.

Carex densa L. Bailey. DENSE SEDGE. Common, marshes, wet fields, low elevation: 6 mi. w. of Occidental on Coleman Valley Road, *Best* (1981); 100 ft. elev. Bodega, *Rose* (1939); Occidental, *Howell* (1936); 3 mi. e. of Valley Ford on Petaluma-Valley Ford Road, *Howell* (1930); Pitkin Marsh, *Howell* (1938), *Rubtzoff* (1952); Cunningham Marsh, 1 mi. sw. of Cunningham Station, *Howell* (1937), *Baker* (1947); near Wright School s. of Hwy. 12, *Baker* (1936 NCC); 2 sites w. of Santa Rosa in Laguna de Santa Rosa, *Harrison* (1978 NCC); 125 ft. elev. w. side of Hwy. 116 s. of Hessel Road, *J. & B. Guggolz* (1991 NCC); Stony Point, *Howell* (1938); s. of El Rose Drive and w. of D St., Petaluma, *B. Anderson* (1987); near Kenwood, *Howell* (1935); Mark West Cr., *Bigelow* (type 1854 U); Pepperwood; Fulton, *Eastwood & Howell* (1936); 1 3/4 mi. n. of Windsor, *Howell* (1943); near Asti, *Eastwood & Howell* (1936); Caldwell Pines, *Jokerst* (1985); 2 mi. n. of Cloverdale, *Eastwood & Howell* (1936). *Carex densa*, if interpreted in broader characterization, might include *C. dudleyi*, which is also listed here. < 5000 ft. Early Mar. —*C. breviligulata* Mackenzie, in part. *C. vicaria* L. Bailey.

Carex deweyana Schweinitz ssp. **leptopoda** (Mackenzie) Calder & Taylor. SHORT-SCALED SEDGE. Occasional, shaded places, near marshes and streams: Gualala River, *Abrams* (1920); n. end of Bodega Harbor, *Rose* (1964); near Camp Meeker, *Hoover* (1941); Joy Road 1.8 mi. from jct. with Bodega Hwy. (s. of Occidental), *Ferris & Lorraine* (1961); Pitkin Marsh, *Rubtzoff* (1951); edge of Santa Rosa Cr. just below Montgomery Drive bridge, *Wagnon* (1989 NCC); 2860 ft. elev. Pine Flat Road, acc. Clifton (1984); 2190 ft. elev. Squaw Cr., acc. Clifton (1984); Caldwell Ranch, *Jokerst* (1985). < 8000 ft. Mid May. —*C. leptopoda* Mackenzie.

Carex dudleyi L. Bailey. DUDLEY'S SEDGE. Common, wet ground, open or wooded places: mouth of Salmon Cr., *Baker* (1932); Bodega, *Rose* (1939); Occidental, *Howell* (1936); Pitkin Marsh, *Howell & Stacey* (1938); Sebastopol, *Howell & Stacy* (1937); near Wright School s. of Hwy. 12 w. of Santa Rosa, *Baker* (1936); Llano Road (near Todd Road Preserve), *Baker* (1934 NCC); Fairfield Osborn (H); Stony Point, *Howell* (1938); near El Verano, *Best* (1980); Glen Ellen, *Bioletti* (1893); Fulton, *Eastwood & Howell* (1936); Pepperwood, near Petrified Forest, *Esau* (1935), *Howell* (1937); 2190 ft. elev. Squaw Cr., acc. Clifton (1984); Caldwell Ranch, *Jokerst* (1985). Closely related to *C. densa* and intergrading with that species when the two grow together, as in Pitkin Marsh, acc. *Rubtzoff, p. 357,* (1953). < 3000 ft.

Carex echinata Murray ssp. **phyllomanica** (W. Boott) Reznicek. Swampy areas: Pitkin Marsh, *Howell* (1945), *Rubtzoff* (1951); acc. *Mackenzie, p. 35,* (1922), a specimen by Bigelow in "California" was probably from Santa Rosa (i. e., Laguna de Santa Rosa). However, the Bigelow collection could have been made in Marin Co. when Bigelow made his historic trip from Corte de Madera to Punta de los Reyes, April 16-20, 1854 (cf. *Howell, p. 30,* [1949]). < 700 ft. Late Apr. —*C. p.* W. Boott.

Carex feta L. Bailey. GREEN-SHEATHED SEDGE. Occasional, moist or wet soil: Pitkin Marsh, *Rubtzoff* (1951); laguna near Sebastopol, *Howell & Stacey* (1937); Fairfield Osborn (H); Kenwood, *Bioletti* (1893); Mark West Cr., *Bigelow* (1854); Caldwell Ranch, *Jokerst* (1985). Type locality: Cloverdale, *Bolander* (1864 U). < 8000 ft. Mid Apr.

Carex globosa Boott. Common, woods, brushy places: Cazadero, *Congdon* (1895); Willow Creek; near Monte Rio, *Howell & Stacey* (1936); Neeley Hill area s. of Guerneville, *Rubtzoff* (1953); 3 mi. from Guerneville on Pocket Can. Road, *C. P. Smith* (1908); Green Valley Road w. of Graton near Harrison Grade, *Ferris & Lorraine* (1961); Vine Hill School, *Howell* (1945); Bouverie; Gates Road off Calistoga Road, *Wells et. al.* (1984); Pepperwood; 2800 ft. elev. w. flank Mt. St. Helena, *Grummer* (1981 PUA). < 4300 ft.

Carex gracilior Mackenzie. SLENDER SEDGE. Rather common, grassy and wooded slopes that are wet in spring: Pitkin Marsh, *Howell* (1938), *Rubtzoff* (1951); Cunningham Marsh, *Howell* (1947); 3 mi. e. of Valley Ford on Petaluma-Valley Ford Road, *Howell* (1930); Fairfield Osborn (H); Petaluma, Sonoma, Santa Rosa, *Stacey* (1933); Pepperwood; 3200 ft. elev. Socrates Mine, *Clifton* (1984 PUA); Cloverdale, *Bolander* (type 1864 U). Locations cited by *Mackenzie* (1922): Mark West Cr., *Bolander, Bigelow;* Russian River, *Bolander*. < 3500 ft. Early May.

Carex gynodynama Olney. Rare, moist or wet meadows: 2 mi. s. of Gualala River (now part of The Sea Ranch), *Eastwood, Howell & Stacey* (1937); 249 ft. elev. Salt Point, *Clifton* (1981 PUA); Fort Ross, *Heller* (1903); n. end of Bodega Bay, *Rose* (1964); Bodega Bay ne. of Inn at the Tides, *Connors* (1992 BML); Guerneville, *Rubtzoff* (1954). When *C. gynodynama* and *C. mendocinensis* grow together they may hybridize. < 2000 ft.

Carex harfordii Mackenzie. Rarely detected, damp and marshy ground: The Sea Ranch, *Wells* (1981); Bodega Head, acc. Connors (1991); Pitkin Marsh, *Rubtzoff* (1951); Occidental, *Howell* (1936); acc. *Mackenzie, p. 47,* (1922); Santa Rosa, *Stacey* (1933). < 2000 ft. —*C. montereyensis* Mackenzie.

Carex hassei L. Bailey. Rare, wet places: Laguna de Santa Rosa near Llano Road, *Rubtzoff* (1963). < 9000 ft.

Carex hendersonii L. Bailey. HENDERSON'S SEDGE. Uncommon, shaded areas, moist coastal woods: border of marsh and redwood forest at Duncans Mills, *Rubtzoff* (1960); Guerneville, *Davy*, acc. *Mackenzie, p. 609,* (1922), *Rubtzoff* (1954), the southernmost-known collections. < 3000 ft. Extends n. to s. British Columbia.

Carex lanuginosa Michaux. WOOLLY SEDGE. Rare, marshes: The Sea Ranch; Pitkin Marsh, *Rubtzoff* (1959); Perry Marsh, *Howell* (1945). < 11000 ft. Late May.

Carex mendocinensis Olney. Rare, wet soil, springs, marshes: 2 mi. s. of Gualala River (now part of The Sea Ranch), *Eastwood, Howell & Stacey* (1937); 2 mi. s. of Monte Rio on road to Camp Meeker, *Hoover* (1941). < 5300 ft.

Carex multicaulis L. Bailey. MANY-STEMMED SEDGE. Occasional, rocky outcrops, dry ground, woods, brush: 1200 ft. elev. sw. flank of Mt. Hood ¼ mi. below base of Sugarloaf Peak, *Baker* (1947 NCC); Pepperwood, *Breedlove* (1983); geysers Unit 13, *Quibell* (1981 NCC); 2440 ft. elev. Caldwell Ranch, acc. Clifton (1984); 3000 ft. elev. Caldwell Pines, acc. Clifton (1984). < 7300 ft. Mid Mar.

Carex nebrascensis Dewey. NEBRASKA SEDGE. Infrequent, marshes, moist or dry meadows: sterile plants in Perry Marsh, *Howell* (1945), acc *Rubtzoff, p. 161,* (1953); sterile plants in Cunningham Marsh, *Howell* (1946); 1½ mi. n. of Sebastopol, *Baker* (1946 NCC). < 8300 ft. Late June.

Carex nudata W. Boott. TORRENT SEDGE. Common, rocky beds of perennial brooks and streams: Russian Gulch (n. of Jenner), acc. *MacKenzie* (1922); Dutch Bill Cr. between Occidental and Monte Rio, *Stocking* (1964 NCC); Vacation Beach near Guerneville, *Rubtzoff* (1962); Bouverie; Adobe Can., *Howell* (1933); Mark West Cr., *Bolander* (1864); Pepperwood; Russian River near Healdsburg, *Rubtzoff* (1952); 2860 ft. elev. Pine Flat Road, acc. Clifton (1984); 2190 ft. elev. Squaw Cr., acc. Clifton (1984); 2800 ft. elev. upper Squaw Cr. 1.1 airmiles wnw. of Geyser Rock, acc. Stone (1983); 2840 ft. elev. Bear. Cr. Can., acc. Clifton (1984); The Geysers, *Quibell* (1980 NCC); Sulphur Cr. (between Cloverdale and The Geysers), *Geis* (1902). < 5000 ft. Early Mar.

Carex obnupta L. Bailey. Common, damp swales, flats, seepages, marshes: Gualala River, *Abrams* (1920); The Sea Ranch; s. end of The Sea Ranch ca 7 mi. s. of Gualala River, *J. H. Thomas* (1976); Salt Point; Willow Creek; Bodega Head; ditch along Bodega Road, *Klimansky* (1930 NCC); Sebastopol,

Congdon (1880), acc. *Mackenzie* (1922); Pitkin Marsh, *Howell & Stacey* (1936), *Rubtzoff* (1951), *Quibell* (1988 NCC); Cunningham Marsh, *Baker* (1946 NCC); w. of Santa Rosa in Laguna de Santa Rosa drainage, *Rubtzoff* (1958), *Harrison* (1978 NCC); Button Ranch on Walker Road (U. C. property) between Petaluma-Valley Ford Road and Pepper Road, acc. Connors (1992). < 3000 ft.

Carex ovalis Good. Infrequent, moist or wet banks, grassland: Bodega Head, *Connors* (1991 BML); Pitkin Marsh, *Rubtzoff* (1951); Cunningham Marsh, *Howell* (1947). < 3700 ft. —*C. tracyi* Mackenzie. *C. leporina* Linnaeus, misapplied.

Carex pachystachya Chamisso. THICK-HEADED SEDGE. Probably southernmost point of distribution of a sedge appearing n. to Alaska and e. to the Rocky Mountains: Pitkin Marsh, *Rubtzoff* (1951, identified by Richard Whitkus, Ohio State Univ., 1989). < 11100 ft. Early May.

Carex praegracilis W. Boott. CLUSTERED FIELD SEDGE. Infrequent, moist places, slopes and flats: Sebastopol, *Howell & Stacey* (1937); Cunningham Marsh, *Howell* (1947); Laguna de Santa Rosa, *Rubtzoff* (1957); Petaluma, *Stacey* (1933); 2190 ft. elev. Squaw Cr., acc. Clifton (1984); Pepperwood. < 9000 ft. Late Mar.

Carex saliniformis Mackenzie. Rare, coastal slopes and flats, moist or wet soil: 2 mi. s. of Gualala River, *Eastwood, Howell & Stacey* (1937); between Hwy. 1 and ocean s. of s. end of Meyers Grade, *J. & B. Guggolz* (1991 NCC). < 400 ft.

Carex serratodens W. Boott. SAW-TOOTHED SEDGE. Occasional, springs, seepages, usually on serpentine: ca 1 mi. nw. of Camp Meeker, *J. & B. Guggolz* (1991 NCC); Occidental Marsh, *Rubtzoff* (1957); Los Guilicos Valley 2.5 mi. from Kenwood en route to Santa Rosa, *Bioletti* (1893), acc. *Mackenzie,* (1922), p. 69 as *C. bifida*; Pepperwood; 5 mi. s. of The Geysers, *Cantelow* (1938); 3200 ft. elev. Socrates Mine, acc. Clifton (1984); ca 2000 ft. elev. Anderson Trail on Black Mtn. between Geyser Peak and upper end of Pine Flat Road, acc. J. & B. Guggolz (1992); 2190 ft. elev. Squaw Cr., acc. Clifton (1984); 2720 ft. elev. Caldwell Ranch, acc. Clifton (1984). < 6000 ft. Late Apr.

Carex simulata Mackenzie. SHORT-BEAKED SEDGE. Infrequent but plentiful locally, moist soil of bogs, meadows, streambanks: Kenwood Marsh, *Rubtzoff* (1962). < 11000 ft. Mid-June.

Carex stipata Muhlenberg. STIPED SEDGE. Infrequent, wet or moist ground: Pitkin Marsh, *Quibell* (1987 NCC); Sebastopol, *Howell & Stacey* (1937); Santa Rosa, *Heller* (1902), acc. *Mackenzie*, p. 34, (1922). < 5700 ft. Early June.

Carex subbracteata Mackenzie. Occasional, open grassland, woods: Russian River, *Bolander* (1864 UC); Bodega Head, *Best* (1985); Occidental, *Howell & Stacey* (1936); Pitkin Marsh, *Rubtzoff* (1953); Cunningham Marsh, *Howell* (1947); Blank Road near Gravenstein Hwy., acc. J. & B. Guggolz (1991); Stony Point, *Howell* (1938). < 3000 ft.

Carex subfusca W. Boott. RUSTY SLENDER SEDGE. Occasional, moist grassland, seepages, sometimes on serpentine: Russian River, *Bolander* (1864), acc. *Mackenzie*, p. 40, (1922); 1024 ft. elev. Coleman Valley Road, acc. Clifton (1984); Pepperwood; 3200 ft. elev. Socrates Mine, acc. Clifton (1984); 2190 ft. elev. Squaw Cr., acc. Clifton (1984); 2800 ft. elev. Caldwell Ranch, *Jokerst* (1985). < 10500 ft. Early May. —*C. teneriformis* Mackenzie.

Carex tumulicola Mackenzie. FOOTHILL SEDGE. Rather common, grassy slopes, flats in open or in woodland: Duncans Mills quad.; Bodega Head, acc. Connors (1991); Cunningham Marsh, *Howell* (1947), *Baker* (1947 NCC); near Wright School, *Baker* (1936); w. of Santa Rosa on Hwy. 12, *Baker* (1936 NCC); jct. of Warrington and Petaluma Hill roads, *Thompson* (1975 NCC); Button Ranch on Walker Road (U. C. property) between Petaluma-Valley Ford Road and Pepper Road, acc. Connors (1992); Petaluma, *Congdon* (1880); 5 mi. sw. of Petaluma, *Morgan* (1961); Bouverie; Santa Rosa, *Congdon* (1880), *Heller* (1902); Fulton, *Eastwood & Howell* (1936); Mark West Cr., *Bolander* (1864); Pepperwood. < 4000 ft. Mid-May.

Carex utriculata Boott. BEAKED SEDGE. Uncommon, wet places: Pitkin Marsh, *Rubtzoff* (1971). < 11400 ft. Late Apr. —*C. rostrata* Stokes.

Carex vesicaria Linnaeus var. **major** Boott. Occasional, marshy ground, shallow water: Willow Cr. Marsh, *Utech & Knight* (1986 CM); marsh at Twin Pine Ranch n. of Sebastopol, acc. Baker (1932); Laguna de Santa Rosa near Sebastopol, *Howell & Stacey* (1937). The "Twin Pine Ranch" named by M. S. Baker is another name for the Perry Ranch. Acc. *Rubtzoff*, p. 192, (1953) does not allude to this species as possibly growing at the Perry Marsh, so Baker's record may have been overlooked. < 6000 ft. —*C. exsiccata* L. Bailey.

Cyperus

1. Style bifid; achene lenticular
    2. Plant perennial, inflorescence not umbellate . . . . . *C. niger*
    2: Plant annual; inflorescence umbellate . . . . . *C. bipartitus*

1: Style trifid; achene triangular
    3. Floral rachis not winged
        4. Plant perennial with short rhizomes
            5. Spikelets 10-20 mm long; culms bluntly triangular, smooth
                6. Plants not dwarf . . . . . *C. eragrostis* var. *eragrostis*
                6: Plants dwarf . . . . . *C. e.* var. *compactus*
            5: Spikelets 4-7 mm long; culms angled, striate . . . . . *C. alternifolius*
        4: Plant annual with fibrous roots
            7. Scales acuminate or awned, the tip recurved . . . . . *C. acuminatus*
            7: Scales obtuse . . . . . *C. difformis*
    3: Floral rachis winged with pair of inner hyaline appendages at each node
        8. Plant perennial; achenes obtuse or mucronulate . . . . . *C. esculentus*
        8: Plant annual, robust; achenes distinctly mucronate . . . . . *C. erythrorhizos*

Cyperus acuminatus Torrey & Hooker. Rare or rarely detected, wet ground: bed of Santa Rosa Cr., acc. Baker (1954); Pepperwood. < 1399 ft. June-Oct.

★ **Cyperus alternifolius** Linnaeus. UMBRELLA PLANT. Uncommon, probable escape from cultivation: Freestone Marsh, *Rubtzoff* (1970). Native of Africa.

Cyperus bipartitus Torrey. Rare, streamsides: 2800 ft. elev. Caldwell Ranch, *Clifton* (1984 PUA). < 5000 ft. Aug.-Sept. —*C. niger* Ruiz López & Pavón var. *rivularis* (Kunth) V. Grant.

★ **Cyperus difformis** Linnaeus. Uncommon, weed in wet spots: Russian River near Guerneville, *Rubtzoff* (1956). < 1700 ft. July-Nov. Native of Asia.

Cyperus eragrostis Lamarck var. **compactus** (Desvaux) Kükenthal. Occasional, damp ground: Stoetz Lane e. of Occidental, *Best* (1983); Willow Brook n. of Petaluma, *Knight* (1987 NCC). < 2300 ft. May-Nov.

Cyperus eragrostis Lamarck var. **eragrostis.** NUT-GRASS. Common, weedy, various habitats: The Sea Ranch; Salt Point; Bodega Head; Duncans Mills Marsh, *Rubtzoff* (1957); Pitkin Marsh, *Rubtzoff* (1951); laguna at Sebastopol, *Baker* (1946 NCC); 2 sites w. of Santa Rosa in Laguna de Santa Rosa drainage, *Harrison* (1978 NCC); Todd Road Preserve; Llano Road ca 1 mi. n. of Gravenstein Hwy., *Howell & Knight* (1986); jct. Railroad Ave. and US 101 near Cotati, *Knight* (1963); near Stony Point, *Knight* (1963); Fairfield Osborn (H); Lakeville St. near US 101 at Petaluma, *Knight* (1963); e. of Sonoma towards Napa, *Baker* (1932 NCC); Bouverie; Kenwood, *Baker* (1954 NCC); Ducker Cr. in e. Santa Rosa, *Wells* (1985); Spring Lake; Santa Rosa, *Heller* (1902); Pepperwood; Healdsburg, King acc. *Jepson*, p. 192 (1909); Pine Flat Road, *Knight* (1963); ca 2000 ft. elev. Anderson Trail on Black Mtn. between Geyser Peak and upper end of Pine Flat Road, acc. J. & B. Guggolz (1992); 2190 ft. elev. Squaw Cr., acc. Clifton (1984); 2800 ft. elev. Bear Cr. Can., acc. Clifton (1984). < 3000 ft. May-Nov. —*C. serrulatus* Watson. *C. virens* Michaux. *C. vegetus* Willdenow.

Cyperus erythrorhizos Muhlenberg. Occasional, marshes, streams, reservoir strands: Vacation Beach near Guerneville, *Rubtzoff* (1958); Forest Hills across the Russian River from Hacienda, *Rubtzoff* (1961); Laguna de Santa Rosa ne. of Graton, *Rubtzoff* (1961); Spring Lake; Fitch Mtn. near Healdsburg, *Rubtzoff* (1963). < 1700 ft. July-Oct.

Cyperus esculentus Linnaeus. CHUFA, YELLOW NUT-GRASS. Infrequent, worldwide weed, often in cultivated ground: Vacation Beach near Guerneville, *Rubtzoff* (1956); Forest Hills across the Russian River from Hacienda, *Rubtzoff* (1965); Cloverdale, *Fembrini* (1932). Low elevs. June-Oct.

Cyperus niger Ruiz López & Pavón var. **niger**. Common, wet ground of marshes, low fields, ditches: Vacation Beach near Guerneville, *Rubtzoff* (1955 NCC); Guerneville, Rubtzoff (1955); Forest Hills across the Russian River from Hacienda, *Rubtzoff* (1961); Pitkin Marsh, *Howell & Stacey* (1936), *Rubtzoff* (1951, 1952); Cunningham Marsh, *Rubtzoff* (1961 NCC); 2 mi. n. of Sebastopol, *Baker* (1945 NCC); Laguna de Santa Rosa, *Rubtzoff* (1958); Bennett Valley, acc. Baker; marsh 5 mi. ne. of Penngrove (now Fairfield Osborn), *Rubtzoff* (1964); Perry Marsh, *Baker & Howell* (1945); Laguna de Santa Rosa, *Rubtzoff* (1958); Bennett Valley Road 4½ mi. s. of Santa Rosa, *Baker* (1938); Spring Lake; Fitch Mtn. near Healdsburg, *Rubtzoff* (1963). < 5000 ft. July-Nov. —*C. n.* var. *capitatus* (Britton) O'Neill, currently not recognized. *C. diandrus* Torrey var. *castaneus* Torrey. *C. melanostachyus* Kunth.

Eleocharis

1. Style trifid; achenes plump or 3-angled
    2. Achenes with several longitudinal ridges and many fine transverse lines
        3. Culms spongy, striate; stamens 2; scales straw-colored . . . . . *E. radicans*
        3: Culms exceedingly slender, furrowed; stamens 3; scales brown . . . . . *E. acicularis*
    2: Achenes not longitudinally ridged
        4. 2- to 7-flowered; culms capillary
            5. Culms 2-7 cm tall . . . . . *E. parvula*
            5: Culms 7-14 (-40) cm tall . . . . . *E. pauciflora*
        4: Spikelets 10- to many-flowered; culms coarse . . . . . *E. rostellata*
1: Style bifid; achenes lenticular . . . . . *E. macrostachya*

Eleocharis acicularis (Linnaeus) Roemer & Schultes. SMALL SPIKERUSH. Common, beds of ponds, marshes: Freestone Marsh; *Rubtzoff* (1962) Guerneville Marsh, *Rubtzoff* (1954); Forestville Marsh, *Rubtzoff* (1955); Pitkin Marsh, *Rubtzoff* (1951); 3 sites w. of Santa Rosa in Laguna de Santa Rosa drainage, *Harrison* (1978 NCC); Todd Road Preserve; Bennett Mtn. Lake, *Rubtzoff* (1955). < 8300 ft. May-Aug.

Eleocharis macrostachya Britton. PALE SPIKERUSH. Rather common, marshes, low valley lands wet during the spring: The Sea Ranch; Austin Cr. Marsh, *Rubtzoff* (1954); Willow Creek; 69 ft. elev. Salmon Cr., acc. Clifton (1984); Bodega Head and dunes, acc. Connors (1991); Pitkin Marsh, *Rubtzoff* (1951); Laguna de Santa Rosa n. of Sebastopol, *Rubtzoff* (1958); Atascadero Cr. Marsh w. of Sebastopol, *Rubtzoff* (1956); Cunningham Marsh, *Howell* (1947); 2 sites w. and sw. of Santa Rosa in Laguna de Santa Rosa drainage, *Harrison* (1978 NCC); Todd Road Preserve (H&K); Fairfield Osborn (S)(NCC); Los Guilicos Valley, *Bioletti* (1893); Bennett Mtn. Lake, *Rubtzoff* (1963); Pepperwood; property bordered by Chianti and Dutcher Cr. roads w. of US 101 and w. of Geyserville, acc. J. & B. Guggolz (1991); 2190 ft. elev. Squaw Cr., acc. Clifton (1984). < 8300 ft. —*E. palustris* (Linnaeus) Roemer & Schultes, misapplied, a European species.

★ **Eleocharis parvula** (Roemer & Schultes) Link. Rare, coastal salt marshes: near Salmon Cr. (town), *Rubtzoff* (1969), *Rubtzoff* with *Bunyan* (1973). < 8300 ft. July-Aug. Circumarboreal.

Eleocharis pauciflora (Lightfoot) Link. Infrequent, marshy places, flats that are wet in spring: Forestville Marsh, *Rubtzoff* (1955); Pitkin Marsh, *Rubtzoff* (1951); Laguna de Santa Rosa near Llano Road, *Rubtzoff* (1958); Bennett Mtn. Lake, *Rubtzoff* (1955). < 12300 ft. May-Aug. — *E. quinqueflora* (F. X. Hartmann) Schwartz.

Eleocharis radicans (Poiret) Kunth. Occasional, moist places: Stoetz Lane ne. of Occidental, *Best* (1983); Laguna de Santa Rosa near Laguna Road, *Rubtzoff* (1960); near Llano Road, *Rubtzoff* (1962); 5 mi. n. of Sears Point, *Rubtzoff* (1970); lower Adobe Can. near Kenwood, *Baker* (1939 NCC); Pepperwood. < 5000 ft. May-Aug. —*E. acicularis* var. *r.* (Poiret) Britton, misapplied.

Eleocharis rostellata (Torrey) Torrey. WALKING SEDGE. Occasional, openings on marshy ground: Bodega Marine Reserve, *Connors* (1991 BML); Forestville Marsh, *Rubtzoff* (1955); locally abundant at Pitkin Marsh, *Rubtzoff* (1951 NCC); Laguna de Santa Rosa e. of Cunningham, *Rubtzoff* (1958). < 6700 ft. May-Aug.

Eriophorum gracile Koch. SLENDER COTTON GRASS. Probably now extirpated: swamps near Santa Rosa, *Bigelow* (1854 U), acc. S. Watson, *Botany of CA, vol. 2, p. 220,* (1880). Usually < 6700 ft. May-July.

Rhynchospora

1. Achenes obovate with a subulate tubercle; bristles retrorsely barbed (except *R. californica*)
    2. Scales whitish; bristles 10-12 . . . . . *R. alba*
    2: Scales brown; bristles 5-7
        3. Achenes smooth . . . . . *R. capitellata*
        3: Achenes rugulose . . . . . *R. californica*
1: Achenes subglobose with a compressed tubercle; bristles upwardly barbed . . . . . *R. globularis*

Rhynchospora alba (Linnaeus) Vahl. WHITE BEAKED-RUSH. Rare, open wet ground: Pitkin Marsh, *Howell & Leschke* (1945), *Baker* (1945 NCC), *Howell, Leschke & Baker* (1945), *H. L. Mason* (1946), *Rubtzoff* (1951), *J. & B. Guggolz* (1987 UC). < 6700 ft. July-Aug.

Rhynchospora californica Gale. CALIFORNIA BEAKED-RUSH. Occasional, threatened by encroachment of willows, marshy ground: Pitkin Marsh, *Howell* (type 1936 GH), *Rubtzoff* (1952); Perry Marsh, *Howell & Baker* (1945); Cunningham Marsh 1.2 mi. nw. of Cunningham, *Rubtzoff* (1961). Also known in the Coast Ranges in Marin County, *Howell* (1945). Recently discovered in Butte County by Oswald. < 3300 ft. July.

Rhynchospora capitellata (Michaux) Vahl. Infrequent, wet marshy spots: Pitkin Marsh, *Howell, Leschke & Baker* (1945); Perry Marsh, *Howell & Baker* (1945). < 6000 ft. July-Aug. —*R. glomerata* (Linnaeus) Vahl var. *minor* Britton.

Rhynchospora globularis (Chapman) Small. ROUND-HEADED BEAKED-RUSH. Occasional locally, damp or wet ground: Pitkin Marsh, *Howell & Stacey* (1936), *Howell, Leschke & Baker* (1945), *Rubtzoff* (1952), *J. & B. Guggolz* (1987 UC); Perry Marsh, *Howell & Baker* (1945); Cunningham Marsh, *Howell & Baker* (1947). These marshes are the only stations known for the species in the w. U.S. acc. *Rubtzoff*, p. 156, (1953). < 1700 ft. July-Aug. —*R. g.* var. *recognita* Gale.

"The phytogeographic clustering of 4 species of beaked-rush in 3 almost contiguous marshes in Sonoma Co. is one of the notable anomalies in the California flora." —John Thomas Howell.

Scirpus

1. Involucral leaves 2-5, usually longer than the inflorescence

2. Spikelets small, 0.3-0.6 cm long . . . . . *S. microcarpus*

2: Spikelets larger, 1-4 cm long

3. Leaves usually 8-10 mm broad; secondary rootlets obvious

4. Inflorescence loose-umbellate; achene triangular; style trifid; bristles nearly equalling achene . . . . . *S. fluviatilis*

4: Inflorescence capitate with 1-several elongated rays; achene lenticular; style usually bifid; bristles half as long as achene . . . . . *S. maritimus*

3: Leaves normally not over 6 mm broad; secondary rootlets inconspicuous . . . . . *S. tuberosus*

1: Involucral leaf solitary, often appearing as a continuation of the culm

5. Culms leafy, triangular or subterete; spikelets few 1-12

6. Plants perennial with rhizomes; culms sharply triangular; bristles present

7. Leaf-blades rolled up longitudinally; involucral leaf 3-10 cm long; plant small to 1.1 m tall . . . . . *S. pungens*

7: Leaf-blades flat; involucral leaves 1-3 cm long; plant larger to 2.2 m tall

8. Achenes light brown or gray, the surface minutely pitted . . . . . *S. americanus*

8: Achenes dark brown, the surface horizontally rugose . . . . . *S. mucronatus*

6: Plants annual with fibrous roots; culms filiform; bristles absent

9. Scales only slightly keeled at tip, obtuse; involucral leaves < 0.5 cm long . . . . . *S. cernuus*

9: Scales sharply keeled, acute or acuminate; involucral leaves to 2.5 cm long; (except perennial at Bodega Head) . . . . . *S. koilolepis*

5: Culm leaves reduced to basal sheaths with short blades, to 8 cm long; culms terete and stout; spikelets numerous in umbels

10. Bristles filiform, barbed

11. Scales > > the achene; pith firm . . . . . *S. acutus* var. *occidentalis*

11: Scales equalling or only very slightly > the achene; pith soft . . . . . *S. tabernaemontani*

10: Bristles broad, ciliate or plumose, never barbed . . . . . *S. californicus*

Scirpus acutus Bigelow var. **occidentalis** (S. Watson) Beetle. COMMON TULE. Occasional, open wet ground, shallow water: Russian River Mouth Marsh, *Rubtzoff* (1963); Willow Cr. Marsh, *Rubtzoff* (1959); Lower Salmon Cr. Marsh, *Rubtzoff* (1962); w. of Westside Park at Bodega Head, acc. Connors (1991); Laguna de Santa Rosa ne. of Graton, *Rubtzoff* (1960); Pitkin Marsh, *Rubtzoff* (1951); Atascadero Cr. Marsh w. of Sebastopol, *Rubtzoff* (1958); e. of Cunningham, *Rubtzoff* (1958); Fairfield Osborn (S); Kenwood Marsh, acc. Baker; Spring Lake; Howarth Park, Santa Rosa, *Knight* (1964). < 8300 ft. May-Aug. —*S. rubiginosus* Beetle.

Scirpus acutus × **S. tabernaemontani**: Forestville Marsh, *Rubtzoff* (1955).

Scirpus americanus Persoon. AMERICAN BULRUSH. Occasional, pond edges, damp areas: Stony Point Marsh, *Rubtzoff* (1972); e. of Hwy. 121, 5 mi. n. of Sears Point, *Rubtzoff* (1970); streamlet 0.7 mi. ne. of Arnold Drive, Eldridge, *Rubtzoff* (1970); Kenwood Marsh, *Rubtzoff* (1974); Spring Lake; Pepperwood. < 6700 ft. June-Aug. —*S. olneyi* A. Gray, misapplied.

Scirpus californicus (C. A. Meyer) Steudel. CALIFORNIA TULE. Occasional, tidal flats, edge of lakes: Lower Salmon Cr. Marsh, *Rubtzoff* (1967); marsh at n. end of Bodega Bay, extending "from freshwater part of marsh through the *Salicornia* belt into the water of Bodega Harbor," *Rubtzoff* (1955), *Connors* (1991); Laguna de Santa Rosa ca 1 mi. ese. of Sebastopol, *Rubtzoff* (1966); Button Ranch on Walker Road (U. C. property) between Petaluma-Valley Ford Road and Pepper Road, acc. Connors (1992); 5 mi. n. of Sears Point on Hwy. 121, *Rubtzoff* (1970); Bennett Mtn. Lake, *Rubtzoff* (1963); Spring Lake. < 700 ft. June-Sept.

Scirpus cernuus Vahl. Occasional, wet ground, marshes, seepages, low-growing and easily overlooked: The Sea Ranch, *Knight* (1984); Horseshoe Cove, *Knight* (1963); Salt Point; Duncans Mills, *M. Jones* (1882); Salmon Cr., *Baker* (1930 NCC), Clifton (1984 PUA); Bodega Harbor, *Howell* (1930); Bodega Head, *Clifton* (1980, 1981 PUA); *Best* (1985); Laguna de Santa Rosa ne. of Graton near Laguna Road, *Rubtzoff* (1960); Forestville Marsh, *Rubtzoff* (1955); Pitkin Marsh, *Howell* (1937); e. of Cunningham Marsh, *Rubtzoff & Harrison* (1978); Button Ranch on Walker Road (U. C. property) between Petaluma-Valley Ford Road and Pepper Road, acc. Connors (1992). < 2700 ft. May-Aug. —*S. c.* var. *californicus* (Torrey) Beetle, currently not recognized.

Scirpus fluviatilis (Torrey) A. Gray. RIVER BULRUSH. Occasional, wet places: Lower Salmon Cr. Marsh, *Rubtzoff* (1954); Guerneville Marsh, *Rubtzoff* (1955); Laguna de Santa Rosa near Trenton s. of Woolsey Road, *Rubtzoff* (1962); Pitkin Marsh, *Rubtzoff* (1957); near Guerneville Road at end of Piner Road w. of Santa Rosa, *Rubtzoff* (1958); 0.6 mi. n. of Gravenstein Hwy. under bridge on Llano Road, *Howell & Knight* (1987); Spring Lake. < 4300 ft. June-July.

Scirpus koilolepis (Steudel) Gleason. Occasional, easily overlooked, wet soil: Salt Point, *Wagnon* (1989 NCC); Bodega Head; wet places in laguna of Santa Rosa Cr., *Bigelow* (1854), Pitkin Marsh, *Howell* (1937, 1957), *Rubtzoff* (1952 NCC); 1 site w. of Santa Rosa and 1 site near Todd Road in Laguna de Santa Rosa drainage, *Harrison* (1978 NCC); 4¼ mi. w. of Glen Ellen on Bennett Valley Road, *Rubtzoff & Froebe* (1971). < 2500 ft. Apr.-June. —*S. carinatus* (Hooker & Arnott) A. Gray.

Scirpus maritimus Linnaeus. PRAIRIE BULRUSH. Common, fresh and brackish marshes: Lower Salmon Cr. Marsh, *Rubtzoff* (1954); Bodega Harbor, *Rubtzoff* (1955); Lakeville St., Petaluma, *Knight* (1963); Petaluma River e. of Petaluma, *Rubtzoff* (1963); Gilardi's Resort, Lakeville, *Knight* (1984); mouth of Tolay Cr. near Sears Point, *Mason & Grant* (1946); mouth of Sonoma Cr. on Sears Point Road (Hwy. 37), *Baker* (1946 NCC); Spring Lake; Mark West Cr., *de Nevers* (1981). < 8300 ft. Apr.-Aug. —*S. robustus* Pursh. *S. paludosus* A. Nelson.

Scirpus microcarpus C. Presl. PANICLED BULRUSH. Common, seepages, along streams, in marshes: The Sea Ranch; near Stewarts Point, *Abrams* (1920); Stewarts Point Skaggs Springs Road 3 mi. from w. end, *Best* (1981); Salt Point; Duncans Mills Marsh, *Rubtzoff* (1955); 69 ft. elev. Salmon Cr., acc. Clifton (1984); Bodega Head Marsh, *Rubtzoff* (1954); Bodega Harbor Marsh, *Rubtzoff* (1955); Guerneville Marsh, *Rubtzoff* (1954); Pitkin Marsh, *Rubtzoff* (1951); Laguna de Santa Rosa e. of Sebastopol, *Rubtzoff* (1958); marsh e. of Cunningham, *Rubtzoff* (1958); Penngrove, *Muick* (1975 NCC); Bouverie; Spring Lake; near Burke's Sanitarium on Mark West Cr., *Heller* (1902); Pepperwood; 2720 ft. elev. Caldwell Ranch, acc. Clifton, (1984). < 9700 ft. May-Aug.

★ **Scirpus mucronatus** Linnaeus. Uncommon, shallow water: Hinebaugh Cr., Rohnert Park, *Rubtzoff & Harrison* (1974); confluence of Mark West and Leslie creeks at Mark West Springs, *de Nevers* (1981); e. of Ida Clayton Road n. of Hwy. 128, *Rubtzoff et al.*, (1979). < 2000 ft. July-Aug. Native of Eurasia.

Scirpus pungens Vahl. THREESQUARE. Uncommon, moist or marshy places, alkaline or saline soils: The Sea Ranch; Horseshoe Cove, *Knight* (1963); Salt Point; Duncans Mills, *M. E. Jones* (1882); mouth of Salmon Cr., acc. Baker (1932). < 6600 ft. May-Aug. —*S. americanus* Persoon, misapplied.

Scirpus tabernaemontani C. Gmelin. SOFT-STEM TULE, GREAT BULRUSH. Occasional, marshes, streams: Willow Cr. Marsh, *Rubtzoff* (1959); Bodega Harbor Marsh, *Rubtzoff* (1955); turning basin, Petaluma, *Knight* (1985); lower Russian River from Fitch Mtn. westward at 4 stations, *Rubtzoff*; Russian River n. of Cloverdale, *Heller* (1902). < 700 ft. Aug.-Sept. —*S. validus* Vahl.

★ **Scirpus tuberosus** Desfontaines. Rare, in plant bed in garden, cultivated for waterfowl food: Sonoma Mission, *Knight* (1990 NCC). < 500 ft. Early Apr. Native of Old World.

HYDROCHARITACEAE. FROGBIT OR WATERWEED FAMILY

1. Leaves in whorls of 3 (-4), with inconspicuous, unfringed nodal scales; stamens 9; styles 2-lobed
   2. Flowers solitary in a spathe . . . . . *Elodea*
   2: Flowers generally 3 in a spathe . . . . . *Egeria*
1: Leaves in whorls of 3-8, with fringed nodal scales; stamens 3; styles simple . . . . . *Hydrilla*

★ **Egeria densa** Planchon. BRAZILIAN WATERWEED. Submerged plants forming large masses of elongated branching stems with occasional slender roots: Fairfield Osborn, *Larned* (1985 NCC). < 7300 ft. June-Oct. Native of S. America. —*Elodea densa* (Planchon) Caspary. Cultivated for aquaria.

★ **Elodea canadensis** Richard. WATERWEED. Submerged plants forming large masses of elongated branching stems with occasional slender roots: Fairfield Osborn (S). Suspected to be an escape. < 8700 ft. July-Aug. Native of Rocky Mts. to Quebec, Kentucky, Virginia.

★ **Hydrilla verticillata** (Linnaeus f.) Caspary. Invasive, noxious, aquatic weed, supported by tubers, which clutters ponds: Spring Lake, *Oliver* (1984 NCC). < 700 ft. Eradicated in 1985. Native of Eurasia.

IRIDACEAE. IRIS FAMILY

1. Style-branches petaloid . . . . . *Iris*
1: Style-branches not petaloid
   2. Inflorescence not spicate
      3. Rootstocks short rhizomatous; stem compressed, winged . . . . . *Sisyrinchium*
      3: Rootstock a corm; stem neither compressed nor winged . . . . . *Romulea*
   2: Inflorescence spicate
      4. Flowers orange; stigmas filiform . . . . . *Crocosmia*
      4: Flower color otherwise; stigmas obovate to obcordate
         5. Flowers purplish-pink; leaves 1-2 dm long, ca 1 cm wide . . . . . *Gladiolus*
         5: Flowers bright scarlet; basal leaves 3-6 dm long, 2.5-6 cm wide . . . . . *Watsonia*

★ **Crocosmia × crocosmiiflora** (Burbidge & Dean) N. E. Brown. Probably rooted from garden discards: Bodega Head Marsh, *Rubtzoff* (1955). A hybrid between two S. African species, *C. aurea* (Pappe) Planchon and *C. pottsii* (Baker) N. E. Brown. < 150 ft. —*Montbretia c.* V. Lemoine, misapplied. *Tritonia c.* (V. Lemoine) Nicholson.

★ **Gladiolus italicus** Miller. "Specimens were sent to Alice Eastwood that Mrs. Adeline Crane of Petaluma found growing on her part of the Vallejo grant. She had seen it appear for ten years, coming up in different places...[It] was perhaps introduced by some members of the Vallejo family..." acc. *Leaflets of Western Botany 1(12):131,* (1934). Scattered for at least a mile along E. Washington St. between Ely and Adobe roads in hayfield on n. side of road, acc. Roderick (1987); Adobe Road near Hardin Lane, Petaluma, *Quibell* (1970 NCC); Occidental, *Western Hills Nursery* (1978). < 1700 ft. Native of s. Europe. —*G. segetum* Ker Gawler.

Iris

1. Falls with a beard of multicellular hairs . . . . . *I. germanica*
1: Falls glabrous, or puberulent with unicellular hairs
   2. Rhizome 5-15 mm thick
      3. Leaves odoriferous . . . . . *I. foetidissima*
      3: Leaves not odoriferous
         4. Stigmas truncately flattened to bilobed . . . . . *I. purdyi*
         4: Stigmas triangular, rounded or tongue-shaped
            5. Perianth-tube not exceeding 30 mm long; leaves with pinkish or reddish bases . *I. douglasiana*
            5: Perianth-tube exceeding 30 mm long, usually slender; basal leaves without red (entire plant frequently colored reddish on *I. fernaldi*)

6. Spathe-bracts 4-9 mm wide, linear-lanceolate; perianth-tube without a distinct throat; flower deep purple to golden-yellow . . . . *I. macrosiphon*

6: Spathe-bracts 6-11 mm wide, broadly lanceolate; perianth-tube occasionally showing a distinct throat; flowers creamy-yellow, occasionally darker veined . . . . *I. fernaldii*

2: Rhizome ca 20-30 mm thick

7. Leaves evergreen; flowers blue; dry habitat . . . . *I. longipetala*

7: Leaves not evergreen; flowers yellow; wet habitat . . . . *I. pseudacorus*

Iris douglasiana Herbert. DOUGLAS IRIS. Common, open grassland, meadows: The Sea Ranch, *Wells* (1981); 1½ mi. s. of Horseshoe Cove near Plantation turnoff, *Knight* (1964); Gualala River crossing e. of Stewarts Point, *Heller* (1902); 2 mi. s. of Stewarts Point, *Porter* (1967); 7 mi. s. of Stewarts Point, *Cantelow* (1938); Salt Point; Duncans Mills, *Rubtzoff* (1955); Willow Creek; 3 mi. n. of Bodega Bay, *McNally* (1970 NCC), *Elliott* (1930 NCC); Bodega Bay, *Heller & Brown* (1902), *Rubtzoff* (1962), *Clifton* (1971 PUA); Bodega Head, *Lamb* (1970 NCC); Graton-Occidental Road 3½ mi. from Occidental, *Thomas* (1955); Forestville Marsh, *Rubtzoff* (1955); Pitkin Marsh, *Rubtzoff* (1951); Button Ranch on Walker Road (U. C. property) between Petaluma-Valley Ford Road and Pepper Road, acc. Patterson (1990); Stony Point, *Congdon* (1880); Geyserville and Alexander Valley quads. Natural hybrids are produced with *I. purdyi*, *I. fernaldii* and *I. macrosiphon*. < 3300 ft. Mar.-May.

Iris fernaldii R. C. Foster. Common, ± shaded slopes: Fairfield Osborn (S); Bouverie; near Kenwood, *Baker* (1947); w. base of Mt. Hood, *Heller & Brown* (1902); Gates Road off Calistoga Road e. of Rincon Valley, acc. Wells et al. (1984); 5 mi. w. of Petrified Forest on the Santa Rosa-Calistoga Hwy., *Foster* (type 1937 GH), *Mason* (1939); ca 9 mi. up Mark West Road from US-101, *Koch* (1938); roadside of confluence of Mark West Cr. with Leslie Cr., Pepperwood, *de Nevers* (1981); Petrified Forest, *Lenz & Balls* (1952 RSA); 5 mi. e. of Mark West Springs, *Eastwood & Howell* (1938); 4.4 mi. n. on Franz Valley Road, *Lenz & Balls* (1952 RSA); Geysers-Healdsburg Road at Big Sulphur Cr., *Breslin* (1975 NCC). Natural hybrids are produced with *I. purdyi, I. douglasiana* and *I. macrosiphon*. < 6700 ft. Apr.-May.

★ **Iris foetidissima** Linnaeus. SCARLET-SEEDED IRIS. Uncommon, damp area below spring: Pepperwood. Inconspicuously flowered. Native of w. and s. Europe, n. Africa

★ **Iris germanica** Linnaeus. GERMAN IRIS. Escape from cultivation: Bouverie. Native of se. Europe.

Iris longipetala Herbert. Infrequent, moist open places: Salt Point; Occidental Marsh, *Rubtzoff* (1962); Atascadero Cr. Marsh, *Rubtzoff* (1958); 6 mi. ne. of Penngrove on Lichau Road (now Fairfield Osborn), *Rubtzoff* (1965). < 2000 ft. Apr.-May.

Iris macrosiphon Torrey. Common, sunny, woody to grassy hillsides: se. part of Bodega Harbor (now mostly extirpated by development), *Ott* (1934 NCC); 4 mi. n. of Camp Meeker (probably on Bohemian Hwy. en route to Monte Rio), *Lenz* (1958 RSA); ca 5 mi. n. of Cazadero, acc. Raiche (1984); Stoetz Lane near Occidental on serpentine, *Best* (1982); Armstrong Redwoods State Park near Guerneville, *Balls* (1954 NCC); Camp Meeker, *Ewing* (1964 NCC); Gravenstein Hwy. between Sebastopol and Graton, *Baker* (1929 NCC); Hidden Acres Road between Grange and Sonoma Mtn. roads, acc. Quibell (1991); Button Ranch on Walker Road (U. C. property) between Petaluma-Valley Ford Road and Pepper Road, acc. Patterson (1990); Bouverie; near Windsor, *Heller & Brown* (1902); Felta Cr. area sw. of Healdsburg, *Bice* (1895 NCC); w. of Healdsburg in Dry Cr. area, *Raiche* (1985); chaparral area sw. of Bradford Mtn. 1½ mi. w. of Dry Cr., acc. J. & B. Guggolz (1992); Lake Sonoma area; ca 2000 ft. elev. Anderson Trail on Black Mtn. between Geyser Peak and upper end of Pine Flat Road, acc. J. & B. Guggolz (1992). Natural hybrids are produced with *I. douglasiana, I. purdyi* and *I. fernaldii*, and one involves both *I. douglasiana* and *I. purdyi*. Usually < 3300 ft. Apr.-May.

★ **Iris pseudacorus** Linnaeus. Infrequent, clumps expand by rhizomes and seed, pond margins: Austin Cr. n. of Hwy. 116 on road to Cazadero, *Rubtzoff* (1987); intermittent lake off w. side of Hwy. 1 se. of Salmon Cr. *Rubtzoff* (1971); Forestville Marsh, *Rubtzoff* (1952); Spring Lake. < 1000 ft. Spring. Native of Mediterranean basin

Iris purdyi Eastwood. Sporadically occasional, wooded areas: Berkeley Y.M.C.A. Camp on Wheatfield Fork Gualala River on Stewarts Point-Skaggs Springs Road, *Stocking* (1964 NCC); w. of Skaggs Springs on road to Stewarts Point, *Heller* (1940); Barrel Springs on Mill Cr. Road, *Elliott* (1930 NCC); Lake Sonoma area; 105 ft. elev. Alder Glen Springs n. of Cloverdale, *Aitken* (1974 PUA). Produces natural hybrids with *I. douglasiana* and *I. macrosiphon* and one involving both *I. douglasiana* and *I. macrosiphon*. < 4000 ft. May-June.

★ **Romulea rosea** (Linnaeus) Ecklon var. **australis** (Ewart) De Vos. SATIN FLOWER. Escape: Pepperwood, *de Nevers* (1986); Foothill Reg. Park near Windsor, acc. Wagnon (1992). < 1200 ft. Introduced from S. Africa. —*R. longifolia* (Salisbury) Baker.

Sisyrinchium:

1. Flowers blue; filaments ± completely fused . . . . *S. bellum*
1: Flowers yellow; filaments ± completely free . . . . *S. californicum*

Sisyrinchium bellum S. Watson. BLUE-EYED-GRASS. Widespread, meadows, open hills, mostly in grassland: Salt Point; Todd Road Preserve (H&K); Fairfield Osborn (*); Bouverie; 350-400 ft. elev. Old Skaggs Springs Road, *Utech, Sato & Ohara* (1990 CM); 2040 ft. elev. Squaw Cr., acc. Clifton (1984). Usually < 8000 ft. Mar.-May.

Sisyrinchium californicum (Ker Gawler) Dryander. GOLDEN-EYED-GRASS. Common, damp places: The Sea Ranch; Goat Rock Beach, *Rubtzoff* (1962); near Spud Point Marina at Bodega Harbor, *Clifton* (1984 PUA); 7 ft. elev. Bodega Head, *Clifton* (1980 PUA); Pitkin Marsh, *Stocking* (1968 NCC); 3 mi. from Healdsburg on Dry Cr., *Elliott* (1930 NCC). Usually < 2000 ft. May-June. —*Marica californica* Ker Gawler, type locality: Port Bodega in "northwest coast," *Menzies* (1792 U).

★ **Watsonia bulbillifera** Matthews & Bolus. Gregarious along roadside: e. side of Hwy. 1 at Salt Point, *Best* (1983). Clusters of bulbils at lower nodes of inflorescence. < 200 ft. May-July. Native of S. Africa

JUNCACEAE. RUSH FAMILY

1. Leaf-sheaths open; leaves stiff (terete to flat); stems usually with spongy pith . . . . *Juncus*
1. Leaf-sheaths closed; leaves soft (flat); stems hollow . . . . *Luzula*

Juncus

1. Plants annual
    2. Flowers inserted singly and each subtended by 2 bractlets
        3. Perianth 3-4 mm . . . . *J. bufonius* var. *occidentalis*
        3: Perianth 4-7 mm . . . . *J. b.* var. *bufonius*
    2: Flowers inserted in true heads, the heads (but not the flowers) subtended by bracts
        4. Lowest bract of inflorescence leaflike, extending past the head; capsule half as long as perianth . . . . *J. capitatus*
        4: Lowest bract not so; capsule longer than perianth
            5. Heads often 2-several-flowered; bracts 2, subequal, 1-1.5 mm long . . . . *J. kelloggii*
            5: Heads 1-flowered; bracts 1-2, these unequal and shorter . . . . *H. hemiendytus*
1: Plants perennial
    6. Lowest bract of inflorescence terete and looking like a continuation of the stem
        7. Stamens 3
            8. Perianth 2.5-4.2 mm, firm when fresh, rigid when dry . . . . *J. effusus* var. *pacificus*
            8: Perianth 1.8-3.5 mm, soft when fresh, not rigid when dry . . . . *J. e.* var. *brunneus*
        7: Stamens 6
            9. Foliage blue-green; anthers ca = filaments; stems in tussocks . . . . *J. patens*
            9: Foliage not blue-green; anthers >> filaments; stems not usually in clumps
                10. Perianth 5-6 mm long . . . . *J. lesueurii*
                10: Perianth usually 3.5-4.5 mm long . . . . *J. balticus*
    6: Lowest bract not precisely like a continuation of the stem, or if so, grooved along inner side
        11. Leaf-blades flat without internal septa on surface facing the stem
            12. Flowers inserted singly and each subtended by 2-3 bracts, leaflike and >> inflorescence
                13. Inflorescences congested, mostly < 3 cm; tepals usually at least 4 mm, brownish with a broad green midstripe . . . . *J. occidentalis*
                13: Inflorescences open, usually > 3 cm; tepals often < 4 mm, greenish . . . . . *J. tenuis*
            12: Flowers inserted in true heads; heads subtended by bracts
                14. Perianth-segments < mature capsules 2-3.5 mm long . . . . *J. covillei*

14: Perianth-segments not usually < capsules 5-6 mm long
15. Heads 5-10-flowered; leaves not stiff . . . . *J. orthophyllis*
15: Heads 8-25-flowered; leaves stiff . . . . *J. falcatus*
11: Leaf-blades terete, or if flat, with edge toward stem gladiate with ± discernible internal septa
16. Leaf-blades terete or ± compressed but not gladiate, with septa usually complete
17. Stamens 3 . . . . *J. bolanderi*
17: Stamens 6 . . . . *J. dubius*
16: Leaf-blades gladiate, flat, septa usually incomplete
18. Anthers mostly > the filaments; style at least – in length to ovary at anthesis
19. Perianths 4-5 mm long
20. Inflorescences 1-6 (rarely 8) dark-brown, generally large, many-flowered, mostly spherical to hemispherical heads; leaves bright green . . . . *J. phaeocephalus* var. *phaeocephalus*
20: Inflorescences of ca 10 to ca 40 lighter-brown, smaller, fewer-flowered, hemispherical to < hemispherical heads arranged in a loose to fairly dense panicle; leaves slightly glaucous . . . . *J. p.* var. *paniculatus*
19: Perianths 3-4 mm long . . . . *J. oxymeris*
18: Anthers << the filaments; style short
21. Stamens generally 6; leaves 3-12 mm wide; capsule tapering gradually into elongate beak . . . . *J. xiphioides*
21: Stamens generally 3; leaves 2-5 mm wide; capsule tapering abruptly into a very short mucro . . . . *J. ensifolius*

Juncus balticus Willdenow. WIRE RUSH. Infrequent, open places, moist soil: The Sea Ranch; Pitkin Marsh, *Rubtzoff* (1951); Petaluma, *Davy* (1890s UC); 2190 ft. elev. Squaw Cr., acc. Clifton (1984). Usually < 7300 ft. May-Aug.

Juncus bolanderi Engelmann. BOLANDER'S RUSH. Common, wet or swampy ground: mouth of stream through golf course at The Sea Ranch, *Wells* (1985); Stewarts Point, *Abrams* (1920); Salt Point; Duncans Mills, *M. Jones* (1882); Salmon Cr. Marsh, *Rubtzoff* (1967); Bodega Head Marsh, *Rubtzoff (1954);* Bodega Bay Marsh, *Rubtzoff* (1962); 7 ft. elev. Bodega Harbor, *Clifton* (1980 PUA); Willow Cr. Marsh, *Rubtzoff* (1959); ca 1 mi. nw. of Camp Meeker, acc. J. & B. Guggolz (1991); Occidental Marsh, *Rubtzoff* (1962); Austin Cr. Marsh, *Rubtzoff* (1960); Coleman Marsh (between Occidental and coast on Coleman Valley Road), *Rubtzoff* (1964); Vacation Beach near Guerneville, *Rubtzoff* (1959); Fife Cr. Marsh near Guerneville, *Rubtzoff* (1960); Laguna de Santa Rosa near Llano Road, *Rubtzoff* (1958); Roblar Road near source of Stemple Cr., *Knight* (1963); Fairfield Osborn (S); Button Ranch on Walker Road (U. C. property) between Petaluma-Valley Ford Road and Pepper Road, acc. Connors (1992); Spring Lake; Pepperwood; Fitch Mtn., *Rubtzoff* (1962); Asti (s. of Cloverdale), *Raven* (1963), *Eastwood & Howell* (1936); ca 2000 ft. elev. Anderson Trail on Black Mtn. between Geyser Peak and upper end of Pine Flat Road, acc. J. & B. Guggolz (1992); 2720 ft. elev. Caldwell Ranch, *Clifton* (1984 PUA). < 5300 ft. June-Aug.

Juncus bufonius Linnaeus var. **bufonius**. TOAD RUSH. Common, moist ground, frequently weedy in gardens and irrigated fields: pygmy forest 3/4 mi. s. of Gualala River, *Rhyne, Wells & Madison* (1985); 7 mi. s. of Gualala on The Sea Ranch, *J. H. Thomas* (1976); Salt Point; Bodega Point, *Eastwood* (1915); Bodega Head Marsh; *Rubtzoff* (1954); Willow Cr. Marsh, *Rubtzoff* (1959); Vacation Beach near Guerneville, *Rubtzoff* (1956, 1959); Fife Cr. Marsh at Guerneville, *Rubtzoff* (1960); Pitkin Marsh, *Howell* (1938), *Rubtzoff* (1951); Laguna de Santa Rosa e. of Sebastopol, *Rubtzoff* (1958); Sebastopol, *Dows* (1907); Cunningham Marsh, *Howell* (1947); between Cotati and Santa Rosa, *Eastwood* (1921); many sites w. and sw. of Santa Rosa in Laguna de Santa Rosa drainage, *Harrison* (1978 NCC); Stony Point Marsh, *Rubtzoff* (1955); Fairfield Osborn (H); Bennett Mtn. Lake, *Rubtzoff* (1961); Spring Lake; Santa Rosa Cr. e. of Santa Rosa, *Heller* (1902); Santa Rosa, *Eastwood* (1921); 2 mi. nw. of Windsor on Old Redwood Hwy., *de Nevers & Kaza* (1981); Pepperwood; Fitch Mtn., *Rubtzoff* (1952); 2800 ft. elev. Bear Cr. Can., acc. Clifton (1984); 2190 ft. elev. Squaw Cr., acc. Clifton (1984). < 10600 ft. Apr.-Sept. Cosmopolitan distribution.

Juncus bufonius var. **occidentalis** F. J. Hermann. Uncommon, dry pools: Llano Road e. of Sebastopol, acc. Baker (1932). < 10600 ft. Apr.-Aug. —*J. sphaerocarpus* Nees, misapplied.

★ **Juncus capitatus** Weigel. Occasional, moist slopes, around vernal pools: firebreak at The Sea Ranch, *Wells* (1984); Salt Point; vernal pool area at Santa Rosa, *Harrison* (1978); Pepperwood. < 3300 ft. Apr. Introduced from the Old World.

Juncus covillei Piper. COVILLE'S RUSH. Occasional, moist shady banks, streambeds: The Sea Ranch, *Wells* (1981); near Petrified Forest, *Howell* (1937); Pine Flat in creek bed, *Harrison* (1979 SRJC); 2190 ft. elev. Squaw Cr., acc. Clifton (1984). < 10000 ft. June-July.

Juncus dubius Engelmann. Common, mostly marshes of coastal hills, occasionally inland: Occidental Marsh, *Rubtzoff* (1957); Guerneville Marsh, *Rubtzoff* (1954); Pitkin Marsh, *Howell* (1938), *Rubtzoff* (1951); Forestville Marsh, *Rubtzoff* (1955); Laguna de Santa Rosa, *Rubtzoff* (1958); 5 mi. n. of Sears Point, *Rubtzoff* (1970); Spring Lake, *Rubtzoff* (1971), *Wagnon* (1989 NCC); 2420 and 2800 ft. elev. Caldwell Ranch, *Clifton* (1984 PUA). < 6700 ft. Apr.-Aug.

Juncus effusus Linnaeus var. **brunneus** Engelmann. Fairly common, coastal, swamps, swales: The Sea Ranch, *Wells* (1985); near Stewarts Point, *Abrams* (1920); Lower Salmon Cr. Marsh, *Rubtzoff* (1954); Bodega Point, *Eastwood* (1915). Usually < 400 ft. June-July.

Juncus effusus var. **pacificus** Fernald & Wiegand. Common, often growing in large tufts, springs, stream banks, marshes, pastures: Timber Ridge on The Sea Ranch, *Wells* (1985); Fort Ross gulch, *McKenzie* (1981); Duncans Mills Marsh, *Rubtzoff* (1957); 7 ft. elev. Bay Flat Road near Spud Point Marina at Bodega Harbor, *Clifton* (1980 PUA); Pitkin Marsh, *Howell* 1938, *Rubtzoff* (1951); Cunningham Marsh, *Howell* (1946); Button Ranch on Walker Road (U. C. property) between Petaluma-Valley Ford Road and Pepper Road, acc. Patterson (1990); w. of Sears Point Raceway in Tolay Cr. drainage, acc. Patterson (1990); 3 mi. up Cavedale Road, *Knight* (1984); Annadel, *Wells* (1984); Pepperwood; 2300 ft. elev. w. flank Mt. St. Helena, acc. Grummer (1981); 2860 ft. elev. Pine Flat Road, acc. Clifton (1984); 3200 ft. elev. Socrates Mine, *Clifton* (1984 PUA); ca 2000 ft. elev. Anderson Trail on Black Mtn. between Geyser Peak and upper end of Pine Flat Road, acc. J. & B. Guggolz (1992); 2190 ft. elev. Squaw Cr., acc. Clifton (1984); 2800 ft. elev. Bear Cr. Can., acc. Clifton (1984); 2720 ft. elev. Caldwell Ranch, acc. Clifton (1984); Pine Mtn. Road near Sulphur Cr. Road, *Knight* (1963). < 8300 ft. June-Aug.

Juncus ensifolius Wikström. Uncommon, marshy places, wet meadows: The Sea Ranch on Deer Trail, *Wells* (1984); Salt Point; Spring Lake. < 9300 ft. July-Aug.

Juncus falcatus E. Meyer. To be expected in low places in moist soil along the immediate coast. It has been collected in Marin and Mendocino counties. < 350 ft. May-Aug.

Juncus hemiendytus F. J. Hermann. Uncommon, moist flats, drying rainpools: near summit of Bennett Valley Road, *Eastwood & Howell* (1940). < 11300 ft. June-July.

Juncus kelloggii Engelmann. KELLOGG'S DWARF RUSH. Infrequent, moist banks, clay depressions: hills e. of Agua Caliente in Sonoma Valley, *Howell* (1943); Bennett Valley Road, *Rubtzoff* (1971). < 2700 ft. May-July.

Juncus lesueurii Bolander. COASTAL WIRE RUSH. Occasional, ponds, moist sandy flats, coastal influence: Willow Cr. Marsh, *Rubtzoff* (1959); Coleman Valley Marsh, *Rubtzoff* (1954); vicinity of Bodega Bay, *Eastwood* (1900), (1915), *Sauffer* (1973); Bodega Point, *Eastwood* (1915); 7 ft. elev. Bay Flat Road near Spud Point Marina at Bodega Harbor, *Clifton* (1980 PUA); Americano Cr. e. of Valley Ford, (1961). < 500 ft. May-Aug. —*J. l.* var. *tracyi* Jepson, indistinct.

Juncus occidentalis (Coville) Wiegand. Occasional, meadows and grassy slopes that are wet in the spring and dry in the summer: Deer Trail on The Sea Ranch, *Wells* (1984); 600 ft. elev. between Hwy. 1 and ocean s. of the s. end of Meyers Grade Road, *J. & B. Guggolz* (1991 NCC); 1024 ft. elev. Coleman Valley Road, *Clifton* (1984 PUA); Pitkin Marsh, *Rubtzoff* (1951); Cunningham Marsh, *Howell* (1947); 3 sites w. and sw. of Santa Rosa in Laguna de Santa Rosa drainage, *Harrison* (1978 NCC); ne. corner of junction of Llano and Todd roads, *Howell* & Knight *(1987);* Fairfield Osborn (H); Button Ranch on Walker Road (U. C. property) between Petaluma-Valley Ford Road and Pepper Road, acc. Patterson (1990); El Rose Drive w. of D St. in Petaluma, *B. Anderson* (1987); 960 ft. elev. in Mt. Hood area at quarry on w. side of Buzzard Peak, *J. & B. Guggolz* (1991 NCC); Annadel, *Wells* (1984); Spring Lake; Santa Rosa, *Heller* (1902); Pepperwood; 2040 ft. elev. Squaw Cr., acc. Clifton (1984). < 7700 ft. May-Aug. —*J. tenuis* Willdenow var. *congestus* Engelmann.

Juncus orthophyllus Coville. Infrequent, wet soil, grassland bordering marshes: Pitkin Marsh, *Rubtzoff* (1952); Cunningham Marsh, *Howell* (1947). < 11700 ft. July-Aug.

Juncus oxymeris Engelmann. Locally common, marshes, wet areas: Stewarts Point-Skaggs Springs Road 18 mi. e. of Stewarts Point, *Best* (1981); ca 1 mi. nw. of Camp Meeker, acc. J. & B. Guggolz (1991); Stoetz Lane e. of Occidental, *Best* (1981); Guerneville Marsh, *Rubtzoff* (1954); Todd Road

Preserve, *Hughey* (1992 NCC); Laguna de Santa Rosa e. of Sebastopol, *Rubtzoff* (1969); River Road 3 mi. w. of Fulton, *Rubtzoff* (1969); 3160 ft. elev. Socrates Mine, *Clifton* (1984 PUA); 2800 ft. elev. Bear Cr. Can., *Clifton* (1984 PUA). < 7000 ft. June-Aug.

Juncus patens E. Meyer. Very common, damp meadows, riparian and marsh situations: The Sea Ranch, *Wells* (1985); Duncans Mills Marsh, *Rubtzoff* (1957); Bodega Head, acc. Connors (1991); ca 1 mi. nw. of Camp Meeker, acc. J. & B. Guggolz (1991); Pitkin Marsh, *Rubtzoff* (1951); Laguna de Santa Rosa e. of Cunningham, *Rubtzoff* (1958); Llano Road 0.6 mi. n. of Gravenstein Hwy., *Howell & Knight (1986);* Lafferty Ranch on Sonoma Mtn., acc. Arnold et al. (1971); Fairfield Osborn (H); Button Ranch on Walker Road (U. C. property) between Petaluma-Valley Ford Road and Pepper Road, acc. Patterson (1990); w. of Sears Point Raceway in Tolay Cr. drainage, acc. Patterson (1990); Norrbom Road near Sonoma, *Knight* (1985); Bouverie; Spring Lake; Santa Rosa, *Heller* (1902); w. of Windsor, *Heller* (1902); Pepperwood; Fitch Mtn., *Rubtzoff* (1952); Pine Flat Road, *Knight* (1963); 2480 ft. elev. Bear Cr. Can., acc. Clifton (1984); Caldwell Ranch, *Jokerst* (1985). < 5300 ft. June-July.

Juncus phaeocephalus Engelmann var. **paniculatus** Engelmann. Very common, marshes, damp areas: Willow Cr. Marsh, *Utech & Knight* (1986 CM); Coleman Marsh on Coleman Valley Road between Occidental and coast, *Rubtzoff* (1954); Occidental Marsh, *Rubtzoff* (1957); Freestone Marsh, *Rubtzoff* (1962); Fife Cr. Marsh near Guerneville, *Rubtzoff* (1960); Guerneville Marsh, *Rubtzoff* (1954); Pitkin Marsh, *Rubtzoff* (1951); Cunningham Marsh, *Howell* (1947); Laguna de Santa Rosa e. of Sebastopol, *Rubtzoff* (1958); Americano Cr. near Bloomfield, *Rubtzoff* (1962); 1400 ft. elev. 6½ mi. ne. of Penngrove on Lichau Road, *Rubtzoff* (1968); Button Ranch on Walker Road (U. C. property) between Petaluma-Valley Ford Road and Pepper Road, acc. Connors (1992); Fairfield Osborn (H); Bennett Mtn. Lake, *Rubtzoff* (1955); Pepperwood; Old Redwood Hwy. 2 mi. nw. of Windsor, *de Nevers & Kaza* (1981). < 7300 ft. May-Aug.

Juncus phaeocephalus var. **phaeocephalus**. Common, wet places: The Sea Ranch; Stewarts Point, *Abrams* (1920); Goat Rock Beach, *Best* (1984); Bodega Head, acc. Connors (1991); Coleman Marsh on Coleman Valley Road between Occidental and coast, *Rubtzoff* (1962, 1964); Pitkin Marsh, *Rubtzoff* (1951); Cunningham Marsh, *Howell* (1947); Todd Road Preserve (H&K); Button Ranch on Walker Road (U. C. property) between Petaluma-Valley Ford Road and Pepper Road, acc. Connors (1992); s. of El Rose Drive and w. of D St. in Petaluma, *B. Anderson* (1987). < 500 ft. May-July. The 2 varieties do not seem to tend to intergrade, acc. *Rubtzoff & Gankin*, p. 56 (1988).

Juncus tenuis Willdenow. Occasional, damp places: Pitkin Marsh, *Rubtzoff* (1951); Fairfield Osborn (H); Bouverie; Pepperwood. Usually < 5000 ft. May-Aug.

Juncus xiphioides E. Meyer. Occasional, wet soil, marshes, near springs: Salt Point; Vacation Beach near Guerneville, *Rubtzoff* (1959); Fife Cr. Marsh near Guerneville, *Rubtzoff* (1956, 1961); Fairfield Osborn (S); St. Helena Road, *Knight* (1963); Spring Lake; Pepperwood; Fitch Mtn., *Rubtzoff* (1962); 2420 ft. elev. Caldwell Ranch, *Clifton* (1984 PUA); Cloverdale. < 7000 ft. May-Oct.

Luzula comosa E. Meyer. WOOD RUSH. Common, seasonally moist to wet marshy ground, grassland, brush and woods, sun or shade: s. end of The Sea Ranch, *J. H. Thomas* (1976); 49 ft. elev. Gerstle Cove at Salt Point State Park, *Clifton* (1981 PUA); grade nw. of Cazadero, *Cantelow* (1939); Bodega Point, *Eastwood* (1915); Bodega Bay, *Heller & Brown* (1902); Vacation Beach near Guerneville, *Rubtzoff* (1953); Pitkin Marsh, *Howell* (1937), *Rubtzoff* (1951); sw. of Hwy. 116 and s. of Sebastopol, *J. & B. Guggolz* (1991 NCC); Button Ranch on Walker Road (U. C. property) between Petaluma-Valley Ford Road and Pepper Road, acc. Patterson (1990); Sugarloaf Ridge; Calistoga Road, *Wiggins* (1949); Lake Sonoma quad.; property bordered by Chianti and Dutcher Cr. roads w. of US 101 and w. of Geyserville, acc. J. & B. Guggolz (1991); 2040 ft. elev. Squaw Cr., acc. Clifton (1984). < 10700 ft. May-Aug. —*L. campestris* (Linnaeus) de Candolle. *L. multiflora* (Ehrhart) Lejeune.

JUNCAGINACEAE. ARROW-GRASS FAMILY
[Including Liliaececeae]

1. Perianth present, stamens 3 or 6 . . . . *Triglochin*
1: Perianth none; stamen 1 . . . . *Lilaea*

Lilaea scillioides (Poiret) Hauman. FLOWERING-QUILLWORT. Common, low places where water persists: upper Stemple Cr. Marsh, *Rubtzoff* (1961); Bodega Head Marsh, *Rubtzoff* (1954); near Forestville, *Rubtzoff* (1951); Pitkin Marsh, *Rubtzoff* (1951); 2 mi. w. of Graton, *Knight* (1967); Atascadero Cr. Marsh, *Rubtzoff* (1958); many sites w. and sw. of Santa Rosa in Laguna de Santa Rosa drainage, *Harrison* (1978 NCC); Todd Road Preserve; 3 mi. e. of Valley Ford, *Howell* (1930); near Bloomfield, *Howell* (1930), *Rubtzoff* (1970); Stony Point Marsh, *Rubtzoff* (1955); Game Refuge Road from Petaluma Hill Road, *Koch* (1939 NCC); Button Ranch on Walker Road (U. C. property) between Petaluma-Valley Ford Road and Pepper Road, acc. Patterson (1990); Chileno Valley, *Rubtzoff* (1972); w. of Sears Point Raceway in Tolay Cr. drainage, acc. Patterson (1990); 5 mi. n. of Sears Point, *Rubtzoff* (1971); Fifth St., Sonoma, *Baker* (1952 NCC); Arnold Drive vernal pools at e. edge Sonoma Valley Reg. Park, acc. Quibell & Moran (1991); near Kenwood, *Baker* (1899 NCC); Bennett Mtn. Lake, *Rubtzoff* (1955); Spring Lake; near Santa Rosa, *Baker* (1899 NCC); Pepperwood, *Breedlove* (1983); near Asti, *Raven* (1963), *Rubtzoff* (1963); n. of Palomino Road near Cloverdale, *Guggolz* (1978). < 5700 ft. Mar.-Oct.

Triglochin:

1. Fertile carpels 6; stamens usually 6, occasionally 3
   2. Ligule of the leaf-sheath entire; the stems often densely tufted from a proliferating caudex or oblique rhizome, 1- 10 dm high . . . . *T. maritima*
   2: Ligule of the leaf-sheath 2-lobed or at least deeply emarginate; plants generally not tufted but well-spaced along slender, naked rhizomes, sometimes arising from a stout, oblique, scaly rhizome, 0.4-5 dm high *T. concinna*

1: Fertile carpels 3, alternating with 3 sterile carpels; stamens 1 or 2 or rarely 3; plants from slender rhizomes, 0.5-3 dm high . . . . *T. striata*

Triglochin concinna Burtt Davy. ARROW-GRASS. Uncommon, coastal brackish marshes: Bodega Marine Reserve Salt Marsh, *Connors* (1990 BML). < 8300 ft. Mar.-Aug.

Triglochin maritima Linnaeus. SEASIDE ARROW-GRASS. Infrequent, saline marshes: Lower Salmon Cr. Marsh, *Rubtzoff* (1955); 1st estuary s. of Bodega Harbor, acc. Quibell (1991); Bodega Head Marsh, *Rubtzoff* (1954); near Petaluma, *Congdon* (1880), *Rubtzoff* (1963). < 9300 ft. Apr.-Aug.

Triglochin striata Ruiz López & Pavón. Uncommon, saline and brackish marshes: Lower Salmon Cr. Marsh, *Rubtzoff* (1973); Salmon Cr. Salt Marsh, *Harrison* (1977 SRJC). < 35 ft. May-Sept.

LEMNACEAE. DUCKWEED FAMILY

1. Plants without roots; fronds floating at or just below water surface or part of frond breaking through surface; flowers breaking through upper surface of frond . . . . *Wolffiella*

1: Plants with evident roots; fronds floating on surface of water; flowers developing in the reproductive pouch
   2. Roots solitary on each segment of frond; fronds 1-3 mm long, underside green or streaked with brown . . . . *Lemna*
   2: Roots several from each segment of frond; fronds 3-6 mm long, underside generally reddish . . *Spirodela*

Lemna

1. Vein 1 per plant body (sometimes obscure)
   2. Plant body 1-2.5 mm long, widely elliptic, generally thicker in middle, usually in pairs; vein < 2/3 distance from root node to tip of plant body . . . . *L. minuscula*
   2: Plant body 2-4 mm long, ± oblong, very thin, generally clusters of 4-8; vein up to 3/4 distance from root node to tip of plant body . . . . *L. valdiviana*

1: Veins 3-5 per plant body
   3. Root sheath with 2 obvious, winglike appendages at base . . . . *L. aequinoctialis*
   3: Root sheath not winged
      4. Plant body widely elliptic to round, tip asymmetric, lower surface usually much swollen or rounded . . . . *L. gibba*
      4: Plant body widely elliptic, tip symmetric, lower surface flat . . . . *L. minor*

Lemna aequinoctialis Welwitsch. Uncommon, ± inland: Vacation Beach near Guerneville, *Rubtzoff* (1959). < 700 ft. Aug.-Sept. —*L. perpusilla* Torrey, misapplied.

Lemna gibba Linnaeus. Rare, surface of pools: Willow Cr. Marsh, *Rubtzoff* (1959); Americano Cr. 0.5 mi. e. of Valley Ford, *Rubtzoff* (1961); Fairfield Osborn (S). Usually < 5000 ft.

Lemna minor Linnaeus. Occasional, surface of pools: Pitkin Marsh, *Rubtzoff* (1957 NCC); Lafferty Ranch on Sonoma Mtn., acc. Arnold et al. (1971); Carneros Cr. at Hwy. 27 between Sonoma and Napa near county line, *Raven* (1951); Sugarloaf Ridge; Pepperwood. Worldwide. < 6700 ft. —*L. minima* Philippi.

Lemna minuscula Herter. Rare, wet areas: Cunningham Marsh, *J. & B. Guggolz* (NCC 1991). < 7300 ft. Much like *L. valdiviana*.

Lemna valdiviana Philippi. Occasional, surface of pools: Pitkin Marsh, *Rubtzoff* (1951 NCC); Laguna de Santa Rosa on Llano Road 0.6 mi. from Gravenstein Hwy., *Howell & Knight* (1986); Willow Brook n. of Petaluma, *Knight* (1987). < 5000 ft.

Spirodela polyrrhiza (Linnaeus) Schleiden. GREATER DUCKWEED. Rare, quiet water: Pitkin Marsh, *Rubtzoff* (1951 NCC). < 8300 ft. Cosmopolitan.

Wolffiella lingulata (Hegelmaier) Hegelmaier. MUD-MIDGET. Uncommon, still water: Bennett Mtn. Lake, *Rubtzoff* (1955). < 1000 ft. June.

LILIACEAE. LILY FAMILY
[Including ALLIACEAE and AMARYLLIDACEAE]

1. Inner and outer tepals obviously alike
 2. Foliage leaves basal or mostly so, cauline leaves if present (as in *Clintonia, Xerophyllum* and *Zigadenus*) reduced in size or few
  3. Flowers in racemes or panicles
   4. Styles 3, distinct nearly to base
    5. Basal leaves evergreen, very numerous, rough, elongate-linear . . . . *Xerophyllum*
    5: Basal leaves dying before winter, smooth, linear
     6. Leaves 2-ranked and ± equitant; plants with rhizomes . . . . *Tofieldia*
     6: Leaves neither 2-ranked nor equitant; plants with bulbs . . . . *Zigadenus*
   4: Styles 1, stigmas 3-lobed or 3-parted
    7. Flowers in panicle; tepals white with purplish midvein . . . . *Chlorogalum*
    7: Flowers in a raceme
     8. Flowers nodding, whitish, yellowish, or rose-pink . . . . *Erythronium*
     8: Flowers otherwise
      9. Flowers ± erect, blue . . . . *Camassia*
      9: Flowers above perpendicular to stem, clustered, scapose; flowers below ± pendulous, orange-red . . . . *Kniphofia*
  3: Flowers in umbels, or almost in heads
   10. Flower stems fleshy to 1 m tall . . . . *Amaryllis*
   10: Flower stems slender, < 1 m tall
    11. Umbels terminal and lateral, on a stem bearing 1 or several foliaceous bracts . . *Clintonia*
    11: Umbel solitary, terminating a leafless stem
     12. Tepals united into a distinct tube
      13. Tepal-tube inflated and/or saccate . . . . *Dichelostemma*
      13: Tepal-tube not inflated or saccate
       14. Leaves rounded without a keel . . . . *Brodiaea*
       14: Leaves distinctly keeled . . . . *Triteleia*
     12: Tepals divided nearly or all the way to the base
      15. Plants with herbage of onionlike taste and odor (except *A. paniculatum*) *Allium*
      15: Plants without herbage of onionlike taste and odor
       16. Pedicels subtended by bractlets above the main spathelike bracts *Muilla*
       16: Pedicels not subtended by bractlets above the main spathelike bracts . . . . *Nothoscordum*

2: Foliage leaves cauline
17. Leaves alternate, not whorled or clustered, narrowly to broadly ovate; fruit a berry
18. Leaves few, usually 2 or 3; tepals 4 . . . . *Maianthemum*
18: Leaves numerous; tepals 6
19. Stem branched . . . . *Disporum*
19: Stem unbranched below the inflorescence
20. Tepals 2-7 mm . . . . *Smilacina*
20: Tepals 6-17 mm . . . . *Veratrum*
17: Leaves ± clustered, the stems generally bearing 1 or more obvious whorls, linear-lanceolate to narrowly ovate; fruit a capsule
21. Tepals generally 4 cm long or more; stigma 3-lobed . . . . *Lilium*
21: Tepals < 4 cm long; stigma deeply parted . . . . *Fritillaria*
1: Inner and outer tepals different in size and color
22. Basal foliage leaves none; flowers 1, subtended by 3 broad cauline leaves . . . . *Trillium*
22: Basal foliage leaves 1-3; flowers generally > 1
23. Leaves linear; inner tepals broader than outer; flowers subtended by leaflike bracts . *Calochortus*
23: Leaves elliptic-oblong, mottled; inner tepals much narrower than the outer; flowers solitary on leafless stalks . . . . *Scoliopus*

Allium (Key ± acc. D. Barbe [in part, 1.-4:], and D. W. McNeal [in part, 3-13:], *Hickman* {1993}).
1. Herbage and bulb without smell of onions; cut surface of bulb with odor of raw potato, rapidly discolors orange to dark red on exposure to air . . . . *A. paniculatum*
1: Herbage and bulb with smell of onions; cut surfaces of bulbs do not readily discolor
2. Scape fistulose and inflated . . . . *A. cepa*
2: Scape not so
3. Umbel with bulbils only; no flowers (occasionally *A. vineale* with flowers and bulbils)
4. Leaves 5-20 mm wide, solid, flat . . . . *A. sativum*
4: Leaves 0.2-4 mm wide, fistular, semicylindrical or filiform . . . . *A. vineale*
3: Umbel with flowers
5. Leaves flat, sickle-shaped . . . . *A. falcifolium*
5: Leaves narrowly channeled to ± cylindric, if flat, not sickle-shaped
6. Bulb coat sculpture 0 or obscure
7. Bulb symmetric, present at flowering time; rhizome 0
8. Inflorescence ± hemispheric; flowers erect; perianth parts elliptic, obtuse *A. neapolitanum*
8: Inflorescence open, ± 1-sided; flowers drooping; perianth parts lanceolate, acute . . . . *A. triquetrum*
7: Bulb oblique, gone by flowering time; new bulbs borne on rhizome tips originating from old bulb root pad
9. Rhizomes conspicuous, usually > 3 cm; leaves widely channeled, flat, or keeled; perianth parts ovate, spreading,entire . . . . *A. unifolium*
9: Rhizomes inconspicuous, < 2 cm; leaves ± cylindric; perianth parts ± lanceolate, erect, at least inner finely serrate . . . . *A. bolanderi*
6: Bulb coat sculpture obvious, netlike (with hand lens), meshes prominent, thick-walled
10. Bulb coat sculpture in wavy lateral rows, herringbone pattern indistinct; perianth parts spreading, outer ± = inner . . . . *A. amplectens*
10: Bulb coat sculpture in sharply serrate bilateral rows, herrringbone pattern distinct; perianth parts erect , outer > inner
11. Perianth parts folded over ovary in fruit; flowers and pedicels deciduous *A. serra*
11: Perianth parts erect, rigid in fruit; flowers persistent
12. Leaves 3-6, curved to curled; inflorescence dense, pedicels erect, 0.7-2 × longer than the flowers . . . . *A. dichlamydeum*
12: Leaves 2-3, straight to ± curved; inflorescence open, pedicels spreading, 1-3 × longer than the flowers
13. Leaves curved; flowers 8-12 mm, stigma minute, entire . . . . *A. peninsulare* var. *franciscanum*
13: Leaves straight; flowers 10-15 mm, stigma 3-lobed or headlike . . . . *A. p.* var. *peninsulare*

Allium amplectens Torrey. Common, dry fields, hillsides, sometimes on serpentine: Schellville on road between Arnold Drive and Napa, *Rose* (1931), *Barneby* (1934); Sonoma, *Bigelow* (type 1854 UC); foot of Trinity Road, *Baker* (1940 NCC); above Kenwood Springs, *Baker* (1929 NCC); Adobe Can., *Baker* (1940 NCC); Bennett Mtn. Lake, *Rubtzoff* (1963); Hidden Oaks Road on w. side of w. end of Sonoma Mtn. Road, *Quibell* (1988 NCC); Annadel; Pepperwood; Petrified Forest, *Eastwood* (1915); w. of Dry Cr. Valley, *Baker* (1934 NCC); on serpentine on Dorman Can. Road off W. Dry Cr. Road, acc. Edwards (1990); Lake Sonoma area. < 6000 ft. Mar.-June.

Allium bolanderi S. Watson. Occasional, clayey soil, openings in woods and brush: Hope Valley near Kenwood, acc. Baker (1932); Sweetwater Springs Road sw. of Healdsburg, *Raiche* (1984 JEPS); ca 420 ft. elev. Pine Flat Road 1.6 mi. from Hwy. 128, acc. Raiche (1991); ca 3260 ft. elev. along Rattlesnake Fire Trail in geothermal area, acc. Raiche (1991). < 3400 ft. May-July.

★ **Allium cepa** Linnaeus. ONION. Uncommon, salt marsh: near Petaluma, *Rubtzoff* (1963). Introduced from Asia.

Allium dichlamydeum E. Greene. Common, rocky places, dry clayey soil, coastal: The Sea Ranch; Stewarts Point, *Baker* (1932 NCC); ca 2 mi. s. of Stewarts Point on big rock outcrop on ocean side of Hwy. 1, acc. Roderick (1989); n. from Jenner on Hwy. 1 ca ¼ mi. s. of Russian Gulch parking lot on large rock outcrop, acc. Roderick (1989); near Cazadero, *Gray* (1928); 8½ mi. n. of Bodega Bay, acc. Roderick (1989); ca ¼ mi. n. of Shell Beach entrance on ocean side of road, *Howell* (1930); 1 mi. n. of Salmon Cr., *Keck & Clausen* (1935); 1 mi. up Salmon Cr., *Baker* (1935 NCC). < 500 ft. May-July.

Allium falcifolium Hooker & Arnott. Occasional, rocky or heavy soil, openings in brush, sometimes on outcrops of serpentine: Big Austin Cr. area, acc. Raiche (1985); Cooper Hill at Dry Cr., *L. M. Brewer* (1929 NCC); Lake Sonoma area; 3100 ft. elev. w. flank Mt. St. Helena, *Grummer* (1979 PUA); along ridge between Socrates Mine and Mercuryville in geothermal area, *Dolan & LaPré* (1987 UCR); 3599 ft. elev. Geyser Rock, acc. Clifton (1983); 3058 ft. elev. Squaw Cr., acc. Clifton (1983); 3117 ft. elev. Bear Cr. Can., acc. Clifton (1984). < 7000 ft. Mar.-July.

★ **Allium neapolitanum** Cirillo. Garden escape: Rio Nido, *Boyle* (1941). < 350 ft. Mar.-Apr. Introduced from Mediterranean region.

★ **Allium paniculatum** Linnaeus. PANICLED ONION. Rare, waste places: Geothermal area, acc. *Barbe* (1989). < 2500 ft. Native of se. Europe to the Caucasus. Noxious weed.

Allium peninsulare Lemmon var. **franciscanum** D. McNeal & F. Ownbey. Uncommon, dry, wooded or open hillsides: Petaluma, *Congdon* (1880). < 1000 ft. Mar.-June.

Allium peninsulare var. **peninsulare**. Uncommon, dry, wooded or open hillsides: ½-3/4 mi. up Cavedale Road from Hwy. 12 scattered along rocky ledges sticking out of chaparral, acc. Roderick (1978); 1 mi. sw. of Kenwood in Hope Valley, *Baker* (1950 NCC). < 1000 ft. Mar.-June.

★ **Allium sativum** Linnaeus. GARLIC. Escape: Rincon Valley at Ducker Cr., *Wells* (1985). Introduced from central Asia.

Allium serra D. McNeal & F. Ownbey. Uncommon, grassy meadows, hillsides: Pepperwood. < 4000 ft. Mar.-May. —*A. serratum* S. Watson, misapplied.

★ **Allium triquetrum** Linnaeus. Garden escape: Bodega Head; Sonoma Mtn. Road at jct. of Pressley Road, acc. Quibell (1991); Gates Road off Calistoga Road, acc. Wells et al. (1984). < 500 ft. Native of w. Mediterranean region.

Allium unifolium Kellogg. Occasional, damp, sometimes clayey soil: The Sea Ranch; Salt Point; Freestone Marsh, *Rubtzoff* (1962); Laguna de Santa Rosa, *Rubtzoff* (1958); Todd Road Preserve; Pepperwood; Dry Cr., *Cantelow* (1934); Stewarts Point-Skaggs Springs Road, *Quibell* (1973 NCC); Lake Sonoma area. < 3700 ft. Apr.-June.

★ **Allium vineale** Linnaeus. WILD GARLIC. Occasional, weedy: in a planter at Sonoma Joe's Restaurant, Petaluma, *Howell* (1976); Hidden Oaks Road between Grange and Sonoma Mtn. roads, acc. Quibell (1991); field on s. side of Bennett Valley Road e. of jct. with Yulupa Ave., Santa Rosa, acc. Quibell (1991); near Piner School w. of Santa Rosa, *Harrison* (1991 SRJC); 1 quadrangle in central part of county and 6 quadrangles in e. part of county, acc. *Barbe* (1989). < 1000 ft. Native of most of Europe, Africa, Caucasus Mts. and Lebanon. Noxious weed.

★ **Amaryllis belladonna** Linnaeus. NAKED LADY. Sometimes planted along fence rows or buildings usually in rural areas: The Sea Ranch; ca 1 mi. nw. of Camp Meeker, acc. J. & B. Guggolz (1991); ¼ mi. along s. end of Lohrman Lane just off Bodega Ave. w. of Petaluma, acc. Roderick & Knight (1989); along Pepper Road ca ½ mi. from Stony Point Road, acc. Roderick (1978); Spring Lake. Native to S. Africa.

Brodiaea

1. Flowers 24-38 mm long
   2. Staminodia narrowly linear and slightly involute; perianth-tube splitting in fruit
      3. Filaments with 2 filiform appendages . . . . *B. appendiculata*
      3: Filaments without filiform appendages, (sometimes rudimentary in *B. californica*)
         4. Tepals 6-10 mm wide; ovary 7-11 mm long . . . . *B. californica* var. *californica*
         4: Tepals 4-7 mm wide; ovary 5-7 mm long . . . . *B. c.* var. *leptandra*
   2: Staminodea broad, either strongly involute or plane; perianth-tube remaining intact in fruit
      5. Perianth-tube ovoid to narrowly campanulate; staminodea involute, directed inward toward the stamens . . . . *B. coronaria*
      5: Perianth-tube funnelform; staminodea erect and plane, distant from stamens . . . . *B. elegans*
1: Flowers 14-24 mm long
   6. Staminodea leaning inward, margin semi-involute, apex broad and slightly hooded; tepals campanulate; scape 0.5-5.0 cm high . . . . *B. terrestris*
   6: Staminodea erect, margin strongly involute; tepals rotate; scape 10-30 cm high . . . . *B. stellaris*

Brodiaea appendiculata Hoover. Uncommon, heavy soil, low open hills, plains: s. of Piner School w. of Santa Rosa, *Harrison* (1981 SRJC). < 2000 ft. Apr.-May.

Brodiaea californica Lindley var. **californica** Infrequent, open plains, wooded hills: near Warfield station (old site at Hwy. 12 and Trinity Road), *Baker* (1928 NCC); Adobe Can., *Baker* (1927 NCC); Mt. Hood above Baker Ranch (Maplewood) in burned chaparral, *Baker* (1947 NCC). < 2700 ft. May-July.

Brodiaea californica var. **leptandra** (E. Greene) Hoover. Occasional, open plains, wooded hills: Spring Lake; 5 mi. e. of Mark West Springs on road to Petrified Forest, *Eastwood & Howell* (1938); Cavedale Road, acc. Greenhouse (1984); Adobe Can., *Baker* (1925); Parktrail Drive, Santa Rosa, *Harrison* (1976 SRJC); Pepperwood; Windsor, *Harrison* (1981 SRJC); 2 mi. se. of Healdsburg, *Hoover* (1930); 530 ft. elev. w. side of Dry Cr. Valley, *Raiche* (1985). < 3000 ft. May-July. --*Hookera synandra* A. A.Heller, type locality: Petrified Forest, *Heller* (1902 U).

Brodiaea coronaria (Salisbury) Engler. Common, grassy hillsides, wooded slopes, dry clay or adobe soil, gravelly places: ca 1 mi. nw. of Camp Meeker, acc. J. & B. Guggolz (1991); Temelec at corner of Arnold Drive and Watmaugh Road sw. of Sonoma, acc. Roderick (1965); Rincon Valley, *Brandon* (1927 NCC); between Santa Rosa and Sebastopol, *Brandegee* (1905 UC); Spring Lake; Petrified Forest, *Heller* (1902 UC); Lake Sonoma area. < 5300 ft. May-July.

Brodiaea elegans Hoover. HARVEST BRODIAEA. Common, grassland of brushy or open slopes: Coleman Valley Road, acc. Clifton (1984); ca 1 mi. nw. of Camp Meeker, acc. J. & B. Guggolz (1991); Occidental Road, *Ferris & Lorraine* (1961); Lichau Road (now Fairfield Osborn), *Rubtzoff* (1968); Button Ranch on Walker Road (U. C. property) between Petaluma-Valley Ford Road and Pepper Road, acc. Patterson (1990); Sobre Vista Road 0.9 mi. w. of Arnold Drive, *Rubtzoff* (1971); Bouverie; Sugarloaf Ridge; 4.25 mi. w. of Glen Ellen on Bennett Valley Road, *Rubtzoff* (1971); Spring Lake; Pepperwood; Lake Sonoma area; Alexander Valley quad.; w. flank Mt. St. Helena, *Grummer* (1979 PUA); ca 1900 ft. elev. Geysers Road between Jimtown and Mercuryville, *Utech & Knight* (1986 CM); Socrates Mine, acc. Clifton (1984); ca 2000 ft. elev. Anderson Trail on Black Mtn. between Geyser Peak and upper end of Pine Flat Road, acc. J. & B. Guggolz (1992). < 7300 ft. Apr.-July.

Brodiaea stellaris S. Watson. Occasional, open places, rocky or clayey soil, often on serpentine: near Camp Meeker, *Wolfe* (1928 NCC); Occidental Marsh, *Rubtzoff* (1957); Occidental Road, *Ferris & Lorraine* (1961); Stoetz Lane e. of Occidental, *Best* (1979), *Utech & Knight* (1986 CM); Graton-Occidental Road, *Baker* (1926 NCC); Laguna de Santa Rosa, *Rubtzoff* (1969); Skaggs Springs, *Hawver* (1915); Lake Sonoma area. < 3000 ft. May-July.

Brodiaea terrestris Kellogg. GROUND BRODIAEA. Common, varying soil qualities: The Sea Ranch, *Wells* (1981); Salt Point, *Best* (1983); ½ mi. n. of Moon Rocks Campground at Salt Point State Park, *Huntington* (1989 NCC); ca 1 mi. n. from entrance of Fort Ross State Park, acc. Roderick (1985); Coleman Valley Road ca 2 mi. from Hwy. 1 around rocky outcrop, acc. Roderick (1989); Freestone Marsh, *Rubtzoff* (1962); upper Stemple Cr. Marsh, *Peñalosa* (1961); Harrison Grade Road, acc. *Kruckeberg* (1984); Pitkin Marsh, *Baker* (1928 NCC), *Howell* (1945); Perry Marsh near Molino, *Baker* (1939 NCC); Laguna de Santa Rosa ne. of Graton at end of Piner Road, *Rubtzoff* (1969); Cunningham Marsh, *Baker* (1934 NCC); 5 sites w. and sw. of Santa Rosa in Laguna de Santa Rosa drainage, *Harrison* (1978 NCC); Todd Road Preserve; jct. of Wilfred and Langer roads nw. of Rohnert Park, acc. Quibell

(1991); w. Railroad Ave. between US 101 and Stony Point Road, *Rubtzoff & Bacigalupi* (1971); between Shellville and El Verano, *Cantelow* (1936); Bennett Mtn. Lake, *Rubtzoff* (1961); Piner Road, Santa Rosa, *Ferguson* (1979 NCC); 3 mi. n. of Windsor, *Eastwood & Howell* (1936). < 5000 ft. Apr.-July. —*B. grandiflora* Smith var. *macropoda* (Torrey) Hoover, type locality: Santa Rosa Cr. *Bigelow* (1854 UC). *B. coronaria* (Salisbury) Engler var. *m.* (Torrey) Hoover.

Calochortus

1. Flowers closed-globose or closed-obovate, nodding at maturity
   2. Stem to 3 dm high; flowers globose, deep yellow; outer tepals = or > the inner, inner tepals glabrous; gland pouched, capsule winged . . . . . *C. amabilis*
   2: Stem to 11 dm high; flowers obovate, pale yellow; outer tepals to 2.5 cm long, inner tepals to 4.5 cm long and moderately hairy to tips; gland angled; capsule non-winged . . . . . *C. raichei*

1: Flowers broadly campanulate to bowl-shaped, erect or spreading
   3. Surface of gland naked or nearly so, although the bordering membranes may be fringed and extend over the gland
      4. Petals glabrous or nearly so, not ciliate or bearded; flowers lilac, often with purple spot above gland . . . . . *C. uniflorus*
      4: Petals ± densely bearded, often ciliate or fimbriate; flowers white or cream, or tinged with rose or purple . . . . . *C. tolmiei*
   3: Surface of gland with short hairlike processes; petal blotch red-brown
      5. Shape of gland doubly lunate; petal blotch very large and surrounded by pale yellow; flowers white to purplish . . . . . *C. vestae*
      5: Shape of gland not doubly lunate; flowers deep yellow . . . . . *C. luteus*

Calochortus amabilis Purdy. YELLOW GLOBE LILY, DIOGENES' LANTERN. Common, dry slopes, woods or brush, loamy to rocky soil: Cazadero, *Wood* (1918), *Baker* (1937 NCC); between Camp Meeker and Monte Rio, *Morse* (1981 NCC); Harrison Grade Road, *Ferris & Lorraine* (1961); Graton-Occidental Road, *Hoover* (1965); Bouverie; Cavedale Road, *Roderick* (1967), *Paskorton* (1979 NCC), *Utech & Knight* (1988 CM); lower Cavedale Road at monument, acc. Quibell (1991); 3 mi. up Trinity Road, *Reich* (1974 NCC); Kenwood, *Nelson* (1933); between Santa Rosa and Agua Caliente, *Heller & Brown* (1902); Sugarloaf Ridge; Los Guilicos, *Mitchell* (1967 NCC); Los Alamos Road, *Stocking* (1967 NCC); probably between Santa Rosa and Mt. St. Helena, *Voznesenskii* (1840-1841 LE); 1000 ft. elev. w. of Los Alamos Road, *Newman* (1964); Alpine Road off Calistoga Road, *Stocking* (1967 NCC); Gates Road off Calistoga Road, acc. Wells et al. (1984); Pepperwood; near Petrified Forest, *Thomas* (1966); Franz Valley Road at Kellogg end, *Murray* (1980 SFSU); Ida Clayton Road, *Roderick* (1967), McNally (1970 NCC); Chalk Hill Road 11 mi. e. of Healdsburg, *J. Smith* (1976 NCC); 1598 ft. elev. Bailache Ave. near Russian River near Healdsburg, *Pitcher* (1975 PUA); chaparral area sw. of Bradford Mtn. 1½ mi. w. of Dry Cr., acc. J. & B. Guggolz (1992); Geyser Peak, acc. Knight & Edwards (1986); 350-400 ft. elev. Old Skaggs Springs Road, *Utech, Sato & Ohara* (1990 CM); Skaggs Springs, *Hawver* (1915); Lake Sonoma area; 2900 ft. elev. w. flank Mt. St. Helena, *Grummer* (1979 PUA); near Pine Flat, *Malloch* (1977 PG&E); ca 2000 ft. elev. Anderson Trail on Black Mtn. between Geyser Peak and upper end of Pine Flat Road, acc. J. & B. Guggolz (1992). < 3300 ft, Apr.-June.

Calochortus luteus Lindley. YELLOW MARIPOSA. Common, grassland, open flats or hills: ca 1 mi. nw. of Camp Meeker, acc. J. & B. Guggolz (1991); Todd Road Preserve; Fairfield Osborn (S); Button Ranch on Walker Road (U. C. property) between Petaluma-Valley Ford Road and Pepper Road, acc. Patterson (1990); between Magnolia Ave. and Bodega Ave. just w. of Horn Ave. w. of Petaluma, acc. Roderick (1936); w. of Sears Point Raceway in Tolay Cr. drainage, acc. Patterson (1990); Sonoma Cr., *Rattan* (1885); Agua Caliente in Sonoma Valley, *Lehman* (1975 NCC); Bouverie; 1½ mi. up Cavedale Road from Hwy. 12, *Utech & Knight* (1986 CM); Trinity Road, *Ruckert* (1960); Sugarloaf Ridge; Bennett Valley Road, *Stocking* (1963 NCC); Bennett Mtn. Lake, *Rubtzoff* (1961); Hidden Oaks Road on w. side of w. end of Sonoma Mtn. Road, *Quibell* (1988 NCC); Spring Lake; Pepperwood; Windsor, *Jepson* (1918 JEPS); near Healdsburg, *Abrams* (1918), *Benton* (1929 NCC); mts. w. of Calistoga, *Kraus* (1894); Lake Sonoma area; ca 2000 ft. elev. Anderson Trail on Black Mtn. between Geyser Peak and upper end of Pine Flat Road, acc. J. & B. Guggolz (1992). < 2400 ft. Apr.-June.

Calochortus raichei S. Farwig & V. Girard. CEDARS FAIRY LANTERN. Rare, serpentine: 800 ft. elev. Austin Cr. State Recreation Area headwaters of Big Austin Cr. at The Cedars, *Raiche* (1986); plus 3 other locations in same area, acc. Raiche. < 1000 ft.

Calochortus tolmiei Hooker & Arnott. PUSSY EARS. Common, woods or open grassy slopes, dry, often rocky soil:The Sea Ranch; Salt Point State Park, *Heblack* (1973 NCC); Fort Ross archeological site, *Thorne* (1971 NCC); Fort Ross Road along Ward Cr., *Plichta* (1978 NCC); 1 mi. n. of Fort Ross on headland, *Roderick* (1967); Bodega Bay bluffs, *McNally* (1970 NCC); Larkspur Hill 1 mi. s. of Bodega Bay, *Ott* (1934 NCC); Bodega, *Heller & Brown* (1902); Los Alamos Road, *Stocking* (1967 NCC); w. of Skaggs Springs, *Heller* (1940); Lake Sonoma area; Geyserville quad.; geysers Unit 18, *Malloch* (1977 PG&E). < 6600 ft. Apr.-July.

Calochortus uniflorus Hooker & Arnott. PINK STAR-TULIP. Common, low wet places, often alkaline:The Sea Ranch; Salt Point State Park, *Heblack* (1973 NCC); Fort Ross archeological site, *Thorne* (1971 NCC); Fort Ross Road along Ward Cr., *Plichta* (1978 NCC); 1 mi. n. of Fort Ross on headland, *Roderick* (1967); Bodega Bay bluffs, *McNally* (1970 NCC); Larkspur Hill 1 mi. s. of Bodega Bay, *Ott* (1934 NCC); Bodega, *Heller & Brown* (1902); Los Alamos Road, *Stocking* (1967 NCC); w. of Skaggs Springs, *Heller* (1940); Lake Sonoma area; Geyserville quad.; geysers Unit 18, *Malloch* (1977 PG&E). < 1000 ft. Apr.-June.

Calochortus vestae Purdy. Infrequent, heavy clayey soils: Pepperwood; Chalk Hill Winery, Windsor, *Harrison* (1989 SRJC); 300 ft. from US 101 on Dutcher Cr. Road, *Guggolz* (1977); Stewarts Point-Skaggs Springs Road, *Guggolz* (1976); Lake Sonoma area. < 3000 ft. May-July.

Camassia quamash (Pursh) E. Greene ssp. **quamash**. CAMAS. Occasional, wet meadows: The Sea Ranch; n. of Fort Ross, *Hastings* (1940 NCC); Fort Ross, *Cantelow* (1940); Occidental Marsh, *Rubtzoff* (1957); Pitkin Marsh, *Stocking* (1966 NCC); Sebastopol Road, *Congdon* (1880); inner corner of Todd Road opposite the preserve, *Quibell* (1972 NCC); Todd Road Preserve in swale, *Harrison* (1974 SRJC). < 11000 ft. May-June. —*C. q.* ssp. *linearis* Gould.

Chlorogalum

1. Bulb coats membranous or with few coarse fibers . . . . *C. pomeridianum* var. *minus*
1: Bulb coats with many coarse fibers
  2. Inflorescence nearly prostrate or branches widely spreading from base, < 4 dm high *C. p.* var. *divaricatum*
  2: Inflorescence erect, 5-25 dm high . . . . *C. p.* var. *pomeridianum*

Chlorogalum pomeridianum (de Candolle) Kunth var. **divaricatum** (Lindley) Hoover. Uncommon, coastal bluffs, hills: Bodega Marine Reserve, *Connors* (1991 BML, NCC). Usually < 350 ft. May-July. —*Laothoe divaricata* E. Greene.

Chlorogalum pomeridianum var. **minus** Hoover. Serpentine outcrops in chaparral: Cazadero quad. 4½ airmiles ne. of Guerneville. < 2400 ft.

Chlorogalum pomeridianum var. **pomeridianum**. SOAP PLANT, AMOLE. Common, grassland, brushy or grassy woodland borders: The Sea Ranch; Bodega Head; serpentine at jct. of Bittner and Joy roads sw. of Occidental, *Ferris & Lorraine* (1961); Stoetz Road ne. of Occidental, *Utech & Knight* (1986 CM); Pitkin Marsh, *Rubtzoff* (1957); Laguna de Santa Rosa, *Rubtzoff* (1958); Todd Road Preserve (H&K); Hidden Oaks Road w. of Sonoma Mtn. Road, acc. Quibell (1991); Fairfield Osborn (S); Tresch Ranch, Walker Road near Two Rock, acc. J. & B. Guggolz (1992); w. of Sears Point Raceway in Tolay Cr. drainage, acc. Patterson (1990); Bouverie; Kenwood, *Waitman* (1985 NCC); Sugarloaf Ridge; Bennett Valley, *Leitholt* (1893); Spring Lake; Pepperwood; Starr Road, Windsor, acc. Bunyan (1987); w. of Windsor near Russian River, *Heller* (1902); chaparral area sw. of Bradford Mtn. 1½ mi. w. of Dry Cr., acc. J. & B. Guggolz (1992); Lake Sonoma area; ca 2000 ft. elev. Anderson Trail on Black Mtn. between Geyser Peak and upper end of Pine Flat Road, acc. J. & B. Guggolz (1992); Geyser Peak, acc. Knight & Edwards (1986); 2700 ft. elev. w. flank Mt. St. Helena, *Grummer* (1979 PUA). < 5000 ft. May-Aug. —*Laothoe pomeridiana* Rafinesque.

Clintonia andrewsiana Torrey. Occasional, damp shady woods: mouth of Gualala River, *Mason* (1930 UC); The Sea Ranch, acc. Wells (1986), acc. *Mahaffey* (1990); bank of Gualala River, *Cantelow* (1934); Annapolis, *Sutliffe* (1920); crossing of south fork of Gualala River e. of Stewarts Point, *Heller* (1940 NY); 0.4 mi. e. Stewarts Point, *Wolf* (1934 GH), *Utech* (1972, 1988 CM); near Stewarts Point, *Gifford* (1950

UC); Salt Point, *Utech* (1988 CM); Hwy. 1 n. of Fort Ross, *Hastings* (1940 NCC); Tin Barn Road between Stewarts Point-Skaggs Springs Road and King Ridge Road, *Roderick* (1961 JEPS); 98 ft. elev. Fort Ross, *Clifton* (1981 PUA); 5 mi. from Jenner on River Road, *McNally* (1970 NCC); Cazadero, *Congdon* (1892 MIN), *Bacigalupi* (1926); ca 3 mi. w. of Cazadero, acc. Roderick (1986); Duncans Mills Marsh, *Rubtzoff* (1957); near Duncans Mills, *Baker* (1899 UC); Willow Cr. Road, *Whitlock* (1972 NCC), *Utech* (1988 CM); Bodega, *Krager* (1924 UC); ca 1 mi. nw. of Camp Meeker, acc. J. & B. Guggolz (1991); Guerneville, *Congdon* (1892 MIN), *Syme* (1920); Armstrong Grove n. of Guerneville, *Johnson* (1933), *Utech* (1972 CM); Buckeye Road (this road now only known just out of e. end of Annadel State Park off Lawndale Road which is off of Hwy. 12), *Patchett* (1917 JEPS); 5 mi. nw. of Healdsburg, *Ruckert* (1960). < 1350 ft. May-July.

Dichelostemma

1. Fertile stamens 6 (3 smaller); perianth tube not narrowed above ovary . . . . *D. capitatum*
1: Fertile stamens 3, equal; perianth tube narrowed above ovary
   2. Inflorescence a dense raceme; anther appendages deeply notched . . . . *D. congestum*
   2: Inflorescence an umbel; anther appendages truncate or rounded . . . . *D. multiflorum*

Dichelostemma capitatum A. W. Wood. BLUE DICKS. Common, grassy or rocky slopes of wooded or open hills: The Sea Ranch, acc. Wells (1985), acc. *Mahaffey* (1990); Salt Point; Cazadero quad.; 5 mi. n. of Bodega Bay, *Baker* (1931 NCC); 8 mi. n. of Bodega Bay, *Howell* (1931); Bodega Head; Todd Road Preserve; Rohnert Park, *Otter* (1978 NCC); Button Ranch on Walker Road (U. C. property) between Petaluma-Valley Ford Road and Pepper Road, acc. Pat-terson (1990); Sonoma Mtn. Road off Adobe Road e. of Petaluma, *Lowry* (1983 SFSU); w. of Sears Point Raceway in Tolay Cr. drainage, acc. Patterson (1990); Bouverie; Trinity Road, *Ruckert* (1960); Sugarloaf Ridge; Los Guilicos, *Mitchell* (1967 NCC); Bennett Mtn. Lake, *Rubtzoff* (1963); Spring Lake; Santa Rosa, *Eastwood* (1921); Pepperwood; Petrified Forest, *Heller* (1902); Independence School near Healdsburg, *Abrams* (1918); Lake Sonoma area; ca 1500 ft. elev. Geysers Road between Jimtown and Mercuryville, *Utech & Knight* (1986 CM); chaparral area sw. of Bradford Mtn. 1½ mi. w. of Dry Cr., acc. J. & B. Guggolz (1992); 350-400 ft. elev. Old Skaggs Springs Road, *Utech, Sato & Ohara* (1990 CM); ca 2000 ft. elev. Anderson Trail on Black Mtn. between Geyser Peak and upper end of Pine Flat Road, acc. J. & B. Guggolz (1992); 2400 ft. elev. Little Geysers, *Malloch* (1977 PG&E). < 7600 ft. Mar.-May. —*Brodiaea capitata* Bentham. *B. pulchella* (Salisbury) E. Greene.

Dichelostemma congestum (Smith) Kunth. OOKOW. Common, hills and canyonsides, open woodland, brush: Fairfield Osborn (*); Petaluma, *Eastwood* (1921); Hidden Oaks Road between Grange and Sonoma Mtn. roads, acc. Quibell (1991); Tresch Ranch, Walker Road near Two Rock, acc. J. & B. Guggolz (1992); Bouverie; Sugarloaf Ridge; Los Alamos Road, *Stocking* (1967 NCC); Spring Lake; Santa Rosa, *Stocking* (1969 NCC); s. of Piner School w. of Santa Rosa, *Harrison* (1986 SRJC); Pepperwood; chaparral area sw. of Bradford Mtn. 1½ mi. w. of Dry Cr., acc. J. & B. Guggolz (1992); Lake Sonoma area; ca 2000 ft. elev. Anderson Trail on Black Mtn. between Geyser Peak and upper end of Pine Flat Road, acc. J. & B. Guggolz (1992). < 6600 ft. Apr.-June. —*Brodiaea congesta* Smith.

Dichelostemma multiflorum (Bentham) A. A. Heller. WILD HYATHINTH. Uncommon, plains, hillsides: Maxwell Farms Park e. of Verano Ave., El Verano, acc. Bunyan and Hass (1983); s. of Piner School w. of Santa Rosa, *Harrison* (1986 SRJC); Sonoma Co. airport near Windsor, *Harrison* (1980 SRJC). < 6600 ft. May-June. —*Brodiaea multiflora* Bentham.

Disporum

1. Flowers cylindrical, truncate or slightly swollen below; stigma usually 3-cleft . . . . *D. smithii*
1: Flowers turbinate, narrowing at base; stigma entire . . . . *D. hookeri*

Disporum hookeri (Torrey) Nicholson. FAIRY BELLS. Common, usually away from coast, shady woods: The Sea Ranch, acc. Wells; Bodega Bay, *Heller & Brown* (1902); Duncans Mills area, *Yates* (1934 UC); Willow Creek; near Camp Meeker, *Armstrong* (1934 UC); Monte Rio, *Hoak* (1912 UC); Russian River Station, *Newell* (1900 GH); Armstrong Redwoods State Preserve, acc. Roderick (1989); Forestville to Guerneville Road, *Baker* (1898 UC); Crane Can. Road ½ mi. from Petaluma Hill Road, *McNally* (1970 NCC); 0.1 mi. w. on Sonoma Mtn. Road from jct. with Bennett Valley Road along Graham Cr., *Utech, Sato & Ohara* (1990 CM); Fairfield Osborn (S); n. slope of Sonoma Mtn., Baker (1926 NCC); Button Ranch on Walker Road (U. C. property) between Petaluma-Valley Ford Road and Pepper Road, acc.

Patterson (1990); 6 mi. w. of Petaluma on Rainsville Road, *Ott* (1934 NCC); Bouverie; base of Mt. Hood, *Heller & Brown* (1902); Sonoma Cr. Can. (Adobe Can.), *Baker* (1904 UC); Sugarloaf Ridge; Mt. Hood, *Jepson* (1918 JEPS); Santa Rosa, *Congdon* (1880); Gates Road off Calistoga Road, acc. Wells et al. (1984); Pepperwood; 1000 ft. elev. 14 mi. ese. of Preston and US 101 on Geysers Road, *Utech & Knight* (1986 CM); 350-400 ft. elev. Old Skaggs Springs Road, *Utech, Sato & Ohara* (1990 CM); Lake Sonoma area; Cloverdale, *Rattan* (1877). < 5400 ft. Mar.-May.

Disporum smithii (Hooker) Piper. Occasional, usually coastal influence, shady woods, sometimes inland: The Sea Ranch; Stewarts Point, *Parks* (1925); Stillwater Cove s. of Salt Point, acc. Bunyan (1988); 98 ft. elev. Fort Ross, *Clifton* (1981 PUA); 5 mi. se. of Monte Rio (probably on Bohemian Hwy.), *Miller* (1949 WTU); Sebastopol quad.; ca 1350 ft. elev. Geysers Road between Jimtown and Mercuryville, *Utech & Knight* (1986 CM). < 2700 ft. Mar.-May.

Erythronium

1. Filaments strongly dilated toward base; flowers rose-pink . . . . *E. revolutum*

1: Filaments more filiform

  2. Anthers golden-yellow; flowers white with golden-yellow base, style 12-15 mm long . . . . . . *E. helenae*

  2: Anthers white; flowers white to cream with pale greenish-yellow base . . . . *E. californicum*

Erythronium californicum Purdy. Common, openings in woods, brushy slopes: Gualala River, *Baker* (1898); n. side of Fitch Mtn., *Mullin* (1937 NCC); 4 mi. w. of US 101 on Dry Cr. Road, *Juul* (1973 NCC); along Dry Cr. ca 7 mi. w. of Warm Springs Dam, *Breedlove* (1976); Stewarts Point-Skaggs Springs Road w. of summit sw. of Lake Sonoma, acc. Quibell (1991); Warm Springs Dam at Lake Sonoma, *Lewis* (1985 NCC); near Cloverdale, *Rattan* (1877); 991 ft. elev. near Cloverdale, *Siemens* (1970 PUA); Hwy. 128 ca ¼ mi. from Mendocino Co. line scattered on a north-facing slope, acc. Roderick (1987). < 6300 ft. Mar.-Apr.

Erythronium helenae Applegate. Locally common, brush, woods, damp volcanic soil, leaf mold: Ida Clayton Road near e. end where it joins Western Mines Road, *Lovell* (1978), acc. Quibell (1991); 6 to 7 mi. up Ida Clayton Road, *McNally* (1970 NCC); headwaters of Sulphur Cr., *Applegate* (1934); near Mercuryville between Jimtown and The Geysers, *Knight & Edwards* (1986), *Utech & Knight* (1986 CM); 50 yds. from Lincoln Rock in geothermal area, *Quibell* (1980 NCC); geysers Unit 13, *Quibell* (1981 NCC). < 4000 ft. Mar.-May.

Erythronium revolutum Smith. Rare, margins of swamps, bogs or wooded streams: Duncans Mills quad. < 3300 ft. Mar.-June.

Fritillaria

1. Leaves on lower part of stem just above ground; flowers campanulate, not mottled

  2. Tepals white with green striations, 10-16 mm long . . . . *F. liliacea*

  2: Tepals greenish-brown to truly brown, their lower third to half creamy-greenish; 21-25 mm long .*F. biflora*

1: Leaves on upper stem, the lower part above ground naked; flowers plainly mottled

  3. Flower color scarlet, checkered yellow . . . . *F. recurva*

  3: Flower color brown-purple mottled with yellow, to pale greenish-yellow and faintly mottled with purple . . . . *F. affinis*

Fritillaria affinis (Schultes) Sealy. CHECKER LILY. Very common, mostly among oaks, pines and brush: Bodega Head; Larkspur Rock e. of Bodega Bay, *Baker* (1932 NCC, 1942 NCC), *Eastwood & Howell* (1939); Camp Meeker, *Howell* (1924), *Hagg* (1934 NCC); 2 mi. from Guerneville on Pocket Can. Road, *McNally* (1970 NCC); Hidden Oaks Road between Grange and Sonoma Mtn. roads, acc. Quibell (1991); Fairfield Osborn (*); Tresch Ranch, Walker Road near Two Rock, acc. J. & B. Guggolz (1992); 6 mi. w. of Petaluma at Rainsville Road, *Ott* (1934 NCC); Bouverie; around parking lot at Sugarloaf Ridge State Park, acc. Roderick (1975); Gates Road off Calistoga Road, acc. Wells et al. (1984); Mark West Springs Road, *Hawver* (1915); Pepperwood; chaparral area sw. of Bradford Mtn. 1½ mi. w. of Dry Cr., acc. J. & B. Guggolz (1992); Skaggs Springs, *Hawver* (1915); Lake Sonoma area; Sausal Cr. foothills near Jimtown, *Applegate* (1934); Geysers Road between Jimtown and Mercuryville, *Utech & Knight* (1986 CM); 50 yds. nw. of Lincoln Rock in geothermal area, *Quibell* (1980 NCC); geysers Unit 13, *Quibell* (1981 NCC); 997 ft. elev. near Cloverdale, *Seimens* (1970 PUA). < 6000 ft. Feb.-May. —*F. lanceolata* Pursh.

Fritillaria biflora Lindley. CHOCOLATE LILY, MISSION BELLS. Uncommon, coastal flats: Gualala Point Co. Park, acc. Rhyne (1987); 2 locations at The Sea Ranch, acc. Raiche (1986). < 4000 ft. Feb.-June. —*F. roderickii* W. Knight, in part.

Fritillaria liliacea Lindley. FRAGRANT FRITILLARY. Rare, open hills and fields, heavy soil: 3/4 mi. e. of Occidental, *Hastings* (1940 NCC); near Camp Meeker, acc. Raiche (1986); 1 mi. se. of Camp Meeker, *J. & B. Guggolz* (1987 UC); Penngrove, acc. Baker (1954); N. Burma Trail at Annadel, *Quibell* (1980 NCC); grassland n. of summit of Taylor Mtn. s. of Santa Rosa, *Quibell* (1981 NCC). < 700 ft. Feb.-Apr.

Fritillaria recurva Bentham. SCARLET FRITILLARY. Common, dry hillsides, woods or brush: Larkspur Rock e. of Bodega Bay, *Baker* (1932 NCC, 1942 NCC), *Eastwood & Howell* (1939); Camp Meeker, *Howell* (1924), *Hagg* (1934 NCC); 2 mi. from Guerneville on Pocket Can. Road, *McNally* (1970 NCC); Hidden Oaks Road between Grange and Sonoma Mtn. roads, acc. Quibell (1991); Fairfield Osborn (*); Tresch Ranch, Walker Road near Two Rock, acc. J. & B. Guggolz (1992); 6 mi. w. of Petaluma at Rainsville Road, *Ott* (1934 NCC); Bouverie; around parking lot at Sugarloaf Ridge State Park, acc. Roderick (1975); Gates Road off Calistoga Road, acc. Wells et al. (1984); Mark West Springs Road, *Hawver* (1915); Pepperwood; chaparral area sw. of Bradford Mtn. 1½ mi. w. of Dry Cr., acc. J. & B. Guggolz (1992); Skaggs Springs, *Hawver* (1915); Lake Sonoma area; Sausal Cr. foothills near Jimtown, *Applegate* (1934); Geysers Road between Jimtown and Mercuryville, *Utech & Knight* (1986 CM); 50 yds. nw. of Lincoln Rock in geothermal area, *Quibell* (1980 NCC); geysers Unit 13, *Quibell* (1981 NCC); 997 ft. elev. near Cloverdale, *Seimens* (1970 PUA). < 7300 ft. Mar.-June. —*F. r.* var. *coccinea* E. Greene.

★ **Kniphofia uvaria** Hooker. RED HOT POKER. Probably from garden trash: along main entrance road to Salt Point State Park, acc. Wells (1985); Blank Road near Gravenstein Hwy., acc. J. & B. Guggolz (1991); along Peracca Road e. of Grange Road, acc. Quibell (1991); Pepperwood. Native of S. Africa.

Lilium

1. Plants of rather dry habitats, bulbs ovoid, the scales mostly > 25 mm long, not jointed
   2. Flowers tending to nod, bright reddish-orange, thickly spotted with purple; not fragrant . . . *L. columbianum*
   2: Flowers erect or ascending, pale lilac or nearly white, becoming rose-purple, finely dotted with brown, fragrant . . . *L. rubescens*

1: Plants of wet habitats, rhizomatous, the scales mostly < 25 mm long, and jointed
   3. Flowers horizontal to ascending, deep reddish-orange, tepals not revolute . . . *L. maritimum*
   3: Flowers nodding, bright orange-red with a light orange center, conspicuously spotted with large purple spots on the lower half, tepals strongly revolute to below the middle
      4. Plants usually in large clones; tepals 5-11 cm; anthers 10-17 mm . . . *L. pardalinum* ssp. *pardalinum*
      4: Plants usually in small clones; tepals 3-9 cm; anthers 5-14 mm . . . *L. p.* ssp. *pitkinense*

Lilium columbianum Baker. OREGON LILY, COLUMBIA LILY. Rare, among ferns and brush: 302 ft. elev. Salt Point State Park, *Clifton* (1981 PUA), acc. Hickman (1986). < 4300 ft. June-July.

Lilium maritimum Kellogg. COAST LILY. Occasional, usually sandy soil, woods, brush, occasionally marshy areas: The Sea Ranch; Hwy. 1 near Timber Cove, *Best* (1981); 1¼ mi. from start of Kruse Ranch Road near Plantation, acc. Robinett & Robinett (1988); w. of Plantation in w. edge of Kruse Rhododendron State Reserve, acc. Quibell (1991); 1 mi. w. of Plantation, *Hastings* (1940 NCC); Salt Point along roadside of Hwy. 1, *Utech* (1988 CM). < 500 ft. June-July.

Lilium pardalinum Kellogg ssp. **pardalinum**. LEOPARD LILY. Occasional, springy places, stream banks: The Sea Ranch; s. side of Gualala River just w. of jct. of N. Fork of same, acc. Quibell (1991); Stewarts Point-Skaggs Springs Road at bridge crossing S. Fork of Gualala River, acc. Roderick (1986); Guerneville Marsh, *Rubtzoff* (1955); Pitkin Marsh, *Rubtzoff* (1951); Sebastopol, *Congdon* (1880); Pepperwood; 2190 ft. Squaw Cr., acc. Clifton (1984). < 6700 ft. May-July.

Lilium pardalinum ssp. **pitkinense** (Beane & Vollmer) M. Skinner. PITKIN MARSH LILY. Rare, marshy situations: Pitkin Marsh, *Baker* (1939 NCC), *Howell* (1945), *Rubtzoff* (1951), *Beane & Volmer* (type 1954 U); n. end of Pit-kin Marsh at Pitkin Ranch boundary, acc. Rae (1981); creekside 1 mi. w. of Cunningham in marsh, *Baker* (1961 NCC), *Rubtzoff* (1974), *Neihaus* (1977), *Quibell* (1987 NCC); Cunningham Marsh, *Baker* (1946). < 400 ft. June-July. —*L. p.* Beane & Vollmer.

Lilium rubescens S. Watson. REDWOOD LILY. Occasional, wooded, brushy ridges and slopes: Coleman Valley Road w. of Occidental, acc. Ely (1987); Camp Meeker, *D. Wilson* (1978 JEPS); near Occidental, *Crosby* (1938); near MacNab cypress grove on Cavedale Road, acc. Raiche (1986); Gates Road off

Calistoga Road, acc. Wells et al. (1984); e. edge of Annadel in lower Schultz Can., acc. Stafford (1991); Pepperwood; w. slope of Mt. St. Helena w. of county line, acc. Lovell (1978); Knights Valley, *Eastwood & Howell* (1939); ridge above Palmer Cr. drainage area w. of Healdsburg, acc. Raiche (1986). < 5000 ft. June-July.

Maianthemum dilatatum (Wood) Nelson & J. F. Macbride. FALSE LILY-OF-THE-VALLEY. Uncommon, damp, shaded embankments: The Sea Ranch w. of Hwy. 1; Azalea Trail at The Sea Ranch, acc. Bunyan (1988); along public access trail from coast to Hwy. 1 just s. of The Sea Ranch Golf Course, acc. Quibell (1991); Skaggs Springs, *Hawver* (1915). < 1700 ft. May-June.

Muilla maritima (Torrey) S. Watson. Occasional, sub-alkaline flats, rocky slopes: Todd Road Preserve; Todd Road w. of Llano Road, *Davies* (1975 NCC), *Harrison* (1978); between Cotati and Santa Rosa, *Raven* (1958); Taylor Mtn. s. of Santa Rosa, *Baker* (1924 NCC); El Verano, *Griffin* (1922); 100 yds. n. of jct. of Agua Caliente Road on Hwy. 12, acc. Roderick (1989); Bennett Valley, *Stocking* (1964 NCC); Windsor, *Koch* (1938 NCC). < 7700 ft. Mar.-June.

★ **Nothoscordum inodorum** (Aiton) Nicholson. YELLOW FALSE GARLIC. Rare, waste places: near Schellville, acc. *Barbe* (1990). < 350 ft. Native of n. S. America to se. U. S. —*N. fragrans* (Venten) Kunth.

Scoliopus bigelovii Torrey. SLINK POD, BROWNIES. Common, moist, shady, embankments, slopes and flats: The Sea Ranch, acc. Wells (1982); Stillwater Cove between Timber Cove and Salt Point, acc. Bunyan (1988), *Utech* (1988 CM); Salt Point; e. of Plantation, *Roderick* (1960 JEPS); 2½ mi. s. of Fort Ross, *Gould* (1939 UC); ½ and 3 mi. s. of Fort Ross, *Utech* (1988 CM); Willow Creek; 2.75, 5.75 and 6.0 mi. on Willow Cr. Road from Bridgehaven, *Utech & Knight* (1988 CM); 5-6 mi. inland from Hwy. 1 on Coleman Valley Road, acc. Roderick (1989); Duncans Mills Marsh, *Rubtzoff* (1957); near Guerneville, *Baker* (1898 UC); Neeley Hills s. of Guerneville, *Rubtzoff* (1953); Pocket Can. Road between Guerneville and Forestville, *Baker* (1926 NCC); Harrison Grade Road midway up grade from e. side, acc. Quibell (1991); Sturgeon's Mill (Furlong Road w. of Sebastopol ca ¼ mi. from Occidental Road), *Hagg* (1934 NCC), *Utech & Knight* (1988 CM); w. of Sebastopol, *Heller & Brown* (1902); Wagnon Road, just off Bodega Hwy., *Utech & Knight* (1988 CM). < 3700 ft. Feb.-Mar.

Smilicina

1. Flowers, numerous, in panicles . . . . *S. racemosa*
1: Flowers, few to several, in racemes . . . . *S. stellata*

Smilicina racemosa (Linnaeus) Link. FALSE SOLOMON'S SEAL. Common, shady woods: The Sea Ranch, acc. Wells (1982); Annapolis quad.; Jenner, *Kawahara* (1969); Willow Creek; Bodega Bay, *Heller & Brown* (1902); between Occidental and Monte Rio, *H. Thomas* (1933 NCC); Armstrong Redwoods State Preserve, acc. Roderick (1989); w. side of Sonoma Mtn. on McDuffy Ranch, *Stocking* (1965 NCC); Gates Road off Calistoga Road, acc. Wells et al. (1984); 997 ft. elev. Franz Valley Road, *Klusman* (1970 PUA); Lake Sonoma area. < 6700 ft. Mar.-May. —*S. r.* var. *amplexicaulis* (Nuttall) S. Watson, currently not recognized.

Smilicina stellata (Linnaeus) Desfontaines. Common, moist streambanks, slopes: The Sea Ranch; Willow Creek; Camp Meeker, *Haag* (1934 NCC); between Freestone and Occidental, *H. Thomas* (1933 NCC); 0.1 mi. w. on Sonoma Mtn. Road from jct with Bennett Valley Road along Graham Cr., *Utech, Sato & Ohara*, (1990 CM); Crane Can., *Stocking* (1963 NCC); Lichau Road off Petaluma Hill Road, *Stocking* (1964 NCC); Fairfield Osborn (S); 6 mi. w. of Petaluma on Rainsville Road, *Ott* (1934 NCC); Spring Lake; Santa Rosa Cr., *Heller & Brown* (1902); Pepperwood; mts. w. of Calistoga (probably on road to Petrified Forest), *Krauss* (1894); Lake Sonoma area. < 8000 ft. Apr.-June. —*S. s.* var. *sessilifolia* (Baker) L. Henderson.

Tofieldia occidentalis S. Watson. Infrequent, meadows, boggy areas: Pitkin Marsh, *Rubtzoff* (1951), *Stocking* (1968 NCC), acc. B. Guggolz (1990); Perry Marsh, *Howell* (1945); 2 mi. n. of Sebastopol on Gravenstein Hwy., *Baker* (1945 NCC). < 10300 ft. July-Aug. —*T. glutinosa* Persoon ssp. *o.* (S. Watson) C. L. Hitchcock.

Trillium

1. Flowers pedicellate . . . . *T. ovatum*
1: Flowers sessile
   2. Flowers with portion of the filament (connective) joining the 2 cells of an anther greenish; filament usually green, sometimes purple; anther dehiscence lateral; tepals white (rarely purple basally) to pink *T. albidum*
   2: Flowers with stamens and anthers purple throughout; anther dehiscence turned inward; tepals varying from purple to yellow or white . . . . *T. chloropetalum*

Trillium albidum J. D. Freeman. Occasional, damp shady areas: Cazadero, *Trosper* (1922); near Cooper's Grove on Sonoma Mtn. Road, *Utech & Knight* (1986 CM); 0.1 mi. w. on Sonoma Mtn. Road from jct. with Bennett Valley Road along Graham Cr., *Utech, Sato & Ohara* (1990 CM); Sonoma Cr. at base of Mt. Hood, *Heller & Brown* (1902); Santa Rosa, *Congdon* (1880); 350-400 ft. elev. Old Skaggs Springs Road, *Utech, Sato & Ohara* (1990 CM); Skaggs Springs, *Hawver* (1915); Stewarts Point-Skaggs Springs Road 10 mi. w. of its jct. with Dry Cr. Road, *Freeman* (1967). < 6700 ft. Apr.-May.

Trillium chloropetalum (Torrey) T. J. Howell. GIANT TRILLIUM. Common, wooded or brushy slopes, usually shady areas: Vine Hill School Road, *Ferrari* (1979 NCC); Crane Can., *Stocking* (1964 NCC); Sonoma Mtn. Road, *McNally* (1970 NCC); Lichau Road off Petaluma Hill Road, *Stocking* (1964 NCC), *M. B. Smith* (1971 NCC); Fairfield Osborn (*); 6 mi. w. of Petaluma on Rainsville Road, *Ott* (1934 NCC); Rodgers Cr. 1 mi. e. of Adobe Road, *Backett* (1959 NCC); Bouverie; Youngman Can. 2 mi. e. of Glen Ellen, *Koch* (1938 NCC); Sugarloaf Ridge; Gates Road off Calistoga Road, acc. Wells et al. (1984); Franz Valley Road, *Giampoli* (1968 NCC), *Stocking* (1969 NCC); Rancheria Cr., Lake Sonoma area, *Guggolz* (1976 NCC). Highly variable. < 6700 ft. Feb.-July. —*T. c.* var. *giganteum* (Hooker & Arnott) Munz, currently not recognized.

Trillium ovatum Pursh. WESTERN TRILLIUM, WHITE TRILLIUM. Common, moist, wooded or brushy places: The Sea Ranch, acc. Wells (1984); Annapolis quad.; Salt Point, *Utech* (1988 CM); Austin Cr. near Cazadero, *Rupp* (1970 NCC); River Road between Jenner and Guerneville, *McNally* (1970 NCC); Duncans Mills Marsh, *Rubtzoff* (1957); Willow Creek; Armstrong Redwoods State Preserve, acc. Roderick (1989); hills s. of Guerneville, *Rubtzoff* (1953); Sturgeon's Mill site on Furlong Road ¼ mi. off Occidental Road, *Utech & Knight* (1986 CM); Cooper's Grove on Sonoma Mtn. Road, acc. Quibell (1991); Sugarloaf Ridge. < 5400 ft. Feb.-Apr.

Triteleia

1. Flowers with stamen bases attached alternately as 2 distinct rows of 3
   2. Perianth tube sharply pointed to base; pedicels 0 to 2 × longer than perianth . . . . *T. laxa*
   2: Perianth tube broadly tapering to base; pedicels 2 to 10 × longer than perianth . . . . . . *T. peduncularis*
1: Flowers with all 6 stamen bases attached at same level
   3. Anther tops = in height or nearly so . . . . *T. hyacinthina*
   3: Anther tops alternately long and short in height . . . . *T. lugens*

Triteleia hyacinthina (Lindley) E. Greene. WHITE BRODIAEA. Common, low places, wet in the spring: The Sea Ranch, *Wells* (1984); Freestone Marsh, *Rubtzoff* (1959); 3 sites w. and sw. of Santa Rosa in Laguna de Santa Rosa drainage, *Harrison* (1978 NCC); Hidden Oaks Road between Grange and Sonoma Mtn. roads, acc. Quibell (1991); Penngrove, *Gutermute* (1932 NCC); Lynch Road 0.3 mi. from Adobe Road, *Quibell* (1978 NCC); w. of Schellville between Arnold Drive and Napa, *Cantelow* (1968); Sobre Vista Road 0.9 w. of Arnold Drive; *Rubtzoff* (1971); Sugarloaf Ridge; Spring Lake; nw. of Santa Rosa, *Roderick* (1967); Old Redwood Hwy. and Starr Road, *J. Smith* (1974 NCC); Lake Sonoma area, *Quibell* (1982 NCC); property bordered by Chianti and Dutcher Cr. roads w. of US 101 and w. of Geyserville, acc. J. & B. Guggolz (1991). < 6700 ft. Apr.-Aug. —*Brodiaea h.* (Lindley) Baker.

Triteleia laxa Bentham. ITHURIEL'S SPEAR. Common, dry flats, slopes, brush, open woodland, grassland: s. part of The Sea Ranch, *J. H. Thomas* (1976); Salt Point; 1 mi. e. of Bodega Bay, *Howell* (1930); upper Green Valley Marsh w. of Graton, *Rubtzoff* (1974); Sebastopol, *Dows* (1907); Todd Road Preserve; Hidden Oaks Road between Grange and Sonoma Mtn. roads, acc. Quibell (1991); Button Ranch on Walker Road (U. C. property) between Petaluma-Valley Ford Road and Pepper Road, acc. Patterson (1990); w. of Sears Point Raceway in Tolay Cr. drainage, acc. Patterson (1990); Bouverie; Sugarloaf Ridge; Spring Lake; Fitch Mtn., *Rubtzoff* (1952); Lake Sonoma area; Geysers Road between Jimtown and Mercuryville, *Utech & Knight* (1986 CM); along ridge between Socrates Mine and Mercuryville in geothermal area, *Dolan & LaPré* (1987 UCR); ca 2000 ft. elev. Anderson Trail on Black Mtn. between Geyser Peak and upper end of Pine Flat Road, acc. J. & B. Guggolz (1992). < 5000 ft. Apr.-June. —*Brodiaea l.* (Bentham) S. Watson.

Triteleia lugens E. Greene. Uncommon, along streams: Youngman's Grade, (probably on periphery of Sonoma Valley), *Baker* (1928); Annadel, acc. Lovell (1976); St. Helena Road ca 1 mi. from county line, acc. Lovell (1976). < 3300 ft. —*Brodiaea l.* Baker.

Triteleia peduncularis Lindley. LONG-RAYED BRODIAEA. Common, wet places, low fields, marshes: Bodega, *Chandler* (UC); Occidental Marsh, *Rubtzoff* (1957); Freestone Marsh, *Rubtzoff* (1965); Pitkin Marsh, *Howell* (1937); Cunningham Marsh, *Howell* (1947); Laguna de Santa Rosa se. of Sebastopol, *Rubtzoff* (1964); 2 sites w. and sw. of Santa Rosa in Laguna de Santa Rosa drainage, *Harrison* (1978); Stony Point, *Quibell* (1978 NCC); Pressley Road opposite Crane Cr. Reg. Park ¼ mi. w. of entrance, acc. Quibell (1991); near Kenwood, *Baker* (1935), *Howell* (1935); Santa Rosa, *Congdon* (1880), *Stocking* (1964 NCC); Pepperwood; Socrates Mine, acc. Clifton (1984). < 2700 ft. May-July —*Brodiaea p.* (Lindley) S. Watson

Veratrum fimbriatum A. Gray. FRINGED FALSE HELLEBORE, FRINGED CORN LILY. Infrequent, wet openings, meadows: The Sea Ranch, acc. Knight (1967), acc. Wells (1984); Salt Point State Park in moist places on marine terrace, *McKenzie* (1977); 2 mi. n. of Stewarts Point, *Baker* (1920 NCC); low places on coastal bluffs 7 mi. n. of Fort Ross, *Howell* (1931); Stillwater Cove County Park between Salt Point and Timber Cove, acc. Bunyan (1988); Green Hill Road between Graton and Cherry roads, *Best* (1981). < 600 ft. July-Sept.

Xerophyllum tenax (Pursh) Nuttall. BEAR-GRASS, INDIAN BASKET-GRASS. Common, open dry ridges and slopes: The Sea Ranch timber preserve, acc. Wells (1984); Annapolis quad.; Salt Point; headwaters of Austin Cr. at The Cedars, *Raiche* (1984); Vine Hill area (bloomed in 1990), acc. Quibell; Sonoma Valley, *Zeile* (1938); Sonoma, *Bioletti* (1892 UC); hills above Sonoma, *Buckingham* (1921); Cavedale Road above cypress grove, acc. Knight (1990); 2 mi. from Hwy. 12 on Trinity Road, acc. Roderick (1986); Annadel, acc. Raiche (1986); n. of Crane Cr. on the w. side of Dry Cr. Valley, *Raiche* (1986); Pepperwood; Franz Valley School Road near county line, acc. Roderick (1964); ca 2000 ft. elev. Anderson Trail on Black Mtn. between Geyser Peak and upper end of Pine Flat Road, acc. J. & B. Guggolz (1992). Infrequently blossoms, mostly after a fire. < 7700 ft. May-Aug.

Zigadenus

1. Perianth 8-14 mm long; stamens ca half as long
   2. Plants 3-10 dm high . . . . . *Z. fremontii* var. *fremontii*
   2: Plants 1-2 dm high . . . . . *Z. f.* var. *minor*
1: Perianth smaller, or if 8 mm long, the stamens > ½ as long
   3. Flowers normally racemose, 4-5 mm long . . . . . *Z. micranthus* var. *micranthus*
   3: Flowers normally paniculate, ca 6-8 mm long . . . . . *Z. m.* var. *fontanus*

Zigadenus fremontii (Torrey) S. Watson var. **fremontii**. STAR LILY. Very common, dry, grassy or brushy slopes: The Sea Ranch; Timber Cove, *Lawner* (1980 NCC); Stewarts Point quad.; Salt Point; Stoetz Lane ne. of Occidental, *Utech & Knight* (1986 CM); 2 mi. n. of Cotati, *Baker* (1924); Bouverie; 1.5 mi. up Cavedale Road from Hwy. 12, *Utech & Knight* (1986 CM); Kenwood, *Bioletti* (1893 UC); Baker's home (Maplewood) in Adobe Can., *Baker* (1924 NCC); Sugarloaf Ridge; Green Hill Drive, Santa Rosa, *Stocking* (1965 NCC); Rincon Ridge in Santa Rosa, *H. Thomas* (1933 NCC); w. face of Taylor Mtn. s. of Santa Rosa off Petaluma Hill Road, *Quibell & Lehner* (1975 NCC); Pepperwood; chaparral area sw. of Bradford Mtn. 1½ mi. w. of Dry Cr., acc. J. & B. Guggolz (1992); Lake Sonoma area; Ida Clayton Road quad.; ca 2000 ft. elev. Anderson Trail on Black Mtn. between Geyser Peak and upper end of Pine Flat Road, acc. J. & B. Guggolz (1992); 3200 ft. elev. geysers Unit 17, *Malloch* (1977 PG&E); 2800 ft. elev. geothermal area, *Cooney* (1977 PG&E). < 3300 ft. Mar.-Apr.

Zigadenus fremontii var. **minor** (Hooker & Arnott) Jepson. ZIGADENE. Occasional, open, moist fields: The Sea Ranch, *Wells* (1987); 2 mi. n. of Cotati, *Baker* (1924 NCC); Wilfred Station (now defunct but at present it is Wilfred Ave. area of Rohnert Park), *Focht* (NCC); n. slope of Taylor Mtn. s. of Santa Rosa off Petaluma Hill Road, *Baker* (1939 NCC); near Windsor, *Heller & Brown* (1902); journey between Fort

Ross and Mt. St. Helena, *Voznesenskii* (1841 LE); Taylor Mtn. s. of Santa Rosa off Petaluma Hill Road, *Baker* (1920 JEPS). Feb.-Mar. Frequently not recognized.

Zigadenus micranthus Eastwood var. **fontanus** (Eastwood) D. McNeal. Occasional, often on serpentine, wet places: 1 mi. s. of Gualala River on coast road (now The Sea Ranch), *Wolf* (1927); marshes near coast (ambiguous), *Morse* (1936); Salt Point Ranch, *Kawahara* (1968); confluence of Wheatfield Fork and S. Fork of Gualala River, *Leech* (1964); Pitkin Marsh, *Howell* (1938), *Rubtzoff* (1951); Bennett Mtn. Lake, *Rubtzoff* (1955, 1961, 1963). < 1700 ft. Apr.-June. —*Z. fontanus* Eastwood.

Zigadenus micranthus var. **micranthus.** Occasional, dry flats, hillsides: ca 1 mi. nw. of Camp Meeker, acc. J. & B. Guggolz (1991); Stoetz Lane near Occidental, *Best* (1980); n. slope of Taylor Mtn. off Petaluma Hill Road, *Baker* (1920); 3/4 mi. up Trinity Road, *Rucker* (1960); Mt. Hood, *Heller & Brown* (1902); Spring Lake; near Santa Rosa, *Heller* (1902); mts. w. of Calistoga (probably near Petrified Forest), *Krauss* (1894); near Franz and Knights valleys, *Krauss* (1895); Geyser Peak, *Knight & Edwards* (1986); Lake Sonoma area. < 3300 ft. May-June.

NAJADACEAE. WATER-NYMPH FAMILY
[Sometimes HYDROCHARITACEAE]

Najas guadalupensis (Sprengel) Magnus. GUADALUPE NAJAS, COMMON WATER-NYMPH. Rare, shallow water: Button Ranch on Walker Road (U. C. property) between Petaluma-Valley Ford Road and Pepper Road, *Connors* (1992 BML); Bennett Mtn. Lake, *Rubtzoff* (1955); Pepperwood, *de Nevers* (1991). < 4000 ft. July.

ORCHIDACEAE. ORCHID FAMILY

1. Leaves reduced and scalelike; plants saprophytic, not green
   2. Plant white; perianth 12-15 mm long . . . . . *Cephalanthera*
   2: Plant brown, yellow or purple; perianth mostly shorter . . . . . *Corallorhiza*

1: Leaves foliaceous; plant green
   3. Flowers mostly 1 to few, with leafy bracts if > 1
      4. Leaf 1, basal; flower 1 . . . . . *Calypso*
      4: Leaves 2 to several; flowers often > 1
         5. Lip an inflated pouch; fertile anthers 2 . . . . . *Cypripedium*
         5: Lip saccate at base, but not pouchlike; anther 1 . . . . . *Epipactis*
   3: Flowers many, spicate or racemose, the bracts not foliaceous
      6. Lip spurred or scrotiform
         7. Sepals 1-nerved; leaves usually near base of stem and withering before anthesis . . . . . *Piperia*
         7: Sepals 3-nerved; leaves usually scattered and green at anthesis . . . . . *Platanthera*
      6: Lip not spurred nor scrotiform
         8. Spike twisted, glabrous . . . . . *Spiranthes*
         8: Spike not twisted, glandular-hairy . . . . . *Goodyera*

Calypso bulbosa (Linnaeus) Oakes. CALYPSO ORCHID, FAIRY SLIPPER. Common, in decaying vegetative material, bogs, woods, shade: ca 1 mi. nw. of Camp Meeker, acc. J. & B. Guggolz (1991); n. slope of Taylor Mtn. off Petaluma Hill Road, *Baker* (1920); ¾ mi. up Trinity Road, *Rucker* (1960); Mt. Hood, *Heller & Brown* (1902); Spring Lake; near Santa Rosa, *Heller* (1902); mts. w. of Calistoga (probably near Petrified Forest), *Krauss* (1894); near Franz and Knights valleys, *Krauss* (1895); Geyser Peak, *Knight & Edwards* (1986); Lake Sonoma area. < 6000 ft. Mar.-July.

Cephalanthera austiniae (A. Gray) A. A. Heller. PHANTOM ORCHID. Uncommon, dry woods: trail e. of Schooner Drive at The Sea Ranch, acc. Wells (1973); Stillwater Cove County Park just n. of Timber Cove, acc. Bunyan (1985); 2000 ft. elev. w. flank Mt. St. Helena, *Grummer* (1981 PUA), acc. Raiche (1985). < 7300 ft. May. —*Eburophyton a.* (A. Gray) A. A. Heller.

Corallorhiza

1. Perianth-parts longitudinally striped . . . . *C. striata*
1: Perianth-parts not striped
   2. Lip 3-lobed, spotted . . . . *C. maculata*
   2: Lip entire, not spotted . . . . *C. mertensiana*

Corallorhiza maculata Rafinesque. SPOTTED CORALROOT. Common, deep woods: The Sea Ranch, acc. Wells (1971); Plantation Road, *Baker* (1932); 2 mi. e. of Stewarts Point, *Okada* (1975); Cazadero quad.; Vacation Beach near Guerneville, *Rubtzoff* (1953); Neeley Hill area s. of Guerneville, *Rubtzoff* (1953); Camp Meeker, *Armstrong* (1929 NCC), *Hagg* (1934 NCC); Dutch Bill Cr. along Bohemian Hwy. en route to Monte Rio from Occidental, *Keahey* (1968 NCC); near s. end of I St. Ext. s. of Petaluma under Bay Laurel trees, acc. Roderick (1930s); ca 1500 ft. elev. Geysers Road between Jimtown and Mercuryville, *Utech & Knight* (1986 CM). < 9300 ft. June-Aug.

Corallorhiza mertensiana Bongard. WESTERN CORALROOT. Uncommon, rich, wooded areas: 3 mi. n. of Fort Ross, *L. P. Bolander* (1949); road to Plantation, *Baker* (1932). < 7300 ft. June-July.

Corallorhiza striata Lindley. STRIPED CORALROOT. Occasional, rich, wooded areas: w. slope of Sonoma Mtn., *Stocking* (1965 NCC); near s. end of I St. Ext. s. of Petaluma under Bay Laurel trees (growing with *C. maculata)*, acc. Roderick (1930s); Bouverie; Sugarloaf Ridge State Park, acc. R. Coleman (1989); Spring Lake; confluence of Mark West and Leslie creeks, *de Nevers* (1981); Lake Sonoma area. < 7300 ft. May-July.

Cypripedium

1. Leaves 2, opposite; lip greenish-yellow . . . . *C. fasciculatum*
1: Leaves several, alternate; lip white or pinkish
   2. Sepals 4-6 cm long, dark brown . . . . *C. montanum*
   2: Sepals ca 1.5 cm long, greenish-yellow . . . . *C. californicum*

Cypripedium californicum A. Gray. CALIFORNIA LADY'S-SLIPPER. Rare, wet hillsides and rocky ledges: head-waters of Austin Cr., *Graff* (1928), acc. Raiche (1986), acc. R. Coleman (1989); 5 mi. up Willow Cr. Road from Bridgehaven, acc. Roderick (1955) since extirpated by logging. < 7300 ft. May-June.

Cypripedium fasciculatum S. Watson. CLUSTERED LADY'S-SLIPPER. Listed in C.N.P.S. Inventory 1994 as occurring in the county, but no positive information. < 5700 ft. Apr.-May.

Cypripedium montanum Lindley. MOUNTAIN LADY-SLIPPER. Rare, damp woods: Cazadero, *Graff* (1926); 1 mi. n. of Occidental en route to Camp Meeker, acc. Roderick (1950); near Sebastopol, *Cantelow* (1948); Mill Cr. Can., *Baker* (1930); 350-400 ft. elev. Old Skaggs Springs Road, *Utech, Sato & Ohara* (1990 CM). < 7300 ft. May-Aug.

Epipactis

1. Sepals and petals 12-20 mm, yellowish to greenish, veined red-purple; lip 14-20 mm, grooved to tip, wet places . . . . *E. gigantea*
1: Sepals and petals 8-13 mm, white to greenish, tinged pink-purple; lip 10-12 mm, ± flat above middle; ± dry places . . . . *E. helleborine*

Epipactis gigantea Hooker. STREAM ORCHID. Common, stream banks: s. bank of Gualala River 2 mi. from mouth, acc. Knight & C. Young (1968); Annapolis, *Sutliffe* (1920); purplish-black-leaved form at The Cedars, acc. Raiche (1988); Griffin Ranch at Kenwood, *Armstrong* (1940 NCC); near Berkeley Y.M.C.A. camp on Stewarts Point-Skaggs Springs Road, acc. Roderick (1989); 1-2 mi. from US 101 on Sulphur Cr. Road on rocky roadcut, acc. Vale (1991). < 8700 ft. May-Aug.

★ **Epipactis helleborine** (Linnaeus) Crantz. BROADLEAF HELLEBORINE. Escape from cultivation: redwood area in Jack London State Park near Glen Ellen, *Wagnon* (1989 NCC). < 4300 ft. Native of Eurasia.

Goodyera oblongifolia Rafinesque. RATTLESNAKE-PLANTAIN. Uncommon, dry forest floor: near end of Pilot's Reach at The Sea Ranch, acc. Wells (1985); Stillwater Cove County Park just n. of Timber Cove, acc. Bunyan (1985); Cooper's Grove on Sonoma Mtn. Road, acc. Baker (1954). < 7300 ft. July-Aug. —*Peramium decipiens* (Hooker) Piper.

Piperia Key ± acc. R. Morgan.

1. Spur 1.5-6 mm
   2. Inflorescence usually secund; lateral petals straight, white, midvein green, upper sepal mostly white or with white margins, lower sepals and lip mostly white; lip recurved toward spur, ± obtuse ..... *P. candida*
   2: Inflorescence ± open (sometimes dense on *P. michaelii*); perianth green throughout; lip ± straight to upcurved beyond middle
      3. Lip narrowly lanceolate, pointed forward or down; lateral petals ± linear, spreading or erect, ± recurved; spur 4-7 mm, curved, pointed down ..... *P. leptopetala*
      3: Lip oblong to ± deltate, usually pointed down, tip upcurved; lateral petals ± erect to pointed forward; spur 1.5-5 mm, pointed back or down ..... *P. unalascensis*
1: Spur 6-15 mm
   4. Spur ± straight, perpendicular to inflorescence axis; upper sepal pointed forward; stem usually < 3 mm diameter ..... *P. transversa*
   4: Spur ± curved, parallel to inflorescence axis; upper sepal erect, lip curved down or pointed forward; stem usually > 3 mm diameter
      5. Plants with sepals white with dark green midvein, petals white to pale green ..... *P. elegans*
      5: Plants with perianth green to yellow-green thoughout
         6. Lip triangular; lateral petals ± sickle-shaped, flat; lower sepals ± reflexed ..... *P. elongata*
         6: Lip ± ovate-deltate; lateral petals ± straight, ± concave; lower sepals spreading .. *P. michaelii*

Piperia candida R. Morgan & J. Ackerman. WHITE-FLOWERED PIPERIA. Rare, coniferous forest: Cazadero, *Randall* (1930 UC). < 4300 ft. May-July.

Piperia elegans (Lindley) Rydberg. Frequent, dry woods, bluff tops: tip of Bodega Head, *A. F. Johnson* (1975 BML); Harrison Grade, *Hastings* (1940 NCC); Fairfield Osborn (S); Sonoma Mtn. on Roth Ranch, *Stocking* (1967 NCC); Sonoma, *Armstrong* (1929 NCC); Bouverie; Hidden Oaks Road on w. side of w. end of Sonoma Mtn. Road, *Quibell* (1988 NCC); Sugarloaf Ridge State Park, acc. R. Coleman (1989); Annadel, *Knight & Edwards* (1986); Matanzas Cr., Santa Rosa, *Quibell* (1977 NCC); Pepperwood, *Lindsay* (1982); Leslie Road near e. Windsor, *Hoffman* (1981 NCC); Mill Cr., *Von Grafen* (1929 NCC); Fitch Mtn., *Rubtzoff* (1952); Lake Sonoma area. < 1700 ft. May-Sept. —*Habenaria maritima* (E. Greene) Ames. *H. e.* (Lindley) Bolander.

Piperia elongata Rydberg. Infrequent, wooded flats and slopes, dry to moist soil: Sugarloaf Ridge, *R. Morgan* (1988); Fitch Mtn., *R. Morgan* (1988); Lake Sonoma area; Cloverdale, *Rattan* (1877). < 7300 ft. Apr.-Aug. —*Habenaria unalascensis* (Sprengel) S. Watson var. *elata* (Jepson) Correll.

Piperia leptopetala Rydberg. Uncommon, serpentine: The Cedars, n. of Cazadero, acc. R. Coleman (1989); Harrison Grade, *Hastings* (1940 NCC). < 7300 ft. May-June.

Piperia michaelii (E. Greene) Rydberg. Possibly overlooked, especially along the coast. < 2300 ft. May-Aug.

Piperia transversa Suksdorf. Infrequent, dry woods and hills: ca 1 mi. nw. of Camp Meeker, acc. J. & B. Guggolz (1991); Sonoma, *Armstrong* (1929 NCC); Hope Valley n. of Baker Ranch off Adobe Can., *Baker* (1928 NCC); Sugarloaf Ridge State Park, acc. R. Coleman (1989); Annadel, *Knight & Edwards* (1986); Pepperwood, *de Nevers* (1988); ne. slope of Fitch Mtn., *Rubtzoff* (1952). < 8700 ft. June-Aug.

Piperia unalascensis (Sprengel) Rydberg. Dry to moist soil in chaparral, grassland or oak woodland: Possibly overlooked. < 10000 ft. Apr.-June. —*Habenaria u.* (Sprengel) S. Watson.

Platanthera

1. Flowers white, lip broadly dilated; column small compared to the dorsal sepal; leaves sheath the stem and cluster near bottom ..... *P. leucostachys*
1: Flowers green; lip linear with slight thickening at base; column > ½ of dorsal sepal; leaves alternate on the stalk ..... *P. sparsiflora*

Platanthera leucostachys Lindley. WHITE-FLOWERED BOG-ORCHID. Uncommon, moist meadows and bogs: Pitkin Marsh, *Rubtzoff* (1951 RSA). < 11300 ft. May-Aug. —*Habenaria l.* (Lindley) S. Watson.

Platanthera sparsiflora (S. Watson) Schlechter. SPARSE-FLOWERED BOG-ORCHID. Infrequent, boggy places, along streams: Pitkin Marsh, *Armstrong* (1929 NCC). < 11300 ft. June-Aug. —*Habenaria s.* S. Watson.

Spiranthes

1. Lip not narrowed below the tip, basal callosities evident . . . . *S. porrifolia*
1: Lip definitely narrowed below the tip; basal callosities not evident . . . . *S. romanzoffiana*

Spiranthes porrifolia Lindley. LADIES' TRESSES. Occasional, springy wet areas: Pitkin Marsh, *Stocking* (1967 NCC); Laguna de Santa Rosa ne. of Graton near Laguna Road, *Rubtzoff* (1960); w. of Santa Rosa in Laguna de Santa Rosa drainage, *Harrison* (1978 NCC); Bennett Mtn. Lake, *Rubtzoff* (1961). < 8300 ft. July-Aug.

Spiranthes romanzoffiana Chamisso & Schlechtendal. Common, wet meadows, banks: N. Fire Station, The Sea Ranch, *Wells* (1984); near Salmon Cr. Village, acc. Connors (1992); Pitkin Marsh, *Howell* (1936), *Rubtzoff* (1951), *Stocking* (1966 NCC); Perry Marsh, *Howell* (1945); Cunningham Marsh, *Howell* (1947); Fairfield Osborn (S); near Santa Rosa, *Congdon* (1880); Pepperwood. < 10100 ft. June-Aug.

POACEAE. GRASS FAMILY

(Key ± acc. A. Chase, *Munz* [1959]).

KEY TO TRIBES

1. Spikelets with glumes persistent, the rachilla articulated above them, 1-many-flowered; upper lemmas frequently empty; rachilla often prolonged beyond the upper lemma
   2. Spikelets borne in an open or spikelike raceme or panicle, usually upon distinct pedicels
      3. Spikelets with only 1 fertile lemma
         4. Spikelets with 2 sterile or ♂ lemmas below the fertile lemma; palea 1-nerved . . . . . *Phalarideae*
         4: Spikelets without sterile lemmas below the fertile lemma; palea 2-nerved . . . . *Agrostideae*
      3: Spikelets 2-many fertile lemmas
         5. Lemma usually shorter than the empty glumes; the awn dorsal and usually bent (spikelets sessile in *Gaudinia*) . . . . *Aveneae*
         5: Lemma usually longer than the empty glumes; the awn terminal and straight or none (rarely dorsal in *Bromus*;pedicels very short or lacking in *Brachypodium*) . . . . *Festuceae*
   2: The spikelets borne in 2 rows, sessile or nearly so
      6. Spikelets on 1 side of continuous axis, forming 1-sided spikes; spikes usually > 1 . . . . . *Chlorideae*
      6: Spikelets alternately on opposite sides of axis which is often articulated; spike terminal, single *Hordeae*
1: Spikelets falling from the pedicels entire, articulate below the glumes, naked or enclosed in bristles or burlike involucres, 1-flowered or, if 2-flowered, lower flower ♂, no lemmas empty; rachilla not extending beyond upper lemma
   7. Spikelets very flat, strongly laterally compressed, paniculate; glumes reduced or wanting; lemma and palea subequal, both keeled . . . . *Oryzeae*
   7: Spikelets mostly not strongly flattened; glumes, or at least 1, usually developed
      8. Lemma and palea hyaline, thin, much more delicate in texture than glumes
         9. Spikelets in pairs, 1 sessile, 1 pedicellate . . . . *Andropogoneae*
         9: Spikelets not in pairs . . . . *Agrostideae*
      8: Lemma, at least that of perfect flower, similar in texture to glumes, or thicker and firmer, never hyaline and thin . . . . *Paniceae*

KEY TO GENERA

Agrostideae

1. Rachilla articulate below the glumes, these falling with the spikelet
   2. Glumes long-awned . . . . *Polypogon*
   2: Glumes awnless
      3. Glumes united toward the base, ciliate on the keel; inflorescence not capitate and bracteate *Alopecurus*
      3: Glumes not united, glabrous; inflorescence capitate, in the axils of broad bracts . . . . *Crypsis*
1. Rachilla articulate above the glumes

4. Lemma indurate when mature and very closely embracing the grain; callus usually well-developed, bearded
  5. Awn 3-branched, the lateral branches sometimes obsolete (then no line of demarcation between lemma and awn) . . . . . . *Aristida*
  5: Awn simple, a line of demarcation between the awn and lemma
    6. Awn twisted and bent, persistent, several to many × longer than fruit
      7. Inflorescence wide, branches widely spreading to ± drooping; palea < 1/3 lemma . . *Nasella*
      7: Inflorescence narrow, branches usually ascending; palea > ½ to = lemma . . . *Achnatherum*
    6: Awn not twisted, deciduous, rarely > 3-4 × as long as fruit . . . . . . *Piptatherum*
4: Lemma usually hyaline or membranaceous at maturity; callus not well-developed
  8. Glumes mostly longer than lemma (sometimes not much longer in *Agrostis*)
    9. Panicle feathery, capitate, nearly as broad as long; spikelets woolly . . . . . . *Lagurus*
    9: Panicle not feathery; spikelets not woolly
      10. Glumes compressed-carinate, stiff-ciliate on keel; panicle dense, cylindric or ellipsoid *Phleum*
      10: Glumes not compressed-carinate, not ciliate
        11. Glumes saccate at base; lemma long-awned; panicle contracted, shining . . *Gastridium*
        11: Glumes not saccate at base; lemma awned or not; panicle open or contracted
          12. Floret bearing a tuft of hairs at the base from the short callus; palea well-developed, the rachilla prolonged behind the palea as a hairy bristle . . . . . . *Calamagrostis*
          12: Floret without hairs at base or with short hairs; palea usually small or obsolete . . . . . . *Agrostis*
  8: Glumes not longer than the lemma, usually shorter (except sometimes for the awn-tips)
    13. Floret with a tuft of hairs at the base from the short callus . . . . . . *Ammophila*
    13: Floret without hairs at base . . . . . . *Sporobolus*

Andropogoneae

1. Spikelets all alike, fertile, surrounded by soft hairs; inflorescence a narrow panicle . . . . . . *Imperata*
1: Spikelets unlike, the sessile perfect, the pedicellate ♂ or sterile
  2. Racemes of several joints silky-villous . . . . . . *Andropogon*
  2: Racemes reduced to 1 or few joints, these in a compound panicle, not silky villous . . . . . . *Sorghum*

Aveneae

1. Spikelets sessile appressed to the concave rachis . . . . . . *Gaudinia*
1: Spikelets otherwise
  2. Florets 2, one perfect, the other ♂
    3. Lower floret ♂, the awn twisted, geniculate, exserted . . . . . . *Arrhenatherum*
    3: Lower floret perfect, awnless; upper floret awned . . . . . . *Holcus*
  2: Florets 2 or more, all alike except the reduced upper ones
    4. Articulation below the glumes (above on *T. canescens*), the spikelets falling entire . . . . . . *Trisetum*
    4: Articulation above the glumes, the glumes similar in shape
      5. Lemmas bifid at apex, awned or mucronate between the lobes; spikelets several-flowered *Danthonia*
      5: Lemmas toothed but not bifid and awned or mucronate between the lobes
        6. Glumes 2-3.5 cm long, 7-9-nerved; spikelets 2-flowered, or with a rudimentary 3rd one, pendulous, annual . . . . . . *Avena*
        6: Glumes not > 1 cm long, 1-5-nerved; spikelets not pendulous, mostly perennial
          7. Lemmas keeled, the awn, when present, from above the middle . . . . . . *Koeleria*
          7: Lemmas convex, awned from below the middle
            8. Rachilla prolonged behind the upper floret; lemmas truncate and erose-dentate at summit . . . . . . *Deschampsia*
            8: Rachilla not prolonged; lemmas tapering into 2 slender teeth . . . . . . *Aira*

Chlorideae

1. Spikelets with > 1 perfect floret . . . . . . *Leptochloa*
1: Spikelets with only 1 perfect floret, though occasionally with additional imperfect florets above or below

2. Rachilla articulate below glumes, spikelets falling entire; spikes not digitate, but arranged on an elongate axis
   3. Glumes narrow, unequal . . . . . *Spartina*
   3: Glumes broad and boat-shaped, equal . . . . . *Beckmannia*
2. Rachilla jointed above the glumes; spikes digitate . . . . . *Cynodon*

Festuceae

1. Plants tall stout reeds with large plumelike panicles; lemmas or rachilla with long silky hairs as long as lemmas
   2. Leaves crowded at base of culms . . . . . *Cortaderia*
   2: Leaves distributed on the culm
      3. Lemmas naked; rachilla hairy . . . . . *Phragmites*
      3: Lemmas hairy; rachilla naked . . . . . *Arundo*
1: Plants low or to ca 1.5 m high, not reedlike
   4. Plants dioecious, perennial . . . . . *Distichlis*
   4: Plants not dioecious, except in some spp. of *Poa* (with villous lemmas)
      5. Spikelets of 2 forms, sterile and fertile intermixed; panicle dense, ± secund
         6. Fertile spikelets 2-3-flowered; sterile spikelets with numerous rigid awn-tipped lemmas; panicle dense, spikelike . . . . . *Cynosurus*
         6: Fertile spikelets with 1 perfect floret, long-awned; sterile spikelets with many obtuse sterile lemmas; panicle-branches short, nodding . . . . . *Lamarckia*
      5: Spikelets all alike in the same inflorescence
         7. Lemmas 3-nerved, the nerves prominent . . . . . *Eragrostis*
         7: Lemmas 5- to many-nerved, the nerves sometimes obscure
            8. Lemmas as broad as long, with outspread margins; florets closely imbricate, horizontally spreading . . . . . *Briza*
            8: Lemmas longer than broad, the margins clasping the palea; florets not horizontally spreading
               9. Lemmas keeled on back (sometimes ± rounded in *Poa*)
                  10. Spikelets strongly compressed, crowded in 1-sided clusters at ends of stiff naked panicle-branches . . . . . *Dactylis*
                  10: Spikelets not strongly compressed, not crowded in 1-sided clusters
                     11. Lemmas awned from a minutely bifid apex, or nearly awnless; spikelets large . . . . . *Bromus*
                     11: Lemmas awnless; spikelets small . . . . . *Poa*
               9: Lemmas rounded on back
                  12. Glumes papery; lemmas firm, strongly nerved, scarious-margined; upper florets sterile, often reduced to rudiments infolded by the broad upper lemmas . . . . . *Melica*
                  12: Glumes not papery; upper florets like the others
                     13. Nerves of lemma parallel, not converging at the summit or but slightly so
                        14. Spikelets in long loose racemes . . . . . *Pleuropogon*
                        14: Spikelets in open or contracted panicles
                           15. Sheath closed to near top; upper glume usually 1-veined . *Glyceria*
                           15: Sheath open to base; upper glume usually 3-veined
                              16. Lemmas faintly 5-veined; annual or perennial, rhizomes 0; open, saline or alkaline soils . . . . . *Puccinellia*
                              16: Lemmas clearly 7-9-veined; perennial from rhizomes; near freshwater . . . . . *Torreyochloa*
                     13: Nerves of lemma converging toward the summit, the lemmas narrowed at apex
                        17. Lemmas awned or awn-tipped from a minutely bifid apex (rarely awnless); palea adhering to the caryopsis
                           18. Spikelets in open to contracted panicles; stigmas borne at the sides of the summit of ovary . . . . . *Bromus*
                           18: Spikelets nearly sessile in strict racemes; stigmas terminal on the ovary . . . . . *Brachypodium*
                        17: Lemmas entire, pointed, awnless or awned from tip (*F. elmeri* minutely toothed at tip)

19. Lemmas awned (awnless or scarcely awned in *Festuca pratensis* and *F. viridula*)
    20. Plants perennial; stamens 3 . . . . *Festuca*
    20: Plants annual; stamen usually 1 . . . . *Vulpia*
19: Lemmas awnless (sometimes minutely awned in *F. pratensis* and *F. viridula*)
    21. Spikelets on slender pedicels in compound panicles; perennial *Poa*
    21: Spikelets on thick short pedicels in simple panicles; annual . . . . *Desmazeria*

Hordeae

1. Spikelets solitary at each node of the rachis (rarely 2 in ssp. of *Elytrigia*, but never throughout)
  2. Spikelets 1-flowered, sunken in hollows in the rachis; spikes slender, cylindric; plant low annuals
    3. Lemmas with awns; florets lateral to the rachis . . . . *Scribneria*
    3: Lemmas not awned; florets dorsiventral to the rachis
      4. First glume wanting; spikes 1-2 dm long . . . . *Hainardia*
      4: First glume present; spikes 0.7-1 dm long . . . . *Parapholis*
  2: Spikelets 2-several-flowered, not sunken in the rachis
    5. Spikelets placed edgewise to the rachis, the lateral ones with a single glume . . . . *Lolium*
    5: Spikelets placed flatwise to the rachis
      6. Plants perennial . . . . *Elytrigia*
      6: Plants annual
        7. Spikelets turgid or cylindric . . . . *Aegilops*
        7: Spikelets compressed . . . . *Triticum*
1. Spikelets normally > 1 at each node of the rachis
  8. Spikelets 3 at each rachis-node, 1-flowered, the lateral pair pedicelled, usually reduced to awns *Hordeum*
  8: Spikelets 2 or more (sometimes 1 in *Elymus*) at each node of rachis, alike, 2-6-flowered
    9. Lemmas 3-veined . . . . *Taeniatherum*
    9: Lemmas 5-veined
      10. Glumes ± elliptic to lanceolate; lemma awns 16-30 mm . . . . *Elymus*
      10: Glumes usually awl-like (except *L. mollis*); lemma awns usually 0-5 mm . . . . *Leymus*

Oryzeae

One genus . . . . *Leersia*

Paniceae

1. Spikelets subtended by bristles
  2. Bristles persistent, the spikelets deciduous . . . . *Setaria*
  2: Bristles falling with the spikelets at maturity . . . . *Pennisetum*
1: Spikelets not subtended by bristles
  3. Glumes awned or mucronate; apex of palea not enclosed by lemma . . . . *Echinochloa*
  3: Glumes awnless; apex of palea usually enclosed by lemma
    4. Spikelets in panicles
      5. Plants perennial
        6. Spikelets 1.5-2.5 mm; lower glume 0.5-1 mm . . . . *Dichanthelium*
        6: Spikelets 3-4 mm; lower glume 0; (several racemes = panicle) . . . . . . *Paspalum dilatatum*
      5: Plants annual . . . . *Panicum*
    4: Spikelets in 1-sided spikelike racemes
      7. Racemes slender 3-12 . . . . *Digitaria*
      7: Racemes stout, in pairs . . . . *Paspalum distichum*

Phalarideae

1. Lower floret ♂; spikelets brown, shining . . . . *Hierochloe*
1: Lower florets sterile; spikelets green or yellowish
  2. Lower florets not reduced to small scalelike lemmas
    3. Lower florets consisting of awned hairy lemmas exceeding fertile floret . . . . *Anthoxanthum*
    3: Lower florets consisting of awnless glabrous lemmas enclosing fertile floret . . . . *Ehrharta*
  2: Lower florets reduced to small awnless scalelike lemmas much smaller than the fertile florets . . *Phalaris*

Achnatherum lemmonii (Vasey) Barkworth. LEMMON'S NEEDLEGRASS. Uncommon, forest openings: 2760 ft. elev. Caldwell Ranch, acc. Clifton (1984); 3200 ft. elev. Caldwell Pines, acc. Clifton (1984). < 7700 ft. May-July. —*Stipa l.* (Vasey) Scribner.

★ **Aegilops triuncialis** Linnaeus. BARBED GOATGRASS. Noxious weed, overgrazed pasture land: Spring Lake, *Wells* (1987); confluence of Mark West and Leslie creeks, *de Nevers* (1989); River Road 6 mi. from Cloverdale, *Fuller* (1960); 3 locations in n. and e. part of county, acc. *Barbe* (1991). < 3300 ft. May. Introduced from Eurasia.

Agrostis

1. Plants annual; lemma with slender awn
  2. Spikelets ca 1.5 mm long; lemma awned below the tip . . . . *A. elliottiana*
  2: Spikelets at least 2.5 mm long; lemma awned from middle . . . . *A. microphylla*
1: Plants perennial; lemma awned or awnless
  3. Palea evident, 2-nerved, at least half as long as lemma
    4. Rachilla prolonged beyond the palea as a minute bristle . . . . *A. avenacea*
    4: Rachilla not so prolonged
      5. Glumes scabrous on the keel and on the back; panicle contracted, lobed, the short branches densely verticillate . . . . *A. viridis*
      5: Glumes scabrous on keel only; panicle open, or if contracted, not lobed or with densely verticillate branches
        6. Branches of panicle or some of them floriferous from base; ligule 3-6 mm long *A. stolonifera*
        6: Branches of panicle naked at base; ligule 1-2 mm long . . . . *A. capillaris*
  3: Palea obsolete or a minute nerveless scale, occasionally up to ca 0.5 mm long
    7. Plants spreading by creeping rhizomes
      8. Hairs at base of lemma 1-2 mm long; culms 6-9 dm tall . . . . *A. hallii*
      8: Hairs at base of lemma minute or none; culms mostly lower . . . . *A. pallens*
    7: Plants without rhizomes, stolons occasionally developed
      9. Panicle narrow, contracted, at least some of lower branches bearing spikelets from base
        10. Culms slender, mostly 1-2 dm tall, tufted, with leaves basal; blades not > 7 cm long, < 2 mm wide; panicles mostly < 5 mm wide . . . . *A. blasdalei*
        10: Culms stouter, taller, not in tufts with dense basal leaves; at least some of blades 8-10 cm long, 3-5 mm wide;glumes scabrous on keel
          11. Lemma acute, not toothed; palea minute . . . . *A. exarata*
          11: Lemma minutely 4-toothed; palea 1/4-1/3 as long as lemma . . . . *A. densiflora*
      9. Panicle open, occasionally diffuse; branches very slender, scabrous, lower branches not bearing spikelets at base . . . . *A. oregonensis*

★ **Agrostis avenacea** Gmelin. Uncommon, disturbed wet area, weedy: Bodega Head, *Connors* (1991 BML). < 1000 ft. June-July. Introduced from Polynesia.

Agrostis blasdalei A. Hitchcock. BLASDALE'S BENT GRASS. Occasional, coastal strands or near dunes: The Sea Ranch; 2.3 mi. s. of Stewarts Point, *Frankel* (1964); Duncans Landing, 4.6 mi. n. of Salmon Cr., *Crampton* (1957); Bodega Head, acc. Connors (1991). < 350 ft. May-July.

★ **Agrostis capillaris** Linnaeus. COLONIAL BENT. Occasional, probably escaping from cultivation: The Sea Ranch, *Wells* (1978); Kenwood, *Baker* (1944); Pepperwood, *Breedlove* (1980). < 3300 ft. July-Sept. Native of Europe. —*A. tenuis* Sibthorp.

Agrostis densiflora Vasey. Occasional, edges of open coastal grassland mesas: The Sea Ranch, *Wells* (1978); n. and s. of Stewarts Point, *Crampton* (1957 AHUC); between Goat Rock and Salmon Creek, *Crampton* (1957 AHUC); Bodega Point, *Eastwood* (1915); Bodega Head, acc. Connors (1991). < 700 ft. May-Aug. —*A. clivicola* Crampton. *A. californica* Trinius.

Agrostis elliottiana Schultes. Rare, vernal pool margins: Pepperwood, *D. Smith* (1993). < 1700 ft. Apr.-May. —*A. exigua* Thurber.

Agrostis exarata Trinius. SPIKE BENT. Common, streamsides, damp or wet places: 15 mi. e. of Stewarts Point, *Best* (1981); near Occidental, *Crampton* (1959); Bodega Head; Pitkin Marsh, *Rubtzoff* (1951); Cunningham Marsh, *Howell* (1947); Button Ranch on Walker Road (U. C. property) between Petaluma-Valley Ford Road and Pepper Road, *Connors* (1992 BML); w. of El Verano, *Best* (1980); Bennett Valley, *Heller* (1902), *Baker* (1931 NCC); College Ave. hwy. near top of hill, *Baker* (1939 NCC); 3 mi. n. of Santa Rosa, *Koch* (1938); Pepperwood; w. of Windsor, *Heller* (1902); Sausel Cr. n. of Alexander Valley on serpentine, *Hoover* (1966); Sulphur Cr. Road to The Geysers Resort, *Koch* (1940 NCC); 2300 ft. elev. upper Big Sulphur Cr. drainage at Little Geysers, acc. Stone (1983); 2400 ft. elev. Bear Cr. Can., *Clifton* (1984 PUA). < 6700 ft. June-Aug. —*A. ampla* A. Hitchcock. *A. e.* var. *pacifica* Vasey.

Agrostis hallii Vasey. Occasional, dry woods: Cazadero quad.; 1024 ft. elev. Coleman Valley Road, *Clifton* (1984 PUA); 5 mi. w. of Sebastopol, *Wagnon* (1939); Button Ranch on Walker Road (U. C. property) between Petaluma-Valley Ford Road and Pepper Road, acc. Connors (1992); Santa Rosa Cr. Can., *Baker* (1941 NCC); 3290 ft. elev. Socrates Mine, *Clifton* (1984 PUA). < 5000 ft. May-July.

Agrostis microphylla Steudel. Occasional, vernal situations: 6½ mi. n. of Sonoma in vernal pool, *Crampton* (1966); vernal pool in Bennett Valley, *Wagnon* (1939); 2400 ft. elev. Little Geysers, *Malloch* (1977 PG&E). < 2500 ft. May-July.

Agrostis oregonensis Vasey. OREGON REDTOP. Rare, wet places: Pitkin Marsh, *Rubtzoff* (1951). < 7000 ft. June-Aug.

Agrostis pallens Trinius. THIN GRASS. DUNE BENT. Common, dunes, woods, chaparral, meadows, sometimes on rhyolite tuff: The Sea Ranch; Bodega Head, *Connors* (1991 BML); 2 mi. ne. of Occidental, *Ferris & Lorraine* (1961); Pitkin Marsh, *Howell* (1945); Button Ranch on Walker Road (U. C. property) between Petaluma-Valley Ford Road and Pepper Road, acc. Connors (1992); Cavedale Road, acc. Edwards (1990); ne. slopes of Fitch Mtn. near Healdsburg, *Rubtzoff* (1952). < 11800 ft. Apr.-Aug. —*A. diegoensis* Vasey.

★ **Agrostis stolonifera** Linnaeus. CREEPING BENT. Escape, waste places, meadows, along roads: The Sea Ranch, *Wells* (1985); Bodega Marine Reserve, *Connors* (1991 BML); Freestone, *Huntington* (1987 NCC); 98 ft. Atascadero Cr., *Clifton* (1981 PUA); Vine Hill region, *Baker* (1946 NCC); Pitkin Marsh, *Rubtzoff* (1951 NCC); Cunningham Marsh, *Howell* (1947); 5th St., Petaluma, *Morgan* (1955); Ducker Cr., Rincon Valley, *Wells* (1985); Santa Rosa Jr. Coll., *Wagnon* (1939 NCC); 2530 ft. elev. Caldwell Ranch, *Clifton* (1984 PUA). < 3300 ft. June-Sept. Introduced from Europe. —*A. alba* Linnaeus var. *palustris* (Hudson) Persoon. *A. a.* var. *major* (Gaudin) Farwell.

★ **Agrostis viridis** Gouan. WATER BENT. Common, weedy, low, moist places: Shell Beach ¼ mi. s. of Goat Rock, *Morgan* (1950); Button Ranch on Walker Road (U. C. property) between Petaluma-Valley Ford Road and Pepper Road, acc. Connors (1990); Monte Rio, *Stacy* (1923); along Copeland Cr., Rohnert Park, *Self* (1979 NCC); Spring Lake; 3 mi. n. of Santa Rosa, *Koch* (1939 NCC). < 6700 ft. Native of Europe. —*A. semiverticillata* (Forsskal) C. Christensen.

Aira

1. Panicle dense, spikelike . . . . *A. praecox*

1: Panicle open

  2. Lower floret with awn as long as that of upper floret . . . . *A. caryophyllea*

  2: Lower floret awnless or nearly so . . . . *A. elegantissima*

★ **Aira caryophyllea** Linnaeus. SILVER EUROPEAN HAIRGRASS, SHIVER GRASS. Common, open places: The Sea Ranch; Salt Point, *Wells* (1982); Bodega Head; 1024 ft. elev. Coleman Valley Road, *Clifton* (1984 PUA); Pitkin Marsh, *Rubtzoff* (1951); w. of Santa Rosa in Laguna de Santa Rosa drainage, *Harrison* (1978 NCC); Fairfield Osborn (H); Button Ranch on Walker Road (U. C. property) between Petaluma-Valley Ford Road and Pepper Road, acc. Patterson (1990); w. of Sears Point Raceway in Tolay Cr. drainage, acc. Patterson (1990); Bouverie; Mt. Hood, *Baker* (1947 NCC); Pepperwood; chaparral area sw. of Bradford Mtn. 1½ mi. w. of Dry Cr., acc. J. & B. Guggolz (1992); Lake Sonoma area; 3200 ft. elev. Socrates

Mine, *Clifton* (1984 PUA); ca 2000 ft. elev. Anderson Trail on Black Mtn. between Geyser Peak and upper end of Pine Flat Road, acc. J. & B. Guggolz (1992); 2040 ft. elev. Squaw Cr., acc. Clifton (1984); 2440 ft. elev. Caldwell Ranch, acc. Clifton (1984); 3000 ft. elev. Caldwell Pines, acc. Clifton (1984). < 6300 ft. Apr.-June. Introduced from Europe.

★ **Aira elegantissima** Schur. ELEGANT EUROPEAN HAIRGRASS. Occasional, open ground: w. of Vine Hill School, *Wagnon* (1939); Chileno Valley Road, *Howell* (1944); 7 mi. from Petaluma on road to Marshall, *Koch* (1939 NCC); Bouverie; Spring Lake; Pepperwood. Introduced from Europe. < 1000 ft. April. —*A. capillaris* Host. *A. elegans* Gaudin.

★ **Aira praecox** Linnaeus. Uncommon, sandy places: The Sea Ranch; bluffs at Salt Point, *Wagnon* (1989 NCC); s. side of mouth of Russian River, *Best* (1980); Bodega Head, acc. Connors (1991). < 350 ft. May-July. Introduced from Europe.

Alopecurus (Key ± acc. W. J. Crins, *Hickman,* [1993]).

1. Spikelets excluding awns > 4 mm
    2. Plants perennial; stem usually > 4.5 dm; glumes, lemma acute . . . . *A. pratensis*
    2: Plants annual; stem usually < 4.5 dm; glumes, lemma acute or obtuse . . . . *A. saccatus*

1: Spikelets excluding awns < 4 mm
    3. Awn straight, > lemma by 0-2 mm . . . . *A. aequalis*
    3: Awn bent, > lemma by 1.5-5 mm
        4. Spikelets > 3 mm; upper leaf-sheath obviously inflated; awn > lemma by 3-5 mm . . . *A. saccatus*
        4: Spikelets usually < 3 mm; upper leaf-sheath not inflated; awn > lemma by 1.5-3.5 mm *A. geniculatus*

Alopecurus aequalis Sobolevski var. **aequalis**. SHORT-AWN FOXTAIL. Common, low damp areas: Duncans Mills Marsh, *Rubtzoff* (1955); Guerneville Marsh, *Rubtzoff* (1955); Occidental Marsh, *Rubtzoff* (1957); Freestone Marsh, *Rubtzoff* (1962); Ross Marsh ca 1¼ mi. ssw. of Forestville, *Rubtzoff* (1955); Pitkin Marsh, *Rubtzoff* (1951); Cunningham Marsh n. of Blucher Cr., *Howell* (1947); Bloomfield, *Congdon* (1880); Llano Road near Hwy. 116 nw. of Cotati, acc. Lovell (1977); Bennett Mtn. Lake, *Rubtzoff* (1955). < 11700 ft. May-July. —*A. a.* var. *sonomensis* Rubtzoff, not clearly distinct.

Alopecurus geniculatus Linnaeus. WATER FOXTAIL. Uncommon, damp areas: Willow Cr. Marsh, *Rubtzoff* (1959). < 5300 ft. June-July.

★ **Alopecurus pratensis** Linnaeus. MEADOW FOXTAIL. Uncommon, fields, waste places: The Sea Ranch, *Wells* (1978); Button Ranch on Walker Road (U. C. property) between Petaluma-Valley Ford Road and Pepper Road, *Connors* (1992 BML). < 5700 ft. May-June. Introduced from Eurasia.

Alopecurus saccatus Vasey. Occasional, along borders of vernal ponds in annual grassland: 2 sites w. of Santa Rosa in Laguna de Santa Rosa drainage, *Harrison* (1978 NCC); near Aurora School, *Howell* (1946); 6½ mi. ne. of Penngrove, *Rubtzoff* (1969); Santa Rosa, *Heller* (1902); 850 ft. elev. Shiloh Ranch in valley of small tributary in Mark West Cr. drainage, 2.3 airmiles ne. of jct. of Shiloh and Faught roads, 0.6 mi. w. of Mark West Springs, acc. Stone (1985). < 2300 ft. Mar.-June. —*A. howellii* Vasey.

★ **Ammophila arenaria** (Linnaeus) Link. EUROPEAN BEACHGRASS. Naturalized on dunes where it was planted as a sand binder; outcompeting native dune flora, reducing dunes to floristic redundancy: The Sea Ranch. *J. H. Thomas* (1976); Salt Point; mouth of Russian River, *Best* (1981); Salmon Cr. dunes, *Morgan* (1955); Bodega Dunes, *Connors* (1990 BML). < 100 ft. May-Aug. Introduced from coast of Europe.

Andropogon (Key ± acc. K. W. Allred, *Hickman*, [1993]).

1. Ligule > 1 mm; lower glume-keel scabrous in lower half; leaf-sheath minutely scabrous . . . . *A. glomeratus* var. *scabriglumis*

1: Ligule < 1 mm; lower glume-keel glabrous in lower half; leaf-sheath glabrous . . . . *A. virginicus*

★ **Andropogon glomeratus** (Walter) Britton, Sterns & Poggenburg var. **scabriglumis** C. S. Campbell. BUSHY BEARD GRASS.Occasional, low, moist ground: Guerneville, *Hoffman* (1943); Vine Hill, *Bufford* (1936 NCC); Geyser geothermal area, *Quibell* (1980 NCC); The Geysers Resort, *Wagnon* (1940 NCC). < 2000 ft. Sept.-Mar. Introduced from southern California.

★ **Andropogon virginicus** Linnaeus. BROOMSEDGE, BLUESTEM. Occasional, open, damp areas: Sweetwater Springs Road, acc. Edwards (1986); Pitkin Marsh, *Rubtzoff* (1951); Cunningham Marsh, *Howell* (1947); 2 mi. se. of Forestville, *Yates* (1936); 2300 ft. elev. Little Geysers in Big Sulphur Cr. watershed upstream from The Geysers at edge of hot spring, *Malloch* (1976 PG&E). < 2500 ft. Sept.-Jan. Native to e. U.S.

Anthoxanthum

1. Plants perennial . . . . *A. odoratum*
1: Plants annual . . . . *A. aristatum*

★ **Anthoxanthum aristatum** Boissier. VERNAL GRASS. Occasional, waste places: Duncans Mills quad.; Freestone, *Huntington* (1987 NCC); Piner School w. of Santa Rosa, *Harrison* (1979 SRJC); Pepperwood; 2 mi. n. of Windsor, *Eastwood & Howell* (1936), *Koch* (1940 NCC); half way between Windsor and Healdsburg, *Beane & James* (1949); this grass is very common in vernally wet areas of n. Santa Rosa, and still exists abundantly in Arata Lane vernal pool, which is very possibly the *Beane & James* site of 1949 previously listed, acc. Harrison (1992); 2040 ft. elev. Squaw Cr., acc. Clifton (1984); 2720 ft. elev. Caldwell Ranch, acc. Clifton (1984); 3000 ft. elev. Caldwell Pines, acc. Clifton (1984). < 3300 ft. May-June. Introduced from Europe.

★ **Anthoxanthum odoratum** Linnaeus. SWEET VERNAL GRASS. Common, pastures, lawns, forest openings, waste places: The Sea Ranch; 7 mi. s. of Gualala, *Howell* (1976); Stewarts Point-Skaggs Springs Road 3 mi. e. of Stewarts Point, *Best* (1981); 2 mi. s. of Stewarts Point on Hwy. 1, *Porter* (1967); Salt Point; Duncans Mills Marsh, *Rubtzoff* (1955); between Bodega Bay and Salmon Cr., *Connors* (1991 BML); Hwy. 116 four mi. w. of Monte Rio, *Morgan* (1961); Fife Cr. Marsh near Guerneville, *Rubtzoff* (1960); Blank Road near Gravenstein Hwy., acc. J. & B. Guggolz (1991). < 3700 ft. May-June. Introduced from Eurasia.

Aristida oligantha Michaux. PRAIRIE THREE-AWN, OLDFIELD THREE-AWN. Occasional, dry areas: Pepperwood; e. side of geysers Unit 18, *Malloch* (1977 PG&E). < 3300 ft. Aug.-Sept.

★ **Arrhenatherum elatius** (Linnaeus) J. S. & C. Presl. TALL OAT GRASS. Infrequent, escape from cultivation: Bohemian Hwy. 2 mi. s. of Occidental, *Morgan* (1956). < 6000 ft. May-June. Introduced from Old World.

★ **Arundo donax** Linnaeus. GIANT REED. Occasional, moist places, ditches, streams, seeps: Stump Beach at Salt Point State Park, *Wells* (1982); Adobe Road near jct. of Davis Road e. of Penngrove, acc. Knight (1990); Bouverie; Ducker Cr., Rincon Valley, *Wells* (1985); Spring Lake. < 1700 ft. Mar.-Sept. Introduced from warm areas of the Old World.

Avena

1. Teeth of lemmas awned or setaceous; pedicels capillary . . . . *A. barbata*
1: Teeth of lemmas acute, not setaceous; pedicels stoutish
    2. Lemmas pubescent with long brown hairs; spikelets generally 3-flowered . . . . *A. fatua*
    2: Lemmas subglabrous; spikelets mostly 2-flowered . . . . *A. sativa*

★ **Avena barbata** Link. SLENDER WILD OAT. Common, weed, rock outcrops, wooded areas, waste places: The Sea Ranch; Bodega Head; many sites w. and sw. of Santa Rosa in Laguna de Santa Rosa drainage, *Harrison* (1978 NCC); Todd Road Preserve, *Howell & Knight* (1987); Fairfield Osborn (H); Tresch Ranch, Walker Road near Two Rock, acc. J. & B. Guggolz (1992); w. of Sears Point Raceway in Tolay Cr. drainage, acc. Patterson (1990); Bouverie; Rincon Valley, *Wells* (1984); Spring Lake; Pepperwood; chaparral area sw. of Bradford Mtn. 1½ mi. w. of Dry Cr., acc. J. & B. Guggolz (1992); Lake Sonoma area; Geyserville and Ida Clayton Road quads.; 3200 ft. elev. Socrates Mine, acc. Clifton (1984); ca 2000 ft. elev. Anderson Trail on Black Mtn. between Geyser Peak and upper end of Pine Flat Road, acc. J. & B. Guggolz (1992); 3600 ft. elev. Geyser Rock, acc. Clifton (1983); 2040 ft. elev. Squaw Cr., acc. Clifton (1984). < 4000 ft. Mar.-June. Native of Old World.

★ **Avena fatua** Linnaeus. WILD OAT. Common, waste and cultivated places: The Sea Ranch; Willow Creek; opposite Sebastopol cemetery, *Wagnon* (1939 NCC); 0.6 mi. from Gravenstein Hwy. on Llano Road, *Howell & Knight* (1987); Todd Road Preserve; w. of Sonoma Lane, Rohnert Park, *O'Toole* (1979 NCC); Copeland Cr. at Sonoma State Univ. campus, *Self* (1979 NCC); Fairfield Osborn (H); w. of Sears Point Raceway in Tolay Cr. drainage, acc. Patterson (1990); Lafferty Ranch at Sonoma Mtn., acc. Arnold et al. (1971); Mountain View Ave., Petaluma, *Morgan* (1956); Bouverie; Sugarloaf Ridge; Spring Lake; Pepperwood; 3 mi. s. of Healdsburg, *Heller* (1902); Mill Cr. Road, *Knight* (1964); Lake Sonoma area. < 4100 ft. Apr.-June. Introduced from Europe.

Avena sativa Linnaeus. CULTIVATED OAT. Occasional, escaping from cultivation: The Sea Ranch, *Wells* (1978); Salt Point; E. Cotati Ave. near Sonoma State Univ., *Knight & Howell* (1987); Watmaugh Road ½ mi. w. of Arnold Drive, *Morgan* (1956); Spring Lake. < 3700 ft. Apr.-June. Introduced from Europe.

Beckmannia syzigachne (Steudel) Fernald. AMERICAN SLOUGH GRASS. Common, muddy flats, marshes, streambanks: Cunningham Marsh, *Howell* (1947); upper Stemple Cr. Marsh, *Rubtzoff* (1961); Laguna de Santa Rosa, *Rubtzoff* (1961); Atascadero Cr. Marsh, *Rubtzoff* (1958); 2 mi. e. of Sebastopol, *Bufford* (1936 NCC); 2 sites w. of Santa Rosa in Laguna de Santa Rosa, *Harrison* (1978 NCC); American Cr. s. of Bloomfield, *Rubtzoff* (1962 NCC); Chileno Valley, *Rubtzoff* (1962); 1 mi. s. of Kenwood on Hwy. 12, *Morgan* (1956); Bennett Mtn. Lake, *Rubtzoff* (1961); Wright School, *Baker* (1929 NCC). < 5000 ft. May-July. —*B. erucaeformis* Calif. refs, misapplied.

Brachypodium

1. Plants perennial; from rhizomes; leaf blades 7-30 cm long . . . . *B. phoenicoides*
1: Plants annual; leaf blades 1.5-8 cm long . . . . *B. distachyon*

★ **Brachypodium distachyon** (Linnaeus) Beauvois. Occasional moist disturbed areas: Duncan Mills quad; hills e. of Agua Caliente, *Howell* (1948); Rincon Valley, *Wells* (1984); Spring Lake, *Wagnon* (1988); Pepperwood. <1800 ft. Apr-May. Native of sw. Europe.

★ **Brachypodium phoenicoides** (Linnaeus) Roemer & Schultes. Uncommon: The Sea Ranch; between Salmon Cr. and Bodega Dunes Campground, *Connors* (1991 BML); < 1000 ft. Apr.-May. Native of sw. Europe in the Mediterranean region.

Briza

1. Spikelets 4-5 mm wide, ca 3 mm long; panicle erect . . . . *B. minor*
1: Spikelets 10 mm wide, ca 12 mm long, panicle drooping . . . . *B. maxima*

★ **Briza maxima** Linnaeus. RATTLESNAKE GRASS, BIG QUAKING GRASS. Common, open areas: The Sea Ranch; Salt Point; Bodega Head; Pitkin Marsh, *Rubtzoff* (1952 NCC); Sebastopol, *Bately* (1932 NCC); Vine Hill Schoolhouse, *Baker* (1946 (NCC); w. of Graton, *Davis* (1975); w. of Santa Rosa in Laguna de Santa Rosa drainage, *Harrison* (1978 NCC); Lafferty Ranch at Sonoma Mtn., acc. Arnold et al. (1971); Button Ranch on Walker Road (U. C. property) between Petaluma-Valley Ford Road and Pepper Road, acc. Patterson (1990); w. of Sears Point Raceway in Tolay Cr. drainage, acc. Patterson (1990); Bouverie; 1½ mi. up Cavedale Road, *Utech & Knight* (1986 CM); Spring Lake; Pepperwood; ½ mi. e. of Geyserville on Hwy. 128, *Parsons* 1983 (SFSU); chaparral area sw. of Bradford Mtn. 1½ mi. w. of Dry Cr., acc. J. & B. Guggolz (1992); Lake Sonoma area; ca 2000 ft. elev. Anderson Trail on Black Mtn. between Geyser Peak and upper end of Pine Flat Road, acc. J. & B. Guggolz (1992); 2040 ft. elev. Squaw Cr., acc. Clifton (1984); 2440 ft. elev. Caldwell Ranch, acc. Clifton (1984). < 2700 ft. Apr.-July. Native of Europe.

★ **Briza minor** Linnaeus. LITTLE QUAKING GRASS. Common, waste places, fields: The Sea Ranch; Bodega Head; Freestone, *Huntington* (1987 NCC); Todd Road Preserve; Roblar Road near Stony Point, *Knight* (1963); Fairfield Osborn (H); Lafferty Ranch at Sonoma Mtn., Arnold et al. (1971); Tresch Ranch, Walker Road near Two Rock, acc. J. & B. Guggolz (1992); w. of Sears Point Raceway in Tolay Cr. drainage, acc. Patterson (1990); Bouverie; 1½ mi. up Cavedale Road, *Utech & Knight* (1986 CM); near Kenwood, *Howell* (1935); Sugarloaf Ridge; Spring Lake; Santa Rosa, *R. F. Smith* (1974 NCC); Pepperwood; ½ mi. e. of Geyserville Hwy. 128, *Parsons* (1983 SFSU); chaparral area sw. of Bradford Mtn. 1½ mi. w. of Dry Cr., acc. J. & B. Guggolz (1992); Lake Sonoma area; ca 2000 ft. elev. Anderson Trail on Black Mtn. between Geyser Peak and upper end of Pine Flat Road, acc. J. & B. Guggolz (1992); 2840 ft. elev. Bear Cr. Can., acc. Clifton (1984); 2040 ft. elev. Squaw Cr., acc. Clifton (1984); 2720 ft. elev. Caldwell Ranch, acc. Clifton (1984). < 3000 ft. Apr.-July. Native of Europe.

Bromus (Key ± acc. *Munz* [1959], of partial thesis by H. K. Wagnon, and ± acc. D. H. Wilken and E. L. Painter, *Hickman* [1993], with subsequent reevaluation by Peter G. Connors).

1. Plants annual
  2. Spikelets strongly compressed, the lemmas keeled, terminal teeth not > 0.5 mm long
    3. Lemmas awnless or the awn < 3 mm long (sometimes biennial) . . . . *B. catharticus*
    3: Lemmas with awns 7-15 mm long
      4. Panicle with spreading or drooping branches; lemma 1.6-2 mm wide . *B. carinatus* var. *carinatus*
      4: Panicle with stiff ascending branches; lemmas 2.5 mm wide . . . . *B. stamineus*
  2: Spikelets terete or slightly compressed, the lemmas slightly compressed or with back rounded, their terminal teeth mostly 0.6-5 mm long
    5. Lemmas broad, not acuminate, the teeth mostly < 1 mm long; 1st glume 3-5-nerved
      6. Panicles compact, the spikelets mostly longer than the pedicels; lemmas scarious, prominently nerved, unequal,the lower longer than the upper . . . . *B. hordeaceus*
      6: Panicles more open, spikelets mostly not longer than pedicels; lemmas ± coriaceous, not prominently veined . . . . *B. japonicus*
    5: Lemmas narrow, elongate, tapering at the tip, the teeth 2-5 mm long; 1st glume 1-nerved
      7. Panicle erect, contracted; awn 1-2 cm long
        8. Culms glabrous below the slightly open panicle; sheaths mostly smooth . . . . *B. madritensis* ssp. *madritensis*
        8: Culms pubescent below the dense panicle; sheaths pubescent . . . . *B. m.* ssp. *rubens*
      7: Panicle open, with spreading or drooping branches
        9. Awn 12-14 mm long; lemmas ca 1 cm long; spikelets 1-2 cm long
          10. Spikelets villous . . . . *B. tectorum* var. *tectorum*
          10: Spikelets glabrous . . . . *B. t.* var. *glabratus*
        9: Awns 20-50 mm long; lemmas 1.7-3 cm long; spikelets 2.5-4 cm long
          11. Lemmas 25-30 mm long; awns 3.5-5 cm long; 1st glume ca 16-20 mm long *B. diandrus*
          11: Lemmas 17-20 mm long; awns 2-3 cm long; 1st glume 7-9 mm long . . . . . *B. sterilis*
1: Plants perennial or biennial
  12. 1st glume 3-7-nerved
    13. Spikelets strongly compressed, lemmas keeled
      14. Fruit compressed; lemma wing between outer vein and margin ± 1 mm wide . *B. stamineus*
      14: Fruit cylindric; lemma wing < 0.5 mm wide
        15. Spikelets not crowded or overlapping; panicle branches > spikelets; widespread . . . . *B. carinatus* var. *carinatus*
        15: Spikelets crowded, overlapping; panicle branches < spikelets, coastal . . . . *B. c.* var. *maritimus*
    13: Spikelets terete or lightly compressed; lemmas not keeled
      16. 2nd glume 5-nerved . . . . *B. laevipes*
      16: 2nd glume 3-nerved . . . . *B. orcuttianus*
  12: 1st glume 1-nerved . . . . *B. vulgaris*

Bromus carinatus Hooker & Arnott var. **carinatus**. CALIFORNIA BROME. Common, open, brushy or wooded areas, waste places: The Sea Ranch; Bodega Head; Coleman Valley, *Koch* (1940 NCC); Vacation Beach near Guerneville, *Rubtzoff* (1953); Guerneville Marsh, *Rubtzoff* (1954); Pitkin Marsh, *Rubtzoff* (1952); Tresch Ranch, Walker Road near Two Rock, acc. J. & B. Guggolz (1992); Alexander Valley quad. < 11700 ft. Apr.-July. —*B. marginatus* Nees.

Bromus carinatus var. **maritimus** (Piper) C. Hitchcock. Occasional, dunes, ocean bluffs: Stewarts Point, Duncans Mills and Arched Rock quads.; Bodega Point, *Eastwood* (1915); Bodega Dunes, *Connors* (1990 BML); dunes between Bodega Bay and Bodega Head, *Raven* (1963). < 350 ft. Apr.-July. —*B. m.* (Piper) A. Hitchcock.

★ **Bromus catharticus** Vahl. RESCUE GRASS. Uncommon, dunes, waste places: near Bodega Head, *Best* (1982); Cotati, *Tunison* (1968 NCC); Spring Lake; Pepperwood. < 5000 ft. Apr.-Nov. Native of S. America. —*B. willdenovii* Kunth.

Bromus diandrus Roth. Ripgut Grass. Abundant, weed, many habitats:The Sea Ranch; Stump Beach at Salt Point State Park, *Wells* (1982); Bodega Head; Freestone, *Huntington* (1987 NCC); 4 sites w. and sw. of Santa Rosa in Laguna de Santa Rosa drainage, *Harrison* (1978 NCC); 0.6 mi. n. on Llano Road from Gravenstein Hwy., *Howell & Knight* (1987); Todd Road Preserve (H&K); between Cotati and Santa Rosa, *Rosen* (1950); s. side of Copeland Cr. (near Sonoma State Univ. campus), *Self* (1979 NCC); Fairfield Osborn (H); w. of Sears Point Raceway in Tolay Cr. drainage, acc. Patterson (1990); Bouverie; Sugarloaf Ridge; Gates Road off Calistoga Road, acc. Wells et al. (1984); Rincon Valley, *Wells* (1984); Spring Lake; Santa Rosa, *Heller* (1902); Pepperwood; between Santa Rosa and Calistoga, *J. H. Thomas* (1966); Lake Sonoma area; ca 2000 ft. elev. Anderson Trail on Black Mtn. between Geyser Peak and upper end of Pine Flat Road, acc. J. & B. Guggolz (1992). < 6700 ft. Apr.-June. Naturalized from Europe. —*B. rigidus* Roth misapplied.

★ **Bromus hordeaceus** Linnaeus. SOFT CHESS. Common, weed, waste places, open grassland, about settlements: The Sea Ranch; Salt Point; Willow Creek; Bodega Head; Freestone, *Huntington* (1987 NCC); Fife Cr. Marsh near Guerneville, *Rubtzoff* (1960); 4 mi. w. of Sebastopol, *Morgan* (1961); Todd Road Preserve; ne. corner Llano and Todd roads, *Howell & Knight* (1987); Fairfield Osborn (H); Button Ranch on Walker Road (U. C. property) between Petaluma-Valley Ford Road and Pepper Road, acc. Patterson (1990); w. of Sears Point Raceway in Tolay Cr. drainage, acc. Patterson (1990); Bouverie; Gates Road off Calistoga Road, acc. Wells et al. (1984); Rincon Valley; *Wells* (1984); Santa Rosa Cr., *Baker* (1930 NCC); Spring Lake; Pepperwood; chaparral area sw. of Bradford Mtn. 1½ mi. w. of Dry Cr., acc. J. & B. Guggolz (1992); Lake Sonoma area; 3200 ft. elev. Socrates Mine, acc. Clifton (1984); 3600 ft. elev. Geyser Rock, acc. Clifton (1984); 3300 ft. elev. Squaw Cr., acc. Clifton (1983); 2440 and 2720 ft. elev. Caldwell Ranch, acc. Clifton (1984). < 9000 ft. May-July. Naturalized from Eurasia. —*B. mollis* Linnaeus and *B. racemosus* Linnaeus, both misapplied.

★ **Bromus japonicus** J. A. Murray. HAIRY CHESS. Occasional, grassy meadows, waste ground: Freestone Marsh, *Huntington* (1987 NCC); Atascadero Cr. Marsh, *Rubtzoff* (1962); Cunningham Marsh, *Howell* (1947); Healdsburg quad. < 7700 ft. May-July. Native of Eurasia. —*B. commutatus* Schrader.

Bromus laevipes Shear. Common, sometimes on serpentine outcrops: 2 mi. ne. of Occidental, *Baker* (1947 NCC); Bennett Valley Road near Grange Road jct., *Morgan* (1950); Spring Lake; Pepperwood; Fitch Mtn. near Healdsburg, *Rubtzoff* (1952); Stewarts Point Skaggs Springs Road 15 mi. w. of US 101, *Best* (1981); Geyserville and Alexander Valley quads.; 3200 ft. elev. Socrates Mine, *Clifton* (1984 PUA). < 8600 ft. May-Aug. —*B. pseudolaevipes* Wagnon.

Bromus madritensis Linnaeus ssp. **madritensis**. FOXTAIL CHESS. Occasional, open woods, waste places: Ida Clayton Road and Cazadero quads. < 7300 ft. Apr.-June. Introduced from Europe.

★ **Bromus madritensis** ssp. **rubens** (Linnaeus) Husnot. FOXTAIL CHESS. Common, grassy slopes, waste places: Salt Point; Mt. Hood, *Baker* (1947 NCC); Calistoga Road, *Baker* (1931 NCC); Annadel, *Wells* (1984); hills near Santa Rosa, *Heller* (1902); Spring Lake; Lake Sonoma area; Geyserville, Alexander Valley and Ida Clayton Road quads.; 3100 ft. elev. w. flank Mt. St. Helena, *Grummer* (1981 PUA); 3200 and 3290 ft. elev. Socrates Mine, acc. Clifton (1984); ca 2000 ft. elev. Anderson Trail on Black Mtn. between Geyser Peak and upper end of Pine Flat Road, acc. J. & B. Guggolz (1992); 3200 ft. elev. geysers Unit 17, *Malloch* (1977 PG&E); 2440 ft. elev. Geyser Rock, acc. Clifton (1984); 3120 ft. elev. Bear Cr. Can., acc. Clifton (1984); 3060 ft. elev. Squaw Cr., acc. Clifton (1983); 2440 ft. elev. Caldwell Ranch, acc. Clifton (1984); Sulphur Cr. Road en route to The Geysers, *Koch* (1940 NCC). < 7300 ft. Mar.-June. Naturalized from Europe. —*B. rubens* Linnaeus.

Bromus orcuttianus (Shear) A. Hitchcock. ORCUTT BROME. Rare, open woods: Fort Ross and Healdsburg quads. < 9700 ft. June-Sept.

★ **Bromus stamineus** Desvaux. Common, weed: 125 ft. elev. Goat Rock Beach, *Woodhouse* (1975 PUA); Fife Cr. Marsh near Guerneville, *Rubtzoff* (1962); Odd Fellows Park at Hilton between Guerneville and Hacienda, *Howell* (1959); 1 mi. nw. of Freestone, *Rose* (1952); Pitkin Marsh, *Rubtzoff* (1952 NCC); Sonoma State Univ. campus, *Howell* (1986), *Schneider* (1986 NCC); ½ mi. s. of Sonoma State Univ. Hospital on Arnold Drive, *Morgan* (1956); 1 mi. e. of Schellville railroad station, *Morgan* (1956); 2 mi. se. of Sonoma on road to Napa, *Howell* (1955). < 700 ft. May-July. Introduced from South America.

★ **Bromus sterilis** Linnaeus. STERILE BROME. Occasional, waste places: Cazadero quad.; Fairfield Osborn (H); s. of Sonoma, *Howell* (1949); 4 mi. sw. of Petaluma on D St. Ext., *Morgan* (1956); Spring Lake; confluence of Dry and Galloway creeks, *Howell* (1967); Lake Sonoma area; Geyserville quad.; 2040 ft. elev. Squaw Cr., acc. Clifton (1984). < 3700 ft. Apr.-June. Introduced from Eurasia.

★ **Bromus tectorum** Linnaeus var. **glabratus** Shear. Uncommon, weed: quarry on Wagnon Road, w. of Sebastopol, *Wagnon* (1940 NCC); Fairfield Osborn (H); Spring Lake. < 7300 ft. May-June. Introduced from Europe.

★ **Bromus tectorum** var. **tectorum**. DOWNY CHEAT. Occasional, weedy, waste places: Cazadero quad.; w. of Santa Rosa in Laguna de Santa Rosa drainage, *Harrison* (1978 NCC); 2860 ft. elev. Pine Flat Road, acc. Clifton (1984); 3200 ft. elev. Socrates Mine, acc. Clifton (1984); 3300 ft. elev. Squaw Cr., acc. Clifton (1983). < 7300 ft. May-June. Native of Eurasia.

Bromus vulgaris (Hooker) Shear. Occasional, wooded, shaded places: The Sea Ranch; between Stewarts Point and Annapolis, *Kennedy* (1915); between Plantation and Hwy. 1, *Koch* (1940 NCC); Joy Road 1½-2 mi. from jct. with Bodega Hwy., *Ferris & Lorraine* (1961); s. of Guerneville, *Rubtzoff* (1953); 5 mi. w. of Sebastopol, *Wagnon* (1939 NCC). < 6300 ft. May-Aug.

Calamagrostis

1. Awn longer than the glumes, geniculate
    2. Panicle open, the branches spreading, naked below . . . . *C. bolanderi*
    2: Panicle compact, the branches appressed, floriferous from their base
        3. Glumes ca 10 mm long, gradually long-acuminate; awn nearly 10 mm long above the bend *C. foliosa*
        3: Glumes 6-8 mm long, abruptly acute or acuminate; awn usually < 5 mm long above the bend . . . . *C. ophitidis*

1: Awn not or scarcely longer than the glumes, straight or geniculate
    4. Awn bent, protruding sidewise from glumes; callus-hairs shorter than lemma
        5. Sheaths ± pubescent on the collar . . . . *C. rubescens*
        5: Sheaths glabrous on the collar
            6. Culms stout, mostly > 10 dm high; ligule 3-8 mm long . . . . *C. nutkaensis*
            6: Culms slender, 4-8 dm tall; ligule 2-3 mm long . . . . *C. koelerioides*
    4: Awn straight, included; callus-hairs somewhat shorter than lemma . . . . *C. stricta* ssp. *inexpansa*

Calamagrostis bolanderi Thurber. BOLANDER'S REED GRASS. Occasional, meadows, openings in coniferous forest: Hedgegate Road at The Sea Ranch, *Wells* (1981); Stump Beach in Salt Point State Park, acc. Amme (1989); Stewarts Point, *Crampton* (1957); Joy Road 1.8 mi. from jct. with Bodega Hwy., *Ferris & Lorraine* (1961); Pitkin Marsh, *Rubtzoff* (1951); Wagnon Road 5 mi. w. of Sebastopol, *Wagnon* (1939 NCC); Cunningham Marsh, *Baker* (1952 NCC). < 600 ft. June-Aug.

Calamagrostis foliosa Kearney. LEAFY REED GRASS. Rocky places near the coast: Sonoma to Humboldt cos. acc. *Munz, p. 1517,* (1995), S. Edwards reports it has not been seen south of n. Mendocino Co. < 4000 ft. May-Aug.

Calamagrostis koelerioides Vasey. Uncommon, open areas: region of Monte Rio, *Baker* (1936 NCC); ca 1 mi. nw. of Camp Meeker, acc. J. & B. Guggolz (1991); Llano Road, *Baker* (1930 NCC); Lake Sonoma quad. < 7700 ft. June-Aug.

Calamagrostis nutkaensis (C. Presl) Steudel. PACIFIC REED GRASS. Bunchgrass, mostly moist, coastal areas: Moonraker Road at The Sea Ranch, *Wells* (1981); The Sea Ranch Timber Preserve, *Wells* (1983); Hwy. 1 near Salt Point, *Best* (1981); Gerstle Cove at Salt Point State Park, *Wells* (1982); marsh near mouth of Russian River, *Rubtzoff* (1962); Goat Rock Beach, *Best* (1984); 7 ft. elev. Bodega Head, *Clifton* (1980 PUA); Pitkin Marsh, *Rubtzoff* (1952). < 5700 ft. May-Aug.

Calamagrostis ophitidis (J. T. Howell) Nygren. SERPENTINE REED GRASS. Occasional, areas of serpentine: Bohemian Hwy. 3½ mi. n. of Occidental, *Best* (1981); Harrison Grade Road 3½ mi. from jct. Green Valley Road, *Ferris & Lorraine* (1961); summit of Harrison Grade Road, *Hoover* (1965); 2350-2400 ft. elev. ca 3/4 mi. e. of Mercuryville off of Healdsburg-Geyser Road, acc. Raiche (1991); 3290 ft. elev. Socrates Mine, acc. Clifton (1984); 3060 ft. elev. Squaw Cr., acc. Clifton (1983). < 3500 ft. May-June.

Calamagrostis rubescens Buckley. PINE GRASS. Uncommon, rocky places, open banks: Mt. Hood, collection 83,376 from seed in Regional Parks Botanic Garden, Berkeley. < 3000 ft. June-Sept.

Calamagrostis stricta (Timm) Koeler ssp. **inexpansa** (A. Gray) C. W. Greene. Uncommon, swampy areas: Pitkin Marsh, *Rubtzoff* (1966), acc. J. & B. Guggolz (1988). < 11400 ft. June-Aug. —*C. crassiglumis* Thurber.

★ **Cortaderia jubata** (Lemoine) Stapf. WEEDY PAMPAS GRASS. Rampant invader, roadsides, cliffsides: The Sea Ranch; Annapolis Road between Gualala River bridge and Annapolis, acc. Knight (1990); Jenner, acc. Knight (1990); 2 mi. e. Bodega Bay, acc. Knight (1994); along US 101 between San Antonio Cr. and Petaluma, *Howell* (1981). < 2700 ft. Naturalized from Argentina. Has been known erroneously as *C. atacamensis* (Philippi) Pilger and *C. quila* (Nees) Stapf.

Crypsis

1. Plants prostrate much-branched, forming mats to 3 dm in diameter; sheath-margins hairy . . . . *C. vaginiflora*
1: Plants tufted, branching, erect to spreading 1-3 dm long; sheath-margins glabrous . . . . *C. schoenoides*

Crypsis schoenoides (Linnaeus) Lamarck. SWAMP GRASS. Occasional, vernal alkali flats, waste places: Button Ranch on Walker Road (U. C. property) between Petaluma-Valley Ford Road and Pepper Road, acc. Connors (1992); ca 1 mi. nw. of Camp Meeker, *J. & B. Guggolz* (1991 NCC); shore of small pond on Stoetz Road, *Best* (1981); Odd Fellows Park near Guerneville, *Howell* (1957); Rohnert Park, *Self* (1979 NCC); Willow Brook, *Knight* (1987 NCC); San Antonio Cr., US 101 at Marin-Sonoma co. line, *Rubtzoff* (1972); Spring Lake Village; Spring Lake; Pepperwood. < 1700 ft. May-June. Native of Old World. —*Heleochloa s.* (Linnaeus) Host.

★ **Crypsis vaginiflora** (Forsskal) Opiz. Uncommon, wet areas: Chileno Valley, *Rubtzoff* (1962); Laguna de Santa Rosa, *Rubtzoff* (1960). < 1700 ft. July. Introduced from Eurasia. —*C. niliaca* Figari.

★ **Cynodon dactylon** (Linnaeus) Persoon. BERMUDA GRASS. Common, appearing about communities as an escape from lawns: The Sea Ranch; Todd Road Preserve (H&K); Llano Road near Gravenstein Hwy, *Howell & Knight* (1987); Button Ranch on Walker Road (U. C. property) between Petaluma-Valley Ford Road and Pepper Road, acc. Patterson (1990); Spring Lake; Pepperwood; Sulphur Cr., *Bufford* (1936 NCC). < 3000 ft. June-Aug. Native of Africa.

Cynosurus

1. Annual; panicles subcapitate; awns obvious . . . . *C. echinatus*
1: Perennial; panicles narrow, spikelike; awns not obvious . . . . *C. cristatus*

★ **Cynosurus cristatus** Linnaeus. CRESTED DOGTAIL. Uncommon: The Sea Ranch, *Wells* (1978). < 1000 ft. Native of Europe.

Cynosurus echinatus Linnaeus. HEDGEHOG DOGTAIL. Common, open places, edge of thickets: S. Fork Gualala River at The Sea Ranch, *Wells* (1984); Gerstle Cove at Salt Point State Park, *Wells* (1981); Bodega Road near coast, *Baker* (1930 NCC); near Todd Road in Laguna de Santa Rosa drainage, *Harrison* (1978 NCC); Todd Road Preserve (H&K); Fairfield Osborn (H); Button Ranch on Walker Road (U. C. property) between Petaluma-Valley Ford Road and Pepper Road, acc. Patterson (1990); w. of Sears Point Raceway in Tolay Cr. drainage, acc. Patterson (1990); Bouverie; Kenwood, *Griffin* (1956 NCC); Sugarloaf Ridge; Pepperwood; 11 mi. from Geyserville on road to Kellogg, *Ferris & Lorraine* (1948); chaparral area sw. of Bradford Mtn. 1½ mi. w. of Dry Cr., acc. J. & B. Guggolz (1992); Stewarts Point Skaggs Springs Road, *Best* (1981); Lake Sonoma area; ca 2000 ft. elev. Anderson Trail on Black Mtn. between Geyser Peak and upper end of Pine Flat Road, acc. J. & B. Guggolz (1992); 2840 ft. elev. Bear Cr. Can., acc. Clifton (1984); 2040 ft. elev. Squaw Cr., acc. Clifton (1984); 2440 ft. elev. Caldwell Ranch, acc. Clifton (1984); near The Geysers, *Malloch* (1977 PG&E). < 3300 ft. May-July. Introduced from Europe.

★ **Dactylis glomerata** Linnaeus. ORCHARD GRASS. Common, grassy meadows, waste places: The Sea Ranch, acc. Stone (1980); Stewarts Point Skaggs Springs Road 1 mi. e. of Stewarts Point, *Best* (1981); Salt Point, *Wells* (1982); Bodega Head; 2.8 mi. from Occidental en route to Freestone, *Koch* (1940 NCC); Freestone, *Huntington* (1987 NCC); w.of Sebastopol, *Wagnon* (1939 NCC); w. of Santa Rosa in Laguna de Santa Rosa drainage, *Harrison* (1978 NCC); Fairfield Osborn (H); Bouverie; Sugarloaf Ridge; False Lake Meadow at Annadel, *Wells* (1984); Pepperwood; Alexander Valley quad.; 3200 and 3290 ft. elev. Socrates Mine, acc. Clifton (1984); 2840 ft. elev. Bear Cr. Can., acc. Clifton (1984); 3300 ft. elev. Squaw Cr., acc. Clifton (1983). < 8300 ft. May-Aug. Native of Eurasia, cultivated as a pasture grass.

Danthonia (Key ± acc. K. W. Allred, *Hickman* [1993]).

1. Upper leaf-blades abruptly spreading to reflexed; mature spikelets usually spreading
   2. Leaves, especially sheaths, densely hairy, hairs papillate at base; florets usually 3-5 . . . . *D. californica* var. *americana*
   2: Leaves usually glabrous, hairy near ligule but not papillate; florets usually 4-8 . . . *D. c.* var. *californica*
1: Upper leaf-blades erect to appressed; mature spikelets ascending to erect . . . . *D. pilosa*

Danthonia californica Bolander var. **americana** (Scribner) Hitchcock. Occasional, open areas: Bodega Head, acc. Connors (1991); 4 sites w. and sw. of Santa Rosa in Laguna de Santa Rosa drainage, *Harrison* (1978 NCC); Todd Road Preserve; Button Ranch on Walker Road (U. C. property) between Petaluma-Valley Ford Road and Pepper Road, acc. Connors (1992); 1 mi. e. of Vallejo's Old Adobe, *Koch* (1939 NCC); Sugarloaf Ridge; 6 mi. from Santa Rosa on Road to Sonoma, *Howell* (1935); Spring Lake; behind old Santa Rosa cemetery on Franklin Ave., *Tofsrud* (1941 NCC); field n. of Windsor, *Wagnon* (1940 NCC). < 8000 ft. May-July.

Danthonia californica var. **californica**. CALIFORNIA OATGRASS. Common, open areas; acc. S. Edwards, a dominant bunch grass in some overgrazed pastures: The Sea Ranch; Salt Point, Stewarts Point, Fort Ross, Duncans Mills and Arched Rock quads.; 1025 ft. elev. Coleman Valley Road, acc. Clifton (1984); Fairfield Osborn (S); Button Ranch on Walker Road (U. C. property) between Petaluma-Valley Ford Road and Pepper Road, acc. Patterson (1990); Todd Road Preserve (H&K); Lafferty Ranch at Sonoma Mtn., acc. Arnold et al. (1971); 8 mi. w. of Petaluma on Bodega Hwy., *Howell* (1930); Spring Lake; Pepperwood; Ida Clayton quad.; 2 mi. nw. of Windsor by Old Redwood Hwy., *de Nevers* (1981); 3200 ft. elev. Socrates Mine, acc. Clifton (1984); 2190 ft. elev. Squaw Cr., acc. Clifton (1984). < 6300 ft. May-July.

★ **Danthonia pilosa** R. Brown. HAIRY OATGRASS. Escaped from cultivation: The Sea Ranch; Miller Ridge (e. of The Sea Ranch), *Roos* (1957); 3.1 mi. n. of Fort Ross on Seaview Road, *Frankel* (1964); Stump Beach parking lot, *Best* (1986); Hwy. 1 ca 6 mi. s. of Stewarts Point, *Best* (1986); Bodega Head, acc. Connors (1991). < 2700 ft. May-July. Introduced from Australia.

Deschampsia (Key ± acc. D. H. Wilken, *Hickman* [1993]).

1. Plants annual . . . . *D. danthonioides*
1: Plants perennial
  2. Inflorescence < 1 cm wide; basal leaf-blades ± 1 mm wide . . . . *D. elongata*
  2: Inflorescence > 1 cm wide; basal leaf-blades 1-7 mm wide
    3. Inflorescence open, lower branches spreading to drooping
      4. Glumes 5 mm long or less . . . . *D. cespitosa* ssp. *cespitosa*
      4: Glumes > 5 mm long . . . . *D. c.* ssp. *beringensis*
    3: Inflorescence compact, lower branches ascending to erect . . . . *D. c.* ssp. *holciformis*

Deschampsia cespitosa (Linnaeus) Beauvois ssp. **beringensis** (Hultén) W. E. Lawrence. Uncommon, coastal prairie: The Sea Ranch in Annapolis Road area just below and w. of landing strip, acc. Stone (1980). Many agrostologists consider this indistinct. < 11500 ft. May-July.

Deschampsia cespitosa ssp. **cespitosa**. TUFTED HAIRGRASS. Rare at low elevations, except in wet places on serpentine, and even then quite unusual: Pitkin Marsh, *Howell* (1945); Cunningham Marsh, *Howell* (1947); ca 450 ft. elev. on hilltop e. of Windsor, *Northen* (1990 NCC). < 11500 ft. July-Aug. Cosmopolitan in n. hemisphere.

Deschampsia cespitosa ssp. **holciformis** (C. Presl) W. B. Lawrence. PACIFIC HAIRGRASS. Occasional, open places, wet meadows, sea bluffs, coastal prairie: The Sea Ranch; Gerstle Cove in Salt Point State Park, *Wells* (1984); Goat Rock State Park, *Best* (1982); 4 mi. s. of Jenner on Hwy. 1, *Morgan* (1956); Bodega Head, acc. Connors (1991); quarry on Wagnon Road 5 mi. w. of Sebastopol, *Wagnon* (1939 NCC); Kenwood, *Baker* (1950 NCC); Santa Rosa, *Congdon* (1880). < 3000 ft. May-July.

Deschampsia danthonioides (Trinius) Bentham. ANNUAL HAIRGRASS. Common, open areas: many sites w. and sw. of Santa Rosa in Laguna de Santa Rosa drainage, *Harrison* (1978 NCC); Todd Road Preserve (H&K); Fairfield Osborn (H); Petaluma, *Congdon* (1980); Sonoma Valley, *Jepson* (1911 JEPS); Bouverie; n. of El Verano, *Koch* (1940 NCC); Bennett Mtn. Lake, *Rubtzoff* (1961); vernal pool s. of Bennett Valley summit, *Baker* (1938); Pepperwood. < 9000 ft. Mar.-Apr.

Deschampsia elongata (Hooker) Bentham. SLENDER HAIRGRASS. Common, open, wet places: Gerstle Cove at Salt Point, *Wells* (1982); Stewarts Point-Skaggs Springs Road 1½ mi. e. of Hwy. 1, *Best* (1981); Russian River mouth marsh, *Rubtzoff* (1962); Austin Cr. Marsh, *Rubtzoff* (1960); near Guerneville, *Davy* (1896); 5 mi. w. of Sebastopol, *Wagnon* (1939 NCC); Laguna de Santa Rosa, *Rubtzoff* (1958); Fairfield Osborn (H); Button Ranch on Walker Road (U. C. property) between Petaluma-Valley Ford Road and Pepper Road, *Connors* (1992 BML); Bennett Valley, *Baker* (1931 NCC); Spring Mtn. Road, *Baker* (1941 NCC); ne. slopes of Fitch Mtn., *Rubtzoff* (1952); 3200 ft. elev. Socrates Mine, acc. Clifton (1984); 2440 ft. elev. Caldwell Ranch, acc. Clifton (1984). < 10400 ft. May-Aug.

★ **Desmazeria rigida** (Linnaeus) Tutin. Uncommon, weed: 2nd St., Santa Rosa, *Koch* (1940 NCC); Pepperwood. < 2350 ft. May-Sept. Introduced from Europe. —*Catapodium rigidum* (Linnaeus) C. E. Hubbard. *Scleropoa r.* (Linnaeus) Grisebach.

Dichanthelium acuminatum (Swartz) Gould & Clark var. **acuminatum**. Locally abundant, moist areas, sometimes the only plant growing in the most severe hydrothermal barrens: 2400 ft. elev. Little Geysers, *Malloch* (1977 PG&E); Hot Springs Cr. near confluence with Big Sulphur Cr., *Crampton* (1985 DAV); 2300 ft. elev. upper Big Sulphur Cr. drainage at Little Geysers, acc. Stone (1983). < 8700 ft. June-Aug. —*D. lanuginosum* (Elliott) Gould var. *thermale* (Bolander) Spellenberg. *Panicum thermale* Bolander, The Geysers, *Bolander* (type 1864 U). —*Panicum pacificum* Hitchcock & Chase. *P. acuminatum* Swartz, var. *acuminatum* in part.

Digitaria

1. Sheaths pilose or villous; spikelets 2.5-3.5 mm long; fertile lemma pale . . . . *D. sanguinalis*
1: Sheaths glabrous; spikelets 2 mm long; fertile lemma brown . . . . *D. ischaemum*

★ **Digitaria ischaemum** (Schreber) Muhlenberg. CRABGRASS. Naturalized, waste places: Salt Point; Llano Road 1.6 mi. from Gravenstein Hwy., *Howell & Knight* (1986); campus at Sonoma State Univ., *Howell & Best* (1986 NCC); Petaluma, acc. *Munz* (1959); Spring Lake. < 1000 ft. Sept.-Nov. Native of Europe.
★ **Digitaria sanguinalis** (Linnaeus) Scopoli. CRABGRASS. Common, weed, cultivated areas: sandy beach of Russian River at Odd Fellows Park near Guerneville, *Howell* (1957); Llano Road, *Howell* (1986); ditch along Todd Road ca 1 mi. w. of Llano Road, *Best* (1987); campus at Sonoma State Univ., *Best* (1986); ½ mi. n. of Penngrove on Redwood Hwy., *Morgan* (1960); Ducker Creek, Rincon Valley, *Wells* (1985); Spring Lake; Pepperwood. < 1700 ft. June-Sept. Native of Europe.

Distichlis spicata (Linnaeus) E. Greene. SALTGRASS. Common, saline flats, salt marshes:The Sea Ranch, *Wells* (1981); Stump Beach at Salt Point State Park, *Wells* (1982); Shell Beach, *Morgan* (1959); Salmon Cr. Marsh, *Rubtzoff* (1962); Bodega Head, *Clifton* (1980 PUA); Bodega Bay, *Wagnon* (1940); Llano Road near sewage plant, *Knight & Howell* (1986 NCC); Schollenberger Park, Petaluma, *W., I. & T. Knight* (1989 NCC); Gilardi's Resort, Lakeville, acc. Knight (1989). < 3300 ft. Apr.-July.

★ **Echinochloa crus-galli** (Linnaeus) Palisot de Beauvois. BARNYARD GRASS. Naturalized, damp, cultivated ground, waste places: 5 mi. w. of Sebastopol, *Wagnon* (1939 NCC); Todd Road Preserve, *Howell* (1987); Llano Road between Gravenstein Hwy. and Todd Road near sewage plant, *Knight & Howell* (1986); Copeland Cr., Rohnert Park, *Self* (1979 NCC); Ducker Cr., 1 mi. sw. of Sonoma, *Howell* (1957); Rincon Valley, *Wells* (1985); Bennett Valley, *Knight* (1963); Spring Lake; Pepperwood. < 5000 ft. July-Oct. Native of Old World.

Ehrharta

1. Spikelets 3-3.5 mm; sterile lemmas ± glabrous, awns 0 . . . . *E. erecta*
1: Spikelets 5-8 mm; sterile lemmas soft-hairy, short-awned to pointed . . . . *E. calycina*

★ **Ehrharta calycina** Smith. VELDT GRASS. Rare, damp, saline places: Salmon Cr. beach, *Harrison* (1977 SRJC); Bodega Dunes, *Connors* (1991 BML); several places in dunes near Salmon Cr., *Connors* (1992 BML). < 700 ft. Introduced from S. Africa.
★ **Ehrharta erecta** Lamarck. Uncommon, disturbed soil at edge of dunes: near Salmon Cr., *Connors* (1992 BML). < 700 ft. Mar.-June. Introduced from S. Africa.

Elymus: (Key ± acc. M. E. Barkworth, *Hickman* [1993]).

1. Glumes 0; top of stalk below lower floret ± knotlike, minutely hairy . . . . *E. californicus*
1: Glumes present, lanceolate to subulate
  2. Glumes awnlike to subulate, base usually < 1.5 mm wide, awn 25-100 mm
    3. Glumes divided into 3+ outward-curving awns; sheath-appendages evident 0.5-1.5 mm *E. multisetus*
    3: Glumes entire or divided into 2 outward-curving awns; sheath-appendages obscure < 0.5 (1) mm . . . . *E. elymoides* ssp. *californicus*
  2: Glumes flat, base usually > 1.5 mm wide, awn < 25 mm
    4. Inflorescence with 2-4 spikelets at all or most nodes
      5. Inflorescence nodding; glumes curving outward from base; base of lower florets exposed . . . . *E. canadensis*
      5: Inflorescence erect; glumes not curving outward at base; base of lower florets concealed
        6. Leaf sheaths smooth or rough but not hairy . . . . *E. glaucus* ssp. *glaucus*
        6: Leaf sheaths, blades sparsely to densely hairy, hairs spreading or directed downwards . . . . *E. g.* ssp. *jepsonii*
    4: Inflorescence with 1 spikelet at all or most nodes
      7. Lemma awn > lemma body 8-30 mm . . . . *E. trachycaulus* ssp. *subsecundus*
      7: Lemma awn < lemma body < 7 mm . . . . *E. t.* ssp. *trachycaulus*

Elymus californicus (Bolander) Gould. CALIFORNIA BOTTLE-BRUSH GRASS. Occasional, ± near the coast: 2.5 mi. w. of Bodega on Salmon Cr. Road, *Baker* (1942); Willow Cr. Road, *Utech* (1988 CM). < 1000 ft. June-Aug. —*Hystrix californica* (Bolander) Kuntze.

Elymus canadensis Linnaeus. CANADIAN WILDRYE. Uncommon, creek banks, sandy soil: Cazadero, *Bacigalupi* (1926). < 700 ft. June-Aug. Native to central and eastern U.S.

Elymus elymoides (Rafinesque) Swezey ssp. **californicus** (J. G. Smith) Barkworth. SQUIRRELTAIL. Rare, dry hills: Fairfield Osborn (S); Bennett Valley Road, *Wagnon* (1939); Alexander Valley quad. < 14000 ft. July-Aug. —*Sitanion hystrix* (Nuttall) J. G. Smith.

Elymus glaucus Buckley ssp. **glaucus**. BLUE WILDRYE. Common, wooded or grassy places: The Sea Ranch; Salt Point; Stewarts Point-Skaggs Springs Road 1.8 mi. e. of Stewarts Point, *Best* (1981); Duncans Mills Marsh, *Rubtzoff* (1957); Freestone Marsh, *Rubtzoff* (1959); 5 mi. w. of Sebastopol, *Wagnon* (1939 NCC); Todd Road Preserve (H&K), Fairfield Osborn (H); Tresch Ranch, Walker Road near Two Rock, acc. J. & B. Guggolz (1992); w. of Sears Point Raceway in Tolay Cr. drainage, acc. Patterson (1990); Bouverie; Sugarloaf Ridge; Annadel, *Wells* (1984); Parker Hill Road, Santa Rosa, *Tofsrud* (1941 NCC); Pepperwood; Lake Sonoma, Geyserville and Alexander Valley quads.; 2550 ft. elev. w. flank Mt. St. Helena, *Grummer* (1982 PUA); 2860 ft. elev. Pine Flat Road, acc. Clifton (1984); 3200 ft. elev. Socrates Mine, acc. Clifton (1984); 2040 and 3060 ft. elev. Squaw Cr., acc. Clifton (1984 & 1983); The Geysers, *Koch* (1940 NCC); 2440 ft. elev. Caldwell Ranch, acc. Clifton (1984); 3000 ft. elev. Caldwell Pines, acc. Clifton (1984). On Bodega Head a highly variable long-awned, rhizomatous *Elymus* may be referred to *E. glaucus*, *Connors* (1991 BML). < 8300 ft. June-Aug.

Elymus glaucus Buckley ssp. **jepsonii** (Burtt Davy) Gould. Occasional, open, woodland areas: McDonald Reservoir, Santa Rosa, *Tufsrud* (1941 NCC); Geyser Peak, *Knight & Edwards* (1986). < 7300 ft. June-Aug. —*E. divergens* Burtt Davy, type locality: dry bank near Petaluma, *Davy* (1897 JEPS).

Elymus multisetus (J. G. Smith) Burtt Davy. BIG SQUIRRELTAIL. Common, waste places: Stoetz Road e. of Occidental, *Ferris & Lorraine* (1961); 5 mi. w. of Sebastopol, *Wagnon* (1939 NCC); 3 sites w. of Santa Rosa in Laguna de Santa Rosa drainage, *Harrison* (1978 NCC); Fairfield Osborn (S); Button Ranch on Walker Road (U. C. property) between Petaluma-Valley Ford Road and Pepper Road, acc. Connors (1992); Hwy. 37 near Sears Point Raceway, *Baker* (1946 NCC); Wright School, *Baker* (1931 NCC); Pepperwood; 3000 ft. elev. w. flank Mt. St. Helena, *Grummer* (1981 PUA); along ridge between Socrates Mine and Mercuryville in geothermal area, *Dolan & LaPré* (1987 UCR); 3200, 3280 and 3290 ft. elev. Socrates Mine, acc. Clifton (1983); 3600 ft. elev. Geyser Rock, acc. Clifton (1983); 3620 ft. elev. Anderson Cr. 1 mi. sw. summit of Cobb Mtn., acc. Clifton (1984); 2040 ft. elev. Squaw Cr., acc. Clifton (1984). < 10500 ft. May-July. —*Sitanion jubatum* J. G. Smith.

A sterile hybrid between **Elymus glaucus** and **E. multisetus** = **Elysitanion** × **hansenii** (Scribner) Bowden. HANSEN'S SQUIRRELTAIL. Uncommon, grassland: cemetery on Bennett Valley Road, *Wagnon* (1939 NCC); Spring Lake; Pepperwood, *Breedlove* (1980). June-Aug. —*Sitanion h.* (Scribner) J. G. Smith.

Elymus trachycaulus (Link) Shinners ssp. **subsecundus** (Link) Gould. WHEATGRASS. Uncommon, moist meadows, woods: Salt Point. <11300 ft. June-Aug. —*E. s.* (Lincoln) A. D. Löve. *Agropyron subsecundum* (Link) A. Hitchcock.

Elymus trachycaulus ssp. **trachycaulus**. SLENDER WHEATGRASS. Occasional, dry to moist and rocky areas: seed collected on area of serpentine near Camp Meeker, *Edwards* (1984); 4 mi. e. of Valley Ford, *Howell* (1930); Pepperwood; 3200 ft. elev. Socrates Mine, *Clifton* (1984 PUA). < 11000 ft. June-Aug. —*Agropyron trachycaulum* (Link) Malte. *A. tenerum* Vasey.

Elytrigia (Key ± acc. J. K. Jarvie & M. E. Barkworth, *Hickman* [1993]).

1. Plants strongly rhizomed; glume-tip usually acute to obtuse; glumes keeled
  2. Glumes acute to awn-tipped; blades drooping, distance between veins >> vein widths . . . . *E. repens*
  2: Glumes truncate to obtuse, not awn-tipped; blades stiff, distance between veins < or = vein width . . . . *E. intermedia*
1: Plants without rhizomes; ; glume-tip truncate; glumes weakly keeled . . . . *E. elongata*

★ **Elytrigia elongata** (Host) Nevski. TALL WHEAT GRASS. Perhaps becoming established: Bodega Marine Reserve, *Connors* (1991 BML). < 5000 ft. May-June. Native of Mediterranean region. —*Agropyron elongatum* (Host) Beauvois.

Elytrigia intermedia (Host) Nevski. INTERMEDIATE WHEATGRASS. Rare, adventive: 3350 ft. elev. crest of Mayacmas Mts. 1.2 airmiles nw. of Geyser Rock, acc. Stone (1983). < 7000 ft. May-June. Native of Europe. —*Elymus hispidus* ssp. *barbatus* (Schur) Meldis. *Agropyron trichophorum* (Link) Richter.

★ **Elytrigia repens** (Linnaeus) Nevski. QUACKGRASS. Infrequent, waste places: 2 locations near Cotati, 1 location near Petaluma, acc. *Barbe* (1989). < 6000 ft. June-Aug. Introduced from Eurasia. —*Agropyron r.* (Linnaeus) Beauvois. Noxious weed.

Eragrostis

1. Plants perennial; anthers 0.5-1+ mm; spikelet axis breaking in late maturity tardily above glumes and between florets . . . . *E. curvula*
1: Plants annual; anthers usually 0.2-0.4 mm; spikelet axis persistent, glumes and lemmas deciduous, paleas deciduous or remaining attached
  2. Plants glandular or papillate along the leaf-margins, keel of lemmas, or on panicle branches
    3. Spikelets 2.5-3 mm wide; lemmas ca 2.5 mm long; glands prominent on keels of lemmas *E. cilianensis*
    3: Spikelets ca 2 mm wide; lemmas ca 2 mm long; glands mostly on panicle-branches and leaves *E. minor*
  2: Plants not glandular on lemmas, panicle-branches, or leaf-margins, occasionally so on sheaths and near nodes
    4. Spikelets ca 1 mm wide, slender, linear
      5. Spikelets 3-5 mm long; lemmas 1-1.5 mm long; pedicels mostly longer than the spikelets *E. pilosa*
      5: Spikelets 5-7 mm long; lemmas ca 2 mm long; pedicels mostly shorter than spikelets . . . . *E. mexicana* ssp. *virescens*
    4: Spikelets ca 1.5 mm wide or wider, linear to ovoid
      6. Plants creeping, rooting at the nodes to form mats . . . . *E. hypnoides*
      6: Plants not creeping and forming mats . . . . *E. pectinacea*

★ **Eragrostis cilianensis** (Allioni) Janchen. STINKGRASS. Occasional, weedy, strong-scented annual, waste places: Russian River at e. Guernewood, *Rubtzoff* (1955); Vacation Beach near Guerneville, *Rubtzoff* (1958); Forest Hills across the river from Hacienda, *Rubtzoff* (1961). < 5000 ft. June-Oct. Introduced from Europe. —*E. megastachya* (Koeler) Link.

★ **Eragrostis curvula** (Schrader) Nees. WEEPING LOVEGRASS. Uncommon, waste places: Meacham Road 0.35 mi. s. of Stony Point Road, *Rubtzoff* (1971). < 1700 ft. Aug.-Oct. Introduced from Africa.

Eragrostis hypnoides (Lamarck) Britton, Sterns & Poggenburg. Occasional, sand bars along Russian River: Odd Fellows Park near Guerneville, *Howell* (1955); Forest Hills across the river from Hacienda, *Rubtzoff* (1961); Mirabel Park, *Rubtzoff* (1970). < 1700 ft. July-Sept.

Eragrostis mexicana (Horneman) Link ssp. **virescens** (C. Presl) Koch & E. Sánchez. MEXICAN LOVEGRASS. Uncommon, moist fields, waste places: Vacation Beach near Guerneville, *Rubtzoff* (1958); ½ mi. n. of Penngrove on Redwood Hwy., *Morgan* (1960); Spring Lake Village; Spring Lake; Pepperwood; Russian River at Fitch Mtn. Grove, *Rubtzoff* (1963). < 6700 ft. May-Oct. —*E. orcuttiana* Vasey.

★ **Eragrostis minor** Host. Uncommon, waste places: Vacation Beach near Guerneville on open, wet, sandy bank of Russian River, *Rubtzoff* (1958). < 4700 ft. June-Sept. Native of Europe. —*E. poaeoides* Beauvois.

★ **Eragrostis pectinacea** (Michaux) Nees. LOVEGRASS. Fairly common, weed of lawns, gardens, disturbed areas: Odd Fellows Park along Russian River near Guerneville, *Howell* (1957); Vacation Beach near Guerneville, *Rubtzoff* (1956); Denman Flats near Petaluma, *Morgan* (1960); San Antonio Cr. at Sonoma-Marin co. line, *Best* (1980). <4700 ft. June-Sept. Native in much of N. America, but introduced locally. —*E. diffusa* Buckley.

Eragrostis pilosa (Linnaeus) Palisot de Beauvois. Occasional, weed, waste places: Pepperwood. < 1000 ft. July- Sept. Native of Europe.

★ **Eriochloa villosa** (Thunberg) Kunth. Probably from seed in a bird-feeder: Rincon Valley, Santa Rosa, *Wells* (1985). Native of e. Asia.

Festuca

1. Awns scarcely if at all evident, < 1 mm long
    2. Leaf-blades flat 4-8 mm wide
        3. Panicle 1-2 dm long; lemmas 5-7 mm long . . . . *F. pratensis*
        3: Panicle 1.5-3.5 dm long; lemmas 7-10 mm long . . . . *F. arundinacea*
    2: Leaf-blades involute, narrower . . . . *F. viridula*

1: Awns well-developed, 2-many mm long
    4. Leaf-blades flat, ± soft and lax
        5. Florets long-stipitate, the rachilla appearing to be jointed a short distance below the floret *F. subuliflora*
        5: Florets not stipitate; awn from between 2 teeth . . . . *F. elmeri*
    4: Leaf-blades narrow or involute, not lax
        6. Collar and mouth of sheath villous . . . . *F. californica*
        6: Collar and mouth of sheath not villous
            7. Awn of lemma longer than the body; ovary pubescent at the summit . . . . *F. occidentalis*
            7: Awn of lemma shorter than the body; ovary glabrous at the summit
                8. Culms loosely tufted and decumbent at the base; blades smooth; sheaths red-brown with obvious pale nerves . . . . *F. rubra*
                8: Culms closely tufted, not decumbent; blades scabrous; sheaths not red-brown at base . . . . *F. idahoensis*

★ **Festuca arundinacea** Schreber. TALL FESCUE. Common, disturbed areas: The Sea Ranch; Bodega Head, acc. Connors (1991); Freestone, *Huntington* (1987 NCC); Button Ranch on Walker Road (U. C. property) between Petaluma-Valley Ford Road and Pepper Road, acc. Connors (1992); Roblar Road ½ mi. w. of Stony Point Road, *Morgan* (1956); ne. corner Llano and Todd roads, *Howell & Knight* (1987); ½ mi. from Lakeville on Stage Gulch Road, *Morgan* (1956); 5.4 mi. s. of Santa Rosa city limits on Hwy. 12, *Morgan* (1956); Pepperwood. < 9000 ft. May-June. Native of Europe.

Festuca californica Vasey. CALIFORNIA FESCUE. Common, variety of habitats: Annapolis quad.; 9 mi. s. of Stewarts Point, *Koch* (1940 NCC); Fort Ross quad.; Willow Creek; Fairfield Osborn (H); Button Ranch on Walker Road (U. C. property) between Petaluma-Valley Ford Road and Pepper Road, acc. Patterson (1990); Bouverie; Kenwood, *Tofsrud* (1941 NCC); Sugarloaf Ridge; near summit of Mt. Hood, *Heller* (1902); Annadel, *Wells* (1984); Spring Lake; Pepperwood; Lake Sonoma area; Alexander Valley quad.; 3200 and 3280 ft. elev. Socrates Mine, acc. Clifton (1984); ca 2000 ft. elev. Anderson Trail on Black Mtn. between Geyser Peak and upper end of Pine Flat Road, acc. J. & B. Guggolz (1992); 3060 and 2190 ft. elev. Squaw Cr., acc. Clifton (1983 & 1984); 2440 ft. elev. Caldwell Ranch, acc. Clifton (1984); 3200 ft. elev. Caldwell Pines, acc. Clifton (1984). < 6000 ft. Apr.-July. —*Bromus kalmii* A. Gray var. *aristulatus* Torrey, type locality: Mark West Cr., *Bigelow* (1854 U).

Festuca elmeri Scribner & Merrill. ELMER'S FESCUE. Occasional, wooded areas: quarry on Wagnon Road 5 mi. w. of Sebastopol, *Wagnon* (1939); Button Ranch on Walker Road (U. C. property) between Petaluma-Valley Ford Road and Pepper Road, *Connors* (1992 BML); Pepperwood. < 5000 ft. May-June.

Festuca idahoensis Elmer. BLUE BUNCHGRASS, IDAHO FESCUE. Common, brush-covered areas, grassy slopes, open mead-ows: Button Ranch on Walker Road (U.C. property) between Petaluma-Valley Ford Road and Pepper Road, acc. Connors (1992); 5 mi. w. of Sebastopol, *Wagnon* (1939 NCC); Fairfield

Osborn (H); Mt. Hood, *Heller* (1902); McDonald Reservoir, Santa Rosa, *Tofsrud* (1941 NCC); Spring Mtn. Road, *Baker* (1945 NCC); Lake Sonoma area; Ida Clayton Road quad.; 2400 ft. elev. w. flank Mt. St. Helena, *Grummer* (1982 PUA); Geyser Peak, acc. Edwards (1986). < 6000 ft. May-July.

Festuca occidentalis Hooker. WESTERN FESCUE. Scattered, moist rocky soil, wooded areas, particularly along edge of redwood groves: 6 mi. e. of Stewarts Point at edge of redwoods, *Best* (1981); Guernewood Park, *Bufford* (1935 NCC); Joy Road 1.8 mi. from jct. with Bodega Hwy., *Ferris & Lorraine* (1961); 5 mi. w. of Sebastopol, *Wagnon* (1939 NCC); 2000 ft. elev. Wildhorse Cr., *Clifton* (1984 PUA). < 6500 ft. May-Aug.

★ **Festuca pratensis** Hudson. MEADOW FESCUE. Cultivated for pasture: Hwy. 1 near Salt Point, *Best* (1981); Hwy. 1 on n. side of Bodega Bay, *Best* (1981); near Wright School, *Tufsrud* (1941 NCC); near El Verano, *Best* (1980); Caldwell Ranch, *Jokerst* (1983). < 6700 ft. May-July. Native of Europe. —*F. elatior* Linnaeus.

Festuca rubra Linnaeus. RED FESCUE. Common, hills, meadows, marshes: The Sea Ranch; Russian River, *Kennedy* (1920); Bodega Head, acc. Connors (1991); Pitkin Marsh, *Rubtzoff* 1951); Todd Road Preserve (H&K); Button Ranch on Walker Road (U. C. property) between Petaluma-Valley Ford Road and Pepper Road, acc. Patterson (1990); between Petaluma and Chileno Valley, *Howell* (1944); Pepperwood; Knights Valley, *Breedlove* (1980); Sulphur Cr., *Koch* (1940 NCC); 3620 ft. elev. Anderson Cr., *Clifton* (1984 PUA). < 8500 ft. May-July.

Festuca subuliflora Scribner. CRINKLE-AWN FESCUE. Uncommon, shady canyons: 98 ft. elev. Fort Ross, *Clifton* (1981 PUA). < 3000 ft. June-July.

Festuca viridula Vasey. GREEN FESCUE. Rare, montane, open slopes, meadows; Mt. Hood, *Heller* (1902); The Geysers, *Bolander* (1864). Acc. S. Edwards, this species is usually observed above 6000 ft. elev. and these citations need confirmation. < 9000 ft. July-Aug. —*F. howellii* Hackel.

★ **Gastridium ventricosum** (Gouan) Schinz & Thellung. NIT GRASS. Common, weed, waste places, fields: The Sea Ranch; Fort Ross and Cazadero quads.; Fife Cr. Marsh near Guerneville, *Rubtzoff* (1961); Austin Cr. 5 mi. nw. of Monte Rio, *Morgan* (1955); Vacation Beach on Russian River, *Rubtzoff* (1954); w. of Santa Rosa in Laguna de Santa Rosa drainage, *Harrison* (1978 NCC); Button Ranch on Walker Road (U. C. property) between Petaluma-Valley Ford Road and Pepper Road, acc. Patterson (1990); Trinity Road, *Tofsrud* (1941 NCC); Mt. Hood above Adobe Can., *Baker* (1947 NCC); Bennett Lake, *Rubtzoff* (1963); Santa Rosa, *Congdon* (1880); Spring Lake; Fitch Mtn., *Rubtzoff* (1952); chaparral area sw. of Bradford Mtn. 1½ mi. w. of Dry Cr., acc. J. & B. Guggolz (1992); Lake Sonoma quad.; ca 2000 ft. elev. Anderson Trail on Black Mtn. between Geyser Peak and upper end of Pine Flat Road, acc. J. & B. Guggolz (1992); geothermal area, *Malloch* (1977 PG&E). < 4000 ft. Apr.-May. Introduced from Europe.

★ **Gaudinia fragilis** (Linnaeus) Palisot de Beauvois. Uncommon, roadside: 320 ft. elev. ½ mi. s. of jct. US 101 turnoff to Asti on w. side of road, *J. & B. Guggolz* (1991 NCC). New to California and possibly the United States. < 800 ft. Native of s. Europe.

Glyceria

1. Spikelets linear, mostly 10 mm or longer; panicles narrow, erect
    2. Lemmas ca 3 mm long, rounded at the summit . . . . *G. leptostachya*
    2: Lemmas 4-6 mm long, ± angled at summit . . . . *G. occidentalis*

1: Spikelets ovate or oblong, generally not > 5 mm long; panicles generally nodding . . . . *G. elata*

Glyceria elata (Lamarck) A. Hitchcock. FOWL MANNAGRASS. Infrequent, moist, marshy areas: Pitkin Marsh, *Rubtzoff* (1951); Laguna de Santa Rosa n. of Graton and e. of Laguna Road, *Rubtzoff* (1962). < 8700 ft. July-Aug.

Glyceria leptostachya Buckley. DAVY'S MANNA GRASS. Common, usually damp places: Occidental Marsh, *Rubtzoff* (1957); Pitkin Marsh, *Rubtzoff* (1951); Perry Marsh, *Howell* (1945); Cunningham Marsh, *Howell* (1947); 2 mi. n. of Sebastopol on Gravenstein Hwy., *Baker* (1945 NCC); Fairfield Osborn (S); Button Ranch on Walker Road (U. C. property) between Petaluma-Valley Ford Road and Pepper Road, acc. Connors (1992); Sonoma, *Heller* (1902); Bennett Valley in ditch, *Klimansky* (1931 NCC); marsh on Parker Hill near Santa Rosa, *Tofsrud* (1941 NCC). < 7000 ft. May-June. —*Panicularia davyi* Merrill, type locality: Guerneville, *Davy* (1896 U).

Glyceria occidentalis (Piper) J. C. Nelson. Common, wet places: ditch at Bodega Head, *Best* (1981); 13 ft. elev. Brooks Gulch near Bodega Head, *Clifton* (1984 PUA); Perry Marsh, *Howell* (1945); Todd Road Preserve (H&K); Spring Lake. < 4300 ft. June-Aug.

★ **Hainardia cylindrica** (Willdenow) Greuter. Occasional, moist, subsaline situations, waste places, sometimes hard dry areas: Timber Cove, *Wagnon* (1989 NCC); lower Salmon Cr. Marsh, *Rubtzoff* (1963); Bodega Head, acc. Connors (1991); Button Ranch on Walker Road (U. C. property) between Petaluma-Valley Ford Road and Pepper Road, acc. Connors (1992); 8 mi. w. of Petaluma on Bodega Road, *Howell* (1930), Annadel, *Wagnon* (1989 NCC). < 1000 ft. May-July. Native of Europe. —*Lepturus cylindricus* (Willdenow) Trinius. *Monerma c.* (Willdenow) Cosson & Durieu de Maisonneuve.

Hierochloe occidentalis Buckley. VANILLA GRASS, SWEET GRASS. Abundant, coniferous forests, grassy areas near brush: The Sea Ranch; Stewarts Point-Skaggs Springs Road, *Parker* (1975 NCC); Salt Point; redwood forest 1 mi. n. of Fort Ross, *Howell* (1930); Duncans Mills, *Davy* (1896 JEPS); Duncans Mills Marsh, *Rubtzoff* (1955); Guerneville, *Congdon* (1880); Joy Road 1.8 mi. from jct. with Bodega Hwy., *Ferris & Lorraine* (1961); 6 mi. w. of Sebastopol, *Wagnon* (1939 NCC). < 3100 ft. Jan.-July.

★ **Holcus lanatus** Linnaeus. VELVET GRASS. Common, open places, invasive weed reducing diversity in sea-bluff meadows: The Sea Ranch; s. section of The Sea Ranch, *J. H. Thomas* (1976); Stewarts Point quad.; Salt Point; Duncans Mills Marsh, *Rubtzoff* (1957); Arched Rock quad.; Bodega Head, acc. Connors (1991); 2 sites w. of Santa Rosa in Laguna de Santa Rosa, *Harrison* (1978 NCC); Todd Road Preserve (H&K); ½ mi. n. of Penngrove on Old Redwood Hwy., *Morgan* (1960); Button Ranch on Walker Road (U. C. property) between Petaluma-Valley Ford Road and Pepper Road, acc. Patterson (1990); Spring Lake; Lake Sonoma area. < 8700 ft. June-Aug. Introduced from Europe.

Hordeum

1. Plants perennial
  2. Glumes and awns 1.8-8 cm long . . . . *H. jubatum*
  2: Glumes and awns < 1.8 cm long
    3. Blades pubescent 1.5-5 mm wide; anthers generally 1.5-3 mm long; glumes of central spikelet 1.5-2.5 × as long as palea . . . . *H. brachyantherum* ssp. *californicum*
    3: Blades mostly glabrous 3-9 mm wide; anthers mostly 1-1.5 mm long; glumes of central spikelet often scarcely longer than palea . . . . *H. b.* ssp. *brachyantherum*
1: Plants annual
  4. Glumes of central spikelet and the inner ones of the lateral spikelets with ciliate margins
    5. Spike with 6-8 spikelets to 1 cm of rachis; stamens of central florets included at anthesis . . . . *H. murinum* ssp. *glaucum*
    5: Spike with 3-5 spikelets to 1 cm of rachis; stamens of central florets exserted at anthesis . . . . *H. m.* ssp. *leporinum*
  4: Glumes not ciliate
    6. Auricles long; rachis continuous; all 3 spikelets sessile, fertile . . . . *H. vulgare*
    6: Auricles obsolete; rachis articulate; lateral spikelets pedicellate, usually not fertile
      7. Spike ovate to ovate-oblong, usually < 5 cm long; awns and glumes strongly spreading at maturity; bases of glumes of lateral spikelets prominent above the pedicel . *H. marinum* ssp. *gussoneanum*
      7: Spike linear-oblong, usually > 5 cm long; awns and glumes suberect; bases of glumes of lateral spikelets not prominent above the pedicel . . . . *H. depressum*

Hordeum brachyantherum Nevski ssp. **brachyantherum**. MEADOW BARLEY. Common, moist places: Salt Point; Bodega Head; 4 mi. w. of Sebastopol, *Wagnon* (1939 NCC); many sites w. of Santa Rosa in Laguna de Santa Rosa drainage, *Harrison* (1978 NCC); Todd Road Preserve; Cotati, *Tunison* (1968 NCC); Fairfield Osborn (S); Button Ranch on Walker Road (U. C. property) between Petaluma-Valley Ford Road and Pepper Road, acc. Patterson (1990); w. of Sears Point Raceway in Tolay Cr. drainage, acc. Patterson (1990); Rincon Valley, *Wells* (1984); Cooper's Grove on Sonoma Mtn. Road, *Baker* (1931 NCC); Spring Lake; campus Santa Rosa Jr. Col., *Tofsrud* (1941 NCC); Pepperwood; ca 2000 ft. elev. Anderson Trail on Black Mtn. between Geyser Peak and upper end of Pine Flat Road, acc. J. & B. Guggolz (1992). < 11500 ft. May-Aug. —*H. nodosum* Linnaeus, misapplied.

Hordeum brachyantherum ssp. **californicum** (Covas & Stebbins) Bothmer, N. Jacobsen & O. Seberg. MEADOW BARLEY. Occasional, brushy and grassy places: s. section of The Sea Ranch, *J. H. Thomas* (1976); Stump Beach, *Best* (1986); jct. Occidental Road with Facendini Lane, *Ferris & Lorraine* (1961); Lake Sonoma area; Alexander Valley quad. < 8700 ft. Apr.-Aug. —*H. californicum* Covas & Stebbins.

Hordeum depressum (Scribner & J. G. Smith) Rydberg. Infrequent, open places: Duncans Mills, Arched Rock quad.; Bodega Head; Spring Lake; Asti, *Abrams* (1918). < 6000 ft. Apr.-May.

Hordeum jubatum Linnaeus. FOXTAIL BARLEY. Apparently not commonly collected, waste or moist places: Forestville, *Quibell* (1980 NCC). < 10000 ft. May-July. —*H. j.* var. *caespitosum* (Scribner) Hitchcock, currently sometimes not recognized.

★ **Hordeum marinum** Hudson ssp. **gussoneanum** (Parlatore) Thellung. MEDITERRANEAN BARLEY. Common, damp, waste places: The Sea Ranch, *Wells* (1972); Bodega Head Marsh, *Rubtzoff* (1968); upper Stemple Cr. Marsh, *Rubtzoff* (1961); Freestone, *Huntington* (1987 NCC); 2 mi. s. of Forestville, *Yates* (1936); Pitkin Marsh, *Rubtzoff* (1958); Todd Road Preserve (H&K); Fairfield Osborn (H); Lafferty Ranch near summit Sonoma Mtn., acc. Arnold et al. (1971); Button Ranch on Walker Road (U. C. property) between Petaluma-Valley Ford Road and Pepper Road, *Connors* (1992 BML); 2½ mi. from Petaluma out I St. Ext., *Morgan* (1956); 1 mi. e. of Vallejo's Old Adobe, *Koch* (1939 NCC); Sonoma State Univ. campus, *Tunison* (1968 NCC); Rincon Valley, *Wells* (1988); Santa Rosa, *Heller* (1902); field behind Santa Rosa High School, *Tofsrud* (1941 NCC); Pepperwood; Lake Sonoma area. < 5000 ft. Apr.-June. Native of Europe. —*H. gussoneanum* Parlatore. *H. hystrix* Roth. *H. geniculatum* Allioni.

★ **Hordeum murinum** Linnaeus ssp. **glaucum** (Steudel) Tzvelev. Occasional, open area: The Sea Ranch, *Wells* (1981); Blank Road near Gravenstein Hwy., acc. J. & B. Guggolz (1991); Hessel Road near Gravenstein Hwy., acc. J. & B. Guggolz (1991); Sonoma State Univ. campus, *Schneider* (1986 NCC). < 3300 ft. Apr.-May. Native of Europe. —*H. g.* Steudel. *H. stebbinsii* Covas.

★ **Hordeum murinum** ssp. **leporinum** (Link) Arcangeli. FARMER'S FOXTAIL. Common, weed, fields, waste places: The Sea Ranch; Bodega Head; Bodega Bay, *Heller* (1902); Willow Cr. Marsh, *Rubtzoff* (1959); Laguna de Santa Rosa e. of Graton and Guerneville roads, *Rubtzoff* (1960); Freestone, *Huntington* (1987 NCC); lower Pitkin Marsh, *Rubtzoff* (1951); jct. of Burnside and Gold Ridge Roads 4 mi. from Sebastopol, *Morgan* (1986); near Todd Road in Laguna de Santa Rosa drainage, *Harrison* (1978 NCC); Fairfield Osborn (S); Button Ranch on Walker Road (U. C. property) between Petaluma-Valley Ford Road and Pepper Road, acc. Patterson (1990); Mountain View Ave., Petaluma, *Tofsrud* (1941 NCC); Bouverie; Bennett Valley, *Koch* (1939 NCC); Rincon Valley, *Wells* (1984); Spring Lake; Santa Rosa, *Miller* (1965 NCC); Pepperwood; Lake Sonoma area. < 3300 ft. Apr.-June. Native of Europe. —*H. l.* Link.

★ **Hordeum vulgare** Linnaeus. BARLEY. Occasional, fields, waste places: Crane Can., *Miller* (1965 NCC); 5.4 mi. from Santa Rosa on Hwy. 12, *Morgan* (1956); Santa Rosa, *Tufsrud* (1941 NCC). Various elevations. Apr.-July. Probably originated in Asia.

★ **Imperata brevifolia** Vasey. SATINTAIL. Uncommon, waste places: Windsor area, acc. *Barbe* (1991). < 1700 ft. Sept.-May. Native of s. California and Mexico. Noxious weed.

Koeleria (Key ± acc. D. H. Wilken, *Hickman*, [1993]).

1. Plants perennial; inflorescence-axis puberulent, glumes and lower lemmas glabrous to minutely scabrous on back . . . . *K. macrantha*

1: Plants annual; inflorescence-axis glabrous, glumes and lower lemmas papillate to coarse-hairy on back . . . . *K. phleoides*

Koeleria macrantha (Ledebour) J. A. Schultes, acc. Gould, *Graminae of Texas, p. 126,* (1976). JUNE-GRASS. Common, dry meadows: Bodega Head, acc. Connors (1991); Stoetz Lane near Occidental, *Best* (1981); near Todd Road in Laguna de Santa Rosa drainage, *Harrison* (1978 NCC); Fairfield Osborn (H); Button Ranch on Walker Road (U. C. property) between Petaluma-Valley Ford Road and Pepper Road, acc. Connors (1992); near Sonoma, *Heller* (1902); Bouverie; Adobe Can., *Baker* (1945 NCC); near Wright School, *Tofsrud* (1941 NCC); Monan's Rill n. of St. Helena Road and e. of Calistoga Road, *Wells* (1985); Pepperwood; Fitch Mtn., *Rubtzoff* (1952); Ida Clayton Road quad. < 11700 ft. May-July. —*K. cristata* (Linnaeus) Persoon. *K. pyramidata* (Lamarck) Beauvois, misapplied.

★ **Koeleria phleoides** (Villars) Persoon. Infrequent, serpentine outcrops: Alexander Valley quad.; Hot Springs Road 3 mi. sw. of Cloverdale, *Crampton* (1957). < 1250 ft. Apr.-July. Introduced from Europe.

★ **Lagurus ovatus** Linnaeus. HARE'S TAIL. Occasional, naturalized, coastal: The Sea Ranch; 1 mi. n. of Stewarts Point, *Beetle* (1941); Salt Point; Fort Ross, *Keck* (1948); Goat Rock Beach, *Best* (1984); Bodega Head, acc. Connors (1991). < 700 ft. May-July. Introduced from Mediterranean.

Lamarckia aurea (Linnaeus) Moench. GOLDENTOP. Common, weed, many habitats: Mill Gulch 1½ mi. se. of Fort Ross, *Clifton* (1977 PUA); Cazadero quad.; under bridge on Austin Cr. 3 mi. w. of Monte Rio, *Morgan* (1955); Pepperwood; Dry Cr. Road 0.7 mi. w. of Warm Springs Dam, *Breedlove* (1976); Lake Sonoma area; Ida Clayton Road quad.; ca 2000 ft. elev. Anderson Trail on Black Mtn. between Geyser Peak and upper end of Pine Flat Road, acc. J. & B. Guggolz (1992); Geyser Peak, *Knight & Edwards* (1986); Cloverdale, *McWilliams* (1930 NCC); US 101, 3 mi. n. of Cloverdale at Sonoma-Mendocino Co. line, *Crampton* (1954). < 2400 ft. Feb.-May. Introduced from Mediterranean region.

★ **Leersia oryzoides** (Linnaeus) Swartz. RICE CUTGRASS. Occasional, damp, open places: Russian River near Guerneville, *Hoffman (1943);* along edge of Russian River at Odd Fellows Park, *Howell* (1959); laguna at Sebastopol, *Baker* (1946 NCC); Laguna de Santa Rosa, *Rubtzoff* (1958); Spring Lake; along Franz Cr. near Franz Valley Road, *de Nevers* (1992); Fitch Mtn. grove 2 mi. e. of Healdsburg, *Rubtzoff* (1963). < 2400 ft. Aug.-Oct. Native of Eurasia.

Leptochloa fascicularis (Lamarck) A. Gray. BEARDED SPRANGLETOP. Common, open, moist, occasionally alkaline areas: Russian River bottom at Odd Fellows Park near Guerneville, *Howell* (1957); Vacation Beach near Guerneville, *Rubtzoff* (1961); Russian River at Mirabel Park, *Rubtzoff* (1970); Laguna de Santa Rosa, *Rubtzoff* (1958); jct. of Todd and Llano roads, *Best* (1987); Copeland Cr. near Rohnert Park, *Self* (1979 NCC); 5 mi. n. of Sears Point, *Rubtzoff* (1970); Spring Lake. < 4000 ft. June-Oct. *—Diplachne f.* Beauvois.

Leymus: (Key ± acc. M. E. Barkworth, *Hickman* [1993]).

1. Glumes lanceolate, > 1 mm wide, clearly veined at midlength; plants from rhizomes
   2. Glumes 3-5 mm wide at midlength, back rounded . . . . . *L. mollis*
   2: Glumes 1-2 mm wide at midlength, back keeled . . . . . *L. × vancouverensis*

1: Glumes subulate, not clearly veined at midlength; plants cespitose or from short rhizomes, (rhizomes well-developed in *L. pacificus*, *L. triticoides*)
   3. Plants usually > 12 dm; blades frequently > 10 mm wide . . . . . *L. × multiflorus*
   3: Plants usually < 10 dm; blades < 10 mm wide
      4. Stem 1-6 dm; leaf blade often > inflorescence . . . . . *L. pacificus*
      4: Stem 4.5-13 dm; leaf blade < inflorescence . . . . . *L. triticoides*

Leymus mollis (Trinius) Pilger. Rare, coastal strand: dunes at n. side of mouth of Salmon Cr., *Connors* (1991 BML); Bodega Head, *Connors* (1992 BML). Usually < 40 ft. June-Aug. *—Elymus m.* Trinius.

Leymus × multiflorus (Gould) Barkworth & D. R. Dewey. Uncommon, dunes: Bodega Head, *Connors* (1992 BML). Hybrid between *L. triticoides* and *L. condensatus*. < 1000 ft. June-July. *—Elymus triticoides* Buckley ssp. *m.* Gould.

Leymus pacificus (Gould) D. R. Dewey. Infrequent, sandy areas: The Sea Ranch; Bodega Head and Dunes, *Connors* (1990 BML). < 400 ft. *—Elymus p.* Gould.

Leymus triticoides (Buckley) Pilger. Occcasional, open areas: w. of Santa Rosa in Laguna de Santa Rosa drainage, *Harrison* (1978 NCC); Bodega Road 4 mi. e. of Valley Ford, *Howell* (1930); Button Ranch on Walker Road (U. C. property) between Petaluma-Valley Ford Road and Pepper Road, acc. Patterson (1990); 3 mi. w. of Petaluma, *Morgan* (1959); near Wright School, *Klimansky* (1936 NCC). < 7700 ft. June-July. *—Elymus t.* Buckley.

Leymus × vancouverensis (Vasey) Pilger. Occasional, coastal strand: Hwy. 1 near Salmon Cr. bridge, *Rubtzoff* (1962); beach at Bodega Harbor entrance to Bodega Marine Reserve, *Connors* (1991 BML); Bodega Point, *Eastwood* (1915); Bodega Harbor Marsh, *Rubtzoff* (1960). Hybrid between *L. mollis* and *L. triticoides*. < 400 ft. July-Aug. *—Elymus × v.* Vasey.

Lolium

1. Glume < the spikelet
   2. Lemmas almost or quite awnless; culms subcompressed . . . . . *L. perenne*
   2: Lemmas, at least the upper, awned; culms cylindric . . . . . *L. multiflorum*
1: Glume = or > the spikelet . . . . . *L. temulentum*

★ **Lolium multiflorum** Lamarck. ITALIAN RYEGRASS. Common, escaped, lawn grass: The Sea Ranch; Salt Point; 1 mi. s. of Carmet on Hwy. 1, *Morgan* (1960); Bodega Head; Freestone, *Huntington* (1987 NCC); Pitkin Marsh, *Rubtzoff* (1951); many sites w. of Santa Rosa in Laguna de Santa Rosa drainage, *Harrison* (1978 NCC); Todd Road Preserve (H&K); near Copeland Cr., Rohnert Park, *Self* (1979 NCC); Fairfield Osborn (H); Lafferty Ranch on Sonoma Mtn., acc. Arnold (1971); ¼ mi. n. of Penngrove on Old Redwood Hwy., *Morgan* (1961); 3 mi. from Petaluma on I St. Ext., *Morgan* (1956); Bouverie; Rincon Valley, *Wells* (1984, 1985); Spring Lake; Odd Fellows Cemetery, Santa Rosa, *Tofsrud* (1941 NCC); 7 mi. ne. of Santa Rosa, *Miller* (1955 NCC); Healdsburg, Alexander Valley and Ida Clayton Road quads. < 3300 ft. June-Aug. Native of Europe.

★ **Lolium perenne** Linnaeus. PERENNIAL RYEGRASS. Common, many habitats: The Sea Ranch; Duncans Mills quad.; Bodega Head, acc. Connors (1991); Pitkin Marsh, *Rubtzoff* (1957); many sites w. of Santa Rosa in Laguna de Santa Rosa drainage, *Harrison* (1978 NCC); Llano Road near Gravenstein Hwy., *Howell & Knight* (1987); Fairfield Osborn (H); Spring Lake Village; Spring Lake, Santa Rosa, *Wells* (1987); Rincon Valley, *Wells* (1988); Pepperwood; Geyserville quad. < 3300 ft. May-Sept. Introduced from Europe.

★ **Lolium temulentum** Linnaeus. DARNEL. Occasional, fields, waste places: Sebastopol, *Bufford* (1941 NCC); Bellview community near Todd Road, *Klimansky* (1936 NCC); Blank Road near Gravenstein Hwy., acc. J. & B. Guggolz (1991); ½ mi. w. of jct. of Arnold Drive and Watmaugh Road, *Morgan* (1956); Sugarloaf Ridge; grain field 5 mi. e. of Santa Rosa, *Wagnon* (1939). < 6300 ft. Apr.-June. Introduced from Mediterranean. Toxic.

Melica

1. Lemmas awned
   2. Awn short, mostly < 4 mm long; leaf-blades mostly 1-3 dm long . . . . . *M. harfordii*
   2: Awn 5-12 mm long; leaf-blades 0.6-1.4 dm long . . . . . *M. aristata*
1: Lemmas awnless
   3. Culms bulbous at bases
      4. Lemmas tapering-acuminate, mostly ciliate-pubescent on nerves . . . . . *M. subulata*
      4: Lemmas acute or obtuse, glabrous
         5. Panicle open and broad when mature, with long spreading branches; 1st and 2nd florets > > 0.5 mm long; spikelets ca 16 mm long . . . . . *M. geyeri*
         5: Panicle very narrow, the branches and spikelets appressed; 1st and 2nd florets mostly < 2 mm long; spikelets shorter . . . . . *M. bulbosa*
   3: Culms not bulbous at bases
      6. Fertile florets 1-2
         7. Rudiment 1 mm long, on a stipe 2.5 mm long; lemma pubescent near tip . . . . . *M. torreyana*
         7: Rudiment mostly 2 or more mm long, on a stipe 0.5 mm long; lemma glabrous or scabrous . . . . . *M. imperfecta*
      6: Fertile florets > 2
         8. Lemmas pilose-ciliate on lower margins . . . . . *M. harfordii*
         8: Lemmas glabrous or scabrous . . . . . *M. californica*

Melica aristata Bolander. AWNED MELIC. Rare, open slopes, dry woods: Alexander Valley quad. < 10000 ft. June-Aug.

Melica bulbosa Geyer. ONIONGRASS. Uncommon, rocky hills, woods: 12 mi. from Calistoga on Alexander Valley Road, *Howell* (1934). < 11800 ft. July-Aug.

Melica californica Scribner. WESTERN MELIC, CALIFORNIA MELIC. Common, rocky places, grassy slopes: hills above Russian Gulch, *Tofsrud* (1941 NCC); 3 mi. n. of Occidental, *Morgan* (1961); quarry on Wagnon Road, *Wagnon* (1939 NCC); near Stony Point, *Baker* (1931 NCC); 9 mi. from Petaluma on road to Marshall, *Koch* (1939 NCC); 3 mi. w. of Petaluma on I St. Ext., *Morgan* (1956); near Sonoma,

Heller (1902); hills near Santa Rosa, *Heller* (1902); near railroad cut 3 mi. s. of Healdsburg, *Heller* (1902); s. slopes of Fitch Mtn. near Healdsburg, *Rubtzoff* (1952); 12 mi. from Calistoga on Alexander Valley Road, *Howell* (1934). < 7000 ft. Apr.-May.

Melica geyeri Bolander. GEYER'S ONIONGRASS. Frequent, Redwood-Douglas-fir forests, occasionally in oak woodland: Fort Ross quad.; Bohemian Hwy. 3½ mi. n. of Occidental, *Best* (1981); Sturgeon's Mill on Furlong Road, *Wagnon* (1939); 5 mi. w. of Sebastopol, *Wagnon* (1939 NCC); Tresch Ranch, Walker Road near Two Rock, acc. J. & B. Guggolz (1992); Button Ranch on Walker Road (U.C. property) between Petaluma-Valley Ford Road and Pepper Road, acc. Connors (1992); Fairfield Osborn (H); Bouverie; Bennett Valley, *Koch* (1940 NCC); Pepperwood; Gates Road off Calistoga Road, acc. Wells et al. (1984); ne. slope of Fitch Mtn. near Healdsburg, *Rubtzoff* (1952); Lake Sonoma area; near beginning of Pine Flat Road, acc. Edwards (1989); 2440 ft. elev. Caldwell Ranch, acc. Clifton (1984). < 6700 ft. Apr.-July.

Melica harfordii Bolander. HARFORD'S MELIC. Occasional, shaded slopes, woods: 2 mi. nw. of Mt. View School near Sebastopol, *Yates* (1936); Bouverie; Calistoga Road, *Klimansky* (1930 NCC); 3300 ft. elev. upper Squaw Cr. drainage 1.2 airmiles nw. of Geyser Rock, acc. Stone (1983). < 7400 ft. May-July.

Melica imperfecta Trinius. SMALL-FLOWERED MELIC. Probably poorly collected, brush-covered hillsides: Bodega Dunes, *Connors* (1991 BML); Youngman's Can. on Trinity Road 2½ mi. e. of Glen Ellen, *Wagnon* (1939). < 5000 ft. Apr.-May.

Melica subulata (Grisebach) Scribner. ALASKA ONIONGRASS. Damp areas: Duncans Mills Marsh, *Rubtzoff* (1959); Jonive Cr. 5 mi. w. of Sebastopol, *Wagnon* (1939); Lake Sonoma quad. < 8700 ft. May-July.

Melica torreyana Scribner. TORREY'S MELIC. Common, many habitats: Salt Point; Cazadero quad.; Bohemian Hwy. between Occidental and Monte Rio, *Best* (1981, 1983); Fairfield Osborn (H); 1 mi. n. of Sonoma, *Rose* (1965); Bouverie; Sobre Vista Road between El Verano and Glen Ellen, *Raven* (1950); Trinity Road, *Wagnon* (1939 NCC); w. base of Mt. Hood, *Heller* (1902); summit of Bennett Valley Road, *Koch* (1939 NCC); Annadel, *Wells* (1984); Spring Lake; near Santa Rosa, *Heller* (1902); Monan's Rill e. of Calistoga Road and n. of St, Helena Road, *Wells* (1985); Pepperwood; s. slopes of Fitch Mtn, *Rubtzoff* (1952); Lake Sonoma area; Geyserville, Alexander Valley and Ida Clayton Road quads.; 3600 ft. elev. Geyser Rock, acc Clifton (1983); 2040 ft. elev. Squaw Cr., acc. Clifton (1984); 2440 ft. elev. Caldwell Ranch, acc. Clifton (1984). < 4000 ft. Mar.-June.

Nassella (Key ± acc. M. E. Barkworth, *Hickman* [1993]).

1. Distal segment of awn wavy
   2. Glumes 12-21 mm; awn 45-110 mm . . . . . *N. cernua*
   2: Glumes 4.5-12 mm; awn 20-46 mm . . . . . *N. lepida*

1: Distal segment of awn straight
   3. Lemma glabrous between veins . . . . . *N. formicarum*
   3: Lemma hairy between veins . . . . . *N. pulchra*

Nassella cernua (Stebbins & Löve) Barkworth. NODDING NEEDLEGRASS. Occasional, dry slopes, open areas: Coleman Valley Road w. of Occidental, *Giampaoli* (1968 NCC); Cotati, *P. Watters* (1965 NCC); Cazadero, Geyserville and Alexander Valley quads. < 4700 ft. Apr.-May. —*Stipa c.* Stebbins & Löve. *S. lepida* Hitchcock var. *andersonii* (Vasey) Hitchcock.

★ **Nassella formicarum** (Delessert) Barkworth. Uncommon, maritime exposures: outskirts of Jenner extending n. past Russian Gulch up to the high prairie flat at jct. of Meyers Grade Road and Hwy. 1, *Amme & Edwards* (1988 UT, DAV & JEPS). < 900 ft. Native of Chile. —*Stipa f.* Delessert.

Nassella lepida (A. Hitchcock) Barkworth. FOOTHILL NEEDLEGRASS. Common, bunch grass throughout hilly areas: Salt Point; Cazadero and Duncans Mills quads.; 1024 ft. elev. Coleman Valley Road, acc. Clifton (1984); between Salmon Cr. and Bodega Bay, *Connors* (1991 BML); 5 mi. w. of Sebastopol, *Wagnon* (1939); Button Ranch on Walker Road (U.C. property) between Petaluma-Valley Ford Road and Pepper Road, acc. Connors (1992); Fairfield Osborn (H); Bennett Valley, *Koch* (1940 NCC); summit Spring Mtn. Road at co. line, *Baker* (1945); Pepperwood; s. slopes of Fitch Mtn. near Healdsburg, *Rubtzoff* (1952); Lake Sonoma area; Geyserville and Alexander Valley quads.; 2860 ft. elev. Pine Flat Road, acc. Clifton (1984). < 5700 ft. Mar.-May. —*Stipa l.* A. Hitchcock.

Nassella pulchra (A. Hitchcock) Barkworth. PURPLE NEEDLEGRASS. Very common, bunchgrass, open flats, slopes: The Sea Ranch; 5 mi. w. of Sebastopol, *Wagnon* (1939); Todd Road Preserve (H&K); Fairfield Osborn (H); Tresch Ranch, Walker Road near Two Rock, acc. J. & B. Guggolz (1992); Button Ranch on Walker Road (U. C. property) between Petaluma-Valley Ford Road and Pepper Road, *Connors* (1992 BML); 7 mi. out Petaluma-Marshall Road, *Koch* (1939 NCC); w. of Petaluma on D St. Ext., *Morgan* (1956); w. of Sears Point Raceway in Tolay Cr. drainage, acc. Patterson (1990); Sonoma Valley, *Jepson* (1911 JEPS); near Sonoma, *Heller* (1902); Bouverie; Maplewood 2 mi. n. of Kenwood, *Baker* (1945); Annadel, *Wells* (1984); Rincon Ridge, *Baker* (1946 NCC); Spring Lake; Santa Rosa, *Giampaoli* (1970 NCC); Monan's Rill e. of Calistoga Road and n. of St. Helena Road, *Wells* (1985); Mark West Springs Road, *S. Miller* (1965 NCC); Pepperwood; 3 mi. s. of Healdsburg on US 101, *Crampton* (1957); Healdsburg, *Heller* (type 1902 UC); 2040 ft. elev. Squaw Cr., acc. Clifton (1984). The California State grass. < 5000 ft. Mar.-May. —*Stipa p.* A. Hitchcock.

Panicum

1. First glume not > ¼ as long as spikelet, truncate or broadly triangular; sheaths smooth . *P. dichotomiflorum*
1: First glume as much as ½ as long as spikelet, acute or acuminate; sheaths hispid
    2. Panicle erect; spikelets not > 4 mm long
        3. Fruit without a scar at base . . . . *P. capillare*
        3: Fruit with a lunate scar at base . . . . *P. hillmanii*
    2: Panicle drooping; spikelets 4.5-5 mm long . . . . *P. miliaceum*

Panicum capillare Linnaeus. WITCHGRASS. Occasional, damp, open ground: beach on Russian River at Odd Fellows Park near Guerneville, *Howell* (1957); Vacation Beach near Guerneville, *Rubtzoff* (1956); beach at Forest Hills near Hacienda, *Rubtzoff* (1971); Bouverie; along Channel Drive ca ½ mi. from Spring Lake Regional Park, *Wagnon* (1989 NCC). < 5000 ft. July-Sept. —*P. c.* var. *occidentale* Rydberg, currently not recognized.

★ **Panicum dichotomiflorum** Michaux. Uncommon, moist places as a weed: Todd Road Preserve, *Howell* (1987). < 1700 ft. July-Oct. Native of e. U. S.

★ **Panicum hillmanii** Chase. Infrequent, around habitations, rural areas: 1 mi. n. of Graton on Hwy. 116, *Thomas* (1969); w. of Santa Rosa in Laguna de Santa Rosa drainage, *Harrison* (1978 NCC); Meacham Road 0.35 mi. s. of Stony Point Road, *Rubtzoff* (1971); firehouse at Penngrove, *Howell & Knight* (1986 NCC); Pepperwood. < 1500 ft. July. Native of Great Plains, U.S.A.

★ **Panicum miliaceum** Linnaeus. BROOM CORN MILLET. Infrequent, found in bird seed: Rincon Valley, *Wells* (1985); Spring Lake Village. < 3300 ft. Aug.-Oct. Native of Eurasia.

★ **Parapholis incurva** (Linnaeus) C. E. Hubbard. SICKLE GRASS. Uncommon, subsaline places: s. side of mouth of Russian River, *Best* (1981); Bodega Head, acc. Connors (1991). < 400 ft. Apr.-June. Native of Europe. —*Pholiurus incurvus* Schinz & Thellung.

Paspalum

1. Racemes a pair at the summit of the culm . . . . *P. distichum*
1: Racemes several to many, forming a panicle . . . . *P. dilatatum*

★ **Paspalum dilatatum** Poiret. DALLIS GRASS. Common, ditches, waste places, roadsides: The Sea Ranch; between Salmon Cr. and Bodega Bay, *Connors* (1991 BML); 1½ mi. w. of Cunningham, *Baker* (1946 NCC); Cunningham Marsh, *Howell* (1946), *Rubtzoff* (1961); w. of Santa Rosa in Laguna de Santa Rosa drainage, *Harrison* (1978 NCC); near sewage treatment plant on Llano Road, *Knight & Howell* (1986); Todd Road Preserve; Butler Road off Stony Point Road, *Wells* (1984); s. side of Copeland Cr., Rohnert Park, *Self* (1979 NCC); Ducker Cr., Rincon Valley, *Wells* (1985); Spring Lake, *Aldrich* (1981); Pepperwood. < 1400 ft. May-Nov. Native of S. America

Paspalum distichum Linnaeus. KNOT GRASS. Common, low, moist areas, ditches: Gualala Point State Park near bridge, *Knight* (1985); Guerneville Marsh, *Rubtzoff* (1954); Vacation Beach near Guerneville (1954); Pitkin Marsh, *Rubtzoff* (1951); Laguna de Santa Rosa, *Rubtzoff* (1958); 5 mi. w. of Sebastopol, *Wagnon* (1940 NCC); Atascadero Cr. Marsh, *Rubtzoff* (1958); Todd Road Preserve; Rohnert Park, *Self* (1979 NCC); Spring Lake; Pepperwood. < 5700 ft. June-Oct.

★ **Pennisetum clandestinum** Chiovenda. KIKUYU GRASS. Uncommon, troublesome weed of gardens and orchards: Bodega Marine Lab. housing, *Connors* (1991 BML); Spring Lake, *Rubtzoff* (1971). < 400 ft. Introduced from Africa. Noxious weed.

Phalaris

1. Spikelets in groups of 7, 1 fertile surrounded by 6 sterile, the group falling entire
   2. Outer glumes of all sterile spikelets clavate instead of only lower part of panicles having such . . . . . *P. paradoxa* var. *praemorsa*
   2: Outer glumes of all sterile spikelets not so . . . . . *P. p.* var. *paradoxa*
1: Spikelets all alike, not falling entire in groups
   3. Plants perennial
      4. Rhizomes lacking; panicle 2-5 cm long, dense, ovate or short-oblong . . . . . *P. californica*
      4: Rhizomes present; panicle 5-18 cm long, narrow
         5. Lemmas of sterile spikelets both developed, ca 1 mm long . . . . . *P. arundinacea*
         5: Lemmas of sterile spikelets solitary, ca 1.5 mm long . . . . . *P. aquatica*
   3: Plants annual
      6. Glumes broadly winged; panicle ovate to short-oblong
         7. Lemmas of sterile spikelets solitary; lemmas of fertile spikelets 3 mm long . . . . . *P. minor*
         7: Sterile lemmas 2; lemmas of fertile spikelets 4-6 mm long . . . . . *P. canariensis*
      6: Glumes wingless or nearly so; panicles linear or oblong
         8. Glumes wingless, acuminate; lemmas of fertile spikelets turgid, apex smooth, acuminate . . . . . *P. lemmonii*
         8: Glumes narrowly winged toward the summit, abruptly pointed or acute; lemmas of fertile spikelets less turgid, villous to the acute apex . . . . . *P. angusta*

Phalaris angusta Nees. Occasional, wet places: Freestone Marsh, *Rubtzoff* (1962); 3 sites w. of Santa Rosa in Laguna de Santa Rosa drainage, *Harrison* (1978 NCC); Cotati, *Tunison* (1968 NCC); Sulphur Cr. at The Geysers, *Koch* (1940 NCC). < 3300 ft. May-June.

★ **Phalaris aquatica** Linnaeus. HARDING GRASS. Fairly common, summer forage for livestock, open places: The Sea Ranch; Stewarts Point-Skaggs Springs Road 2 mi. e. of Stewarts Point, *Best* (1981); Bodega Head, acc. Connors (1991); Freestone, *Huntington* (1987 NCC); Llano Road near sewage treatment plant, *Knight & Howell* (1987); Lafferty Ranch at Sonoma Mtn., acc. Arnold et al. (1971); Fairfield Osborn (S); Live Oak Road in Liberty Dist. w. of Petaluma, *Knight* (1963); Annadel, *Wells* (1984); Rincon Valley, *Wells* (1984); Spring Lake; Pepperwood; Geyserville and Alexander Valley quads. < 4000 ft. June-Sept. Native of Old World. —*P. tuberosa* Linnaeus var. *stenoptera* (Hackel) A. Hitchcock.

★ **Phalaris arundinacea** Linnaeus. REED CANARY GRASS. Uncommon, moist soil: Laguna de Santa Rosa near Occidental Road, *Rubtzoff* (1958); Spring Lake. Escape from cultivation. < 5400 ft. May-Aug. Native of much of N. America and Eurasia, but not in Sonoma County.

Phalaris californica Hooker & Arnott. Occasional, damp areas, w. part of county: roadside between Plantation and Hwy. 1, *Koch* (1940 NCC); Stewarts Point-Skaggs Springs Road near Stewarts Point, *Baker* (1948 NCC); Salt Point; Freestone Marsh, *Rose* (1962); 5 mi. w. of Sebastopol, *Wagnon* (1939 NCC). < 2500 ft. May-Nov.

★ **Phalaris canariensis** Linnaeus. CANARY GRASS. Rare, open areas: 5 mi. w. of Sebastopol, *Wagnon* (1939 NCC); Llano Road, *Klimansky* (1930 NCC). < 1000 ft. Apr.-July. Introduced from Mediterranean region.

Phalaris lemmonii Vasey. Common, moist places: Fife Cr. Marsh near Guerneville, *Rubtzoff* (1960); s. of Santa Rosa at Bellview community, *Klimansky* (1931 NCC); Todd Road Preserve (H&K); near Wright School, *Wagnon* (1940 NCC); 4 mi. w. of Santa Rosa on Sebastopol Road, *Howell* (1947); 3 sites w. and sw. of Santa Rosa in Laguna de Santa Rosa drainage, *Harrison* (1978 NCC); Stony Point Road near Gravenstein Hwy., *Morgan* (1961); Pepperwood; Old Redwood Hwy. n. of Windsor, *Baker* (1931 NCC); Windsor vernal pool, *Oden* (1981). < 2600 ft. Apr.-June.

★ **Phalaris minor** Retzius. Occasional, usually damp areas: Willow Cr. Marsh, *Rubtzoff* (1959); Freestone Marsh, *Rubtzoff* (1962); 2 sites w. of Santa Rosa in Laguna de Santa Rosa drainage, *Harrison* (1978 NCC); Lakeville Hwy. e. of Petaluma, *Rubtzoff* (1962); Rohnert Park, *Self* (1979 NCC); 2200 ft. elev. nw. flank of Mt. St. Helena, *Grummer* (1982 PUA). < 2500 ft. Apr.-July. Introduced from Mediterranean region.

★ **Phalaris paradoxa** Linnaeus var. **paradoxa**. Occasional, waste places: Willow Cr. Marsh, *Rubtzoff* (1959); Freestone Marsh, *Rubtzoff* (1962); 2 sites w. of Santa Rosa in Laguna de Santa Rosa drainage, *Harrison* (1978 NCC); Lakeville Hwy. e. of Petaluma, *Rubtzoff* (1962); Rohnert Park, *Self* (1979 NCC); 2200 ft. elev. nw. flank of Mt. St. Helena, *Grummer* (1982 PUA). < 2200 ft. May-Aug. Introduced from Mediterranean region.

★ **Phalaris paradoxa** var. **praemorsa** (Lamarck) Cosson & Durieu. Rare, waste places: Spring Lake. < 1000 ft. May-Aug. Introduced from Mediterranean region.

Phleum

1. Panicle ovoid, not > twice as long as thick; awn of glumes 2 mm long . . . . *P. alpinum*
1: Panicle long-cylindrical; awn of glumes 1 mm long . . . . *P. pratense*

Phleum alpinum Linnaeus. ALPINE TIMOTHY. Occasional, dunes, coastal swales: 5 mi. n. of Jenner, *Keck & Clausen* (1939); Bodega Point, *Eastwood* (1915); s. end of Bodega Bay (now Bodega Harbour development), *Wagnon* (1940 NCC). <11500 ft. July-Aug.

★ **Phleum pratense** Linnaeus. CULTIVATED TIMOTHY. Occasional, forage for livestock, escaped from cultivation: Salt Point; Guerneville, *Davy* (1901 JEPS); Perry Marsh, *Howell* (1945); Atascadero Cr. Marsh, *Rubtzoff* (1958); 5th St., Petaluma, *Morgan* (1955); Glen Ellen, *Davy* (1901 JEPS). < 7000 ft. May-June. Introduced from Eurasia.

Phragmites australis (Cavanilles) Steudel. COMMON REED. Rare, forming canelike thickets, wet places: Pepperwood; 2720 ft. elev. Caldwell Ranch, acc. Clifton (1984). < 5400 ft. July-Nov. —*P. communis* Trinius var. *berlandieri* (Fournier) Fernald.

★ **Piptatherum miliaceum** (Linnaeus) Cosson. SMILO GRASS. Possibly common, waste places: The Sea Ranch, acc. Wells (1982); Spring Lake. < 1000 ft. Apr.-Sept. Introduced from Mediterranean region. —*Oryzopsis m.* (Linnaeus) Ascherson & Schweinfurth.

Pleuropogon

1. Lemma 4-6 mm long; leaf blades mostly 2-4 mm wide . . . . *P. californicus*
1: Lemma 6-9 mm long; leaf blades mostly 5-12 mm wide . . . . *P. hooverianus*

Pleuropogon californicus (Nees) Bentham. SEMAPHORE GRASS. Common, generally low, wet places: Cunningham Marsh, *Howell* (1947); many sites w. of Santa Rosa in Laguna de Santa Rosa drainage, *Harrison* (1978 NCC); Todd Road Preserve (H&K); Llano Road near Todd Road, *Klimansky* (1931 NCC); between Petaluma and Chileno Valley, *Howell* (1944); w. outskirts of Sonoma, *Baker* (1952 NCC); near El Verano, *Best* (1980); Rincon Valley, *Wells* (1984); low spot e. of Santa Rosa fairgrounds, *Wagnon* (1939 NCC); Pepperwood; 2 mi. nw. of Windsor near Old Redwood Hwy., *de Nevers* (1981); property bordered by Chianti and Dutcher Cr. roads w. of US 101 and w. of Geyserville, acc. J. & B. Guggolz (1991). < 2400 ft. Mar.-May.

Pleuropogon hooverianus (L. Benson) J. T. Howell. NORTH COAST SEMAPHORE GRASS. Occasional, meadows: Salmon Cr., *Clifton* (1981 PUA); Freestone area on road to Occidental in ditch on n. side of Freestone Flat Road, acc. Lovell (1977); ½ mi. w. of Llano Road, 1½ mi. e. of Hwy. 116 on Todd Road n. of right angle turn of Todd Road, acc. Rae (1982); marshy area near Pressley Road, acc. Lovell (1977). < 1700 ft.

Poa (Key ± acc. D. D. Keck, *Munz* [1959]).

1. Plants annual
    2. Lemmas pubescent, but not webbed at base . . . . *P. annua*
    2: Lemmas glabrous or pubescent, with a tuft of cobwebby hairs at base . . . . *P. howellii*
1: Plants perennial (*P. kelloggii* sometimes behaving as an annual)
    3. Creeping rhizomes present
        4. Plants dioecious, the ♂ and ♀ florets morphologically similar . . . . *P. douglasii*
        4: Plants not dioecious, the florets perfect
            5. Panicle open, pyramidal, its elongated lower branches lax and only in outer ½ floriferous; lemmas with a tuft of cobwebby hairs at base

6. Lemmas pilose on keel and marginal nerves; strongly perennial . . . . *P. pratensis*
6: Lemmas glabrous except for the web; annual or weakly perennial . . . . *P. kelloggii*
5: Panicle more contracted, oblong, its short lower branches not lax; lemmas not webbed at base, although sometimes vestigially webbed . . . . *P. compressa*
3: Creeping rhizomes wanting
7. Florets mostly age into dark purple bulblets; culm with bulblike base . . . . *P. bulbosa*
7: Florets normal, green; culms not bulblike at base
8. Lemmas with a tuft of cobwebby hairs at base
9. Lemmas glabrous, or the keel slightly pubescent . . . . *P. trivialis*
9: Lemmas pubescent on keel and marginal nerves . . . . *P. palustris*
8: Lemmas not webbed at base; lemmas ± hairy on back, keel or nerves, at least towards base, sometimes obscurely so, occasionally glabrous in *P. secunda*
10. Spikelets distinctly compressed; lemmas keeled, lemmas puberulent towards base only . . . . *P. unilateralis*
10: Spikelets little compressed, elongate; lemmas rounded on back, crisp puberulent toward base, the keel obscure . . . . *P. secunda*

★ **Poa annua** Linnaeus. ANNUAL BLUEGRASS. Common, cultivated fields, moist, disturbed areas, lawns, gardens: The Sea Ranch; Fort Ross-Russian River-Bodega areas, *Voznesenskii* (1840-1841 LE); Willow Creek; Larkspur Rock at Bodega Bay, *Bufford* (1936 NCC); 2 sites w. of Santa Rosa in Laguna de Santa Rosa drainage, *Harrison* (1978 NCC); Todd Road Preserve; Fairfield Osborn (H); gardens and lawns at Royal Oak Mobile Home Park, Petaluma, acc. Knight (1989); ¼ mi. n. of Agua Caliente in Sonoma Valley, *Baker* (1952 NCC); Sugarloaf Ridge; Spring Lake Village: Spring Lake; Pepperwood; Lake Sonoma area; 2190 ft. elev. Squaw Cr., acc. Clifton (1984). < 6700 ft. Winter and spring; continuously when moisture permits. Introduced from Europe.

★ **Poa bulbosa** Linnaeus. BULBOUS BLUEGRASS. Common, waste places: Button Ranch on Walker Road (U. C. property) between Petaluma-Valley Ford Road and Pepper Road, *Connors* (1992 BML); plaza in Cotati, *Morgan* (1961); Bouverie; Lake Sonoma area. < 6700 ft. Spring. Introduced from Europe.

★ **Poa compressa** Linnaeus. CANADIAN BLUEGRASS. Rare, open or partially shaded places: Watson School on Bodega Hwy. 3 mi. from Freestone, *Best* (1984). < 6000 ft. Summer. Introduced from Europe.

Poa douglasii Nees. SAND-DUNE BLUEGRASS. Occasional, dunes, coastal bluffs: The Sea Ranch (at that time Olson Ranch), *P. B. Kennedy* (1914); Salt Point; s. side of mouth of Russian River, *Best* (1981); near mouth of Salmon Cr., *Baker* (1931); Bodega Bay, *Heller* (1902 JEPS); Bodega Head. Usually < 400 ft. Early summer.

Poa howellii Vasey & Scribner. HOWELL'S BLUEGRASS. Occasional, open places: quarry on Wagnon Road near Sebastopol, *Wagnon* (1939); w. slope of Mt. Hood, *Baker* (1947 NCC); summit of St. Helena Road, *Baker* (1942 NCC); Little Sulphur Cr. near bridge, *Koch* (1940 NCC). < 3300 ft. Spring. —*P. bolanderi* Vasey ssp. *h.* (Vasey & Scribner) Keck.

Poa kelloggii Vasey. Occasional, shady places, moist woods, coastal influence: The Sea Ranch (at that time Olson Ranch), *P. B. Kennedy* (1919); Stewarts Point-Skaggs Springs Road 1 mi. e. of Stewarts Point, *Best* (1981); 249 ft. elev. Salt Point, *Clifton* (1981 PUA). < 1700 ft. Spring.

Poa palustris Linnaeus. FOWL BLUEGRASS. Open ground, moist ditches and meadows: 2550 ft. elev. nw. flank Mt. St. Helena, *Grummer* (1982 PUA); 2000 ft. elev. geysers Unit 18, *Malloch* (1977 PG&E). < 6700 ft. Spring to early summer. Introduced from Europe.

★ **Poa pratensis** Linnaeus. KENTUCKY BLUEGRASS. Common, open areas, meadows, damp places: Duncans Mills Marsh, *Rubtzoff* (1957); Austin Cr. Marsh, *Rubtzoff* (1960); Fife Cr. Marsh near Guerneville, *Rubtzoff* (1960); 1024 ft. elev. Coleman Valley Road, acc. Clifton (1984); upper Stemple Cr., *Rubtzoff* (1961); Bodega Head, *Rubtzoff* (1960); 2 mi. ne. of Occidental, *Best* (1981); Pitkin Marsh, *Rubtzoff* (1951); Laguna de Santa Rosa, *Rubtzoff* (1958); 2 sites w. of Santa Rosa in Laguna de Santa Rosa drainage, *Harrison* (1978 NCC); Petaluma, *Morgan* (1961); Royal Oaks Mobile Home Park at lawn margins, acc. Knight (1989); near Santa Rosa, *Tofsrud* (1941 NCC); 3200 ft. elev. Socrates Mine, acc. Clifton (1984); 2190 ft. elev. acc. Clifton (1984); 2800 ft. elev. Caldwell Ranch, acc. Clifton (1984). Commonly cultivated for lawns. < 11700 ft. Spring to early summer. Introduced from Europe.

Poa secunda J. S. Presl. PINE BLUEGRASS, PACIFIC BLUEGRASS. Common, open, rocky slopes: Freestone, *Huntington* (1987 NCC); 2 mi. w. of Occidental, *Baker & Bolton* (1936 NCC), *Wagnon* (1939 NCC); Stoetz Lane ne. of Occidental, *Best* (1981); 5 mi. w. of Sebastopol, *Wagnon* (1940 NCC); Button

Ranch on Walker Road (U. C. property) between Petaluma-Valley Ford Road and Pepper Road, *Connors* (1992 BML); Fairfield Osborn (H); near El Verano, *Koch* (1940 NCC), *Best* (1980); Bouverie; Adobe Can. near Kenwood, *Brewer* (1930 NCC), *Baker* (1950 NCC); Calistoga Road, *Klimansky* (1934 NCC); Monan's Rill n. of St. Helena Road and e. of Calistoga Road, acc. Wells (1985); Santa Rosa Cr. Can., *Koch* (1941 NCC); e. of Santa Rosa Cemetery *Koch* (1941 NCC); Larkspur Rock near Bodega Bay, *Baker* (1935 NCC); Pepperwood; Geysers geothermal area, *Quibell* (1980 NCC); Lake Sonoma area; 2440 ft. elev. Caldwell Ranch, acc. Clifton (1984). < 12700 ft. Spring and early summer. —*P. scabrella* (Thurber) Vasey.

★ **Poa trivialis** Linnaeus. ROUGH BLUEGRASS. Common, moist places: Willow Cr. Road ½ mi. from Bridgehaven, *Ferris & Lorraine* (1961); Willow Cr. Marsh, *Rubtzoff* (1959); Duncans Mills Marsh, *Rubtzoff* (1955); Fife Cr. Marsh near Guerneville, *Rubtzoff* (1960); Freestone, *Huntington* (1987 NCC); Guerneville Marsh, *Rubtzoff* 1954; Pitkin Marsh, *Rubtzoff* (1951); Laguna de Santa Rosa, *Rubtzoff* (1960); laguna just e. of Sebastopol, *Howell* (1951); Fairfield Osborn (S). < 2400 ft. Spring and early summer. Native of Europe.

Poa unilateralis Vasey. OCEAN-BLUFF BLUEGRASS. Occasional, rocky meadows, cliffs and bluffs at the coast: The Sea Ranch; ocean cliffs at Salt Point, *Best* (1984); rocky slope 8 mi. n. of Bodega Bay, *Howell* (1931); Bodega Bay, *Heller* (1902 JEPS); Bodega Head, *Best* (1985). < 700 ft. Spring.

Polypogon

1. Plants annual
   2. Glumes slightly lobed, the lobes not ciliate . . . . *P. monspeliensis*
   2: Glumes prominently lobed, the lobes ciliate-fringed . . . . *P. maritimus*

1: Plants perennial
   3. Awns ± stiff and straight; ligules 3-8 (13) mm long; culms tufted . . . . *P. interruptus*
   3: Awns delicate, flexuous; ligules 2-3 mm long; culms erect and not tufted . . . . *P. australis*

★ **Polypogon australis** Brongniart. CHILEAN BEARD GRASS. Common, coastal wet areas: The Sea Ranch, *Wells* (1978); Duncans Mills Marsh, *Rubtzoff* (1955); Willow Cr. Marsh, *Rubtzoff* (1959); Bodega Bay Marsh, *Rubtzoff* (1962); Pitkin Marsh, *Rubtzoff* (1951). < 3300 ft. May-June. Introduced from Chile and Argentina.

★ **Polypogon interruptus** Kunth. DITCH BEARD GRASS. Common, low places, fields, meadows, roadside ditches: Stump Beach at Salt Point State Park, *Wells* (1982); 69 ft. elev. Salmon Cr., acc. Clifton (1984); Bodega Marine Reserve Freshwater Marsh, *Connors* (1991 BML); ca 1 mi. nw. of Camp Meeker, *J. & B. Guggolz* (1991 NCC); Vacation Beach on Russian River, *Rubtzoff* (1956); Forestville Marsh, *Rubtzoff* (1955); 5 mi. w. of Sebastopol, *Wagnon* (1939 NCC); Laguna de Santa Rosa, *Rubtzoff* (1960); marsh near Todd Road Preserve, *Howell & Knight* (1986); Copeland Cr., Rohnert Park, *Self* (1979 NCC); Fairfield Osborn (S); Pepperwood; Lake Sonoma area; 3160 ft. elev. Socrates Mine, *Clifton* (1984 PUA); 2190 ft. elev. Squaw Cr., acc. Clifton (1984). < 4400 ft. May-Aug. Native of South America.

★ **Polypogon maritimus** Willdenow. MEDITERRANEAN BEARD GRASS. Occasional, low, damp areas: Austin Cr. Marsh, *Rubtzoff* (1960); Rus-sian River bottom at Forest Knolls (near Hacienda), *Rubtzoff* (1969); Vacation Beach near Guerneville, *Rubtzoff* (1962); 3 sites w. and sw. of Santa Rosa in Laguna de Santa Rosa drainage, *Harrison* (1978 NCC); 6.5 mi. ne. of Penngrove on Lichau Road, *Rubtzoff* (1968); Hwy. 37 near Sears Point Raceway, *Baker* (1946 NCC); Bennett Mtn. Lake, *Rubtzoff* (1961); Spring Lake, *Wells* (1987); Pepperwood, *Breedlove* (1980). < 2000 ft. June. Introduced from Mediterranean region.

★ **Polypogon monspeliensis** (Linnaeus) Desfontaines. RABBITFOOT GRASS, ANNUAL BEARD GRASS. Very common, low places where water accumulates during the spring: The Sea Ranch; Stump Beach at Salt Point State Park, *Wells* (1983); 4 mi. w. of Sebastopol, *Wagnon* (1939 NCC); 4 sites w. and sw. of Santa Rosa in Laguna de Santa Rosa drainage, *Harrison* (1978 NCC); Todd Road Preserve (H&K); Copeland Cr. ¼ mi. e. of US 101, *Self* (1979 NCC); Fairfield Osborn (S); Button Ranch on Walker Road (U. C. property) between Petaluma-Valley Ford Road and Pepper Road, acc. Patterson (1990); w. of Sears Point Raceway in Tolay Cr. drainage, acc. Patterson (1990); Griffin Road, Kenwood, *Baker* (1931 NCC); Rincon Valley, *Klimansky* (1930 NCC); Spring Lake; Mark West Springs Road, *Baker & Wagnon* (1957 NCC); Pepperwood; Circle 8 Ranch on Pine Flat Road, *Knight* (1963); 3290 ft. elev. Socrates Mine, acc. Clifton (1984); 2800 ft. elev. Bear Cr. Can., acc. Clifton (1984); 2800 ft. elev. Caldwell Ranch, acc. Clifton (1984). < 7000 ft. Apr.-Aug. Introduced from s. & w. Europe.

Puccinellia nutkaensis (J. S. Presl) Fernald & Weatherby. ALASKA ALKALI GRASS. Uncommon, salt marsh: Bodega Bay area, *Clifton* (1980 PUA); Bodega Marine Reserve Salt Marsh, *Connors* (1991 BMC). < 40 ft. June-Aug. —*P. grandis* Swallen. *P. lucida* Fernald & Weatherby.

Scribneria bolanderi (Thurber) Hackel. Occasional, thin soil, rocky areas: Russian River Valley, *Bolander* (1864); Stoetz Lane 2 mi. ne. of Occidental, Best (1985); 5 mi. w. of Sebastopol, *Wagnon* (1940 NCC); 3 mi. se. of Sonoma, *Howell* (1965); Hwy. 12, ¼ mi. n. of Agua Caliente, Sonoma Valley, *Baker* (1952 NCC); hills e. of Agua Caliente, *Howell* (1948). < 9000 ft. Mar.-June. —*Lepturus b.* Thurber.

Setaria

1. Bristles below each spikelet numerous, at least > 5
    2. Plants annual . . . . *S. pumila*
    2: Plants perennial . . . . *S. gracilis*

1: Bristles below each spikelet 1 or 3
    3. Bristles ± retrorsely scabrous . . . . *S. verticillata*
    3: Bristles not retrorsely scabrous . . . . *S. viridis*

Setaria gracilis Kunth. KNOTROOT BRISTLE GRASS. Occasional, open ground, cultivated soil, waste places: Jonive Road near Bodega Hwy., *Best* (1983); Todd Road near Stony Point Road, *Rubtzoff* (1971); 0.6 mi. on Llano Road from Gravenstein Hwy., *Howell & Knight* (1986). 1400 ft. May-Sept. —*S. geniculata* (Lamarck) Beauvois, misapplied.

★ **Setaria pumila** (Poiret) Roemer & Schultes. Occasional, waste places, cultivated soil: between Salmon Cr. and Bodega Bay, *Connors* (1991 BML); Odd Fellows Park near Guerneville, *Howell* (1957); Forest Hills near Hacienda, *Rubtzoff* (1968); Meacham Road 0.35 mi. s. of Stony Point Road, *Rubtzoff* (1971); US 101 at Sonoma-Marin co. line, *Morgan* (1958). < 4000 ft. June-Oct. Introduced from Europe. —*S. lutescens* (Weigel) C. E. Hubbard. *S. glauca* (Linnaeus) P. Beauvois.

★ **Setaria verticillata** (Linnaeus) P. Beauvois. BUR BRISTLE GRASS. Uncommon, waste places, cultivated soil: near Twin Oaks s. of Penngrove, *Knight* (1963); Pepperwood. < 1200 ft. May-July. Native of Europe.

★ **Setaria viridis** (Linnaeus) Beauvois. BRISTLE GRASS. Occasional, waste places, cultivated soil: Todd Road Preserve (H&K); 0.6 mi. from Gravenstein Hwy. on Llano Road, *Howell & Knight* (1987); jct. of Todd and Llano roads, *Best* (1987); Franz Valley, Mark West Cr. and Pepperwood, *de Nevers* (1981). < 1200 ft. June-Aug. Introduced from Eurasia.

Sorghum: (Key ± acc. K. W. Allred, *Hickman* [1993]).

1. Plants perennial with rhizomes . . . . *S. halepense*

1: Plants annual . . . . *S. bicolor*

★ **Sorghum bicolor** (Linnaeus) Moench. SUDAN GRASS, MILO, SORGHUM. Up until 1940 it was fed to chickens as "milo" and plants around Petaluma are probably derived from this importation: banks of Russian River at Summerhome Park, *Oden* (1981); along River Road between Fulton and Mirabel Park, *Oden* (1981); Marshall Ave. w. of Petaluma, *Knight* (1963); Pepperwood. Milo and Sudan grass are entirely different looking. Sudan grass looks just like *S. halepense*. < 2000 ft. Aug.-Sept. Native of Africa —*S. sudanense* (Piper) Stapf.

★ **Sorghum halepense** (Linnaeus) Persoon. JOHNSONGRASS. Common, escaped from cultivation, becoming a troublesome weed: Summerhome Park on Russian River, *de Nevers* (1981); Lindberg Lane and Lakeville Hwy., Petaluma, *Knight* (1963); Stage Gulch Road 2½ mi. from Lakeville, *Morgan* (1956); Bennett Valley Road, *Stocking* (1963 NCC); Rincon Valley, *Wells* (1984); Russian River near Cloverdale, *Filippi* (1964 NCC). > 2600 ft. May-Aug. Native of Mediterranean region.

Spartina foliosa Trinius. CALIFORNIA CORD GRASS. Dominant in lower parts of salt marshes: Bodega Bay, acc. Connors (1991); Petaluma, *Congdon* (1880); Petaluma Salt Marsh, acc. Strong (1991) of BML. < 30 ft. July-Nov.

★ **Sporobolus cryptandrus** (Torrey) A. Gray. SAND DROPSEED. Infrequent, weed: 0.6 mi. n. of Gravenstein Hwy. on Llano Road, *Howell & Knight* (1987). Acc. D. Wilken, probably not native; widely distributed by Caltrans for roadside landscaping. < 10000 ft. May-Aug. Native only to sw. CA deserts and eastward.

★ **Taeniatherum caput-medusae** (Linnaeus) Nevski. MEDUSAHEAD. Occasional, weed, grassland: near Todd Road in Laguna de Santa Rosa drainage, *Harrison* (1978 NCC); Todd Road Preserve (H&K); Button Ranch on Walker Road (U. C. property) between Petaluma-Valley Ford Road and Pepper Road, acc. Connors (1992); Fairfield Osborn (S); Spring Lake; Knights Valley, *Breedlove* 1980; Pepperwood; Hot Springs Road sw. of Cloverdale, *Howell* 1967; Alexander Valley and Ida Clayton Road quads. < 3300 ft. June-July. Introduced from Europe. —*Elymus c.-m.* Linnaeus. Noxious weed.

Torreyochloa pallida (J. S. Presl) Church var. **pauciflora** (J. S. Presl) J. I. Davis. ALKALI GRASS, WEAK MANNAGRASS. Common, shallow water and low wet areas: Duncans Mills Marsh, *Rubtzoff* (1955); Willow Cr. Marsh, *Rubtzoff* (1959); Bodega Head, *Rubtzoff* (1955), *Connors* (1991 BML); Guerneville Marsh; *Rubtzoff* (1954); Forestville Marsh, *Rubtzoff* (1955); Atascadero Cr. Marsh, *Rubtzoff* (1958); 2 mi. n. of Sebastopol, *Howell* (1945), *Rubtzoff* (1958); Laguna de Santa Rosa, *Rubtzoff* (1958); Bennett Mtn. Lake, *Rubtzoff* (1961). < 11700 ft. July-Sept. —*Glyceria pauciflora* J. S. Presl.

Trisetum canescens Buckley. TALL TRISETUM. Common, moist areas: Stump Beach at Salt Point State Park, *Best* (1985); 9 mi. e. of Stewarts Point on Stewarts Point-Skaggs Springs Road, *Best* (1981); Annapolis and Fort Ross quads.; Joy Road w. of Occidental, *Ferris & Lorraine* (1961); near Guernewood Park, *Klimansky* (1931 NCC); 5 mi. w. of Sebastopol, *Wagnon* (1939); Pitkin Marsh, *Rubtzoff* (1951); Fairfield Osborn (H); Button Ranch on Walker Road (U. C. property) between Petaluma-Valley Ford Road and Pepper Road, *Connors* (1992 BML); Trinity Road, *Baker* (1941 NCC); Annadel; McDonald Reservoir (Lake Ralphine), Santa Rosa, *Baker* (1941 NCC); Pepperwood; Fitch Mtn. near Healdsburg, *Rubtzoff* (1952); Lake Sonoma area; Alexander Valley quad. < 9000 ft. May-Aug. —*T. cernuum* Buckley var. *c.* (Buckley) Beal.

★ **Triticum aestivum** Linnaeus. WHEAT. Occasional, vernal flats: jct. Crane Can. and Petaluma Hill roads, *C. Jones* (1980 NCC); 5.4 mi. from Santa Rosa on Hwy. 12 en route to Kenwood, *Morgan* (1956); Rincon Valley, *Baker* (1941 NCC); Rincon Ridge, *Raven* (1950); Mark West Cr. Road, *Baker* (1941 NCC); Pepperwood. < 4700 ft. Apr.-July. Native of Old World. —*T. vulgare* Villars. *T. compactum* Host.

Vulpia

1. Spikelets densely 5-12-flowered; lemmas without scarious margin . . . . . *V. octoflora*
1: Spikelets loosely 1-5-flowered; lemmas with narrow scarious margin
  2. Inflorescence narrow, branches ascending or appressed-ascending
    3. Lemmas conspicuously long-ciliate toward apex . . . . . *V. myuros* var. *hirsuta*
    3: Lemmas not ciliate
      4. Lower glume 2/3-3/4 as long as the 2nd . . . . . *V. bromoides*
      4: Lower glume < ½ as long as the 2nd . . . . . *V. myuros* var. *myuros*
  2: Inflorescence broader, the principal branches spreading
    5. Spikelets glabrous, or ± scabrous, but not with long well-developed hairs *V. microstachys* var. *pauciflora*
    5: Spikelets pubescent with long well-developed hairs
      6. Pedicels appressed or slightly spreading; lower branches of panicle usually spreading or reflexed
        7. Lemmas glabrous; glumes pubescent . . . . . *V. m.* var. *confusa*
        7: Lemmas hirsute; glumes glabrous or pubescent . . . . . *V. m.* var. *ciliata*
      6: Pedicels and panicle-branches all finally spreading or reflexed . . . . . *V. m.* var. *microstachys*

★ **Vulpia bromoides** (Linnaeus) S. F. Gray. SIX-WEEKS FESCUE. Common, marshes, moist flats: The Sea Ranch, *Wells* (1978); Fort Ross, Cazadero and Duncans Mills quads.; Willow Cr. Marsh. *Rubtzoff* (1959); Bodega Point, *Eastwood* (1915); Bodega Head Marsh, *Rubtzoff* (1960); Occidental Marsh border, *Rubtzoff* (1957); Laguna de Santa Rosa e. of Cunningham near Llano Road, Rubtzoff (1958); Laguna de Santa Rosa e. of Graton between Occidental and Guerneville roads, *Rubtzoff* (1960; Forestville Marsh, *Rubtzoff* (1955);

Pitkin Marsh, *Rubtzoff* (1951); 5 mi. w. of Sebastopol, *Wagnon* (1940); Cunningham Marsh, *Howell* (1947); Todd Road Preserve (H&K 1987); Stony Point Marsh, *Rubtzoff* (1955); Button Ranch on Walker Road (U. C. property) between Petaluma-Valley Ford Road and Pepper Road, acc. Patterson (1990); w. of Sears Point Raceway in Tolay Cr. drainage, acc. Patterson (1990); w. slope Mt. Hood, *Baker* (1947 NCC); Spring Lake; Santa Rosa, *Tufsrud* (1941 NCC); Monan's Rill e. of Calistoga Road and n. of St. Helena Road, *Wells* (1985); Pepperwood; ne. slope Fitch Mtn. near Healdsburg, *Rubtzoff* (1952); Lake Sonoma, Geyserville, Alexander Valley and Ida Clayton Road quads.; Sulphur Cr., *Koch* (1940 NCC); The Geysers, *Koch* (1940 NCC). Usually < 5000 ft. Apr.-June. Native of Europe. —*Festuca b.* Linnaeus, misapplied. *F. dertonensis* (Allioni) Acherson & Graebner.

Vulpia microstachys (Nuttall) Bentham var. **ciliata** (Beal) Lonard & Gould. Occasional, open places: hills e. of Agua Caliente, *Howell* (1943); sw. slope of Mt. Hood, *Baker* (1947 NCC); The Geysers, *Koch* (1940 NCC). Usually < 3300 ft. Apr.-June. —*Festuca eastwoodiae* Piper. *F. grayi* (Abrams) Piper.

Vulpia microstachys var. **confusa** (Piper) Lonard & Gould. Rare, dry hillsides: The Geysers, *Koch* (1940 NCC). Usually < 3300 ft. Mar.-June. —*Festuca c.* Piper.

Vulpia microstachys var. **microstachys**. NUTTALL'S FESCUE. Occasional, usually chaparral, grassy slopes: Pepperwood; Lake Sonoma area; Alexander Valley quad.; 3200 ft. elev. Socrates Mine, acc. Clifton (1984). Usually < 3300 ft. Apr.-June. —*Festuca m.* (Nuttall).

Vulpia microstachys var. **pauciflora** (Beal) Lonard & Gould. Common, around rocks and brush, woods and open slopes: Harrison Grade, *Wagnon* (1989 NCC); 5 mi. w. of Sebastopol, *Wagnon* (1940 NCC); 7 mi. from Petaluma to Marshall, *Koch* (1939 NCC); Kenwood, *Baker* (1941); sw. slope of Mt. Hood, *Baker* (1947 NCC); Mark West Springs, *Baker* (1934 NCC); Pepperwood; Sulphur Cr., *Koch* (1940 NCC); Lake Sonoma area; Alexander Valley quad.; 3090 ft. elev. Socrates Mine, acc. Clifton (1984); 2400 ft. elev. Little Geysers, *Malloch* (1977 PG&E); 2440 ft. elev. Caldwell Ranch, acc. Clifton, (1984). Usually < 5000 ft. Apr.-June. —*Festuca pacifica* Piper. *F. reflexa* Buckley.

★ **Vulpia myuros** (Linnaeus) C. Gmelin var. **hirsuta** (Hackel) Ascherson & Graebner. FOXTAIL FESCUE. Common, various habitats: Salt Point; Gold Ridge Road 4 mi. w. of Sebastopol, *Morgan* (1961); Fairfield Osborn (S); Bennett Mtn. Lake, *Rubtzoff* (1963); Lake Sonoma and Alexander Valley quads. Usually < 5000 ft. Apr.-June. Native of Europe. —*V. megalura* (Nuttall) Rydberg. *Festuca megalura* Nuttall.

★ **Vulpia myuros** var. **myuros**. RATTAIL FESCUE. Common, openings of chaparral, open slopes, disturbed areas Stewarts Point, Fort Ross and Cazadero quads.; Fairfield Osborn (S); Kenwood, *Baker* (1952 NCC); Adobe Can. Road near Kenwood, *Baker* (1945 NCC); Rincon Valley, *Wells* 1987; Spring Lake; Lake Sonoma area; Geyserville and Alexander Valley quads.; 3200 ft. elev. Socrates Mine, acc. Clifton (1984). Usually < 5000 ft. Mar.-May. Native of Europe. —*Festuca m.* Linnaeus.

Vulpia octoflora (Walter) Rydberg. SIX-WEEKS FESCUE. Uncommon, openings or burns, chaparral or brushy hillsides: Spring Mtn. Road on summit, *Baker* (1942 NCC); 3200 and 3290 ft. elev. Socrates Mine, acc. Clifton (1984). < 6600 ft. Apr.-June. —*Festuca o.* Walter.

POTAMOGETONACEAE. PONDWEED FAMILY
[Including RUPPIACEAE]

1. Flowers axillary or terminal in spikes or heads . . . . . *Potamogeton*
1: Flowers axillary and solitary, or very few in a group . . . . . *Ruppia*

Potamogeton

1. Stipules adnate to leaf-base, forming a sheath enfolding stem; leaf appearing as though arising from top of sheath
  2. Plants with submerged leaves only; fruiting spikes slender . . . . . *P. pectinatus*
  2: Plants with long-petioled, floating leaves and sessile, linear, submersed leaves, if floating leaves absent the fruiting spikes of the submersed parts capitate and sessile or on very short peduncles . . . *P. diversifolius*
1: Stipules axillary and free from the leaf-base, either free from one another or united to form a cylinder around stem
  3. Leaves all submersed and essentially alike; petioles short or absent
    4. Leaves lanceolate, oblong or ovate
      5. Leaves broadly lanceolate-attenuate, large, serrulate only at tip . . . . . *P. illinoensis*
      5: Leaves oblong and crisped, serrulate throughout, rounded at tip . . . . . *P. crispus*
    4: Leaves filiform or setaceous < 2 mm. broad
      6. Keel of fruit obviously dentate-winged; spikes capitate . . . . . *P. foliosus*

6: Keel of fruit not dentate; spikes short-cylindric . . . . *P. pusillus*

3: Leaves of two kinds, submersed and floating, the floating leaves with broad blades and long petioles

7. Submersed leaves lanceolate, flat, on long petioles . . . . *P. nodosus*

7: Submersed leaves linear, usually bladeless and filiform, 0.8-2 mm wide, blade when present linear-lanceolate and on a very long petiole; floating leaves broad, many-veined, base of blade subcordate . . . . *P. natans*

★ **Potamogeton crispus** Linnaeus. CRISPATE-LEAVED PONDWEED. Occasional, fresh water ponds, slow-running streams: Coleman Marsh submerged in Salmon Cr., *Rubtzoff* (1964); Bodega Head; Atascadero Cr. Marsh, *Rubtzoff* (1962 NCC); Button Ranch on Walker Road (U. C. property) between Petaluma-Valley Ford Road and Pepper Road, *Connors* (1992 BML); larger lake at Sonoma State Univ. campus, *Stuart* (1987 NCC). < 7000 ft. July-Sept. Native of Europe.

Potamogeton diversifolius Rafinesque. RAFINESQUE'S DIVERSE-LEAVED PONDWEED. Uncommon, still water, shallow ponds, slow streams: Todd Road Preserve; shallow area of Bennett Mtn. Lake, *Rubtzoff* (NCC 1962). < 8300 ft. May-June.

Potamogeton foliosus Rafinesque. LEAFY PONDWEED. Occasional, ± shallow, still, or slowly running water: Button Ranch on Walker road (U. C. property) between Petaluma-Valley Ford Road and Pepper Road in a metal stock-watering tank, *Connors* (1990 BML); Vacation Beach, *Rubtzoff* (1955 NCC); near Guerneville on Russian River, *Rattan* (1877). < 7700 ft. July-Oct.

Potamogeton illinoensis Morong. SHINING PONDWEED. Rare, cold, deep water: near Sebastopol, *Baker* (1899 NCC). < 7000 ft. June-Aug.

Potamogeton natans Linnaeus. FLOATING-LEAVED PONDWEED. Rare, marshy lakes, ponds: Duncans Mills Marsh, *Rubtzoff* (1955). < 9000 ft. June-Aug.

Potamogeton nodosus Poiret. LONG-LEAVED PONDWEED. Common, ponds, streams: Guerneville Marsh, *Rubtzoff* (1954 NCC); Button Ranch along Walker Road (U. C. property) between Petaluma-Valley Ford Road and Pepper Road, *Connors* (1992 BML); 1200 ft. elev., in 6 in. of water in Coney Lake near Kenwood, *Baker* (1954 NCC); Bennett Mtn. Lake, *Rubtzoff* (1954 NCC); Pepperwood; banks of Russian River n. of Cloverdale, *Heller* (1902). < 9200 ft. May-Aug.

Potamogeton pectinatus Linnaeus. FENNEL-LEAF PONDWEED. Occasional, still, brackish or fresh water: lower Salmon Cr. Marsh, *Rubtzoff* (1962); Bodega Head Marsh, *Clifton* (1980 PUA); e. of Aurora School, *Best & Follette* (1981); Button Ranch on Walker Road (U. C. property) between Petaluma-Valley Ford Road and Pepper Road, *Connors* (1992 BML); Russian River near Cloverdale, *Heller* (1902). < 8000 ft. May-July.

Potamogeton pusillus Linnaeus. SMALL PONDWEED. Common, slow-running streams and ponds: near mouth of Russian River, *Baker* (1899); Duncans Mills Marsh, *Rubtzoff* (1963 NCC); upper Stemple Cr. Marsh, *Rubtzoff* (1961); near Sebastopol, *Baker* (1899); Button Ranch on Walker Road (U. C. property) between Petaluma-Valley Ford Road and Pepper Road, *Connors* (1992 BML); 1 mi. w. of Schellville, *Raven* (1955); Pepperwood, *Breedlove* (1980). < 9000 ft. May-June.

Ruppia

1. Peduncles 0.2-2.5 cm long, not spiraling; leaves obtusish . . . . *R. maritima*

1: Peduncles 3-30 cm long, spiraling or flexuous; leaves acute . . . . *R. cirrhosa*

Ruppia cirrhosa (Petagna) Grande. Uncommon, brackish water: Lower Salmon Cr. Marsh, *Rubtzoff* (1962). Cosmopolitan. < 6800 ft. Mar.-Aug. —*R. spiralis* Linnaeus.

Ruppia maritima Linnaeus. WIGEON GRASS, DITCH-GRASS. Uncommon, brackish water: Bodega Marine Reserve, *Connors* (1991 BML); Hwy. 37 between Sears Point Raceway and Sonoma Cr., *Baker* (1946 NCC). < 400 ft. Apr.-July.

SPARGANIACEAE. BUR-REED FAMILY

[Sometimes treated as part of TYPHACEAE]

Sparganium

1. Stigmas 2; mature achenes ± truncate at apex . . . . *S. eurycarpum*

1: Stigma 1; mature achenes tapered at apex

2\. Leaves rounded on back 1.5-4 mm wide, floating, the middle and upper enlarged and sheathing at the base . . . . *S. angustifolium*

2: Leaves mostly flat 5-12 mm wide, usually erect, scarcely inflated at base . . *S. erectum* ssp. *stoloniferum*

Sparganium angustifolium Michaux. NARROW-LEAVED BUR-REED. Uncommon, slow-moving water, ponds: n. half of lower Pitkin Marsh, *Rubtzoff* (1952 NCC). < 11500 ft. July-Aug.

Sparganium erectum Linnaeus ssp. **stoloniferum** (Graebner) C. Cook & M. S. Nicholls. Common, fresh water, swamps, ponds: The Sea Ranch pond e. of Timber Ridge Road and n. of Annapolis Road, *Wells* (1984); swamp at upper end of Miller Cr. (e. of The Sea Ranch), *Wells* (1984); bog 2 mi. e. of Stewarts Point, *Nobs & Smith* (1949); Salt Point; Duncans Mills Marsh, *Rubtzoff* (1955); Guerneville Marsh, *Rubtzoff* (1954); Pitkin Marsh, *Rubtzoff* (1951, 1952); Atascadero Cr. Marsh w. of Sebastopol, *Rubtzoff* (1972); Knights Valley 1½ mi. w. of Franz Valley Road and 1 mi. sw. of Hwy. 128, *Harrison* (1979 NCC). < 3100 ft. June-Aug. —*S. multipedunculatum* (Morong) Rydberg. *S. emersum* Rehmann var. *m.* (Morong) Rydberg.

Sparganium eurycarpum Engelmann. Common, along streams, swamps: ravine between Gualala Point Regional Park and The Sea Ranch Lodge, *Knight* (1984); marsh at jct. of Hwy. 1 and Hwy. 116 s. from Jenner, acc. Clifton (1981); Bodega Harbor Marsh, *Rubtzoff* (1955); Bodega Head Marsh, *Rubtzoff* (1954); mouth of Salmon Cr., *Ferris* (1937); laguna n. of Sebastopol at Occidental Road crossing, *Howell* (1946); Laguna de Santa Rosa, *Rubtzoff* (1958); Cunningham Marsh, *Howell* (1947); 1 mi. se. of Bloomfield on Two Rock Road, *Howell* (1955); Healdsburg, *Rattan* (1877). < 500 ft. Apr.-Aug. —*S. greenei.*

TYPHACEAE. CATTAIL FAMILY

Typha

1\. Leaves mostly 5 mm wide, strongly convex on back, dark green; ♀ spikes much overtopped by leaves . . . . *T. angustifolia*

1: Leaves 6-15 mm wide; ♀ spikes not or ± overtopped by leaves

- 2\. Leaf-sheaths auriculate, the blades moderately convex on back, blue-green . . . . *T.* × *glauca*
- 2: Leaf-sheaths tapered to subtruncate, not auriculate
  - 3\. Leaf-blades moderately convex on back; interval between ♂ and ♀ spikes 1-4 cm long *T. domingensis*
  - 3: Leaf-blades flat; no interval between ♂ and ♀ spikes . . . . *T. latifolia*

Typha angustifolia Linnaeus. NAIL ROD, NARROW-LEAVED CATTAIL. Common, brackish or fresh water: Bodega Head; ca 1 mi. nw. of Camp Meeker, acc. J. & B. Guggolz (1991); Sebastopol-Santa Rosa Hwy. near Wright Road, *Rubtzoff* (1971); Roth Ranch on Sonoma Mtn., *Stocking* (1966 NCC); marsh on e. side of Hwy. 121 5 mi. n. of Sears Point, *Rubtzoff* (1971); Stewarts Point-Skaggs Springs Road, *Quibell* (1973 NCC); Lake Sonoma area. < 6600 ft. June.

Typha angustifolia × **T. domingensis**: Spring Cr. Reservoir, Santa Rosa, *Rubtzoff* (1976).

Typha domingensis Persoon. COMMON CATTAIL, SOUTHERN CATTAIL. Common, brackish and freshwater marsh habitats: Bodega Head; ditch near Freestone Marsh, *Rubtzoff* (1970); Russian River at Vacation Beach, *Rubtzoff* (1955); Laguna de Santa Rosa ne. of Graton, *Rubtzoff* (1966); Laguna de Santa Rosa e. of Cunningham, *Rubtzoff* (1970); Fairfield Osborn (S); San Antonio Road w. of US 101 along the county line, *Knight* (1964); e. side of Hwy. 121 5 mi. n. of Sears Point, *Rubtzoff* (1970); Spring Lake, *Stocking* (1969 NCC); Spring Cr. Reservoir, *Rubtzoff* (1971); Lake Sonoma area. < 5000 ft. June-July.

Typha × glauca Godron. Uncommon, freshwater situations: San Antonio Road west of US 101 along county line, *Knight* (1964). Acc. *Munz Supplement, p. 175,* (1968), *T. glauca* represents hybrids between *T. angustifolia* and *T. latifolia*, *T. domingensis* and *T. latifolia*, and trihybrids; however, currently not recognized. June.

Typha latifolia Linnaeus. SOFT-FLAG, BROAD-LEAVED CATTAIL. Common, freshwater marshes: The Sea Ranch; Salt Point; Duncans Mills Marsh, *Rubtzoff* (1957); Guerneville Marsh, *Rubtzoff* (1955); Forestville Marsh, *Rubtzoff* (1955); Pitkin Marsh, *Rubtzoff* (1962); Todd Road Preserve; Fairfield Osborn (S), (NCC); Button Ranch on Walker Road (U. C. property) between Petaluma-Valley Ford Road and Pepper Road, acc. Patterson (1990); w. of Sears Point Raceway in Tolay Cr. drainage, acc. Patterson (1990); Spring Lake; Lake Sonoma area. < 6600 ft. June-July.

ZANNICHELLIACEAE. HORNED-PONDWEED FAMILY

Zannichellia palustris Linnaeus. HORNED-PONDWEED. Occasional, pools, sluggish streams: near mouth of Russian River, *Baker* (1899 NCC); lower Salmon Cr. Marsh, *Rubtzoff* (1955); Russian River at Vacation Beach, *Rubtzoff* (1956); Button Ranch on Walker Road (U. C. property) between Petaluma-Valley Ford Road and Pepper Road, *Connors* (1992 BML). < 7400 ft. Oct.-Nov.

ZOSTERACEAE. EEL-GRASS FAMILY

1. Plants dioecious; leaves not > 3-6 mm broad; ovary sagittate-cordate; tide pools along the ocean *Phyllospadix*
1: Plants monoecious; leaves 8-12 mm broad; ovary ovoid; bays and estuaries . . . . *Zostera*

Phyllospadix

1. Inflorescence cauline on elongate peduncles, > 10 cm long; spadices several . . . . *P. torreyi*
1: Inflorescence basal or nearly so on short peduncles, 1-6 cm long; spadices usually solitary . . . . . *P. scouleri*

Phyllospadix scouleri Hooker. SURF-GRASS. Uncommon, intertidal rocky shore: Bodega Head, *Connors* (1993 BML), acc. Fishlyn & Phillips, *Biological Bulletin, vol. 158, pp. 34-48* (1980). < 500 ft. May-July.

Phyllospadix torreyi S. Watson. Uncommon, beach near the low tide level: Shell Beach, which is 1st beach n. of Wright's Beach, *Oxford* (1964 NCC); Bodega Head, *Connors* (1993 BML), acc. Fishlyn & Phillips, *Biological Bulletin, vol. 158, pp. 34-48* (1980). < 10 ft. May-Nov.

Zostera marina Linnaeus. EEL-GRASS. Archibald Menzies reported in his log (1792) that this plant was seen floating on debris around Kotzebue's ship *Predriiatiie* along the coast. There is little doubt, after reading the log, that this was in the vicinity of Bodega Head, locally common: Bodega Harbor, *Connors* (1991 BML). < -2 ft. June-Sept.

GLOSSARY

Abaxial. On the side away from the axis; dorsal.
Abortive. Defective; barren; imperfectly developed.
Abscission. Shedding by cutting off.
Acaulescent. Stemless or nearly so.
Accrescent. Enlarging with age, as often with the calyx after flowering.
Achene. (Akene). Dry, indehiscent, one-celled, one seeded fruit, often appearing to be, and referred to, as a seed.
Acicular. Needlelike.
Actinomorphic. Having radial symmetry, as a regular flower.
Aculeate. Covered with prickles.
Acumen. A tapering point. (Adj.: acuminate).
Acute. Sharp-pointed, tapering to a point, the sides straight or ± convex.
Adaxial. On the side nearest the axis; ventral. See abaxial.
Adherent: Quality of holding firmly.
Adnate. Fusion of unlike parts, as stamens to petals.
Aestivation. The arrangement of the parts in a flowering bud.
Alveolate. Honeycombed.
Ament. A usually flexuous, scaly-bracted, spikelike inflorescence of cymules; e.g., in oaks, birches and willows; catkin.
Amphitropous. Inverted (of ovule), with the attachment near the middle of one side.
Amplexicaul. Clasping the stem, as the base of certain leaves.
Anastomosing. Netted; especially applied to veins so connected by cross veins as to form a network.
Androgynous. Having staminate and pistillate flowers in the same inflorescence, or in *Carex* in the same spikelet, the former above the latter.
Angiosperm. Having seeds in closed ovary; contains 2 subclasses: monocotyledons and dicotyledons.
Angulate. Of angular form; angled. (Adv.: angulately).
Annual. (Noun) A plant that completes its life cycle, producing flowers and seeds, within one year or one growing season.
Annular. Forming a ring.
Annulus. In ferns, the thick-walled ring that surrounds the sporangium and by contraction causes it to rupture when mature.
Anthesis. Flowering or time of flowering.
Anthocyanins. Blue to red pigments of petals and other plant parts. (Adj.: anthocyanous).
Antisepalous. Having stamens opposite, or in front of, the sepals.
Apetalous. Having no petals.
Apical. At or of the apex or tip.
Apiculate. Ended abruptly in a little point.
Appendage. Any attached supplementary or secondary part.
Appressed. Pressed flat against another organ.
Arachnoid. Cobwebby, with soft and slender, entangled hairs.
Arcuate. Somewhat curved, as if bent like a bow.
Areole. Areola. A little area defined on a surface as the angular space between vein reticulations; e.g., in nutlets of *Plagiobothrys*.
Aril. An exterior covering or appendage of certain seeds.
Arista. An awn, or bristlelike tip. (Adj.: aristate).
Articulate. Jointed; having a place for natural separation with a clean-cut scar.
Ascending. Rising obliquely or curving upward; approaching erect.
Asymmetric. Not symmetrical.
Attenuate. Slenderly prolonged or tapering; more gradual than acuminate.
Auricle. An ear-shaped appendage. (Adj.: auriculate, bearing auricles).
Awn. A bristlelike appendage, as at the tip of glumes or lemmas of many grasses.
Axil. Upper angle between a leaf or branch and the stem that bears it. (Adj.: axile, axillary).
Axis. Stem or imaginary line around which organs like leaves or petals are arranged. (Adj.: axial). (Pl.: axes).
Banner. Upper petal of a papilionaceous flower, e.g., the pea family.
Barbellate. With short, usually stiff, hairs.
Basifixed. Attached at the base, as anther terminal on filament.
Bearded. Bearing long stiff hairs.
Berry. A multiseeded, indehiscent fruit in which the pericarp is fleshy throughout, i.e. tomato or grape.
Bi- Two.
Biconvex. Convex on both sides.
Biennial. A plant normally requiring two years to complete its life cycle.
Bifid. Two-cleft about to middle, as some petals or leaves.
Bifurcate. Two-forked, as in some Y-shaped hairs, stigmas or styles.
Bilabiate. Two-lipped; e.g., calyx, or corolla of snapdragon.
Bilateral. [bilaterally symmetrical]. Divisible into two equal halves by only one plane; zygomorphic; as in flowers of *Scrophulariaceae*.
Bilocular. Two-celled, as an ovary.
Bipinnatifid. Twice pinnately cleft.
Biradial. Divisible into mirror-image halves in two ways.
Bisexual. Having both stamens and pistils; possessing perfect flowers.
Bladdery. Thin and inflated.
Blade. The flat expanded portion of a leaf or petal.
Bract. A reduced leaf subtending a flower, usually associated with an inflorescence.
Bracteate. Bearing bracts.
Bracteole. A bract borne on a secondary axis, as on the pedicel or even on the petiole.
Bractlet. A secondary bract born on a pedicel instead of subtending it; sepaloid organs subtending the sepals in many *Rosaceae*.
Branchlet. Small branch.
Bristle. A stiff, short hair of a plant.
Bulb. A modified underground stem with thickened scales or coats as an onion.
Bulbil. The miniature bulb or plantlet produced in leaf axils of some bulbous plants or in leaf sinuses.
Burl. A dome-shaped growth on or at the base of the trunk of a tree or large shrub.
Caducous. Falling off prematurely or very early.
Callus. (Callosity). Hard protuberance or thickening. (Pl.: calli).
Calyculate. Calyxlike; bearing a part resembling a calyx; more usually, provided with bracts against or beneath the calyx and looking like a supplementary calyx.
Calyx. Outer whorl of flower parts, usually green leaf- or bractlike structures, called sepals. (Pl.: calyces).
Campanulate. Bell-shaped, as flowers of *Campanula*.
Canescent. Gray-pubescent; hoary or becoming so.
Capillary. Very slender; hairlike.
Capitate. Formed like a head; in heads.
Capsule. A dry dehiscent fruit composed of more than one carpel.
Carinate. Keeled.
Carpel. A simple pistil or element of a compound pistil.
Carpophore. A prolongation of the floral axis between the carpels, as that which supports the pendulous fruit; e.g., of the parsley family.

Cartilaginous. Hard and tough, but not bony.
Caruncle. A protuberance or appendage of a seed, at or near the hilum (the scar at the point of attachment of an ovule or seed). (Adj.: carunculate).
Caryopsis. The fruit of grasses.
Catkin. A usually flexuous, spikelike inflorescence of cymules; e.g. in oaks, birches and willows; ament.
Caudate. Tailed or bearing a tail-like appendage as in leaves of some members of the arum family.
Caudex. Woody base of an otherwise herbaceous perennial. (Pl.: caudices).
Caulescent. Having a conspicuous leggy stem above ground.
Cauline. Belonging to the stem.
Cell. The mostly microscopic particle of protoplasm constituting the structural unit of organisms; this in plants enclosed by walls; also the locule or locules of an anther, ovary, or fruit.
Centripetal. Growing from without toward the center; an indeterminate inflorescence.
Cernous. Nodding,drooping.
Cespitose. Growing in dense clumps or tufts sometimes forming cushion-like growth forms.
Chaff. Thin, dry scales. (Adj.: chaffy).
Chartaceous. Of papery or tissuelike texture and not usually green.
Ciliate. Fringed with hairs (cilia). Ciliolate: fringed with fine hairs.
Cincinnus (-i). Used in the plural for the branches of a unilateral scorpioid cyme, e.g., *Crassulaceae*.
Cinereous. Ash-colored.
Circinate. Coiled into a circle or partially so; e.g., characteristic of some fern fronds.
Circumscissile. Dehiscing by a circular line around the fruit so that a lid comes off, as in pods of *Anagallis*.
Cismontane. W. of the main Sierran crest, as opposed to the deserts.
Clavate. Club-shaped, or like a baseball bat. (Clavellate: the diminutive).
Claw. The narrow petiole-like base of some petals and sepals.
Cleft. Cut about halfway down to a midrib.
Cleistogamous. Closed and self-fertilizing, as in flowers of some violets.
Clone. (1) Genetically identical individuals resulting from asexual reproduction such as by rhizomes, stolons or branchlets of a single individual. (2) Individuals of horticultural value (e.g., the so-called garden "varieties" of rose, iris, gladiolus, etc.).
Coherent. Congenitally united with another organ of the same kind (coalescent), or of another kind (adnate).
Collar. Outer side of the grass leaf at the junction of sheath and blade.
Coma. Tuft of hairs, or resembling such. (Adj.: comose).
Commissure. The face by which two carpels cohere, as in *Apiaceae*.
Compound. Two or more similar parts in one organ.
Compressed. Flattened laterally but fruits of *Apiaceae* and spikelets of *Poaceae* can be compressed either "laterally" or "dorsally," not necessarily used as opposite of obcompressed.
Concave. Shallowly hollowed; saucerlike.
Concolorous. Of uniform color.
Conduplicate. Folded together lengthwise, as the leaves of many grasses.
Confluent. State of blending together.
Conic. Cone-shaped.
Connate. United congenitally; fusion of lilea parts, such as petals with petals. See adnate.
Connivent. Coming together or converging but not fused.
Constricted. Tightened or drawn together.
Convex. Rounded on the surface.
Cordate. Heartshaped, with a broad and notched base.
Coriaceous. Leathery, tough.
Corm. A short, thick, firm, fleshy, bulblike subterranean stem, usually broader than high, producing roots from base and leaves and flowering stems from top; e.g., *Gladiolus*. (Without scales or coats as with an onion.)
Corolla. The petals of a flower.
Corona. The crownlike cup found at the orifice of the corolla tube as in a daffodil.
Coroniform. Crown-shaped.
Cortex. Rind or bark.
Corymb. A short and broad, more or less flat-topped inflorescence with the lateral flowers usually on longer pedicels and opening before the central flowers; contrast with cyme. (Adjs.: corymbose, corymbiform).
Cosmopolitan. Widely distributed over the globe.
Costa. A rib. Costate. Ribbed, having longitudinal elevations.
Cotyledons. The seed leaves or first leaves of the embryo; one in the monocotyledons, two in the dicotyledons.
Crenate. Edged with scallops or rounded teeth, as a petal or leafblade. (Diminutive: crenulate).
Crest. Elevated appendage or ridge at summit of an organ.
Crisped. Irregularly curled, e.g. leaf-margins, petal-margins, hairs.
Cucullus. A hood, as on some seeds.
Culm. The type of pithy or hollow slender stem found in grasses and sedges.
Cuneate. Wedge-shaped.
Cupulate. Cup-shaped, as the cup (involucre) of an acorn.
Cusp. A sharp and rigid point. (Adj. cuspidate.)
Cyathium. Inflorescence of *Euphorbia*, consisting of a cuplike involucre.
Cyme. A flat-topped flower cluster in which the central flower blooms first.
Deciduous. Falling, as the leaves of non-evergreen trees.
Decumbent. Lying or reclining on the ground, but with tip ascending.
Decurrent. Extending down and adnate to the stem; e.g., the leaf-bases of *Helenium puberulum*.
Decurved. Curved downward.
Decussate. Opposite, with successive pairs at right angles to each other; e.g., leaves of *Euphorbia lathyris*.
Dehiscence. Process of opening when ripe to discharge the contents, as an anther or seed vessel. (Adj.: dehiscing).
Deliquescent. Melting away or dissolving, as a trunk disappearing into branches.
Deltate. Triangular with the sides almost equal.
Deltoid. Equilaterally triangular.
Dendritic. Treelike; e.g., the branching hairs of some *Brassicaceae*.
Dentate. With sharp teeth pointing outward instead of toward tip. See serrate.
Denticulate. Slightly and finely dentate.
Depauperate. Lacking in the natural size or development.
Dicot. A plant having two cotyledons, or seed leaves.
Dichotomous. A system of branching in which the main axis forks repeatedly into two branches.
Diffuse. Scattered, widely spread.
Digitate. Fingered; the members arising from one point like fingers of an open hand; e.g., the leaflets on *Aesculus californica*.
Dilated. Flattened and broadened, as an expanded filament.
Dimorphic. Dimorphous. Having two unlike forms.
Dioecious. Having ♂ and ♀ flowers on different plants.
Disarticulating. Separating joint from joint at maturity.
Discoid. Lacking ray flowers; in a flower head in the *Asteraceae*.

Dissected. Deeply divided into numerous fine segments.
Distal. Opposite the point of attachment, apical, away from the axis.
Distended. Expanded; increased in size.
Divaricate. Widely divergent.
Divergent. Extending away from each other by degrees.
Dorsal. Attached or pertaining to the back; abaxial. See ventral.
Dorsiventral. Having distinct dorsal and ventral surfaces.
Drupe. A stone fruit; e.g., a plum, in which the outer part is fleshy and the inner part hard. (Diminutive: drupelet, as an element in a blackberry).
Eccentric. One-sided; abaxial; out of the center.
Eciliate. Not having marginal hairs, as on a leaf-blade.
Eglandular. Not having glands.
Ellipsoid. An elliptic solid, as a compressed sphere.
Elliptic. In the shape of a flattened circle more than twice as long as broad.
Elongate. Drawn out in length.
Emarginate. Notched at the apex.
Emersed. Rising above the water instead of floating on it.
Endemic. Confined to a given locality.
Endocarp. The inner layer of the pericarp or fruit wall.
Entire. Undivided; the margin continuous, not incised or toothed.
Epappose. Not pappus-bearing.
Ephemeral. Lasting for a day or less.
Equitant. Overlapping in two ranks, each folded leaf astride the inner younger ones; e.g., leaves of *Iris*.
Erose. Having margin irregular as if gnawed.
Erucic. (Acid). Pertaining to a crystalline acid found in *Eruca* and *Brassica*.
Estuarine. Found in estuaries.
Evanescent. Soon disappearing.
Exserted. Protruding as stamens projecting beyond the corolla; not included.
Exocarp. The outer layer of a pericarp or fruit wall.
Explanate. Spread out flat.
Exserted. Produced or projecting beyond some enclosing organ or part, e.g., a flower with exserted stamens.
Extirpated. Wholly destroyed.
Exudate. A discharge of moisture, juice or gum as through pores or incisions.
Falcate. Sickle-shaped.
Falls. Outer whorl or series of perianth parts of an iridaceous flower, often broader than those of the inner series (the standards) and, in some *Iris*, drooping or flexuous.
Farinose. Covered with a meal-like powder.
Fascicle. A close cluster or bundle of flowers or leaves; the bundle of needlelike leaves of pines. (Adj.: fasciculate).
Fastigiate. With branches erect and more or less appressed.
Felted. Having hairs closely woven or matted together.
Fibrous. (1) Composed of fine or slender structures. (2) Having a root system composed of many roots similar in length and thickness, e.g., as some members of *Ranunculaceae*.
Filament. Thread; especially the stalk of a stamen (terminated by the anther).
Filiform. Threadlike.
Fimbriate. Fringed with long or coarse hairs.
Fistulose. Hollow.
Flabelliform. Fan-shaped; broadly wedge-shaped.
Flaccid. Weak, limp, soft or flabby.
Flexuous. Zigzag, having alternate opposite curvatures.
Floccose. Covered with tufts of soft woolly hairs that generally rub off readily.
Floret. Lemma + palea + flower of grasses. One of the small individual flowers that make up the head of a composite flower, or member of the Sunflower Family.
Floriferous. Bearing or producing flowers.
Foliaceous. Resembling a leaf in shape or texture.
Follicle. A simple 1-celled pod opening by the inner suture.
Foveolate. Diminutively pitted; honeycombed.
Frond. The leaf of a fern; occasionally used in the same sense of foliage, especially of palms.
Fruit. Seed-bearing organ, a matured pistil or pistils; e.g., achene, berry, capsule, drupe, follicle, nut, pome, samara, silicle, silique, utricle, etc.
Frutescent. Fruticose. Shrubby or bushy in the sense of being woody.
Funnelform. With the tube gradually widening upward and passing into the limb; e.g., flowers of *Convolvulus*.
Furcate. Forked.
Furfuraceous. Covered with branlike scales; scurfy.
Fuscous. Grayish brown.
Fusiform. Spindle-shaped; narrowed both ways from a swollen mid-area.
Galea. A structure shaped like a helmet, often drawn out like a beak; e.g., corolla of *Castilleja*. (Adj.: galeate).
Gamophyllous. With segments ± united; as in involucres, calyces and corollas.
Geminate. In pairs; twin.
Geniculate. Bent like a knee.
Gibbous. Swollen on one side.
Glabrate. (Glabrescent). Nearly glabrous, or becoming glabrous with maturity or age.
Glabrescent. Becoming glabrous, or somewhat so.
Glabrous. Not hairy.
Gladiate. Sword-shaped, but may be curved or straight.
Gland. A small, generally round body on the epidermis or at the tip of a hair, often secreting a usually sticky substance; present on stems, leaves or inflorescences of some plants; e.g., the glandular hairs of tarweeds.
Glandular. With glands.
Glaucescent. Nearly glaucous, or becoming so.
Glaucous. Covered with a whitish, waxy coating or bloom that rubs off.
Globose. Round; spherical.
Glomerate. Crowded, as in a head-like inflorescence.
Glomerule. A compact, capitate cyme.
Glumes. A pair of bracts at the base of a spikelet in grasses.
Granular. Having a structure or texture appearing to consist of grains or granules.
Gregarious. Growing in groups or colonies.
Grit cell. A stone cell, especially one occurring in leaves or fleshy fruits, as pears and quinces.
Gymnosperm. A class including plants having the seeds naked, or not enclosed in an ovary.
Gynaecandrous. Having staminate and pistillate flowers in the same spikelet, the latter above the former.
Gynobase. An enlargement of the receptacle bearing the pistil; e.g., in many borages.
Gynoecium. The female component of an angiosperm flower.
Gynophore. The prolonged stipe of a pistil.
Hastate. Shaped like an arrowhead but with the basal lobes turned outward.
Head. A dense, usually rounded inflorescence with a very short axis.
Hemispheric. Shaped like half a sphere.
Herbaceous. Pertaining to an herb; opposed to woody; having the texture or color of a foliage leaf; dying to the ground each year.
Herbage. Succulent parts of herbaceous plants, especially the foliage and young stems.

Hermaphroditic. Having stamens and pistils in the same flower.
Heterogamous. Producing two kinds of flowers in same head.
Heteromorphic. Of more than one kind of form. See homomorphic.
Hilum. The scar or "eye" of a seed, marking the point of attachment.
Hirsute. With moderately stiff hairs of medium length. (Diminutive: hirtellous).
Hispid. Bristly, bearing dense, erect, straight, harshly stiff (but ± flexuous) hairs.(Diminutive: hispidulous).
Homogamous. With flowers of head all of one kind, with stigmas becoming receptive at the time pollen is shed in the same flower.
Homomorphic. Alike in form or size.
Husk. The dry external covering of certain fruits or seeds.
Hyaline. Translucent.
Hypanthium. An enlarged cuplike receptacle below the calyx.
Hypogynous. Free from and inserted beneath the pistil.
Imbricate. Overlapping as shingles on a roof.
Imperfect. Referring to a flower lacking either male or female sex organs.
Incised. Cut ± sharply and deeply.
Included. Not projecting beyond the mouth of the corolla; said of stamens and pistils.
Indehiscent. Not opening to release contents.
Indurate. Hardened.
Indusium. An epidermal outgrowth covering the sorus of ferns. (Pl.: indusia).
Inferior. Growing below, as an ovary adnate to the hypanthium and so below the free part of the calyx and corolla, e.g., members of *Onagraceae* such as *Clarkia* and *Epilobium*. See superior.
Inflated. Blown up, bladdery.
Inflexed. Turned abruptly inward.
Inflorescence. Flower cluster; also the way flowers are arranged; e.g., ament, catkin, corymb, cyme, panicle, raceme, spadix, spike, umbel.
Infraspecific. Lower in rank than specific.
Innovation. A lateral branch or offshoot from the stem, generally restricted to new growth; a new shoot.
Internode. The part of a stem between two nodes or joints.
Interval. Space between ridges.
Invaginated. Infold so that an outer becomes an inner surface.
Involucel. A secondary involucre, with an umbellet of the *Apiaceae*.
Involucrate. Having an involucre.
Involucre. A ring or imbricate series of bracts around a flower cluster or flower.
Involute. Rolled inwards from the edges, the dorsal surface outermost.
Irregular. Showing a lack of uniformity; asymmetric, as a zygomorphic flower.
Isotype. A duplicate (part of a single gathering made by a collector at one time) of the holotype; it is always a specimen.
Keel. A projecting longitudinal ridge, or like the keel of a boat; e.g., fruits of *Plagiobothrys*, or a pea flower.
Labiate. Lipped; e.g., as in corolla of the *Lamiaceae*.
Lacerate. With the margin irregularly cut about one-half to two-thirds the distance to the midvein, the margins of segments so formed entire or otherwise; appearing as if torn.
Laguna. A large diffusion of stagnant water, marsh. An uneven country, full of marshes.
Lamina. The expanded portion of a leaf or floral segment.
Lanceolate. In leaves, much longer than broad; lance-shaped.
Lateral. Belonging to the side.
Leaflet. A division of a compound leaf.
Legume. A superior one-celled fruit usually dehiscent into two valves, having the seeds attached along the ventral suture — commonly called *pod*, and restricted to fruits of the pea family.
Lemma. In grasses, the lower and generally larger of the pair of bracts (lemma and palea) that enclose a flower. See palea.
Lenticular. Lens-shaped.
Lepidote. Covered with small scurfy scales.
Ligulate. Strap-shaped.
Ligule. In *Asteraceae*, the straplike outer portion of the corolla in ligulate or ray flowers. In *Poaceae*, and other grasslike plants, the appendage between the sheath and blade.
Limb. The uppermost, usually expanded, part of a gamophyllous calyx or corolla, as distinguished from the tube and throat.
Linear. Long and narrow, as of a leaf, with sides parallel or nearly so.
Lineate. Marked with parallel lines, striped.
Linguiform. Tongue-shaped.
Lip. Upper or lower of two parts of unequally divided calyx or corolla; in orchids, one petal unlike the others, usually larger.
Littoral. A shore and its adjacent areas. (Adj.: of a shore or coastal region).
Lobe. A (usually ± rounded) projection of a deeply indented leaf or other organ. (Adj.: lobed).
Locule. Cell or chamber in ovary. (Adj.: locular, as 2-locular or bilocular or 2-celled).
Loculicidal. In dehiscence of capsule, splitting down the back of the cell and so through the locule. See septicidal.
Lunate. Crescent-shaped.
Lunulate. Resembling a small crescent.
Lyrate. Pinnately lobed, with the terminal lobe the largest and laterals becoming progressively smaller toward the base.
Marcescent. When withered not falling off, especially leaves and corollas.
Maritime. Of or pertaining to the sea.
Membranaceous. Membranous; thin and soft in texture and ± translucent.
Mericarp. One of the two carpels which split apart like separate fruits in *Apiaceae*. Also in *Limnanthaceae, Aceraceae* and *Geraniaceae*.
-merous. Parted; suffix, as in flowers 3-merous, with parts such as petals and stamens in threes, as in lilies.
Micropyle. A minute scar (not always present) indicative of the opening into the ovule.
Monadelphous. Having the stamens united into one group or tube.
Monocot. Any seed plant having a single cotyledon.
Monoecious. With the male and female elements in the same individual, but in separate unisexual flowers.
Monostele. The solid central cylinder in the stems and roots of vascular plants, consisting of the vascular bundles collectively; e.g., ferns.
Monotypic. Having only a single type or representative, e.g., only one genus in a family.
Mucro. An abrupt sharp terminal point. (Adj. mucronate).
Mucronulate. Minutely mucronate.
Multicellular. As referring to hairs of many cells giving a ± prismatic appearance.
Muricate. Roughened by short firm often sharp projections.
Mycoparasite. A fungal parasite.
Naturalized. Adaptation of a foreign entity (plant, etc.) to the environment of an area.
Nectary. A gland secreting nectar usually at the base of the corolla or petals, or in the spur of such flowers as larkspurs and violets.

Neutral. Devoid of functional stamens or gynoecium.
Node. The joint of a stem, that part of the stem which bears a leaf or leaves. The space between two nodes is an internode. (Adj.: nodal).
Nut. A dry fruit from a compound ovary, with a hard wall not splitting open.
Nutlet. A small nut.
Obcompressed. Flattened the other way; opposite to the usual way; e.g., *Layia* ray-achenes are flattened front to back, with the flat part toward the center of the flower.
Obconic. Inversely conical, with the point of attachment at the small end.
Obcordate. Inversely heart-shaped, with the notch at the apex.
Oblique. With sides markedly unequal and base asymmetric; e.g., leaves of *Ulmus*.
Obovate. Inversely ovate.
Obovoid. Having the form of an inverted egg.
Obpyriform. Stem on wide end of pear-shaped fruit.
Obsolete. Not evident, vestigal.
Obtuse. Rounded or blunt at the end.
Ochroleucous. Yellowish white to buff-colored.
Ocrea. Sheath around stem above leaf-base especially in some *Polygonaceae*. (Pl.: ocreae).
Offset. A short lateral shoot by which certain plants are propagated.
Operculate. Furnished with a lid.
Orbicular. Circular or nearly so; e.g., leaves of garden nasturtium.
Orifice. Opening of a tubular corolla, sometimes constricted.
Orthotropous. Erect and straight (of an ovule or seed, with micropyle at apex and hilum at base).
Ovary. Lower part of pistil containing ovule or ovules.
Ovate. Egg-shaped in outline, in two dimensions, the widest part below the middle, as a leaf.
Ovoid. Egg-shaped, in three dimensions, as a fruit.
Ovule. The immature seed in the ovary.
Palate. The projecting part of the lower lip into the throat of a two-lipped corolla.
Palea. The upper of the two bracts enclosing the flower in *Poaceae*.
Paleaceous. Chaffy or chafflike in texture; paleate.
Palmate. Palmatifid. Handlike, with the fingers spread. Of leaf venation: with main veins spreading like fingers from base of blade. Of a lobed or compound leaf: with lobes or leaflets similarly spreading. See pinnate.
Panicle. An often large and much-branched inflorescence, like a compound raceme. (Adj.: paniculate).
Pannose. Covered with a felt of woolly hairs.
Papillae. Small nipple-shaped protuberances. (Sing.: papilla; adj.: papillate, papillose).
Papilionaceous. Butterfly-like, as in the corolla of the pea, with banner, wings and keel.
Pappose. Pappus-bearing.
Pappus. Crown of bristlelike structures, hairs, awns or scales, on the summit of the achene, in some members of *Asteraceae*.
Papule. A small solid, usually conical, elevation.
Parietal. Attached to the wall of the ovary, instead of to the axis; e.g., ovules or a placenta.
Pectinate. Divided like the teeth of a comb.
Pedicel. The stalk supporting a single flower (Adj.: pedicellate).
Peduncle. The stalk of a single flower or of a flower cluster. (Adj.: pedunculate).
Peltate. Shield-shaped; in leaves, with petiole attached on the underside of blade at or near middle.
Pelviform. Basin-shaped.
Pendulous. Inclined to hang downward; drooping.
Penicillate. Tipped with a tuft of fine hairs, as the stigmas of some grasses.
Pepo. A berry with a leathery rind; derived from an inferior ovary; use often restricted to the fruit of the *Cucurbitaceae*.
Perennial. A plant that lives for more than two years or two growing seasons. (Also an adj.).
Perfect. A flower having both stamens and pistils.
Perfoliate. In leaves, with a base that extends around the stem; e.g., *Claytonia perfoliata*.
Perianth. The floral envelopes, the calyx or corolla, or both, sometimes used when they are not distinguishable. See tepal.
Pericarp. The ripened wall of the ovary, referring to a fruit.
Perigynium. The saclike or bottle-like structure that encloses the pistillate flower and fruit in *Carex*. (Pl.: perigynia).
Perigynous. Literally, around the ovary; usually said of organs adnate to the perianth.
Petal. One of the segments of the inner floral envelope or corolla.
Petiole. Stalk of a leaf. (Adj.: petiolate).
Petiolule. Stalk of a leaflet. (Adj.: petiolulate).
Phalange. A fascicle of stamens united by their filaments.
Phyllary. A single bract of the involucre subtending the flower heads of members of *Asteraceae*.
Phyllode. Phyllodium. A modified and expanded, leaflike petiole, resembling a leaf-blade as in many Australian acacias. (Pl.: phyllodia).
Pilose. Hairy; bearing soft, straight spreading hairs.
Pinna. Primary division of a pinnately compound leaf or frond. **Pinnae** (pl.) may be further divided into pinules.
Pinnate. Featherlike. Of a compound leaf: with leaflets spreading from the sides of an elongate rachis.
Pinnatifid. Cleft pinnately into lobes not quite reaching the midrib.
Pinnatisect. Pinnately divided to the midrib.
Pistil. Female reproductive organ in the flower, consisting of an ovary, style and stigma.
Placenta. The ovule-bearing surface in the ovary. (Pl.: placentae).
Plane. Surface flat and even, not curved.
Plano-convex. Plane or flat on one side, and convex on the other.
Plicate. Folded, as in a fan, or approaching this condition.
Plumose. Provided with featherlike compound hairs.
Pollen. Mostly fine dustlike grains, formed in the anthers; the male element.
Pollinia. Pollen masses; e.g., orchids and milkweeds.
Polygonal. Having many angles and, hence many sides.
Polymorphic. Taking various forms.
Polystelic. Having an axial cylinder with more than one stele (consisting of vascular bundles collectively); e.g., roots and stems in *Perideridia*.
Pome. Represented by the fruit of the apple, pear, cotoneaster, etc., in the *Rosaceae*.
Prehensile. Adapted for grasping by wrapping around as referring to tendrils; e.g., in *Lathyrus*.
Prickle. A small, sharp point, as on the bark of a plant.
Prismatic. With flat faces separated by angles.
Process. (Noun). A natural outgrowth, projection or appendage.
Procumbent. Trailing on the ground but not rooting.
Proliferating. Abnormal or redundant development.
Proximal. Designating that end of a limb or other part which is nearest to the point of attachment; opposed to *distal*.

Puberulent. Minutely pubescent.
Pubescent. Clothed with hair, the hair usually fine and soft.
Pulverulent. Dusted as with fine powder.
Punctate. Marked with dots, depressions, or translucent glands. (Diminutive: puncticulate).
Pungent. (1) Ending in a rigid sharp point or prickle. (2) Acrid to the taste or smell.
Pustulate. Bearing blisterlike swellings or *pustules*, mostly at the bases of hairs; e.g., in *Cryptantha, Plagiobothrys*.
Pyramidal. Pyramidlike.
Pyriform. Pear-shaped.
Quadrate. Square or nearly square.
Quadrangular. Having four angles.
Raceme. An unbranched inflorescence with flowers borne on pedicels along a main axis and usually opening from below upward.
Racemose. In racemes or resembling a raceme.
Rachilla. The axis that bears the florets in the spikelet of grasses.
Rachis. The axis of a spike or raceme, or of a petiole through a compound leaf.
Radial. Forming uniformly around a central axis; pertaining to a ray or ray flower.
Radiate. Having rays or ray flowers; spreading from a common center; e.g., chicory.
Radicle. The embryonic root of a germinating seed.
Ray. One of the branches of an umbel; one of the outer (± strap-shaped) flowers in heads of *Asteraceae*.
Receptacle. That portion of the floral axis upon which the flower parts are borne, or, in *Asteraceae*, that which bears the florets, in the head. (Adj.: receptacular).
Reflexed. Recurved; bent backwards.
Regular. Said of a flower having radial symmetry, with the parts in each series alike.
Remote. Distantly spaced.
Reniform. Kidney-shaped.
Repand. With an undulating margin, less strongly wavy than sinuate.
Repent. Creeping (prostrate and rooting).
Resinous. Containing or producing resin, said of bud scales when coated with a sticky exudate of resin.
Resupinate. Twisted 180°, as the ovary (and flowers) of most orchids; upside down; also sometimes leaves along prostrate stems of *Arctostaphylos uva-ursi*.
Reticulate. Netted.
Retrorse. Bent or turned downward or backward.
Retuse. In leaves, with a rounded apex very slightly notched at the end of the midvein; less extreme than emarginate, and much less than obcordate.
Revolute. Rolled toward the underside from both margins.
Rhizome. A subterranean rootlike stem, usually producing roots from its lower surface and shoots from its upper surface. (Adj.: rhizomatous, rhizomal).
Rhomboidal. Obliquely 4-sided. Like an equilateral parallelogram with the corners not right-angled. (Synonym: rhombic).
Rosette. A circular cluster of leaves usually close to the ground. (Adj.: rosulate).
Rostrate. Beaked.
Rosulate. Forming a rosette.
Rotate. Wheel-shaped; having the parts flat and spreading or radiating like those of a wheel, e.g., the corolla of *Solanum*.
Ruderal. Growing among rubbish or in waste places.
Rudiment. A part imperfectly developed in size or structure.
Rufous. Reddish brown.
Rugose. Wrinkled. (Dimunitive; rugulose).
Saccate. Pouchy; bag-shaped.
Sagittate. Arrowhead-shaped; with a pair of basal lobes (or ears) turned downwards.
Salient. Projecting or pointing outward.
Salverform. With a long corolla-tube and a shorter limb at right angles to it; e.g., some *Nicotiana*.
Samara. An indehiscent winged fruit as in maple and ash.
Saprophyte. A plant lacking chlorophyll living on dead organic matter.
Scabrid. Somewhat rough.
Scabrous. Rough to the touch.
Scale. Any small, usually dry, appressed leaf or bract, often only vestigial.
Scandent. Climbing.
Scape. A leafless flower stalk arising from the ground. (Adj.: scapose).
Scarious. Thin, dry and membranaceous, not green.
Scorpioid. With the axis of the inflorescence coiled at the tip like the tail of a scorpion; e.g., some in *Boraginaceae*.
Scrotiform. Pouch-shaped.
Scurfy. With small branlike scales on the epidermis.
Secund. One-sided, or seemingly so.
Seed. Structure developing from the ovule after fertilization and containing the embryo of the plant of the next generation.
Segment. One of the divisions of a bract, sepal, petal or leaf that is divided but not compound.
Sepal. A division or segment of a calyx. (Adj.: sepaloid = sepal-like).
Septum. (Pl.: septa). Partition between cavities; e.g., in leaves of *Juncus xiphioides*. (Adj.: septal, septate).
Septicidal. In dehiscence of capsule, splitting through the septa and between the locules. See loculicidal.
Septifragal. In dehiscence of capsule, a breaking away of the valves from the septa.
Sericeous. Silky with straight soft hairs.
Serrate. With sharp rather coarse sawlike teeth that point toward tip. See dentate.
Serrulate. A margin representing the diminutive of serrate, finely serrate.
Sessile. Not stalked, as a leaf without a petiole.
Seta. A bristle, or a rigid, sharp-pointed bristlelike element.
Setaceous. Bristly or bristlelike.
Setose. Beset with bristles.
Setulose. Bearing minute bristles.
Sheath. A tubular structure enclosing the stem, as the lower part of a grass leaf.
Sickle-shaped. Like a curved, hooklike blade.
Sigmoid. Like the letter S, doubly curved.
Silicle. A short fruit of certain *Brassicaceae*, usually not more than twice as long as broad.
Silique. A long slender fruit of certain *Brassicaceae*.
Simple. Unbranched, as a stem or hair; uncompounded, as a leaf; single, as a pistil of one carpel.
Sinuate. With a margin strongly wavy by turning inwards and outwards but too shallow to be lobed.
Sinus. The space or recess between two lobes or divisions of a leaf or other expanded organ.
Sinus-plait. A separation by an attachment between the lobes of a corolla, e.g., *Gentiana*. (Synonym: sinus appendage).
Sordid. Of a dull or dirty hue.
Sorus. A cluster of sporangia in ferns. (Pl.: sori).
Spadix. A spike with a fleshy axis enveloped in a spathe; e.g., *Calla*. (Pl.: spadices).
Spathaceous. Resembling or having a spathe.
Spathe. A large showy bract enclosing or subtending a spadix or flower cluster; e.g., *Lysichiton*.
Spatulate. Spoon- or spatula-shaped.
Spheric. Spherical. Like a ball.

Spheroidal. Approximately spherical.
Spicate. Spikelike or arranged in a spike. (Diminutive: spiculate).
Spiciform. Shaped like a spike.
Spike. An unbranched inflorescence in which the flowers are sessile and nearly always open from bottom to top.
Spikelet. Glumes + floret(s) of grass.
Spine. Any stiff, sharp pointed woody process, distinguished from a thorn by the absence of vascular tissue; occasionally a whole leaf may be modified into a spine.
Spinescent. More or less spiny, spine-tipped.
Spinose. Bearing spines.
Sporangium. Small sac or case containing spores. (Pl.: sporangia).
Spore. A single microscopic cell, with a thick wall; the reproductive body of ferns and other non-flowering plants; analagous to seed of flowering plants.
Sporocarp. A body produced by certain fern allies which contains clusters of sporangia or spores.
Sporogenous. Producing or developing spores.
Sporophyll. A spore-bearing leaf (as in ferns); a leaflike organ bearing reproductive parts or organs.
Spurred. Having a hollow and slender extension of some part of the flower, usually nectariferous as the calyx of larkspur or the corolla of the violet.
Squarrose. Spreading rigidly at right angles or more, as the tips of bracts.
Stamen. In flowering plants, the male, pollen-producing organ of a flower. (Adj.: staminal, stamineal).
Staminal or stamineal. Pertaining to, or having or consisting of, stamens.
Staminode. Staminodium. A sterile stamen (lacking an anther), or what corresponds to a stamen; e.g., *Brodiaea elegans*. (Pl.: staminodia).
Stele. The central cylinder or cylinders of vascular and related tissue in the stem, root, petiole, leaf.
Stellate. Starlike; stellate hairs have 3 to many branches radiating from one point; e.g., *Sidalcea*.
Sterile. Infertile; Lacking any functional sex organs.
Stipe. The stalk beneath an ovary; the stalk of a fern frond. (Adj.: stipitate).
Stipule. Appendage, usually paired, at the base of petiole in leaves of some plants; e.g., in *Fabaceae*. (Adj.: stipular).
Stolon. A horizontal stem below or at the surface of the ground that gives rise to adventitious roots and new plants usually at the tip or at nodes.
Stoloniferous. Having stolons.
Stoma or stomate. Breathing pore in a leaf or stem. (Pl.: stomata).
Striate. Marked with fine longitudinal lines or furrows.
Strict. Very straight and upright, not at all lax or spreading.
Strigose. With stiff, sharp. straight, appressed bristles or hairs. (Diminutive: strigulose).
Strobilus. Conelike cluster of sporophylls. (Pl.: strobili).
Strophiole. An appendage at the hilum of certain seeds. (Adj.: strophiolate).
Style. The slender part of a pistil between the ovary and stigma.
Stylopodium. A disclike enlargement at the base of the style; as in *Apiaceae*.
Sub- A prefix generally signifying rather, slightly, or somewhat.
Submersed. Submerged.
Subtend. Under or supporting, as a leaf or bract subtends a flower when borne in its axil. (Adj.: subtended).
Subulate. Awl-shaped.
Succulent. Juicy; fleshy and soft.
Suckering. Sending out suckers (shoots) from a subterranean stem or root.
Suffrutescent. Pertaining to a perennial plant, slightly woody only at its base.
Suffruticose. Pertaining to a perennial plant with a woody stem several cm above ground; more woody than suffrutescent, but very low.
Sulcate. Furrowed with longitudinal grooves; e.g., scapes of *Plantago lanceolata*.
Superior. Growing above, as an ovary free from other floral parts, e.g., roses, blackberries. See inferior.
Swale. A moist meadowy area.
Suture. The line of dehiscence of fruits or anthers; the line of natural union or division between coherent parts.
Sympetalous. With united petals.
Taxon. Any taxonomic unit, as an order, genus, species, etc. (Pl.: taxa).
Tepal. Sepals and petals of a perianth that are alike.
Terete. Round in cross-section; cylindrical.
Terminal. Situated at the end; an apical growth; e.g., a cone at the end of a branch.
Ternate. In threes.
Tessillate. Checkered by depressions or color patterns, or having the appearance of a mosaic.
Tetraploid. Characterized by having four basic chromosome sets.
Throat. The ± expanded upper part of a gamophyllous calyx or corolla between the tube and the limb.
Thyrse, thyrsus. A compact, ovate panicle; strictly, with main axis indeterminate, but with other axes cymose.
Tomentum. A covering of short densely matted woolly hairs. (Adj.: tomentose; diminutive: tomentulose).
Torose. Knobby, as a rhizome, cylindric but swollen and contracted at intervals. (Diminutive: torulose).
Torus. That part of the axis of a flower that bears the petals; the receptacle; e.g., *Eschscholzia californica*.
Trifid. Three-cleft.
Trichome. A bristle or hair.
Triquetrous. Sharply three-angled, the sides concave; e.g., some sedges.
Truncate. Nearly straight across, as if cut across.
Tuber. A thick, solid, short, underground stem with many buds.
Tubercle. Warty swelling or a little tuber. (Adj.: tuberculate, tuberous).
Tubular. Shaped like a hollow cylinder.
Tunicate. Having thin coats, as a bulb; e.g., *Allium*.
Turbinate. Top-shaped, inversely conical.
Turgid. Inflated; swollen.
Turion. A scaly shoot or sucker, e.g., in some *Epilobium*.
Tussock. A dense cluster or tuft of grass or sedge.
Umbel. A flat or convex inflorescence whose flowers are on pedicels radiating from a common point at the apex of the peduncle; e.g., in *Apiaceae*. (Adj.: umbellate).
Umbelliform. Umbel-shaped.
Uncinate. Hooked at the tip.
Undulate. Wavy; repand; with less pronounced "waves" than sinuate.
Urceolate. Urn-shaped.
Utricle. A bladdery, small one-seeded fruit.
Valvate. Opening as if by valves, as most capsules and some anthers; in aestivation, meeting at the edges without overlapping. (Noun: valve).
Vernal. Of or pertaining to spring.
Ventral. Opposed to the dorsal; attached or pertaining to the front; adaxial.
Verticil. A whorl; a ring of similar organs radiating from a node, as in some inflorescences of *Lupinus*. (Adj.: verticillate).
Vestigial. Reduced to a trace of a more perfectly developed part or organ.
Villous. Bearing long and soft and not matted hairs; shaggy.

Viscid. Sticky.
Whorl. A verticil; circular arrangement of similar parts emanating from ca the same point on an axis; e.g., leaves of *Galium*.
Winged. Having a membranous expansion attached to an organ.
Xerophytic. Requiring only a small amount of moisture.
Zygomorphic. Bilaterally symmetrical; divisible by only one plane into two equal halves; as corollas of *Scrophulariaceae*.

REFERENCES

Abrams, L. 1923. *Illustrated Flora of the Pacific States*, Vol. 1; Vol. 2, 1944; Vol. 3, 1951. Stanford Univ. Press, Stanford, CA.

Abrams, L. and R. S. Ferris. 1960. *Illustrated Flora of the Pacific States*, Vol. 4. Stanford Univ. Press, Stanford, CA.

Axelrod, D. I. 1944. *The Sonoma Flora*. Carnegie Instit. of Washington, Washington, DC.

Baker, M. S. 1954. *A Partial List of Seed Plants of the North Coast Ranges of California*. North Coast Herbarium. Santa Rosa Junior College, Santa Rosa, CA.

Barbe, G. D. 1991. *Noxious Weeds of California*. Dept. of Food & Agriculture, Sacramento, CA.

Barbour, M. G. 1970. *The Flora and Plant Communities of Bodega Head, CA*. Madroño 20:239-313. San Francisco, CA.

Barbour, M. G., R. B. Craig, F. R. Drysdale and M. T. Ghiselin. 1977. *The Coastal Ecology of Bodega Head*. U. C. Press, Berkeley, CA.

Behr, H. H. 1888. *Flora of the Vicinity of San Francisco*. Publisher unknown. (From Sonoma to Santa Clara and from Niles to the Pacific Ocean.)

Chaney, R. W. 1951. *Geologic Guidebook of the S. F. Bay Counties*, Bulletin #154, State of CA, Dept. of Nat. Resources, Division of Mines.

Correll, D. S. 1950. *Native Orchids of North America North of Mexico*. Revised 1978 by Stanford Univ. Press. Issued 1950 by Chronica Botanica Co.

Dempster, L. T. 1979. *Rubiaceae. A Flora of California* (Jepson), vol. 4, part 2. Jeps. Herbarium & Library, Univ. of CA., Berkeley, CA.

de Nevers, G. 1985. *Pepperwood Flora*. Calif. Acad. of Sci., San Francisco, CA.

Dolan, R. W., & L. F. LaPré. 1987. *Streptanthus morrisonii* complex. Tierra Madre Consultants, Riverside, CA.

Ellman, P. 1986. *Bouverie Audubon Preserve plant list* (unpublished).

Ferlatte, W. J. 1974. *A Flora of the Trinity Alps of Northern California*. U. C. Press, Berkeley, CA.

Holland, R. F. (with coauthor Jain) 1977, 1981. See Jokerst, J. D. 1990.

Greene, E. L. 1894. *Manual of the Botany of the Region of the San Francisco Bay*. Cubery & Co., San Francisco, CA.

Griffin, J. R. and W. B. Critchfield. 1972. *The Distribution of Forest Trees in California*. USDA For. Serv. Res. Paper PSW-82.

Guggolz, J. 1982. *Vascular plants of the Warm Springs area*. Unpublished list as a project for U. S. Army Corps of Engineers.

Hannum, W. T. 1951. *Geologic Guidebook of the San Francisco Bay Counties*, Bulletin 154. Calif. Div. of Mines, San Francisco, CA.

Harrison, N. 1976. *Plants and Plant Communities of Sonoma Co.*, Santa Rosa Jr. Coll., Santa Rosa, CA.

Harrison, N. 1991. *A Flora of Selected Sites of the Laguna de Santa Rosa*. Santa Rosa Jr. College, Santa Rosa, CA.

Heiser, Jr., C. B. 1969. *Nightshades, The Paradoxical Plants*. W. H. Freeman & Co., San Francisco, CA.

Hektner, M. M. and T. C. Foin.1977. *Vegetation Analysis of a Northern Coastal Prairie: The Sea Ranch, Sonoma Co., CA*. Madroño 24:83-103. (W. of Hwy. 1).

Hektner, M. M. and T. C. Foin. 1978. *A Flora of the Coastal Terraces of The Sea Ranch, Sonoma Co., CA*. The Wasmann Journal of Biology, San Francisco, CA. (W. of Hwy. 1).

Hektner, M. M. and M. P. Wells. 1975-1978. *Grasses of The Sea Ranch, CA*, (unpublished).

Hendry, G. W. 1931. *The Adobe Brick as a Historical Source*. Agricultural History. 5(3): 110-127.

Hendry, G. W. and M. K. Bellue. 1936. *An Approach to Southwestern Agricultural History through Adobe Brick Analysis*, pp. 1-8. The Univ. of New Mexico Bulletin, Albuquerque, NM.

Hendry, G. W. and M. P. Kelly. 1925. *The plant content of adobe bricks, with a note on adobe brick making*. 4(4): 360-374. California Historical Society Quarterly.

Hickman, J. C. 1989. *Introduction to The Jepson Manual*. U. C. Press, Berkeley, CA.

Hickman, J. C., Editor. 1993. *The Jepson Manual, Higher Plants of California*. U. C. Press, Berkeley, CA.

Hitchcock, A. S. 1950. *Manual of the Grasses of the United States*. U. S. Government Printing Office, Washington D. C.

Hitchcock, C. L. and A. Cronquist. 1961. *Vascular Plants of the Pacific Northwest*, Part 3. Univ. of Washington Press, Seattle, WN.

Hoover, R. F. 1970. *The Vascular Plants of San Luis Obispo County, CA*. U. C. Press, Berkeley, CA.

Howell, J. T. 1937. *A Russian Collection of Californian plants*. Leaf. West. Bot. 2:2 pp. 17-20, San Francisco, CA.

Howell, J. T. 1949. *Marin Flora*. U. C. Press, Berkeley, CA.

Howell, J. T., et al. 1981. *Fairfield Osborn Preserve plant list* (unpublished).

Jepson, W. L. 1909-1943. *Flora of California*. U. C. Press, Berkeley, CA.

Jepson, W. L. 1925. *Manual of the Flowering Plants of California*. Sather Gate Bookstore, Berkeley, CA.

Jokerst, J. D. 1990. *Floristic Analysis of Volcanic Mudflow Vernal Pools*. This paper was a part of Vernal Pool Plants — Their Habitat and Biology. Chico State Univ., #8, Chico, CA.

Kartesz, J. T. and R. Kartesz. 1980. *A Synonymized Checklist of the Vascular Flora of the United States, Canada, and Greenland*. The Univ. of N. Carolina Press, Chapel Hill, NC.

Knight, W. and I. et al. 1988. *Willow Brook list* (unpublished).

Kozloff, E. N. and L. H. Beidleman. 1994. *Plants of the S. F. Bay Region — Mendocino to Monterey*. Sagen Press, Pacific Grove, CA.

Kruckeberg, A. R. 1984. *California Serpentines: Flora, Vegetation, Geology, Soils and Management Problems*. U. C. Press, Berkeley, CA.

Lellinger, D. B. 1985. *A Field Manual of the Ferns & Fern Allies of the United States & Canada*. Smithsonian Inst., Washington, DC.

Lovell, B. 1978. *Franz Valley study*. Sonoma Co. Planning Dept., Santa Rosa, CA.

Mackenzie, K. K. 1906-1923. Bound collection of articles from various journals. Calif. Acad. of Sci., San Francisco, CA.

Mackenzie, K. K. 1922. *A Monograph of the California Species of the Genus* Carex. Erythea, 8:1-95. Berkeley, CA.

Mahaffey, E. 1990. *Wildflowers of The Sea Ranch*. Bookcrafters, Chelsea, MI.

Mason, H. L. 1957. *A Flora of the Marshes of California*. U. C. Press, Berkeley, CA.

McMinn, H. E. 1951. *Studies in the Genus* Diplacus (Scrophulariaceae). Madroño 11(2) pp. 33-127.

Mertens, T. R. and P. H. Raven. 1965. *Taxonomy of* Polygonum, *section* Polygonum (Avicularia) in N. Amer. Madroño 18(3) pp. 85-92.

Munz, P. A. 1959. *A California Flora*. U. C. Press, Berkeley, CA. Also a *Supplement* (1968).

Muth, G. 1985. *Napa Flora*. Pacific Union College, Angwin, CA. (Unpublished).

Neilson, J. A. 1981. *Flora of the Mayacmas Mountains*. Ecoview Environmental Consultants, Napa, CA.

Pacific Gas & Electric Co. *Report of Geothermal Area Rare and Endangered Plants*. 1985. San Francisco, CA.

Patterson, C. 1989. *Survey for Alternative Land Fill Sites for Sonoma Co*. Woodward-Clyde Consultants, Oakland, CA. Unpublished.

Powell, W. R. 1968. *Some Vascular Plants in Sonoma County*. State Cooperative Soil-Vegetation Survey. Berkeley, CA.

Radford, A. E., W. C. Dickison, J. R. Massey, C. R. Bell. 1974. *Vascular Plant Systematics*, Harper & Row, NY.

Rubtzoff, P. 1953. *A Phytogeographical Analysis of the Pitkin Marsh*. The Wasmann Journ. Biol. 11:129-219. San Francisco, CA.

Rubtzoff, P. 1958. Iris longipetala *in Sonoma County, CA*. Leafl. West. Bot. 8:272. San Francisco, CA.

Rubtzoff, P. and R. Gankin. 1988. *Botany of the Saddle Area wetlands, San Bruno Mts., San Mateo Co., CA*. The Four Seasons 8 (2): pp. 36-57. Oakland, CA.

Rattan, V. 1900. *Elementary Exercises and West Coast Botany*. The Whitaker & Ray Co., San Francisco, CA.

Self, D. W. and D. M. Taniguchi. 1982. *Todd Preserve plant list* (unpublished).

Serpa, L. 1979-1980. *Fairfield Osborn Preserve Plant Llist* (unpublished). Specimens on file NCC.

Skinner, M. W. and B. M. Pavlik. 1994. *Inventory of Rare and Endangered Vascular Plants of California, 54th edition*. C.N.P.S., Sacramento, CA.

Smith, J. P., Jr. and K. Berg. 1988. *Inventory of Rare and Endangered Vascular Plants of California, 4th edition*. C.N.P.S., Berkeley, CA.

Stocking, K. (no date) *Sugarloaf Ridge State Park Plant List* (unpublished).

Tunison, T. 1977. *Annadel State Park Plant List* (unpublished).

Tutin, T. G., V. H. Heywood, N. A. Burgess, D. H. Valentine, S. M. Walters, D. A. Webb, editors. 1964-1980. *Flora Europaea*. 5 volumes. Cambridge Univ. Press, Cambridge, U. K.

Weinreb, E. 1979. *The Vegetation of Willow Creek, Sonoma Co., CA*, (self-published).

Wells, F. G. 1951. *Geologic Guidebook*, U. S. Division of Mines, San Francisco, CA.

Wells, M. P. 1983. *Collections from Timber Preserve adjacent to The Sea Ranch, CA*.

Wells, M. P. 1984. *Species List for The Sea Ranch as a Whole*, (unpublished; material from the Hektner & Foin lists included).

Wolf, C. B. and W. E. Wagener. 1948. *The New World Cypresses*. El Aliso, vol. 1. Rancho Santa Ana Botanic Garden, Anaheim, CA.

SOIL-VEGETATION SURVEY 1947-1964

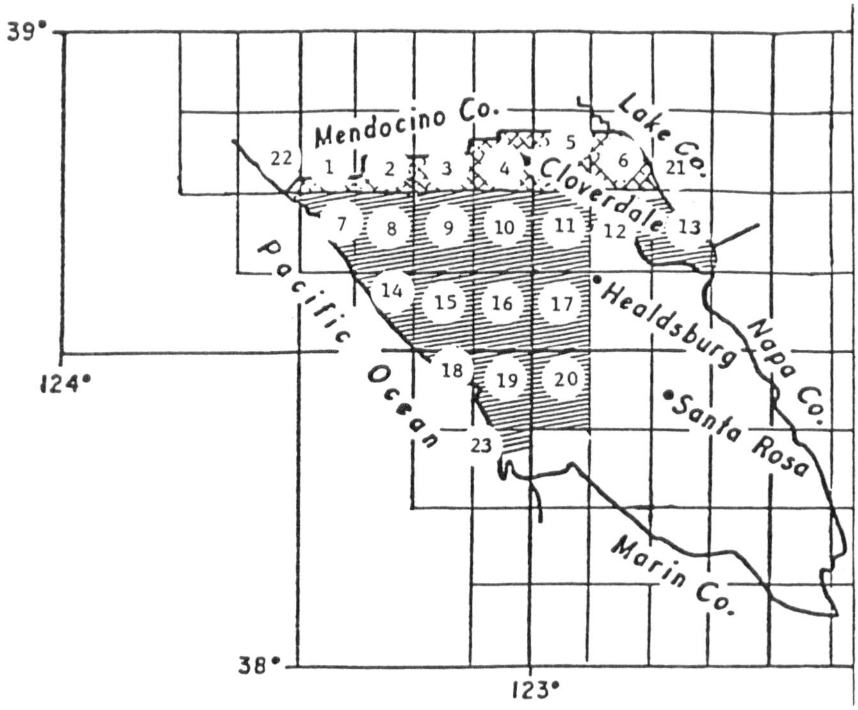

Information was compiled from the records of the State Cooperative Soil-Vegetation Survey of Sonoma County. The northern tier of quadrangles was mapped from 1947 through 1950, and the remainder from 1961 through 1964. About 554,000 acres were mapped. These data are from a paper published by State Cooperative Soil-Vegetation Survey, February 1968, authored by W. Robert Powell.

The locations are all included on the Sonoma and Napa counties AAA map.

Some locations have been titled in bold face in the text, along with quadrangle names, due to better descriptive geographic quality of Sonoma County sites. Otherwise some names of quadrangles would be Mendocino and Lake county designations.

Ornbaum Valley Quadrangle 1 - **Rockpile Creek** area in the extreme nw. part of the county.
Ornbaum Valley Quadrangle 2 - **North Fork Buckeye Creek** area.
Hopland Quadrangle 3 - **Mt. Tom**/Galloway Creek area.
Hopland Quadrangle 4 - Cherry Creek/**Cloverdale** area.
Kelseyville Quadrangle 5 - **Big Sulphur Creek**, Pine Mtn. Road area.
Kelseyville Quadrangle 6 - The Geysers/Cobb Creek/**Geothermal** area.
Stewarts Point Quadrangle 7 - The Sea Ranch/Stewarts Point area, Annapolis area.
Annapolis Quadrangle 8 - Fuller Mtn./Berkeley YMCA area.
Tombs Creek Quadrangle 9 - Hayfield Hill/Big Mtn. area.
Skaggs Springs Quadrangle 10 - **Lake Sonoma** area.

Healdsburg Quadrangle 11 - **Geyserville** area.
Healdsburg Quadrangle 12 - **Alexander Valley** area.
Calistoga Quadrangle 13 - **Ida Clayton Road** area.
Plantation Quadrangle 14 - **Salt Point**, Timber Cove area.
Fort Ross Quadrangle 15 - Seaview/ Meyers Grade area.
Cazadero Quadrangle 16 - Guerneville/Armstrong Grove area.
Healdsburg Quadrangle 17 - Hacienda/Rio Nido area.
Arched Rock Quadrangle 18 - East Branch Russian Glade area.
Duncans Mills Quadrangle 19 - Monte Rio area.
Sebastopol Quadrangle 20 - Vine Hill area.
Lower Lake Quadrangle 21 - Very small part of quadrangle between Anderson Springs and **Socrates Mine**.
Point Arena Quadrangle 22 - Very small part of quadrangle at **mouth of Gualala River**. Most of the area is in Mendocino County, which is irrelevant to this study.

Bodega Head Quadrangle 23 - The extreme n. part of the quadrangle and not of Bodega Head itself. This quadrangle was not mapped by the Soil-Vegetation Survey, but data from another source is included in their study.

Voucher specimens are on file at University of California, Davis, some mounted in the herbarium and some remaining unmounted in Dr. Powell's office.

In the text, quadrangles will be mentioned by name which are coded to the map at the beginning of this Soil-Vegetation survey section.

MARSHES IN SOUTHERN PART OF SONOMA COUNTY

(1) **Russian River Mouth Marsh.** S. bank of Russian River ca ½ mi. upstream from the mouth opposite Jenner and Penny Island).
(2) **Willow Creek Marsh**. Farther up the Russian River, along lower Willow Cr. (a southern tributary to the Russian River), 0.6 mi. ne. of Hwy. 1 crossing of the Russian River.
(3) **Lower Salmon Creek Marsh**. At mouth of Salmon Cr., near town of Salmon Cr. on Hwy. 1.
(4) **Coleman Marsh**. On n. side of Salmon Cr. ca 1 mi. upstream from Hwy. 1 crossing of the creek. So named because of Coleman Hill and Coleman Valley road in the vicinity.
(5) **Bodega Bay Marsh**. Just n. of the town of Bodega Bay on Hwy. 1.
(6) **Bodega Harbor Marsh**. At the head of Bodega Harbor 3/4 mi. w. of town of Bodega Bay.
(7) **Bodega Head Marsh**.On Bodega Head, 1 mi. airline and 2 mi. on road sw. town of Bodega Bay.
(8) **Brooks Gulch Marsh.** Lower portion of Brooks Gulch, 1 mi. se. town of Bodega Bay, Hwy. 1.
(9) **Duncans Mills Marsh**. Just n. of town of Duncans Mills on Hwy. 12 along lower Orrs Cr.
(10) **Austin Creek Marsh.** Ne. of the preceding on Hwy. 12, near lower Austin Cr.
(11) **Fife Creek Marsh.** Along lower Fife Cr., just n. of Guerneville.
(12) **Guerneville Marsh.** Ca ½ mi. se. of Guerneville, in the lower part of Mays Can. and the adjacent parts of Pocket Can.
(13) **Occidental Marsh.** Ca 3/4 mi. ne. of Occidental on the Occidental-Graton Road.
(14) **Freestone Marsh.** Ca 1 mi. nw. of Freestone, on Bohemian Hwy.
(15) **Forestville Marsh.** Ca ½ mi. e. of Forestville.
(16) **Pitkin Marsh.** Ca 1½ mi. se. of Forestville, e. of Hwy. 12.
(17) **Ross Marsh.** Near the Ross Station Road crossing of the Green Valley Cr., 1 mi. sw. of Hwy. 12, ½ mi. sw. of Ross.
(18) **Atascadero Creek Marsh.** Ca 1 mi. nw. of Graton along lower Atascadero Cr., a tributary of Green Valley Cr. Also in Ragle Ranch Regional Park.
(19) **Perry Marsh**. Near Molino, 2 mi. nw. of Sebastopol.
(20) **Cunningham Marsh.** Ca 3 mi. airline ssw. of Sebastopol or 1 mi. sw. of Cunningham.
(21) **Stony Point Marsh.** On Stony Point Road ca ¼ mi. n. of Stony Point (ca. 2 mi. airline sw. of Cotati).
(22) **Upper Stemple Creek Marsh.** Ca 2 mi. s. of the preceding, along upper Stemple Cr.

(23) **Bennett Mtn. Lake.** Shallow lake with marshy shores located 3 mi. airline w. of Kenwood in a flat on the e. slope of Bennett Mtn. at 1180 ft. elev. It is accessible from Lawndale (ca 2 mi. w. of Kenwood) on a trail 2 mi. long up the hill. Now known as **Ledson Marsh** within Annadel State Park.
(24) **Kenwood Marsh.** At Kenwood 0.3 mi. w. of Hwy 12.

Americano Cr. area. There are some marshy spots along American Cr. in the Valley Ford-Bloomfield region Marsh near Enterprise Road w. of Glen Ellen on Bennett Valley Road. (Not on map).
Aurora School Marsh. Ca 3 mi. from Tomales en route to Two Rock Church jct.
Laguna de Santa Rosa. This is a sluggish stream flowing in a nw. direction through the middle of the area shown on the map. It empties into Mark West Cr., a tributary of the Russian River. The Laguna de Santa Rosa is accompanied in its course by a series of marshy areas. Because the section of the stream is ca 12 mi. long, the different locations along it have been specified more exactly, such as Laguna de Santa Rosa near Trenton, ne. of Graton, n. of Sebastopol, e. of Sebastopol, e. of Cunningham, etc.
Sears Point Marsh. Ca 0.5 mi. n. of jct. of Hwys. 37 and 121 in Tolay Cr. drainage. (Not on map).

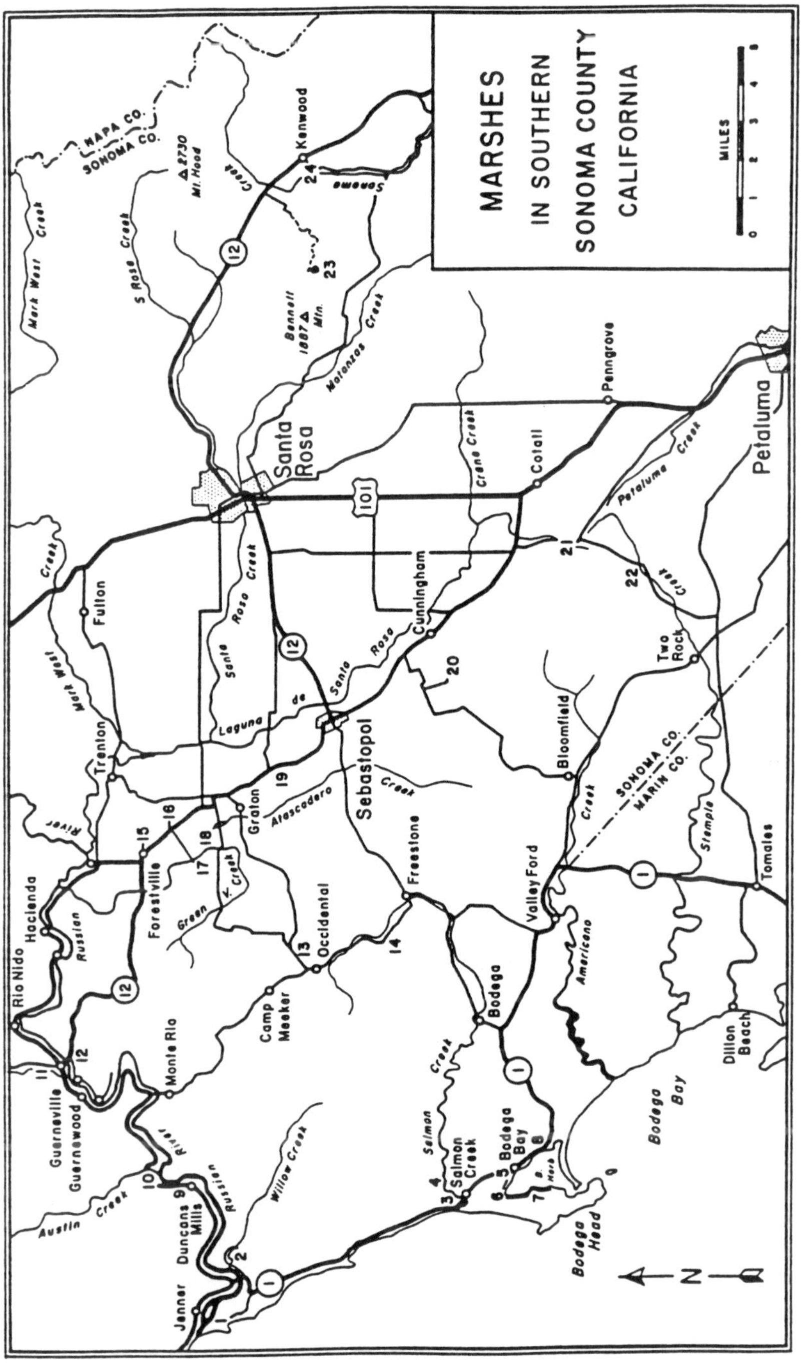
MARSHES
IN SOUTHERN
SONOMA COUNTY
CALIFORNIA
MILES
NAPA CO.
SONOMA CO.
MARIN CO.
Kenwood
Mt. Hood
Bennett Mtn.
Santa Rosa
Petaluma
Penngrove
Cotati
Cunningham
Fulton
Trenton
Sebastopol
Graton
Forestville
Occidental
Freestone
Bloomfield
Two Rock
Valley Ford
Tomales
Bodega
Camp Meeker
Monte Rio
Rio Nido
Hacienda
Guerneville
Guernewood
Duncans Mills
Jenner
Salmon Creek
Bodega Bay
Bodega Head
Dillon Beach
Russian River
Austin Creek
Willow Creek
Salmon Creek
Americano
Stemple
Atascadero Creek
Laguna de Santa Rosa
Santa Rosa Creek
Mark West Creek
Matanzas Creek
Crane Creek
Petaluma Creek
Sonoma Creek
Green V. Creek
Bodega Bay

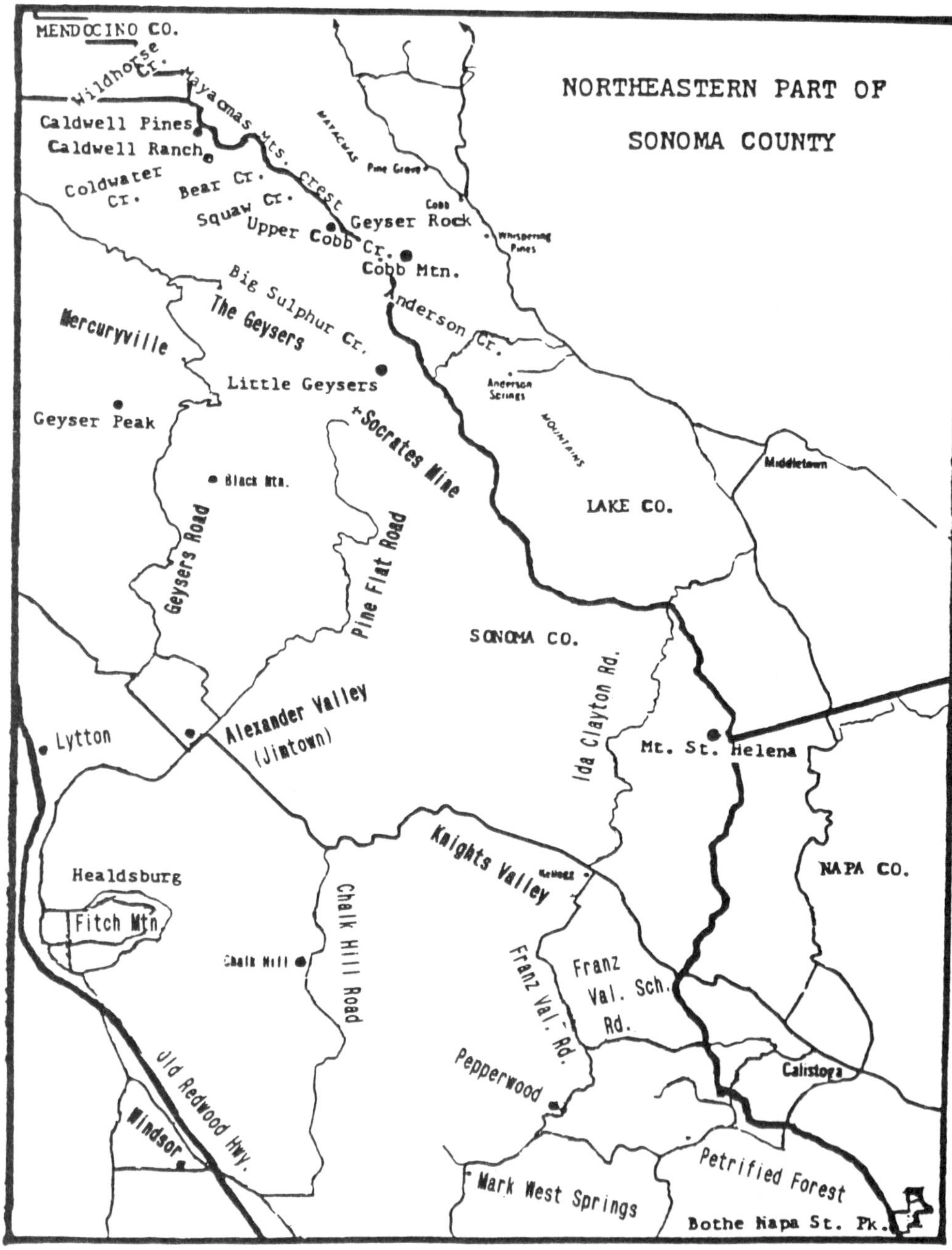
NORTHEASTERN PART OF SONOMA COUNTY
MENDOCINO CO.
Wildhorse Cr.
Mayacmas Mts. Crest
Caldwell Pines
Caldwell Ranch
Coldwater Cr.
Bear Cr.
Squaw Cr.
Upper Cobb Cr.
Geyser Rock
Whispering Pines
Pine Grove
Cobb Mtn.
Big Sulphur Cr.
Anderson Cr.
Anderson Springs
The Geysers
Mercuryville
Little Geysers
Geyser Peak
Socrates Mine
Black Mtn.
LAKE CO.
Middletown
Geysers Road
Pine Flat Road
SONOMA CO.
Ida Clayton Rd.
Mt. St. Helena
Alexander Valley (Jimtown)
Lytton
NAPA CO.
Healdsburg
Fitch Mtn
Chalk Hill
Chalk Hill Road
Knights Valley
Franz Val. Rd.
Franz Val. Sch. Rd.
Calistoga
Pepperwood
Old Redwood Hwy.
Windsor
Mark West Springs
Petrified Forest
Bothe Napa St. Pk.

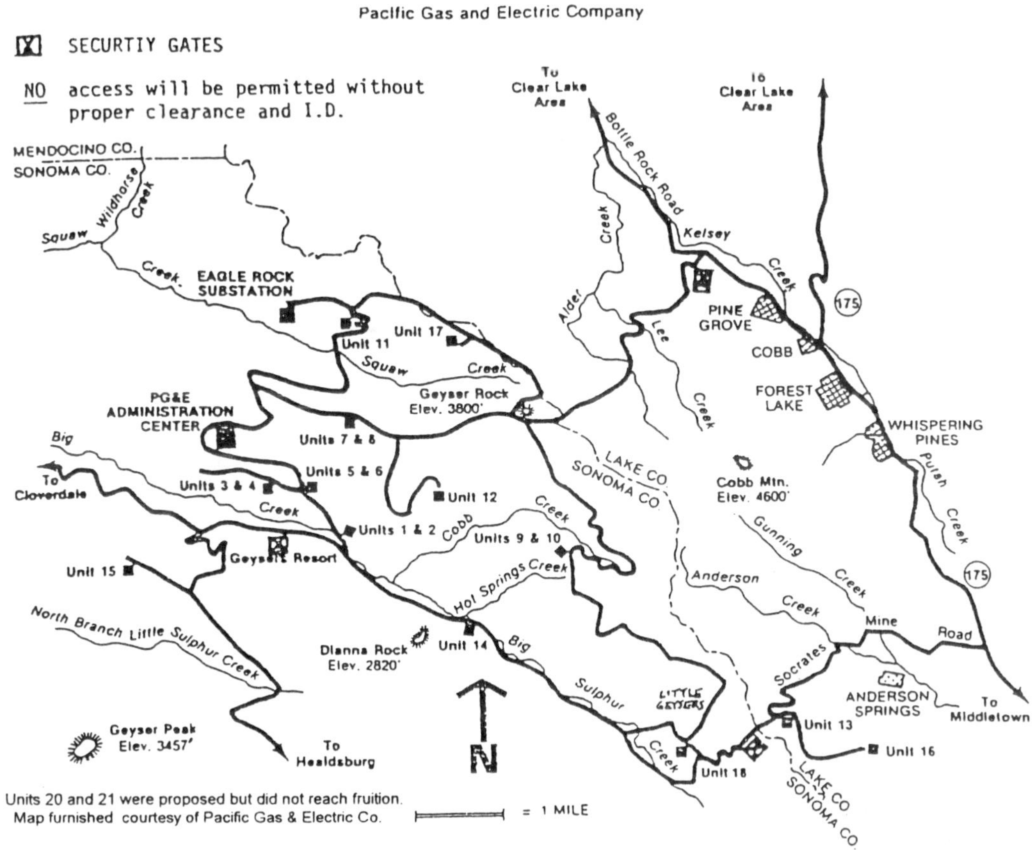

Units 20 and 21 were proposed but did not reach fruition.
Map furnished courtesy of Pacific Gas & Electric Co.

COLLECTION SITES

1. Adobe Canyon M-7
2. Alexander Valley J-3
3. Annadel L-7
4. Annapolis C-3
5. Arnold Drive N-9
6. Atascadero Cr I-8
7. Aurora School J-10
8. Austin Creek F-6
9. Bennett Peak L-7
10. Bennett Valley L-8
11. Bloomfield J-9
12. Bodega H-9
13. Bodega Bay G-9
14. Bodega Head (Point) F-9
15. Bouverie N-8
16. Calistoga Rd L-6
17. Camp Meeker H-7
18. Cavedale Rd N-8
19. Cazadero F-6
20. Chileno Valley K-11
21. Cloverdale G-1
22. Cobb Mtn K-2
23. Coleman Valley Rd G-8
24. Cotati K-9
25. Crane Can L-8
26. Cunningham Marsh J-8
27. Dry Cr M-3, M-4
28. Duncans Mills F-7
29. El Verano N-9
30. Estero Americano H-9
31. Fairfield Osborn L-9
32. Fitch Mtn I-4
33. Forestville I-6
34. Fort Ross D-6
35. Franz Valley Rd K-4
36. Freestone I-8
37. Fulton J-6
38. Geyser Pk (Sulphur Pk) I-2
39. Geyserville I-3
40. Glen Ellen M-8
41. Goat Rock E-7
42. Grange Rd L-8
43. Graton I-7
44. Gualala River B-2, C-4
45. Guerneville G-6
46. Hacienda H-6
47. Harrison Grade H-8
48. Healdsburg I-4
49. Horseshoe Cove B-5
50. Ida Clayton Rd L-4
51. Jenner F-7
52. Jimtown J-3
53. Kenwood M-7
54. Knights Valley K-4
55. Laguna de Santa Rosa J-7
56. Lake Sonoma area F-3
57. Lakeville N-11
58. Los Alamos Rd L-6
59. Los Guilicos M-7
60. Mark West Springs K-5
61. Mill Cr Rd H-4
62. Monte Rio G-7
63. Mt. Hood M-6
64. Mt. St. Helena L-3
65. Ne. Sonoma Co p. 342
66. Occidental H-8
67. Penngrove L-9
68. Pepperwood K-5
69. Perry Marsh I-7
70. Petaluma L-10
71. Petaluma Hill Rd L-9
72. Pet.-Marshall Rd K-10
73. Petrified Forest L-5
74. Pine Flat Rd J-2
75. Pine Mtn Rd H-1
76. Pitkin Marsh I-7
77. Plantation D-5
78. Pressley Rd L-9
79. Quads p. 338
80. Rincon Valley M-7
81. Rincon Ridge K-7
82. Rockpile Rd E-2
83. Russian River I-2, I-5
84. St. Helena Rd L-6
85. Salmon Cr G-8
86. Salt Point C-5
87. San Antonio Rd L-11
88. Santa Rosa K-7
89. Sears Point O-12
90. Sebastopol J-8
91. Skaggs Springs G-3
92. Sonoma N-9
93. Sonoma Mtn M-0
94. Sonoma Mtn Rd (II) M-9
95. Sonoma State Univ L-9
96. Sonoma Valley N-8, N-9
97. Spring Lake L-7
98. Stemple Cr H-10
99. Stewarts Point B-4
100. Stw Pt-Skgs Spgs Rd O-3
101. Stony Point K-9
102. Sugarloaf Ridge M-7
103. Sulphur Cr I-1
104. Sweetwater Spgs Rd H-5
105. The Cedars G-5
106. The Geysers J-2
107. The Sea Ranch B-3
108. Todd Rd Preserve J-8
109. Trinity Rd N-8
110. Two Rock J-10
111. Units-PG&E p. 343
112. Valley Ford H-9
113. Vine Hill J-7
114. Willow Brook L-10
115. Willow Cr G-7
116. Windsor I-5
117. Wright School J-8
118. Wright's Beach F-8

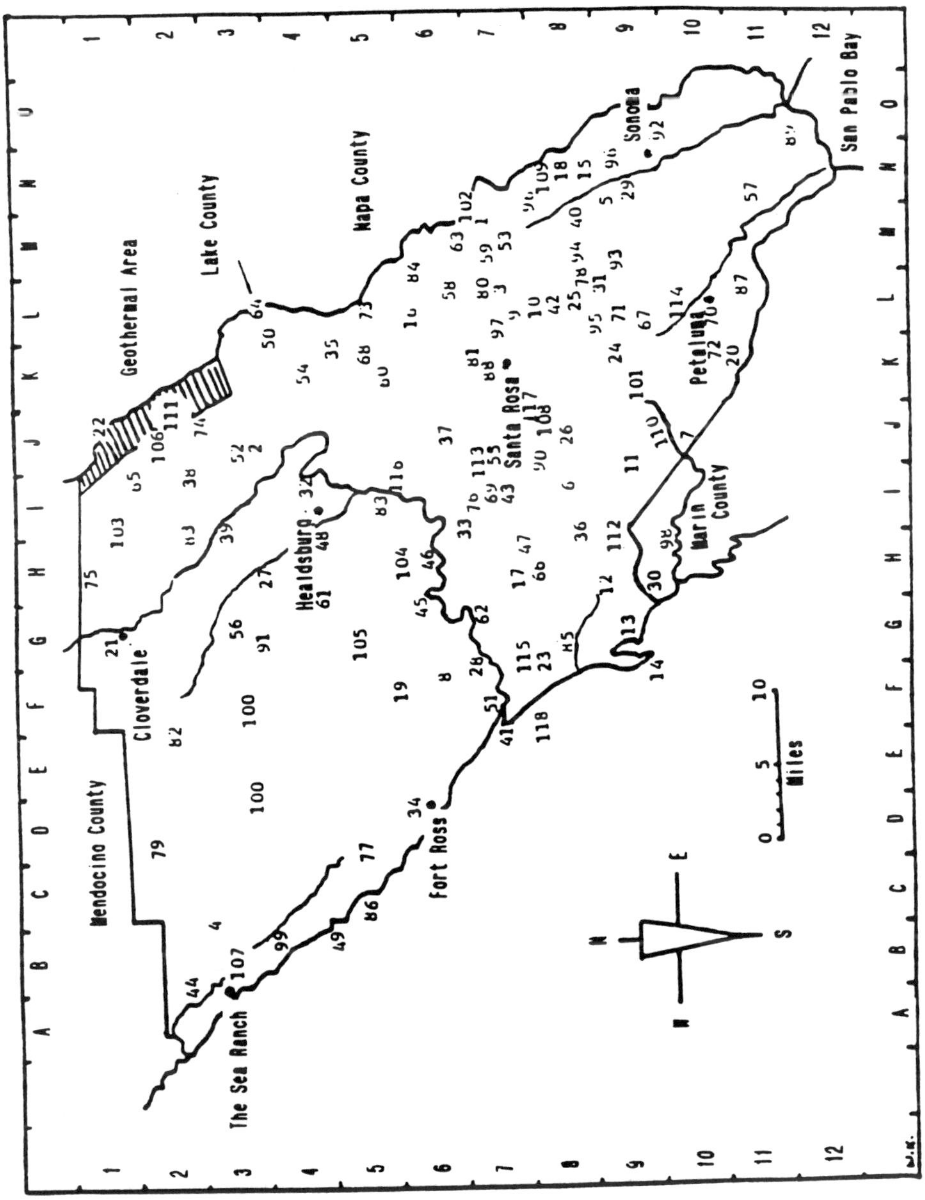
Mendocino County
Geothermal Area
Lake County
Napa County
Cloverdale
Healdsburg
Santa Rosa
Sonoma
Petaluma
Marin County
Fort Ross
The Sea Ranch
San Pablo Bay
Miles

INDEX to Plant Families and Genera

INDEX to Common Names